Springer Tracts in Additive Manufacturing

The book series aims to recognise the innovative nature of additive manufacturing and all its related processes and materials and applications to present current and future developments. The book series will cover a wide scope, comprising new technologies, processes, methods, materials, hardware and software systems, and applications within the field of additive manufacturing and related topics ranging from data processing (design tools, data formats, numerical simulations), materials and multi-materials, new processes or combination of processes, new testing methods for AM parts, process monitoring, standardization, combination of digital and physical fabrication technologies and direct digital fabrication.

Khalid Zarbane · Zitouni Beidouri
Editors

Proceedings of CASICAM 2022

Springer

Editors
Khalid Zarbane
Advanced Mechanics and Smart Factory Team, Laboratory of Advanced Research in Industrial and Logistics Engineering (LARILE)
Hassan II University of Casablanca
Casablanca, Morocco

Zitouni Beidouri
Advanced Mechanics and Smart Factory Team, Laboratory of Advanced Research in Industrial and Logistics Engineering (LARILE)
Hassan II University of Casablanca
Casablanca, Morocco

ISSN 2730-9576 ISSN 2730-9584 (electronic)
Springer Tracts in Additive Manufacturing
ISBN 978-3-031-32926-5 ISBN 978-3-031-32927-2 (eBook)
https://doi.org/10.1007/978-3-031-32927-2

This Springer imprint is published by the registered company Springer Nature Switzerland AG
The registered company address is: Gewerbestrasse 11, 6330 Cham, Switzerland

Organization

Committees

Honorary Chairs

Pr. Houssine Azeddoug, President of Hassan II University of Casablanca, Morocco.
Pr. Badreddine Benameur, Director of ENSEM, Casablanca, Morocco.
Pr. Abdelmajid Badri, Director of ESTC, Casablanca, Morocco.

General Chair

Khalid Zarbane, ESTC, Hassan II University of Casablanca, Morocco.

Technical and Program Committee Chair

Zitouni Beidouri, ESTC, Hassan II University of Casablanca, Morocco.

Technical and Program Committee

Ali Gökhan Demir, Politecnico di Milano, Italy.
Aissa Oubellouch, ESTC, Hassan II University of Casablanca, Morocco.
Allan E. W. Rennie, Lancaster University, UK.
Christoph Klahn, Research Institute Inspire, Switzerland.

Mustapha Ouardouz, Faculty of Sciences and Technologies (FST), Abdelmalek Essaâdi University, Morocco.
Younes Abouliatim, ESTC, Hassan II University of Casablanca, Morocco.

Organizing Committee Chair

Mohammed Nassraoui, ESTC, Hassan II University of Casablanca, Morocco.

Organizing Committee

Abdelhafid Essadki, ESTC, Hassan II University of Casablanca, Morocco.
Elhassan Boudaia, ESTC, Hassan II University of Casablanca, Morocco.
Eujin Pei, University of Brunel, UK.
Fathia Alkelae, Tokyo University of Science, Japan.
Jamal Benhra, ENSEM, Hassan II University of Casablanca, Morocco.
Mohamed Eloumami, ESTC, Hassan II University of Casablanca, Morocco.
Mohamed Hattabi, ENSEM, Hassan II University of Casablanca, Morocco.
Mustapha Ouardouz, FST, Abdelmalek Essaâdi University, Tetouan, Morocco.
Zoulal Mansouri, ESTC, Hassan II University of Casablanca, Morocco.

Scientific Committee

Abdelkarim Chouaf, Hassan II University of Casablanca, Morocco.
Abdelkhalak ElHami, Insitut National des Sciences Appliquées (INSA), Rouen, France.
Abdessamed Bernoussi, Abdelmalek Essaâdi University, Morocco.
Adil Akkouch, Western Michigan University Homer Stryker M.D. School of Medicine, USA.
Ahmed Kadhim, Hussein Babylon University, Iraq.
Alain Bernard, Ecole Centrale de Nantes, France.
Allan E. W. Rennie, Lancaster University, UK.
Ana María Camacho, National University of Distance Education, Spain.
Anton du Plessis, Stellenbosch University, South Africa.
Bandar Abdullah Aloyaydi, Qassim University, Saudi Arabia.
Benallal Laila, Abdelmalek Essaâdi University, Morocco.
Bouchaib Radi, Hassan I University, Morocco.
Christoph Klahn, Research Institute Inspire, Switzerland.

Partners

Hassan II University of Casablanca, Morocco.
ENSEM, Hassan II University of Casablanca, Morocco.
ESTC, Hassan II University of Casablanca, Morocco.
Association Marocaine de Fabrication Additive et d'Impression 3D (AMFAIM3D).
Laboratory of Advanced Research in Industrial and Logistics Engineering (LARILE).
Centre National pour la Recherche Scientifique et Technique (CNRST).
Académie Hassan II des Sciences et Techniques.

Sponsors

FACTORY 3D

Preface

Additive manufacturing is a rapidly evolving field of interdisciplinary research and investigation. Today, additive manufacturing technologies are capable of bringing disruptive transformation not only to the way products are designed and manufactured but also to all phases of the product life cycle. These technologies have invaded the industrial world and are becoming a pillar of Industry 4.0. This book highlights the most recent advances in additive manufacturing, leading researchers from Morocco, Europe and beyond, brought out at the second edition of the Casablanca International Conference On Additive Manufacturing (CASICAM'22) held in Morocco on November 23 and 24, 2022, co-organized by the *Ecole Nationale Supérieure d'Electricité et de Mécanique* and the *Ecole Supérieure de Technologie de Casablanca*, and in collaboration with the Moroccan Association of Additive Manufacturing and 3D Printing (AMFAIM'3D). CASICAM'22 was an opportunity for researchers, scientists, representatives of R&D organizations, doctors, engineers, industrialists, architects, managers, and entrepreneurs to interact and share their research activities, ideas, developments, and emerging applications of additive manufacturing as well as related technologies. The two-day conference also provided a precious networking time and a valuable platform to meet renowned scholars and make new contacts in the field of additive manufacturing. Over a hundred participants from eight countries, including 60 Ph.D. students, attended the conference, which gave rise to more than 36 presentations (oral and posters). All the submissions underwent at least two rigorous review processes.

This book is structured into six parts and covers topics of particular interest, namely:

- FDM Process Parameter Control and Optimization
- Laser-Based Additive Manufacturing (SLM and SLS) Process Parameter Optimization
- Topology Optimization
- AM Process Simulation
- AM Part Quality Control
- Potential of Additive Manufacturing

We hope that this book will contribute to raising interest among young scientists and Ph.D. students, and that it will be useful to the scientific community active in the field of additive manufacturing.

Casablanca, Morocco

Khalid Zarbane
Zitouni Beidouri

Contents

Part I
FDM Process Parameter Control and Optimization

Chapter 1
Experimental Investigation of Recycled Pet Materials Fdm Process Parameters Using Taguchi Analysis

Oğuz Çolak and Anar Abbasov

Abstract This article is a experimental study of recycled PET in the additive manufacturing (AM) of polymers. Parametric optimisation of fused deposition modelling (FDM) process using Raw PET and Recycled Pet materials is compared in present research. Experiments were designed for both materials according to the L9 Taguchi technique for three factors at three levels each–Printing Speed (PS), Extrusion Temperature (ET), filling ratio (FR). Each experimental ASTM D638 test part manufactured by FDM methodology is compared Mechanical properties and dimensional accuracy by using Analysis of Variance (ANOVA) method.

Keywords Additive manufacturing · Polymer · Recycling · Material extrusion Additive manufacturing · Taguchi DOE

1.1 Introduction

Additive manufacturing with the FDM method is the layer-by-layer production of three-dimensional objects from a digital model. As the name suggests, this process works by adding material. In the method, a piece is formed by using filament or granular raw material [1]. The advantages of the FDM method are its ease of application, low cost, ability to produce large parts, and preparation of raw materials according to the application. The parts produced with this method include 5 steps until the production stage. First, the CAD model of the part is created. Later, this model is converted to STL format and after the G codes are created in the slicing program for the 3D printer to work, it is printed. [2]. The materials used in the FDM method are polymer-based, and polymers, polymer composites, and recycled polymers can be used here. The interest in the availability of recycled materials is increasing day by day, especially in terms of low cost and environmental cleanliness. Among these materials, the increase in polyethylene terephthalate (PET) consumption, especially

O. Çolak (✉) · A. Abbasov
Mechanical Engineering Department Eskişehir, Eskisehir Technical University, Eskisehir, Turkey
e-mail: oguzcolak@eskisehir.edu.tr

K. Zarbane and Z. Beidouri (eds.), *Proceedings of CASICAM 2022*, Springer Tracts in Additive Manufacturing, https://doi.org/10.1007/978-3-031-32927-2_1

with the increase in the world population, has also become a problem. PET demand in 2016 is 8400 kilotonnes. For the forecast period 2017–2025, it is expected to increase by 6.9% [3]. Despite the beneficial applications of PET, which contributes to our daily lives, the natural degradation of the final product obtained from PET creates serious problems as it takes approximately 300–450 years to be disposed of in the environment [4]. Currently, waste PET in the environment accounts for 12% by volume and 8% by weight of total solid waste worldwide [5]. A proven scientific method to manage this waste is to recycle PET, thus minimizing environmental issues such as greenhouse effects.

1.2 Materials and Methods

In this work, tensile tests of the samples produced using recycled PET (rPET) material by FDM method were performed and their mechanical properties were investigated. Afterwards, samples from the same material obtained by normal methods were produced and subjected to tensile tests. The mechanical properties of recycled and pure polymers were compared. At the end, Taguchi and Variance (Anova) analyzes were performed for both materials.

The polymers with the trade names Ultrafuse PET and Ultrafuse rPET used in the study were produced by "Basaran Yenilikci Teknolojiler San. ve Tic. Ltd. Sti". It was supplied from the company. Recycled PET polymer with the trade name Ultrafuse rPET has a natural bluish appearance, while Ultrafuse PET has a transparent white appearance. Mechanical properties of PET and rPET polymer filament are given in Table 1.1.

Tensile specimens were produced with the Ultimaker S5 FDM based additive manufacturing system in accordance with ASTM D638 standards.

Taguchi experimental design method was used to determine the number of test samples (Table 1.2). Accordingly, a total of 18 test samples, 9 PET and 9 rPET, were produced and their mechanical properties were examined by tensile testing. For this purpose, 3 different printing speeds, 3 different extrusion temperatures and 3 different filling ratios were selected by keeping some parameters constant (Table 1.3). Print speeds used in 3D printers; V = 30 mm/sec, V = 45 mm/sec and V =

Table 1.1 Mechanical properties of PET and rPET filament

Parameter	PET	rPET
Tensile strength	33.4 MPa	38.6 MPa
Elongation at break	2.7%	4.3%
Young's modulus	1933 MPa	1640 MPa
Flexural strength	66.7 MPa	66.9 MPa
Flexural modulus	2063 MPa	1662 MPa
Flexural strain at break	4.6%	5.5%

Table 1.2 L9 Taguchi technique

	Printing speed (mm/s)	Extrusion temperature (C°)	Filling ratio (%)
1	30	220	30
2	30	230	60
3	30	250	100
4	45	220	60
5	45	230	100
6	45	250	30
7	60	220	100
8	60	230	30
9	60	250	60

Table 1.3 Fixed manufacturing parameters

Parameter	Value
Nozzele diameter	0.4 mm
Layer thickness	0.1 mm
Boundry wall layer	5'
Infill texture	Linear
First layer thickness	0.1 mm
Wall thickness	2 mm
Infill orientation	45/–45
Build platform temperature	75 °C

60 mm/sec. Print temperature 220, 230 and 250 °C. Filling ratios were determined as 30, 60 and 100%.

After sample production was completed, tensile tests were performed on ASTM D638 tensile samples. Among the obtained data were the force, elongation and strain values applied according to time. According to these values, the tensile strength, strain rates and young's modulus of the samples were found. Afterwards, Taguchi and Anova analyzes were performed for both PET and rPET materials using the Minitab application. The production results obtained from the Taguchi method are evaluated by converting them to the Signal/Noise (S/N) ratio. Analysis of the effect of each control factor (printing speed, extrusion temperature, filling ratio) on the tensile strength in the test results is performed with an S/N response table.

1.3 Results and Discussion

The processing parameters selected within the scope of the study are printing speed, extrusion temperature and filling ratio. Considering these parameters, a total of 18 samples were produced as 9 PET and 9 rPET materials. Afterwards, the samples were subjected to tensile tests. The tensile test results made with the Instron brand test device were obtained in the form of "Raw Data". These results constitute 4 different data in themselves. These data are time (s), elongation (mm), load (kgf) and tensile strain (mm/mm). First, these data were transferred to the Excel program. The kgf unit was then converted to the N unit. The reason for this is that the load must be expressed with N. For this, the values given in kgf are multiplied by 9.81.

$$N = 9.81 \times kgf \tag{1.1}$$

After that, the stress to which the specimen was subjected to time was calculated. For this, the forces in N units are divided by 39. Here 39 is the critical cut area of the samples.

$$\sigma = \frac{P}{A} \tag{1.2}$$

Using the aforementioned data, stress–strain graphs were drawn in Excel and the young's modulus was calculated. The tensile test results were collected in a table (Table 1.4) to better display the values on the graph and the parameters with which these values belong to the sample produced. In the mentioned table, extrusion temperature, printing speed, filling ratio, tensile strength, strain and young's modulus are given. Samples given with A in the table belong to PET, and samples given with B belong to rPET samples.

The elastic modulus values given in the table are obtained from the stress–strain ratio graph. In the stress–strain plot, the line up to the tensile stress at break is plotted at separation. The function of the obtained curve is taken as y = kx + b by Excel. Here, the value of k represents the modulus of elasticity. Thus, the modulus of elasticity was obtained using the same method for each of the 18 samples.

Taguchi analysis was performed to find the optimal control levels to obtain the highest tensile strength in the produced samples, and Anova analysis to find the most influential among the 3 parameters. First, analyzes were made for the PET sample. Table 1.5 gives the S/N response table for tensile strength for PET samples. This table, taken using the Taguchi method, gives optimal control levels to get the highest tensile strength. The parameters given in bold in the table show the ideal levels of control factors for tensile strength.

The level values of the control factors for the tensile strength of the PET sample shown in Table 1.5 are given in graphic form in Fig. 1.1. The process parameters of the control factors required to maximize the tensile strength can be easily determined using this chart.

Table 1.4 L9 Taguchi technique results

Sample	Printing speed(mm/s)	Extrusion temperature (°C)	Filling ratio (%)	Tensile strength (MPa)	Strain rate (%)	Young's modulus (MPa)
1A	30	220	30	28	0.012	2433.1
1B	30	220	30	28.5	0.022	1578.4
2A	30	230	60	45.2	0.018	2972.3
2B	30	230	60	31.3	0.021	1773.1
3A	30	250	100	21.5	0.0078	2985.7
3B	30	250	100	38	0.018	2360.9
4A	45	220	60	43.7	0.017	2910.3
4B	45	220	60	30.2	0.02	1691.5
5A	45	230	100	53.5	0.016	3664.2
5B	45	230	100	34.4	0.02	1986.8
6A	45	250	30	39.2	0.017	2727.7
6B	45	250	30	29.7	0.022	1603.4
7A	60	220	100	53.24	0.019	3353.5
7B	60	220	100	30.2	0.02	1726.1
8A	60	230	30	41.3	0.018	2654.7
8B	60	230	30	27.2	0.022	1519.4
9A	60	250	60	38.95	0.014	3146.6
9B	60	250	60	32.7	0.021	1822.8

Table 1.5 Table of S/N ratios for tensile strength in PET samples

Level	Printing speed	Extrusion temperature	Filling ratio
1	29.56	32.09	31.04
2	33.08 (45 mm/s)	33.33 (230 °C)	32.57 (60%)
3	32.88	30.11	31.91
Delta	3.52	3.22	1.53
Rank	1	2	3

As can be seen from the graph of S/N ratios in Fig. 1.1, these parameters to be used as printing speed 45 mm/s, extrusion temperature 230 °C and filling ratio 60% are expected to give the highest value for tensile strength in the experiments.

Then, Anova analysis was performed via Minitab program to find out which of the parameters determined to get the highest tensile strength of PET samples was the most effective (Table 1.6).

From the Anova results, it is seen that the most important parameter affecting the tensile strength for PET material is the printing speed, and the least affecting the filling ratio.

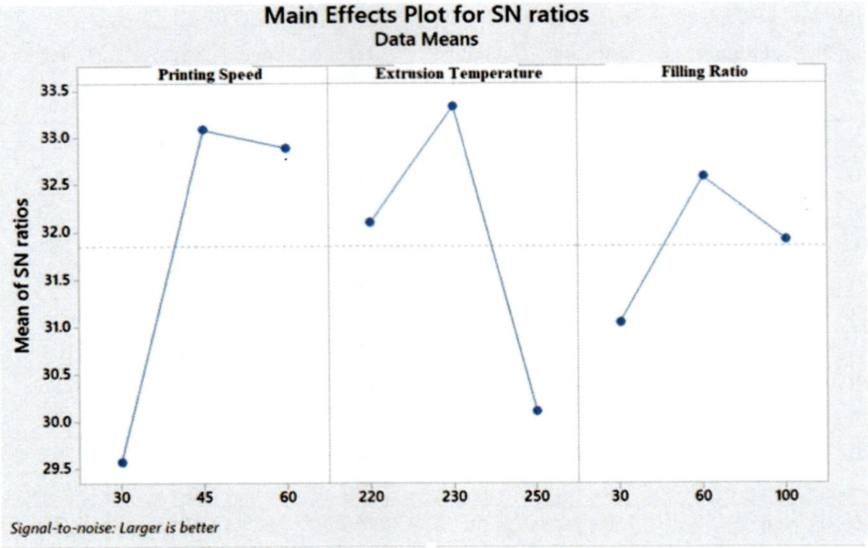

Fig. 1.1 Graph of S/N ratios for tensile strength in PET samples

Table 1.6 Tensile Strength Anova analysis for PET sample

Source	DF	Seq SS	Contribution (%)	Adj SS	Adj MS	F-Value	P-Value
Printing speed	2	361.34	40.80	361.34	180.67	2.23	0.310
Extrusion temperature	2	277.17	31.30	277.17	138.58	1.71	0.369
Filling ratio	2	84.92	9.59	84.92	42.46	0.52	0.656
Error	2	162.20	18.31	162.20	81.10		
Total	8	885.62	100.00				

The regression equation determined for the manufacturing parameters for the Tensile strength of PET material is given in Eq. (1.3).

$$\begin{aligned}\text{Tensile strength } &= 94.1 + 0.431 \text{ Printing speed} - 0.337 \text{ Extrusion temperature} \\ &+ 0.089 \text{ Filling ratio}\end{aligned} \tag{1.3}$$

Then, analyzes were made for the rPET material in the same way as the Minitab application. Table 1.7 gives the S/N response table for the tensile strength of rPET samples.

Figure 1.2 shows the graph of S/N ratios for tensile strength in rPET samples. As can be seen from the graph, these parameters to be used as printing speed of 30 mm/

Table 1.7 Table of S/N ratios for tensile strength in rPET samples

Level	Printing speed	Extrusion temperature	Filling ratio
1	30.20 (30 mm/s)	29.43	29.08
2	29.93	29.78	29.93
3	29.53	30.45 (250 °C)	30.64 (100%)
Delta	0.67	1.01	1.56
Rank	3	2	1

s, extrusion temperature of 250 °C and filling ratio of 100% are expected to give the highest value for tensile strength in the experiments.

Afterwards, Anova analysis was performed to find out which of the parameters determined to get the highest tensile strength for rPET samples was the most effective (Table 1.8).

From the Anova results, it is seen that the most important parameter affecting the tensile strength for rPET material is the filling density ratio, and the least affecting is the printing speed.

The regression equation determined for the manufacturing parameters for Tensile strength in rPET material is given in Eq. (6.2).

$$\text{Tensile strength} = 0.34 - 0.0856\ \text{Printing speed} + 0.1274\ \text{Extrusion temperature}$$

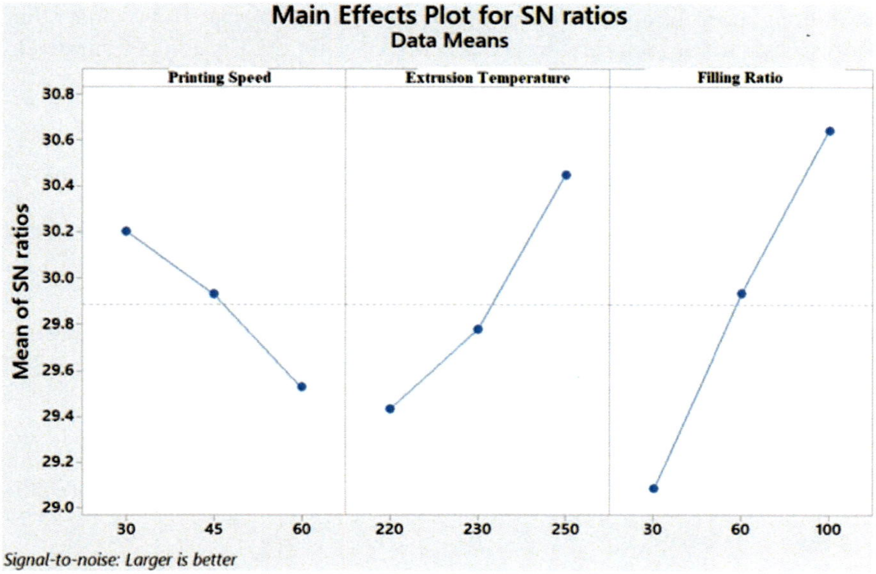

Fig. 1.2 Graph of S/N ratios for tensile strength in rPET samples

Table 1.8 Tensile Strength Anova analysis for rPET sample

Source	DF	Seq SS	Contribution (%)	Adj SS	Adj MS	F-Value	P-Value
Printing speed	1	9.882	11.48	9.882	9.8817	10.74	0.022
Extrusion temperature	1	22.716	26.40	22.716	22.7163	24.70	0.004
Filling ratio	1	48.865	56.78	48.865	48.8651	53.12	0.001
Error	5	4.599	5.34	4.599	0.9198		
Total	8	86.062	100.00				

$$+ \ 0.0813 \text{ Filling ratio} \tag{1.4}$$

1.4 Conclusions

Taguchi and Anova analyzes were performed for both PET and rPET materials in FDM technology. According to the results of Taguchi analysis, the production parameters are 45 mm/s, printing temperature is 230 °C and filling density ratio is 60%, in rPET material printing speed is 30 mm/s, printing temperature is 250 °C and filling density ratio should be set to 100%. According to the results of Anova analysis, the most important parameters affecting the tensile strength for PET material are printing speed, printing temperature and filling density, respectively, while filling density ratio, printing temperature and printing speed for rPET material, respectively.

References

1. Bryll, K., Piesowicz, E., and Szymanski, P.: Sl aczka, W.; Pijanowski, M. In Polymer Composite Manufacturing by FDM 3D Printing Technology. Matec Web Conf. **237**, 02006 (2018)
2. Kumar, L.J., Pandey, P.M.: Wimpenny DI. 3D printing and additive manufacturing technologies Springer (2019)
3. Lopez, Y.M., Paes, J.B., Gustave, D., Gonçalves, F.G., Méndez, F.C., Nantet, A.C.T.: Production of wood-plastic composites using cedrela odorata sawdust waste and recycled thermoplastics mixture from post-consumer products-A sustainable approach for cleaner production in Cuba. J. Clean. Prod. **244**, 118723 (2020)
4. Javier, C.S., Sergio, A.R., Roberto, Z.G., Jorge, D.D.: Optimization of the tensile and flexural strength of a wood-pet composite. Ingeniería, investigación y tecnología **16**(1), 105–112 (2015)
5. Corradini, E., Ito, E.N., Marconcini, J.M., Rios, C.T., Agnelli, J.A., Mattoso, L.: Interfacial behavior of composites of recycled poly (ethyelene terephthalate) and sugarcane bagasse fiber. **28**(2), 183–187 (2009)

Chapter 2
Multi-Response Optimization of Tensile Behavior of 3D Printed Polyethylene Material Using Response Surface Methodology

Anouar El Magri and Sébastien Vaudreuil

Abstract Fused deposition modeling (FDM) is among the most used additive manufacturing technique to create 3D components from thermoplastics materials. FDM has been utilized in many applications, ranging from testing models, lightweight tools to final functional parts. To become an effective processing technology, FDM must overcome many obstacles, including weak and anisotropic mechanical properties. Achieving good properties for a given material requires a proper control of printing parameters, including layer thickness as well as printing speed and temperature. While many studies are available for commonly used FDM materials, this is not the case for polyethylene (PE) which is a notoriously hard to print material. This work aims to improve Young's modulus and tensile strength of 3D printed PE through control and optimization of printing temperature, printing speed, and layer thickness. The statistical significance of each printing parameter was defined by the analyses of variance (ANOVA). Results indicate that printing temperature is the most dominant contributor to elastic modulus and tensile strength. A response surface methodology (RSM) was also applied to analyze the results and optimize tensile properties. According to this latter methodology, the optimum factor levels were achieved at 50 mm s^{-1}, 204 °C printing temperature and 0.24 mm layer thickness.

Keywords Polyethelene · FDM · Printing parameters · Mechanical properties

A. E. Magri (✉) · S. Vaudreuil
Euromed Polytechnic School, Euromed Research Center, Euromed University of Fes, Route de Meknès (Rond Point Bensouda), 30 000, Fès-Morocco, Morocco
e-mail: a.elmagri@ueuromed.org

S. Vaudreuil
e-mail: s.vaudreuil@ueuromed.org

K. Zarbane and Z. Beidouri (eds.), *Proceedings of CASICAM 2022*, Springer Tracts in Additive Manufacturing, https://doi.org/10.1007/978-3-031-32927-2_2

2.1 Introduction

Additive manufacturing (AM) technology, also known as 3D printing, has shown significant potential in manufacturing hard and complex geometries impossible to achieve using more traditional manufacturing processes [1]. The AM is a process for manufacturing 3D objects layer by layer directly from a 3D model data, something opposite to subtractive manufacturing technologies where material is removed from feedstock. For multiple reasons, including ease of operation, Fused deposition modeling (FDM) has become the most widespread AM technique [2, 3]. In this process, a thermoplastic material is continuously fed through a heated nozzle for extrusion in its viscous state. This extruded plastic is then deposited layer by layer in a precise manner to form a 3D printed part.

Various thermoplastics are currently used to 3D-print products satisfactorily by FDM. Among them are polylactic acid (PLA), acrylonitrile butadiene styrene (ABS), polycarbonate (PC), polyphenyl sulfone (PPSF) and polyetheretherketone (PEEK) [4–7]. Parts made with these thermoplastic usually exhibits lower mechanical properties when 3D Printed than when obtained by injection molding. To overcome this limitation, a large number of FDM parameters can be controlled to generate 3D products with enhanced quality and mechanical properties. Among critical printing parameters identified in the literature are the printing temperature, printing speed, layer thickness, infill density, chamber temperature and raster angle. [8, 9]. Printing temperature will, for example, directly influence polymer viscosity and fluidity, which will in turn affect bonding between layers. El Magri et al. [10] evaluated the effects of printing temperature on the tensile properties of 3D printed Poly-(phenylene sulfide) when printed in the [320–350 °C] range. It was found that crystallinity and tensile properties were strongly affected by printing temperature, with higher nozzle temperature generating more porosity that negatively influences tensile properties and part quality. Layer thickness is considered by many as a controversial parameter in FDM technology. Sood et al. [11] found that tensile strength of their samples increased with thicker layers, which was attributed to a stronger diffusion between adjacent layers due to high temperature. Work by Tymark et al. [11] showed that parts with thicker layers exhibited higher elastic modulus, while thin layers samples had the highest tensile strength. El Magri et al. [13] have investigated the effect of layer thickness on tensile and thermal performances of PEEK parts made by 3D printing. They found that layer thickness has a considerable impact on the final tensile properties of the parts. Increasing thickness from 0.1 to 0.15 mm enhanced by 10% the Young's modulus, but was accompanied by a decrease of 1.6 and 6% in tensile strength and elongation at break, respectively. This was investigated using SEM images of fractured specimen surfaces. Results showed that increasing layer thickness led a poorer adhesion, with voids between adjacent layers. Investigating the temperature evolution and layer thickness of deposited materials is nevertheless a priority, as it could be controlled by accounting for the printing speed parameters [14]. Several works studied the effects of printing speed on the 3D parts' final characteristic [13–16]. Vanaei et al. [17] have shown that printing speed is an important factor that can lead to poor interface bonding if using a high printing rate, while gravity will generate deformation at low printing rate. Abeykoon et al. [18] claimed that printing speed, in addition to nozzle temperature, should be considered for the control of material fluidity and solidification.

Published literature shows that FDM is a complex process because of the extensive list of adjustable parameters that influences the final properties of 3D printed parts. Using Taguchi method is, for example, an easy way to systematical perform design optimization. This offers several advantages, such as the simplification and minimization of the number of experimental tests [19–21]. Most studies have used analysis of variance (ANOVA) to determine which input printing parameter is statistically significant to the design output. With this analysis method, the optimum control parameter combination can be defined for a specific process [20]. Many authors have studied the effects of FDM printing parameters on the final mechanical properties of 3D printed parts using the Taguchi method [22, 23]. Results showed that printing orientation, layer thickness, infill density, and printing temperature affected the final mechanical properties of 3D printed parts. By using Design of experiments (DOE) methodology, Durão et al. [24] identified which printing parameters in FDM process affect mostly the manufacturing time, dimensional accuracy, and material consumption for low-cost. They found that printing speed and the number of contours are the most influencing printing parameters for the part quality manufactured using FDM process. El Magri et al. [15] used the response surface methodology (RSM) to identify the statistical significance and effects on tensile and structural properties of nozzle temperature, printing speed, and layer thickness when 3D printing PPS. Results showed that layer thickness is the most influential printing parameter on Young's modulus and degree of crystallinity. This RSM yielded optimum factor levels of 338 °C nozzle temperature, a 30 mm s^{-1} printing speed and the use of 0.17 mm layer thickness.

The present study investigates the relationship between the most important printing parameters for 3d printing polyethylene, using response surface mythology. The effects of printing temperature, layer thickness and printing speed on Young's modulus and tensile strength will be measured, the goal being to identify the optimum printing conditions based on graphical multi-response optimization.

2.2 Material and Samples Fabrication

A polyethylene (*PE*) filament was purchased from 3DXTECH (USA), having a diameter of 1.75 ± 0.02 mm, a density of 0.905 g cm^{-3} and a melting temperature of 130 °C. The filament was dried at 80 °C for 4 h prior to printing. The FDM system used for printing *PE* is a VOLUMIC 3D STREAM-MK2. Tensile specimens were printed flat on the glass build platform (XY surface) to achieve a good first layer adhesion. A 100% infill was used to obtain the solid-like samples required for mechanical testing. Table 2.1 summarize the major printing parameters. Slicing of the

Table 2.1 Fused filament fabrication printing parameter

Printing parameters	Value	Unit
Printing temperature	170–220	(°C)
Printing speed	20–50	(mm s−1)
Bed temperature	90	(°C)
Chamber temperature	30	(°C)
Infill line directions [relative to the long axis of the test bar]	[45/-45]	(°)
Layer height	0.1–0.30	(mm)
Infill pattern	Lines	
Line width	0.4	(mm)
Infill density	100	%
Number of bottom/top layers	2/2	Layer
Number of contours	1	wall

3D-model into individual layers was performed using the SIMPLIFY 3D software version 4.1.2.

2.3 Mechanical Testing

Tensile testing was performed per ASTM D638–14 "Standard Test Method for tensile Properties of Plastics". This standard is used for testing the tensile strength and elastic modulus of our polymer and for calculating their mechanical properties, and outlines accuracy requirements for the test frames and accessories used. A Criterion C45.105 electromechanical universal testing machine (MTS, USA) equipped with a 10 kN load cell and self-tightening jaws was used for testing. A crosshead displacement speed of 5 mm min^{-1} was used. The shape and dimensions of the specimens were defined in accordance with the D638 standard (Fig. 2.1), and six samples of each series were additive manufactured:

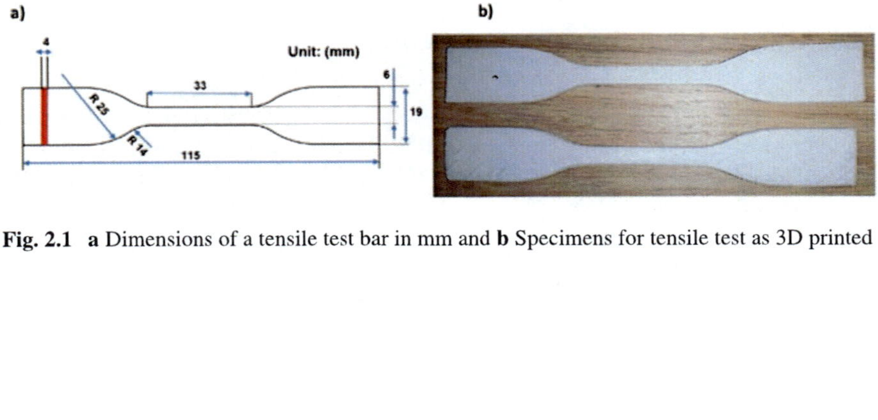

Fig. 2.1 **a** Dimensions of a tensile test bar in mm and **b** Specimens for tensile test as 3D printed

2.4 Design of Experiments and Response Surface Methodology

The use of Design of experiments (*DOE*) is of great help in efficiently planning experiments and analyzing the obtained results. For this purpose, a central composite design (CCD) was used to establish the number of experiments required to optimize three selected FDM manufacturing parameters, namely nozzle temperature (T), printing speed (S), and layer thickness (L) for two responses, namely elastic modulus (E) and tensile strength (σ_t).

The correlation between the measured response (E and σ_t) and experimental factors (T, S, and L) can be evaluated by response surface methodology (RSM). The experimental data is fitted using a second-order polynomial response surface model as expressed in Eq. (2.1).

$$Y = \beta_0 + \sum_{i=1}^{N} \beta_i X_i + \sum_{i=1}^{N} \beta_{ii} {X_i}^2 + \sum_{i \neq j}^{N} \beta_{ij} X_i X_j + \varepsilon \tag{2.1}$$

where Y is the predicted response (E and σ_t), β_0, β_i, β_{ii}, β_{ij} are respectively the regression coefficients for the intercept, linear, quadratic and interaction terms. X_i and X_j represent the coded printing parameters (T, S, and L). The experimental error is represented by ε.

The applied parameters for the CCD and their ranges are listed in Table 2.2. Levels limits have been chosen according to observations and testing in a preliminary experimental phase. This phase yielded values of nozzle temperature ranges in the [170–220 °C] range, printing speed in the [20–50 mm s^{-1}] range and layer thickness in the [0.1–0.3 mm] range. The printing parameters are presented into three levels (low, basal and high) with coded value (−1, 0, 1) and starting points of ± 1.68 for ± α in the CCD pattern.

The Minitab version 18 software was used in this study to generate the statistical model and plot the response surface of parameters optimization. ANOVA is also used to develop the regression model, to assess and to compare the differences between the effects of each printing parameter on the tensile properties.

Table 2.2 Ranges of tested printing parameters in the CDD

Factor	Code	Level					Unit
		−1.68	−1	0	+1	+1.68	
Printing temperature	T	169.77	180	195	210	220.22	(°C)
Printing speed	S	18.18	25	35	45	51.81	(mm s^{-1})
Layer height	L	0.11	0.15	0.20	0.25	0.29	(mm)

Table 2.3 Experimental results according to the CDD

	Factor			*Response*	
# Run	T (°C)	S (mm s^{-1})	L (mm)	*Tensile strength (MPa)*	*Elastic modulus (MPa)*
1	210	25	0.15	16.41 ± 1.22	365.2 ± 7.4
2	210	45	0.25	15.62 ± 1.11	356.9 ± 7.2
3	195	35	0.11	14.97 ± 1.12	324.0 ± 5.6
4	180	45	0.15	13.20 ± 1.14	351.2 ± 5.4
5	195	35	0.2	13.32 ± 1.09	393.3 ± 9.7
6	195	51.8	0.2	13.38 ± 1.18	412.6 ± 10.5
7	195	18.1	0.2	13.42 ± 1.36	359.6 ± 5.6
8	180	25	0.15	13.50 ± 1.22	353.8 ± 4.8
9	180	25	0.25	12.49 ± 1.17	319.5 ± 3.4
10	220	35	0.20	13.40 ± 1.19	384.8 ± 7.8
11	210	45	0.15	13.47 ± 1.14	359.3 ± 5.2
12	180	45	0.25	12.49 ± 1.43	359.4 ± 5.1
13	210	25	0.25	13.02 ± 1.26	360.2 ± 5.3
14	169	35	0.2	12.45 ± 1.05	346.1 ± 5.1
15	195	35	0.28	13.39 ± 1.33	± 8.7

2.5 Results and Discussion

2.5.1 *Tensile Properties*

Tensile properties of samples prepared under identified CCD printing conditions were evaluated, with a focus on Young's modulus and tensile strength. All specimens were printed according to an orientation of [45/−45°] relative to the long axis of the test bar to reduce layer delamination [13, 25]. Table 2.3 summarizes the mechanical properties of the various *PE* test specimens printed. This table shows the impact of combined factors on elastic modulus and tensile strength properties. The highest elastic modulus was registered at 195 °C nozzle temperature, a 51.81 mm s^{-1} printing speed and with 0.2 mm thick layer. The highest tensile strength was achieved for samples printed at 210 °C nozzle temperature, 25 mm s^{-1} printing speed and 0.15 mm layer thickness.

2.5.2 *Analysis of Variance (ANOVA)*

The statistical significance levels of each manufacturing parameter (T, S and L) on various response parameters was established through an analysis of variance (ANOVA). This statistical significance is described by a probability p value that should be lower than the *alpha* value set at 0.15. The Pareto chart indicates the main effects generated by ANOVA and possible interactions. This has the advantage to identify the standardized effects of linear, quadratic and interacted terms versus the significance bar, with standardized value of 1.523 (see Fig. 2.2). Any effect beyond this standardized value line is statistically significant in the fitted model of the measured response.

From the Pareto charts (Fig. 2.2), the quadratic effect of layer thickness parameter ($L \times L$) can be considered as having a significant effect on the elastic modulus and tensile strength responses, while its linear effect (L) impacts significantly the tensile strength response only (p value = 0.023). Nozzle temperature has a significant influence on the elastic modulus response as indicated by both square and linear effects crossing the reference line at a p value of 0.092 and 0.020 respectively. Nozzle temperature also has a superior impact on the tensile strength response (p

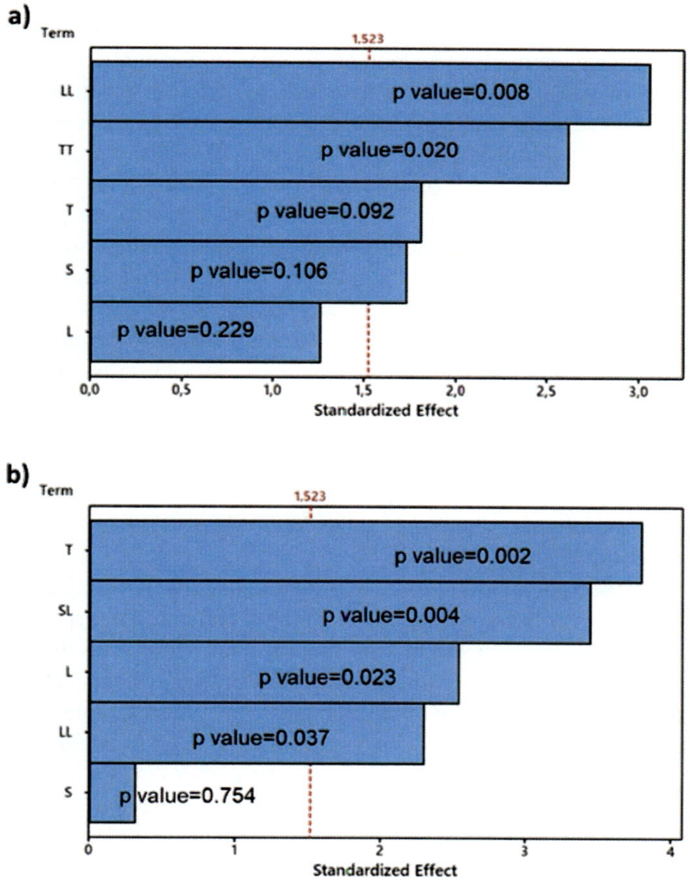

Fig. 2.2 Pareto chart of the standardized effects of: **a** Elastic modulus and **b** Tensile strength

value = 0.002). While printing speed has a statistically significant impact on the elastic modulus, with a p value of 0.106, it is not considered as being significant for tensile strength as its p value is superior to 0.15. The interaction of printing speed and layer thickness ($S \times L$) was however found to affect the tensile strength response (p value = 0.004).

The ANOVA approach was used to examine differences among level means for manufacturing parameters. There is a main effect when various levels of a parameter alter differently the response. The plot graphs of the main effects to the response for each factor level are connected by a line. We have used main effects to study differences between level means in the case of printing temperature, printing speed and layer thickness. As shown in Fig. 2.3a, the lines are not horizontal, meaning that different degrees alter differently the elastic modulus response. The same tendency was observed for printing temperature and layer thickness on tensile strength response. Figure 2.3b shows however that the main effect for printing speed is close to horizontal, indicating the lack of effect on tensile strength properties. In this instance, each level of printing speed changes the responses in the same manner and the mean tensile strength is pseudo-similar across all printing speed levels.

In the case of the elastic modulus response, all printing parameters present a significant curve variation with a significant p-values. Results from the mean effect plot (Fig. 2.3a) show that printing at 200 °C with a 50 mm s^{-1} printing speed and 0.20 mm layer thickness gives a rigid behavior to *PE*, with higher elastic modulus. Printing with high speed, 50 mm s^{-1} in this case, could reduce extrusion defects while increasing the extruded *PE* density as the extrusion speed generally affects the melt pressure inside the nozzle. This will affects both surface quality and adhesion of the extruded filament [13, 16].

It can be seen in Fig. 2.3b that tensile strength response is affected only by printing temperature and layer thickness parameters, as their influence is not linear and have p value of 0.002 and 0.023 respectively. Printing *PE* material with high nozzle temperature (220 °C) promotes *PE* specimens with high tensile strength.

High nozzle temperature impacts directly the fluidity and solidification of extruded filaments, while influencing adhesion between deposited rasters [26]. High tensile properties and a strong inter-bonding between printed layers were achieved at high nozzle temperature [27].

The mean effect plot results Fig. 2.3b show that using very thin layers (0.1 mm) results in specimens with improved tensile strength compared to printed specimens of higher layer thickness. The use of very thin layers (0.1 mm) results in specimens having more layers, resulting in higher cohesion between them while increasing the surface contact between individual filaments. This yield samples having lesser porosity and gaps between layers, which could be attributed to better heat transport that favor a neck growth along the printed polymer's lines [28].

On the other side, higher layer thickness interrupts the adhesion and attachment of adjacent layers. As confirmed by El Magri et al. [29], the increase in layer thickness for printed PEEK and PEI materials caused bigger gaps which increased porosity in the specimen's cross section.

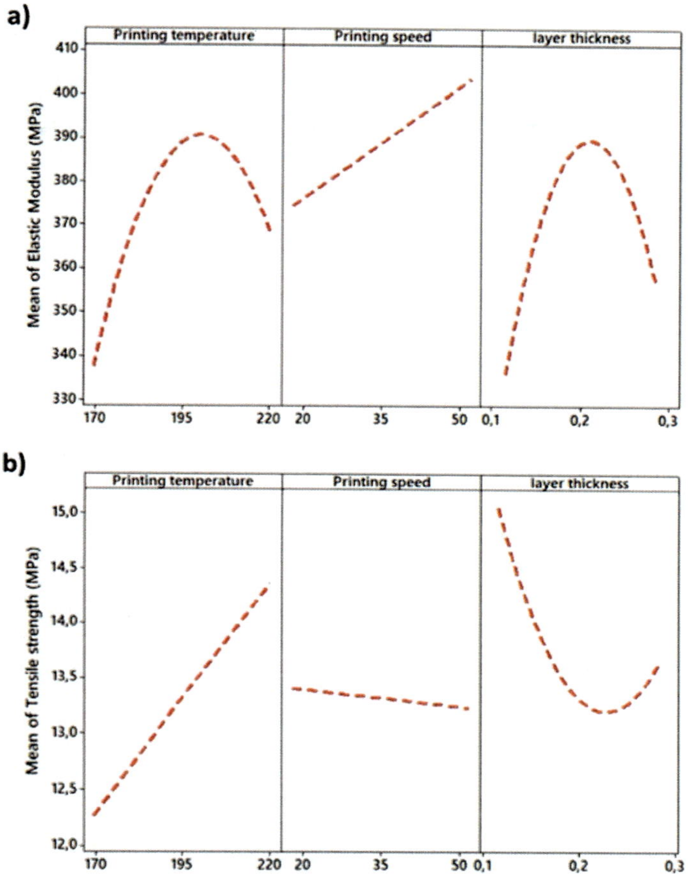

Fig. 2.3 Main effects plot for the response variable **a** Elastic modulus, **b** Tensile strength

2.5.3 *Responses Optimization*

Determining the optimum conditions of printing parameters that maximize all responses (Tensile properties) is the main objective of RSM. Three-dimensional response surface are graphical depictions of the stored regression models to help determine the responses' optimum values. They also help to better understand the interaction between tested manufacturing parameters within the range studied. The expected results of three-dimensional elastic modulus and tensile strength as response surfaces variable for printing temperature and layer thickness variables are shown in Fig. 2.4. Elastic modulus and tensile strength responses are curved in these cases because the models contain statistically significant quadratic terms.

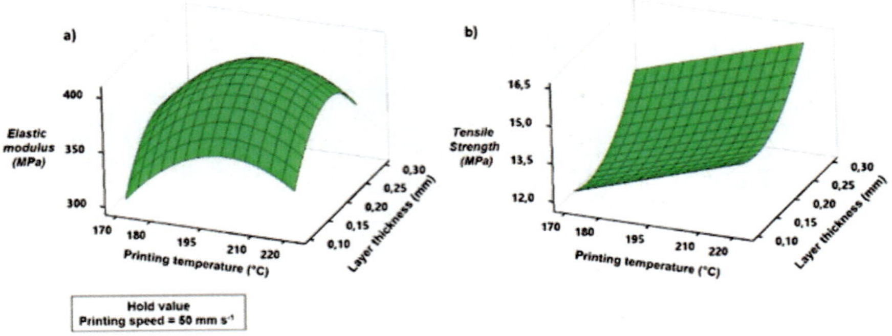

Fig. 2.4 Response surface of: **a** Elastic modulus, **b** Tensile strength

A graphical multi-response optimization technique was then applied to establish the best combination of manufacturing parameters maximizing as much as possible the tensile properties of 3D-printed *PE* material. This is performed by using overlaid contour plot to visually detect the feasible area in which all response variables are in an acceptable range [15, 30]. This feasible area is formed by printing temperature and layer thickness variables, for given holding values of printing speed, such that the fitted values for each response are between their respective contours. A printing speed of 50 mm s^{-1} was kept as this yielded the maximum elastic modulus. Each set of contours defines the limits of acceptable ranges for the fitted response, where the lower and upper limits are respectively the solid and dotted contours. A different contour color is used to display each response (see Fig. 2.5).

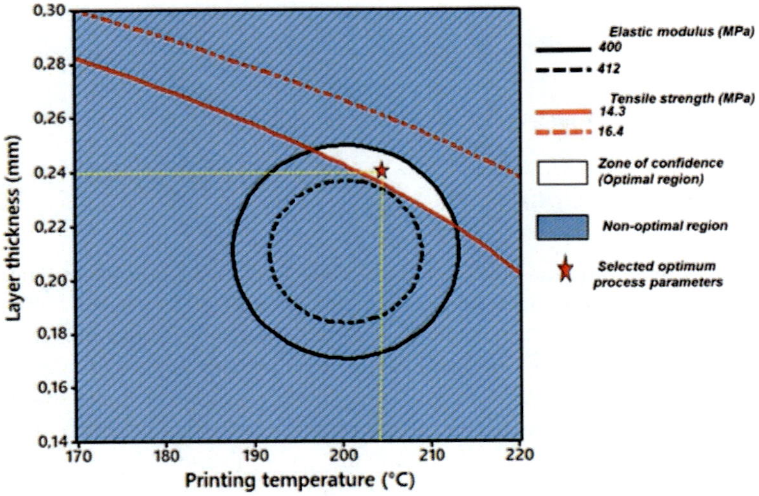

Fig. 2.5 Graphical multi-response optimization

The overlaid contour plots were used in this work to identify a compromise between the elastic modulus and tensile strength responses. They are also used to find the best settings for each printing parameters and to identify a single response yielding good tensile properties. Figure 2.5 shows the contour plot of all superimposed responses and the region (in white) which satisfies all response criteria imposed. The criteria used for constraint optimization were elastic modulus ($400 < E\ (MPa) < 412$) and tensile strength ($14.3 < \sigma_t(MPa) < 16.4$).

The superimposed contour plot exhibits an optimal region where all criteria are met. From this plot, the optimum printing parameters found for a printing speed of 50 mm s^{-1} are a printing temperature of 204 °C and a 0.24 mm layer thickness. Specimens printed with these optimum printing parameters show good agreement, with values of 408 MPa achieved for Young's modulus and 15.8 MPa for tensile strength.

2.6 Conclusion

This study investigated the impacts of three FDM printing parameters on the tensile properties of 3D printed polyethylene (*PE*) pieces. The printing parameters chosen in this analysis are recognized for their critical influence on final properties, and are the printing temperature, printing speed, and layer thickness. A response surface methodology (RSM) was used to analyze the effects of these parameters on the elastic modulus and tensile strength properties. Results indicate that printing temperature is the dominant contributor to Young's modulus and tensile strength responses, with a significant p value. RSM was also used to enhance the responses to the chosen input factors, leading to establish optimum levels of temperature, printing speed, and layer thickness.

References

1. Ahn, S., Montero, M., Odell, D., Roundy, S., Wright, P.K.: Anisotropic material properties of fused deposition modeling ABS. Rapid Prototyp. J. **8**(19), 248–257 (2002). https://doi.org/10.1108/13552540210441166
2. Martínez, J., Diéguez, J.L., Ares, E., Pereira, A., Hernández, P., Pérez, J.A.: Comparative between FEM models for FDM parts and their approach to a real mechanical behaviour. Procedia Engineering **63**, 878–884 (2013). https://doi.org/10.1016/j.proeng.2013.08.230
3. Wu, W., Geng, P., Li, G., Zhao, D., Zhang, H., Zhao, J.: Influence of layer thickness and raster angle on the mechanical properties of 3d-printed peek and a comparative mechanical study between PEEK and ABS, pp. 5834–5846, (2015). https://doi.org/10.3390/ma8095271
4. Schirmeister, C.G., Hees, T., Licht, E.H., Mülhaupt, R.: 3D printing of high density polyethylene by fused filament fabrication. Addit. Manuf. **28**, 152–159 (2019). https://doi.org/10.1016/j.addma.2019.05.003

5. Peterson, A.M.: Review of acrylonitrile butadiene styrene in fused filament fabrication: A plastics engineering-focused perspective. Addit. Manuf. (2019). https://doi.org/10.1016/j.addma.2019.03.030
6. Parandoush, P., Lin, D.: A review on additive manufacturing of polymer-fiber composites. Compos. Struct. **182**, 36–53 (2017). https://doi.org/10.1016/j.compstruct.2017.08.088
7. Coey, J.M.D.: Permanent magnets: Plugging the gap. Scripta Mater. **67**(6), 524–529 (2012). https://doi.org/10.1016/j.scriptamat.2012.04.036
8. Deng, X., Zeng, Z., Peng, B., Yan, S., Ke, W.: Mechanical properties optimization of poly-ether-ether-ketone via fused deposition modeling. Materials (Basel) (2018). https://doi.org/10.3390/ma11020216
9. Türk, D.A., Brenni, F., Zogg, M., Meboldt, M.: Mechanical characterization of 3D printed polymers for fiber reinforced polymers processing. Mater. Des. **118**, 256–265 (2017). https://doi.org/10.1016/j.matdes.2017.01.050
10. El Magri, A., Vaudreuil, S., El Mabrouk, K., Touhami, M. E.: Printing temperature effects on the structural and mechanical performances of 3D printed Poly-(phenylene sulfide) material, (2020). https://doi.org/10.1088/1757-899X/783/1/012001
11. Sood, A.K., Ohdar, R.K., Mahapatra, S.S.: Parametric appraisal of mechanical property of fused deposition modelling processed parts. Mater. Des. **31**(1), 287–295 (2010). https://doi.org/10.1016/j.matdes.2009.06.016
12. Tymrak, B.M., Kreiger, M., Pearce, J.M.: Mechanical properties of components fabricated with open-source 3-D printers under realistic environmental conditions. Mater. Des. (2014). https://doi.org/10.1016/j.matdes.2014.02.038
13. El Magri, A., El Mabrouk, K., Vaudreuil, S., Chibane, H., Touhami, M. E.: Optimization of printing parameters for improvement of mechanical and thermal performances of 3D printed poly(ether ether ketone) parts. J. Appl. Polym. Sci., 49087, (2020). https://doi.org/10.1002/app.49087
14. Vanaei, H.R., et al.: Toward the understanding of temperature effect on bonding strength, dimensions and geometry of 3D-printed parts. J. Mater. Sci. (2020). https://doi.org/10.1007/s10853-020-05057-9
15. El Magri, A., El Mabrouk, K., Vaudreuil, S., Ebn Touhami, M.: Experimental investigation and optimization of printing parameters of <scp>3D</scp> printed polyphenylene sulfide through response surface methodology. J. Appl. Polym. Sci., 49625, (2020). https://doi.org/10.1002/app.49625
16. Geng, P., et al.: Effects of extrusion speed and printing speed on the 3D printing stability of extruded PEEK filament. J. Manuf. Process. (2019). https://doi.org/10.1016/j.jmapro.2018.11.023
17. Vanaei, H.R., et al.: A comparative in-process monitoring of temperature profile in fused filament fabrication. Polym. Eng. Sci. (2020). https://doi.org/10.1002/pen.25555
18. Abeykoon, C., Sri-Amphorn, P., Fernando, A.: Optimization of fused deposition modeling parameters for improved PLA and ABS 3D printed structures. Int. J. Light. Mater. Manuf. (2020). https://doi.org/10.1016/j.ijlmm.2020.03.003
19. Lee, B.H., Abdullah, J., Khan, Z.A.: Optimization of rapid prototyping parameters for production of flexible ABS object. J. Mater. Process. Technol. **169**(1), 54–61 (2005). https://doi.org/10.1016/J.JMATPROTEC.2005.02.259
20. Ghani, J., Choudhury, I., Hassan, H.: Application of Taguchi method in the optimization of end milling parameters. J. Mater. Process. Technol. **145**(1), 84–92 (2004). https://doi.org/10.1016/S0924-0136(03)00865-3
21. Anitha, R., Arunachalam, S., Radhakrishnan, P.: Critical parameters influencing the quality of prototypes in fused deposition modelling. J. Mater. Process. Technol. **118**(1–3), 385–388 (2001). https://doi.org/10.1016/S0924-0136(01)00980-3
22. Torres, J., Cotelo, J., Karl, J., Gordon, A.P.: Mechanical property optimization of FDM PLA in shear with multiple objectives. JOM (2015). https://doi.org/10.1007/s11837-015-1367-y
23. Nancharaiah, T., et al.: An experimental investigation on surface quality and dimensional accuracy of FDM components. Int. J. Emerg. Technol., (2010)a

24. Durão, L.F.C.S., Barkoczy, R., Zancul, E., Lee Ho, L., Bonnard, R.: Optimizing additive manufacturing parameters for the fused deposition modeling technology using a design of experiments. Prog. Addit. Manuf., (2019). https://doi.org/10.1007/s40964-019-00075-9
25. El Magri, A., El Mabrouk, K., Vaudreuil, S., Ebn Touhami, M.: Mechanical properties of CF-reinforced PLA parts manufactured by fused deposition modeling. J. Thermoplast. Compos. Mater., 089270571984724, (2019). https://doi.org/10.1177/0892705719847244
26. Ning, F., Cong, W., Qiu, J., Wei, J., Wang, S.: Additive manufacturing of carbon fiber reinforced thermoplastic composites using fused deposition modeling. Compos. Mater. **80**, 369–378 (2015). https://doi.org/10.1016/j.compositesb.2015.06.013
27. Ning, F., Cong, W., Hu, Y., Wang, H.: Additive manufacturing of carbon fiber-reinforced plastic composites using fused deposition modeling: Effects of process parameters on tensile properties. J. Compos. Mater. (2016). https://doi.org/10.1177/0021998316646169
28. Sood, A.K., Ohdar, R.K., Mahapatra, S.S.: Experimental investigation and empirical modelling of FDM process for compressive strength improvement. J. Adv. Res. **3**(1), 81–90 (2012). https://doi.org/10.1016/j.jare.2011.05.001
29. El Magri, A., Vanaei, S., Vaudreuil, S.: An overview on the influence of process parameters through the characteristic of 3D-printed PEEK and PEI parts. High Perform. Polym., 095400832110099, (2021). https://doi.org/10.1177/09540083211009961
30. Minitab 18 Support—Minitab. https://support.minitab.com/en-us/minitab/18/ . Accessed 09 Dec 2019.

Chapter 3
Impact of Support Structures on the Mechanical Behaviour of Components Produced by Extrusion-Based Additive Manufacturing

Joana Martins, Mário S. Correia, Henrique A. Almeida, and Joel C. Vasco

Abstract Additive manufacturing plays an important role in the current manufacturing context since it enables to create functional parts without any limits of geometric design, as well as building parts of multi-material or material functionally graded. In some additive manufacturing systems, support structures are generated to assist in the building process and has a high importance in the time and material consumed during building and in the process cost of the final part. In some situations, these support structures are inserted into the generated physical model itself. The main objective of this research work is the mechanical characterization of models/components that contain support structures that were generated during the construction of the desired final piece. In this exploratory study, from the data obtained, it is possible to conclude that the support material does indeed assists during the building process but compromises the mechanical resistance of the produced parts.

Keywords Extrusion-based Additive manufacturing · Support structures · Mechanical Behaviour

---

J. Martins · M. S. Correia · H. A. Almeida (✉) · J. C. Vasco
School of Technology and Management, Polytechnic Institute of Leiria, Leiria, Portugal
e-mail: henrique.almeida@ipleiria.pt

M. S. Correia
Centre for Mechanical Engineering, Materials and Processes, University of Coimbra, Coimbra, Portugal

H. A. Almeida
Computer Science and Communication Research Centre, Polytechnic Institute of Leiria, Leiria, Portugal

J. C. Vasco
Institute for Polymers and Composites, University of Minho, Guimarães, Portugal

K. Zarbane and Z. Beidouri (eds.), *Proceedings of CASICAM 2022*, Springer Tracts in Additive Manufacturing, https://doi.org/10.1007/978-3-031-32927-2_3

3.1 Introduction

The current market demands are for better quality, better efficiency and cost reduction, as well as the ability to meet environmental and recycling objectives, with the consequent speed in product development and reduction of the time it takes to reach the market. In particular, Rapid Prototyping (RP) technology for producing prototypes or functional models to aid in the final product design process to reduce product development time and final manufacturing cost has emerged, which has the potential to dramatically change the design process [1].

Models and prototypes can be manufactured with RP technology not only for visualization purposes, but also to build functional parts. "Rapid Tooling" (RT) is a natural extension of RP and emerged from the need to evaluate RP models in terms of their performance. To allow validation of performance, functional prototypes must be produced using the same material as will be used in full-scale production. Furthermore, these processes are not exclusively used for prototyping and new opportunities and applications for these manufacturing processes are emerging. Thus, the RT process complements the RP options, because they are able to supply larger quantities of models in a wide variety of materials [1].

In most cases, additive manufacturing processes need to generate supports for the structures to be manufactured, this is because when they are built in an additive way, some geometries cannot be supported only by the final model that is intended throughout the construction [2].

Support structures are typically hollow or cellular structures that are removed as soon as the object is built, so these represent a waste in the additive manufacturing process. The manufacture of these supports that will later be removed requires time, energy and material consumption, thus the component manufacturing time increases [3]. Thus, the amount of material spent to build the support structures can significantly affect production costs. The presence of support structures increases both the time required to manufacture the part and the time and complexity of post-production operations [4]. In fact, support removal and surface finishing are generally carried out by hand polishing which is a time consuming process, very specialized and expensive. Minimizing the amount of supported surfaces can make this operation shorter as it improves post-processing efficiency [5]. Consequently, the material efficiency of the supporting structures in terms of design and optimization is highly important to improve the sustainability and efficiency of metallic additive production [6].

Taking into account the role of the support material, it is also necessary to study its importance in the mechanical behaviour of the manufactured part. It is important to evaluate the building parameters so that the process is faster and cheaper but without compromising the mechanical strength intended for the prototype [3, 7–10]. This is the focus of this work, which aims to evaluate the mechanical contribution of support structures in components produced by an additive process.

In this study commercial FDM equipment was used to produce specimens that were evaluated by compression testing. Different geometries were elaborated, with

Table 3.1 Infinity features and applications (adapted from [12])

Features	Applications
Support material	
It dissolves in water	Geometries not previously possible on printers
Insurance for domestic sewage	Malleable figurines
Allows for complexity	Bumps/impressions within impressions
Allows movement	

different combinations of nylon and support material, in order to be able to compare and relate the contribution of support structures during production to mechanical behaviour of the produced parts.

3.2 Materials, Methods and Methodology

3.2.1 Materials

For the production of specimens, Nylon for CubePro and Infinity™ Rinse-away Support Material were used as construction materials, with the second corresponding to support material.

Nylons are members of the polyamide (PA) family and are mostly crystalline in structure. Nylon is one of the main materials used in additive manufacturing by laser sintering technologies and also by FDM [11].

For the support material, the material called Infinity™ Rinse-away Support Material was used, whose main characteristic is the fact that it is easily removable after the production of the desired model.

The company 3D Systems [12] presents some characteristics and applications of these materials that are described in Tables 3.1 and 3.2.

3.2.2 Geometry of the Test Specimens

According to ASTM D695, the specimens will have the dimensions shown in Fig. 1a.

In order for the software to admit the support material, it was considered that the specimens would have to be hollow on the inside and also that they would have a suspended structure on top of it so that it was possible through the software control to generate the necessary support material, as can be seen in Fig. 1b. Also in this figure it is possible to observe (through the arrow indication) the construction direction that

Table 3.2 Features and applications of Nylon (adapted from [12])

Features	Applications
Thermoplastic	Functional and durable prototypes and models
Exceptionally durable	Aerospace, automotive and motorsports parts
High rigidity, hardness and resistance	Gears, accessories and bearings
High heat resistance	Boxes and casings;
High resistance to weight	Drivers and connectors;
Excellent abrasion resistance	Sports consumer goods;
Excellent wear and corrosion resistance	Fluid reserves and gas tanks;
High chemical and fuel oil resistance	Intake brackets and manifolds;
Resistance to insects, fungi and mildew	Parts that require secondary operations such as painting, milling or adhesive bonding
Superior adhesion between print layers	

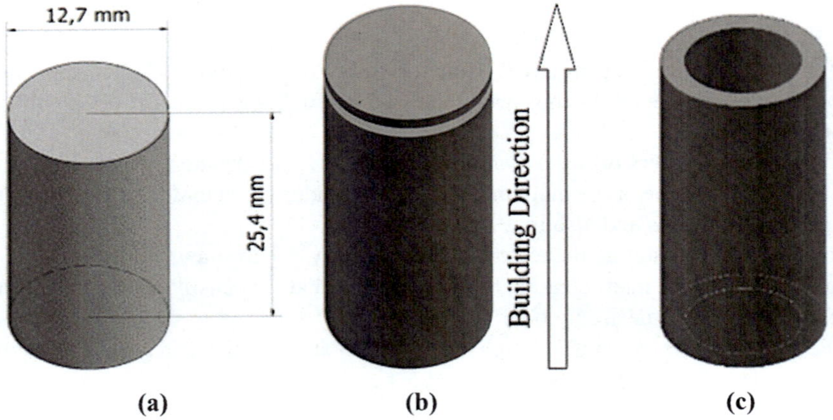

Fig. 3.1 **a** initial dimensions of test pieces and configuration of test pieces for: **b** manufacture; **c** tests

was defined. After manufacturing, it was necessary to remove the suspended structure in order to obtain the specimen in the desired final shape as shown in Fig. 1c.

In order to be able to evaluate the influence of the wall thickness through the solid volume in relation to the volume of support material, five types of geometries that differ in the amount of Nylon material were considered. Within these geometries it is still necessary to add the variable with and without support material in order to obtain data in the tests to mechanically characterize this material, so the following configurations were defined:

- 100% solid volume;
- Solid volume 75% (with and without support);
- Solid volume 50% (with and without support);
- Solid volume 25% (with and without support);
- Support structure only.

The volume and internal radius values were calculated in order to respect these percentages.

The CAD Inventor 2016 system was used to create the STL files which were interpreted and manipulated by the CubePro 2016 software developed by the machine brand.

3.2.3 *Production of Test Specimens*

In this work, the CubePro Duo rapid prototyping machine (Fig. 3.2) from 3D Systems was used. This equipment is available in the Rapid Prototyping and Reverse Engineering laboratory of the Mechanical Engineering Department of the School of Technology and Management of the Polytechnic Institute of Leiria.

The number of specimens to be produced, taking into account the indications of the standard, is five specimens of each configuration, i.e. 5 × 8 configurations = 40 specimens.

Initially, it is necessary to provide the printer with information about the material in which the specimens would be produced so that it can automatically define its temperature parameters and degree of material flow. The material definition is as follows, Nylon Black for the component and Infinity for the support structures.

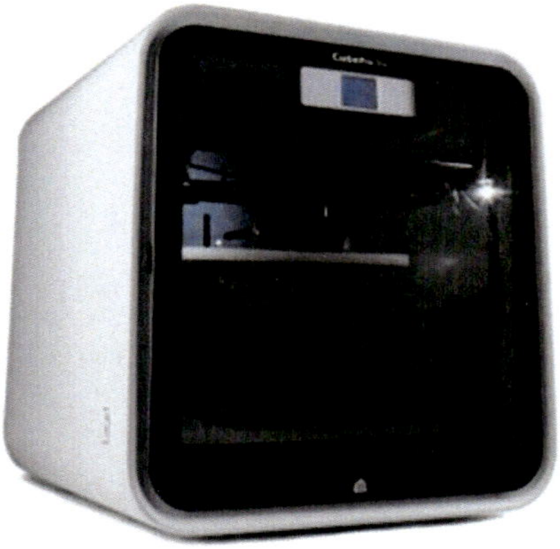

Fig. 3.2 CubePro Duo [12]

Table 3.3 Software manufacturing parameters

Print pattern	Print strength
Diamonds honeycomb	Strong Almost solid Solid

These specimens will have different compositions due to differences in the volume of existing material, however, the layer thickness (Layer Resolution), the type of interior fill (Print Strength) and the type of interior fill (Print Pattern) were defined with the same parameters in order to be able to characterize the support material more effectively.

Thus, among the parameters that the software allows us to choose, as can be seen in Table 3.3, the type of printing "Diamonds" was defined, which corresponds to a strong printing pattern with three directions of support in a cross and the printing force "Solid" which fills the layer with no empty spaces.

The definition of these parameters was based on the parameters that the software of the equipment allows to change, for this material it is not possible to modify the layer thickness (Layer Resolution) which is fixed for all specimens. In Fig. 3.3 is possible to observe the menu where is selected the desired printing parameters.

Thus, using the CubePro software, it is possible to simulate the construction of the model, simulate the support structures, as well as access the forecast of the time needed for its construction and the amount of material to be used in the process.

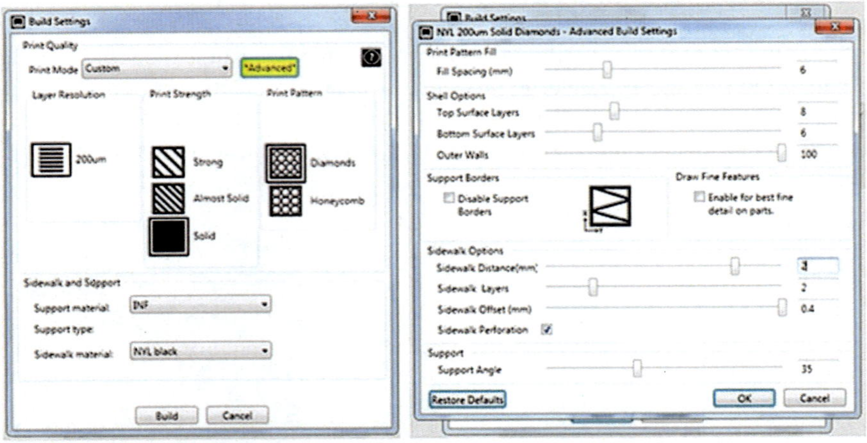

Fig. 3.3 Selection of printing parameters in the software

Fig. 3.4 Machine for compression tests

3.2.4 Compressive Tests

The mechanical compression tests were performed on the universal testing machine ZWICK Z100 (Fig. 3.4) of the Materials Laboratory of the School of Technology and Management in Polytechnic of Leiria.

An analytical balance was also used to obtain the mass values of each of the produced specimens.

These compression tests were performed according to ASTM D695 at a deformation speed of 1 mm/min and at room temperature. Each specimen was tested individually until it reached its breaking point, then the values of force [N] and displacement [mm] were analyzed for each case in order to obtain the force curve [N]—displacement [mm] as example from Fig. 3.5.

3.3 Results and Discussion

3.3.1 Production of Samples

Figure 3.6 shows the simulation of the fabrication of the specimens in two different views and in Fig. 3.7 the result actually obtained after their production. In this image it is possible to observe the specimens with 50% solid volume with support. Figures 3.8 and 3.9 show the specimens obtained, after production and post-processing, without support and with support, respectively. The specimens were constructed five at a time, that is, five of each type at a time.

In Fig. 3.10, it is possible to observe the specimens obtained with support material (top row) and those obtained without support material (bottom row).

As for the mass of the pieces produced, we can see in Fig. 3.11 that, as expected, the supported specimens have greater mass and the maximum difference in mass

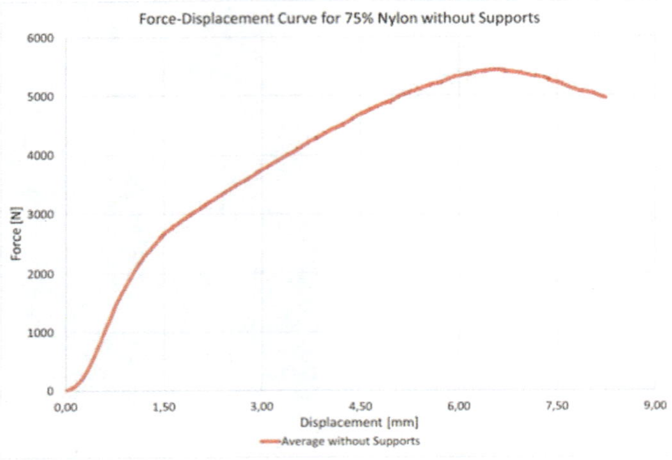

Fig. 3.5 Example of obtained force–displacement curve

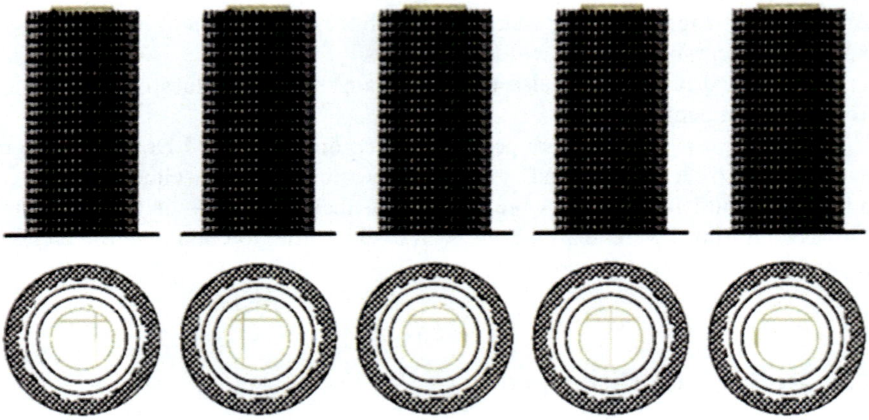

Fig. 3.6 Results in the software of the 50% nylon specimen with support

obtained between the supported and unsupported models was 0.4484 g and was verified in the 25% nylon specimens.

3.3.2 Compressive Tests

In Fig. 3.12 the appearance of the deformations that occur during the compression test is shown. In Fig. 3.13 and in Fig. 3.14, the force–displacement relationships for

Fig. 3.7 Result obtained after fabrication and post-processing of the 50% nylon specimens with support

Fig. 3.8 Obtained unsupported test pieces

the unsupported and supported 25% nylon specimens respectively from which the average curve was taken are shown.

By analyzing the average force–displacement curves and their comparisons between specimens with and without support material present in Fig. 3.15 to Fig. 3.17, it was possible, first of all, to verify that the mechanical behavior of nylon and infinity are very identical, since the force–displacement plots presented have very similar curves along the force that was applied, although with values for mechanical behavior very different. It was possible to observe that both materials have a

Fig. 3.9 Supported test pieces obtained

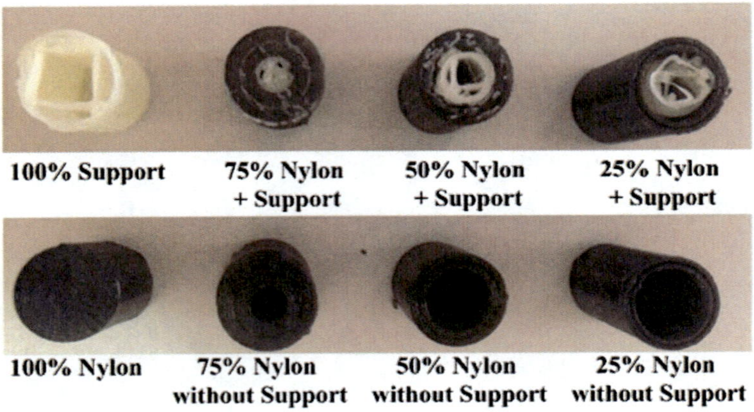

Fig. 3.10 Test pieces obtained from all types of configurations

high elastic domain and that this does not vary considerably with the introduction of support structures in the part.

In these figures, as the percentage of material (nylon) increases, the contribution of the support structure is to reduce the component's ability to resist compressive loads.

Figures 3.18 and 3.19 show the Force–Displacement curves for specimens with 100% Nylon and for 100% infinity respectively. It should be noted that the dispersion of results for the infinity specimens is very high. This fact is due to the construction of the test pieces in Nylon being a more dense structure than those observed for infinity. In the case of the construction of infinity specimens, the structure is much more porous and its behavior is greatly influenced by the geometric shape of this structure.

It can also be seen that the greater the volume of nylon relative to the volume of infinity, the more resistant is the tested part, as can be seen in Fig. 3.20 for

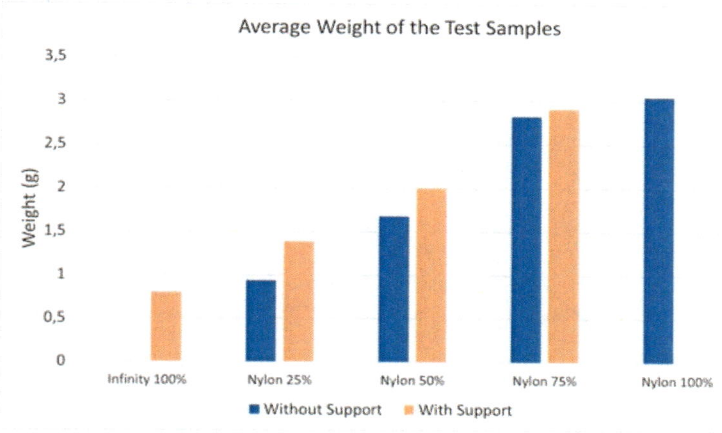

Fig. 3.11 Comparison of the masses of test pieces with and without support

Fig. 3.12 Deformations observed during the compression tests

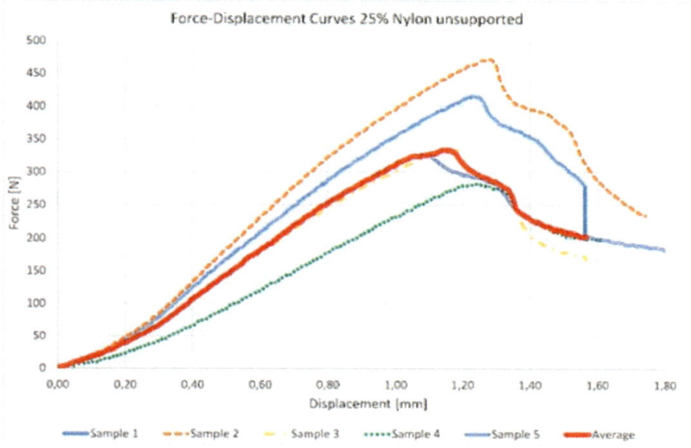

Fig. 3.13 Force–Displacement curves 25% Nylon unsupported

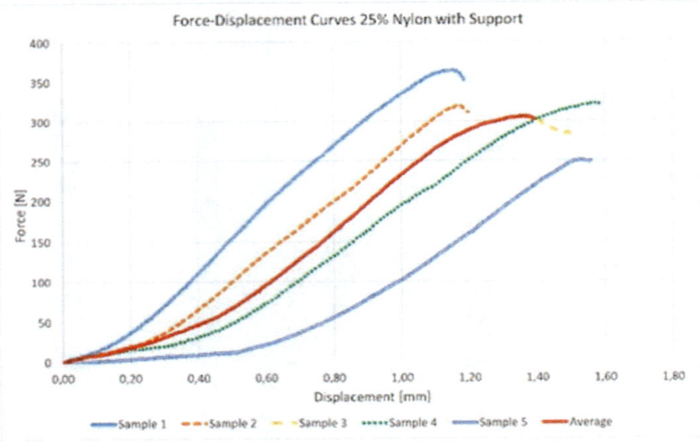

Fig. 3.14 Supported 25% Nylon force–Displacement curves

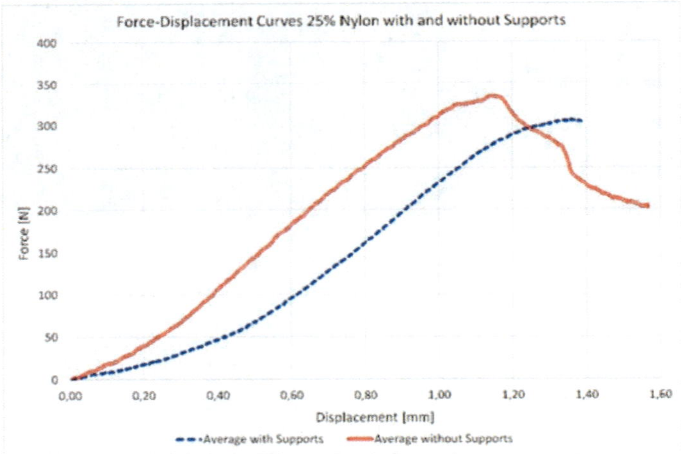

Fig. 3.15 Comparison average force–Displacement curves 25% Nylon with/without support

unsupported specimens and in Fig. 3.21 for supported specimens. This allows to confirm that nylon has higher mechanical strength when constructed autonomously than when constructed with supporting material structures.

As for the maximum load that the components resist, in Fig. 3.22 it is possible to visualize the differences between the maximum forces for all specimen configurations.

In the curves obtained in the compression tests, it was also possible to verify that, in most cases studied, the specimens with a support structure showed a lower compressive strength than the specimens with the same volume of nylon but without internal

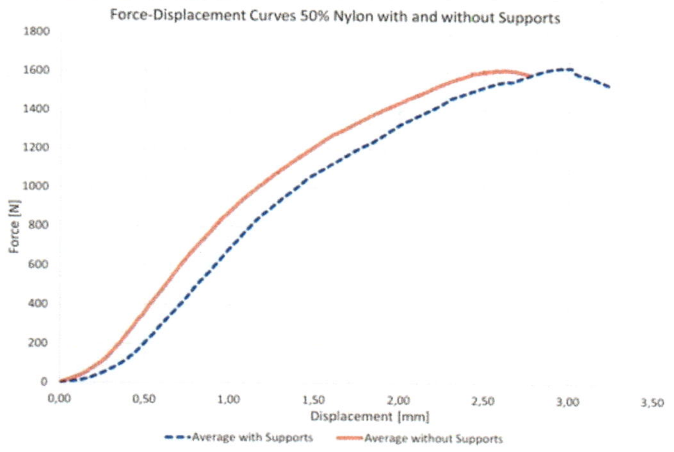

Fig. 3.16 Comparison average force–Displacement curves 50% Nylon with and without support

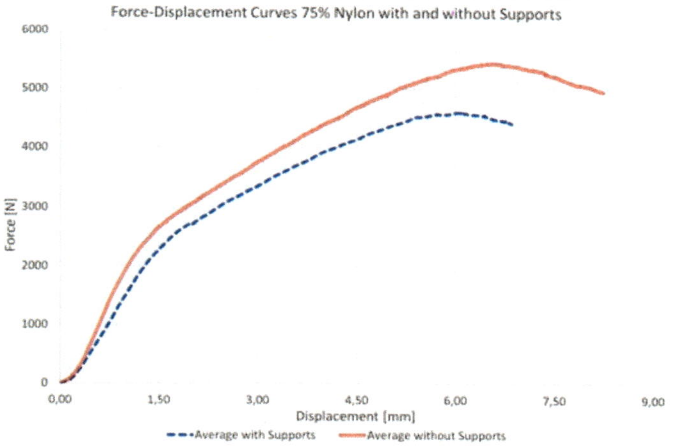

Fig. 3.17 Comparison average force–Displacement curves 75% Nylon with and without support

support. One of the possible causes for these results is the fact that the construction time of the supported specimens compared to the unsupported specimens is always higher; that is, each layer of construction is running longer due to the fact that two materials are used in the same layer. In this way, the first material deposited drops in temperature as support material is being deposited on the same layer, this can cause this layer to decrease the necessary adhesion to the layer that will be built later.

According to these data, it is possible to conclude that when creating a part produced by this process it is very important to predict whether or not the part

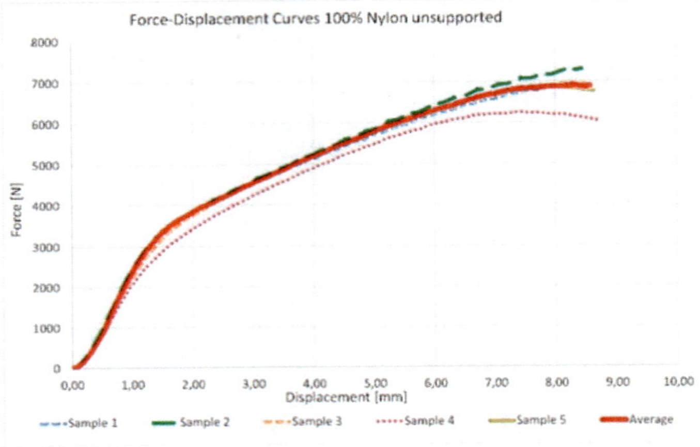

Fig. 3.18 100% Nylon force–Displacement curves

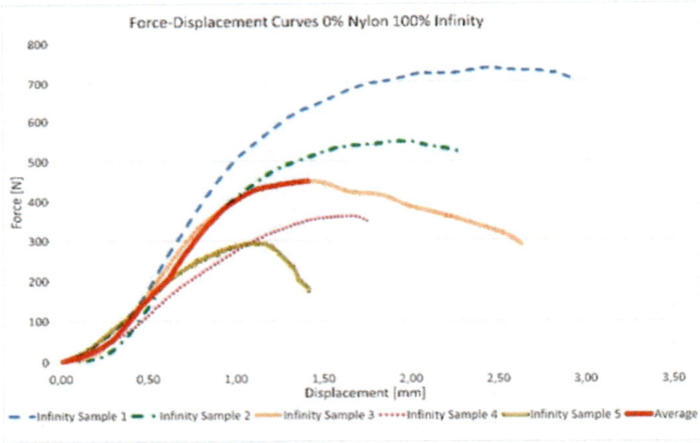

Fig. 3.19 Force–Displacement curves 0% Nylon 100% infinity

actually needs support because, as previously evaluated, if there is support material, the piece will take longer to produce, and this factor will have a negative impact on the mechanical strength of the final part produced.

To validate this evaluation, a new analysis was performed, Fig. 3.23, in which the maximum compressive strength was compared with the mass of each specimen. This comparison arose from the fact that it was not possible to calculate the yield and rupture stress of the specimens. This inability is due, as discussed earlier, to the way in which the support structure is generated that is not possible to control and thus it is not possible to effectively determine the area of the resistant section.

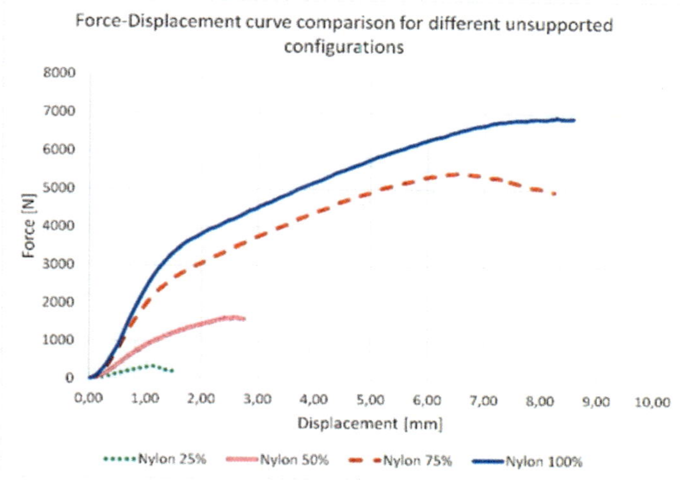

Fig. 3.20 Force–Displacement curve comparison for different unsupported configurations

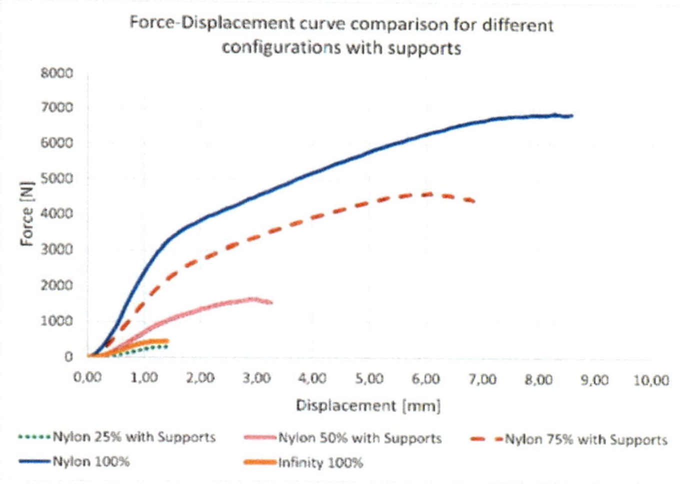

Fig. 3.21 Force–Displacement curve comparison for different supported configurations

It is to be noted in Fig. 3.23 that the compressive strength decreases by about 9% from Nylon 25% specimens to supported Nylon 25% specimens, while the F/m ratio decreases by 38%. Comparing the case of Nylon 50% specimens without and with support, although the resistant force increases residually with the existence of support (approximately 1%), the F/m ratio decreases by 15%. Finally, in the case of Nylon 75%, the resistant strength decreases by 15% while the F/m ratio decreases by 17%.

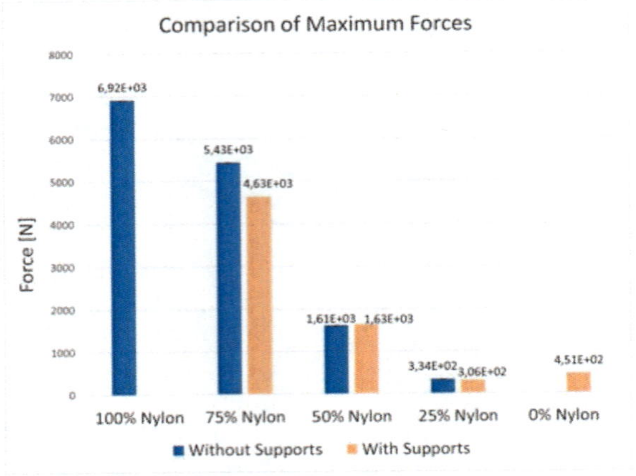

Fig. 3.22 Result of the maximum forces obtained

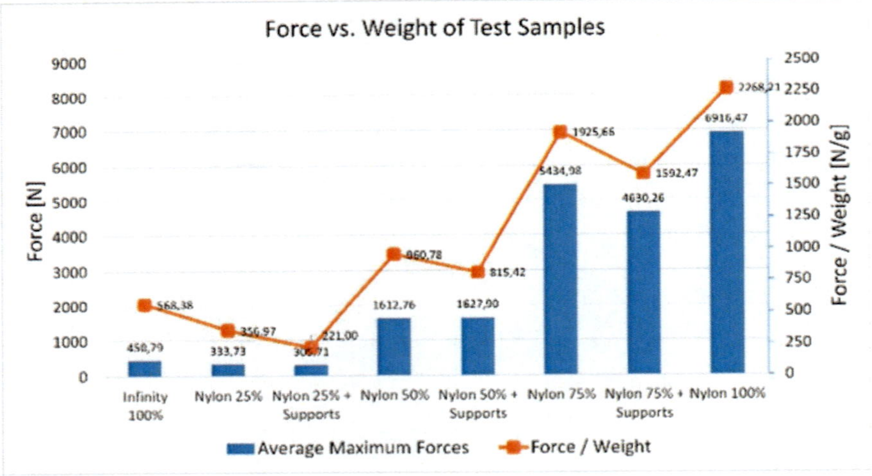

Fig. 3.23 Comparison of maximum force data with the specimen's weight

With this analysis it is clear that the contribution of the existence of supporting structures in components is negative for the mechanical strength of the components, due to the construction methodology using two materials and with two different extrusion heads that extrude material and alternate support. It is then suggested that construction strategies be adopted that eliminate the use of support structures, knowing that in some cases only using them the components are properly constructed and geometrically correct.

3.4 Conclusions

With this work it was possible to mechanically evaluate components produced by the FDM process with and without support structures. For this, several configurations of specimens without support material and others with support material were developed.

It was possible to analyse the additive production process, namely the FDM. In this process, it was verified that the mechanical resistance of the final parts produced is negatively influenced by the support structures. It is concluded that the support generated to successfully produce the final part tends to decrease the compressive strength of the component. As a cause of this decrease, the construction time may be longer due to the alternation of use of extrusion heads from one material (construction) to another (support), causing a decrease in adhesion between the constructed layers.

According to the data obtained, it can be concluded that the production planning of the parts produced by FDM technology is very important in terms of time reduction and final quality. Knowing that some geometries, in FDM, can only be generated using support structures. It is worth mentioning for this study the increase in component manufacturing time and the decrease in mechanical strength with the existence of this type of structures.

References

1. Galantucci, L.M.: Study of compression properties of topologically optimized FDM made structured parts. CIRP Annals—Manufacturing technology. (2008)
2. Chua, C.K., Leong, K.F.: 3D printing and additive manufacturing—principles and applications, 4th edn. World Scientific Publishing (2014)
3. Almeida, H.A., Correia, M.S.: Sustainability impact evaluation of support structures in the production of extrusion based parts. In: Muthu, S.S., Savalani, M.M. (Eds.), Handbook of Sustainability in Additive Manufacturing, Vol. 1, pp. 7–30. Springer (2016). ISBN 978–981–10–0549–7
4. Gibson, I., Rosen, D., Stucker, B.: Additive Manufacturing technologies—3D printing, rapid prototyping and direct digital manufacturing, 2nd edn. Springer, New York (2015)
5. Rosen, D.: What are the principles for design for additive manufacturing?. In: Proceedings of the 1st International conference on progress in additive manufacturing (Pro-AM2014), C.C. Kai et al (eds.), pp. 85–90. Research Publishing Services (2014)
6. Strano, G., Hao, L., Everson, R., Evans, K.: A new approach to the design and optimisation of support struture in additive manufacturing. Int. J. Adv. Manuf. Technol
7. Kotlinski, J.: Mechanical properties of commercial rapid prototyping materials. Rapid Prototyp. J. **20**(6), 499–510 (2014)
8. Sood, A.K., Ohdar, R.K., Mahapatra, S.S.: Experimental investigation and empirical modelling of FDM process for compressive strength improvement. J. Adv. Res. **3**(1), 81–90 (2012)
9. Majewski, C., Hopkinson, N.: Effect of section thickness and build orientation on tensile properties and material characteristics of laser sintered nylon-12 parts. Rapid Prototyp. J. **17**(3), 176–180 (2011)
10. Quintana, R., Choi, J.W., Puebla, K., Wicker, R.: Effects of build orientation on tensile strength for stereolithography-manufactured ASTM D-638 type I specimens. Int. J. Adv. Manuf. Technol. **46**, 201–215 (2010)

11. Kamrani, A.K., Nasr, E.A.: Rapid prototyping, theory and pratice. Springer, New York (2006)
12. 3D Systems.: 3D Systems. (2016) [Online] Available at http://www.3dsystems.com/ Accessed on 25 Aug 2016

Chapter 4
Mechanical Performance of Cellular Structures in Additive Manufacturing by Fused Deposition Modeling

A. Eljihad, M. Nassraoui, and O. Bouksour

Abstract The polylactic acid (PLA) that is one of the materials on which the technology (FDM) is based is mostly used in the additive manufacturing to fabricating different mechanical structures for various 3D printing specifications applications. In this paper, The polylactic acid (PLA) material is chosen to design three cellular structures in the form of lattice which are the diagonal form for two size cases 6*6*6 and 12*12*12, the octet form for two size cases 6*6*6 and 12*12*12, and the centered form for two size cases 6*6*6 and 12*12*12, these three cellular structures of are designed, optimized and simulated by Ansys software based of the finite element method for examining the its mechanical properties for the different sizes by utilizing the Gibson Ashby method and Hook's law. The mechanical properties of the proposed lattice structure sizes are evaluated by the tow mechanical parameters which are the modulus of elasticity and maximum stress by applying a numerical compression test. As result, the obtained results of the simulation show that the octet cell offer the good performance mechanical properties with high stiffness and low deformation, as well the minimize weight.

Keywords Additive manufacturing · Topological optimization · Mechanical properties

A. Eljihad (✉)
National High School of Electricity and Mechanics, University Hassan II of Casablanca, Casablanca, Morocco
e-mail: Abdelghani.eljihad-etu@etu.univh2c.ma

A. Eljihad · M. Nassraoui · O. Bouksour
Laboratory of Mechanics, Production, and Industrial Engineering, High School of Technology of Casablanca, University Hassan II of Casablanca, Casablanca, Morocco
e-mail: mohammed.Nassraoui@univh2c.ma

O. Bouksour
e-mail: Otmane.Bouksour@univh2c.ma

K. Zarbane and Z. Beidouri (eds.), *Proceedings of CASICAM 2022*, Springer Tracts in Additive Manufacturing, https://doi.org/10.1007/978-3-031-32927-2_4

4.1 Introduction

Additive manufacturing (AM) or 3D printing is a forming technology for producing three-dimensional parts. The object manufactured as a finished product is created by adding the material line by line, layer by layer, surface by surface or part by part [1]. This shaping technology has had a strong influence on industries such as automotive, aerospace, motorsports, energy and medical [2]. This innovation has totally changed the design and manufacturing paradigms of the company. Products can be customized and designed into complex shapes using one of the additive manufacturing technologies that offer unlimited design flexibility. The design for additive manufacturing aims to improve productivity in a way that the cost and manufacturing time should be minimal. while maximizing the intrinsic functionality parameters of additively manufactured parts [3]. The materials used for the deposition of layers on one of the technologies are metals, ceramics and synthetic polymers, with extreme temperature conditions. Each layer built is controlled by a computer-aided design (CAD) modeler that measures the initial parameters such as the trajectory of the tool of each product supports and orientation. The 3D printer manufactures thin layers this layer harden in the selected area using a thermal or chemical source by ultraviolet laser beam, solvent jet or binder to be controlled by computer [1].

Topological optimization is a computer-aided design method that automatically generates structural products with better characteristics and performance. Topological optimization is based on the distribution of materials in the discretized design domain [4]. The optimized product is not limited to its initial topology and can achieve improved performance compared to the initial state [4].

Topological optimization leverages the best design as an advantage and potential, and offers a broad design perspective for additive manufacturing, but it faces some challenges such as performance characterization, scaling effect in lattice structures, anisotropy, and material fatigue [5].

This paper focuses on characterization, scaling effect and cell types for lattice structures. Firstly, by simulating each optimized structure in a static way. Secondly, the elastic modulus of each model (octet, diagonal, centered) and scale (6*6*6 and 12*12*12) must be calculated with Hook's law. Finally, we determined the modulus of elasticity by the model of Gibson Ashby which is the closest for the lattice structures with its characteristics of form and scale [4].

This paper is organized as follows. Section 4.2 presents the methodology and analysis process of the proposed cellular structures by using the polylactic acid (PLA) material. In Sect. 4.3, the simulated results of mechanical properties of the lattice structure are presented and discussed. The conclusion is given in last Section.

4.2 Design Approach

The Design for Additive Manufacturing (DFAM) approach that aims to improve the properties of additively manufactured structures with minimization of manufacturing time and material waste, which influences the cost of the manufactured product. The latter favors the realization of functional and lightweight structures, such as lattice structures have explored the main production constraints depending on the technology used [3].

4.2.1 Methodology

The flowchart as shown in the Fig. 4.1 represents the different steps of the methodology that is used to the development of the cellular structures design [3]. details of each step are explained as following,

The design of structure with the prediction of the analytical of the model analysis is based to generate exactly the behavior of the lattice structures and the 3D printed of the different models. These designed specimens are realized in order to verify it's with tensile test by holding grips while one is fixed and other is moved. If these obtained results are significant, otherwise, the geometry parameters will modify.

Lattices have stochastic arrangements in the form of irregular cell geometry and non-periodic this structure can be plastically deformed until obtaining a high deformation as shock absorbers, or non-stochastic unit cells of periodic shape this type has various applications such as aeronautics. It has a very good strength/lightness ratio and meet the requirement of custom functionality. The lattice structures based on lattice have a better capacity to resist the load and adaptable mechanical property.

4.2.2 Prediction of the Mechanical Response

The behavioral law was developed by advanced estimation methods to predict the mechanical characteristics of lattice structures based on the unit cells used.

The normal case hook law of the equation below will be used:

$$E_{Hook} = \frac{\sigma}{\varepsilon} = \left(\frac{F/_{S}}{\Delta l/_{l}} \right) \tag{4.1}$$

with F the force in (N), S is the area in (mm^2), Δl (mm) is the elastic deformation of the samples and l (mm) is the initial length of the sample before deformation [3].

But the Gibson-Ashby model is the most notable for giving an adequate mechanical response [6].

The calculation of the modulus of elasticity includes the Gibson-Ashby scaling law as well as other parameters such as: the unit cell, the mechanical property of solid structures and the relative density. It is different to the simple elasticity model by the behavior law, because the properties of porous structures depend on the mechanical response dominated by bending or stretching [3].

$$E_{lat} = C E_{sol} \left(\frac{\rho_{cell}}{\rho_{sol}} \right)^n \tag{4.2}$$

E_{lat} and E_{sol} are the elastic moduli of the optimized structure the overall solid structure. ρ_{cell}/ρ_{sol} are the relative densities of the lattice structure, n is a coefficient dependent on the cellular structures if dominated by the stress of a bending or stretching load.

The Gibson-Ashby scaling law uses the practical case of elastic force legs after the correlation coefficients have been properly determined. These are calculated as a function of the structure and material used. The methodology used is a non-linear regression method to identify the different constants 'c' and 'n' of the equation.

if we replace the relative modulus of elasticity (E_{lat}/E_{sol}) and density (ρ_{cell}/ρ_{sol}) by the variables Y and X respectively, this equation becomes of the following form:

$$Y = C X^n \tag{4.3}$$

using a numerical gauss–newton method, we obtain the best fit values as follows:
n = 2 and c = 1.15 for octet lattice structures.
n = 1 and c = 0.5 for Diagonal lattice structures.
n = 2 and c = 0.67 for centered lattice structures.

4.2.3 Cellular Structure Analysis Process

Lattices have stochastic arrangements in the form of irregular cell geometry and non-periodic this structure can be plastically deformed until obtaining a high deformation as shock absorbers, or non-stochastic unit cells of periodic shape this type has various applications such as aeronautics. It has a very good strength/lightness ratio and meet the requirement of custom functionality. The lattice structures based on lattice have a better capacity to resist the load and adaptable mechanical properties. Samples of cubic lattice of 24 × 24 × 24 must be manufactured by FDM technology with polyactic filament (PLA). On the table below have six configurations on which the lattice structure of the samples digitized by ANSYS software calculates the mechanical parameters: stresses and strains of geometries [7]. Based on a compression test using the static structural modulus, the lattice structures are meshed by adaptive mesh elements. According to the predefined displacement of the movable crossbeam by 10 mm and fix the other side of the cube by the fixed crossbeam.

4.3 Results and Discussion

The modulus of elasticity or Young's modulus determines the stiffness of the lattice structure under load. For this purpose, calculating this modulus by hook's law and Gibson-Ashby's law to compare the modulus of elasticity of these lattice structures (octet–diagonal–centered).

Based on the mechanical properties of the PLA material in Tables 4.1 and 4.2 to verify and compare with the results obtained by the theoretical method.

The figure below clearly shows the difference between the six structures:

PLA is the most widely used material in melt deposition technology, hence the need to characterize this material for 3D printing applications in the various fields.

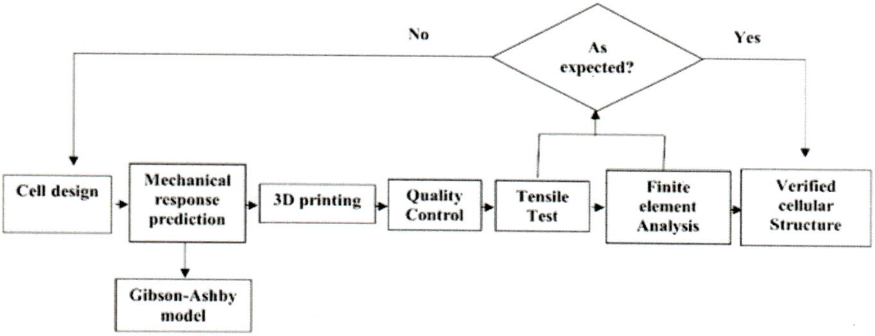

Fig. 4.1 Stages of design and development of cellular structures

Cellular structures with predefined PLA material were designed with two scales 6*6*6 and 12*12*12 of structures (diagonal—octet -centered) to evaluate with the theoretical method by Hook's law and Gibson Ashby's law.

From the results obtained, which generate complex structures but with adequate mechanical properties according to the desired optimization of mass [4]. The Ashby method is proportional to te elasticity model of the PLA material, which illustrates the crucial parameters of lattice structures. The simulation method used in this work demonstrated: the maximum stress and strain of Von mis. The results obtained by the calculations based on Hook's law will be compared to those predicted by Gibson's Ashby model which is most like the elastic modulus of the material [9].

The results detailed in Fig. 4.2 and Table 4.1 clearly show that the mechanical properties of the samples with PLA material modeled by ANSYS software at 10^{-3} m displacement are completely different due to the type of structure used and the orientation of the unit cell [10].

The diagonal cell of size 12*12*12 withstands higher stress than other types of cells. Therefore, octet structures have very high stiffness with low deformation. So, we choose the octet structure in case the stiffness/weight ratio is required.

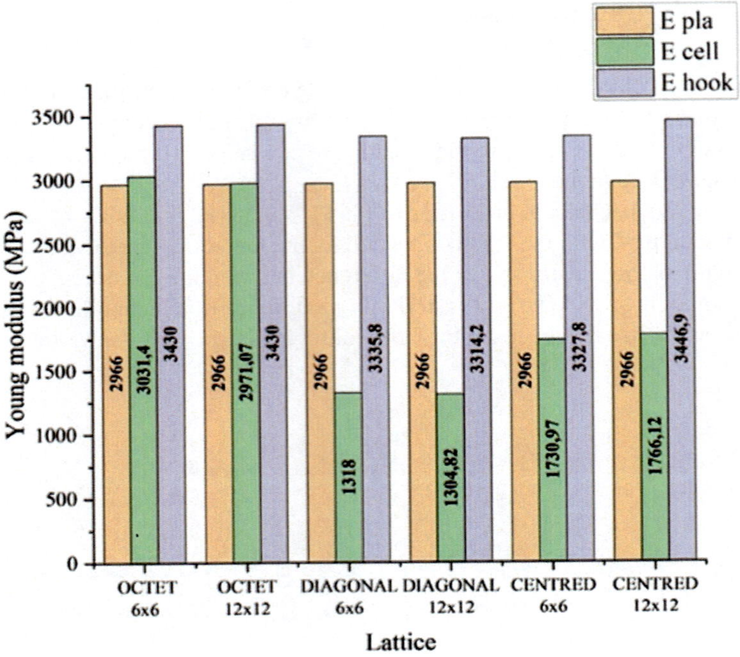

Fig. 4.2 Comparison the modulus elasticity according to the Gibson-Ashby and Hooke models

Finally, it is found that the elastic modulus varies with the type of cell, the dimensions of the cells and the ratio of the desired mechanical properties.

4.4 Conclusions

In this paper, the PLA polymer-based lattice structures are proposed, studied, designed and simulated for considering as design resources for additive manufacturing. Octet, diagonal, centered type of the lattice structures have been designed, simulated by Ansys software based on the finite element method. The simulation of the three types which is based on the numerical study and the Hook's law and the Gibson-Ashby model have been done to characterize the mechanical properties as the modulus of elasticity and maximum stresses. The discussion and analysis of the obtained results confirm the cell type and its own size having a strong influence on the mechanical properties such as the modulus of elasticity. According to the values of the modulus of elasticity, the Octet cell has been chosen as the most rigid with miniature size. The proposed octet cell fabricated with the polylactic acid (PLA) with good mechanical characteristic's will be a very attractive for the technology (FDM) that haves a strong influence on industries such as domains automotive, aerospace,

Table 4.1 Comparison of lattice structures by mechanical characteristics and mass

Cell type	Unit cell size	Cellular model	Shear stresses (Pa)	Total deformation (m/m)	Mass (gr)
 Octet	6× 6×6	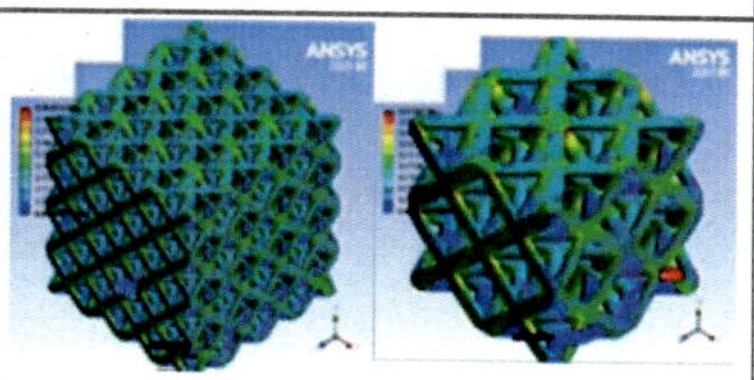	Min: 4,3353e + 5 Max: 2,403e + 8	Min: 2,7807e-4 Max: 6,9934e-2	4,4352
	12 × 12 × 12	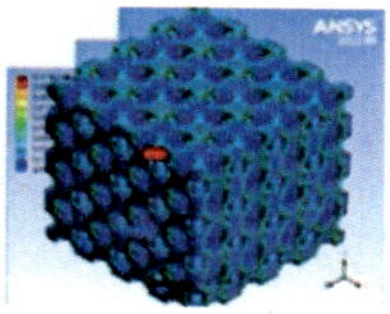	Min: 3,0194e + 5 Max: 1,9913e + 8	Min: 1,8623e-4 Max: 5,792e-2	4,4341
Centered 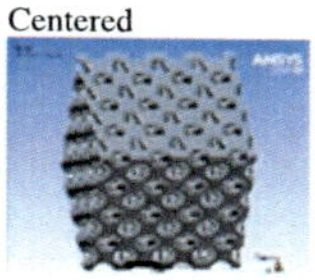	6 × 6×6	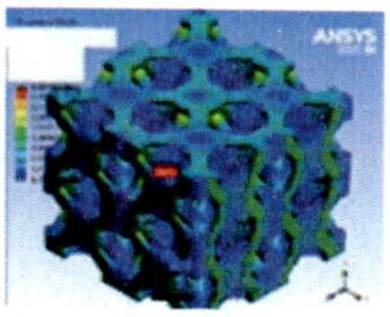	Min: 5593,2 Max: 3,8573e + 8	Min: 3,0178e-6 Max: 0,11,563	7,5132
Centered 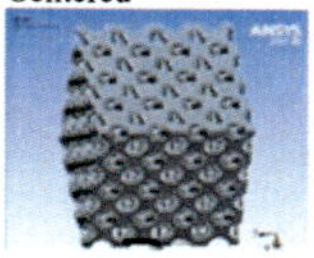	12× 12× 12	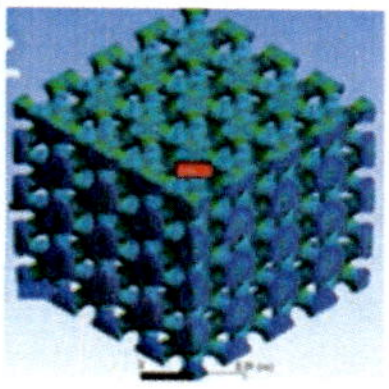	Min: 9589,8 Max: 3,0459e + 8	Min: 5,1507e-6 Max: 9,1902e-2	7,5129
Diagonal 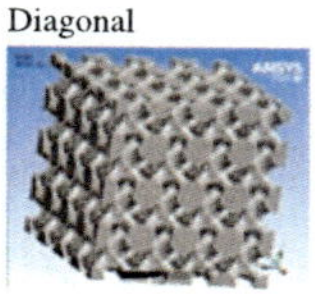	6× 6× 6		Min: 1,3273e + 6 Max: 3,8889e + 8	Min: 1,0042e-3 Max: 0,11,686	7,542
	12 × 12× 12	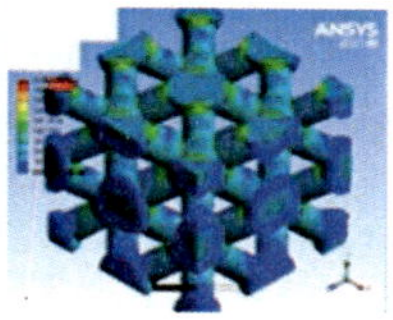	Min: 8,4373e + 5 Max: 3,6703e + 8	Min: 2,4456e-4 Max: 0,10,648	7,2091

Table 4.2 Mechanical properties of PLA material [8]

Property	Standard	Unit	Value
Tensile Strength	ASTM D638	MPa	26–45
Elongation at break	ASTM D638	%	1,0–2,5
Modulus of elasticity	ASTM D638	MPa	2539–3039
Flexural strength	ASTM D790	MPa	45–84
Heat Deflection	ASTM D648	∘C	53

motorsports, energy and medical. As the future work, each type of the lattice structures which are octet, diagonal and centered will be fabricated for the size of 6*6 and 12*12, the experimental results will be tested with compression and compared with other research already published.

References

1. Sandhu, G.S., Boparai, K.S, Sandhu, K.S.: Influence of slicing parameters on selected mechanical properties of fused deposition modeling prints. In: Mater. Today Proc. Elsevier Ltd, pp 1378–1382 (2021)
2. Ouhsti, M., El Haddadi, B., Belhouideg, S.: Effect of printing parameters on the mechanical properties of parts fabricated with open-source 3D printers in PLA by fused deposition modeling. Mech Mech Eng **22**, 895–907 (2018)
3. Eren, O., Sezer, H.K., Yalçın, N.: Effect of lattice design on mechanical response of PolyJet additively manufactured cellular structures. J Manuf Process **75**, 1175–1188 (2022)
4. Wu, J., Sigmund, O., Groen, J.P.: Topology optimization of multi-scale structures: a review. Struct.. Multidiscip. Optim **63**, 1455–1480 (2021)
5. Boyard, N. : Méthodologie de conception pour la réalisation de pièces en Fabrication Additive
6. du Plessis, A., Razavi, S.M.J., Benedetti, M., Murchio, S., Leary, M., Watson, M., Bhate, D., Berto, F.: Properties and applications of additively manufactured metallic cellular materials: A review. Prog Mater Sci **125**, 100918 (2022)
7. Sun, Z.P., Guo, Y.B., Shim, V.P.W.: Deformation and energy absorption characteristics of additively-manufactured polymeric lattice structures—Effects of cell topology and material anisotropy. Thin-Walled Struct **169**, 108420 (2021)
8. Stratasys Inc.: A Global Leader in Applied Additive Technology Solutions, (2017)
9. du Plessis, A., Yadroitsava, I., Yadroitsev, I., le Roux, S.G., Blaine, D.C.: Numerical comparison of lattice unit cell designs for medical implants by additive manufacturing. Virtual. Phys. Prototyp **13**, 266–281 (2018)
10. Salem, H., Abouchadi, H., Elbikri, K.: PLA Mechanical Performance Before and After 3D Printing. Int J Adv Comput Sci Appl **13**, 324–330 (2022)

Chapter 5
Effects of Build Orientation and Raster Angle on Surface Roughness and Mechanical Strength of FDM Printed ABS

Adil El Azzouzi, Hamid Zaghar, Mohammed Sallaou, and Larbi Lasri

Abstract Build orientation and raster angle are critical factors that significantly influence mechanical behavior and surface roughness. Fused deposition modeling (FDM) is used to construct a standard specimen on acrylonitrile butadiene styrene (ABS) filament in order to examine the effects of this parameter. This included considering three construction orientations and three raster angles. As a consequence, the findings indicate that the mechanical strength increases with decreasing raster angle and by aligning from upright to flat orientation due to the fracture mechanism and loading direction. Additionally, there was a big difference in the surface roughness depending on the manufacturing orientation and raster angle; perpendicular measurements increase surface roughness values. The purpose of current work is to complete tensile experiments to investigate the impact of build orientation and raster angle on surface quality and mechanical strength. Our knowledge of appropriate printing angles and construction orientation will improve as a result of the findings of this inquiry.

Keywords FDM · Build orientation · Raster angle · Mechanical strength · Surface roughness

---

A. E. Azzouzi (✉) · M. Sallaou · L. Lasri
Moulay Ismail University, Ensam Meknes, Morocco
e-mail: elazzouziadil1@gmail.com

M. Sallaou
e-mail: mosallaou@yahoo.fr; m.sallaou@ensam.umi.ac.ma

L. Lasri
e-mail: l.lasri@ensam.umi.ac.ma

H. Zaghar
Faculty of Sciences and Techniques, Sidi Mohammed Ben Abdellah University, Fez, Morocco
e-mail: hamid.zaghar@usmba.ac.ma

K. Zarbane and Z. Beidouri (eds.), *Proceedings of CASICAM 2022*, Springer Tracts in Additive Manufacturing, https://doi.org/10.1007/978-3-031-32927-2_5

5.1 Introduction

Additive manufacturing is a rapidly growing technology in a number of industries. Therefore, FDM is the most widely used technology for manufacturing products. In addition, the materials used by this process are characterized by important mechanical properties and don't change with time or intoxication [1]. However, the uncertainty of the printing parameters and the printing process influence the quality of the mechanical properties of the printed parts. At this stage, the materials and the process used have not been studied in a systematic way in order to obtain functional parts, with adjusted mechanical properties, or with the objective of achieving a competitive production time/cost (for small and medium production series) [1–3].

Additionally, the materials employed in this technique are distinguished by significant mechanical qualities and do not alter with time or exposure to the environment [4]. A 3D printer first transforms a computer model created with CAD software into a real item [5]. It is possible to utilize a support material if the sections have shapes that hang over the edge. A strong digital chain, including the many steps from design to additive manufacturing of the items, is followed in the development of the components processed by the FDM technique or other AM techniques. This chain includes planning, getting ready and cutting, simulating the procedure, etc. FDM technique is a thermal process that uses a heating system to melt the raw material (plastic material).After having been melted; the material is deposited by a nozzle of extrusion on the mobile printing table. And since the extruded filament is viscous, the operation of bonding it to the preceding layer becomes easy. Thus, the exposure of the material to the ambient temperature allows its cooling which leads to the manufacture of the final geometry of the product [6, 7]. The printer of the FDM technique admits several parameters having dominant roles in obtaining a good product, the key elements of which are: the layer thickness, the dimensions of the deposited filaments, the orientation of the parts… Optimal values of these parameters is a crucial step to ensure product quality, improve mechanical and aesthetic properties and reduce printing time and cost. In terms of feedstock, the most assessed materials in the available literature are ABS and PLA thermoplastic materials. Moreover, the impact of layer thickness and percent infill on tensile characteristics is generally recognized in the literature, but for build direction and raster angle varied results have been found. Raut et al. [8] investigated the impact of build orientation over the tensile, flexural strength and total cost. The optimal tensile strength and total cost could be obtained at Y-axis 0° build orientation while the optimal flexural strength could be reached at X-axis 0° orientation. Hernandez et al. [9] examined the effects of various orientations (build angles). The results showed that print 90° in the XY plane provided the highest tensile strength. It was also found that print 0° in the XY plane enhanced significantly the compressive and flexural strengths in comparison to other orientations. Lanzotti et al. affirm that tensile strength reduce as printed angle augment for ABS [10, 11]. 45° raster orientation gives a strongest material behavior according to Lecther [12]. The impact of raster orientation in relation to building orientation for ABS filament was also examined by Durgun et al. [13]. Their research demonstrated that for flat and on-edge directions, 0° raster angle revealed the best mechanical qualities, however upright orientation displayed conflicting findings compared to those of the other two construction orientations. Hernandez et al. came to the opposite conclusion, claiming that changing the raster orientation had no discernible impact on the tensile strength of ABS [14]. Also Naveed [16] confirmed that the 45° raster orientation produced a strongest specimen with the average ultimate tensile is 55.45 MPA and highest elongation of 4.24%. Moreover, Ashtankar et al. [17] showed that when the samples were aligned from 0° to 90°, the compressive strength and tensile strength of the ABS parts decreased. The most significant factors used to evaluate the quality of the chosen direction are surface. Studies show that the most influential parameter on the surface finish is the layer thickness followed by the width of the deposited filaments and the orientation of the part [18], while the latter parameter has a great impact on the dimensional accuracy [19, 20]. Low values of layer thickness as well as wide filament widths allow for a decrease in roughness while moderate values of raster width and 0° orientation reduce dimensional errors. The void between the deposited filaments is also a problem in the final part where it has more impact on the dimensional accuracy than the surface finish.

Because of the variation in results and the major limitation of the methods described in the literature to choose the optimal build direction, this study examined the impact of build orientations with three raster angles (0°, 45°, 90) on the mechanical properties and surface roughness in order to choose the optimal orientation and raster angle.

5.2 Fused Deposition Modeling

The fused deposition modeling is a thermal process that uses a heating system to melt the raw material (plastic) to form filaments (Fig. 5.1) after being melted; the material is deposited by a nozzle. Extrusion on the mobile printing table. And since the extruded filament is viscous, the operation of gluing it to the preceding layer becomes easy. Thus, exposing the material to room temperature allows it to cool, which leads to the fabrication of the final geometry of the product.

5.3 Materials and Methods

ABS (acrylonitrile butadiene styrene) is the substance utilized in this investigation; more information about ABS can be found in Table 5.1. Additionally, we utilized the Ultimaker 3 device to create the components (Fig. 5.2).To get a mean value for the observed traits and attributes, such as ultimate tensile strength and surface roughness, three identical tensile specimens are evaluated with each raster. Table 5.2 displays the printing parameters applied in this experimental attempt.

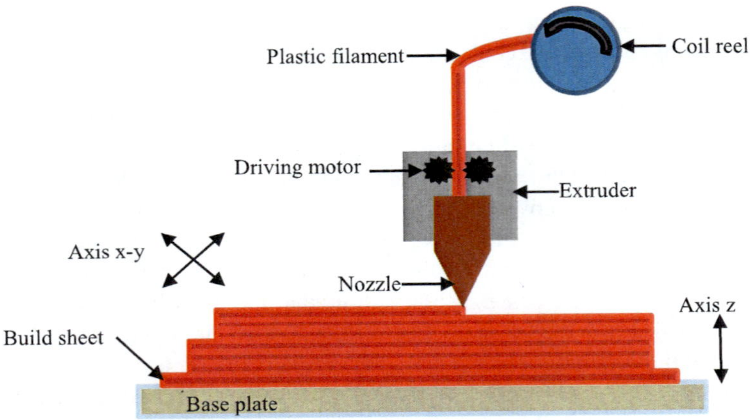

Fig. 5.1 FDM process illustration

Table 5.1 Printing temperatures and filament densities

Material	Volume (cm3)	Mass (g)	Nozzle temperature [°C]	Bed temperature [°C]	Apparent density (g/cm3)
ABS filament	5.27	5.8	190	55	1.10

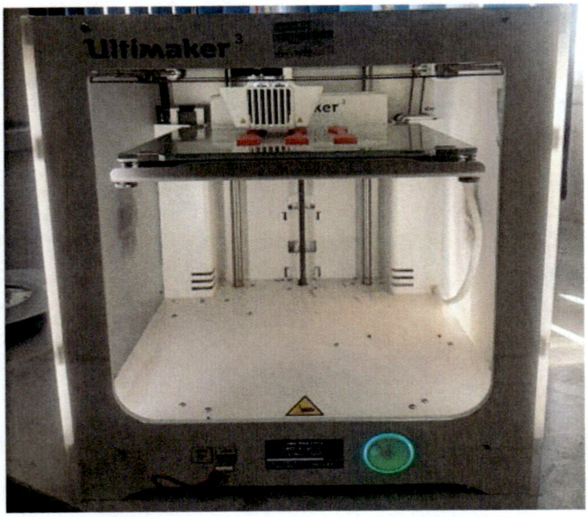

Fig. 5.2 Ultimaker 3 3D printer

Table 5.2 Printing parameters

Parameters	Value
Layer thickness	0.3 mm
Extrusion width	0.5 mm
Nozzle diameter	0.4 mm
Printing speed	60 mm/s

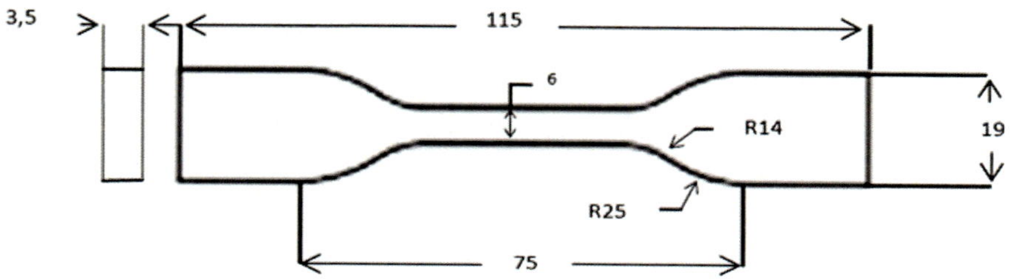

Fig. 5.3 Nominal dimensions according to the ASTM D638 standard

The specimen's geometry matched the requirements of ASTM D638 for the stress components [15], which is published by the American Society for Testing and Materials. The appropriate millimeter dimensions are shown in Fig. 5.3.

The test specimen's virtual three-dimensional geometry was designed using the SolidWorks 3D CAD program, translated to STL format, and then prepared for 3D printing using Cura. Table 5.3, Fig. 5.4 give parameters values used to the samples; for each direction and raster angle specimens.

For the tensile test, DETLAB equipment was used. The Test specimens were run at 5 mm/min. EM506 software was used to set up specimens for this test machine. As part of this software, displacement (mm) and force (N) values are also recorded in order to generate stress–strain curves for each test. Figure 5.5 shows an arrangement for performing the tensile test on 3D-printed specimens. For each sample, surface roughness measurements were taken parallel and perpendicular to the tensile stress direction.RT-90G roughness testers were also used to measure roughness.

Table 5.3 Sample processing parameters specification

Materiel	Build orientation	Raster angle		
ABS	Flat	0°	45°	90°
	On edge	0°	45°	90°
	Upright	0°	45°	90°

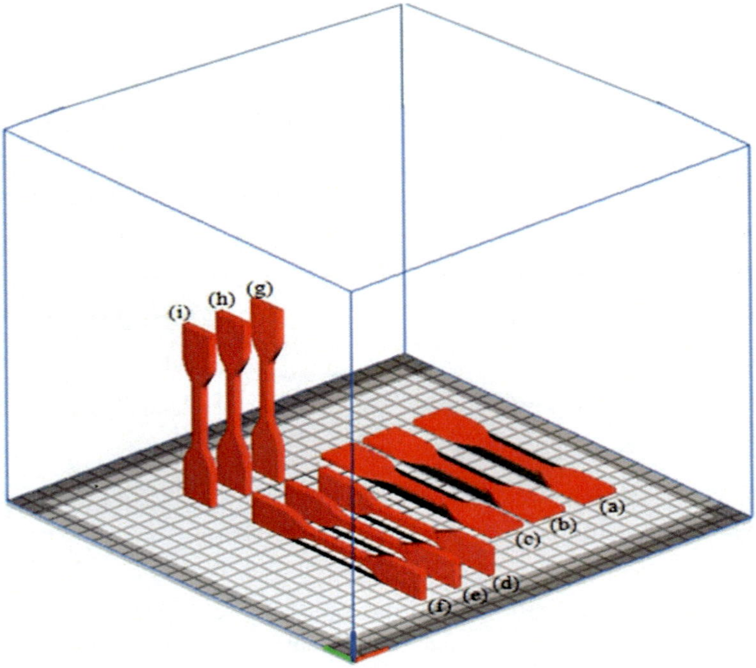

Fig. 5.4 Tensile samples with different construction orientations and printing angles **a** flat 0°, **b** flat 45°, **c** flat 90°, **d** on edge 0°, **e** on edge 45°, **f** on edge 90°, **g** upright 0°, **h** upright 45°, **i** upright 90°

Fig. 5.5 An example of a tensile test set up for 3D-printed specimens

5.4 Results and Discussion

5.4.1 Tensile Test Results

Figure 5.6 presents engineering stress–strain curves for each filament for the tree construction orientations of flat, on edge, upright, and 0°, 45°, and 90° raster. The average values of the mechanical parameters, stress and elongation, were shown by each curve.

As seen in Fig. 5.6, ABS has the maximum tensile strength (45 MPA) when oriented flat (0°), whereas ABS has the lowest tensile strength when oriented upright (17 MPa). Because the tensile strength of FDM 3D-printed items becomes optimum when the pieces are entirely oriented in the direction of loading stress. Furthermore, the largest UTS were observed in 0° raster orientation and it continuously decreased when the angle was raised in terms of print speed. The filament itself can bear the tensile stress at 0°, despite the bond strength decreasing as the raster orientation increases from 0° to 90°.Although raster angle fluctuation was also seen in elastic modules, it was not as noticeable as UTS and fracture strain. One may claim that raster angles at zero degrees displayed more elastic modules. It may be noted that ABS was determined to be favorable with optimum values and its simplicity of fabrication when all mechanical features were considered. The mechanical performance of ABS parts depends on the breaking mode, which comes in two varieties: interlayer fracture and interlayer fracture.

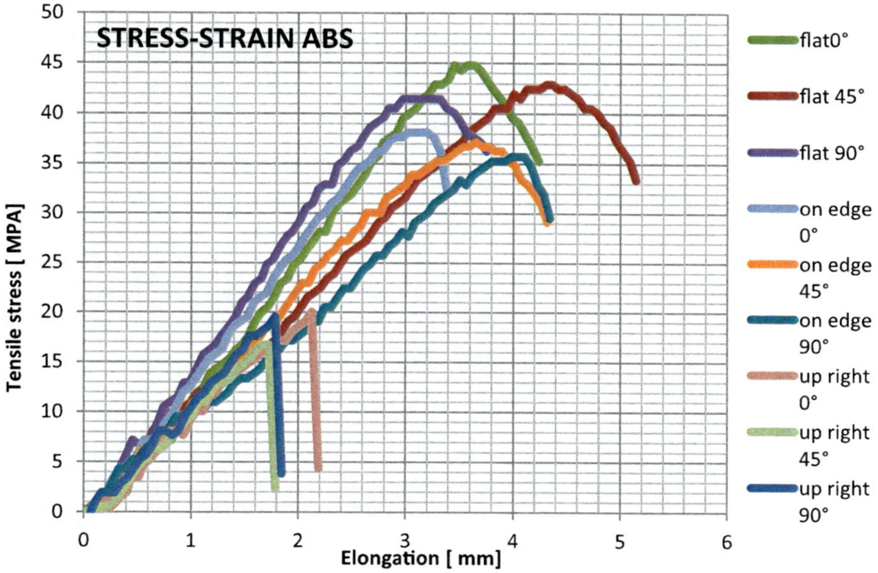

Fig. 5.6 The stress–strain curves for ABS with different building orientations and raster angle

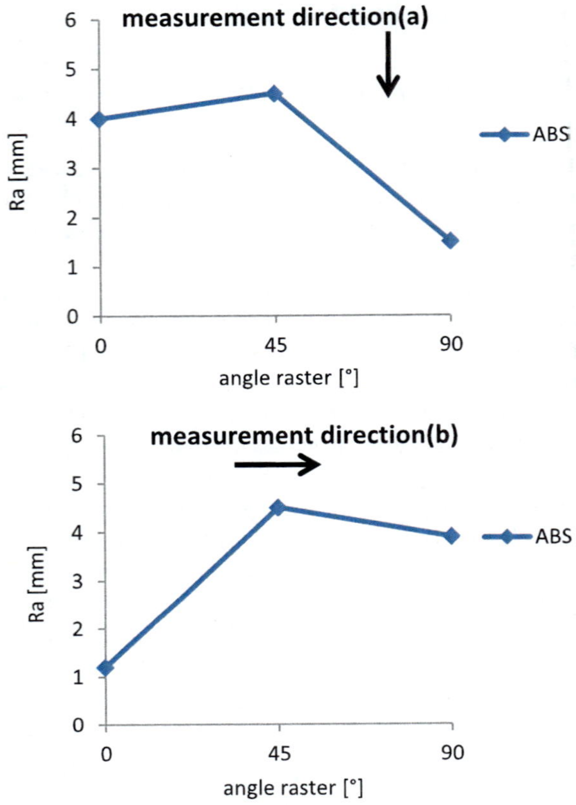

Fig. 5.7 Measurement of surface roughness R_a: **a** Perpendicular direction, **b** Parallel direction

5.4.2 *Surface Roughness Result*

The influence of raster angle on surface roughness was evaluated. Figure 5.7 illustrates these values in terms of surface roughness. It can be seen from this figure that surface roughness Ra values where lowest when the measurement direction was parallel to tensile loading. Consequently 0° and 90° raster angle give lowest values for parallel and perpendicular direction respectively. Also, the maximum value is obtained at 45°.

5.5 Conclusion

The results showed that when the construction orientation was changed from horizontal (flat) to vertical (upright) and the raster angle was changed from 0° to 90°, the tensile strength reduced. Surface roughness levels were at their lowest when the measurement direction was parallel to the raster direction or 45°. A multi-objective optimization approach is required to determine the appropriate construction direction that offers the best trade-off between tensile strength, and surface roughness.

References

1. Available: https://www.statista.com/statistics/560304/worldwide-survey-3d-printing-top-technologies/. Most used 3D printing technologies 2020 Statista. [Online]. Accessed 20 Jan 2021
2. Altan, M., Eryildiz, M., Gumus, B., Kahraman, Y.: Effects of process parameters on the quality of PLA products fabricated by fused deposition modeling (FDM): surface roughness and tensile strength. Mater. Test. **60**(4), 471–477 (2018). https://doi.org/10.3139/120.111178
3. Bardiya, S., Jerald, J.: Satheeshkumar, V, Effect of process pa-rameters on the impact strengthof fused filament fabricated (FFF) polylactic acid (PLA) parts. Mater. Today: Proc. (2020). https://doi.org/10.1016/j.matpr.2020.08.066.Bardiya,S.,Jerald,J.,Satheeshkumar,V
4. Mohamed, O.A., et al.: Optimization of fused deposition modeling process parameters: a review of current research and future prospects. Adv. Manuf. 40–54 (2015). https://doi.org/10.1007/s40436-014-0097-7
5. Utela, B., Storti, D., Anderson, R., et al.: A review of process development steps for new material systems in three dimensional printing (3DP). J. Manuf. Processes. **10**(2), 94–107 (2008). https://doi.org/10.1016/j.jmapro.2009.03.002
6. Guessasma, S., et al.: Challenges of additive manufacturing technologies from an optimization perspective. Int. J. Simul. Multisci. Des. Optim. (2015). https://doi.org/10.1051/smdo/2016001
7. Turner et al.: A review of melt extrusion additive manufacturing processes: II. Materials, dimensional accuracy, and surface roughness, Rapid Prototyping J., 245–265 (2015). BN. https://doi.org/10.1108/RPJ-02-2013-0017
8. Raut, S., Jatti, V.S., Khedkar, N. K., Singh, T. P.: Investigation of the effect of built orientation on mechanical properties and total cost of FDM parts. MSPRO, **6**(Icmpc), 1625–1630 (2014). https://doi.org/10.1016/j.mspro.2014.07.146
9. Hernandez, R., Slaughter, D., Whaley, D., Tate, J., Asiabanpour, B.: Analyzing the tensile,compressive, and flexural properties of 3D printed ABS P430 plastic based on printing orientation using fused deposition modeling. An Addit. Manuf. Conf. SFF 2016, pp. 939–950, (2016). https://hdl.handle.net/2152/89645
10. Es-Said, O.S., Foyos, J., Noorani, R., Mendelson, M., Marloth, R., Pregger, B.A.: Effect of layer orientation on mechanical properties of rapid prototyped samples. Mater. Manuf. Process. **15** (1), 105–125 (2000)
11. Lanzotti, A., Grasso, M., Staiano, G., Martorelli, M.: The impact of process parameters on mechanical properties of parts fabricated in PLA with an open-source 3-D printer. Rapid Prototyp. J. **21**(5), 603–618 (2015). https://doi.org/10.1108/RPJ-09-2014-0135
12. Material property testing of 3D printed specimen in PLA on an entry level 3D printer. In: Proc. of the ASME 2014, International mechanical engineering congress exposition IMECE2014, pp. 1–8. Montreal, Quebec, Canada (2014) T. Letcher and M. Waytashek. https://doi.org/10.1115/IMECE2014-39379
13. Durgun, I., Ertan, R.: Experimental investigation of FDM process for improvement of mechanical properties and production cost. Rapid Prototyp. J. **20**(3), 228–235 (2014). https://doi.org/10.1108/RPJ-10-2012-0091
14. Hernandez, R., Slaughter, D., Whaley, D., Tate, J. and Asiabanpour, B.: Analysing the tensile, compressive, and flexural properties of 3D printed ABS P430 plastic based on printing orientation using fused deposition modelling. In: Proc. of the 26thAnnual International solid free form fabrication symposium-an additive manufacturing conference, pp. 939–950. Austin, Texas, USA (2016). https://doi.org/10.1007/s12206-021-0708-8

15. ASTM Standard D638.: Standard test methods for tensile properties of plastics. ASTM International, West Conshohocken, PA (2010)
16. Naveed, N.: Investigate the effects of process parameters on material properties and microstructural changes of 3D-printed specimens using fused deposition modelling (FDM). Materials Technology. Adv. Perform. Mater., 12 (2020). https://doi.org/10.1080/10667857.2020.1758475
17. Ashtankar, K.M., Kuthe, A.M., Rathour, B.S.: Effect of build orientation on mechanical properties of rapid prototyping (Fused Deposition Modeling) made Acrylonitrile Butadiene Styrene (ABS) Parts. In: ASME 2013 International mechanical engineering congress and exposition, vol. 11, pp. 1–7. https://doi.org/10.1115/IMECE2013-6314
18. Anitha, R., al.: Critical parameters influencing the quality of prototypes in fused deposition modelling. J. Mater. Process. Technol. **118**(1–3), 385–388 2001. https://doi.org/10.1016/S0924-0136(01)00980-3
19. Sood, A.K., Ohdar, R.K. and Mahapatra, S.S.: Improving dimensional accuracy of fused deposition modeling processed part using grey Taguchi method. Mater. Des. **30**(10), 4243–4252 (2009). https://doi.org/10.1016/j.matdes.2009.04.030
20. Gurrala, P.K, Regalla, S.P.: DOE based parametric study of volumetric change of FDM parts. Proc. Mater. Sci. **6**, 354–360, (2014). https://doi.org/10.1016/j.mspro.2014.07.045

Part II
Laser-Based Additive Manufacturing (SLM and SLS) Process Parameter Optimization

Chapter 6
A Review on Selective Laser Sintering 3D Printing Technology for Polymer Materials

Fatima-ezzahrae Jabri, Aissa Ouballouch, Larbi Lasri, and Rachid EL Alaiji

Abstract This review analyzes different approaches used to study selective laser sintering (SLS) technology of polymer materials. These main approaches concern: thermal behavior and fatigue. Regarding the first behavior, researchers studied extensively the effect of process parameters such as laser power, scan speed, and laser energy density on the thermal behavior of 3D printed parts. Numerical and experimental analyses are used to conduct process parameters evaluations. Scan speed and laser power are found to be the most significant parameters on the laser energy density. For the second, according to test protocols and quantitative analysis performed, the authors concluded that the tensile analysis in different environments showed that testing in the water reduced fatigue life of polymer samples. The impact of process parameters on the mechanical properties of 3D parts is also analyzed. All these investigations have made it possible to determine the optimal process conditions to ensure higher quality and better fatigue strength. Therefore, this review shows that researchers can focus on creating a combination of both approaches to expand the use of this process for industrial part production.

Keywords Selective laser sintering (SLS) · Polymer · Thermal behavior · Fatigue · Mechanical properties · Process parameters

F. Jabri (✉) · R. EL Alaiji
Abdelmalek Essaadi University, B.P. 1818, Tangier 90000, Morocco
e-mail: fatimaezzahrae.jabri1@etu.uae.ac.ma

R. EL Alaiji
e-mail: relalaiji@uae.ac.ma

A. Ouballouch
Hassan II University, B.P. 8012, Casablanca, Morocco
e-mail: a.ouballouch@gmail.com

L. Lasri
Moulay Ismail University, B.P. 15290, Meknes, Morocco
e-mail: Lasrilarbi@yahoo.fr

K. Zarbane and Z. Beidouri (eds.), *Proceedings of CASICAM 2022*, Springer Tracts in Additive Manufacturing, https://doi.org/10.1007/978-3-031-32927-2_6

6.1 Introduction

Additive manufacturing (AM) or 3D printing is a process of creating 3D printed objects by following a series of operations. Layer by layer, until obtaining the final part, from a numerical model created using computer-aided design (CAD) software. Polymers are the most widespread materials used for 3D printing. Particularly, for fused deposition modeling (FDM), stereolithography (SLA) and selective laser sintering (SLS) [1, 2]. Principally, selective laser sintering received particular attention. Due to its relativity to high dimensional accuracy, ease modification, design changes, and flexibility in the type of materials [3]. Two critical factors must be considered to provide a high-quality and mechanical strength of 3D printed product. First, control of thermal distribution to ensure optimized process parameters. Because it's related to heat distribution, radiation and thermal convection phenomena provided in the SLS built chamber [4]. For that, thermal behavior is considered as a key factor governing the transformation of microstructure. As result, thermal phenomena have a significant influence on surface quality, dimensional accuracy and material properties (microstructure and mechanical strength) [4–6]. Hence, different aspects are studied for a thorough understanding of the effect of various process parameters and material characteristics on the quality of 3D printed parts. The second factor is the ability of 3D printed parts to withstand the mechanical and environmental stresses. These considerations subsequently cause fatigue crack propagation and material failure. Since, polymers are susceptible to fatigue at applied stresses below yield [7]. Moreover, there are factors that decrease the mechanical strength at the fatigue analysis. This involved process parameters, geometrical and environmental considerations. So, it is critical to understand and analyzed the fatigue behavior to improve the final parts durability and reliability [8]. Consequently, this paper gives an overview of thermal and fatigue behavior of 3D printed polymers. Figure 6.1 illustrates the main parameters that have an impact on these two behaviors. The main objective is to determine whether there are models that can shed light on the printing parameters that produce the best fatigue resistance. Also, to pick optimized parameters and physicochemical properties of powder that will result in a high quality product. This review is structured as follows. Sections 6.2 provide a description of the impact of process parameters, particle morphology, melt pool and temperature distribution on product quality. Section 6.3 identifies the parameters that have a significant effect on fatigue life and mechanical properties. These comprehend process parameters, powder bed density, geometric and environmental considerations, and material properties. The review's conclusion is reached at the end.

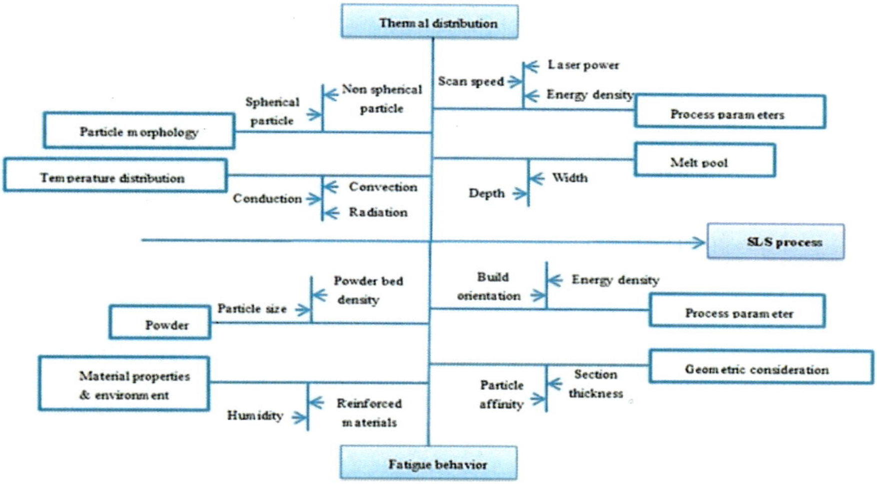

Fig. 6.1 Factors influencing the thermal distribution and fatigue behavior of 3D parts produced by selective laser sintering

6.2 Thermal Behavior

In this section, through numerical simulations and experimental tests. The researchers examined the influence of process parameters variations, particle morphology on part quality and mechanical properties. Also, they analyzed the impact of process parameters on melt pool and temperature distribution. Additionally, it is shown how particle size affects temperature variation.

6.2.1 *Effect of Process Parameters and Particle Morphology*

The thermal behavior is considered as a link between the process parameters and the product quality. In this regard, process parameters such as laser power, scan speed, and energy density. These parameters have been widely reported in literature to have varying effects on surface qualities of parts produced by SLS. In fact, to reduce build times for industrial manufacturing, [9] investigated the powder bed fusion (PBF) at high scan speeds. In this work, they claimed that high-speed processing decreases the density of printed parts and particle fusion. Moreover, [10] revealed from his experimental analysis that with non-constant values of the scan speed during the sintering process. Local overheating and notable deformations are obtained, which also influenced the mechanical performance of 3D parts. In order to explain the combined influence of laser power and scan speed on energy density [11]. Concluded that with low laser power and high scan speed, the energy density is insufficient to melt a single layer. As consequence, it leads to low part density and poor mechanical properties. Additionally, the powder may partially decompose with relatively high laser power and low scan speed, due to the high energy density generated. In this context, [11] noted that high energy could result in undesirable outcomes during the part's construction and might deteriorate the properties of the polymer. Moreover, researchers reported that particle shape is considered as a crucial element of the PBF sub-functions namely recoating and coalescence [12]. In this regard, [13]discovered that spherical particles improves the dimensional accuracy of 3D printed parts. In turn it contributes on the creation of a denser powder bed. Also [14] showed in their study that a non-spherical particle decreases the properties of 3D printed parts.

6.2.2 *Melt Pool*

According to literature, the melt pool is affected by the process parameters. For the reason that, the melt pool generates descriptive relationships between process parameters, mechanical properties and surface performance [17]. In the opinion of [18], both the physical properties and the laser power have an impact on the melt pool. Furthermore, [19] related that the maximum temperature, width and depth of the melt pool all increase as the laser power increases while the scan speed decreases. As well as, [11] stated numerous experimental and numerical simulation analyses for the polyamide 6 (PA6) processing. In this study, they concluded that the melt depth increases with increasing laser power because of the effect of higher heat penetration. Moreover, the depth decreases with increasing scan speed due to inappropriate heat input. [11] Observed also that the bond strength between the powder layers was directly affected by the depth of melt pool. A higher melt depth leads to a higher sintering conditions and better mechanical properties. As a result, the optimized parameters are in the region where the powder can be remelted.

6.2.3 *Temperature Distribution*

Scientists, users, and even machine manufacturers have remarked that the temperature distribution is irregular on some commercial machines. This conduce to inconsistencies in the mechanical properties of the parts [20]. In this reference [11] found that the maximum temperature of the melt pool increased as the laser moved, which is attributed to the influence of the heat stored in the first path. Besides, [18] proposed a three-dimensional finite element (FE) method by COMSOL Multiphysics software. In order to calculate the maximum temperature at the center of the laser spot. In this simulation, the phenomena of convection and thermal radiation are considered. They concluded that the preheating power and the application of temperature with a small beam from the laser caused a large temperature change on the surfaces of the polyamide layer. Moreover, to figure the impact of process parameters on temperature distribution. [11] proved that as the three-dimensional and maximum melting temperature increases, the laser power raises and the scanning speed decreases. In addition, to describe the impact of particle size on temperature variation. [21] demonstrated in his work. As the particle size decreases, the sintering initiation temperature decreases and densification increases independent of temperature. They also showed that high heating kinetics frequently induce surface diffusion and limit the growth of grain contact.

6.3 Fatigue Behavior

The main disadvantages of polymer SLS parts are porosity, shrinkage/warpaging caused by thermal distortion, and low strength in the Z-direction [3]. These failures can lead to cracking or fracture. Especially when the structure is subjected to repeated critical loads, resulting in fatigue damage [22]. Hence, fatigue behavior is a long-term factor that has an impact on mechanical properties such as tensile strength and yield strength. Therefore, the effects of process parameters including energy density, powder bed density and printing orientation are first introduced. Then, the impact of section thickness and particle affinity on fatigue life are discussed in Sect. 3.2. Finally, the influences of materials properties and environment considerations on mechanical properties are examined in details.

6.3.1 Process Parameters and the Powder Bed Density

Several experiments were conducted to analyze the behavior of the structure during fatigue tests. In this fact, [21] proved that the density of the powder sample is an important parameter for the sintering quality. A high value results in good agglomeration and gives more particle-to-particle contacts and a smaller pore size. This leads in turn to faster sintering. However, a low density makes the material difficult to consolidate. Furthermore, by increasing the powder bed density for both nylon-12 and commercially available thermoplastic polyurethane (TPU) powder. Due to better packing and flowability of the powder, [23] observed an increase in printed density, an increase in ultimate tensile properties and a decrease in surface roughness. In this context, [24] concluded that as the energy input increases, the penetration of laser increases, resulting in a denser part morphology, increased elongation at break, and tensile strength. In addition, [24] focused on particle sizes to see how they affected fatigue life. They discovered that a combination of smaller particles and large laser energy densities results in a higher degree of sintering and improved fatigue life. Another factor affecting mechanical properties of produced parts is printing orientation. In this regard, [25] showed that tensile testing of components created in the Z-axis orientation revealed the lowest strength and stiffness values. While tensile testing of parts built in the y-axis orientation revealed the highest mechanical property values. Furthermore, [26] found that the highest strength and stiffness appeared in parts built in the x-axis orientation. The influence of anisotropy [27] on samples plotted in perpendicular (90°) and parallel (0°) directions for both PA12 and thermoplastic polyurethane configurations (TPU). The semi-brittle failure behavior of the SLS-printed TPUs was particularly noticeable for printing directions perpendicular to the stress (90°). They also asserted that the influence of the printing orientations, or more specifically, the weak interfaces between layers, had a mitigating effect on the smooth notch. In addition, [27, 28] concluded that the maximum fatigue strength was found for the 0° printing orientations specimens. While the lowest fatigue strength was found for the 90° printing orientations specimens. Overall, the specimen printing direction induced anisotropy is a crucial consideration for further strength investigation.

6.3.2 Geometric Consideration

In addition to process parameters, geometric considerations such as section thickness and particle affinity have a substantial impact on fatigue life. Studies by [30] examined the effect of geometry on fatigue life by altering section thickness (2, 4, and 6 mm) of PA12 dog bone specimens. These specimens are subjected to uniaxial tension-tension and tension–compression loading. In both cases, fatigue life increased with section thickness. This is mainly due to the fact that the temperature increases faster in small section thicknesses than in large ones. According to literature researches, if there is a poor affinity between the powder particles, the part begins to lose its rigidity and degrade more quickly [22]. In this regard, studies conducted by [31] explored the relationship between fatigue life and particle affinities. [31] created rectilinear boxes from PA12 and polybutylene terephthalate (PBT) from BASF. The 90/10 blend has the longest fatigue life. Moreover, the 90/10 combination did not harden or soften during cyclic loading because the stress change remained essentially constant.

6.3.3 Material Properties and Environment

According to the literature, the fatigue life of SLS polymers seems to be affected more by material characteristics and ambient factors than by process parameters and geometric considerations. In [32], compact tension specimens ($50 \times 48 \times 10\ mm^3$) made of pure PA12 (PA12) and short glass fiber PA12 (PA12-f) were examined in a dry atmosphere at 23 and −50 °C. Both materials have similar characteristics at 23 and −50 °C. Both materials have comparable fatigue life at 23 °C, while PA12-f has a much better fatigue life at −50 °C. Similarly, [33] compact tensile specimens ($50 \times 48 \times 10 mm^3$) were fabricated from PA12 and PA11 and tested in three different environments: dry at 23 °C, dry at 50 °C, and wet at 23 °C. From the results, PA11 exhibited better fatigue resistance at low temperature than at room temperature. It also showed better fatigue crack propagation behavior than PA12. Moreover, they also claimed that the water test decreased the fatigue life of both samples, with a reduction in fatigue parameters, especially for PA12. Therefore, the toughness, stiffness, and strength of polyamides decrease with humidity while their ductility increases.

6.4 Conclusions

This review article summarizes the fundamental aspect of fatigue and thermal behavior of SLS 3D printing for polymers materials. The effect of process parameters on temperature distribution, melt pool, product quality and mechanical properties has been investigated in the part of thermal behavior. In the third section, the fatigue life and mechanical properties are studied in relation to process parameters, material characteristics and geometrical and environmental determinations. Outcomes of this review are listed as follow:

- The two factors that have a significant effect on laser energy density are scan speed and laser power.
- Higher melting depth leads to improved sintering conditions and mechanical properties.
- The optimized values are situated in the remelt area of the powder.
- Temperature is more affected by laser power; scan speed and density 3D printed part.
- The use of spherical particles increases the dimensional accuracy of 3D printed components.
- The fatigue life of the parts was affected by the energy density and the powder bed density.
- The thickness of the section improves the fatigue resistance for polymer materials.
- The water test decreased the fatigue life for polymer material.

More investigations and experiments are required to find the parameters and conditions that would provide long fatigue life and robust mechanical properties, resulting in high-quality of 3D printed parts.

References

1. Safaee, S., Schock, M., Joyee, E.B., Pan, Y., Chen, R.K.: Field-assisted additive manufacturing of polymeric composites. Addit. Manuf. **51**, 102642 (2022). https://doi.org/10.1016/j.addma.2022.102642
2. Karaş, B., Smith, P.J., Fairclough, J.P.A., Mumtaz, K.: Additive manufacturing of high density carbon fibre reinforced polymer composites. Addit. Manuf. **58**, 103044 (2022). https://doi.org/10.1016/j.addma.2022.103044

3. Nouri, A., Rohani Shirvan, A., Li, Y., Wen, C.: Additive manufacturing of metallic and polymeric load-bearing biomaterials using laser powder bed fusion: A review. J. Mater. Sci. Technol. **94**, 196–215 (2021). https://doi.org/10.1016/j.jmst.2021.03.058
4. Yan, Z., Liu, W., Tang, Z., Liu, X., Zhang, N., Li, M., Zhang, H.: Review on thermal analysis in laser-based additive manufacturing. Opt. Laser Technol. **106**, 427–441 (2018). https://doi.org/10.1016/j.optlastec.2018.04.034
5. Lupone, F., Padovano, E., Casamento, F., Badini, C.: Process phenomena and material properties in selective laser sintering of polymers: a review. Materials. **15**, 183 (2021). https://doi.org/10.3390/ma15010183
6. Olakanmi, E.O., Cochrane, R.F., Dalgarno, K.W.: A review on selective laser sintering/melting (SLS/SLM) of aluminium alloy powders: Processing, microstructure, and properties. Prog. Mater Sci. **74**, 401–477 (2015). https://doi.org/10.1016/j.pmatsci.2015.03.002
7. Sauer, J.A., Hara, M.: Effect of molecular variables on crazing and fatigue of polymers. In: Kausch, H.-H. (ed.) Crazing in Polymers, Vol. 2. pp. 69–118. Springer Berlin Heidelberg, Berlin, Heidelberg (1990). https://doi.org/10.1007/BFb0018019.
8. Shanmugam, V., Das, O., Babu, K., Marimuthu, U., Veerasimman, A., Johnson, D.J., Neisiany, R.E., Hedenqvist, M.S., Ramakrishna, S., Berto, F.: Fatigue behaviour of FDM-3D printed polymers, polymeric composites and architected cellular materials. Int. J. Fatigue **143**, 106007 (2021). https://doi.org/10.1016/j.ijfatigue.2020.106007
9. Lexow, M.M., Drexler, M., Drummer, D.: Fundamental investigation of part properties at accelerated beam speeds in the selective laser sintering process. RPJ. **23**, 1099–1106 (2017). https://doi.org/10.1108/RPJ-04-2016-0060
10. Peyre, P., Rouchausse, Y., Defauchy, D., Régnier, G.: Experimental and numerical analysis of the selective laser sintering (SLS) of PA12 and PEKK semi-crystalline polymers. J. Mater. Process. Technol. **225**, 326–336 (2015). https://doi.org/10.1016/j.jmatprotec.2015.04.030
11. Li, M., Han, Y., Zhou, M., Chen, P., Gao, H., Zhang, Y., Zhou, H.: Experimental investigating and numerical simulations of the thermal behavior and process optimization for selective laser sintering of PA6. J. Manuf. Process. **56**, 271–279 (2020). https://doi.org/10.1016/j.jmapro.2020.04.080
12. Berretta, S., Wang, Y., Davies, R., Ghita, O.R.: Polymer viscosity, particle coalescence and mechanical performance in high-temperature laser sintering. J Mater Sci. **51**, 4778–4794 (2016). https://doi.org/10.1007/s10853-016-9761-6
13. Chatham, C.A., Long, T.E., Williams, C.B.: A review of the process physics and material screening methods for polymer powder bed fusion additive manufacturing. Prog. Polym. Sci. **93**, 68–95 (2019). https://doi.org/10.1016/j.progpolymsci.2019.03.003
14. Schmid, M., Amado, A., Wegener, K.: Polymer powders for selective laser sintering (SLS). In: Presented at the Proceedings of PPS-30: The 30th International Conference of the Polymer Processing Society—Conference Papers. Cleveland, Ohio, USA (2015). https://doi.org/10.1063/1.4918516
15. Grossin, D., Montón, A., Navarrete-Segado, P., Özmen, E., Urruth, G., Maury, F., Maury, D., Frances, C., Tourbin, M., Lenormand, P., Bertrand, G.: A review of additive manufacturing of ceramics by powder bed selective laser processing (sintering / melting): Calcium phosphate, silicon carbide, zirconia, alumina, and their composites. Open Ceramics. **5**, 100073 (2021). https://doi.org/10.1016/j.oceram.2021.100073
16. Van den Eynde, M., Verbelen, L., Van Puyvelde, P.: Assessing polymer powder flow for the application of laser sintering. Powder Technol. **286**, 151–155 (2015). https://doi.org/10.1016/j.powtec.2015.08.004
17. Li, J., Yuan, S., Zhu, J., Zhang, W., Li, S., Wang, C.: Numerical investigation of novel process planning in the polymeric powder bed fusion. J. Manuf. Process. **67**, 195–211 (2021). https://doi.org/10.1016/j.jmapro.2021.04.060
18. Yaagoubi, H., Abouchadi, H., Taha Janan, M.: Numerical simulation of heat transfer in the selective laser sintering process of Polyamide12. Energy Rep. **7**, 189–199 (2021). https://doi.org/10.1016/j.egyr.2021.08.089

19. Shen, F., Yuan, S., Chua, C.K., Zhou, K.: Development of process efficiency maps for selective laser sintering of polymeric composite powders: Modeling and experimental testing. J. Mater. Process. Technol. **254**, 52–59 (2018). https://doi.org/10.1016/j.jmatprotec.2017.11.027
20. Goodridge, R.D., Tuck, C.J., Hague, R.J.M.: Laser sintering of polyamides and other polymers. Prog. Mater Sci. **57**, 229–267 (2012). https://doi.org/10.1016/j.pmatsci.2011.04.001
21. Mokrane, A.: Modélisation numérique du frittage laser des polymères en poudre, (2019). http://theses.insa-lyon.fr/publication/2019LYSEI062/these.pdf
22. Safai, L., Cuellar, J.S., Smit, G., Zadpoor, A.A.: A review of the fatigue behavior of 3D printed polymers. Addit. Manuf. **28**, 87–97 (2019). https://doi.org/10.1016/j.addma.2019.03.023
23. Ziegelmeier, S., Christou, P., Wöllecke, F., Tuck, C., Goodridge, R., Hague, R., Krampe, E., Wintermantel, E.: An experimental study into the effects of bulk and flow behaviour of laser sintering polymer powders on resulting part properties. J. Mater. Process. Technol. **215**, 239–250 (2015). https://doi.org/10.1016/j.jmatprotec.2014.07.029
24. Salmoria, G.V., Hotza, D., Klauss, P., Kanis, L.A., Roesler, C.R.M.: Manufacturing of Porous Polycaprolactone Prepared with Different Particle Sizes and Infrared Laser Sintering Conditions: Microstructure and Mechanical Properties. Adv. Mech. Eng. **6**, 640496 (2014). https://doi.org/10.1155/2014/640496
25. Koo, J.H., Ortiz, R., Ong, B., Wu, H.: Polymer nanocomposites for laser additive manufacturing. In: Laser Additive Manufacturing. pp. 205–235. Elsevier (2017). https://doi.org/10.1016/B978-0-08-100433-3.00008-7
26. Dizon, J.R.C., Espera, A.H., Chen, Q., Advincula, R.C.: Mechanical characterization of 3D-printed polymers. Addit. Manuf. **20**, 44–67 (2018). https://doi.org/10.1016/j.addma.2017.12.002
27. Major, Z., Lackner, M., Hössinger-Kalteis, A., Lück, T.: Characterization of the Fatigue Behavior of SLS Thermoplastics. **8**, (2021)
28. Salazar, A., Cano, A.J., Rodríguez, J.: Mechanical and fatigue behaviour of polyamide 12 processed via injection moulding and selective laser sintering. Analysis based on Kitagawa-Takahashi diagrams. Eng. Fract. Mech. **275**, 108825 (2022). https://doi.org/10.1016/j.engfracmech.2022.108825
29. Ajoku, U., Saleh, N., Hopkinson, N., Hague, R., Erasenthiran, P.: Investigating mechanical anisotropy and end-of-vector effect in laser-sintered nylon parts. Proceedings of the Institution of Mechanical Engineers, Part B: Journal of Engineering Manufacture. **220**, 1077–1086 (2006). https://doi.org/10.1243/09544054JEM537
30. Amel, H., Rongong, J., Moztarzadeh, H., Hopkinson, N.: Effect of section thickness on fatigue performance of laser sintered nylon 12. Polym. Testing **53**, 204–210 (2016). https://doi.org/10.1016/j.polymertesting.2016.05.027
31. Salmoria, G.V., Lauth, V.R., Cardenuto, M.R., Magnago, R.F.: Characterization of PA12/PBT specimens prepared by selective laser sintering. Opt. Laser Technol. **98**, 92–96 (2018). https://doi.org/10.1016/j.optlastec.2017.07.044
32. Salazar, A., Rico, A., Rodríguez, J., Segurado Escudero, J., Seltzer, R., Martin de la Escalera Cutillas, F.: Fatigue crack growth of SLS polyamide 12: Effect of reinforcement and temperature. Compos. Part B: Eng., **59**, 285–292 (2014). https://doi.org/10.1016/j.compositesb.2013.12.017
33. Salazar, A., Rico, A., Rodríguez, J., Segurado Escudero, J., Seltzer, R., Martin de la Escalera Cutillas, F.: Monotonic loading and fatigue response of a bio-based polyamide PA11 and a petrol-based polyamide PA12 manufactured by selective laser sintering. Eur. Polym. Journal. **59**, 36–45 (2014). https://doi.org/10.1016/j.eurpolymj.2014.07.016

Chapter 7
Effects of 316L Steel Powder Recycling on Manufactured Parts by Selective Laser Melting Process: A Review

Kaoutar Fellah, Meriem Hayani Mechkouri, Hamid Azzouzi, and Kamal Reklaoui

Abstract Additive manufacturing (AM) is a process of shaping materials in full expansion in several industrial sectors such as aeronautics, space and automotive. The additive manufacturing allows developing optimized parts with complex geometries impossible to achieve with conventional processes. When using powder beds for additive manufacturing, the quality of the powder used is important to the efficiency of the process and the characteristics of the finished product. Despite the fact that this is a universally acknowledged truth, the relationship between powder property and part property is still not fully understood. However, there have been a lot of studies published recently that concentrate on certain powder qualities and how they relate to part properties. This review's objective is to give a broad overview of the state of the art at the moment. The information presented enables better decision-making about the recycling and reuse of powder for particular additive manufacturing procedures.

Keywords Additive manufacturing · Selective laser melting · Recycled powder · Stainless steel 316L

7.1 Introduction

Selective laser melting (SLM) is a powder bed fusion additive manufacturing (AM) process that permits the manufacture of intricate and useful metallic objects. The metal powder is deposition in very thin layers over the build platform, and this AM technique uses a highly concentrated laser beam to selectively melt the metal powder. The powder plays an important role in obtaining competitive mechanical properties. The relationship between the quality of the manufactured parts and several process variables, such as laser power, layer thickness, and construction orientation, has been

---

K. Fellah (✉) · M. H. Mechkouri · H. Azzouzi · K. Reklaoui
Engineering, Innovation and Management of Industrial Systems Laboratory, Faculty of Science and Technology, Tangier, Morocco
e-mail: Kaoutar.fellah1@etu.uae.ac.ma

K. Zarbane and Z. Beidouri (eds.), *Proceedings of CASICAM 2022*, Springer Tracts in Additive Manufacturing, https://doi.org/10.1007/978-3-031-32927-2_7

the subject of extensive investigation. Other important aspects like powder properties and powder reuse have received less attention. Reusing powder material that hasn't been eaten is one of the SLM process most significant advantages because it makes the process more economical and environmentally friendly. Powder that is not melted for integration into a component is gathered in a container beneath the build chamber. Separate investigations have looked into some of the stainless steel 316L powder's useful qualities. After being utilized numerous times in the selective laser melting process, stainless steel 316L powder has been described. Studies have been done on powder composition, morphology, and particle size distribution. With an emphasis on the impact of powder quality in the selective laser melting process on part qualities, the present review attempts to provide an overview of previous research initiatives.

7.2 Influence of Powder Characteristics

The microstructure of metal parts developed by additive manufacturing is influenced by characteristics of raw material. This includes, but is not limited to, the powder manufacturing process, its size, morphology, chemical composition, and other characteristics (castability, density) [1]. The determination of optimum characteristics intrinsic to powder for additive manufacturing applications has been subject of numerous studies [2–7]. The objective is limited to obtaining most dense powder bed possible. The thickness of powder bed varies between 20 and 70 μm depending on type of machine and material used. In what concerns 3D machine (or 3D printer), power and diameter of laser spot affect powder thickness.

7.2.1 Powder Elaboration

The powders used in selective laser melting process are usually atomized by gas (nitrogen or argon) and sometimes by water. Powders atomized by water have a lower cost than powders atomized by gas. However, they are characterized by a large agglomeration of powder particles, a large number of satellites on surface of particles and irregular shapes [4]. A satellite is formed when a smaller particle sticks to a larger particle during solidification. Their oxygen level is also higher than that of powders atomized by gas [3], which can promote the formation of oxide in parts developed by selective laser melting.

With identical selective laser melting parameters, parts produced by selective laser melting from gas atomized powders have a better relative density than those bound from water atomized powders [6]. Given these results, gas-atomized powders are widely preferred for the selective laser melting process [7].

7.2.2 Apparent Density and Castability of Powder

Apparent Density. Apparent density is one of powder characteristics reflecting densification of powder bed. The density should be as high as possible [2, 3].

Castability. Castability refers to the ability of a powder to flow freely and consistently as individual particles. The castability of powder affects homogeneity of coating on entire construction tray before laser beam passes. If castability is not good, powder will be difficult to spread on construction tray. This will result in an inhomogeneous powder bed throughout construction stage [8]. It is influenced by several characteristics of powder [8].

Shape. Castability is improved when there are more spherical powder particles,

Amount of oxides on surface of particles. The higher density of oxides and poorer castability,

Powder moisture. Surface moisture increases interparticle forces, which degrades castability of particles,

Particle size. Very fine powder particles (<20 μm) affect castability of the powder. Inter-particle forces of the Van der Waals type are stronger for these fine particles. This leads to formation of agglomerates and degrades the castability of powder.

The castability and apparent density of powder could have a significant influence on the properties of the parts developed by SLM. Baitimerov et al. [9] demonstrate, from the study of 3 powder batches, that the densest samples are obtained for the batch with best combination of castability/apparent density. For this study, the powder is supplied by gravity and a scraper is used to spread powder on construction tray.

The powders with good castability are recommended for the SLM process.

It is also important to point out that measurement and relevance of criterion of castability are to be considered according to the coating system (characteristics of the scraper, roller, etc.) [10].

7.2.3 Particle Size and Morphology of the Powder

For a good castability of powder and a good densification of powder bed before passing laser beam, the particle size distribution of powder should ideally be Gaussian and the powder should be as spherical as possible [11–13].

Grain size. Powder particles used for the selective laser melting process have a d90 (diameter of which 90% of the number of grains are smaller) less than 50 μm [3] regardless of the thickness of powder bed defined. The size distributions of particles have a strong influence on castability of powder. Small particles (diameters less than 5 μm) will tend to form clusters and thus prevent uniform powder deposition on construction tray [14].

A mixture of small and large powder particles is preferred for selective laser melting process, where small particles percolate through larger particles to fill inter-particulate pores [15, 16] and achieve optimum powder bed density.

The particle size of powder has an influence on the final density of parts developed by SLM. Indeed, Chen et al. [17] compared influence of two powder batches of different sizes on density of the final parts with fixed SLM parameters. The batch containing medium-sized particles of less than 16 μm resulted in denser parts than the 48 μm medium-sized batch.

At fixed SLM parameters, powders of different sizes will lead to formation of pieces of different densities. However, powders of different particle sizes can be used in the same SLM printer, but the process parameters must be adjusted and optimized for each batch of powder [18]. This is the case of study of Spierings et al. [2] which obtains three powder batches of different sizes (d50 = 15, 28 and 37 μm) dense parts at more than 99% by adjusting only power and scanning speed of SLM process. Similarly, Kamath et al. [19] develop 316L steel parts from two powder batches of different particle sizes. They obtain more than 99% dense parts for two powder batches using a laser power greater than 250 W. The energy deposited is sufficient to fuse all powder particles and lead to a dense material.

Morphology. The shape of powder particles has an influence on final density of the parts developed by SLM. Olakanmi et al. [12] demonstrate that density of parts is decreased when powder particles are of heterogeneous shapes. Spherical particles facilitate densification of layers because they generate little inter-particle friction and have better mobility [20].

Gorji et al. [21] performed scanning electron microscopy (SEM) to survey the change in size, grains, surface contamination, morphology and microstructure of the powder particles for virgin feedstock and recycled stainless steel 316 L powders (over 10 times reused).

The Image processing results shows that the size of the recycled powder has not change and the particles have mostly retained their spherical shape.

Some particles in the recycled powder now show a smoother surface than the virgin powder. Heiden et al. suggested that the heat affects some particles dur- ing the melting and solidification that shatters their rough surface [22].

Some recycled powder particles show noticeable signs of porosity and cracks on the surface. Conversely, Heiden et al. reported a reduction in both surface and internal pores in reused powder particles suggesting that pores break up with under the laser's heat and the porous particles are consumed by time [22].

7.3 Microstructures of 316L Stainless Steel

7.3.1 Generality

The 316 stainless steel has many variations in terms of chemical composition.

Table 7.1 Specified chemical composition of the virgin 316L stainless steel powder feedstock, given in wt. % [23]

Element	Fe	Cr	Ni	Mo	Mn	Si	P	C	S
wt. %	Bal	16.5–18.5	10.0–13.0	2.0–2.5	< 2.0	< 1.0	< 0.045	< 0.030	< 0.030

Table 7.2 Tensile properties of 316L stainless steel specimens made with new powder and recycled powder

Source	Conditions	Rm (MPa)		Rp 0.2 (MPa)		Arupture (%)		Machine
		//Z	⊥Z	//Z	⊥Z	//Z	⊥Z	
Liu et al. [24]	Powder new	665	–	489	–	29	–	DiMetal-100
	Recycled 5 times without sieving	500	–	436	–	14	–	
Heiden et al. [22]	Powder new	600	–	450	–	43	–	Prox DMP 200
	Recycled 30 times with sieving	587	–	449	–	50	–	

The 316 steel is an iron-based alloy composed mainly of chromium and nickel (Table 7.1). The latter is an ϒ-genetic element, essential for stabilizing the austenitic structure of steel at room temperature.

The 316L steel ("L" for "Low Carbon") has a carbon content of less than 300 ppm, which allows it to significantly reduce the formation of chromium carbides and thus avoid the phenomena of decromization at grain joints. This gives it greater resistance to intergranular corrosion. Austenitic Steel 316L is a grade of austenitic stainless steel that retains a cubic austenitic structure with centered faces at room temperature. This material is used in various industries such as oil, chemical, pharmaceutical, and nuclear industries. It has a very good corrosion resistance thanks to its high chromium content (>16%) allowing the formation of a protective layer of Cr_2O_3 oxide on the surface while maintaining high mechanical properties [11].

7.4 Mechanical Properties of 316L Stainless Steel Developed by SLM

7.4.1 Influence of Powder

The recycling of non fused powder has been the subject of several studies available in literature. Liu et al. [24] compared mechanical tensile properties of 316L stainless steel specimens constructed with new powder and recycled powder 5 times without sieving and with the same SLM parameters. The mechanical properties of parts made with recycled powder are significantly lower than those of parts made with new powder (Table 7.2).

Heiden et al. [22] compared the properties of two 316L steels made from new and recycled powders 30 times with sieving between each cycle. They see a decrease in the final density of part made with the recycled powder. However, the ductility, yield strength and maximum mechanical strength are stable regardless of powder used (Table 7.2). Sieving filtered out the "bad" particles and "impurities" from reused powders that could have led to degradation of mechanical properties. This variation in properties can be explained by differences in characteristics between new and recycled powders. Slotwinski et al. [25] demonstrated that 8 times recycled stainless steel powder has different characteristics than the initial powder. The particle size distribution increases with number of cycles. Recycled powder has many more satellites and fusion-bound particles.

Heiden et al. [22] found little change between new powder and recycled powder 30 times. The particle size distribution and chemical composition are similar. However, the number of irregularly shaped particles, surface oxygen content and apparent density increase after recycling.

These results demonstrate an influence of powder characteristics on the final properties of 316L steel parts developed by SLM [26].

7.5 Conclusions

The powder properties are useful for defining the final properties of 316L steel parts developed by Selective Laser Melting process. In the examinations on the association between powder and part quality that were the subject of this review, stainless steel 316L powders were utilized a number of occasions. Consequently, it is challenging to draw quantitative conclusions. But despite all of that, the data can be summarized to provide a qualitative assessment and a basic plan of action for the creation of a thorough powder qualification procedure.

References

1. DebRoy, T., Wei, H.L., Zuback, J.S., Mukherjee, T., Elmer, J.W., Milewski, J.O., Beese, A.M., Wilson-Heid, A., De, A., Zhang, W.: Additive manufacturing of metallic components – Process, structure and properties. Prog. Mater. Sci. **92**, 112–224 (2018). https://doi.org/10.1016/j.pmatsci.2017.10.001
2. Spierings, A.B., Levy, G.: Comparison of density of stainlesssteel 316L parts produced with selective laser melting using different powder grades. In: 20th Annu. Int. Solid Free. Fabr. Symp. SFF 2009. (2009)
3. Schade, C.T., Murphy, T.F., Walton, C.: Development of atomizedpowders for additive manufacturing. In : Powder metal. Word congr. (2014). Accessed 2 July 2014

4. Fayazfar, H., Salarian, M., Rogalsky, A., Sarker, D., Russo, P., Paserin, V., Toyserkani, E.: A critical review of powder-based additive manufacturing of ferrous alloys: Process parameters, microstructure and mechanical properties. Mater. Des. **144**, 98–128 (2018). https://doi.org/10.1016/j.matdes.2018.02.018
5. Hedberg, Y., Norell, M., Linhardt, P., Bergqvist, H., Odnevall Wallinder, I.: Influence of surface oxide characteristics and speciation on corrosion, electrochemical properties and metal release of atomized 316L stainless steel powders. Int. J. Electrochem. Sci. **7**, 11655–11677(2012)
6. Cacace, S., Demir, A.G., Semeraro, Q.: Densification mechanism for different types of stainless steel powders in selective laser melting. Procedia CIRP. **62**, 475–480 (2017). https://doi.org/10.1016/j.procir.2016.06.010
7. Li, R., Shi, Y., Wang, Z., Wang, L., Liu, J., Jiang, W.: Densification behavior of gas and water atomized 316L stainless steel powder during selective laser melting. Appl. Surf. Sci. **256**, 4350–4356 (2010). https://doi.org/10.1016/j.apsusc.2010.02.030
8. Dietrich, S., Wunderer, M., Huissel, A., Zaeh, M.F.: A new approach for a flexible powder production for additive manufacturing. Procedia Manuf. **6**, 88–95 (2016). https://doi.org/10.1016/j.promfg.2016.11.012
9. Baitimerov, R., Lykov, P., Zherebtsov, D., Radionova, L., Shultc, A., Prashanth, K.G.: Influence of powder characteristics on processability of AlSi12 alloy fabricated by selective laser melting. Mater. Basel Switz. **11**, 742 (2018). https://doi.org/10.3390/ma11050742
10. Spierings, A.B., Voegtlin, M., Bauer, T., Wegener, K.: Powder flowability characterisation methodology for powder-bed-based metal additive manufacturing. Prog. Addit. Manuf. **1**, 9–20 (2016). https://doi.org/10.1007/s40964-015-0001-4
11. Pillot, S.: Fusion laser sélective de lit de poudresmétalliques. Technique de l'ingénieur, (2016)
12. Olakanmi, E.O.: Selective laser sintering/melting (SLS/SLM) of pure Al, Al–Mg, and Al–Si powders: Effect of processing conditions and powder properties. J. Mater. Process. Technol. **213**, 1387–1405 (2013). https://doi.org/10.1016/j.jmatprotec.2013.03.009
13. Boisselier, D., Sankaré, S.: Influence of powder characteristics in laser direct metal deposition of ss316l for metallic parts manufacturing. Phys. Procedia. **39**, 455–463 (2012). https://doi.org/10.1016/j.phpro.2012.10.061
14. McGEARY, R.K.: Mechanical packing of spherical particles. J. Am. Ceram. Soc. **44**, 513–522 (1961)
15. German, R.M.: Mathematical relations in particulate materials processing ceramics, powder metals, cermets, carbides, hard materials, and minerals/, Wiley, Hoboken, NJ: c, (2008). https://doi.org/10.1002/9780470370087.ch24.
16. Zhanga, S., Weib, Q., Lin, G., Zhao, X., Shi, Y.: Effects of powder characteristics on selective laser melting of 316l stainless steel powder. Adv. Mater. Res. **189–193**, 3664–3667 (2011)
17. https://doi.org/10.4028/www.scientific.net/AMR.189-193.3664
18. Chen, W., Yin, G., Feng, Z., Liao, X.: Effect of powder feedstock on microstructure and mechanical properties of the 316l stainless steel fabricated by selective laser melting. Metals. **8**, 729 (2018). https://doi.org/10.3390/met8090729
19. Zhukov, A., Deev, A., Kuznetsov, P.: Effect of alloying on the 316L and 321 Steels samples obtained by selective laser melting. Phys. Procedia. **89**, 172–178 (2017). https://doi.org/10.1016/j.phpro.2017.08.010
20. Kamath, C., El-dasher, B., Gallegos, G.F., King, W.E., Sisto, A.: Density of additively-manufactured, 316L SS parts using laser powder-bed fusion at powers up to 400 W. Int. J. Adv. Manuf. Technol. **74**, 65–78 (2014). https://doi.org/10.1007/s00170-014-5954-9
21. Egger, G., Gygax, P.E., Glardon, R., Karapatis, N.P.: Optimization of powder layer density in selective laser sintering. Solid Free. Fabr. Proc. (1999), 255–263 (1999)
22. Gorji, N E., Gorji, N.E., O'Connor, R., Mussatto, A., Snelgrove, M., González, P.M., Brabazon, D.: Recyclability of stainless steel (316 L) powder within the additive manufacturing process. Materialia **8**, 100489 (2019). https://doi.org/10.1016/j.mtla.2019.100489
23. Heiden, M.J., Deibler, L.A., Rodelas, J.M., Koepke, J.R., Tung, D.J., Saiz, D.J., Jared, B.H.: Evolution of 316L stainless steel feedstock due to laser powderbed fusion process. Addit. Manuf. **25**, 84–103 (2019). https://doi.org/10.1016/j.addma.2018.10.019

24. Pinto, F.C., Souza Filho, I.R., Sandim, M.J.R., Sandim, H.R.Z.: Defects in parts manufactured by selective laser melting caused by δ-ferrite in reused 316L steel powder feedstock. Addit. Manuf. **31**, 100979 (2020). https://doi.org/10.1016/j.addma.2019.100979
25. Liu, Y., Yang, Y., Mai, S., Wang, D., Song, C.: Investigation into spatter behavior during selective laser melting of AISI 316L stainless steel powder. Mater. Des. **87**, 797–806 (2015). https://doi.org/10.1016/j.matdes.2015.08.086
26. Slotwinski, J.A., Garboczi, E.J., Stutzman, P.E., Ferraris, C.F., Watson, S.S., Peltz, M.A.: Characterization of metal powders used for additive manufacturing. J. Res. Natl. Inst. Stand. Technol. **119**, 460–493 (2014). https://doi.org/10.6028/jres.119.018
27. Chniouel, A.: Study of the production of 316L stainless steel by selective laser melting on a powder bed: influence of the process parameters, the characteristics of the powder, and the heat treatments on the microstructure and the mechanical properties, Doctoral thesis from Paris-Saclay University prepared at Paris-Sud University, Doctoral School n°579 Mechanical and energy sciences, materials and geosciences - Doctoral specialty: Materials Science, (2019).

Chapter 8
Optimization of the Roughness and Dimensional Accuracy of PA12 Parts Produced by Selective Laser Sintering

Zainab Faraj, Smail Zaki, Mohamed Aboussaleh, and Hamid Abouchadi

Abstract In the selective laser sintering process, the surface finish and dimensional accuracy of the printed parts are of great importance, as they have a direct impact on their quality. Part roughness and dimensional accuracy are highly dependent on the manufacturing parameters, as well as the orientation of the parts in the build plate. A critical study and analysis of the influence of process parameters on the roughness and dimensional accuracy of SLS printed parts was conducted. The material selected for the study was PA12. The main process parameters selected for the study were: laser power, layer thickness, scan spacing and scan speed. The tests were performed according to Taguchi's Table T9. The experiments resulted in a sample with a roughness of less than 11 μm. The results show that the thickness of the layer is the parameter that has the most influence on the roughness of the parts, however the dimensional accuracy is more impacted by the laser power than by the other parameters.

Keywords Additive manufacturing · Selective laser sintering · Roughness · Taguchi method · Dimensional accuracy

8.1 Introduction

Additive manufacturing (AM) is a process that allows the layer by layer creation of physical objects from a three-dimensional CAD model [1, 2]. This technology and unlike conventional technologies has allowed the design of parts with quite

Z. Faraj (✉) · S. Zaki · M. Aboussaleh
Engineering of Structures and Complex Systems (ESCS), Innovation and Engineering Systems Laboratory (IES), ENSAM, Moulay Ismail University, Meknes, Morocco

H. Abouchadi
Laboratory of Applied Mechanics andTechnologies (LAMAT), ENSAM, STIS Research Center Mohammed V University, Rabat, Morocco
e-mail: h.abouchadi@um5r.ac.ma

K. Zarbane and Z. Beidouri (eds.), *Proceedings of CASICAM 2022*, Springer Tracts in Additive Manufacturing, https://doi.org/10.1007/978-3-031-32927-2_8

complex features and geometries [3–6]. Selective laser sintering (SLS) is a process of powder-based additive manufacturing that has seen great development and is currently applied in several industries [7]. Moreover, for the construction of solid components, this process uses a high-powered carbon dioxide (CO_2) laser, which by heating allows to selectively fuse a polymer layer. In a first step, the polymer powder is brought to a processing temperature close to the melting temperature of the polymer in question [8], then the polymer layers are deposited by means of a powder distribution roller, a reloading blade or a wiper. As soon as a new layer of about 0.08–0.2 mm is deposited, it is selectively fused by a CO_2 laser according to the profile information of each layer. Each time a layer of powder is applied, the construction platform is lowered by the applied powder thickness. These steps are repeated until the model is built layer by layer. The peripheral powder set that has not been melted reinforces the produced melt and acts as support structures.

The roughness of parts produced by the SLS process is considerably more important than that of parts produced by classical methods. Previous studies on the roughness of SLS-produced parts have revealed surface roughness values of SLS-produced parts in the range of $5-25\ \mu m$, depending on the processing parameters [8], but the surface roughness of SLS-produced parts remains one of the most important problems of this technology. Sachdeva et al. [9] chose Dura form polyamide as a study material to analyze the surface roughness of SLS-produced samples, they found that process parameters are the main factor controlling the surface quality of parts made by 3D SLS technology. Mavoori and Vekatesh [10] evaluated the surface roughness of sintered PA22 parts, they tried to determine the proper values of the variables using the Taguchi method. Calignano et al. [11] also employed the Taguchi method to evaluate the influence of input variables on the surface roughness of SLS 3D printed parts. Taguchi. Negi et al. [12] fabricated glass-filled polyamide specimens and found that scan spacing followed by scan speed and laser power have a significant influence on the surface roughness in the SLS process.

The parameter of dimensional accuracy is of major importance in many application areas. To produce parts with a strict tolerance, certain combinations of suitable process parameters must be used. For this reason, it is important to know the impact of process parameters on dimensional accuracy during production [13–17]. Dimensional inaccuracy is often the result of shrinkage caused by the thermal energy exerted during the sintering process [18]. This non-uniform internal shrinkage causes deformation of the produced parts. The properties of the powder and the manufacturing parameters have a very high impact on the shrinkage of the parts produced by SLS, which is why an optimization of the process parameters is essential.

This work analyzes the surface roughness and the dimensional accuracy of the parts produced by the SLS printing technology by varying the parameters and by printing different plates according to the Taguchi T9 table. It also presents the influence of each of the parameters on the roughness and the dimensional accuracy.

8.2 Equipment and Materials

8.2.1 Equipment

To perform this study, the printing of the samples by laser sintering technology was performed on the ProX™ 500 machine, this 3D Systems printer is equipped with a construction space whose volume is 381 × 330 × 460 mm, as well as a scanning speed that can reach the up to 305 mm/s. To perform the various tests, the manufactured samples geometry was designed according to ASTM D638 standards, all dimensions of manufactured specimens are presented in Fig. 8.1.

Selection of Process Parameters

Before starting to print the samples, the choice of the parameters that have the most influence on the roughness of the parts and their dimensional accuracy is a very important step. According to experiments [19–21], the process parameters that significantly influence the surface finish and dimensional accuracy of the sintered parts are the laser power, layer thickness, scan spacing and scan speed. The individual scans were performed following Taguchi's T9 4 ^ 3 table. For each configuration, five specimens were printed to have statistically valid results. Table 8.2 lists the parameters studied and their different levels.

Fig. 8.1 Specimen dimensions,

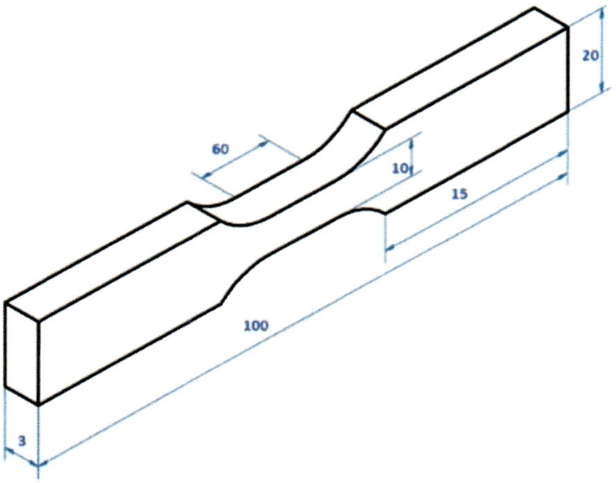

Table 8.2 Process parameters with their different levels

	Level 1	Level 2	Level 3
Laser power (W)	57	62	67
Scan speed (mm/s)	180	200	220
Layer thickness (mm)	0,1	0,11	0,12
Scan spacing (mm)	0,16	0,20	0,24

8.2.2 *Surface Roughness Measurements*

The measurement of surface roughness of samples printed by SLS technology was performed using the Mitutoyo Surftest SJ-410 roughness meter. This instrument is known for its high certainty specification, equipped with a wide range, high resolution detector and drive unit that provide high accuracy measurements, and is also equipped with 46 roughness parameters that comply with the latest ISO, DIN, ANSI and JIS norms.

8.2.3 *Dimensional Accuracy Measurements*

The dimensional accuracy of the realized samples is represented by S, and is calculated by the following Eq. (8.1) [22].

$$S = \frac{LD - LM}{LD} \times 100 \tag{8.1}$$

where LD is the predicted size in the CAD model and LM is the actual size measured after printing. The dimensional accuracy was calculated after measuring the length, width and height of the samples and then their average was taken as final data. The different measurements were made with a digital caliper.

8.3 Result and Discussion

As soon as the manufacturing process is completed, the printed parts were recovered for cleaning before their roughness and dimensional accuracy could be measured. Figure 8.2 shows the final state of the parts produced. The roughness and dimensional accuracy results are grouped in Table 8.3 and will be used later to study the influence of different input variables on the roughness as well as the dimensional accuracy of the parts produced by the selective laser melting technology.

Fig. 8.2 SLS manufactured samples

Table 8.3 Experimental design matrix and collected data

Laser power	Scan speed	Layer thickness	Scan spacing	S	Roughness (Ra) (μm)
57	180	0,10	0,16	3,004	10,2304
57	200	0,11	0,20	3,062	11,2200
57	220	0,12	0,24	2,523	12,8254
62	180	0,11	0,24	3,032	12,4668
62	200	0,12	0,16	3,215	13,4234
62	220	0,10	0,20	2,628	11,7002
67	180	0,12	0,20	2,922	13,2544
67	200	0,10	0,24	3,29	10,7938
67	220	0,11	0,16	3,733	11,5558

8.3.1 Effect of Process Parameters on Dimensional Accuracy

Table 8.4 shows the average dimensional accuracy values calculated for each parameter at each level. Figure 8.3 shows the variation curve of the average dimensional accuracy value as a function of each parameter, which will help determine the influence of each factor on the dimensional accuracy of parts produced by SLS technology.

Table 8.4 Mean of the responses for dimensional accuracy

Niveau	Laser power (W)	Scan speed (mm/s)	Layer thickness (mm)	Scan spacing (mm)
1	2,863	2,986	2,974	3,317
2	2,958	3,189	3,276	2,871
3	3,315	2,961	2,887	2,948
Delta	0,452	0,228	0,389	0,447
Rang	1	4	3	2

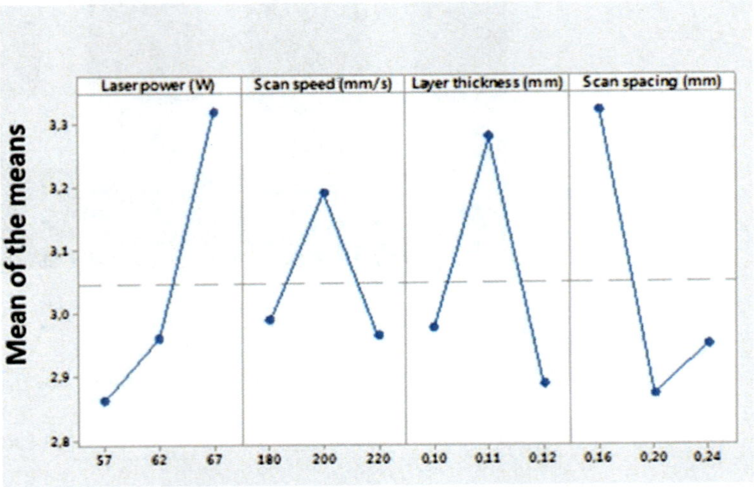

Fig. 8.3 Effect of process parameters on the dimensional accuracy of SLS manufactured parts

Figure 8.3 shows that as the laser power increases, the average dimensional accuracy increases at two different rates, reaching its maximum when the laser power reaches 67W. When the laser power is too high, it is able to sinter the previous powder layer, so the sintering is uneven in some places, which makes the product uneven.

The dimensional accuracy varies in a narrow range with increasing scan speed, it reaches its maximum value when the scan speed is 200 mm/s, and reaches the minimum value when the scan speed is 220 mm/s. As for the layer thickness, when the thickness is increased, the layer dimensions are reduced. This is because when the layer thickness is small, the high power laser is able to sinter the previous powder layer and therefore the dimensional accuracy is affected.

The dimensional accuracy improves with increasing scan spacing and reaches its minimum value when the scan spacing is 0.20 mm. A scan spacing that is too small increases the area of overlap between adjacent scan lines and therefore causes an overconcentration of energy, which in turn leads to overheating, thus affecting the dimensional accuracy of the printed parts.

8.3.2 *Effect of Process Parameters on Roughness*

Table 8.5 shows the average roughness values measured for each parameter at each level. Figure 8.4 shows the variation curve of the average roughness value according to each parameter, thus allowing to determine the influence of each factor on the roughness of the parts printed by the SLS technology.

When the laser power is high, a variation in the surface condition of the parts is noticed. In fact, the roughness increases and reaches its peak when the laser power passes the 62W value and then drops when the laser power reaches the 67W value. In effect, when the laser power is high, it is capable of fusing the previous layer, which improves the surface quality of the printed parts. The roughness of the parts changes with different scanning speeds, reaching its minimum when the scanning speed reaches the value of 200 mm/s. Because when the scanning speed is insufficient, the sintering of the powder particles is done in a heterogeneous way, which can cause the degradation of the roughness of the printed part. With the increase of the layer thickness, the roughness presents an increasing curve at different speeds, since the higher the layer thickness, the more energy is needed to sinter grains, which can

Table 8.5 Mean of the responses for roughness

Niveau	Laser power (W)	Scan speed (mm/s)	Layer thickness (mm)	Scan spacing (mm)
1	11,43	11,98	10,91	11,74
2	12,53	11,81	11,75	12,06
3	11,87	12,03	13,17	12,03
Delta	1,10	0,21	2,26	0,32
Rang	2	4	1	3

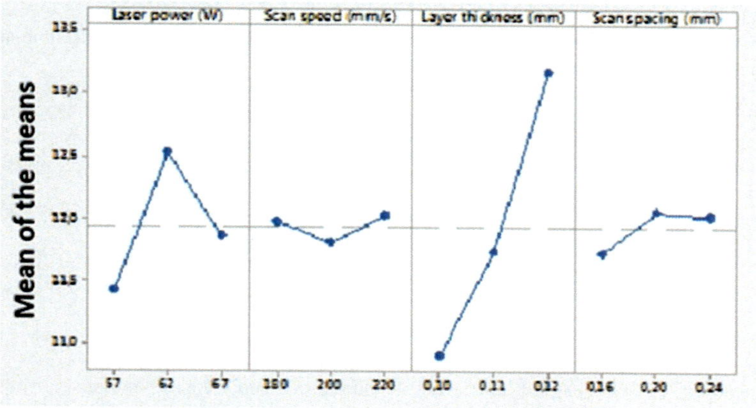

Fig. 8.4 Effect of process parameters on the roughness of SLS manufactured parts

influence the surface appearance. The roughness varies with the increase of the scan spacing value, the minimum value is obtained for the scan spacing level 1.

8.4 Conclusion

The objective of this study was to evaluate the effect of laser power, scan speed, layer thickness and scan spacing on the dimensional accuracy and surface quality of PA12 parts printed by SLS technology. The conclusions of this study are as follows:

- The parameter that has the most influence on the dimensional accuracy is the laser power and the one that has the least impact is the scan speed.
- Among the four parameters studied, the most critical and the one with the greatest impact on the surface condition of the samples is the thickness of the layer, while the one with negligible impact is the scanning speed.
- The lowest roughness value was obtained for the following combination: laser power of 57W, scan speed of 180 mm/s, layer thickness of 0.10 mm and scan spacing of 0.16 mm.
- The highest dimensional deviations were obtained for parts printed with the following combination: laser power of 67W, scan speed of 220 mm/s, layer thickness of 0.11 mm and scan spacing of 0.16 mm.

References

1. Zhang, X., Zheng, Y., Suresh, V., Wang, S., Li, Q., Li, B., Qin, H.: Correlation approach for quality assurance of additive manufactured parts based on optical metrology. J. Manuf. Process. **1**(53), 310–317 (2020)
2. Li, X., Chen, Y.: Micro-scale feature fabrication using immersed surface accumulation. J. Manuf. Process. **1**(28), 531–540 (2017)
3. Li, L., Anand, S.: Hatch pattern based inherent strain prediction using neural networks for powder bed fusion additive manufacturing. J. Manuf. Process. **29**(56), 1344–1352 (2020)
4. Song, X., Chen, Y., Lee, T.W., Wu, S., Cheng, L.: Ceramic fabrication using mask-image-projection-based stereolithography integrated with tape-casting. J. Manuf. Process. **1**(20), 456–464 (2015)
5. Solis, D.M., Silva, A.V., Volpato, N., Berti, L.F.: Reaction-bonding of aluminum oxide processed by binder jetting. J. Manuf. Process. **1**(41), 267–272 (2019)
6. Hasanov, S., Gupta, A., Nasirov, A., Fidan, I.: Mechanical characterization of functionally graded materials produced by the fused filament fabrication process. J. Manuf. Process. **1**(58), 923–935 (2020)
7. Tiwari, S.K., Pande, S., Agrawal, S., Bobade, S.M.: Selection of selective laser sintering materials for different applications. Rapid Prototyp J **21**(6), 630–648 (2015)
8. Bacchewar, P.B., Singhal, S.K., Pandey, P.M.: Statistical modelling and optimization of surface roughness in the selective laser sintering process, Proc. IMechE, Pa. Birds: J. Eng. Manuf. **221**, 35 (2006)
9. Sachdeva, A., Singh, S., Sharma, V.S.: Investigating surface roughness of parts produced by SLS process. Int. J. Adv. Manuf. Technol. **64**(9–12), 1505–1516 (2013)

10. Mavoori, N.K., Vekatesh, S.: Investigation on surface roughness of sintered PA2200 prototypes using Taguchi method. Rapid Prototyp J **25**(3), 454–461 (2019)
11. Calignano, F., Manfredi, D., Ambrosio, E.P., Iuliano, L., Fino, P.: Influence of process parameters on surface roughness of aluminum parts produced by DMLS. Int J Adv Manuf Technol **67**(9–12), 2743–2751 (2013)
12. Negi, S., Dhiman, S., Sharma, R.K.: Investigating the surface roughness of SLS fabricated glass-filled polyamide parts using response surface methodology. Arab J Sci Eng **39**(12), 9161–9179 (2014)
13. Chen, X., Wang, C., Ye, X., Xiao, Y., Huang, S.: "Direct slicing from Power SHAPE models for rapid prototyping. Int. J. Adv. Manuf. **17**(7), 543–547 (2001)
14. Cai, C., ShianTey, W., Chen, J., Zhu, W., Liu, X., Liu, T., Zhao, L., Zhou, K.: Comparative study on 3D printing of polyamide 12 by selective laser sintering and multi jet fusion. J. Mater. Process. Technol. **288**, 116882 (2020)
15. Ma, F., Hua Zhang, K.K.B., Hon, Q.G.: An optimization approach of selective laser sintering considering energy consumption and material cost, J. Cleaner Prod. **199** (20), 529–537 (2018)
16. Fina, F., Goyanes, A., Gaisford, S., Basit, A.W.: Selective laser sintering (SLS) 3D printing of medicines. Int. J. Pharm. **529**(1–2), 285–293 (2017)
17. AtheerAwad, F., Goyanes, A., Gaisford, S., Basit, A.W.: 3D printing: Principles and pharmaceutical applications of selective laser sintering. Int. J. Pharm. **586**(30), 119594 (2020)
18. Zheng, H.Z., Zhang, J., Lu, S.Q., Wang, G.H., Xu, Z.F.: Effect of core–shell composite particles on the sintering behavior and properties of nano-Al2O3/polystyrene composite prepared by SLS. Mater. Lett. **60**(9–10), 1219–1223 (2006)
19. Pham, D.: Rapid manufacturing, 'The technologies & applications of rapid prototyping & rapid tooling. Springer, (2000)
20. Rong-Ji, W., Wang, L.: Influence of process parameters on part shrinkage in SLS. Springer, (2006)
21. Rong-Ji., Xin-hua.: Optimizing process parameters based on neural networks and genetic algorithm. Springer (2008)
22. Ning, Y., Wong, Y.S., Fuh, J.Y.: Effect and control of hatch length on material properties in the direct metal laser sintering process. Proc Inst Mech Eng Part B J Eng Manuf **219**(1), 15–25 (2005)

Chapter 9
Density Improvement of Alsi10mg0.6 Parts Manufactured by Selective Laser Melting Manufacturing Process: Literature Review, Challenges, and Research Opportunities

El-Mehdi Kiass, Khalid Zarbane, and Zitouni Beidouri

Abstract In a highly competitive market where the Selective Laser Melting manufacturing process is used, the improvement of part quality is necessary to ensure business continuity and competitiveness. The present study aims to present a literature review on the methods explored to improve the quality of Alsi10mg0, 6 parts manufactured with Selective Laser Melting technology by reducing the porosity level. A detailed description of the Selective Laser Melting process has been presented in the present work trying to highlight the complexity of this process and the parameters that can impact the porosity rate. A summary of multiple previous works and the studied parameters has been presented in this paper. Multiple research opportunities have been presented and studying the correlation between fused area surface and porosity rate has been proposed as a promising study outlook.

Keywords Selective laser melting · Porosity · AlSi10Mg0 · 6. Laser powder bed fusion · Density improvement · Density improvement

E.-M. Kiass (✉)
National Higher School of Electricity and Mechanics (ENSEM), Hassan II University of Casablanca, B.P. 8118 Oasis, Casablanca, Morocco
e-mail: elmehdi.kaiss1-etu@etu.univh2c.ma

K. Zarbane · Z. Beidouri
Laboratory of Advanced Research on Industrial and Logistics Engineering (LARILE), ENSEM, Hassan II University of Casablanca, B.P. 8112 Oasis, Casablanca, Morocco
e-mail: khalid.zarbane@univh2c.ma

Z. Beidouri
e-mail: zitouni.beidouri@univh2c.ma

K. Zarbane and Z. Beidouri (eds.), *Proceedings of CASICAM 2022*, Springer Tracts in Additive Manufacturing, https://doi.org/10.1007/978-3-031-32927-2_9

9.1 Introduction

Additive Manufacturing (AM) is one of the most promising manufacturing processes and relatively new technology; Gartner's Hype Cycle for emerging technologies relative to the year 2015 [1] indicated that the use of this technology in the enterprise will reach a plateau of productivity by two to five years from 2015, something that has been obtained by the introduction of multiple important parts manufacturers in a serial production mode.

Additive Manufacturing is a generic term that groups several manufacturing processes characterized by their special method which consists in producing parts by adding the material layer by layer, unlike subtractive manufacturing processes like milling or turning where the part is created by using a full form of material and by removing specific areas of this block to obtain the desired part. Although a fundamental difference between the two approaches, they are used in a complementary way to exploit the advantages and compensate for the inconvenience of each technology. Vega et al. [2] listed seven main categories in Additive Manufacturing: Material Extrusion, Directed Energy Deposition (DED), Vat Polymerization, Binder Jetting, Material Jetting, Sheet Lamination, Powder Bed Fusion (PBF), which covers Electron Beam Melting (EBM) Selective Laser Sintering (SLS) and Selective Laser Melting (SLM).

Selective Laser Melting (SLM) also known as Laser Powder Bed Fusion (L-PBF) is a relatively new technology considered a branch of Additive Manufacturing (AM) processes. SLM technology is the most common additive manufacturing method, which was developed by Fraunhofer Institute for Laser Technology (ILT) [3]. Compared to other Additive Manufacturing processes SLM presents the following advantages: the ability to manufacture parts with high mechanical properties and relatively good dimensional accuracy, the use of a large range of materials, and the production of ready-to-use parts by getting a near net shape. As the main disadvantage, we find the need to create an important quantity of supports necessary for the production and a time and material-consuming operation for removing these supports.

Selective Laser Melting technology uses Computer-Aided Design (CAD) models to accumulate material in form of fine powder in a layer-by-layer way to manufacture complex-shape [4] and highly precise parts with outstanding properties using a high-energy laser beam that selectively fuses specific areas of the layers to form the complete part [5–8]. SLM process has several advantages such as the possibility to form extremely complex parts such as waveguides, lightweight structures in the aerospace field, assembly-free heat exchangers, and molds with internal flow channels. Deploying SLM technology has also the advantage to be an ecological method since it contributes to saving resources as the waste can approach zero [9]. Due to the possibility to manufacture complex-shape parts, SLM gives more freedom to parts designers [4].

Multiple material powders can be processed using SLM technology. The most common alloys are Ti6Al4V, Inconel718, 316L, and AlSi10mg0, 6 that are used in fields like aerospace, dental, aeronautic, automotive, robotic and medical.

A material alloy in form of powder with specific characteristics and generally a granulation between 10 and 60 μm [10] is used to obtain high-quality final parts. The raw material can be obtained by different powder atomization processes such as air atomization, inert gas atomization, or plasma atomization.

To build the part in a layer-based approach, a fine powder is spread in form of layers with a thickness that generally varies from 20 to 150 μm. Different types of powder spreading systems are used; the most common ones are flexible carbon fiber brush blades used in EOS [11] machines and rigid rollers used in Addup machines [24]. Figure 9.1 represents an example of layering cinematics in SLM machines.

Describing one of the challenges related to processing AlSi10mg0,6 alloy using SLM technology, Aboulkhair et al. [9] indicated that AlSi10mg0,6 powder has poor flowability rate compared to other powder alloys as described in Table 9.1; this means that the layering of this specific alloy is more difficult and requires more attention to avoid layering irregularities.

After deposition, the thin powder layers are melted selectively based on a CAD model using one or multiple high-energy laser beams with a power that can reach 1000 W in some industrial machines. The laser beams are directed with extreme precision using a scanner head composed of multiple mirrors as described in Fig. 9.2.

Another impacting characteristic was described by Aboulkhair et al. [9]; AlSi10mg0,6 powder has high reflectivity along with high thermal conductivity

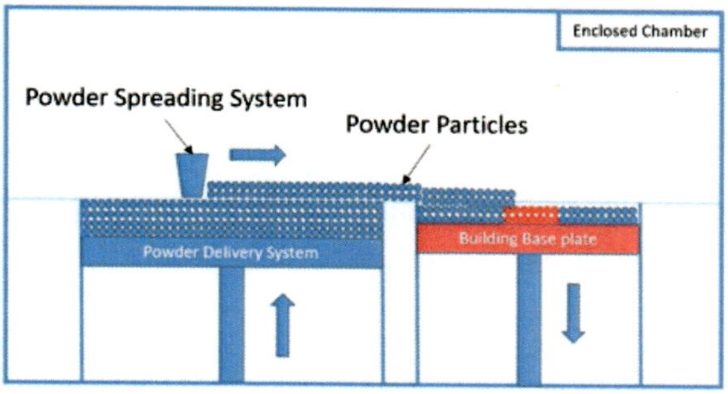

Fig. 9.1 Example of layering cinematics in SLM machines

Table 9.1 Comparing the flowability of different SLM candidate materials [9]

Powder material	Flowability (s/50 gm)
Ti64	47
Stainless steel 316	14.6
Al6061	77
AlSi10Mg	No flow

Fig. 9.2 Laser beam direction by mirrors

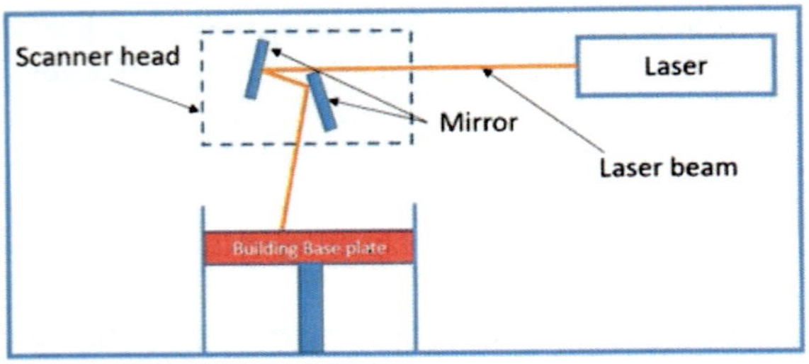

compared to other powder alloys as described in Table 9.2; this means that high laser power is required to compensate the thermal conductivity and reflectivity effect to ensure a correct melting.

Due to its lightweight and high corrosion resistance [13], AlSi10Mg is extensively used in the aeronautic and aerospace industries. However, in a high-temperature situation, the alloy is highly reactive to oxygen and hydrogen and hence the parts are built using an inert gas such as argon in the build chamber with a controlled flow rate [14]. As described in Fig. 9.3 inert gas circulation has three important functions; a part construction without oxidation obtained by filling the build chamber with inert gas, preventing the pollution of the optical system due to the formed fumes, and preventing the pollution of powder bed by the formed spatters both using inert gas flow 1 and 2.

Table 9.2 Comparing thermal conductivity and reflectivity of different SLM candidate materials [9]

Powder material	Thermal conductivity (W/(m K))	Reflectivity (%)
Ti64	6.7	53–59
Stainless steel 316	21.4	60
Al6061	172	91
AlSi10Mg	146	91

Fig. 9.3 Inert gas flow circulation in the SLM printing chamber

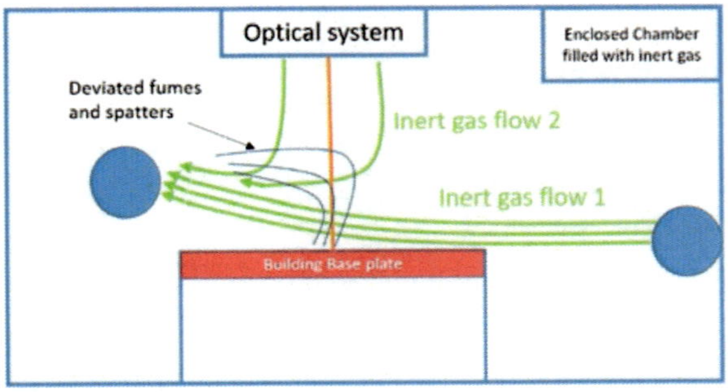

After each production, the remaining powder that has undergone an alteration of its characteristics, especially its particle size distribution due to the thermal effect of the laser, is recycled using a sieving system to retrieve its initial characteristics.

Regarding different reasons mentioned above the melt flow is hard to control and can lead to several defects such as porosity.

The rest of the document is composed of three sections: a first section that provides a literature review of the work already done to reduce the rate of porosity in parts produced in SLM, a second section that will present ways of improvement to reduce the rate of porosity and finally a conclusion to address the next step of our study.

9.2 Porosity Improvement Challenges of Slm Parts: Literature Review

The main object behind this study is to present a literature review including several previous studies performed and having as a target the reduction of porosity level in parts manufactured with Selective Laser Melting technology.

Selective Laser Melting can represent a great benefit when applied in fields such as aerospace, medicine, or aeronautics. These fields have a very high-quality requirement allowing to obtain final products meeting the needs of the end-users efficiently.

We can limit aspects defining the quality of a part manufactured in SLM by five important aspects that are: level of porosity or density, mechanical characteristics, roughness, hardness, and dimensional accuracy. The main challenge nowadays is to obtain a high level of quality at a reasonable cost and time.

Porosity is one of the common phenomena encountered in parts produced in SLM technology. In some applications in the medical biology field, for example, porosity may be considered an advantage; it can allow the manufacturing of biocompatible scaffolds containing internal and external porosity within which bone tissue can adhere perfectly. However, in some fields like aeronautics, aerospace, and automotive, where high mechanical properties are required, porosity is considered a defect and must be reduced. It has been reported that porosity has an important effect on mechanical characteristics [15]. The pores in the parts produced with this technology affect their density which leads to an alteration of the mechanical characteristics and therefore of the functioning of the mechanical systems.

Porosity percentage is defined as the fraction of voids in material over the total volume. Porosity can be quantified in two ways, based on tomography measurement in a volume-based approach and based on the surface area of accessible pores in a surface-based approach. Figure 9.4 shows an example of porosity encountered in Selective Laser Melting parts measured with the second method.

Yanga et al., studying porosity formation mechanisms and fatigue response in Al–Si–Mg alloys made by selective laser melting [16], observed that there are three types of porosity: large irregular-shaped porosities also named lack of fusion porosities

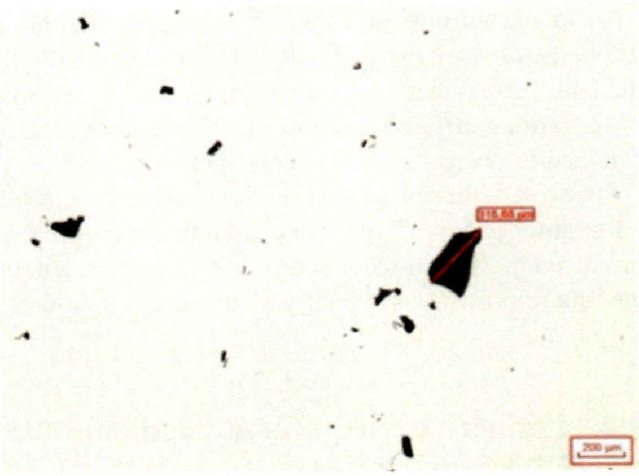

Fig. 9.4 Example of porosities encountered in parts produced in Selective Laser Melting technology and measured based on the accessible surface, the pores are shown in black color and the full material is shown in white color

which are caused by a not sufficient density of energy, large round porosities named keyhole porosities and caused by the use of a high density of energy and gas porosities with a size below 5 μm caused by the presence of hydrogen within the surface of powder grains. The form of porosity can reveal the cause of its formation. This conclusion can allow industrial manufacturers to save a lot of time when trying to resolve a quality problem by trying to detect the root cause of defects related to porosity formation; it can give a clear idea about the value of used energy to melt the powder, if it's low or high energy and also can give an idea about the quality of powder and the impact of hydrogen.

Ferrar et al. [14], reported that there are more than 130 parameters that may influence the SLM process, with approximately 13 of these parameters being crucial to the quality attributes of the final manufactured parts.

Several researchers ([10, 16, 17, 19, 21]) have studied the effect of parameters such as laser power, scan speed, hatching, layer thickness, and multiple other parameters. Gibson et al. [8] listed a group of process parameters influencing the porosity level in AlSi10Mg Selective Laser Melting parts. These process parameters are grouped into four categories, which are Laser-related, Scan-related, Powder-related and Temperature-related.

Most studies focused their research on four main process parameters that are laser power, scan speed, hatching, layer thickness, and volumic energy density (VED) which is a combination of these four parameters, VED [J/mm3] is defined as:

$$VED = \frac{P}{v * h * t} \tag{9.1}$$

where P [W] is the laser power, v [mm/s] is the scan speed, h [mm] is the hatch spacing, and t [mm] is the layer thickness

Chen et al. [17] found that ~ 44.53 J/mm3 of VED was suitable to achieve a dense AlSi10Mg alloy. In addition, Rao et al. [18] demonstrated that whether with a volumic energy density equal to 50 or 100 J/mm3 we obtain the same level of porosity. Giovagnoli et al. [19] and Pal et al. [10] confirmed the limitation of the volumic energy density-based approach on porosity estimation.

Cha et al. demonstrated that there is no significant effect of build atmosphere on the porosity level of AlSi10Mg parts fabricated through selective laser melting [20].

Jiang et al., by performing a factor analysis of selective laser melting process parameters with normalized quantities and the Taguchi method applied on 316L stainless steel powder, demonstrated that the parameter that most influences the porosity is the laser power, followed by the scan speed, then the hatching, after comes to the laser power scan speed interaction and finally the scan speed hatching interaction [21].

Multiple studies ([10, 16, 17, 19, 21]) tried to focus their work on the optimization of the five principal key parameters and they demonstrated that it is possible to reach a low rate of porosity based on those parameters.

In a different approach, the impact of powder storage conditions and characteristics was studied to reduce the porosity level. Weingarten et al. [22] tested different methods to dry aluminum powder to reduce the quantity of hydrogen attached to the surface of powder grains which can reduce the porosity level, they tried to dry powder inside the machine using a low value of laser power and high scan speed, they also demonstrated that a drying temperature of 90 °C using an oven can reduce the hydrogen quantity by 35% and a drying temperature of 200 °C can reduce it by 50%, in this study they found that drying powder outside the machine gives better results in terms of porosity reduction.

Fiegl et al. [23] when studying the effect of AlSi10Mg0.4 long-term reused powder in selective laser melting on the mechanical properties demonstrated that the oxygen and hydrogen content rises significantly during recirculation of the powder. In this study they compared the oxygen and hydrogen level of virgin powder and used powder during a period of 30 months, the content of oxygen during this period increased from 0.05 to 0.12% and the hydrogen content rises from 80 to 150 ppm. They concluded that this powder degradation leads to a density decrease; the porosity in specimens made of reused powder is about 4 times higher than the one in virgin powder samples.

9.3 Research Opportunities

Several authors ([4–6, 9, 10, 16–20, 22, 23]) when studying methods to improve AlSi10mg0.6 parts produced by Selective Laser Melting technology density, focused their research on the variation of standard parameters.

Rao et al. [18] studied the effect of build plate preheating as an influencing parameter on porosity formation. They tested two values of preheating temperature; 35 and 200 °C and proved that cooling rate variation could affect the pores present in the melt pool. One of the parameters that may also affect the cooling rate and therefore the porosity formation is the percentage of build plate filling by parts; having more fused volume within the same base plate will generate more heat and will affect the cooling rate. This represents an interesting research opportunity and can be organized following different approaches:

1. The first approach is a macroscopic one and consists of finding a correlation between the build plate filling rate BPFR defined as Eq. (9.2) and the porosity level of built parts:

$$BPFR = \frac{PV}{BPV} \tag{9.2}$$

where PV is the total volume of different parts present on the baseplate and BPV is the build plate's total volume.

2. The second approach consists of studying the correlation between the surface average of total fused sections by layer and the porosity level.
3. In another approach, we can explore the possibility to find a correlation between the fused surface within a layer and the porosity level within the same layer.

The last approach and the most complete one, can consist of taking into consideration all the precedent approaches and also taking into consideration fused surface variation during layer construction.

Due to the demonstrated effect of cooling rate on porosity level in part produced using selective laser melting technology, another promising research opportunity consists of studying the correlation between the four following parameters and their effect on porosity level:

1. Artificial cooling time, which represents the time allowed for the precedent layer to cool before fusing the current one
2. The height of the layer, which represents the distance between the current layer and the baseplate
3. The position of the part in the baseplate

This approach may permit to define of a minimum artificial cooling time allowing the manufacturing of parts with the desired porosity level independently of layer height or part position with high productivity.

Cha et al. [20] studied the effect of change in chamber gas from argon to nitrogen on porosity level and demonstrated that there is no significant difference between the two gases. But an important research opportunity may be to study the effect of injected inert gas temperature on porosity level. This temperature may affect the heat dissipation behavior and therefore the cooling rate of parts during manufacturing.

The study of Fiegl et al. [23], demonstrated with an experimental qualitative approach that the powder reuse in relation to the recycling cycle count that was applied to the powder impacts the density level of the produced part. An interesting study perspective may be to elaborate a model that links the recycling cycle count and the life cycle history of the powder with the porosity level. The need to include the life cycle history of powder as an influencing parameter is imposed by the fact that a previously charged production may affect the oxygen content of the powder more than a relatively less charged production.

9.4 Conclusion and Outlook

This review of the literature demonstrates that the effect of standard parameters, such as laser power, scan speed, hatching, layer thickness, and powder characteristics, on the porosity level of parts manufactured with SLM technology was studied by multiple authors who have demonstrated that we can obtain an important reduction of porosity by acting only on these parameters. Moreover, it would be important to explore more ways to obtain more reduction of porosity level.

As the next way of optimization, it would be interesting to study the dependencies between the total fused surface and the porosity rate in the same section. Exploring the existence of this correlation can be studied by manufacturing multiple samples with different surface sizes and measuring the porosity level in this section.

References

1. Gartner hype cycle for emerging technologies. Available online https://www.gartner.com/smarterwithgartner/whats-new-in-gartners-hype-cycle-for-emerging-technologies-2015. Last Accessed 26 Mar 2022
2. Vega, G., Paz, R., Gleadall, A., Monzón, M., Alemán-Domínguez, M.E.: Comparison of CAD and Voxel-based modelling methodologies for the mechanical simulation of extrusion-based 3D printed scaffolds. Materials **14**, 5670 (2021)
3. Schleifenbaum, H., Diatlov, A., Hinke, C., Bült-mannn, J., Voswinckel, H.: Direct photonic production: towards high speed additive manufacturing of individual goods. Prod engineering—Research Development. **5**, 359–371 (2011)
4. Zhang, B., Liao, H., Coddet, C.: Effects of processing parameters on properties of selective laser melting Mg–9% Al powder mixture. Mater Design. **8**(34), 753 (2012)
5. Kempen, K., Thijs, L., Van Humbeeck, J., Kruth, J.P.: Mechanical properties of AlSi10Mg produced by selective laser melting. Phys Procedia. **46**(39), 439 (2012)
6. Bartkowiak, K., Ullrich, S., Frick, T., Schmidt, M.: New developments of laser processing aluminium alloys via additive manufacturing technique. Phys Procedia. A **12**, 393–401 (2011)
7. Louvis, E., Fox, P., Sutcliffe, C.J.: Selective laser melting of aluminium components. J. Mater. Process. Technol. **84**(211), 275 (2011)
8. Gibson, I., Rosen, D.W., Stucker, B.: Additive manufacturing technologies. Springer, New York, USA (2010)
9. Aboulkhair, T.N., Everitt, M.N., Ashcroft, I., Tuck, C.: Reducing porosity in AlSi10Mg parts processed by selective laser melt-ing. Addit. Manuf. **1–4**, 77–86 (2014)
10. Pal, S., Lojen, G., Kokol, V., Drstvenšek, I.: Reducing porosity at the starting layers above supporting bars of the parts made by selective laser melting. Powder. Technol. **355**, 268–277 (2019)
11. Peter N., Pitts Z., Thompson, S., Saharan, A.: Benchmarking build simulation software for laser powder bed fusion of metals. Addit. Manuf., **36**, (2020)
12. ASTM Standard B213–13.: Standard test methods for flow rate of metal powders using the hall flowmeter funnel. West Conshohocken, PA: ASTM International. (2013). http://dx.doi.org/10.1520/B0213-13. www.astm.org
13. Martinez MG.: PFC: AlSi10Mg parts produced by Selective Laser Melting (SLM). p. 71, (2013)
14. Ferrar, B., Mullen, L., Jones, R., Stamp, R., Sutcliffe, C.J.: Gas flow effects on selective laser melting (SLM) manufacturing performance. J. Mater. Process. Technol. **212**(2), 355–364 (2012)

15. Yasa, E., Kruth, J.P.: Microstructural investigation of selective laser melting 316L stainless steel parts exposed to laser re-melting. Procedia. Eng. **19**, 389–395 (2011)
16. Yang, K.V., Rometsch, P., Jarvis, T., Rao, J., Cao, S., Davies, C., Wu, X.: Porosity formation mechanisms and fatigue response in Al-Si-Mg alloys made by selective laser melting. Mater. Sci. Eng., A **712**, 166–174 (2018)
17. Chen, J., Hou, W., Wang, X., Chu, S., Yang, Z.: Microstructure, porosity and mechanical properties of selective laser melted AlSi10Mg. Chin. J. Aeronaut. **33**, 2043–2054 (2020)
18. Rao, H., Giet, S., Yang, K., Wu, X., Davies, C.H.J.: The influence of processing parameters on aluminium alloy A357 manufactured by selective laser melting. Mater. Des. **109**, 334–346 (2016)
19. Giovagnoli, M., Silvi, G., Merlin, M., Di Giovanni, M.T.: Optimisation of process parameters for an additively manufactured AlSi10Mg alloy: Limitations of the energy density-based approach on porosity and mechanical properties estimation. Mater. Sci. & Eng. A **802**, 140613 (2021)
20. Rakesh Cha, S., Priyankab, N., Ra, J., J Vasaa, N.: Effect of build atmosphere on the mechanical properties of AlSi10Mg produced by selective laser melting. Mater. Today: Proc. **5**, 17231–17238 (2018)
21. Jiang, H.Z., Li, Z.H., Feng, T., Wu, P.Y., Chen, Q.S., Feng, Y.L., Li, S.W., Gao, H., Xu, H.J.: Factor analysis of selective laser melting process parameters with normalised quantities and Taguchi method. Opt. Laser Technol. **119**, 105592 (2019)
22. Weingarten, C., Buchbinder, D., Pirch, N., Meiners, W., Wissenbach, K., Poprawe, R.: Formation and reduction of hydrogen porosity during selective laser melting of AlSi10Mg. J. Mater. Process. Technol. **221**, 112–120 (2015)
23. Fiegl, T., Frank, M., Raza, A., Hryha, E., Körner, C.: Effect of AlSi10Mg0.4 long-term reused powder in PBF-LB/M on the mechanical properties. Mater. & Des. **212**, 110176 (2021)
24. FormUp® 350.: Available online https://addupsolutions.com/machines/pbf/formup-350/. Last accessed 27 Mar 2022

Part III
Topology Optimization

Chapter 10
The Impact of Topology Optimization Parameters in the Shape and the Strength of the Structure

A. Ait Ouchaoui, M. Nassraoui, and B. Radi

Abstract The additive manufacturing techniques create parts by the deposition of successive layers which help engineers to overcome the limits of traditional manufacturing techniques and give them the freedom to design complexes parts. The association of additive manufacturing and topology optimization allow to take full advantages of this manufacturing techniques, Especially, to design the complexes lightweight parts. This paper gives a synthesis and recapitulation of topology optimization approaches and their algorithms. It also presents the results analysis by the MATLAB code of four topology optimization methods. This code provides the final shape of the optimized structure and generate Standard Tessellation Language (STL) files. One numerical example is presented to exhibit the impact of topology optimization approaches in the shape and the strength of the structure. Eventually, an evaluation of the stress and strain of the optimized structures results using numerical simulation software. The results of the mentioned analysis and codes are discussed and compared with the literature results.

Keywords Additive manufacturing · Topology optimization · SIMP · RAMP · Level Set · BESO

A. A. Ouchaoui (✉)
National High School of Electricity and Mechanics, University Hassan II, Casablanca, Morocco
e-mail: ahmed.aitouchaoui-etu@etu.univh2c.ma

A. A. Ouchaoui · M. Nassraoui
Laboratory of Mechanics, Productics, and Industrial Engineering, EST, University Hassan II, Casablanca, Morocco
e-mail: mohammed.nassraoui@univh2c.ma

B. Radi
Laboratory IMII, FST, University Hassan I, Settat, Morocco

K. Zarbane and Z. Beidouri (eds.), *Proceedings of CASICAM 2022*, Springer Tracts in Additive Manufacturing, https://doi.org/10.1007/978-3-031-32927-2_10

10.1 Introduction

Additive manufacturing (AM), known under the name of rapid prototyping and 3D printing in the nineties, is the process of creating three-dimensional parts by adding material layer-by-layer from STL files. This manufacturing technology is widely used in rapid prototyping, and it can create almost any shape from many materials such as metallic materials, polymers, composites, and biomaterials [1]. AM reduces the time and cost of manufacturing, especially for prototyping and batch production, and minimize the assembly components [2], and helps to surmount the limits of traditional manufacturing techniques. AM plays a crucial part in modern manufacturing industries and has a wide field of applications, including aeronautics, aerospace, mechatronics, medicine. The first AM technology in France and the USA in 1984 is stereolithography (SLA) technology [3]. The AM technologies are classified into seven categories: Material extrusion [4], powder bed fusion [5], vat photopolymerization [6], material jetting [7], binder jetting [8], sheet lamination [9], and direct energy deposition [10]. The investigation of the advantages and inconvenient of additive manufacturing technologies are described in the paper of [11]. Besides the capabilities of AM, it still has many limitations such as accuracy, surface roughness, and material characterization.

Topology optimization is a progressed structural design method to reach the best material distribution satisfying defined load and boundary conditions via the right material distribution [12]. Since the presentation of the first optimization method by [13] it become the most widely used structural design technique in the aeronautical, aerospace, automotive, and architectural fields to optimize weight and improve mechanical properties. The structures obtained by topology optimization are characterized by complex geometry, which makes their manufacture difficult by conventional manufacturing techniques. The first method of topology optimization introduced by [13] is the homogenization method. The idea of the first method has been developed in various way namely: the density-based methods such as solid isotropic material penalization (SIMP) [14], Rational Approximation of Material Properties (RAMP) [15], the level set method (LSM) [16], the evolutionary methods such as evolutionary structural optimization (ESO) [17], Bi directional. evolutionary structural optimization (BESO) [18], Feature driven optimization (FDO) [19], Moving morphable components MMC [20], Moving morphable voids MMV [21].

The educational topology optimization codes for MATLAB, python and Mathematica, are cited in the paper of [22] including Femlab, and FreeFem ++, Struc program for level set method. The SIMP methods algorithms such as ToPy, 99-line program, ParetoOptimalTracing, PolyTop, 88-line program. The paper of [23] stated the theoretical aspects of the topology optimization methods, the general algorithm of this methods and the comparison of the methods with regard to the settings and computation cost.

The aim of this work is to examine the results of the four significantly applied topology optimization approaches: SIMP, RAMP, BESO, and LSM. First, the discussion the effect of the topology optimization method parameters in the shape of the structures. Second, the comparison of the static analysis results for the four methods to expose the impact of topology optimization in the strength of the structure. Finally, a conclusion of the optimal parameters of topology optimization methods according to the suitable mechanical performance.

10.2 Materials and Methods

The problem formulation for minimum compliance with volume constrain and the MATLAB codes are explained in the paper of [24] for the four topology optimization methods: SIMP, RAMP, Level set, and BESO. The results of stress and strains are given by the static analysis which is performed with numerical analysis software using the equilibrium equation given by the generalized Hooke's low. The Table 10.1 present the static analysis settings and the mechanical characteristics of polylactic acid (PLA) from the engineering data sources of the numerical simulation software.

The study case is the topology optimization of a cantilever beam with Sect. 120 × 40 × 4 as shown in the Fig. 10.1.

In this work, the four topology optimization methods, SIMP, RAMP, Level set et BESO, are conducted by the MATLAB codes stated in the paper of [24]. The domain in the Image-based Initialization and Post-Processing (IbiPP) code is defined by the image format which represent the design problem according to RGB values to identify the features such as load, pressure, boundary conditions, design and non-design domain. This part of the code transforms the image into useful input of the topology optimization code. The output of the image post-processing is a matrix with

Table 10.1 Static analysis settings and mechanical characteristics of PLA

Young modulus Ex (MPa)	Poisson's ratio v	Density (kg/m^3)	Element type	Element size	Force
3450	0.39	1250	Tetrahedrons	6 mm	100N

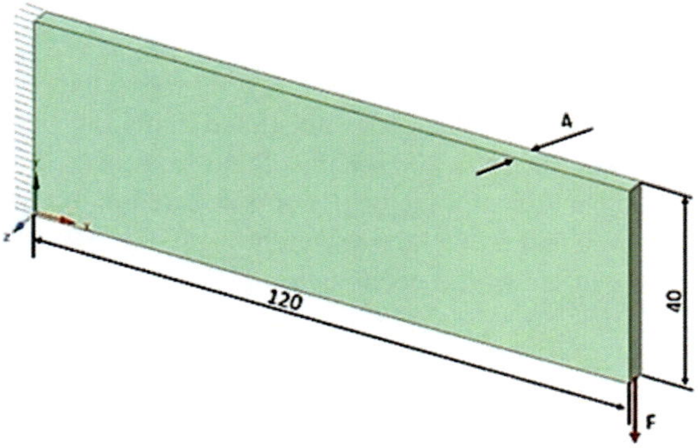

Fig. 10.1 Boundary conditions for the original cantilever beam model

Table 10.2 Topology optimization parameters of the MATLAB code

nelx	r_{min}	P	q	Volfrac	Ft	v	F	E0
200	2	3	3	0.4	2	0.39	1	1

size equal to the pixel size of the image file. Although, the image post-processing is not used in this paper. The image post-processing section in the IbiPP code is replaced with a matrix. Firstly, the domain matrix is one's matrix with size equal to (length of the beam by the element size, width of the beam by the element size). Then, the boundary conditions and the load are defined by the attribution of value 51 to define the fixed support nodes and 20 for the load node.

The effect of optimization parameters is investigated by the following cases: Case 1: effect of penalization factors (p) and (q) and the number of elements (nelx) in the results of SIMP and RAMP methods. Case 2: The effect of the filter minimal radius (r_{min}) in the results of SIMP and RAMP methods. Case 3: the effect of evolutionary ratio (er) in the BESO optimization results. Case 4: The effect of the parameter the regularization factor (tau) in the results of the level set method. The Table 10.2 present the topology optimization parameters where ft is the size of the filter, v the Poisson ratio, F the load and E0 the Young modulus of the solid.

10.3 MATLAB Code Results

The MATLAB codes of the four topology optimization methods create images and STL files for the result of the topology optimization. These STL files shape is repaired by reverse engineering to fix the surface defects. The MATLAB codes results are introduced in this section based on the four study cases.

10.3.1 Case 1: Effect of P and Q and Number of Elements Nelx in the Results of SIMP and RAMP Methods

The effect of penalization factors p for five number of elements in x direction in the topology of the structure is illustrated in the Fig. 10.2. The structure is clear of gray zones from nelx = 100 for p = 2 and from nelx = 50 for p > 2. The shape depends on the penalization factor p. For p = 2 and p = 3 the shape is consistent. However, for p = 4 and p = 5 he is different from the first shape of p = 2 and p = 3.

The effect of the penalization factor q in the RAMP method is presented in the Fig. 10.3. For q = 2 the optimized structure is not clear from gray zones even for nelx = 200. It became clearer in q = 3 after nelx = 200, nelx = 150 for q = 4 and nelx = 100 for q = 5. According to these results, the shape is not very affected by the penalization factor q.

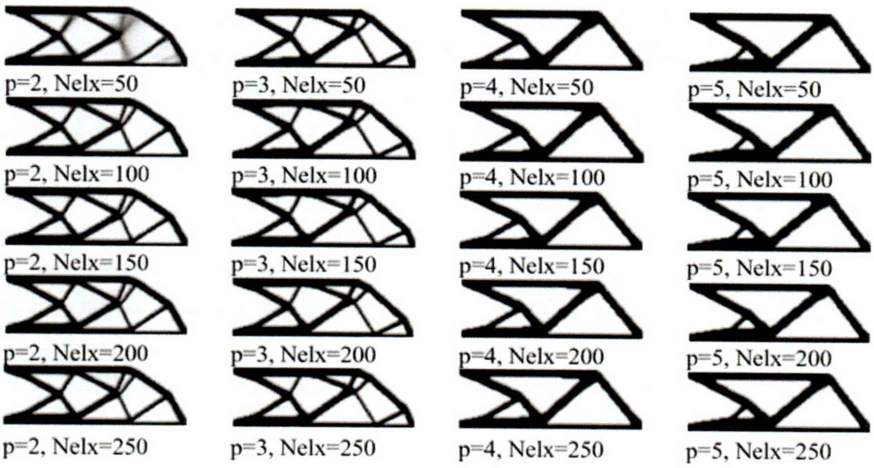

Fig. 10.2 The effect of p and nelx in the shape of SIMP method results

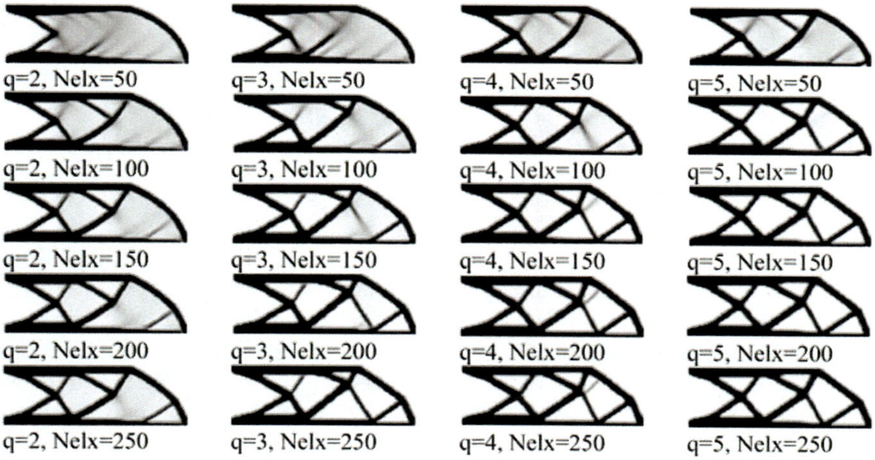

Fig. 10.3 The effect of q and nelx in the shape of RAMP method results

10.3.2 Case 2: The Effect of the Filter Minimal Radius r_{min} in the SIMP and RAMP Methods Results

The effect of the filter minimal radius r_{min} of SIMP and RAMP methods in the shape is investigated by the variation of the filter minimal radius from 1 to 5. The r_{min} variation results for SIMP optimization are presented in the Fig. 10.4. The shape changes and become unclear from $r_{min} > 2$. For $r_{min} = 1$ many elements are not connected. Corresponding to these results, the optimal filter minimal radius is $r_{min} = 2$.

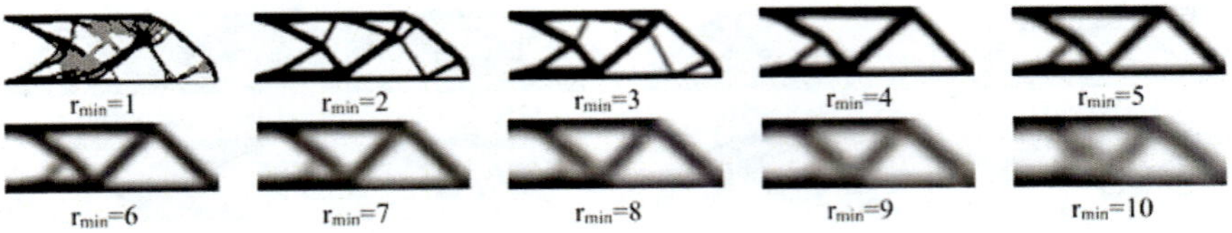

Fig. 10.4 The effect of r_{min} in the shape of SIMP optimization results

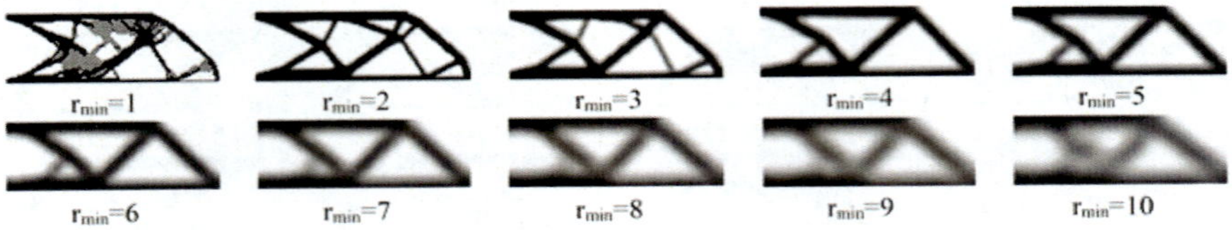

Fig. 10.5 The effect of r_{min} in the shape of RAMP optimization results

The effect of r_{min} in the result of RAMP optimization is more important than the SIMP results as showed in Fig. 10.5. The clear result is the optimization RAMP with $r_{min} = 2$.

10.3.3 *Case 3: The Effect of Evolutionary Ratio in the BESO Optimization Results*

The analysis of the effect of the evolutionary ratio er in the BESO optimization results by the modification of the evolutionary ratio er from 0.15 to 0.75. The effect of er in the shape of BESO optimization results are exposed in the Fig. 10.6. The shape is clearly different from the SIMP and RAMP optimization results and undergoes modifications depending on the er value. For er > 0.45 results the shape is the elements are strangely distributed which make the results inacceptable. The optimal evolutionary ratio is decided by the results of static analysis.

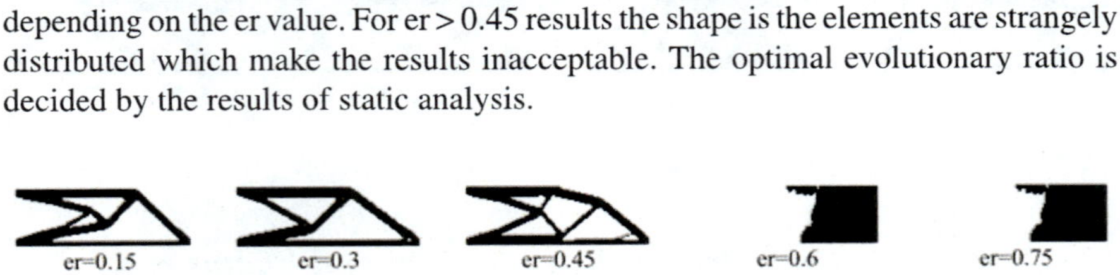

Fig. 10.6 The effect of er in the shape of BESO optimization results

Fig. 10.7 The effect of tau in the shape of level set optimization results

10.3.4 Case 4: The Effect of the Parameter Tau in the Level Set Optimization Results

The Study of the impact of the parameter tau in the Level set optimization results by the changing the parameter tau from 1.10^{-4} to 8.10^{-4}. The impact of tau in the shape of Level Set optimization results are presented in the Fig. 10.7. The shape is similar to the result of SIMP optimization with $p = 4$ and subject to small modifications rely on the tau values. The optimal parameter tau is not obvious in this stage it depends on the static analysis results.

10.4 Numerical Static Analysis Results

The impact of the optimization code parameters in the strength of the optimized structures is demonstrated by the evaluation of the stress and elastic strain results for the geometry of the optimization methods. The original cantilever beam strength is also evaluated to compare it results with the optimized geometry results. As well, the choice of the best topology optimization parameters depends on the static analysis results which are presented in Table 10.3. The static analysis is performed using numerical simulation with numerical simulation software. The geometry used in the static analysis is created by CAD software from the STL files created by the MATLAB code for the four topology optimization methods cited previously. The stress and strain ratio are calculated to exhibit the difference between the strength of the original and the optimized structures.

The topology optimization parameters used in the following figures are explained in Table 10.4.

The Fig. 10.8 demonstrate the effect of the optimization parameter in the equivalent Von Mises stress. The BESO method result are superior to other methods. Meanwhile, the level set method results are inferior to the other methods.

The Fig. 10.9 shows the impact of the topology optimization parameters in the equivalent elastic strain. The superior and inferior graphs are just as the equivalent Von Mises stress.

Table 10.3 Static analysis results for the study cases

Method	Parameter's value	V (10^{-6} m^3)	Volfrac (%)	σ (MPa)	σ ratio	ε (10^{-3} mm/mm)	ε ratio
Original	–	19.20	100	20.965	1.00	6.4865	1.00
SIMP (p)	2	8.470	44	42.964	2.05	12.665	1.95
	3	8.030	42	30.438	1.45	9.2078	1.42
	4	7.940	41	31.944	1.52	9.5120	1.47
	5	8.190	43	22.121	1.06	6.8778	1.47
SIMP (r_{min})	2	8.030	42	30.438	1.45	9.2078	1.42
	3	8.491	44	29.168	1.39	8.8039	1.36
	4	8.660	45	33.151	1.58	9.7967	1.51
RAMP (p)	2	7.728	40	33.184	1.58	10.043	1.55
	3	7.793	41	36.176	1.73	10.657	1.64
	4	7.843	41	26.443	1.26	7.9545	1.23
	5	7.916	41	31.859	1.52	9.5668	1.47
RAMP (r_{min})	2	7.793	41	36.176	1.73	10.657	1.64
	3	7.324	38	52.765	2.52	15.294	2.36
BESO (er)	0.15	7.440	39	45.963	2.19	13.473	2.08
	0.30	7.711	40	46.683	2.23	13.531	2.09
	0.45	7.551	39	59.886	2.86	17.358	2.68
Level Settau x10^{-4}	2	7.676	40	30.111	1.44	9.1520	1.41
	4	7.891	41	23.581	1.12	7.3529	1.13
	6	8.295	43	23.038	1.10	6.9842	6.9842

Table 10.4 Identification of the parameter used in the following figures

Parameter	SIMP (p)	RAMP (q)	SIMP (r_{min})	RAMP (r_{min})	BESO	Level Set
2	P = 2	q = 2	$r_{min} = 2$	$r_{min} = 2$	er = 0.15	tau = 2.10^{-4}
3	P = 3	q = 3	$r_{min} = 3$	$r_{min} = 3$	er = 0.3	tau = 4.10^{-4}
4	P = 4	q = 4	$r_{min} = 4$	–	er = 0.45	tau = 6.10^{-4}
5	P = 5	q = 5	–	–	–	–

10.5 Discussion

The equivalent Von Mises stress are greater than the original beam for all the optimization results. The minimal findings of the volume fraction is 38% for $r_{min} = 3$ in the RAMP(r_{min}) case. The minimal equivalent Von Mises stress is 22.121 MPa and the minimal equivalent elastic strain is 6.877 10^{-3} mm/mm for p = 5 in the SIMP(p) case.

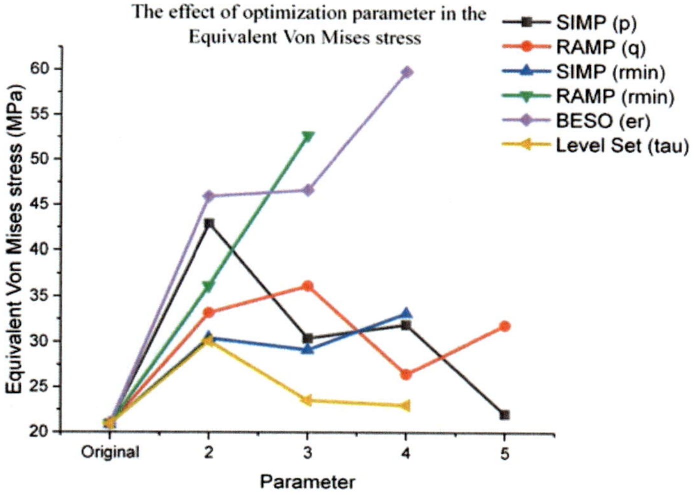

Fig. 10.8 The effect of topology optimization parameters in the equivalent Von Mises stress

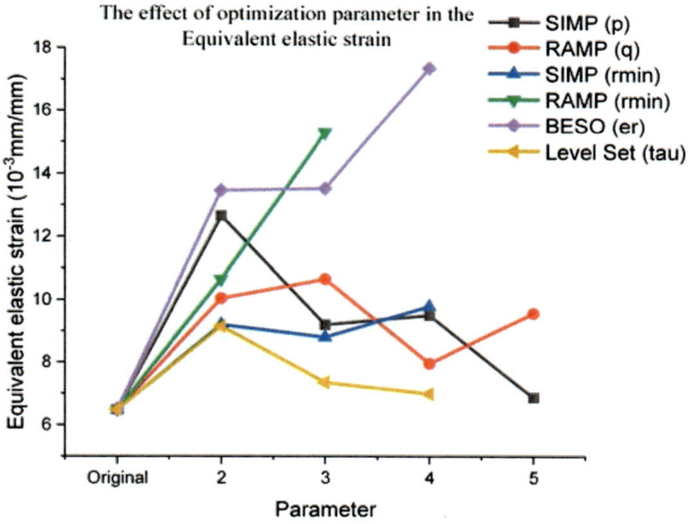

Fig. 10.9 The effect of topology optimization parameters in the equivalent elastic strain

10.6 Conclusion

The findings of the static analysis show that the stress and strains are proportional to the volume fraction. The minimal results of the stress and strain of the optimized beams results which are close to the values of the stress and strain of the original beam are the result of SIMP optimization where the penalization factor p = 5 and filter minimal radius $r_{min} = 3$. The results of the Level set optimization results where the parameter tau = 6.10^{-4} are close from these minimal values. The difference is 0.91 MPa for the stress and 0.11 for the strain. The BESO optimization results are twice greater than the original beam values.

References

1. Badiru AB, Valencia V v, Liu D (2017) Additive manufacturing handbook: product development for the defense industry. CRC Press
2. Zhu, J., Zhou, H., Wang, C., Zhou, L., Yuan, S., Zhang, W.: A review of topology optimization for additive manufacturing: Status and challenges. Chin. J. Aeronaut. **34**, 91–110 (2021)
3. Wohlers, T., Gornet, T.: History of additive manufacturing. Wohlers report **24**, 118 (2014)
4. Taylor, A.C., Beirne, S., Alici, G., Wallace, G.G.: System and process development for coaxial extrusion in fused deposition modelling. Rapid Prototyp J **23**, 543–550 (2017)
5. Yadroitsev, I., Yadroitsava, I., du Plessis, A., MacDonald, E.: Fundamentals of Laser Powder Bed Fusion of Metals. Elsevier (2021)
6. Zheng, Y., Wang, Y., Chen, R.K., Deshpande, S., Nelson, N.S., Buchman, S.R., Shih, A.J.: Tissue transformation mold design and stereolithography fabrication. Rapid Prototyp J **23**, 162–168 (2017)
7. Muthuram, N., Sriram Madhav, P., Keerthi Vasan, D., Mohan, M.E., Prajeeth, G.: A review of recent literatures in poly jet printing process. Mater Today Proc (2022). https://doi.org/10.1016/j.matpr.2022.08.090
8. Ziaee, M., Crane, N.B.: Binder jetting: A review of process, materials, and methods. Addit Manuf **28**, 781–801 (2019)
9. Pilipović A (2022) Sheet lamination. Polymers for 3D Printing: Methods, Properties, and Characteristics 127–136
10. Dass A, Moridi A (2019) State of the Art in Directed Energy Deposition: From Additive Manufacturing to Materials Design. Coatings: Vol 9. Page **418**(9), 418 (2019)
11. Lkadi, O., Nassraoui, M., Bouksour, O.: Aperçu sur la fabrication additive : technologies, matériaux, applications An Overview on Additive Manufacturing : Technologies, Materials and Applications. Incertitudes et fiabilité des systèmes multiphysiques **6**, 1–15 (2022)
12. Sigmund O, Maute K (2013) Topology optimization approaches. A comparative review. Structural and Multidisciplinary Optimization 2013 48:6 48:1031–1055
13. Bendsoe, M.P., Kikuchi, N.: Generating optimal topologies in structural design using a homogenization method. Comput Methods Appl Mech Eng **71**, 197–224 (1988)
14. Bendsoe, M.P., Sigmund, O.: Material interpolation schemes in topology optimization. Arch. Appl. Mech. **69**, 635–654 (1999)
15. Stolpe, M., Svanberg, K.: An alternative interpolation scheme for minimum compliance topology optimization. Struct. Multidiscip. Optim. **22**, 116–124 (2001)
16. Dai, X., Tang, P., Cheng, X., Wu, M.: A variational binary level set method for structural topology optimization. Commun Comput Phys **13**, 1292–1308 (2013)
17. Xie, Y.M., Steven, G.P.: A simple evolutionary procedure for structural optimization. Comput Struct **49**, 885–896 (1993)
18. Xia, L., Xia, Q., Huang, X., Xie, Y.M.: Bi-directional Evolutionary Structural Optimization on Advanced Structures and Materials: A Comprehensive Review. Archives of Computational Methods in Engineering **25**, 437–478 (2018)
19. Zhang, W., Zhou, Y., Zhu, J.: A comprehensive study of feature definitions with solids and voids for topology optimization. Comput Methods Appl Mech Engrg **325**, 289–313 (2017)

20. Guo X (2014) Doing topology optimization explicitly and geometrically: A new moving morphable components based framework. Frontiers in Applied Mechanics 31–32
21. Zhang, W., Chen, J., Zhu, X., Zhou, J., Xue, D., Lei, X., Guo, X.: Explicit three dimensional topology optimization via Moving Morphable Void (MMV) approach. Comput Methods Appl Mech Eng **322**, 590–614 (2017)
22. Liu, K., Tovar, A.: An efficient 3D topology optimization code written in Matlab. Struct. Multidiscip. Optim. **50**, 1175–1196 (2014)
23. Yago, D., Cante, J., Lloberas-Valls, O., Oliver, J.: Topology Optimization Methods for 3D Structural Problems: A Comparative Study. Archives of Computational Methods in Engineering (2022). https://doi.org/10.1007/s11831-021-09626-2
24. Ibhadode, O., Zhang, Z., Bonakdar, A., Toyserkani, E.: IbIPP for topology optimization—An Image-based Initialization and Post-Processing code written in MATLAB. SoftwareX **14**, 100701 (2021)

Chapter 11
Numerical Study of Mechanical Behavior of the Topologically Optimized Part Produced Virtually by Fused Deposition Modeling

Intissar Antar, Mourad Othmani, Khalid Zarbane, Mohamed El Oumami, and Zitouni Beidouri

Abstract Additive manufacturing (AM) has the advantage of produce parts with complex geometries compared to conventional manufacturing processes. Through topological optimization, these parts can then be optimized for weight and cost savings. Indeed, this method strives to find the optimal distribution of the material representing the structure on the basis of a given set of loads. However, the mechanical behavior of these optimized parts can be affected by various process printing parameters used. To be used as supporting or loaded structures (or under a certain loading), appropriate mechanical properties are required. Therefore, this research aims to numerically analyze the mechanical behavior of the digital 3D Messerschmitt-Bolkow-Blohm (MBB) beams printed in correspondence with Fused Deposition Modeling (FDM) process in a numerical environment. To this end, a digital model was created using a developed numerical approach based on the G-code. This model was then topologically optimized using the Solid Isotropic Material with Penalization (SIMP) method according to two configurations. The mechanical

I. Antar (✉) · K. Zarbane · M. El Oumami · Z. Beidouri
Laboratory of Advanced Research on Industrial and Logistics Engineering, National Higher School of Electricity and Mechanics, Hassan II University of Casablanca, B.P. 8118 Oasis, Casablanca, Morocco
e-mail: intissar.antar@ensem.ac.ma

K. Zarbane
e-mail: khalid.zarbane@univh2c.ma

M. El Oumami
e-mail: mohoumami@gmail.com

Z. Beidouri
e-mail: zitouni.beidouri@univh2c.ma

M. Othmani
Laboratory of Mechanics Engineering and Innovation National Higher School of Electricity and Mechanics, Hassan II University of Casablanca, B.P. 8118 Oasis, Casablanca, Morocco
e-mail: mourad.othmani@ensem.ac.ma

K. Zarbane and Z. Beidouri (eds.), *Proceedings of CASICAM 2022*, Springer Tracts in Additive Manufacturing, https://doi.org/10.1007/978-3-031-32927-2_11

behavior of these parts was predicted and compared. The results showed the impact of raster angle on mechanical behavior of the MBB optimized beam.

Keywords Additive manufacturing · FDM · Mechanical behavior · Topology optimization · SIMP

11.1 Introduction

Topology optimization (TO) has been identified as an important structural optimization tools, it is widespread used in several fields including aerospace and automobile to produce a lightweight structure. It strives to seek in the design space, the optimal distribution of material in the design domain of the structure for a specified set of loads and boundary conditions [1]. A range of standard methods of TO were proposed, like solid Isotropic Material with Penalization (SIMP). Rational Approximation of Material Properties (RAMP) [2], Evolutionary approach Structural Optimization (ESO) which considered as a hard -kill method [3]. And Level Set method (LSM) [4, 5]. The simplicity of the SIMP method, it becomes the most popular mathematical methods used in TO. For solving the TO problems, optimization algorithms are needed, the commonly used are: the Optimality Criteria method (OC) [6], the Method of Moving Asymptotes (MMA) [7] and the Sequential Linear Programming method (SLP) [8]. Therefore, additive manufacturing (AM) techniques are an obvious choice for produce this sort of structures, as they are able to manufacture structures faster, cheaper and more accurately than conventional methods [9]. There are numerous AM processes such as Selective Laser Sintering (SLS), Fused Deposition Modeling (FDM), selective laser melting (SLM). Hence, the mechanical behavior of the optimized parts printed by AM can be affected by various process printing parameters. In this regard, many authors had been studied the mechanical behavior of topological optimized structures printed by different AM processes. Hoglund et al[10] presented a novel approach which it a modified SIMP approach that determines the optimal distribution of an orthotropic material with fixed material orientation to design fiber-reinforced FDM parts. The results agree with the isotropic model computed by the SIMP method. Rashid et al. [11] studied the Bi-directional Evolutionary Structural Optimization (BESO) algorithm to generate topologically optimized lattice unit cells printed in two different unit cell arrangements using Selective Laser Melting (SLM) process, his results based on the bending test showed that the performance of BESO lattice unit beams were impeded by the brittle failure of SLM-printed parts. Garcia-Granada et al. [12] focuses in his research on structural optimization of additive manufactured parts of thermoplastic parts based on analysis of the stiffness/weight (mass) ratio for a beam subjected to a three-point bending load. The test samples printed by FDM according to two angles 0 and 45 are compared with those printed by Poly jet 3D printer using TO software. The results showed that the obtained stiffness to mass ratio varies with printing orientation. The study of Mohan and Simhambhatla [13] attempts to refine the outcome from TO with AM-specific considerations by introducing a neighborhood density function. The effectiveness of their method is validated subject on the three-point bending test to analyze the structural behavior on Messerschmitt-Bolkow-Blohm (MBB) beam using Poly Jet 3D printing. The structural performance of the AMTO geometry is marginally lower compared to the conventional TO geometry. Bo Li et al. [14] provides in his work a structurally heterogeneous lattice design method suitable for the quasi-static stress state of 3D printed parts and he demonstrated that the heterogeneous lattice part exhibited better comprehensive mechanical performance than the uniform lattice.

To sum up, several studies have been conducted to investigate the mechanical behavior of the structures printed by various AM process using different TO methods, however, less researches had been carried out on the optimization of the topology oriented to the FDM printing technique. The present work provides a novel numerical approach to investigate the mechanical behavior of the topological optimized structures. This research is organized as follow: in Sect. 11.2 the mathematical model of the SIMP method and numerical simulation of topology optimization in software ABAQUS are presented. Section 11.3 is devoted to the proposed numerical approach to investigate the mechanical behavior of the optimized part and printed virtually by FDM. The results are discussed in Sect. 11.4. The conclusions of the present research are deduced in Sect. 11.5.

11.2 Formulation and Numerical Simulation of Topology Optimization

11.2.1 Mathematical Formulation of SIMP Method

The most common optimization technique called Solid Isotropic Material with Penalization (SIMP) uses the density-based approach [15] to reach the optimal distribution of a material in a given space. In SIMP method, each design domain is discretized into a number of finite elements and each finite element is connected with a density ρ_e which evaluated continuously between 0–1, where 0 indicates the material is removed and 1 indicates the material is required. To avoid the singularity in finite element matrix, the density is given a lower limit ρ_{min} The ratio between the Young's modulus and the density can be written as:

$$E_e(\rho_e) = \rho_e^p E_0 \tag{11.1}$$

where E_0 is the young's modulus of the solid material.

Topology optimization problem for minimizing the compliance of the structure can be written as:

$$Minimize : C(\rho) = F^t U = \sum_{e=1}^{N} (\rho_e)^p \{u_e\}^t [K_e] \{u_e\} \tag{11.2}$$

$$\begin{cases} \frac{V(\rho)}{V_0} = V_f \\ [K]\{u\} = \{F\} \\ 0 < \rho_{min} \leq \rho_e \leq 1 \end{cases} \tag{11.3}$$

where U and F are the global displacement and force vector,K is the global stiffness matrix, u_e and K_e are the element displacement vector and stiffness matrix, respectively. ρ_{min} is a vector of minimum relative densities which is set to $\rho_{min} = 0.001$. N is the number of elements used to discretize the design domain,p is the penalization power. This parameter penalizes intermediate densities. $V(\rho)$ and V_0 is the material volume and design domain volume. V_f prescribes the volume fraction [16].

Equation (11.2) represents the objective function which is minimizing the compliance of structure with the density as a variable subject to the constraint of the volume fraction.

11.2.2 Topology Optimization Simulation Using Software ABAQUS

In this study, ABAQUS 2022 from Dassault Systems was used to perform topology optimization for minimizing the compliance and the test case chosen is the so-called Messerschmitt-Bolkow-Blohm (MBB) beam benchmark problem. It is considered one of the most popular tests problems in topology optimization is optimize according to the SIMP method. A schematic view of the MBB beam is shown in Fig. 11.1. The dimensions and loading conditions are depicted in Table 11.1.

After modeling the MBB beam, the elastic constants of ABS are introduced respectively Young's Modulus $E = 2200 MPa$ Poisson's ratio $\nu = 0.33$. Thus, a

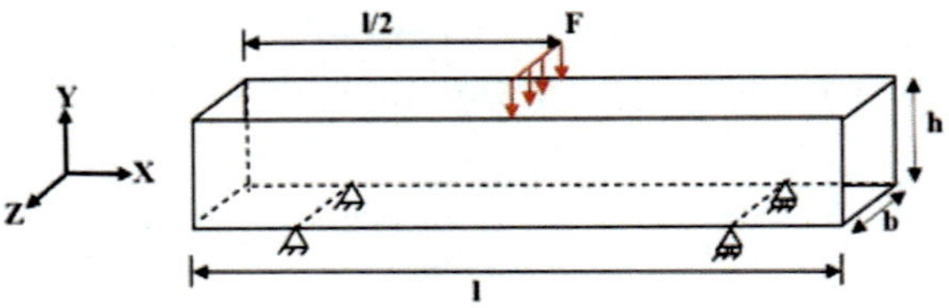

Fig. 11.1 MBB beam design domain and its boundary conditions

Table 11.1 Parameter values of MBB beam

Parameter	Value (mm)
Length (l)	80
Width (b)	20
Height (h)	4

Table 11.2 Configuration of the topology optimization for MBB beam

Objective function	Minimize strain energy
Optimization algorithm	Optimality Criteria (OC)
Constraint	Volume Fraction $V_f \leq 30\%$
Force	12N
SIMP factor	3
Filter radius	0.6 mm

linear and isotropic solid element type is applied. The parameters necessary for topology optimization are tabulated in Table.11.2. Linear hexahedral elements C3D8R are used in mesh type.

Following the optimization process in Fig. 11.2 topology optimization of the MBB beam was performed to achieve the minimum compliance under a static load, through the minimization of the strain energy subject to the volume fraction, which determines the preserved material at the end of optimization, considering that the material in all design remains in the elastic zone. The load and boundary conditions regions are frozen. The minimum density is set to 0.001 and the optimization problem is solved by the method Optimality Criteria (OC).

The final topological optimization design is shown in Fig. 11.3. The optimization process needed 42 iterations to achieve the optimal design. Convergence history for the MBB beam is given in Fig. 11.4. The minimum compliance obtained for 30% of the volume preserved is 1.23 N.m.

11.3 Proposed Method

In order to study the mechanical behavior of the FDM printed parts, a numerical approach has been developed and presented. This approach enables to obtain a digital optimized design similar to printed shape. A script using Python programming language is developed in which all printing parameters are taken into consideration. This code runs in ABAQUS 2022 (Dassault Systems) in a manner to execute the G-code extracted from 'Slic3r', then draws the tool path and the raster section [17].

11.3.1 Generation of Virtual Model

The virtual optimized model is obtained in several steps summarized in Fig. 11.5. Initially, CAD model of the optimized MBB beam mentioned above is saved in Stereo Lithography (STL) format, then, its G-code is extracted from "Slic3r" in order to reproduce the optimized structure considering the printing and filament settings

Fig. 11.2 Flowchart of the optimization process in ABAQUS

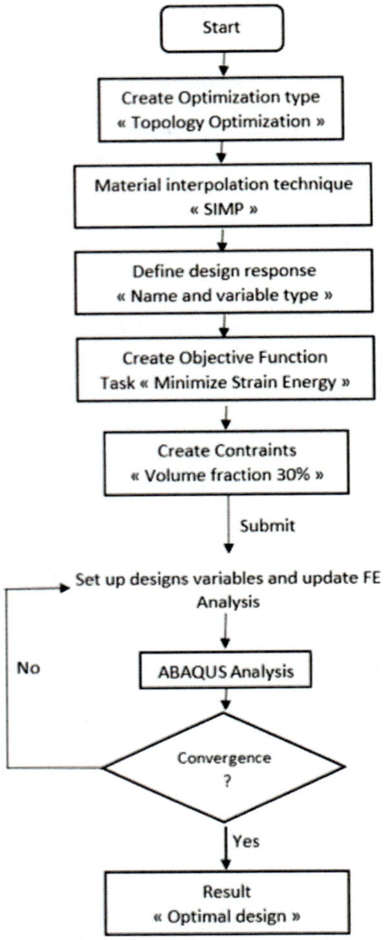

Fig. 11.3 Optimal design of MBB beam

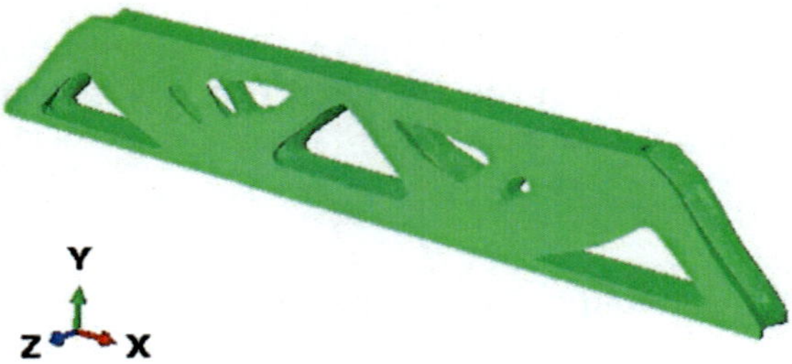

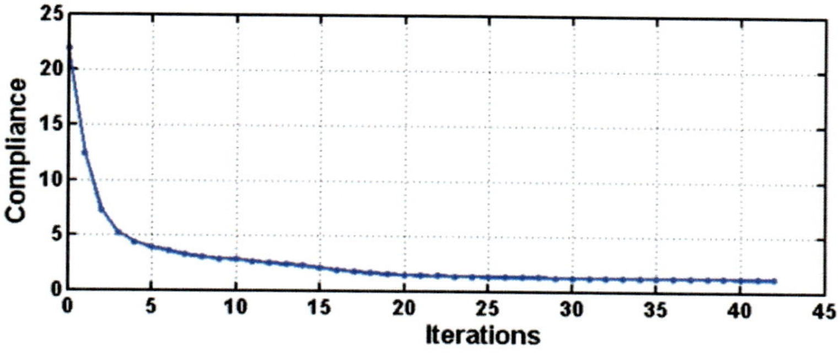

Fig. 11.4 Evolution history of the objective function

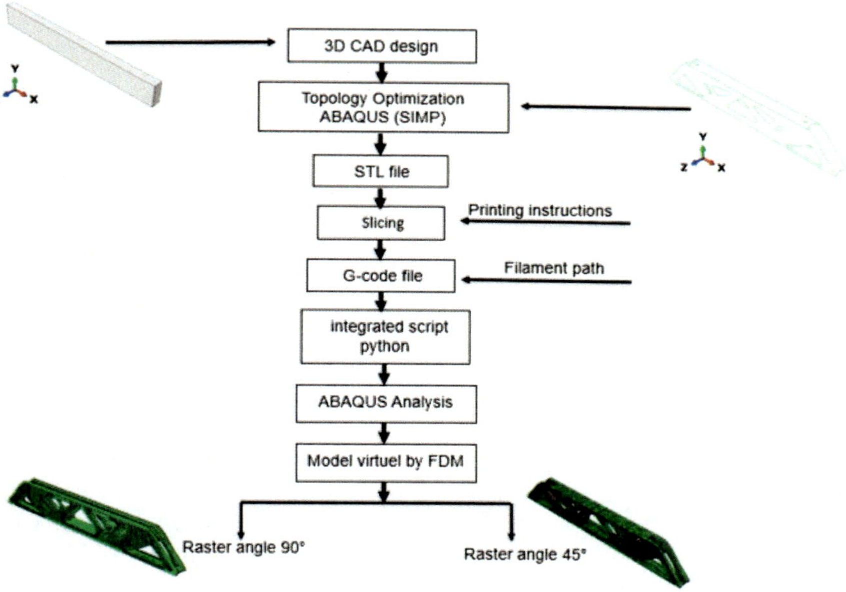

Fig. 11.5 Computational simulation of the proposed approach

stated in Table 11.3. Secondly then, a script in "Python" based on the G-code file has been developed and integrated in the "ABAQUS 2022." At the end of the script generation, the virtual model has been obtained with two raster angles presented in Fig. 11.6.

Table 11.3 Printing setting of the digital models

Properties	Value
Layer thickness	0.4 mm
Raster angle	90–45°
Raster width	0.8 mm
Overlap	5%
Infill patern	Aligned rectilinear
First layer speed	30 mm/s
Print speed of other layers	60 mm/s
Number of perimeters	1
Temperature	240°

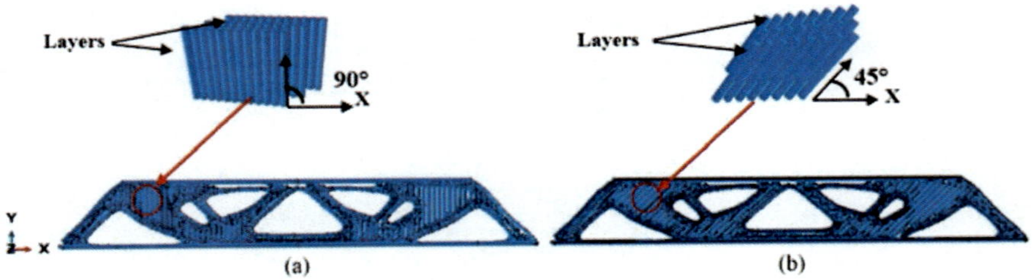

Fig. 11.6 Virtual models of MBB beam according to: **a** Raster angles 90°. **b** Raster angle 45°

11.3.2 Analysis

Once the generation of digital models is completed, the elastic constants of ABS are introduced [17]. Material is assumed to be homogeneous and isotropic in the analysis. Independent instance type is selected in the 'Assembly' module, and tie type interaction contact is created between the contour and infill aiming to stick the filament between them to determine which part of the surface of the model comes into contact during the deformation [18]. The digital models were submitted the similar boundary conditions to the previous optimized MBB beam. Linear 3D tetrahedral elements C3D4 (4 nodes linear tetrahedron) are used for meshing the virtual models. The boundary conditions and the mesh of the virtual models are shown in Fig. 11.7.

11.4 Results and Discussion

The results of the finite element analysis of mechanical properties of the virtual printed models are presented in Table 11.4. For the digital printed model of 90°, the maximum principal strain is 0.47(%), maximum displacement is 2.26 mm, and the

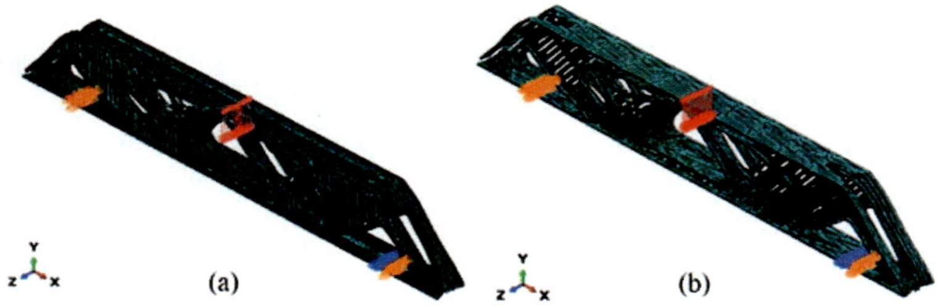

Fig. 11.7 Mesh of the virtual models: **a** raster angle of 90°, **b** raster angle of 45°

Table 11.4 Mechanical properties of the virtual structures

Raster angle	Max Principal strain (%)	Max von Mises stress (GPa)	Max displacement (mm)
90°	0.47	14.4	2.26
45°	3.67	38.7	9.69

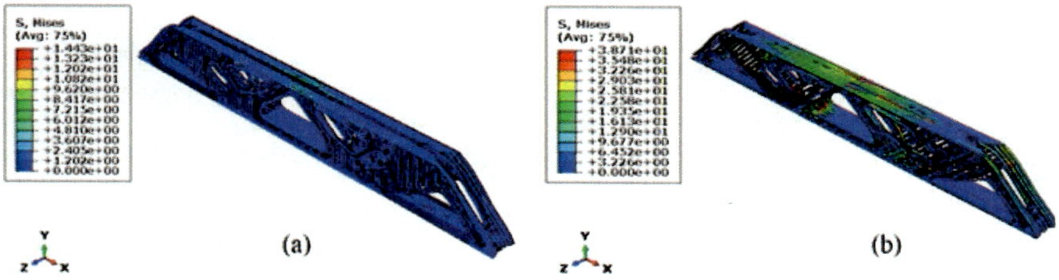

Fig. 11.8 FEA results of the optimal typologies (virtual structures): **a** raster angle of 90°, **b** raster angle of 45°

maximum von Mises stress is $\sigma_{max}^{90^\circ} = 14,4$ GPa. However, the maximum principal strain for 45° is 3.67(%), the maximum displacement is 9.69(mm), and the maximum von Mises stress is $\sigma_{max}^{45^\circ} = 38.7$ GPa. The von Mises stress distribution presented in Fig. 11.8 appears clearly in the 45° numerical model compared to the 90°model.

11.5 Conclusion

In this study, a novel numerical approach to examine the mechanical behavior of the 3D printed optimized structures has been proposed. This method allowed us to create the digital models taking into consideration the printing parameters through a script developed in Python based on the G-code file. The digital models were created in a numerical environment according to 90° and 45° raster angles. This method is applied on a MBB beam. The numerical results based on the distribution of von Mises stress demonstrate that the topological optimized MBB beam of 90° has significantly better mechanical behavior than 45°. We concluded from our results that the filaments parallel to the load provides greater strength than the inclined filaments. The future work includes developing an experimental approach to investigate the mechanical behavior of the optimized structures.

References

1. Sigmund, O., Bondsgc, M.P.: Topology optimization. In: State-of-the-Art and Future Perspectives, Copenhagen: Technical University of Denmark (DTU) (2003)
2. Stolpe, Mathias, Svanberg, Krister: An alternative interpolation scheme for minimum compliance topology optimization. Struct. Multidiscip. Optim. **22**(2), 116–124 (2001)
3. Xie, Y.M., Steven, G.P.: Evolutionary structural optimization for dynamic problems. Comput. Struct. **58**(6), 1067–1073 (1996)
4. Allaire, Grégoire., Jouve, François, Toader, Anca-Maria.: Structural optimization using sensitivity analysis and a level-set method. J. Comput. Phys. **194**(1), 363–393 (2004)
5. Wang, M.Y., Wang, X., Guo, D.: A level set method for structural topology optimization. Comput. Methods Appl. Mech. Eng. **192**(1–2), 227–246 (2003)
6. Bendsøe, M.P.: Optimization of Structural Topology, Shape, and Material, vol. 414. Springer (1995)
7. Svanberg, Krister: The method of moving asymptotes-a new method for structural optimization. Int. J. Numer. Methods Eng. **24**(2), 359–373 (1987)
8. Dunning, P.D., Kim, H.A.: Introducing the sequential linear programming level-set method for topology optimization. Struct. Multidiscip. Optim. **51**(3), 631–643 (2015)
9. Berrio Bernal, J.D., Silva, E.C.N., Montealegre Rubio, W.: Characterization of effective young's modulus for fused deposition modeling manufactured topology optimization designs. Int. J. Adv. Manuf. Technol. **103**(5), 2879–2892 (2019)
10. Hoglund, R., Smith, D.E.: Non-isotropic material distribution topology optimization for fused deposition modeling products. In: 2015 International Solid Freeform Fabrication Symposium. University of Texas at Austin (2015)
11. Rashid, R., Masood, S.H., Ruan, D., Palanisamy, S., Huang, X., Rahman Rashid, R.A.: Topology optimisation of additively manufactured lattice beams for three-point bending test. In: 2018 International Solid Freeform Fabrication Symposium. University of Texas at Austin (2018)
12. Garcia-Granada, A.A., Catafal-Pedragosa, J., Lemu, H.G.: Topology optimization through stiffness/weight ratio analysis for a three-point bending test of additive manufactured parts. In: IOP Conference Series: Materials Science and Engineering, vol. 700, p. 012012. IOP Publishing (2019)
13. Shashi Ranjan Mohan and Suryakumar Simhambhatla: Adopting feature resolution and material distribution constraints into topology optimisation of additive manufacturing components. Virtual Phys. Prototyp. **14**(1), 79–91 (2019)
14. Li, Bo., Shen, Ciming: Solid stress-distribution-oriented design and topology optimization of 3d-printed heterogeneous lattice structures with light weight and high specific rigidity. Polymers **14**(14), 2807 (2022)
15. Sigmund, Ole, Maute, Kurt: Topology optimization approaches. Struct. Multidiscip. Optim. **48**(6), 1031–1055 (2013)
16. Sigmund, Ole: A 99 line topology optimization code written in Matlab. Struct. Multidiscip. Optim. **21**(2), 120–127 (2001)

17. Othmani, M., Zarbane, K., Chouaf, A.: Enhanced mesostructural modeling and prediction of the mechanical behavior of acrylonitrile butadiene styrene parts manufactured by fused deposition modeling. Int. Rev. Mech. Eng. **14**(4)
18. Antar, I., Othmani, M., Zarbane, K., El Oumami, M., Beidouri, Z.: Topology optimization of a 3D part virtually printed by FDM. J. Achiev. Mater. Manuf. Eng. **21**(2), 120–127 (2022)

Chapter 12
Topological Optimization for Fused Deposition Modeling (FDM) Process

Abderrazak Boualaoui, Driss Sarsri, and Mohammed Lamrhari

Abstract The emergence of additive manufacturing processes has changed our knowledge not only in terms of manufacturing but also in terms of design. 3D printing by deposition of fused material is better known as FDM or Fused Deposition Modeling which is one of the additive manufacturing processes for plastic parts. Topological optimization for the FDM process is an asset that reduces the total weight of a part. It aims to distribute the material in an optimal way based on a set of loads, volume limits of the structure, or user conditions. By combining topological optimization with the FDM process, the field of possibilities widens considerably: • On the one hand, topological optimization makes it possible to design complex geometries that traditional machining cannot achieve. • On the other hand, additive manufacturing makes it possible to create equally complex shapes that cannot be reproduced by conventional processes. 3D printing and topological optimization therefore make it possible to accelerate and improve the creation of high-performance and complex products, without increasing their design and manufacturing costs. We are therefore in a medium-term research project which is to develop design tools and methods for plastic additive manufacturing.

Keywords Topology optimization · Additive manufacturing · FDM

A. Boualaoui (✉)
ENSA, Tanger, Morocco
e-mail: BOUALAOUI7374@GMAIL.COM

D. Sarsri
ENSA, Tanger, Morocco

M. Lamrhari
ENSAM, Meknes, Morocco

K. Zarbane and Z. Beidouri (eds.), *Proceedings of CASICAM 2022*, Springer Tracts in Additive Manufacturing, https://doi.org/10.1007/978-3-031-32927-2_12

12.1 Introduction

Additive manufacturing (AM) is the general term for technologies that successively join material to create physical objects as specified by data from a 3D model [1]. These technologies are currently used in various industrial engineering applications as well as in other sectors of society, such as medicine, education, architecture, cartography, toys and entertainment [2].

FDM (Fused Deposition Modeling) technology is a 3D printing technique that creates parts additively using filament [3]. The process consists of depositing molten wire layer by layer, from a 3D model. It is a perfect option, whether for prototyping or mass production of functional parts. It is indeed an affordable 3D printing solution for mechanical and functional parts. In addition, FDM technology offers a wide choice of materials, from plastic to metal.

With design tools (CAE: Computer Aided Engineering), simulation on a 3D digital model can recover the spatial distribution field of stresses and strains on a loaded part. The distribution of the stresses makes it possible to identify the places more or less loaded. Based on the spatial distribution of the load, it is then possible to optimize the distribution of the material partly by reinforcing the highly loaded places and relieving the volume with a low load [4].

Several topological optimization methods have appeared, including we will quote in a non-exhaustive way:

- The methods of penalizing the material, the most widespread of which is the SIMP method (Solid Isotropic Microstructure with Penalization) which defines the density of the material as a design variable of the mesh domain, [Rozvany 92, Bendsøe 99, Bendsøe 89].
- ESO (Evolutionary Structural Optimization) methods where the topology optimal is achieved by gradually removing the least desired material (least efficient, least worked, etc.) from the design domain, [Tanskanen 02, Zuo 14].
- Methods based on the topological derivative which measures the influence of the (infinitesimal) creation of small holes or cracks in the objective function, [Burger 04, Norato 07, Novotny 13].
- The level-line or Level-Set methods which track, model and simulate the dynamic interface of the structure by using a function Level-Set (Lipshitzian) [Osher 88].

The optimization of structures is one of the essential concerns for the design of systems in the mechanical industry (civil engineering, aeronautics, automotive, etc.). Design offices not only improve the mechanical performance of the parts they design, but also seek to optimize their weight, size and production cost. The problem that interests us here, crucial in many industrial applications, is to find a form of structure having the best compromise between its resistance and its mass [5].

This work presents the optimization procedure of a plastic component (gate release handle) which is mechanically loaded. Therefore, the research results can be applied to the design of similar products produced by the FDM method.

12.2 Materials and Methods

12.2.1 Primary CAD Model

The starting point for additive manufacturing is a model that digitally represents the object in 3 dimensions, sliced. This information will be sent to the 3D printer which will manufacture the part by adding successive layers (Fig. 12.1).

The primary model is produced on ANSYS 2022R2 under SpaceClaim subject to the rules and recommendations for the creation of digital models. The topological optimization is ensured by the creation in the model of the structural elements to define the supports, the loads and the boundary conditions. The steps implemented in the text are carried out by the ANSYS tool.

For the optimization procedure a model of the release handle is used. It is a component of the garage door control system. The calculation by the finite element method is ensured by the CAD data of the primary model.

The boundary conditions represent the loading forces in the release handle bearing. The main model is shown in Fig. 12.2. A tetrahedral finite element lattice of size 4 mm is defined to perform the finite element calculation and the properties of the ABS material are affected, in particular the material properties the Young's modulus, which characterizes the deformation of the part under a given load. The FEM network definition of the network, load, and storage is shown in Fig. 12.3. The identical material is then used to define the material characteristics for topology optimization (Fig. 12.4).

The parameters of the optimization with the SIMP method are:

_ A tetrahedral finite element lattice of size 4 mm
_Penalty factor is 3.
_Objective: to minimize compliance.
_Constraint on the response: 50% of the mass.
_Manufacturing constraint: 45° overhang.

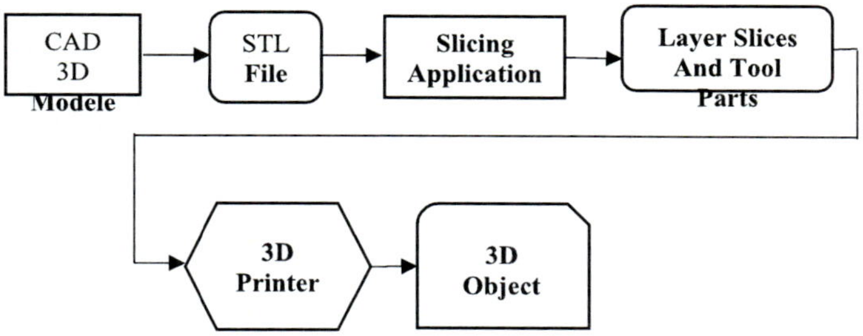

Fig. 12.1 3D printing process

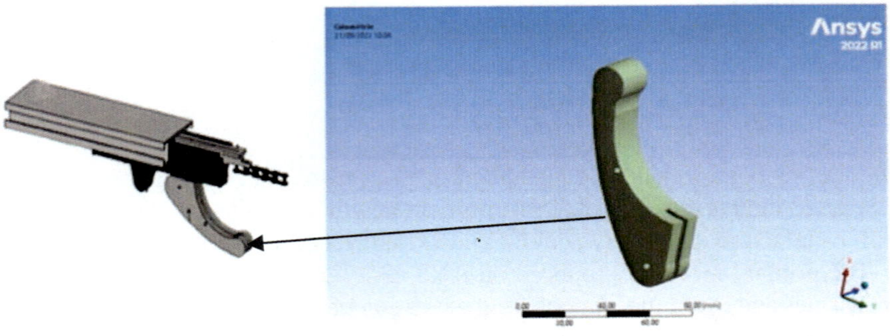

Fig. 12.2 Initial handle for garage door

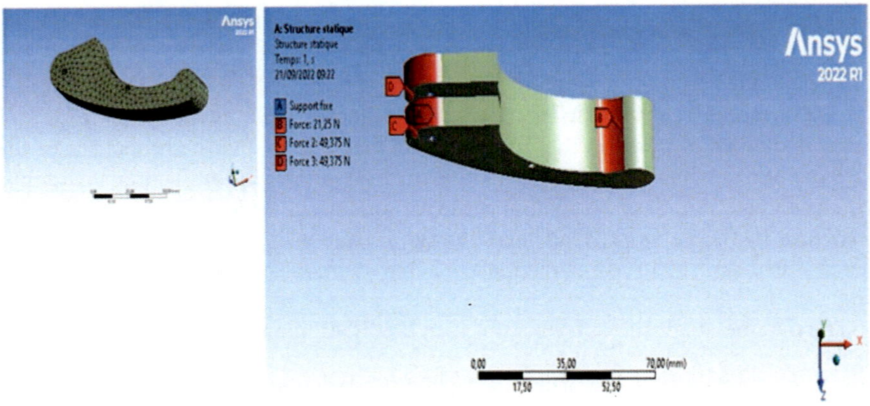

(a) (b)

Fig. 12.3 **a** Initial handle with FEM. **b** Loads (21,25N sur B and 49,375N sur D et C)

(b) Results of the simulation

The simulation is started to analyze the stress distribution in the part at the defined loads. To evaluate the functional behavior of the loaded handle the reduced von-mises constraint is chosen. The handle in ABS material whose characteristics according to Table 12.1

(b-1) Results of the simulation before optimization (Figs. 12.5 and 12.6).

(b-2) Results of the simulation with the manufacturing orientation along the Y direction after optimization (Fig. 12.7).

On Fig. 12.8, almost the same deformation on the whole part except with a maximum at the hole.

On Fig. 12.9, we notice a balancing of the stress on all the volume but with a peak (84,367MPA) which largely exceeds the maximum stress before optimization.

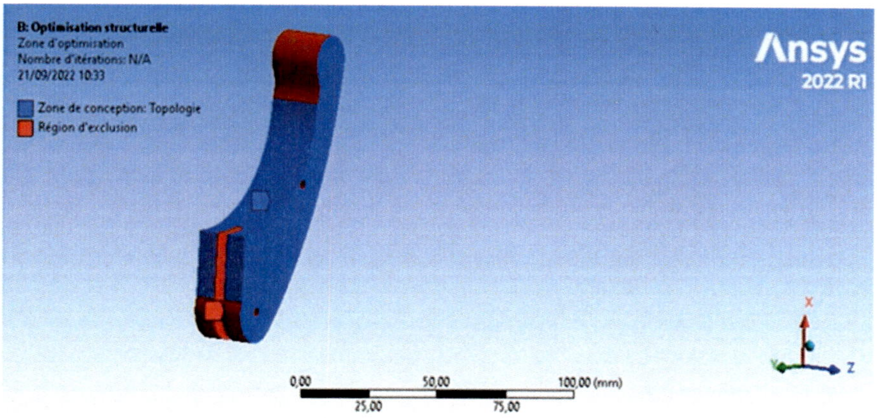

Fig. 12.4 Optimization areas

Table 12.1 Physical and mechanical characteristics of the handle

Material	ABS plastic
Density	1040 kg/m^3
Young's module	2390 MPa
Volume	38,37 cm^3
Mass	0,04 kg

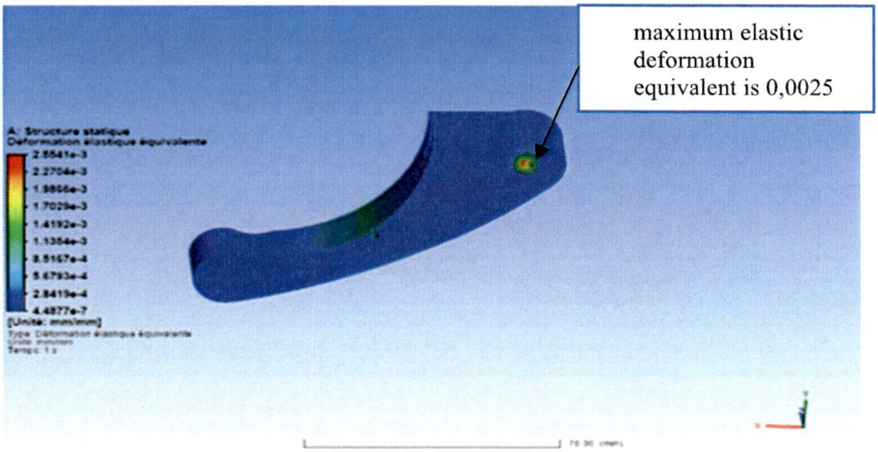

Fig. 12.5 Equivalent elastic deformation

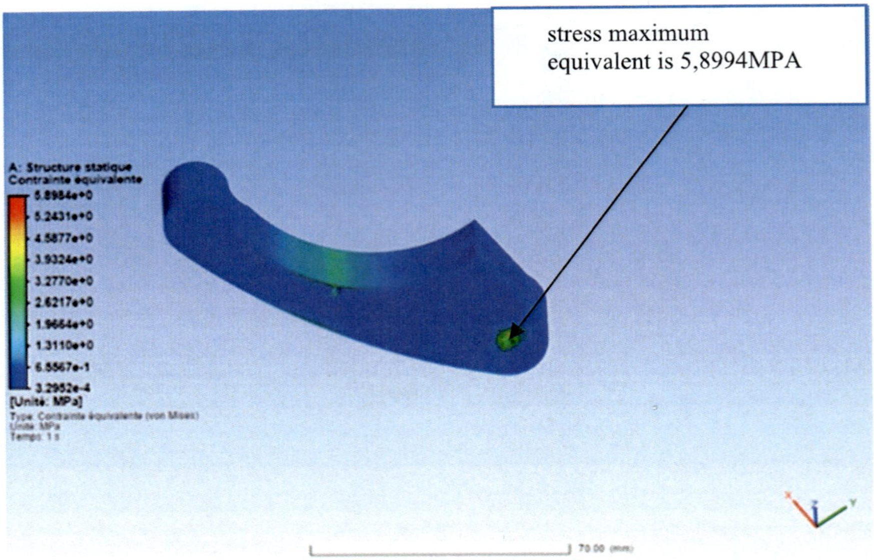

Fig. 12.6 Equivalent stress

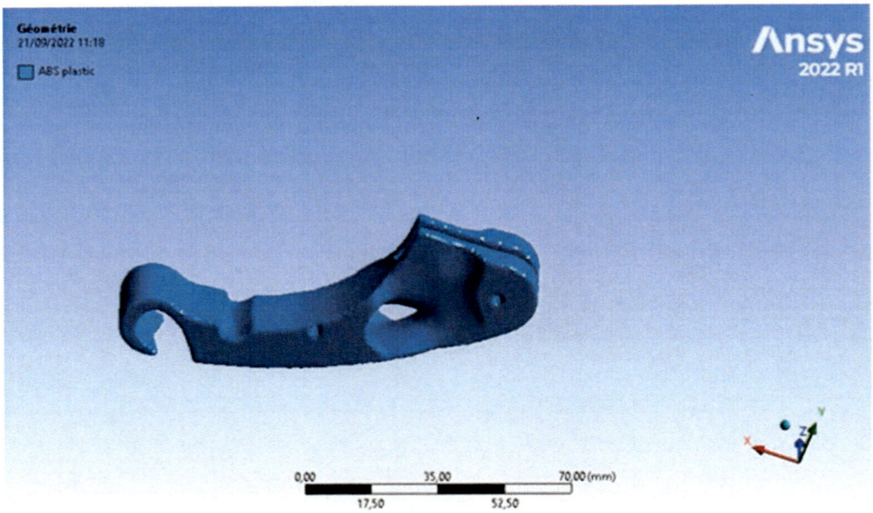

Fig. 12.7 Optimized and smoothed handle

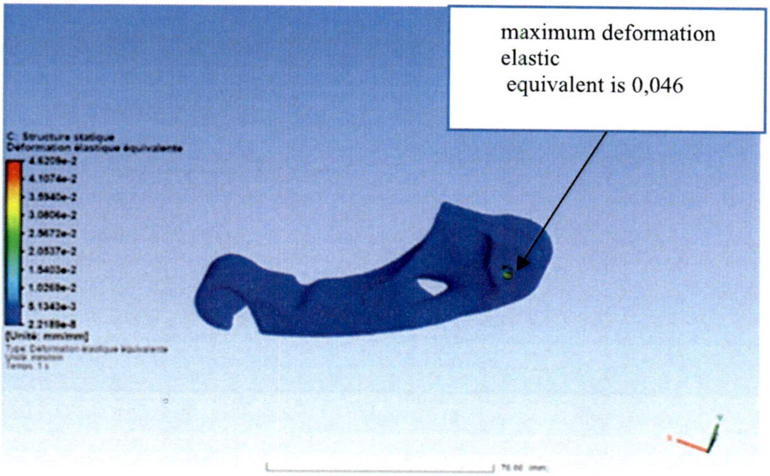

Fig. 12.8 Deformation elastic equivalent after optimization

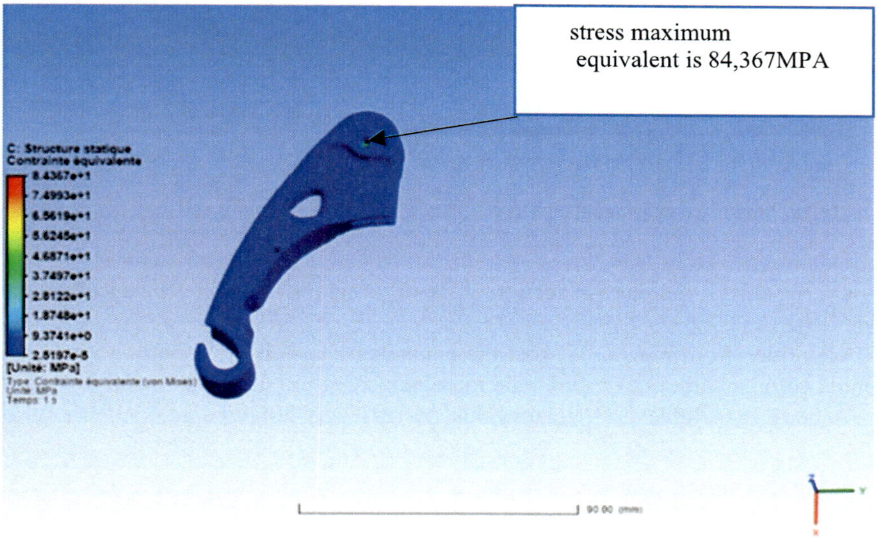

Fig. 12.9 Equivalent stress after optimization

According to Table 12.2, we note a decrease in volume and also in mass of 50%.

(b-2) Results of the simulation with the manufacturing orientation along the Z direction (Fig. 12.10).

Table 12.2 Physical and mechanical characteristics of the optimized handle with the manufacturing orientation along the Y direction

Material	ABS plastic
Density	1040 kg/m^3
Young's module	2390 MPa
Volume	26,17 cm^3
Mass	0,0272 kg

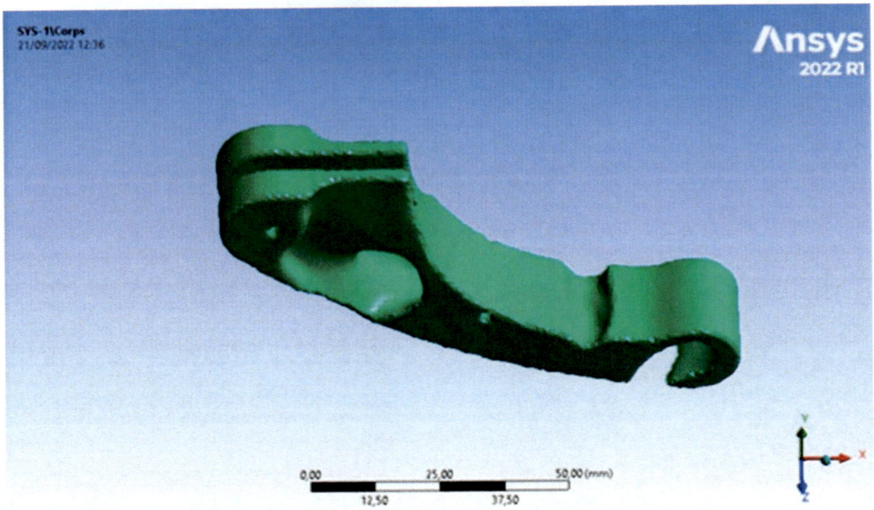

Fig. 12.10 Optimized and smoothed handle

On Fig. 12.11, there is a reduction of the deformation and that is balanced on the whole part.

We notice from Fig. 12.12, that the equivalent stress is well distributed on the whole volume without exceeding the maximum stress on the initial handle.

According to Table 12.3, the mass has decreased by 50%.

12.3 Discussion

In both optimization cases, the mass is minimized by 50%. For the optimization following Y, we notice a remarkable overshoot of the initial maximum handle stress. While in the case of the optimization along Z, the deformation is minimized and the stress is well balanced on the whole volume and without concentration of critical stresses.

So, with topological optimization we can optimize the stresses and reduce the mass by choosing the right manufacturing direction.

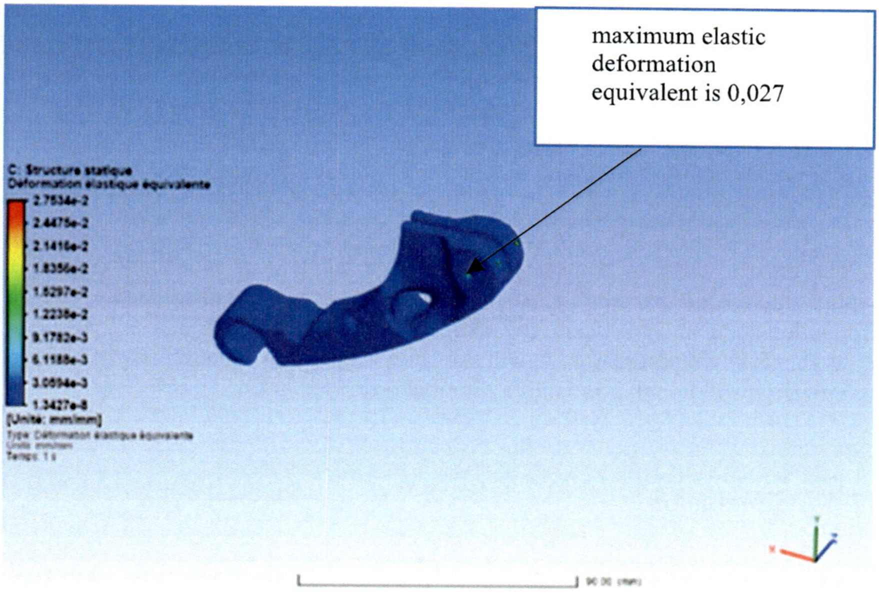

Fig. 12.11 Deformation after optimization

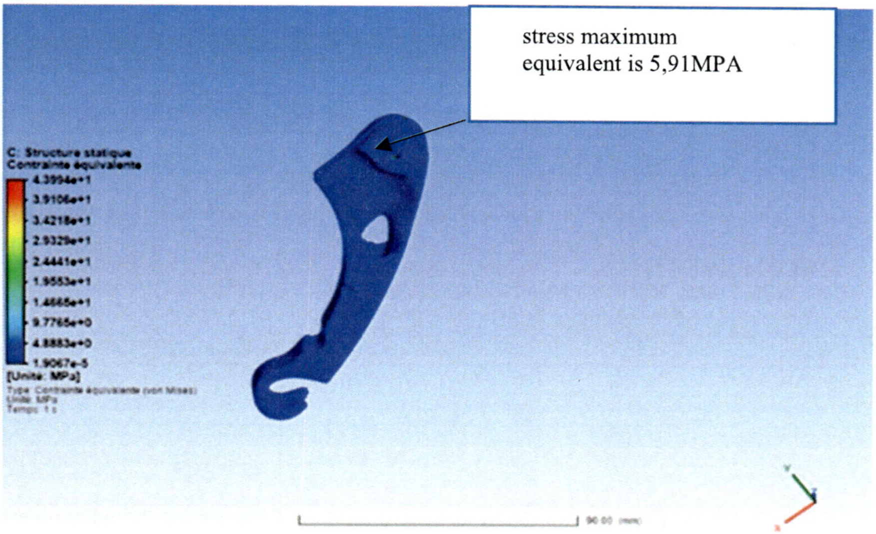

Fig. 12.12 Equivalent stress after optimization

Table 12.3 Physical and mechanical characteristics of the optimized handle with the manufacturing orientation along the Z direction

Material	ABS plastic
Density	1040 kg/m^3
Young's module	2390 MPa
Volume	26,173 cm^3
Mass	0,027 kg

The results stated on the sample part can be applied in equivalent or similar applications of the design of a structural solution of a constrained part, where the production perspective is one of the additive technologies.

In the following research, I will see how to choose an optimal topological optimization solution that satisfies a minimal cost.

12.4 Conclusion

Topological optimization makes it possible to orient designers very early on towards the optimal shapes in line with the objectives set. It is increasingly used in the design of industrial parts to optimize their design according to one or more functional constraints (mechanical resistance, weight reduction, etc.). This new methodology saves time (reduction of calculation iterations) and optimizes product design by avoiding traditional design, a long trial and error process.

References

1. Yu, H. et al.: Topology optimization for multipatch fused deposition modelling 3D printing 10(3) (2020)
2. ISO/ASTM 52900:2021(fr)
3. Dvorakova, J.: Int J Mech Eng Robot Res 10(2) (2021)
4. Rozvany: Aims. Struct Multidisc **21**, 90–108 (2001)
5. Ngo, T.D. et al.: Additive manufacturing (3D printing): a review of materials, methods, applications and challenges **143**, 172–196 (2018)

Part IV
AM Process Simulation

Chapter 13
Local Structural Anisotropy in Particle Simulations of Powder Spreading in Additive Manufacturing

Sudeshna Roy, Hongyi Xiao, Mohamad Yousef Shaheen, and Thorsten Pöschel

Abstract Producing consistent and homogeneous packing structure in powder layer deposition for cohesive raw materials under varying thermal conditions is challenging for additive manufacturing. Interparticle cohesion and thermal parameters play key roles on the structure of powder layer deposited on the substrate in additive manufacturing. In this work, we characterize the structural anisotropy of the deposited powder layer and quantify the packing structure on the particle-level using a threshold-free local packing anisotropy based on Voronoi tessellation. Based on the statistics of the local anisotropy, we observe a transition in the structure of the deposited powder layer from homogeneous to heterogeneous for cohesive materials at $Bo = 10$. Including an idealized temperature-dependence of the normal contact force does not influence the structure of the deposited layer in powder spreading.

Keywords Discrete element method · Powder spreading · Cohesion · Thermal · Homogeneity

S. Roy and H. Xiao: These authors have contributed equally to this work.

S. Roy (✉) · H. Xiao · T. Pöschel
Institute for Multiscale Simulation, Friedrich-Alexander-Universität Erlangen-Nürnberg, Cauerstrasse 3, 91058 Erlangen, Germany
e-mail: sudeshna.roy@fau.de

H. Xiao
e-mail: hongyi.xiao@fau.de

T. Pöschel
e-mail: thorsten.poeschel@fau.de

M. Y. Shaheen
Multi Scales Mechanics, University of Twente, Drienerlolaan 5, 7522 Enschede, Netherlands
e-mail: m.y.shaheen@utwente.nl

K. Zarbane and Z. Beidouri (eds.), *Proceedings of CASICAM 2022*, Springer Tracts in Additive Manufacturing, https://doi.org/10.1007/978-3-031-32927-2_13

13.1 Introduction

Powder-based additive manufacturing, e.g., powder bed fusion, has attracted significant attention [1–3] due to its advantages of fast prototyping, superior design, and geometrical flexibility. This technique allows near-net-shape production, enabling quick production of exact and customized parts, saving time and reducing waste. However, a more extensive usage of this manufacturing technology is still missing due to limitations in process quality control, e.g. non-uniform powder packing during spreading and the limited range of powder materials commercially available. Various types of structural defects are observed which strongly correlates to deficit in quality of sintered parts [4, 5]. Discrete Element Method (DEM) simulation is a particle-based technique to examine the structure of layers in powder spreading. Using DEM simulations, Nasato et al. found that small frequency and amplitude of a vibrating recoating mechanism lead to a smaller powder bed porosity [6]. The recoating velocity also plays a critical role on the effects of particle shape influencing the powder bed porosity [7]. Shaheen et al. showed that powder layer defects are more likely to occur for higher particle rolling and sliding friction [8]. He et al. investigated the effects of particle size and cohesion on powder spreadability, and investigated their effects on the local packing density and surface roughness and the DEM simulations are validated with qualitative agreements with the corresponding experiments [9].

The key aspect of simulation-driven optimization of the layer quality is a good characterization of the packing structure. Despite local packing density being widely used as a measure for characterizing granular packing, this parameter alone is not a good indicator to identify structural defects in disordered packing [10] since many packing arrangements correspond to the same local density. On the other hand, local structural anisotropy is an inherent property of non-crystalline packing and is associated with important mechanical properties in disordered packing, e.g. jamming [11], plasticity [10], shear band formation [12–14]. Therefore, in this study, we use local structural anisotropy as a measure to characterize structures of the deposited particles.

13.2 Numerical Model

13.2.1 Numerical Setup

The spreading processes are simulated using an open-source code MercuryDPM [15]. We use a simple linear elastic adhesive force to study the effect of cohesion and a temperature dependent contact model to study the thermal effect. We simulate a small part of the powder bed of width 1 mm using periodic boundary condition in the lateral direction y. The spreading tool blade is shown in Fig. 13.1, moving at a constant velocity v_T along the spreading direction x. The substrate is assumed to be flat with the wall-particle friction coefficients same as that of the particle-particle.

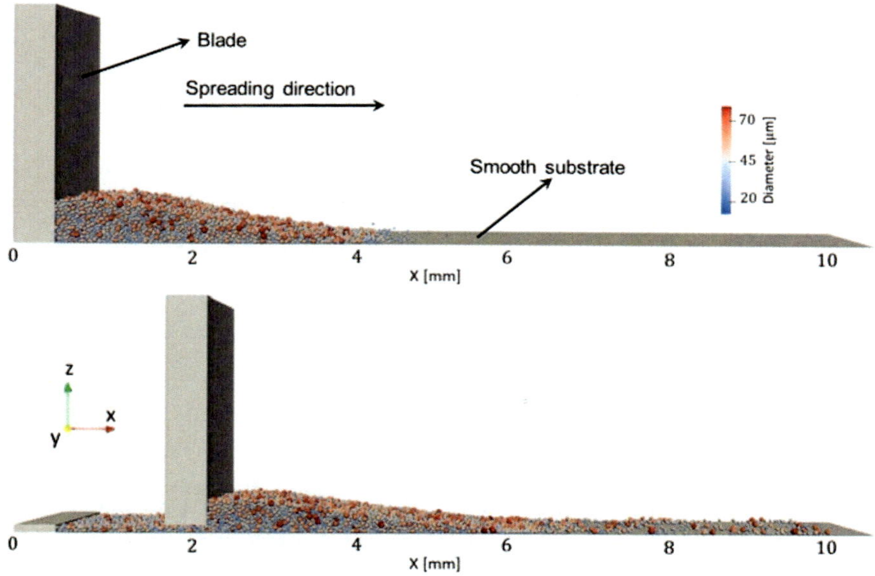

Fig. 13.1 Numerical setup for powder spreading on a smooth substrate, before and during spreading for non-cohesive particles, $\mu_r = 0.005$ and $\mu_s = 0.1$ s

We insert particles of diameter $D_{50} = 37$ μm, $D_{\text{min}} = 12$ μm and $D_{\text{max}} = 79$ μm in front of the spreader tool, at $(x, y, z) \in [0.5, 2.5]$ mm $\times[0, 1]$ mm $\times[0, h]$ mm until the total bulk particle volume equals 0.75 mm^3 which is sufficient to create a powder layer of 7 mm length, 1 mm width with a tool gap of $H = 100$ μm. After the particles settle down and relax, the simulation of the spreading process starts with velocity $v_T = 10$ mm/s. Particles reaching the end of the powder bed ($x = 10$ mm) are not considered for the analysis. The spreading process ends when the blade arrives at the end and the simulation ends after time 1.5 s when the system is static, i.e., the kinetic energy is sufficiently low.

13.2.2 Cohesive Contact Model

Many models exist in DEM to describe dry cohesion of small particles by including an attractive force due to van der Waals interaction between particles close to each other or in contact. A linear law for the adhesive force is used, which was shown to have the same bulk rheology as the more realistic non-linear models [16]. The normal force is composed of a linear elastic, dissipative and a linear adhesive force:

$$f_{ij}^{adh} = \begin{cases} -f_{max}^{adh}, & \text{if } \delta_{ij}^n \geq 0 \\ -(f_{max}^{adh} + k^{adh}\delta_{ij}^n), & \text{if } -\frac{f_{max}^{adh}}{k^{adh}} \leq \delta_{ij}^n < 0 \\ 0, & \text{otherwise} \end{cases} \tag{13.1}$$

where k^{adh} is the adhesion stiffness during loading and unloading. The maximum adhesion force is defined as $f_{max}^{adh} = \frac{3}{2}\pi\gamma\frac{D_{eff}}{2}$, where γ is the surface energy and $D_{eff} = \frac{D_i D_j}{D_i + D_j)}$ is the effective diameter of the two interacting particles i and j.

13.2.3 *Thermal Contact Model*

A recently developed thermal discrete element model is used in this work that simulates the powder spreading at elevated temperature. The model is developed for the application of Laser Powder Bed Fusion (LPBF) on polymers. Here, as an initial attempt, we use this contact model for analyzing the effect of powder temperature on the structure of deposited powder layer in the powder spreading process. Simulations are initiated with a homogeneous temperature $T_{part} = 430$ K, which is higher than the ambient temperature $T_{amb} = 298.15$ K but are below the melting temperature $T_m = 451.15$ K. The wall-particle contacts are assumed to be adiabatic. During spreading, the particle temperatures evolve as heat, Q_i is transferred between the particles in contact through conduction as Q_i^{cond} and with the ambient through thermal convection as Q_i^{conv} and through radiation as Q_i^{rad},

$$\begin{aligned} Q_i &= Q_i^{cond} + Q_i^{conv} + Q_i^{rad} \\ &= \sum k_s \frac{s_{ij}}{r_{ij}}(T_j - T_i) + h_{conv}A_i(T_i - T_{amb}) + \epsilon\sigma_{SB}A_i(T_i^4 - T_{amb}^4) \end{aligned} \tag{13.2}$$

where k_s is the thermal conductivity coefficient, s_{ij} is the contact area, r_{ij} is the interparticle distance, h_{conv} is the convective heat transfer coefficient, A_i is the particle's projected area, ϵ is the material emissivity and $\sigma_{SB} = 5.67 \times 10^{-8}$ W/(m^{2}K^4) is the Stefan-Boltzmann constant. Temperature-dependent elastic modulus and damping coefficients are used in this study to capture the thermal effects. If no melting occurs (particles are below their melting temperature, $T_m = 451.15$), a purely repulsive Hertzian force is assumed. Note that the temperature dependence of elastic modulus is taken from Luding et al. [17] which can be matched to real materials in the future. A detailed description of the thermal contact model is described by Shaheen et al. [18].

13.2.4 *Material Parameters*

To study the effect of particle cohesion, we simulate the powder spreading process for varying surface energy γ from 0 to 0.8 mJ/m^2 with an interval of 0.1 mJ/m^2.

To characterize the ratio between the interparticle cohesion and particle gravity, the Bond number is calculated as $Bo = \frac{9\gamma}{4\rho D_{50}^2 g}$. Note that $Bo = 4$ corresponds to the surface energy of the Ti-Al-4V powder with $\gamma = 0.1$ mJ/m^2 [19]. We also perform a simulation with thermal particles and compare the results with athermal particles simulation. We use PA12 powder material parameters for the thermal model presented in this work while keeping other contact parameters the same for the ease of comparison of our results.

13.3 Structural Characteristics of Cohesive Particles

Figure 13.2 shows the deposited layer packing for the two extreme cases; non-cohesive particles ($Bo = 0$) and highly cohesive particles ($Bo = 30$). The structures are widely varying between the two cases and have clearly distinguishable features. The particle packing is dense and homogeneously distributed for $Bo = 0$. In contrast, the cohesive particles show a highly heterogeneous structure with regions of locally dense and dilute packing. In the following subsections, we illustrate the method of characterizing the structural inhomogeneity using a threshold-free measure based on the local structural anisotropy calculated from Voronoi tessellation (Table 13.1).

Table 13.1 DEM simulation parameters

Variable	Unit	Values
Particle density (ρ)	kg/m^3	4430
Normal stiffness (k_n)	kg/s^2	2.2
Normal dissipation (η_n)	kg/s	3.3×10^{-6}
Tangential stiffness (k_t)	kg/s^2	$2/7k_n$
Tangential dissipation (η_t)	kg/s	$2/7\eta_n$
Sliding friction coeff. (μ_s)	–	0.1
Rolling friction coeff. (μ_r)	–	0.005
Particle diameter (d_p)	μm	12–79
Additional parameters for cohesive model		
Surface energy (γ)	mJ/m^2	0–0.8
Adhesion stiffness (k^{adh})	kg/s^2	$0.5k_n$
Additional parameters for thermal model		
Solid heat capacity (c_{solid})	J/kgK	1200
Thermal cond. coeff. (k_s)	W/mK	0.10–0.50
Thermal conv. coeff. (h_{conv})	W/m^{2}K	150
Thermal emmissivity (ϵ)	–	0.9

13.3.1 Anisotropy Vector and Divergence

The measure of the local structural anisotropy is developed by Rieser et al. from the observation that the center of a particle deviates from the centroid of its Voronoi cell in a disordered packing [10]. Any two particles with a shared Voronoi cell face are defined as neighbors; from this a Delaunay triangulation is generated by connecting groups of three mutual neighbors into triangles. Figure 13.2a includes a schematic illustration of the Voronoi tessellation calculated based on the projections of the particle positions on the xy-plane, which is plotted below the particle packing, the Delaunay triangulation k and the anisotropy vector $\overrightarrow{C}$ pointing from particle centers to corresponding Voronoi cell centroids. We mathematically quantify the extent to which the vectors $\overrightarrow{C}$ at the three vertices of the Delauney triangles points inward or outward in a given space of the triangle. We quantify this by locally measuring the divergence of the vectors, taken over a Delaunay triangle k with area A_k, which is calculated based on the concept of constant strain triangle of finite element analysis [20]. $\bar{A}$ is the average of all A_k within the packing. The local structural anisotropy, Q_k^o, calculated from the divergence is defined as:

$$Q_k^o \equiv (\nabla \cdot \overrightarrow{C})(A_k/\bar{A}) \tag{13.3}$$

By construction, Q_k^o is dimensionless with a mean near zero. It is sensitive to the local structural anisotropy and has a geometrical significance: positive (negative) values correspond to overpacked (underpacked) regions. The distribution of Q_k^o values over

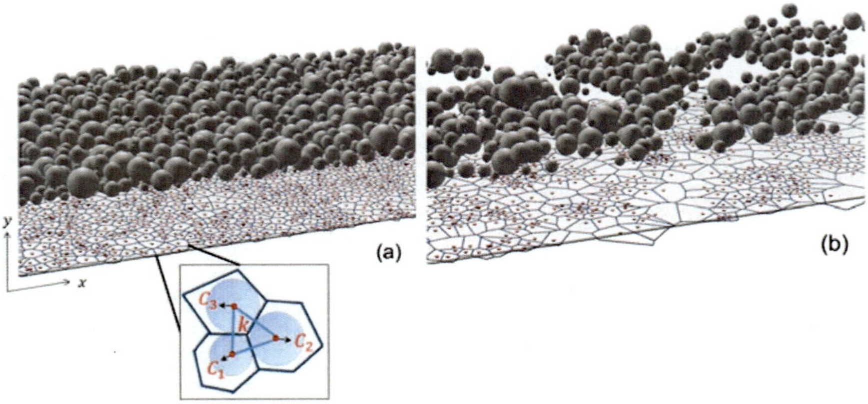

Fig. 13.2 Granular packing of powder layer for $Bo = 0$ (**a**) and $Bo = 30$ (**b**). The points represent the projections of particle centers on the xy plane and the corresponding 2D Voronoi tessellations are also shown on the plane. Also shown in **a**, a schematic representation of particle packing with superimposed Voronoi tessellation (dark blue) and Delaunay triangles (blue). Also shown are vectors C_p (black) that point from each particle center to the centroid of its Voronoi cell

a packing is nearly Gaussian [11]; hence, it is well described by just the standard deviation and the skewness.

13.3.2 Modification for the Quasi-2D Deposited Layer

The challenging aspect here is to extend the 2D calculation, explained in condition (13.2), to a quasi-2D thin free-surface layer of height 2 to $3d_p$ which is typical to powder spreading. Since the deposited layer is thin and many regions only contain a single layer of particles, we use the 2D projections of the particles on the xy plane for calculating the anisotropy. While the 2D projection is informative, the simplification of using 2D projections could lead to erroneous contributions of highly positive Q_k^o, especially in non-cohesive packing, due to the vertically aligned particles in a quasi-2D layer. To reduce the contribution of vertically aligned particle, we scale Q_k^o with the ratio of projected area of overlapping particles, A_p, on the xy plane and the sum of the real area of the particles, $A_r = \sum \pi r_i^2$. Therefore, the new definition of Q_k is given as follows:

$$Q_k \equiv Q_k^o(A_p/A_r) \equiv (\nabla \cdot \vec{C})(A_k/\bar{A})(A_p/A_r) \tag{13.4}$$

For highly dense packings of vertically aligned particles, $A_p/A_r < 1$ and the anisotropy is reduced. For dilute packings, $A_p/A_r = 1$ and thus Q_k value in this case is same as the original definition without correction. In dense and uniform regions, Q_k fluctuates randomly, giving a Gaussian peak [11]. In anisotropic regions, one can observe that the highly positive and negative Q_k values show up together, suggesting that the distribution of Q_k is rather a good indicator of the structural characterization.

13.3.3 Divergence Fields and Distributions

Figure 13.3a and b show the divergence map for $Bo = 0$ and $Bo = 30$, respectively. The Delaunay triangles are colored according to the corresponding values of Q_k. The value of each individual triangle is determined by its immediate surroundings, rather than the overall packing density. Thus, the dense and uniform regions (dark triangles in both $Bo = 0$ and $Bo = 30$) are not indicated as overpacked and only anisotropic regions show extreme values, which mostly exist in $Bo = 30$.

The distribution of Q_k is a strong structural indicator that is associated with important mechanical properties of a disordered packing, including jamming and shear band formation [11–14]. Figure 13.4a shows the distribution of Q_k for different Bo. The majority of the Q_k resides in the region around zero with a Gaussian distribution, which is made clear by fitting a Gaussian distribution using values between $\bar{Q}_k - 0.5$ to $\bar{Q}_k + 0.5$ for each dataset, where $\bar{Q}_k$ is the mean of the distribution. The solid

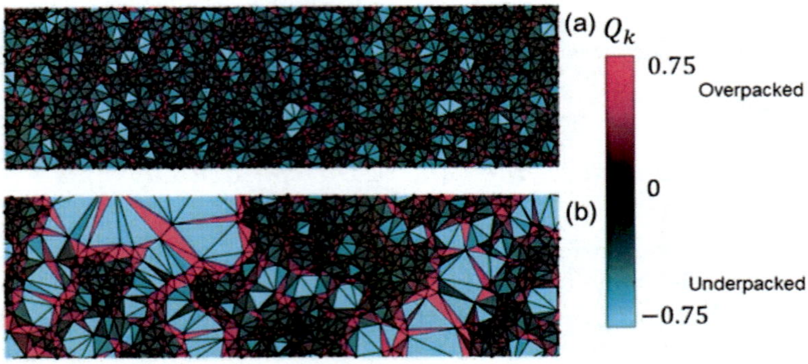

Fig. 13.3 Delaunay triangles with color representing local values of Q_k for **a** $Bo = 0$, and **b** $Bo = 30$ for a section of deposited layer $x \in (4\ldots6)$ mm and $y \in (0.2\ldots8)$ mm

curves are Gaussian fits for different Bo. For lower Bo, we observe a consistent slope of the distribution throughout the range $Q_k < 0$ which is an indication of homogeneous structure throughout the packing. In contrast, a transition of the slope of the distribution at $Q_k = -1$ is clearly distinct for higher Bo, suggesting the coexistence of dense uniform regions and dilute anisotropic regions for cohesive materials. For $Q_k < -1$, the distribution deviates from Gaussian and becomes exponential-like for higher Bo. This exponential tail corresponds to the existence of highly underpacked sites distributed sparsely in the packing. For $-1 < Q_k < 1$, the distribution is narrower with increasing Bo, indicating the existence of locally uniform packing. This variation in structure is also evident in cohesive system from the experimental studies by Xiao et al. [12] and it is lacking in non-cohesive particles system as distinguished by Harrington et al. [14] for disordered particles.

We show the results of the standard deviation and the skewness of Q_k distribution, respectively, in Fig. 13.4b and observe a transition behavior of these parameters from low to high Bo. The standard deviation (skewness) remains constant up to $Bo = 10$ and a monotonic increase (decrease) for Bo>10. Thus, $Bo = 10$ shows a transition in structural anisotropy for cohesive materials. For higher cohesion ($Bo > 10$), particles cannot fully relax during spreading, leading to a more anisotropic structure. We observe fluctuations in the trend for higher Bo which is a result of the high anisotropy of the structures. We also compared the standard deviations and skewness of the distributions with the anisotropy calculated from condition (13.3) for Q_k^o and condition (13.4) for Q_k, respectively. The divergence Q_k^o from condition (13.3) gives slightly higher values for both the standard deviations and skewness but qualitatively shows the same behavior as Q_k.

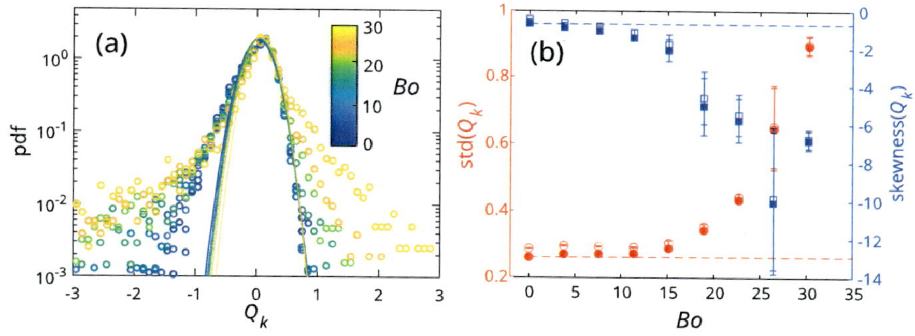

Fig. 13.4 **a** Probability density of normalized divergence of center-to-centroid vectors for the quasi-2D packing of powder deposited for different Bo. $Q_k > 0$ regions are more densely packed than their surroundings; hence, we call these regions overpacked. $Q_k < 0$ regions are more loosely packed than their surroundings, and are therefore labeled underpacked. **b** Standard deviations (red circles) and skewness (blue squares) Vs Bo. We also compared the standard deviations and skewness of the distributions with divergence calculated from condition (13.3) (Q_k^o, hollow points) and condition (13.4) (Q_k, solid points). The dashed lines indicate the standard deviations and skewness of Q_k distribution for $Bo = 0$

13.4 Structural Characteristics of Thermal Particles

To investigate the temperature dependence, simulations of particles with temperature-dependent contact models are performed. Since particles are below their melting temperature, the normal forces due to viscosity of the melt and sintering effects are not active in the thermal contact model. Figure 13.5a shows a packing of deposited quasi-2D layer of thermal particles with the particles colored according to their temperature after deposition. The qualitative structure shows that particles are densely packed similar to the case of athermal non-cohesive particles. Interesting features emerge through our numerical implementation, such as, the larger particles show less temperature drop compared to smaller particles. Figure 13.5b shows a quantitative comparison of structures showing the Q_k distributions of athermal particles and thermal particles. The distributions are overlapping, indicating that the structure of the deposited layer is not sensitive to the initial particle temperature under the idealized temperature dependence of the stiffness, which varies in the range of 6×10^7 to 8×10^7. Note that inter-particle cohesion such as Van der Waals forces could be temperature dependent as well, which is not included in the interaction of thermal particles but will be included in our studies in future.

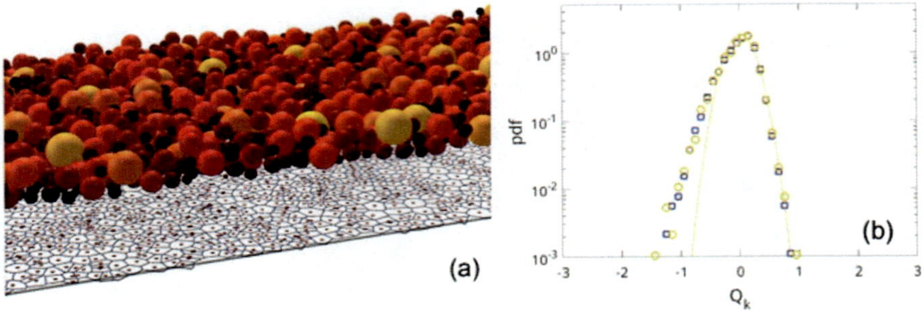

Fig. 13.5 **a** Granular packing of powder layer for thermal particles. The particles are colored according to their temperature with color scale varying from 330 K (black) to 380 K (yellow). **b** Probability density of the normalized divergence of center-to-centroid vectors for quasi-2D packing of powder deposited for athermal particles (blue squares) and thermal particles (yellow circles)

13.5 Conclusions

The concept of the divergence of anisotropy vector, Q_k, is extended as a geometrical measure of structural anisotropy to our example of a quasi-2D system of a thin free-surface layer of deposited powder. In particular, Q_k with zero mean, by construction, is easy to interpret in terms of overpacked and underpacked regions in the powder layer packing. Additionally, the transition of the slope of Q_k distribution at $Q_k = -1$ displays a strong signature coexistence of local sites of dense regions with uniform and isotropic structures as well as dilute regions with highly anisotropic structures for cohesive materials. The standard deviation and the skewness of the Q_k distributions show a monotonic increase for $Bo > 10$, indicating a transition in structural anisotropy of deposited layer at $Bo = 10$. An analysis of the structures of the thermal particles deposited on the substrate reveals that features start to emerge, such as, the larger particles show less temperature loss through heat transfer, and their implications on the packing structure and the sintering process will be investigated in future studies.

Acknowledgements We gratefully acknowledge Deutsche Forschungsgemeinschaft (DFG, German Research Foundation) for funding the Collaborative Research Center 814 (CRC 814), Project Number 61375930-SFB 814 'Additive Manufacturing', sub-project B1. We also thank Humboldt Research Foundation for granting the 'Humboldt Research Fellowship'. The work was supported by the Interdisciplinary Center for Nanostructured Films (IZNF), the Central Institute for Scientific Computing (ZISC), and the Interdisciplinary Center for Functional Particle Systems (FPS) at Friedrich-Alexander-Universität Erlangen-Nürnberg.

References

1. Bhavar, V., Kattire, P., Patil, V., Khot, S., Gujar, K., Singh, R.: A review on powder bed fusion technology of additive manufacturing. Addit. Manuf. Handb. **3**, 251–253 (2017)
2. Vock, S., Kloeden, B., Weissgaerber, T., Kieback, B.: Powders for powder bed fusion: a review. Progress Addit. Manuf. **4**, 383–397 (2019)
3. Chen, H., Sun, Y., Yuan, W., Pang, S., Yan, W., Shi, Y.: A review on discrete element method simulation in laser powder bed fusion additive manufacturing. Chin. J. Mech. Eng.: Addit. Manuf. Front. **1**, 100017 (2022)
4. Cunningham, R., Nicolas, A., Madsen, J., Fodran, E., Anagnostou, E., Sangid, M.D., Rollett, A.D.: Analyzing the effects of powder and post-processing on porosity and properties of electron beam melted Ti-6Al-4V. Mater. Res. Lett. **5**, 516–525 (2017)
5. Gong, H., Rafi, K., Gu, H., Janaki Ram, G.D., Starr, T., Stucker, B.: Influence of defects on mechanical properties of Ti-6Al-4V components produced by selective laser melting and electron beam melting. Mater. Des. **86**, 545–554 (2015)
6. Nasato, D., Briesen, H., Pöschel, T.: Influence of vibrating recoating mechanism for the deposition of powders in additive manufacturing: Discrete element simulations of polyamide 12. Addit. Manuf. **48**, 102248 (2021)
7. Nasato, D., Pöschel, T.: Influence of particle shape in additive manufacturing: discrete element simulations of polyamide 11 and polyamide 12. Addit. Manuf. **36**, 101421 (2020)
8. Shaheen, M.Y., Luding, S., Thornton, A.R., Weinhart, T.: The influence of material and process parameters on powder spreading in additive manufacturing. Powder Technol. **383**, 564–583 (2021)
9. He, Y., Hassanpour, A., Bayly, A.E.: Combined effect of particle size and surface cohesiveness on powder spreadability in additive manufacturing. Powder Technol. **392**, 191–203 (2021)
10. Richard, D., Ozawa, M., Patinet, S., Stanifer, E., Shang, B., Ridout, S.A., Xu, B., Zhang, G., Morse, P.K., Barrat, J.-L., Berthier, L., Falk, M.L., Guan, P., Liu, A.J., Martens, K., Sastry, S., Vandembroucq, D., Lerner, E., Manning, M.L.: Predicting plasticity in disordered solids from structural indicators. Phys. Rev. Mater. **4**, 113609 (2020)
11. Rieser, J.M., Goodrich, C.P., Liu, A.J., Durian, D.J.: Divergence of Voronoi cell anisotropy vector: a threshold-free characterization of local structure in amorphous materials. Phys. Rev. Lett. **116**, 088001 (2016)
12. Xiao, H., Ivancic, R.J.S., Durian, D.J.: Strain localization and failure of disordered particle rafts with tunable ductility during tensile deformation. Soft Matter **35**, 8226–8236 (2020)
13. Harrington, M., Durian, D.J.: Anisotropic particles strengthen granular pillars under compression. Phys. Rev. E. **97**, 012904 (2018)
14. Harrington, M., Xiao, H., Durian, D.J.: Stagnant zone formation in a 2D bed of circular and elongated grains under penetration. Granul. Matter **22**, 1–9 (2020)
15. MercuryDPM. https://www.mercurydpm.org/
16. Roy, S., Singh, A., Luding, S., Weinhart, T.: Micro-macro transition and simplified contact models for wet granular materials. Comput. Part. Mech. **3**, 449–462 (2016)
17. Luding, S., Manetsberger, K.: Müllers: a discrete model for long term sintering. J. Mech. Phys. Solids **53**, 455–491 (2005)
18. Shaheen, M.Y., Luding, S., Thornton, A.R., Weinhart, T.: Thermal discrete particle model of powder consolidation in additive manufacturing (in preparation) (2023)
19. Meier, C., Weissbach, R., Weinberg, J., Wall, W.A., Hart, A.J.: Modeling and characterization of cohesion in fine metal powders with a focus on additive manufacturing process simulations. Powder Technol. **343**, 855–866 (2019)
20. Cook, R.D.: Concepts and Applications of Finite Element Analysis. Wiley, New York (2007)

Chapter 14
An Open-Source Discrete Element Model for SS316L Alloy Powder Characterization Using a Virtual Hall-Flow Meter to Study the Flowability in Powder Bed Fusion Additive Manufacturing

Bouabbou Abdelkrim and Sébastien Vaudreuil

Abstract Powder-bed fusion is the most widely used process for additively manufacturing metallic parts. It has drawn a lot of attention because of its remarkable flexibility, especially from high-demand industries like aerospace and automotive. Improving this process requires a greater understanding of metal alloy powder behaviour, something that can be accomplished by using numerical modelling as a low-cost alternative to in-situ testing. Establishing a connection between process variables and the quality of the powder layer is made easier by examining the effects of powder flowability, humidity, porosity, and density. The present work studies the flowability of metallic powder using Discrete Element Method (DEM) to investigate the intrinsic properties for virtual metallic powders generation. The use of a high-fidelity particle-scale model to capture the dynamics of metal particles interactions in a virtual Hall-flow meter enabled us to obtain suitable sliding and rolling coefficients. This was achieved by quantifying the flow rate and angle of repose from an experimental apparatus combined with DEM simulations.

Keywords Discrete element method · Flowability · Hall flow · Metal powder · SS316L · LIGGGHTS

B. Abdelkrim (✉) · S. Vaudreuil
Euro-Mediteranean University of Fez, Fez, Morocco
e-mail: Abdelkrim.bouabbou@ueuromed.org

S. Vaudreuil
e-mail: S.vaudreuil@ueuromed.org

K. Zarbane and Z. Beidouri (eds.), *Proceedings of CASICAM 2022*, Springer Tracts in Additive Manufacturing, https://doi.org/10.1007/978-3-031-32927-2_14

14.1 Introduction

Like any other additive manufacturing (AM) technology, Laser powder bed fusion (LPBF) uses the same basic principle to manufacture high-quality parts. The LPBF process begins by slicing a 3D CAD model into thin slices. Once the virtual model is sliced into layers, the physical building process will rely on the deposition of a thin powder layer spread by a Raking or Rolling system. This is followed by selectively melting (ISO/ASTM: PBF-LB/M) each powder layer using a laser beam. After lowering the build platen, this process is then repeated layer by layer until the part is finished. The quality of the built part is affected by several factors, such as build parameters, metal powder composition, and powder bed properties. The powder bed in return is influenced by powder flowability, packing density, morphology and particle size distribution. Each of these properties can have a substantial impact on the quality of the produced metallic part. For instance, a low-quality powder-bed can have unfavourable consequences such as melt pool instability, partial melting, and so forth. To achieve a better quality, faster manufacturing, and at lower cost, it is imperative to study the influence of the different process parameters on the deposited powder layer. Spreading parameters and variability in powder properties due to different metal powder suppliers, powder storage conditions, and recycling, renders this task very complex. Due to the interest in powder granulometry in AM, many authors explored the effects of power morphology distribution, and density, flowability, and thermal properties on the part's density, surface quality and other mechanical properties. Authors like [1] used Hausner ratio and Avalanche angle from revolution powder analyser (RPA) as the main metric of metallic powder flowability. However, relying on experimental assessments of powder flowability alone cannot be used to characterize the powder bed spreading process owing to the different dynamic conditions used. Modelling and numerical simulation constitute an ideal component that helps understand the associated physical phenomena at different scales while providing additional information that can't be easily obtained experimentally. The powder bed is a discrete domain, the discrete element method is a perfect match for modelling the complex particle's behaviour under the spreading process. For instance, using DEM modelling, the authors of [2] have determined that the predominant factor governing the powder bed properties is the particle size distribution rather than the particle sphericity. Several researchers have attempted to study powder's flowability to optimise the spreading process, [3] showed that a high count number of fine-sized particles fraction, generate superior powder bed quality in terms of the smooth finished surface and high mass density. However, an excessive fine fraction in the powder bed cause major flowability issues.

The accuracy of any DEM model depends on the input values assigned to the parameters as determined through a calibration process [4]. For this reason, authors like [5] attempted to study powder flowability using DEM simulation validated by an experimental assessment, these authors used revolution powder analyser, where they have shown that the rolling motion of non-spherical powders was found to cause an instantaneous disturbance to the flow of neighboring powder particles. Others studied flowability by simulating a Hall-flow meter, another flowability characterization experimental technique. Through similar work, [6] calibrated their DEM model for hall-flow simulation using experimental angle of repose and mass-flow rate measurements. Our work provides a unique approach for characterizing metal powders at the micro-scale, with a focus on AM powder bed flowability. It combines DEM simulations of the Hall-flow meter characterization technique and in-situ experimental measurements to calibrate the relative DEM metal powder model parameters for the stainless steel 316L metal alloy. With the developed model in this study, we are able to gain an in-depth understanding of the effects DEM system parameters have on simulated powder flow behaviour and enable extraction of relevant friction coefficients for the appropriate powder flow properties usually used in powder-bed spreading DEM models.

14.2 Materials and Methods

14.2.1 Metallic Powder Alloys and Characterization Method

The flowability is not an inherent property of metallic alloys, because it is affected by not only the physical properties (shape, particle size, humidity, etc.) but also on the used experimental apparatus and the relative measurement method, namely Hall flow meter [7], Revolution powder analyser [5] and shear cell test [8]. This work investigates Stainless steel 316L metal powder flowability using a Hall-flow meter. The experimental apparatus is presented in (Fig. 14.1). In the test, the time for 50 g of powder to flow through the funnel is recorded and called the Hall Flow value.

14.2.2 DEM and Simulation Setup

Discrete element method DEM is a Lagrangian-based method that's able to model with a high-fidelity the granular materials. An approach capable of modelling small particle interactions, after which their subsequent motion is computed using initial conditions, contact models, and other forces such as $\vec{F}_i^{wall}$ forces applied by surrounding walls, $\vec{F}_i^{bond}$ particle bonds, drag $\vec{F}_i^{drag}$ and $\vec{F}_i^{em}$ electro-magnetic forces as shown in the Eq. (14.1).

$$m_i \ddot{r}_i = \vec{F}_i^{Contact} + \vec{F}_i^{wall} + \vec{F}_i^{bond} + \vec{F}_i^{drag} + \vec{F}_i^{em} \tag{14.1}$$

$$\vec{F}_{ij}^{Contact} = \mathcal{F}\left(\|r_i - r_j\|, R_i, R_j, \text{materialparameters}\right) \tag{14.2}$$

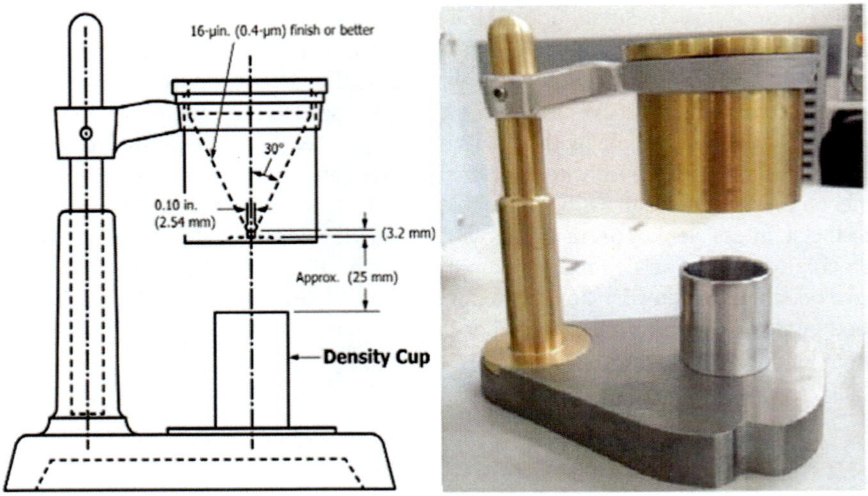

Fig. 14.1 Experimental Hall-flow meter (Standard ASTM B12 -17)

where m_i and r_i are respectively the mass and position vector for each ith particle. R_i, R_j are two radiuses of two particles **i** and **j**, the contact between these particles is activated when $\|r_i - r_j\| \le R_i + R_j$. Particle contact mechanics used in this work capture normal F^n_{ij}, tangential F^t_{ij} and angular contact related to normal interactions and angular velocity between particles which helps determine a directional torque proportional to the normal contact force [9].

$$F^n_{ij} = (k_n \delta_n - \gamma_n V_{n,ij}) n_{ij}, F^t_{ij} = (k_t \delta_t - \gamma_t V_{t,ij}) t_{ij} \tag{14.3}$$

where $\delta_n n_{ij}$ and $\delta_t t_{ij}$ are respectively the normal and tangential overlap. Also, $V_{n,ij} n_{ij}$ and $V_{t,ij} t_{ij}$ are respectively normal and tangential relative velocity. k_n, k_t, γ_n and γ_t are model coefficients respectively computed based on Young and shear modulus. A Hertzian contact model without cohesion was used to compute the particle–particle and particle–wall interaction forces. The forces and torques due to gravity, collisions, and static and rolling friction are considered in this work. The governing equations were solved in the open-source framework LIGGGHTS [10]. The funnel geometry is based on the specification of [7] and the simulation domain is illustrated Fig. 14.2.

For this model, we consider a highly spherical stainless steel SS316L metal alloy where the parameters used in our DEM model are presented in (Table 14.1). Therefore, the modelled morphology is spherical and the particle size distribution was derived from the volume-based to the number-based particle size distribution (Fig. 14.2) with which we have used to generate the initial particle cloud. All of the simulated particles are under the influence of gravity only and without considering drag forces from the surrounding air. Determining sliding and rolling friction was carried out on multiple values between 0.1 and 0.5 to find out the parameter values

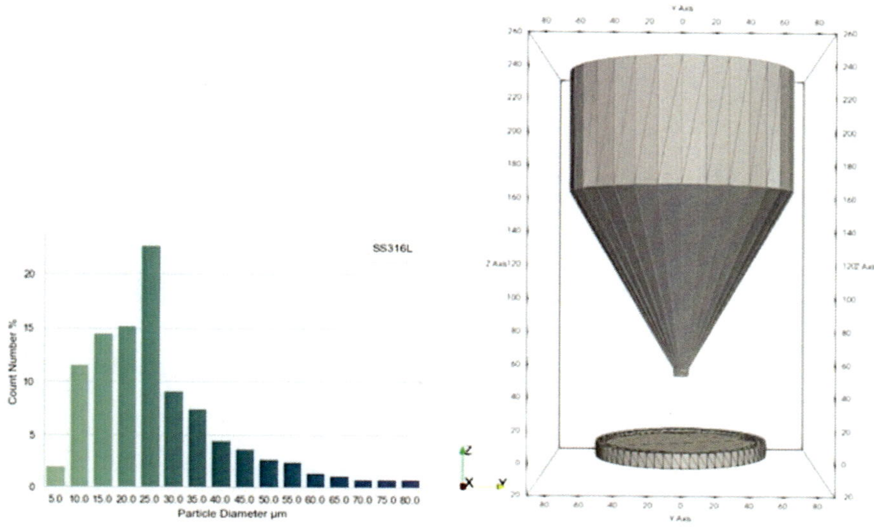

Fig. 14.2 Simulation setup and Particle size distribution (PSD) of the modelled powder

Table 14.1 SS316L powder model parameters for DEM Simulation

Material properties			
Young modulus (GPa)	208	Rolling friction	[0.1–0.5]
Density (kg/m^3)	7980	Sliding friction	[0.1–0.5]
Poisson ratio	0.267	Coefficient of restitution	0.3
Cumulative Volume % [D_{10}, D_{50}, D_{90}]	[8,22,45]	Time step (s)	10^{-6}

matching experimental AOR. However, for a Hall-flow characterization, this is not enough to model flowability, this is why a Mass flow assessment is needed to choose the appropriate parameters matching both AOR and Mass flow.

14.3 Results and Discussion

The present results are acquired by simulating 30 k particles with a 10^{-6} time step. This choice took into account both Hertz and Rayleigh Time Limit [11]. Furthermore, the computing power needed to model full scale is a major constraint for modelling metal powder particles using DEM at a micro-scale. The simulation trials were carried out in parallel with 18 CPUs on an IntelXeon workstation. For this reason, particle size is often scaled up, the scaling down of the simulation domain was carried out using Beverloo law [12] which facilitate a reliable estimation of the mass flow rate through the scaled down funnel geometry. In our model, two scaling factors were

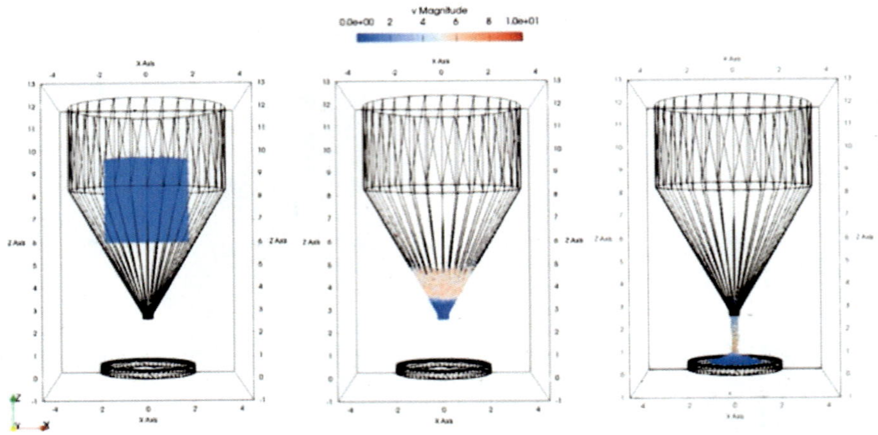

Fig. 14.3 Simulation steps cloud generation, free fall and funnel flow (Left to right)

employed, the particles were scaled up for 10^3 and the geometry scaled down for 10^2. We have used Paraview for rendering the simulated data (Fig. 14.3) along with Python for data analysis. This simulation comprises 3 steps, Cloud generation, Free fall and Funnel flow at the orifice ($D_0 = 2.5$ mm).

The first two steps Cloud generation and free fall step, where the powder settles at the base of the funnel, took less than 1 s. This model uses sliding friction SF, rolling friction RF, and instead of cohesive forces, we have employed a constant directional torque (CDT) adding torque contribution to the modeled particles. The computed results acquired are velocity magnitude, forces, and angular velocity presented in (Fig. 14.4). At the orifice, depending on sliding friction, the modelled particles interact with the wall of the funnel which decreases their velocity.

At the end of Funnel flow step, the simulation particle forms a heap of which we take to measure the angle of repose as seen in (Fig. 14.5).

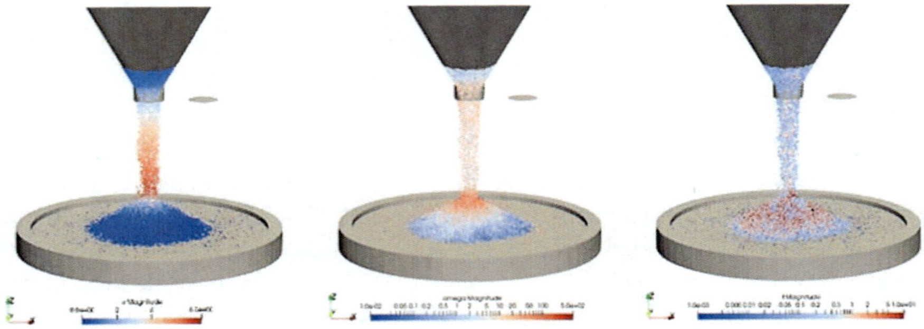

Fig. 14.4 Particle velocity (m/s), omega the angular velocity (rad/s) and contact forces magnitude (N)

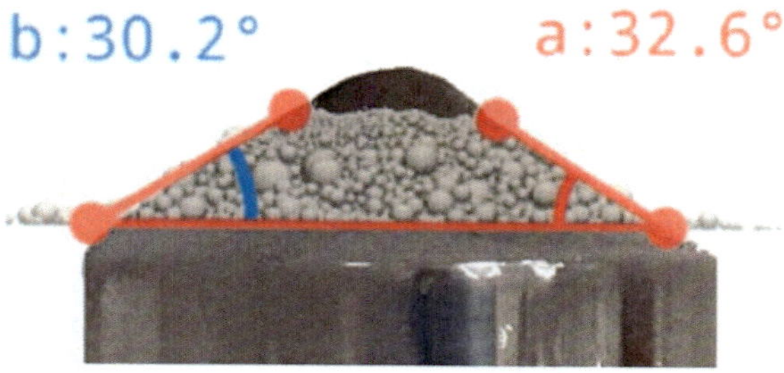

Fig. 14.5 Simulated powder's Angle of repose acquired by our DEM model super imposed to experimental powder heap

In the work of [13], four different types of particle size distributions for SS316L powders were measured using the Hall flow meter. A PSD with more fine-sized particles decreases the powder flowability than that of the coarse-sized particle distribution. This is reflected on both the mass flow (0,27 g/s) for coarse and 0,58 g/s for fine-sized powder, and the angle of repose 29,5° for coarse and 32,0° for fine-sized. As we have modelled a fine-sized particle distribution of SS316L, these results helped to determine the sliding and rolling coefficients needed to carry out DEM in our model parameters.

Different combinations of sliding and rolling frictions can sometimes result in similar AOR values. Furthermore, static AOR alone cannot reflect the actual flowability of the powder. Therefore, in order to find the model parameters that describe best the flowability of the simulated powder, our simulation need to generate a similar mass flow at the orifice (Hall value) taken by the in-situ measurements in combination with the experimental AOR using the specific sliding and rolling friction. As seen in (Fig. 14.5) the simulated AOR is between 30,2° and 32,6° acquired for 0,5 rolling resistance and 0,5 sliding resistance.

Figure 14.6 on the other hand, presents the flow measurements as a function of time, the simulated data taken at the orifice of our 3D funnel at 10^{-3} s resolution are, the total mass passed through the orifice, the number of particles and their respective flow rates, the mass flow rate and particle number per unit of time.

The flow regime ends at 4.378 s and by that time, 2.5 g of mass had exited the orifice. This computes to a mean mass flow rate of 0.571 g/s. This verifies the acquired Hall flow value of 29.03 s/50 g for the desired experiment AOR. We can also notice in the number of particles (Fig. 14.6) in green, it exhibits an oscillating behaviour where a peak is recorded before tailing off to zero. This peak is attributed to the large number of small-sized particles lagging behind due to the small travel distance in each contact iteration. Moreover, the effect of rolling friction on Mass flow rate are not so drastic. However, the more rolling friction, the more the simulated Angle of repose is increased. Conversely, the sliding friction decreases the Mass flow rate drastically when increased and, along with it, an increase in the angle of repose. For this SS316L model, the acquired results were generated at 0.5 SF and 0.5 RF.

Other parameters like Young modulus and Restitution coefficients had no effects on the bulk flow of the modelled particles. However, the execution time increases drastically for higher values, and the simulation diverges as Rayleigh and Hertz's

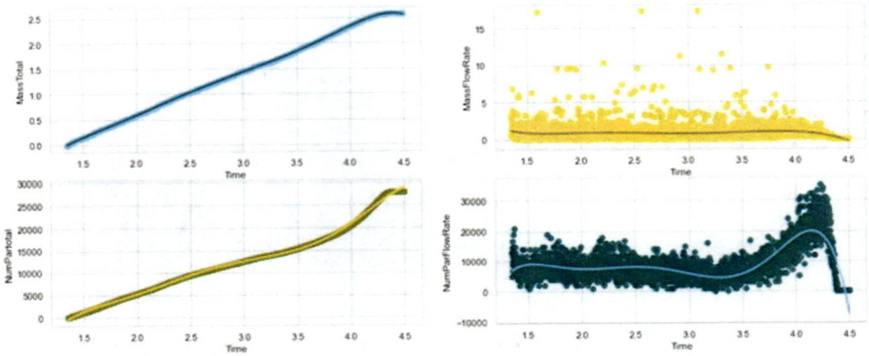

Fig. 14.6 Simulated Mass-flow at the orifice results acquired by our DEM model

Time values are well above 20%. Therefore, a smaller time step is required and the maximum number of simulated particles should be constrained to reach iterative convergence while keeping the computational time at a minimum.

14.4 Conclusion

Owing to the versatility of DEM modeling and the high throughput simulation it provided, this approach presents itself as an ideal candidate to model metal powder flowability. The virtual powder parameters (e.g., PSD, Friction coefficients) inferred from the powder's intrinsic physical properties taken by in-situ measurements, control the resultant flowability of the simulated powder particles. Our approach helped model, through a virtual static characterization, the flowability of metal powder validated by the simulated mass flow rate that results in the appropriate angle of repose obtained experimentally.

This approach will help researchers investigating powder bed spreading to incorporate the appropriate particle flow behaviour into their spreading process model, this also can be replicated for other different materials, having different particle size distributions, while taking into account the effects of different morphologies using friction coefficients and additional torque provided by a constant directional torque (CDT) contact model.

Acknowledgements This work was funded by the Académie Hassan II des Sciences et Techniques under the project Thales3D « Effet du recyclage de la poudre AlSi7Mg0.6 sur les pièces fabriquées par "Selective Laser Melting" (SLM) ». We would like to extend our thanks to our Lab engineers, Mr Zakaria Mohammed and Mr Yassine El Ansary for their contribution in Manufacturing the Hall Flow meter.

References

1. Riener, K., et al.: Influence of particle size distribution and morphology on the properties of the powder feedstock as well as of AlSi10Mg parts produced by laser powder bed fusion (LPBF). Addit. Manuf. **34**, 101286 (2020). https://doi.org/10.1016/j.addma.2020.101286
2. Yim, S., Bian, H., Aoyagi, K., Yamanaka, K., Chiba, A.: Spreading behavior of Ti48Al2Cr2Nb powders in powder bed fusion additive manufacturing process: experimental and discrete element method study. Addit. Manuf. **49**, 102489 (2022). https://doi.org/10.1016/j.addma.2021.102489
3. Ma, Y., Evans, T.M., Philips, N., Cunningham, N.: Numerical simulation of the effect of fine fraction on the flowability of powders in additive manufacturing. Powder Technol. **360**, 608–621 (2020). https://doi.org/10.1016/j.powtec.2019.10.041
4. Bouabbou, A., Vaudreuil, S.: Understanding laser-metal interaction in selective laser melting additive manufacturing through numerical modelling and simulation: a review. Virtual Phys. Prototyp. **17**(3), 543–562 (2022). https://doi.org/10.1080/17452759.2022.2052488
5. Dai, L., Chan, Y.R., Vastola, G., Zhang, Y.W.: Discrete element simulation of powder flow in revolution powder analyser: effects of shape factor, friction and adhesion. Powder Technol. **408**, 117790 (2022). https://doi.org/10.1016/j.powtec.2022.117790
6. Phua, A., Doblin, C., Owen, P., Davies, C.H.J., Delaney, G.W.: The effect of recoater geometry and speed on granular convection and size segregation in powder bed fusion. Powder Technol. **394**, 632–644 (2021). https://doi.org/10.1016/j.powtec.2021.08.058
7. ASTM B213-17: Test methods for flow rate of metal powders using the hall flowmeter funnel (2020). https://doi.org/10.1520/B0213-17
8. Zegzulka, J., Gelnar, D., Jezerska, L., Prokes, R., Rozbroj, J.: Characterization and flowability methods for metal powders. Sci. Rep. **10**(1) (2020). https://doi.org/10.1038/s41598-020-77974-3
9. Angus, A., et al.: Calibrating friction coefficients in discrete element method simulations with shear-cell experiments. Powder Technol. **372**, 290–304 (2020). https://doi.org/10.1016/j.powtec.2020.05.079
10. Kloss, C., Goniva, C.: LIGGGHTS – open source discrete element simulations of granular materials based on lammps. In: Supplemental Proceedings, pp. 781–788. Wiley (2011). https://doi.org/10.1002/9781118062142.ch94
11. Burns, S.J., Piiroinen, P.T., Hanley, K.J.: Critical time step for DEM simulations of dynamic systems using a Hertzian contact model. Int. J. Numer. Methods Eng. **119**(5), 432–451 (2019). https://doi.org/10.1002/nme.6056
12. Mankoc, C. et al.: The flow rate of granular materials through an orifice. July 31, 2007. http://arxiv.org/abs/0707.4550. Accessed 03 Oct 2022
13. Du, K., Li, S., Jie, S., Gao, X., Yu, Y.: Effect of 316L stainless steel powder size distribution on selective laser melting process. J. Phys. Conf. Ser. **1347**(1), 012121 (2019). https://doi.org/10.1088/1742-6596/1347/1/012121

Chapter 15
Temperature Gradients as a Source of Balling and Humping in Laser Processing of Titanium

Michael Blank and Thorsten Pöschel

Abstract The spheroidization of the melt in laser melting processes deteriorates the mechanical properties of the welding seam or the manufactured part. This phenomenon is called humping or balling. To improve the reliability of the product quality a thorough understanding of the occurring physical phenomena is indispensable. By comparison of three-dimensional Smoothed Particle Hydrodynamics simulations of single-line laser tracks of titanium with experiments, we show that high-temperature gradients reduce the wetting forces which act at the three-phase contact line which in turn promotes the fragmentation of the cylindrical melt.

Keywords Humping · Balling · Laser processing · Smoothed particle hydrodynamic

15.1 Introduction

A liquid tends to spheroidize to minimize its surface energy. In laser melting processes the melt pool may fragment into coarsened spheres in case of inappropriate processing conditions. This phenomenon is called balling in powder-based additive manufacturing processes [33] and can be related to insufficient wetting of the solid substrate or droplet splashing during the manufacturing process. This impairs the uniform deposition of fresh powder, which favors the formation of pores and thermal stresses, delamination due to weak interlayer bonding, and deteriorates the part geometry [24, 35]. Similarly, the spheroidization of the welding seam in laser welding is called humping. Several theories have been developed to explain the humping phenomenon ranging from Rayleigh's capillary instability, Kelvin-Helmholtz instability, and hydraulic jump to the fluid flow, e.g., due to Marangoni forces [29]. In this work, we investigate the influence of the temperature difference across the solid-liquid

M. Blank (✉) · T. Pöschel
Institute for Multiscale Simulation, Friedrich-Alexander-Universität Erlangen-Nürnberg, Cauerstraße 3, 91058 Erlangen, Germany
e-mail: michael.u.blank@fau.de
URL: http://www.mss.cbi.fau.de/

K. Zarbane and Z. Beidouri (eds.), *Proceedings of CASICAM 2022*, Springer Tracts in Additive Manufacturing, https://doi.org/10.1007/978-3-031-32927-2_15

interface on the surface geometry of the resolidified melt obtained from single-line laser tracks. For this reason, we simulate the three-dimensional laser melting process of titanium using the Smoothed Particle Hydrodynamics (SPH) method where the ambient gas phase is modeled by boundary conditions. Moreover, we employ a ray tracing approach the model the laser-metal interaction.

15.2 Material Properties of Titanium

The material properties of titanium are modeled temperature-dependent. Data for transition temperatures and enthalpies, dynamic viscosity, and emissivity can be found in [3, 7, 9–11]. The specific heat capacity, c_p, and density, ρ, is modeled as given by [26]. The thermal conductivity of titanium, k, is obtained by interpolation of the experimental data given by [20] and employing the Wiedemann-Franz law using the experimental data of the electric resistivity given by [22]. The modeled density, heat capacity, and thermal conductivity are shown in Fig. 15.1.

The surface tension coefficient, σ, and the enthalpy of evaporation, h_{vap} are provided in [14, 34] The recoil pressure, p_{recoil}, which acts on the evaporating metal surface is modeled following [14]. To model titanium's reflectivity as a function of temperature and a laser wavelength of 1.08 µm, we follow the work of Ujihara [28] and Siegel [23] which are based on the Drude theory. The required material properties for these models can be found in [12, 13, 18, 31].

The permittivity measured by Werner et al. [31] is used to model the refractive indices of atomically clean titanium in this work. When the titanium surface is rough, oxidized, or contaminated with chemicals, the reflectivity of electromagnetic radiation is significantly decreased. To model such an aged titanium surface we employ the experimental data of Johnson and Christy [13] and compute the plasma and collision frequency required by [23, 28] by substitution of the real into the imaginary part of the permittivity. Using Fresnel laws, the power reflectance of titanium can be computed [32]. Figure 15.2 shows the power reflectance of the titanium surface for light with a wavelength of 1.08 µm at normal incidence.

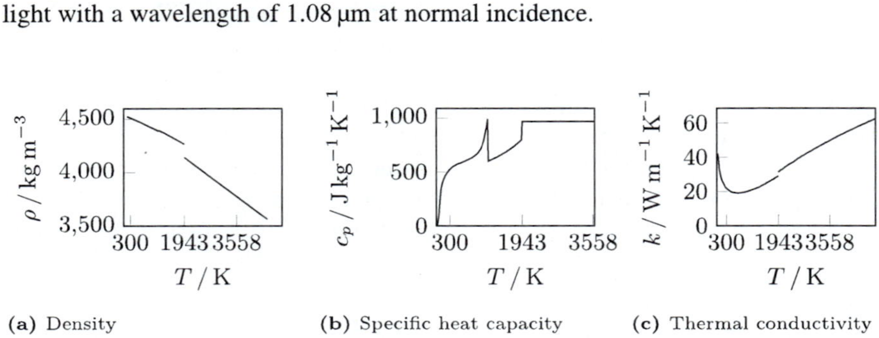

Fig. 15.1 Modeled temperature-dependent material properties of titanium

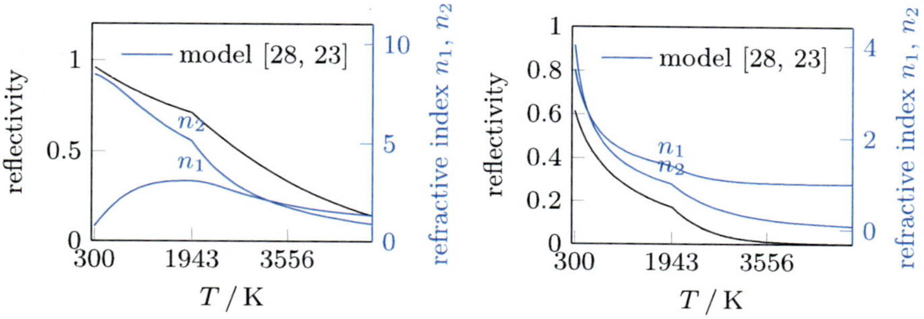

(a) Clean titanium surface. (b) Aged titanium surface.

Fig. 15.2 Modeled real and imaginary parts of the refractive index and corresponding reflectivity of titanium as a function temperature of an aged and atomically clean titanium surface for light with a wavelength of $\lambda = 1.08\,\mu\text{m}$ at normal incidence

15.3 Simulation Method

To simulate the laser melting process of titanium we employ SPH. The ambient gas phase is not represented by SPH particles, instead appropriate boundary conditions are applied at the liquid-gas interface. Moreover, titanium is assumed to be incompressible and Newtonian. In the simulation, we distinguish between the subset of solid, Ω^{s}, wall, Ω^{w}, and liquid SPH particles, Ω^{l}. The whole domain containing all SPH particles is denoted as Ω. To identify the subsets of SPH particles we compute the Shepard filters computed from all, S_a, only liquid, S_a^{l}, only solid, S_a^{s}, and only surface SPH particles, S_a^{v} by

$$S_a = \sum_b \frac{m_b}{\rho_b} W_{ab}\,, \quad S_a^{\text{l}} = \sum_{b \in \Omega^{\text{l}}} \frac{m_b}{\rho_b} W_{ab}\,, \quad S_a^{\text{s}} = \sum_{b \in \Omega^{\text{s}}} \frac{m_b}{\rho_b} W_{ab}\,, \quad S_a^{\text{v}} = \sum_{b \in \Omega^{\text{v}}} \frac{m_b}{\rho_b} W_{ab}\,. \tag{15.1}$$

Here, m_b and ρ_b are the mass and density assigned to particle b and W_{ab} is the Wendland C2 kernel function [30] evaluated for the particle pair a and b. The subscripts a and b denote that a property is defined at the position $\mathbf{r}_a$ or $\mathbf{r}_b$ of an SPH. Surface particles are identified as vertices of the surface mesh which is computed in each time step using the CGAL-Library [25] and which is required for the applied ray tracing algorithm. We use a threshold of $S_a < 0.95$ to identify SPH particles in the vicinity of the free surface and $S_a \geq 0.95$ to identify SPH particles located in the bulk with sufficient neighbor particle support. The total acceleration of a liquid SPH particle is

$$\frac{\mathrm{D}\mathbf{u}_a}{\mathrm{D}t} = \mathbf{f}_a^{\text{p}} + \mathbf{f}_a^{\text{v}} + \mathbf{f}_a^{\text{s}} + \mathbf{f}_a^{\text{b}}\,, \tag{15.2}$$

where $\mathbf{u}_a$ is the velocity of particle a, t is the time, and $\mathbf{f}_a^{\text{p}}$, $\mathbf{f}_a^{\text{v}}$, $\mathbf{f}_a^{\text{s}}$, $\mathbf{f}_a^{\text{b}}$ are the acceleration of particle a due to pressure, viscous, surface tension, and body forces. The acceleration of a liquid SPH particle due to pressure forces in the vicinity of the free surface ($S_a \geq 0.95$) is obtained using the symmetric pressure gradient approximation given by [15]. The solid SPH particles are static and do not move in the simulations. We solve a Poisson Equation for pressure (PPE) to obtain a truly incompressible material. The computed pressure field is imposed on the velocity field to achieve incompressibility using a predictor-corrector integration scheme [6]. The PPE is modified for liquid SPH particles near the liquid-gas interface ($S_a < 0.95$) with insufficient neighbor particle support using the approach by [17]. Here, the gas phase is modeled by applying a Dirichlet pressure boundary condition to the PPE. Analogously to the derivation of the PPE in [17], the symmetric pressure gradient approximation is modified to

$$\mathbf{f}_a^{\mathrm{p}} = \sum_b m_b \left(\frac{p_a - p_b}{\rho_b^2} \right) \nabla W_{ab} \,, \quad \text{if } S_a < 0.95 \,, \tag{15.3}$$

where p_b and p_a is the pressure of particle b and the ambient pressure experienced by particle a. In this work $p_a = \Delta \tilde{p}_{\mathrm{recoil},a}$, where

$$\Delta \tilde{p}_{\mathrm{recoil},a} = \frac{1}{S_a^{\mathrm{v}}} \sum_{b \in \Omega^{\mathrm{v}}} \frac{m_b}{\rho_b} \Delta p_{\mathrm{recoil},b} W_{ab} \,. \tag{15.4}$$

Here, $\Delta p_{\mathrm{recoil},b}$ is the recoil pressure acting on particle b given by [14]. To avoid the penetration of liquid SPH particles into a solid wall we identify SPH wall particles by $a \in \Omega^{\mathrm{s}} \wedge S_a^{\mathrm{l}} > 10^{-3}$. The pressure of these SPH wall particles is computed implicitly by solving the PPE as

$$\sum_{b \in \Omega^{\mathrm{l}}} \frac{m_b}{\rho_b} \frac{4}{\rho_a + \rho_b} (p_a - p_b) F_{ab} = \sum_{b \in \Omega^{\mathrm{l}}} -\frac{m_b}{\rho_b} \frac{\mathbf{u}_{ab} \cdot \nabla W_{ab}}{\Delta t} \,, \qquad \mathbf{u}_{ab} = \mathbf{u}_a - \mathbf{u}_b\}, \,, \tag{15.5}$$

where

$$F_{ab} = \frac{\mathbf{r}_{ab} \cdot \nabla W_{ab}}{r_{ab}^2 + 0.01 h^2} \,, \qquad \mathbf{r}_{ab} = \mathbf{r}_a - \mathbf{r}_b \,, \qquad r_{ab} = |\mathbf{r}_a - \mathbf{r}_b| \,. \tag{15.6}$$

Here, ρ_a and p_a is the density and the pressure of particle a, $0.01h^2$ is a small number to avoid division by zero if $r_{ab} \to 0$, and $\mathbf{r}_a$ and $\mathbf{r}_b$ is the position of a and b, respectively. The viscous acceleration, $\mathbf{f}_a^{\mathrm{v}}$, is given by [16] and the body force in this work is $\mathbf{f}_a^{\mathrm{b}} = (0,\ 0,\ -9.98)\ \mathrm{m\,s^{-2}}$ in Cartesian coordinates. To ensure the no-slip boundary condition at walls, the approach by [1] is employed. Surface tension is modeled using the Continuum Surface Force (CSF) approach [2] applied to free surfaces

$$\mathbf{f}_a^{\mathrm{s}} = \mathbf{f}_a^{\perp} + \mathbf{f}_a^{\parallel} \,, \qquad \mathbf{f}_a^{\perp} = \frac{1}{\rho_a} 2\sigma_0 \bar{\kappa}_a \hat{\bar{\mathbf{n}}}_a \left| \mathbf{n}_a^{\mathrm{lg}} \right| \,, \qquad \mathbf{f}_a^{\parallel} = \frac{1}{\rho_a} \nabla_{\mathrm{s}} \sigma_0 / \Delta x \,, \tag{15.7}$$

Here, $\mathbf{f}_a^{\perp}$ and $\mathbf{f}_a^{\parallel}$, is the acceleration acting in normal or tangential direction to liquid-gas interface, σ_0, $\bar{\kappa}_a$, $\hat{\mathbf{n}}^{\mathrm{lg}}$, and $\hat{\tilde{\mathbf{n}}}_a^{\mathrm{lg}}$ are the surface tension coefficient, mean curvature, normal vector smoothed normalized normal vector, and surface gradient of surface tension at the position of particle a, respectively. Moreover, Δx is the SPH discretization length representing the edge length of the cubic volume of an SPH particle. The normal vector, smoothed normal vector, and smoothed normalized normal vector are given by

$$\mathbf{n}_a^{\mathrm{lg}} = -\sum_b \frac{m_b}{\rho_b} \nabla W_{ab}\,, \qquad \tilde{\mathbf{n}}_a = \frac{1}{S_a} \sum_b \frac{m_b}{\rho_b} \mathbf{n}_b W_{ab}\,, \qquad \hat{\tilde{\mathbf{n}}}_a^{\mathrm{lg}} = \frac{\tilde{\mathbf{n}}_a^{\mathrm{lg}}}{\left|\tilde{\mathbf{n}}_a^{\mathrm{lg}}\right|}\,. \tag{15.8}$$

Normal vectors whose magnitude is smaller than $0.1/h$ are discarded from the curvature computation. To include wetting forces into CSF, we employ the smoothed normal correction scheme by [4] which modifies the orientation of normal vectors of liquid SPH particles as a function of the particle's distance to the solid wall and the desired equilibrium contact angle, Θ_∞. We define the unit normal vectors of liquid SPH particles as

$$\hat{\tilde{\mathbf{n}}}_a^{\mathrm{lg}} = \begin{cases} \hat{\tilde{\mathbf{n}}}_a^{\mathrm{slg}} & \text{if} a \in \Omega^{\mathrm{slg}} \\ \hat{\tilde{\mathbf{n}}}_a^{\mathrm{lg}} & \text{else} \end{cases}, \quad a \in \Omega^{\mathrm{slg}} \text{ if} S_a^{\mathrm{s}} > 0 \wedge \left|\mathbf{n}_a^{\mathrm{lg}}\right| > 0.1/h\,, \tag{15.9}$$

where $\hat{\tilde{\mathbf{n}}}_a^{\mathrm{slg}}$ is the corrected normal vector of a particle a in vicinity of the three-phase contact line, Ω^{slg}. Since the gas phase is not discretized by SPH particles we additionally modify the normal vectors of selected solid SPH particles and include the resulting normal vectors into the curvature computation of the liquid SPH particles to increase its accuracy in regions with low neighbor particle support, in particular for low desired equilibrium contact angles. The orientation of normal vectors of solid particles b in the neighborhood of a liquid, central SPH particle a, by

$$\hat{\mathbf{n}}_{\infty,b}^{\mathrm{lg,s}} = \begin{cases} \hat{\mathbf{t}}_a^{\mathrm{sf}} \sin\Theta_\infty^{\mathrm{s}} - \hat{\tilde{\mathbf{n}}}_a^{\mathrm{sf}} \cos\Theta_\infty^{\mathrm{s}} & \text{if } a \in \Omega^{\mathrm{slg}} \wedge \mathbf{r}_{ab} \cdot \hat{\tilde{\mathbf{n}}}_a^{\mathrm{lg}} > 0 \\ 0 & \text{else} \end{cases}, \tag{15.10}$$

Here, $\Theta_\infty^{\mathrm{s}}$ is the equilibrium contact angle used to adjust the solid normal vectors, $\hat{\tilde{\mathbf{n}}}_a^{\mathrm{sf}}$ is the normalized normal vector of particle a pointing outwards of the solid phase

$$\mathbf{n}_a^{\mathrm{sf}} = -\sum_{b\in\Omega^{\mathrm{s}}} \frac{m_b}{\rho_b} \nabla W_{ab}\,, \qquad \tilde{\mathbf{n}}_a^{\mathrm{sf}} \frac{1}{S_a^{\mathrm{s}}} \sum_{b\in\Omega^{\mathrm{s}}} \frac{m_b}{\rho_b} \mathbf{n}_a^{\mathrm{sf}} W_{ab}\,, \qquad \hat{\tilde{\mathbf{n}}}_a^{\mathrm{sf}} = \frac{\tilde{\mathbf{n}}_a^{\mathrm{sf}}}{\left|\tilde{\mathbf{n}}_a^{\mathrm{sf}}\right|}\,, \tag{15.11}$$

and $\hat{\mathbf{t}}^{\mathrm{sf}}$ is the normalized tangent to the solid phase

$$\mathbf{t}_a^{\mathrm{sf}} = \hat{\tilde{\mathbf{n}}}_a^{\mathrm{lg}} - \left(\hat{\tilde{\mathbf{n}}}_a^{\mathrm{lg}} \cdot \hat{\tilde{\mathbf{n}}}_a^{\mathrm{sf}}\right) \hat{\tilde{\mathbf{n}}}_a^{\mathrm{sf}}, \qquad \hat{\mathbf{t}}_a^{\mathrm{sf}} = \frac{\mathbf{t}_a^{\mathrm{sf}}}{\left|\mathbf{t}_a^{\mathrm{sf}}\right|}. \tag{15.12}$$

The condition $\hat{\tilde{\mathbf{n}}}_a^{\mathrm{lg}} > 0$ in Eq. (15.10) identifies all neighboring solid SPH particles which are located on the liquid side of the tangent plane at the position of the liquid particle a defined by the normal vector of the liquid particle a. Solid SPH particles that are located on the gas side of the tangent plane are not considered in the curvature computation of the liquid SPH particle. Moreover, $\Theta_\infty^{\mathrm{s}}$ is a free parameter to calibrate the wetting forces. By setting $\Theta_\infty^{\mathrm{s}} > \Theta_\infty$, the underestimated curvatures are corrected to larger values, which prevents the continuous spreading of the liquid on the solid substrate. Finally, the mean curvature of a liquid SPH particle ($a \in \Omega^{\mathrm{l}}$) is computed by

$$\bar{\kappa}_a = -0.5 \sum_{b\in\Omega^{\mathrm{l}}} \frac{m_b}{\rho_b} \left(\hat{\tilde{\mathbf{n}}}_a^{\mathrm{lg}} - \hat{\tilde{\mathbf{n}}}_b^{\mathrm{lg}}\right) \hat{\nabla} W_{ab} - 0.5 \sum_{b\in\Omega^{\mathrm{s}}} \frac{m_b}{\rho_b} \left(\hat{\tilde{\mathbf{n}}}_a^{\mathrm{lg}} - \hat{\mathbf{n}}_{\infty,b}^{\mathrm{lg,s}}\right) \hat{\nabla} W_{ab}, \tag{15.13}$$

where, $\hat{\nabla} W_{ab}$ is the normalized kernel gradient [19]. Substituting the linear dependence of surface tension by [27] into $f_a^{\|}$ in Eq. (15.7) yields

$$\nabla_{\mathrm{s}} T_a = \nabla T_a - \left(\nabla T_a \cdot \hat{\tilde{\mathbf{n}}}_a^{\mathrm{lg}}\right) \hat{\tilde{\mathbf{n}}}_a^{\mathrm{lg}}, \qquad \nabla T_a = -\sum_b \frac{m_b}{\rho_b} (T_a - T_b) \hat{\nabla} W_{ab}, \tag{15.14}$$

where T_a and T_b is temperature of particle a and b. Thus, the acceleration of particle a due to the tangential component of surface tension is

$$\mathbf{f}_a^{\|} = \frac{1}{\rho_a \Delta x} \sigma_T \left(\nabla T_a - \left(\nabla T_a \cdot \hat{\tilde{\mathbf{n}}}_a^{\mathrm{lg}}\right) \hat{\tilde{\mathbf{n}}}_a^{\mathrm{lg}}\right) \text{ if } S_a^{\mathrm{s}} \leq 0.5. \tag{15.15}$$

The condition $S_a^{\mathrm{s}} \leq 0.5$ is required to avoid the penetration of liquid SPH particles into the wall, which could lead to unstable simulations.

The energy equation for each SPH particle a is given by [5]

$$\rho_a \tilde{c}_{p,a} \frac{\mathrm{D}T_a}{\mathrm{D}t} = \sum_b \frac{m_b}{\rho_b} \frac{4 k_a k_b}{k_a + k_b} (T_a - T_b) F_{ab} + \frac{\dot{q}_a^{\mathrm{G}} + \dot{q}_a^{\mathrm{rad}} + \dot{q}_a^{\mathrm{vap}}}{\Delta x}. \tag{15.16}$$

Here, $\tilde{c}_p$ is the effective heat capacity which incorporates the effect of the latent heat of crystallization and melting [8], k_a and k_b are the thermal conductivities assigned to particle a and b, and $\dot{q}_a^{\mathrm{G}}$, $\dot{q}_a^{\mathrm{rad}}$, and $\dot{q}_a^{\mathrm{vap}}$ are the energy absorbed due to laser irradiation, energy lost to the ambiance due to heat radiation and vaporization given by

$$I = \sum_{i=0}^{N_{\mathrm{ray}}} \frac{2 P_0}{\pi w_0^2} \exp\left(\frac{-2 r_{\mathrm{ray}}^2}{w_0^2}\right), \quad \dot{q}_a^{\mathrm{rad}} = \sigma_{\mathrm{SB}} \epsilon_{\mathrm{p}} \left(T^4 - T_0^4\right), \quad \dot{q}_{,a}^{\mathrm{vap}} = \dot{j}_a^{\mathrm{net}} h_{\mathrm{vap},a}(T), \tag{15.17}$$

where P_0 is the total laser power, w_0 is the beam waist radius, r_{ray} is the radial distance of a ray to the beam center axis, $\sigma_{SB} = 5.67 \cdot 10^{-8}\, text W m^{-2} K^{-4}$ is the Stefan-Boltzmann constant, $T_0 = 300\,K$ is the ambient temperature, j^{net} is the net evaporation flux [14]. In the ray tracing algorithm, we compute a surface mesh on the positions of the SPH particles using the CGAL-Library [25]. Upon intersection of a ray with the surface mesh, the transmitted energy from the ray to the material is evaluated using Fresnel laws [32] and the temperature-dependent refractive indices given in Sect. 15.2. The temperature of a triangle is computed as the mean of the SPH particle temperatures which constitute the triangle. Moreover, the absorbed energy by the triangle is distributed among the three SPH particles depending on their distance to the intersection point. The rays are traced until they left the simulation domain or transmitted 99.9 % of their initial energy. Equation (15.16) is integrated using the explicit Euler scheme.

15.4 Results and Discussion

To validate the developed numerical model the Institute of Photonic Technologies (LPT) at the Friedrich-Alexander-Universität Erlangen-Nürnberg (FAU) provided microscopy images of welding seam cross-sections obtained by single-line laser tracks on titanium grade 2 metal sheets and different processing parameters. In the performed experiments, an Aconity mini from Aconity GmbH, a redPower Qube fiber laser from SPI Lasers Ltd. with a wavelength of 1.08 µm, and an Axialscan 30 from Raylase is used. The laser is operated in continuous wave mode with Gaussian intensity distribution and the 86.5 % spot radius is adjusted to 75 µm or 120 µm in this work. Argon at atmospheric pressure is used as inert gas. The obtained cross-sections through the welding seams are polished with OP-S and H_2O_2, and etched according to Kroll. Images of the welding seam were taken using a Zeiss microscope with a polarization filter.

The simulations are performed using a smoothing length of the kernel function of $h = 2\Delta x$. Table 15.1 shows the adjusted laser power and scanning velocity, u_l /mm s^{-1}, of experiment and simulation, the dimensions of the simulated domain with length, L, height, H, and depth, D, as well as the discretization length and the total number of SPH particles, N_{SPH}.

The simulated and experimentally measured melt depth and melt width are shown in Fig. 15.3 as a function of the energy density, $E_A = \frac{P_0}{2 w_0 u_l}$. Exp. 1 and Exp. 2 represent the experimental measurements. Sim. 1 denotes simulation results where atomically clean titanium is assumed and sim. 2 denotes the simulation results where the initial surface layer of SPH particles is modeled as an aged titanium surface. The simulated melt geometry agrees well with the experiments if an aged titanium surface is assumed whereas the greater reflectivity of the clean titanium leads to underestimated melt depth and width. This indicates the importance of the correct modeling of the metal surface properties.

Table 15.1 Domain dimensions, applied spatial discretization as well as total amount of simulated SPH particles as a function of the laser properties and $w_0 = 120\,\mu\text{m}$

P_0/W	u_1/mm s^{-1}	Δx/μm	L/μm	D/μm	H/μm	N_{SPH}/1000
100	900	5.0	1415	400	100	452.80
100	600	7.5	1415	450	250	374.22
100	400	7.5	1415	450	250	374.22
100	250	7.5	1415	600	250	498.96
100	100	10.0	1415	500	350	248.50

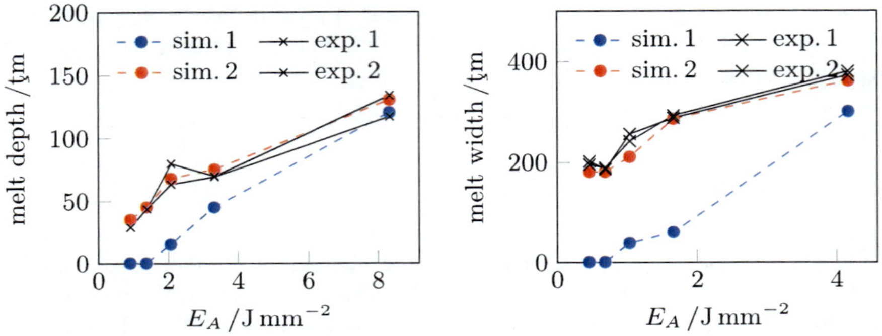

Fig. 15.3 Comparison of simulated melt depth (top) and melt width (bottom) with experiments for $P_0 = 100\,\text{W}$ and $w_0 = 120\,\text{tm}$. Experimental data provided by Florian Huber, LPT, FAU

Table 15.2 Processing parameters and simulated domain sizes of single-line laser tracks of titanium

set	P_0 W	u_1 mm s^{-1}	w_0 μm	E_A J mm^{-2}	Δx μm	L μm	H μm	D μm	N_{SPH}/1000
1	100	600	120	0.69	4.0	1415	225	100	938.10
2	350	500	120	2.92	10	3415	600	300	615.60
3	500	1500	120	3.47	7.5	3415	700	225	1272.24
4	600	3000	120	0.83	7.5	3415	600	225	1272.24
. 5	350	2000	75	1.17	5.0	1665	400	175	932.40

In the next simulations we calibrate the wetting forces by adjusting Θ_∞ and Θ_∞^{s} for the parameters given in Table 15.2 by comparing the geometry of the simulated weld cross section with the experiments.

This procedure is shown in Fig. 15.4 for parameter set 1 and is repeated for all parameter sets. As a result, the wetting forces are adjusted to $\Theta_\infty = 25°$ and $\Theta_\infty^{\text{s}} = 27°$ for parameter set 1 and $\Theta_\infty = 60°$ and $\Theta_\infty^{\text{s}} = 65°$ for parameter sets 2–5. The spreading of a material on its kind is not well investigated. According to [21] homologous wetting depends on the Stefan number. Therefore, we compute the

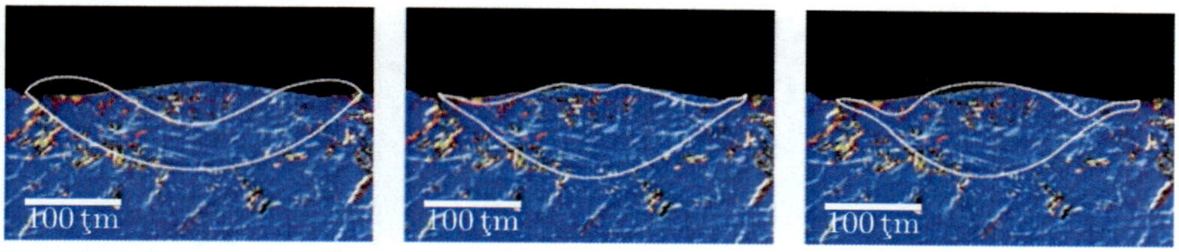

(a) $\Theta_\infty = 20°$, $\Theta^{\mathrm{s}}_\infty = 22°$ (b) $\Theta_\infty = 30°$, $\Theta^{\mathrm{s}}_\infty = 32°$ (c) $\Theta_\infty = 40°$, $\Theta^{\mathrm{s}}_\infty = 45°$

Fig. 15.4 Comparison of the weld geometries obtained by experiments and simulations with varying wetting forces for parameter set 1 ($w_0 = 120\,\mu\mathrm{m}$, $P_0 = 100\,\mathrm{W}$, $u_1 = 600\,\mathrm{mm\,s^{-1}}$). Experimental images provided by Florian Huber, LPT, FAU.

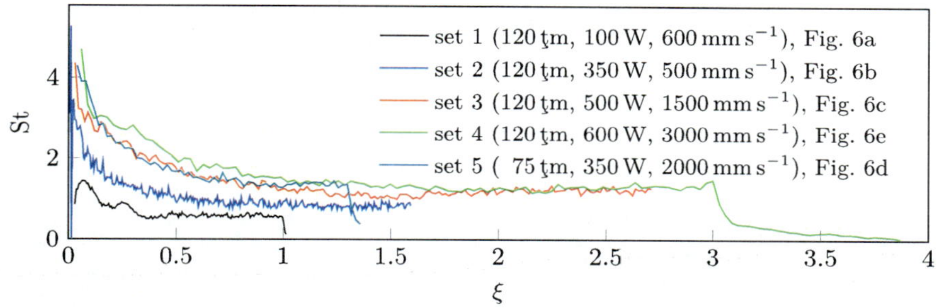

Fig. 15.5 Stefan numbers as a function of the laser position obtained by the simulations with parameter set 1–5

mean Stefan number from all liquid SPH particles, $N^{\mathrm{l}}_{\mathrm{SPH}}$, that are located in a normal distance of 10–12 μm to the solid-liquid interface by

$$\mathrm{St} = \frac{c_p\left(T^{\mathrm{l}} - T_{\mathrm{m}}\right)}{h_{\mathrm{m}}}, \qquad T^{\mathrm{l}} = \frac{1}{N^{\mathrm{l}}_{\mathrm{SPH}}} \sum_{b\in\Omega^{\mathrm{l}}} T_b\,. \tag{15.18}$$

Here, $c_p = 881.72\,\mathrm{J\,kg^{-1}}$ is the mean value of the heat capacities of the solid and liquid phase at melting temperature. Figure 15.5 shows the computed Stefan numbers for parameter sets 1 to 5 as a function of the laser position $\xi = u_1/1\,\mathrm{mm}$.

The computed Stefan number is affected by the adjusted laser processing parameters. For example, Fig. 15.7 shows a snapshot of the simulation with parameters set 4 where humping occurred. When comparing the Stefan numbers to the corresponding experimental images of the weld shape in Fig. 15.6, it can be seen that higher Stefan numbers are associated with more elevated welding seams. Moreover, greater Stefan numbers are related to larger temperature differences between liquid SPH particles near the solid-liquid interface particles and the melting temperature. This implies a greater temperature gradient across the solid-liquid interface, which may influence the resultant wetting forces.

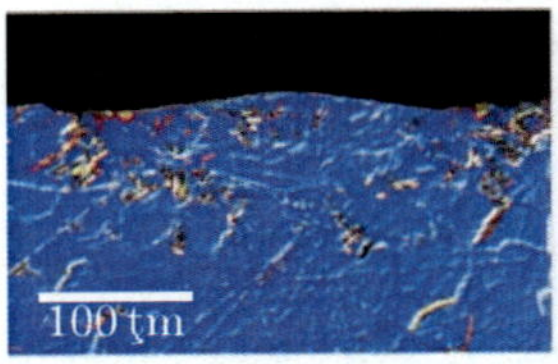

(a) Parameter set 1 ($w_0 = 120\,\mu m$, $P_0 = 100\,W$, $u_l = 600\,mm\,s^{-1}$).

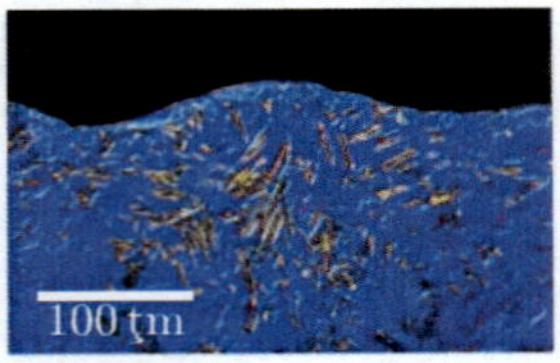

(b) Parameter set 2 ($w_0 = 120\,\mu m$, $P_0 = 350\,W$, $u_l = 500\,mm\,s^{-1}$).

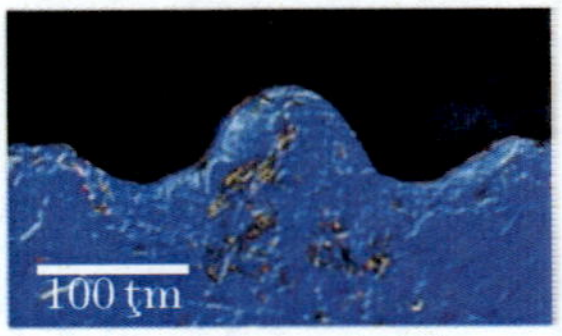

(c) Parameter set 3 ($w_0 = 120\,\mu m$, $P_0 = 500\,W$, $u_l = 1500\,mm\,s^{-1}$).

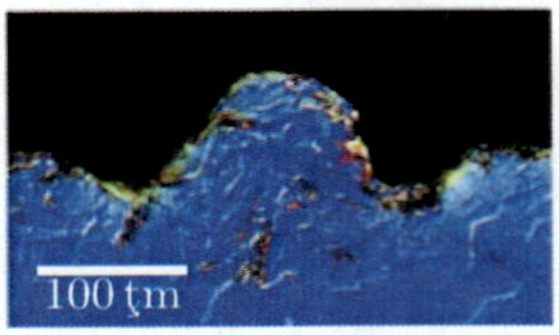

(d) Parameter set 5 ($w_0 = 120\,\mu m$, $P_0 = 350\,W$, $u_l = 2000\,mm\,s^{-1}$).

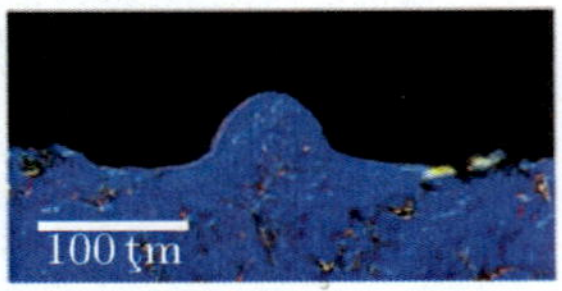

(e) Parameter set 4 ($w_0 = 120\,\mu m$, $P_0 = 600\,W$, $u_l = 3000\,mm\,s^{-1}$).

Fig. 15.6 Microscopy images of the welding seam cross-sections provided by Florian Huber, LPT, FAU

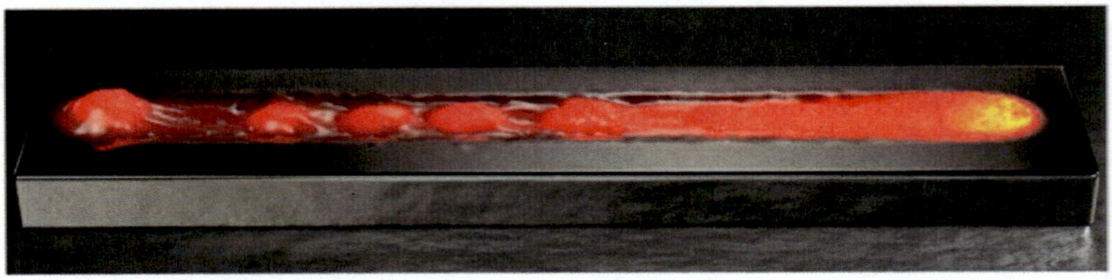

Fig. 15.7 Simulation snapshot of the laser welding process of titanium using parameter set 4 exhibiting humping

Acknowledgements Funded by the Deutsche Forschungsgemeinschaft (DFG, German Research Foundation)-Project-ID 61375930-SFB 814 "Additive Manufacturing" TP B1. We thank the Gauss Centre for Supercomputing for providing computer time through the John von Neumann Institute for Computing on the GCS Supercomputer JUWELS at Jülich Supercomputing Centre. The work was also supported by the Interdisciplinary Center for Nanostructured Films (IZNF), the Competence Unit for Scientific Computing (CSC), and the Interdisciplinary Center for Functional Particle Systems (FPS) at Friedrich-Alexander University Erlangen-Nürnberg.

References

1. Adami, S., Hu, X.Y., Adams, N.A.: A generalized wall boundary condition for smoothed particle hydrodynamics. J. Comput. Phys. **231**(21), 7057–7075 (2012)
2. Brackbill, J., Kothe, D.B., Zemach, C.: A continuum method for modeling surface tension. J. Comput. Phys. **100**(2), 335–354 (1992)

3. Bradshaw, F.J.: The optical emissivity of titanium and zirconium. Proc. Phys. Soc. Lond. Sect. A **63**(8), 573–577 (1950)
4. Breinlinger, T., Polfer, P., Hashibon, A., Kraft, T.: Surface tension and wetting effects with smoothed particle hydrodynamics. J. Comput. Phys. **243**, 14–27 (2013)
5. Cleary, P.W., Monaghan, J.J.: Conduction modelling using smoothed particle hydrodynamics. J. Comput. Phys. **148**(1), 227–264 (1999)
6. Cummins, S.J., Rudman, M.: An SPH projection method. J. Comput. Phys. **152**(2), 584–607 (1999)
7. Desai, P.: Thermodynamic properties of titanium. Int. J. Thermophys. **8**(6), 781–794 (1987)
8. Farrokhpanah, A., Bussmann, M., Mostaghimi, J.: New smoothed particle hydrodynamics (SPH) formulation for modeling heat conduction with solidification and melting (2016)
9. Fortov, V.E., Dremin, A.N., Leont'ev, A.A.: Evaluation of the parameters of the critical point. Teplofiz Vys Temp **13**(5), 1072–1080 (1975)
10. Iida, T., Guthrie, R.I.L.: The Physical Properties of Liquid Metals. Oxford University Press, Oxford; New York, Clarendon Press (1988)
11. Ishikawa, T., Paradis, P.-F., Okada, J.T., Watanabe, Y.: Viscosity measurements of molten refractory metals using an electrostatic levitator. Meas. Sci. Technol. **23**(2), 025305 (2012)
12. Jacobson, M.I.: An investigation of the lattice parameters, electrical resistivity, and magnetic susceptibility of some titanium alloys: proposal of an electronic band structure. Ph.D. thesis, The Ohio State University (1958)
13. Johnson, P.B., Christy, R.W.: Optical constants of transition metals: Ti, V, Cr, Mn, Fe Co, Ni, and Pd. Phys. Rev. B **9**(12), 5056–5070 (1974)
14. Klassen, A.: Simulation of evaporation phenomena in selective electron beam melting. doctoral thesis, FAU University Press (2018)
15. Monaghan, J.J.: Smoothed particle hydrodynamics. Rep. Prog. Phys. **68**(8), 1703 (2005)
16. Morris, J.P., Fox, P.J., Zhu, Y.: Modeling low Reynolds number incompressible flows using SPH. J. Comput. Phys. **136**(1), 214–226 (1997)
17. Nair, P.: Modeling free surface flows and fluid structure interactions using smoothed particle hydrodynamics. Ph.D. thesis, Department of Mechanical Engineering, Indian Institute of Science, Bangalore (2015)
18. Naito, S., Arai, T., Yokoyama, I., Waseda, Y.: The long-wavelength limit of the structure factor of liquid 3d transition metals using one-component plasma model. High Temp. Mater. Process. **8**(2), 125–134 (1989)
19. Oger, G., Doring, M., Alessandrini, B., Ferrant, P.: An improved SPH method: towards higher order convergence. J. Comput. Phys. **225**(2), 1472–1492 (2007)
20. Powell, R.W., Ho, C.Y., Liley, P.E.: Thermal conductivity of selected materials, part 2. National Bureau of Standards, U.S. Department of Commerce (1968)
21. Schiaffino, S., Sonin, A.A.: Motion and arrest of a molten contact line on a cold surface: an experimental study. Phys. Fluids **9**(8), 2217–2226 (1997)
22. Seydel, U., Fucke, W.: Electrical resistivity of liquid Ti, V, Mo and W. J. Phys. F: Met. Phys **10**(8), L203–L206 (1980)
23. Siegel, E.: Optical reflectivity of liquid metals at their melting temperatures. Phys. Chem. Liquids **5**(1), 9–27 (1976)
24. Simchi, A., Petzoldt, F., Pohl, H.: Direct metal laser sintering: material considerations and mechanisms of particle bonding. Int. J. Powder Metall. **37**(2), 49–62 (2001)
25. The CGAL Project: CGAL User and Reference Manual. CGAL Editorial Board, 5.4 edition (2022)
26. Thurnay, K.: Thermal properties of transition metals. Technical report 0947-8620, Germany (1998)
27. Tong, M., Browne, D.J.: An incompressible multi-phase smoothed particle hydrodynamics (SPH) method for modelling thermocapillary flow. Int. J. Heat Mass Transf. **73**, 284–292 (2014)
28. Ujihara, K.: Reflectivity of metals at high temperatures. J. Appl. Phys. **43**(5), 2376–2383 (1972)
29. Wei, P.: Thermal science of weld bead defects: a review. J. Heat Transf. **133**, 031005 (2011)

30. Wendland, H.: Piecewise polynomial, positive definite and compactly supported radial functions of minimal degree. Adv. Comput. Math. **4**(1), 389–396 (1995)
31. Werner, W.S.M., Glantschnig, K., Ambrosch-Draxl, C.: Optical constants and inelastic electron-scattering data for 17 elemental metals. J. Phys. Chem. Ref. Data **38**(4), 1013–1092 (2009)
32. YU, P., Cardona, M.: Fundamentals of Semiconductors: Physics and Materials Properties. Springer Science & Business Media
33. Zhang, J., Song, B., Wei, Q., Bourell, D., Shi, Y.: A review of selective laser melting of aluminum alloys: processing, microstructure, property and developing trends. J. Mater. Sci. Technol. **35**(2), 270–284 (2019)
34. Zhou, K., Wang, H.P., Chang, J., Wei, B.: Experimental study of surface tension, specific heat and thermal diffusivity of liquid and solid titanium. Chem. Phys. Lett. **639**, 105–108 (2015)
35. Zhu, H.H., Lu, L., Fuh, J.Y.H.: Study on shrinkage behaviour of direct laser sintering metallic powder. Proc. Inst. Mech. Eng. B: J. Eng. Manuf. **220**(2), 183–190 (2006)

Chapter 16
CFD Modelling of Mortar Extrusion and Path Planning Strategy at the Corner for 3D Concrete Printing

Khalid El Abbaoui, Issam Al Korachi, Md Tusher Mollah, and Jon Spangenberg

Abstract Computational Fluid Dynamics (CFD) modelling has become an important tool for accelerating different costly and time-consuming processes for 3D concrete printing. In this study, a CFD model is developed to simulate the mortar extrusion and deposition flow in order to investigate the path planning at the corner. The mortar is characterized by a viscoplastic Bingham fluid. The printhead moves according to a prescribed speed to print trajectory with a turn of 90° angle. The model solves the Navier Stokes equations and uses the volume of fluid (VOF) technique. This work investigates the effects of changing printhead direction in the geometrical conformity and the process precision. The direction change is carried out by the acceleration steps and jerk. A smoothing factor is provided to smooth the toolpath. Several simulations were performed by varying the smoothing factor and jerk. Overfill has occurred at the sharp corner when applying a jerk due to the constant extrusion rate. Dependance between the acceleration steps, the acceleration time limit and the jerk has been found.

Keywords 3D concrete printing · CFD modelling · Path-planning

16.1 Introduction

The automated construction process, 3D concrete printing (3DCP), has gained significant attention over recent years due to its ability for structural customization and printing scale. This process consists of depositing successive layers of concrete

K. El Abbaoui (✉) · I. Al Korachi
Euromed Polytechnic School, Euromed Research Center, Euromed University of Fez, 30000 Fez, Morocco
e-mail: k.elabbaoui@ueuromed.org

M. T. Mollah · J. Spangenberg
Department of Civil and Mechanical Engineering, Technical University of Denmark, 2800 Kgs, Lyngby, Denmark

K. Zarbane and Z. Beidouri (eds.), *Proceedings of CASICAM 2022*, Springer Tracts in Additive Manufacturing, https://doi.org/10.1007/978-3-031-32927-2_16

contouring an object without formwork, i.e., by direct concrete placement. Therefore, 3DCP is a free form buildup [1] and was proved on the rationalization of ecological materials [2, 3]. The basic physics of the full of manufacturing steps of 3DCP was detailed in the reference [4]. Despite its enormous progress, the process is still accompanied by several errors and failures. A specific constraint is that an angular toolpath needs to be implemented at the corner, which is typically rounded.

The trajectory planner sits between the G-code and the servomotor execution of the toolpath travel. The G-code file may contain a million lines in it because of how much path error is permissible, how long it is going to take, how smooth desired, the axes coupled and dynamic computation. Printing objects comes with the understanding of the extruder dynamics of the 3D printer and how planning the path [5]. Fleck et al. [6] observed an overflow at the corner because of trapezoidal velocity profiles employed through Marlin firmware [7]. This firmware involves a slam velocity at turn. Research work has been done on material extrusion-based additive manufacturing with investigating the print precision of cornering [8, 9]. Giberti et al. [10] proposed an algorithm for path planning to inspect the material extrusion rate strained by velocity pursuant of a toolpath over the execution of 3D printing process. Jin et al. [11] tried to smooth paths by decreasing the number of sharp corners using an implicit algorithm. Bos et al. [12] mentioned that square nozzles should be tangent to the toolpath when moving the printhead to overcome the twisting of filament. AlSakka et al. [13] identified the optimal deposition path of the 3DCP which reduce the weak bonds and preserve the buildability of the extruded material by using a discrete event model.

In complement to the physical printing, CFD modelling is able to reproduce the extrusion-deposition of the printable concrete and investigate some specific constraints to the physical print [14–17]. The angular toolpath planning in 3DCP is in its infancy. However, several CFD modelling studies have been performed in other 3D printing technologies, like fused deposition modeling. [18, 19] developed CFD models to investigate the underfill and overflow areas in the deposited corners, and investigated to smoothen the corners. Furthermore, Mollah et al. [20] investigated corner deposition for Bowden and Direct-drive extruders, where the Direct-drive extruder controls the amount of material extrusion near the corner. These models were developed using the Newtonian fluid model.

In the present study, a CFD model is developed to solve the Navier Stokes equations with the Bingham rheological model to simulate the mortar extrusion through a cylindrical nozzle. The simulations predict the evolution of the layer shape and the overfill area of a deposited mortar at a turn of 90° angle. Different printing ways are simulated to inquire the consequences of smoothing the extruder nozzle with a constant extrusion rate. The structure of the study is as follows. Section 16.2 describes the methodology of the study with the theoretical details, CFD model, and path planning strategy. The results are discussed in Sect. 16.3. Section 16.4 concludes the study.

16.2 Methodology

16.2.1 Governing Equations

The flow dynamics of fresh mortar is modelled by the Navier Stokes' equations for the non-Newtonian incompressible fluid with a constant density, formed by the following continuity and the momentum equations:

$$\nabla \cdot u = 0 \tag{16.1}$$

$$\rho\left(\frac{\partial \mathbf{u}}{\partial t} + \mathbf{u} \cdot \nabla \mathbf{u}\right) = -\nabla p + \rho \mathbf{g} + \nabla \cdot \boldsymbol{\sigma} \tag{16.2}$$

where $\boldsymbol{u}$ is the velocity vector field, $\rho = 2100$ kg/m^3 is the density, p is the pressure, $\boldsymbol{g}$ is the gravitational vector field acting downwards, σ is the constitutive stress tensor, and t is the time variable. The rheology of the fresh mortar can be modelled by the viscoplastic Bingham model [21]. This type of fluids flows when the shear stress exceeds the yield stress. The evolution of the constitutive stress tensor can be approximated with the Generalized Newtonian Fluid constitutive model:

$$\boldsymbol{\sigma} = 2\eta_{app}(\dot{\gamma})\mathbf{D} \tag{16.3}$$

where $\mathbf{D} = (\nabla \mathbf{u} + \nabla^T \mathbf{u})/2$ is the strain rate tensor, $\dot{\gamma} = \sqrt{2tr(\mathbf{D}^2)}$ is the shear rate, and $\eta_{app}(\dot{\gamma})$ is the apparent viscosity of the viscoplastic Bingham model. However, the apparent viscosity of the material increases to infinity when the applied stress is below the yield stress. To overcome this singularity issue of the Bingham fluid flow in the numerical simulation, the apparent viscosity is smoothed with the bi-viscosity method:

$$\eta_{app}(\dot{\gamma}) = \begin{cases} \eta_{app}^{MAX} & \text{for } \dot{\gamma} < \dot{\gamma}_c \\ \frac{\tau_Y}{\dot{\gamma}} + \mu_p & \text{for } \dot{\gamma} \geq \dot{\gamma}_c \end{cases} \tag{16.4}$$

where $\tau_Y = 630$ Pa is the yield stress, $\mu_p = 7.5$ Pa.s is the plastic viscosity, $\eta_{app}^{\mathrm{MAX}} = \tau_Y/\dot{\gamma}_c + \mu_p$ is the maximum value of the smoothed apparent viscosity and $\dot{\gamma}_c = 0.01$ s^{-1} is the critical threshold of the shear rate [14]. For more details on the regularization methods of the Bingham constitutive equation, please see [22].

16.2.2 CFD Model

The physical problem was numerically simulated with the software FLOW-3D®, version 12.0 [23]. The Eqs. (16.1) and (16.2) are discretized by the finite volume

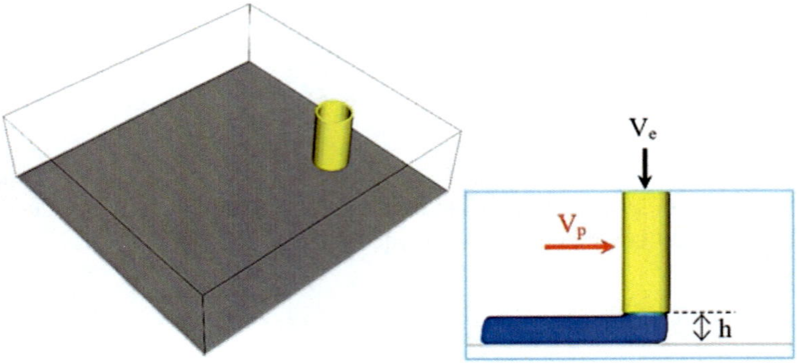

Fig. 16.1 Geometry of the CFD model (left) and a picture during the simulation (right)

method. The volume of fluid method [24] was used to capture the extrudate mortar sharp interfaces with the split Lagrangian method. The numerical domain was discretized with a uniform Cartesian non-conforming mesh. The cell size is 1 mm for each direction. A no slip boundary condition was applied to the nozzle wall and the build plate. The fresh mortar begins to flow through the extruder nozzle with a constant velocity and exit with a fully developed flow profile. The other domain boundaries are set as continuative boundary conditions. Intel® Xeon® W-2155 CPU 3.30 GHz 20 cores is used to execute all the numerical simulations. Each simulation takes around 2 days and 4.5 h. The geometry of the model consists of a cylindrical nozzle and a build plate. The nozzle comes with an inner diameter D = 25 mm and a thickness e = 2 mm, the numerical domain occupies a volume of 240 × 230 × 65 mm^3. The distance between the nozzle outlet and the build space is fixed at h = 12.5 mm. The printhead starts to move with a prescribed velocity when the mortar reaches the build plate (see Fig. 16.1).

16.2.3 Path Planning Strategy

The path planning of deposited corners consists of printing two segments with 90° turn without stopping the printhead. The servomotor of the 3D printer controls the displacement through a finite jerk and a linear acceleration limit. The jerk is the velocity leap that occurs instantly when the printhead begins an acceleration and a deceleration. Five depositing strategies of mortar layer with sharp and smooth corner have been simulated (see Fig. 16.2). The printhead moves with a maximum velocity Vp = 50 mm/s, decelerates and accelerates at the corner with a maximum value A_{max} = 30–50 mm/s^2. The deceleration time $\delta t = (t_{(s_1)out} - t_{(s_1)in}) = 1$ s and the extrusion velocity (V_e = 33.4 mm/s) are kept constant along the path except in the case (a) of the most used printing strategy in 3DCP, where the jerk is infinite because of the very small servo time (0.0001 s) at the acceleration transitions. A smoothing factor χ was used to smooth the printhead trajectory around the corner defined as the ration between the acceleration time limit $(t_{(s_1)out} - t_{(s_2)in})$ and the deceleration time δt required to decrease the velocity of the printhead until it stops in the first direction (X axis in our cases). S_1 and S_2 refer to the two segments respectively. The following Table 16.1 presents the path planning strategies that are numerically simulated.

16.3 Results and Discussion

The simulated results of the deposition flow are post-processed using FlowSight, version 12.0 and presented in Fig. 16.3. In the left side of the figure, we illustrate a perspective view of the different simulated strategies. The right side shows the ideal widths and the maximum flow depth of the different deposited layers. To exactly execute a toolpath with the G-code of the 3D printer, the extruder nozzle has to stop every time reaching a corner because in each segment of the toolpath, the extruder nozzle moves at a fixed ratio between the motors. When coming to a stop at the corner, the flow rate drew from the extruder nozzle and it will take a long time to deposit. Figure 16.3.a, b and c correspond to the sharp corner cases. The first strategy corresponds to an ideal model with an infinite jerk where we have to be aware because it could cause vibration and slam to the printhead and as a result a poor deposition could be obtained. It is observed that the selected jerk did not improve the print quality of the corner (see Fig. 16.3c). Furthermore, the mortar overflowed and crushed at the 90° turn. This is occurring because of the feed rate that has been kept constant and the increase of the extrusion pressure due to the slowing of the printhead while changing direction. In order to smooth corners, it is needed to act on the velocity, acceleration and jerk limits of the actuator of the 3D printer. Figure 16.3d shows an overflow at the exterior of the turn due to the corner shape and the acceleration time limit of 0.3 s. The accuracy print is obtained when the acceleration and the deceleration steps are blended instantly. Figure 16.3e shows a uniform corner bounded by an elliptical arc through the endpoints with a minimal overflow at the exterior of the turn despite of the smoothing of the toolpath. The observation is that when the printhead goes through the corner, the velocity is momentarily decelerated and accelerated. Therefore, a curvilinear path at the corner exterior is obtained.

The overflow and the underflow amount at the inside and the outside of the corner with 90° angle regarding to the ideal model (a) were quantified to analyze the deposition behavior when the printhead initiate a turn for the studied printing strategies (b), (c), (d), (e), and a printing strategy with a proportional extrusion velocity (see Fig. 16.4), where the deceleration/acceleration phases are blended (see Fig. 16.5).

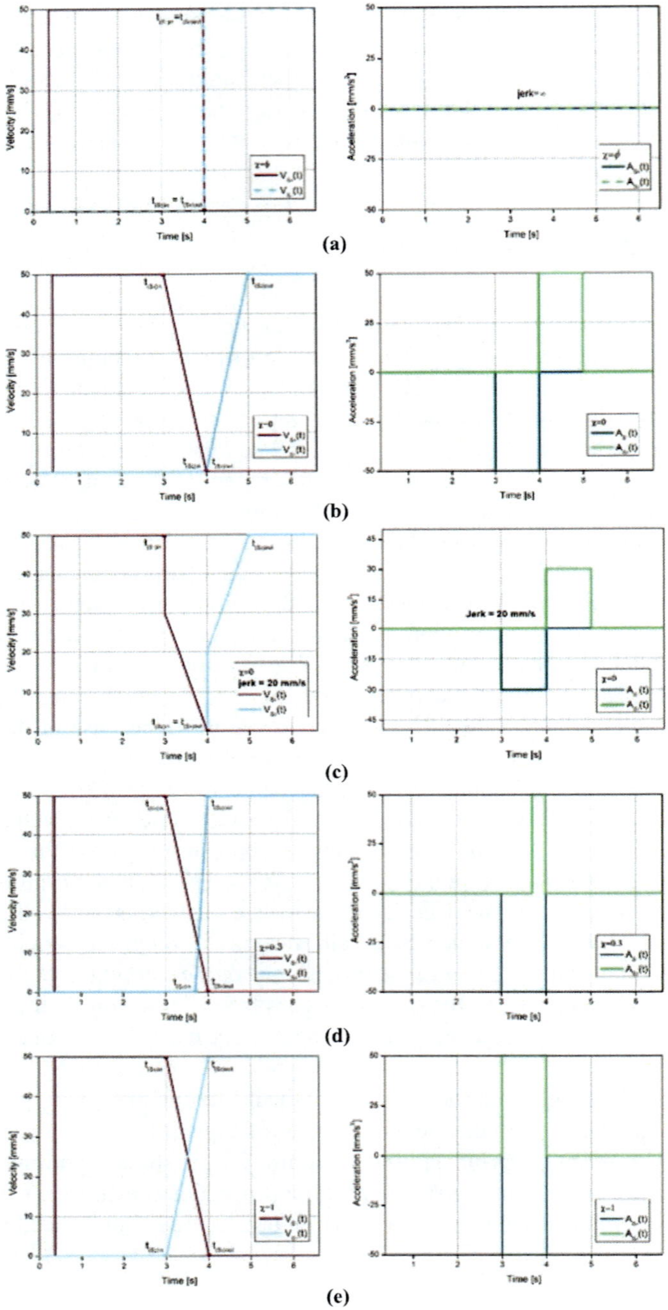

Fig. 16.2 Per-axis velocity profiles and there corresponding acceleration steps with different smoothing factor

Table 16.1 Path planning strategies cases numerically simulated

Corner type	Smoothing factor χ
Sharp corner	0 (infinite jerk)
	0 (without jerk)
	0 (jerk = 20 mm/s)
Smooth corner	0.3
	1

The blue bars correspond to a comparison between the case (c) and the printing strategy applying a jerk of 20 mm/s and a proportional extrusion velocity. The virtual printed layer for the sharp corner strategy applying a jerk of 20 mm/s was noticeably improved because of the synchronization of the extrusion rate at the direction change of the printhead (blue bar at the top). As illustrated on the graphs, the deposited material largely overflowed at the exterior side of the turn compared to the interior side for all the toolpath planning scenarios. This is because the double deposition occurs at the inner side of the turn, one from the pre-corner strand and the other from the post-corner layer. In addition, the toolpath of the printhead was fitted at the turn to have a curvature line without reaching the sharp corner. Therefore, the precision of the print quality depends on the way we program the print settings.

16.4 Conclusions

A CFD model was developed to investigate different toolpath strategies to assess the corner precision in 3DCP. The rheology of the printable mortar is approximated with the Generalized Newtonian constitutive model using the Bingham rheological fluid model. A 90° turn was investigated. It was found that the printhead should be planned to not slam or slow to a stop for good print quality of corners. It was observed that the tested case involving a jerk of 20 mm/s and an extrusion rate synchronized with the nozzle speed reduces the swelling of the strand at the corner. In fact, a commensurate extrusion rate with the printing velocity should generate a uniform corner in the case of the 90° turn; however, it could create a major swelling of the strand in the case of smaller angles. The CFD model provides a powerful tool that could be utilized to investigate different angular turns in 3DCP. This will minimize the number of prototypes and experimental tests required to implement toolpath strategies for corners with high precision.

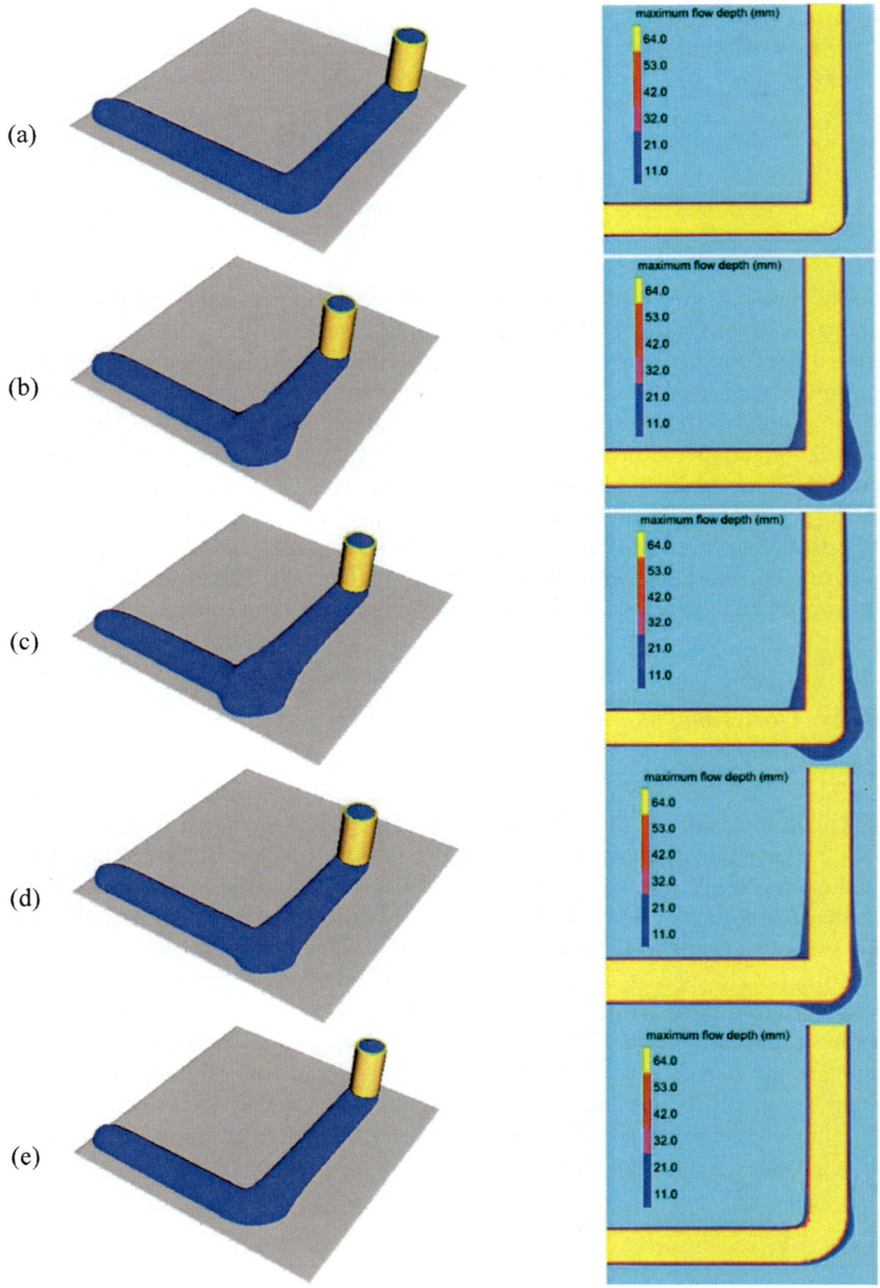

Fig. 16.3 Simulations of mortar extrusion along a toolpath with direction change of 90° using five deposition strategies

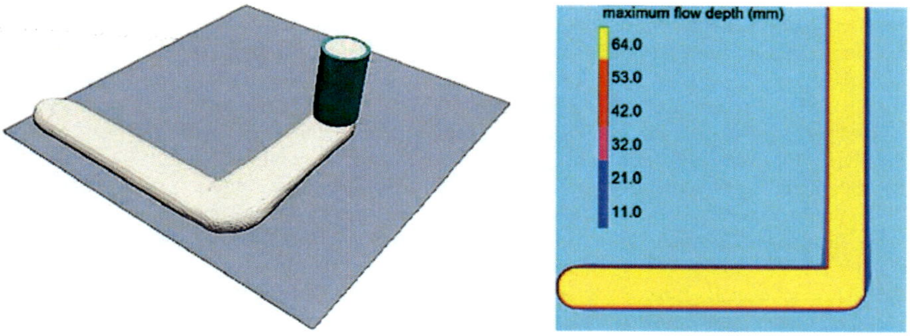

Fig. 16.4 Printing strategy applying a jerk of 20 mm/s and a synchronized extrusion velocity with the printing veocity

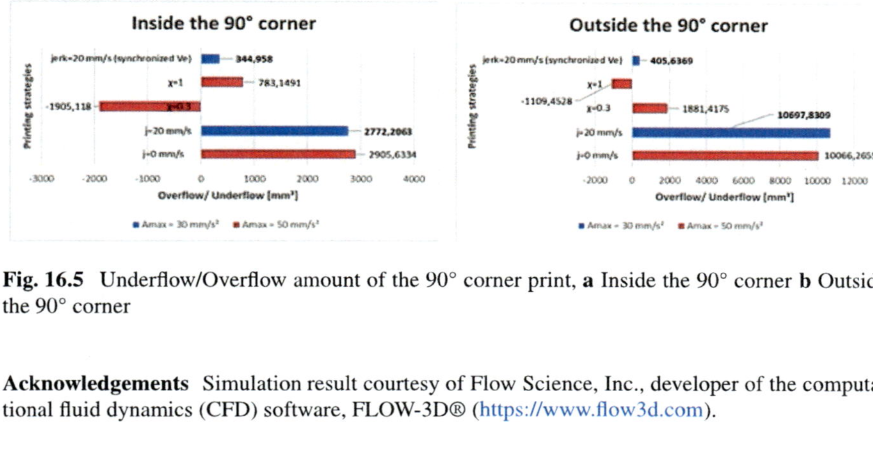

Fig. 16.5 Underflow/Overflow amount of the 90° corner print, **a** Inside the 90° corner **b** Outside the 90° corner

Acknowledgements Simulation result courtesy of Flow Science, Inc., developer of the computational fluid dynamics (CFD) software, FLOW-3D® (https://www.flow3d.com).

References

1. Pegna, J.: Exploratory investigation of solid freeform construction. Autom. Constr. **5**(5), 427–437 (1997)
2. De Schutter, G., Lesage, K., Mechtcherine, V., Nerella, V.N., Habert, G., Agusti-Juan, I.: Vision of 3D printing with concrete—Technical, economic and environmental potentials. Cem. Concr. Res. 112:25-36 (2018)
3. Dixit, M.: 3-D printing in building construction: a literature review of opportunities and challenges of reducing life cycle energy and carbon of buildings. In: IOP Conference Series: Earth and Environmental Science, vol 1. IOP Publishing, p 012012 (2019)
4. Mechtcherine, V., Bos, F.P., Perrot, A., da Silva, W.L., Nerella, V., Fataei, S., Wolfs, R.J., Sonebi, M., Roussel, N.: Extrusion-based additive manufacturing with cement-based materials–production steps, processes, and their underlying physics: a review. Cem. Concr. Res. **132**, 106037 (2020)

5. Go, J., Schiffres, S.N., Stevens, A.G., Hart, A.J.: Rate limits of additive manufacturing by fused filament fabrication and guidelines for high-throughput system design. Addit. Manuf. **16**, 1–11 (2017)
6. Fleck, T.J., McCaw, J.C., Son, S.F., Gunduz, I.E., Rhoads, J.F.: Characterizing the vibration-assisted printing of high viscosity clay material. Addit. Manuf. **47**, 102256 (2021)
7. Marlin-Firmware. https://marlinfw.org/meta/gcode/. Accessed 6 Oct 2022
8. Akhoundi, B., Nabipour, M., Kordi, O., Hajami, F.: Calculating printing speed in order to correctly print PLA/continuous glass fiber composites via fused filament fabrication 3D printer. J. Thermoplast. Compos. Mater. **0**(0), 0892705721997534. https://doi.org/10.1177/0892705721997534
9. Li, L., McGuan, R., Isaac, R., Kavehpour, P., Candler, R.: Improving precision of material extrusion 3D printing by in-situ monitoring & predicting 3D geometric deviation using conditional adversarial networks. Addit. Manuf. **38**, 101695 (2021)
10. Giberti, H., Sbaglia, L., Urgo, M.: A path planning algorithm for industrial processes under velocity constraints with an application to additive manufacturing. J. Manuf. Syst. **43**, 160–167 (2017)
11. Jin, Y., Du, J., Ma, Z., Liu, A., He, Y.: An optimization approach for path planning of high-quality and uniform additive manufacturing. Int. J. Adv. Manuf. Technol. **92**(1), 651–662 (2017)
12. Bos, F., Wolfs, R., Ahmed, Z., Salet, T.: Additive manufacturing of concrete in construction: potentials and challenges of 3D concrete printing. Virtual Phys. Prototyp. **11**(3), 209–225 (2016). https://doi.org/10.1080/17452759.2016.1209867
13. AlSakka, F., Senan, M.H., Abou Yassin, A., Hamzeh, F.: Path optimization in 3D concrete printing to minimize weak bonds formation. Period. Polytech. Arch. **50**(2), 163–168 (2019)
14. Comminal, R., Leal da Silva, W.R., Andersen, T.J., Stang, H., Spangenberg, J.: Modelling of 3D concrete printing based on computational fluid dynamics. Cem. Concr. Res. **138**, 106256 (2020). https://doi.org/10.1016/j.cemconres.2020.106256
15. Roussel, N., Spangenberg, J., Wallevik, J., Wolfs, R.: Numerical simulations of concrete processing: from standard formative casting to additive manufacturing. Cem. Concr. Res. **135**, 106075 (2020)
16. Spangenberg, J., da Silva, W.R.L., Comminal, R., Mollah, M.T., Andersen, T.J., Stang, H.: Numerical simulation of multi-layer 3D concrete printing. RILEM Tech. Lett. **6**, 119–123 (2021)
17. Spangenberg, J., Leal da Silva, W.R., Mollah, M.T., Comminal, R., Juul Andersen, T., Stang, H.: Integrating reinforcement with 3D concrete printing: experiments and numerical modelling. In: RILEM International Conference on Concrete and Digital Fabrication, pp. 379–384. Springer (2022)
18. Comminal, R., Serdeczny, M., Pedersen, D., Spangenberg, J.: Numerical modeling of the material deposition and contouring precision in fused deposition modeling. In: Proceedings of the 29th Annual International Solid Freeform Fabrication Symposium, Austin, TX, USA, pp 13–15 (2018)
19. Comminal, R., Serdeczny, M.P., Pedersen, D.B., Spangenberg, J.: Motion planning and numerical simulation of material deposition at corners in extrusion additive manufacturing. Addit. Manuf. **29**, 100753 (2019)
20. Mollah, M.T., Moetazedian, A., Gleadall, A., Yan, J., Alphonso, W.E., Comminal, R.B., Seta, B., Lock, T., Spangenberg, J.: Investigation on corner precision at different corner angles in material extrusion additive manufacturing: an experimental and computational fluid dynamics analysis. In: Solid Freeform Fabrication Symposium 2022: 33rd Annual Meeting. The University of Texas at Austin, pp. 872–881 (2022)
21. Roussel, N.: Rheological requirements for printable concretes. Cem. Concr. Res. **112**, 76–85 (2018)

22. Tu, Y., Hassan, A., Siadat, A., Yang, G., Chen, Z.: Numerical simulation and experimental validation of deposited corners of any angle in direct ink writing (2022)
23. Flow Science, I.: FLOW-3D, Version 12.0. (2019)
24. Hirt, C.W., Nichols, B.D.: Volume of fluid (VOF) method for the dynamics of free boundaries. J. Comput. Phys. **39**(1), 201–225 (1981). https://doi.org/10.1016/0021-9991(81)90145-5

Chapter 17
Effect of Raster Width on the Strength of the Cohesive Zone Between ABS Filaments from a Printed Digital CT

Oumaima Aourik, Mourad Othmani, and Abdelkerim Chouaf

Abstract Considering the industrial requirements on the mechanical performances requested in the printed parts, many research axes have been developed. Among these requirements, the resistance to breakage of this type of structures printed by FDM is very complex and little developed. The objective of this study is to analyze and understand the effect of raster widths on the mechanical performance and crack propagation in an ABS (Acrylonitrile–Butadiene–Styrene) sample obtained by FDM (Fused Deposition Modeling). A numerical approach has been developed to link this printing parameter (the raster width) to the crack propagation phenomenon. And more particularly our study was oriented towards the analysis of the cohesion between filaments, for that, we treated only two cases of width of the raster ($l = 0.42$ and $l = 0.56$ mm). In order to analyze the results obtained from the concepts of the energetic theory of the rupture.

Keywords FDM · Raster width · ABS · CT specimen · Crack · Numerical simulation

17.1 Introduction

Today, additive manufacturing covers practically all industrial sectors [1]. Indeed, thanks to the maturity of its technique, it is possible to produce parts with very satisfactory appearance performances [2]. However, performance in terms of mechanical behaviour has been little developed, given the complexity of the printed structures

O. Aourik (✉) · M. Othmani · A. Chouaf
Laboratory of Mechanics, Engineering and Innovation, Hassan II University, National School of Electricity and Mechanics, Casablanca, Morocco
e-mail: oumaima.aourik@ensem.ac.ma

M. Othmani
e-mail: mourad.othmani@ensem.ac.ma

A. Chouaf
e-mail: a.chouaf@ensem.ac.ma

K. Zarbane and Z. Beidouri (eds.), *Proceedings of CASICAM 2022*, Springer Tracts in Additive Manufacturing, https://doi.org/10.1007/978-3-031-32927-2_17

and their sensitivity to the numerous printing parameters. Indeed, the impact of these printing parameters on the mechanical properties has been the subject of much research.

Ziemian et al. studied the effect of build orientation on the fatigue strength of acrylonitrile butadiene styrene (ABS) parts made by 3D printing [3]. Similar work was carried out by Durgun et al. where the effect of raster angle and orientation on the mechanical properties of ABS printed parts was evaluated using the three-point tensile and bending test [4]. In addition, Sood et al. also investigated the compressive strength of ABS FDM sample build orientation, raster angle and layer thickness, raster width and spacing [5]. Tymrak et al. improved the tensile strength and elastic modulus of printed ABS and PLA FDM parts by changing the build orientation, layer thickness and raster spacing [6]. In addition to orientation and layer thickness, several other parameters have also been evaluated in previous research. Zhang et al. investigated the fracture behaviour of a glass fibre reinforced polymer joint by tensile testing with different temperatures [7]. Vicente et al. tested the influence of the nozzle size on the mechanical properties of 3D printed ABS parts via the tensile test [8].

Despite this very interesting amount of work on the mechanical performance of printed parts, work on the resistance to crack propagation in this type of structure remains to be developed.

Indeed, compared to other techniques, the phenomenon of crack propagation in polymeric structures obtained by FDM is very complicated due to its microstructure which depends on various printing parameters [9]. It is in this sense that we are interested in the damage aspects of parts printed by the FDM (Fused Dipositive Modeling) technique [10, 11]. We recall that this technology is more precise and is a well established digital manufacturing method in the additive technology of polymer parts [12, 13]. FDM technology belongs to modern additive technology, where the three-dimensional model is manufactured layer by layer.

The principle of this FDM process is based on the manufacture of a part layer by layer. During this process, the filaments making up the layer are welded to each other within the same layer and between adjacent layers. This welding takes place on the contact surfaces and along the deposited filaments. The quality of this welding determines the resistance to crack propagation between filaments.

In the present study, we investigated the effect of raster width on the resistance to crack propagation in a 3D printed structure (FDM). In particular, we focused our study on the analysis of the cohesion between filaments. For this purpose, we have developed a numerical approach that simulates the mechanical behaviour of a digital CT specimen. In order to make this digital specimen as close as possible to a printed physical specimen, its microstructure was regenerated using an original approach based on the simulation of the FDM manufacturing process that we developed in our laboratory. Given the complexity of the numerical simulations, we have treated only two cases of raster width ($l = 0.42$ and $l = 0.56$ mm). The results obtained are analysed using the concepts of the energy theory of rupture. The other width cases are under development.

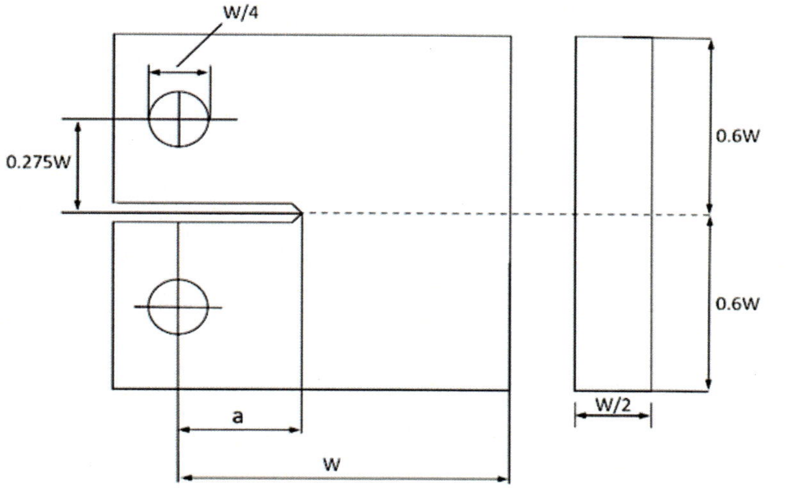

Fig. 17.1 Dimensions of the CT specimen (a = 0.45W and W = 20 mm) ASTM D5045 [14]

17.2 Elaboration of the Numerical CT Samples

To study the effect of raster width on the resistance to crack propagation in FDM printed ABS structures, we considered CT (Compact Tension) specimens. Based on ASTM D5045 [14], the geometrical characteristics we adopted for this specimen are shown in Fig. 17.1. This specimen was modeled on a CAD software. The virtual model was converted into a Stereolithography (STL) file, which was also translated into a machine instruction (G-code) file. This file describes the trajectories of the nozzle that deposits the filler material. Once the file is obtained, the samples will be ready to be made physically and digitally.

In order to produce these CT specimens digitally, a digital model was developed using the FDM manufacturing method and a script in the programming language "Python". This program allows the numerical calculation codes to execute and draw the trajectory instructions written in the G-code file, and to prepare the virtual specimens for the static traction simulation (Fig. 17.2). The adopted printing parameters are shown in Fig. 17.2.

Parameters	Values	Units
Layer thickness	0.2	Mm
First layer printing speed	30	mm/s
Print speed of other layers	150	mm/s
Build orientation	XYZ	
Raster angle	90	Deg
Infill percentage	100	%

(continued)

(continued)

Parameters	Values	Units
Number of perimeters	2	
Temperature	230	°C

We recall that this FDM printing technique is a process of building successive layers that allows us to control and produce samples with different raster widths. For our study, we have considered only two cases of raster width (l = 0.42 and l = 0.56) due to the complexity of the simulations. In Fig. 17.3, we have schematically highlighted the width of the filament.

17.3 Numerical Simulation

After the preparation of the virtual model of the CT specimen, we used a 'Tie' type interaction, to create the contact between the successive filaments, with 3D linear quadratic elements. For the material used (ABS), we considered a Young's modulus equal to 2.0 GPa, a Poisson's ratio equal to 0.3 and a density of 1050 kg/m^3. The boundary conditions applied during the numerical simulation of the static traction in the calculation code "Abaqus Standard" are identical to those applied during a physical test. In addition, we imposed an embedding on the reference point-2, while on the other reference point-1, an imposed displacement u0 along the axis perpendicular to the axis of symmetry of the notch (Fig. 17.4).

17.4 Results and Discussion

Due to the complexity of the microstructure of the numerical CT specimen, which contains a multitude of discontinuities, the simulations take several days. Therefore, we were obliged to simplify our model of the CT specimen, while waiting for the use of a more powerful computing station. The simplified model that we proposed consists in putting in contact at the level of the plane of symmetry of the notch two layers comprising filaments (at 90° with respect to the notch), and the remainder of the part is made up of the continuous ABS (Fig. 17.5).

17.4.1 Stress Distribution

From our simulation results, we have plotted the Von Mises stress distributions for the two configurations of the printed CT specimens (Fig. 17.6). For both cases (l = 0.42 mm and l = 0.56 mm), it can be seen that the shape of the plasticized zone

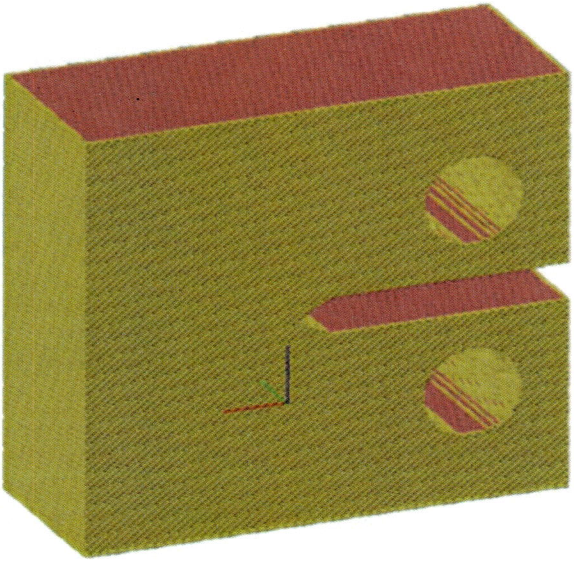

Fig. 17.2 Virtual model of the CT specimen and the considered parameters for its printing

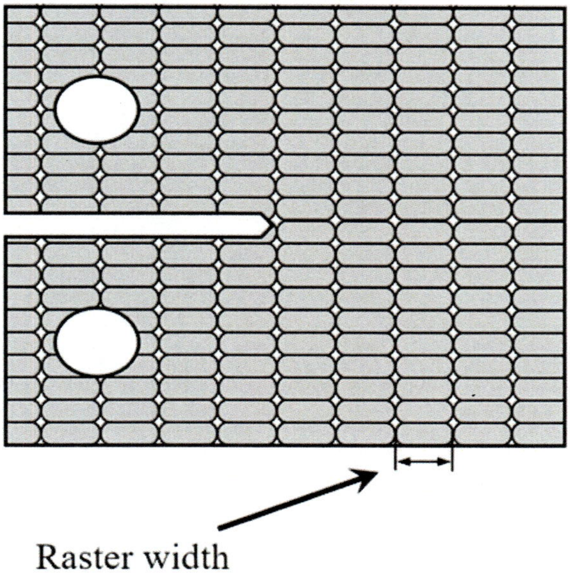

Fig. 17.3 Raster width of constituting the CT sample

at the notch tip is typically that of mode I behavior. However, the difference in the size of this zone is very marked from one case to another. For the case of width l = 0.42 mm, this size is relatively much larger on both sides of the notch axis. On the other hand, for the case l = 0.56 mm the size of the plasticized zone is relatively

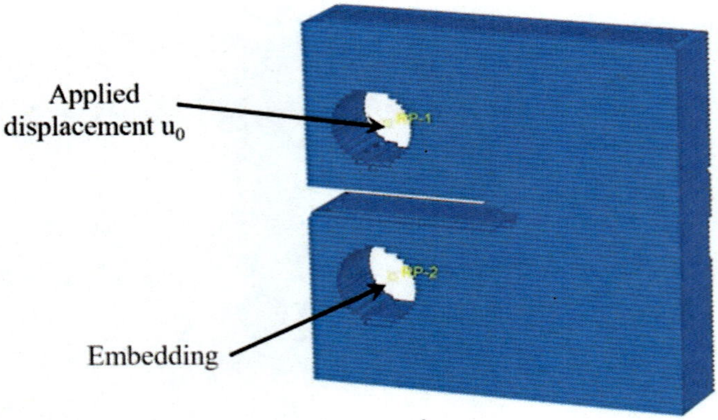

Fig. 17.4 Meshing of the printed CT specimen and boundary conditions

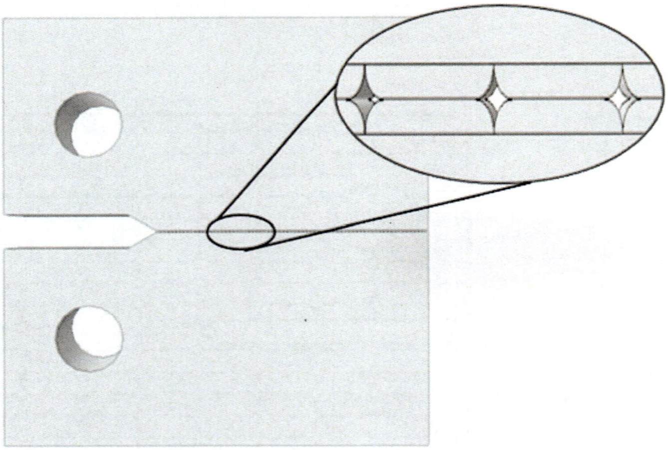

Fig. 17.5 Simplified CT specimen, in the plane of symmetry of the notch, two layers of filaments are in contact

smaller and concentrated at the front of the notch. Indeed, the gap between the lips is easier to produce in the case l = 0.42 mm than in the case l = 0.56 mm. This suggests that with the raster width 0.56 mm, the specimen has a better strength than with the width 0.42 mm.

It is also important to note that the Von Mises stress level reaches a value of 38.7 MPa in both cases. As this value is relatively close to the elastic limit of the studied material ($\sigma_e = 35$ MPa), we can consider that we are in the case of a confined plasticity.

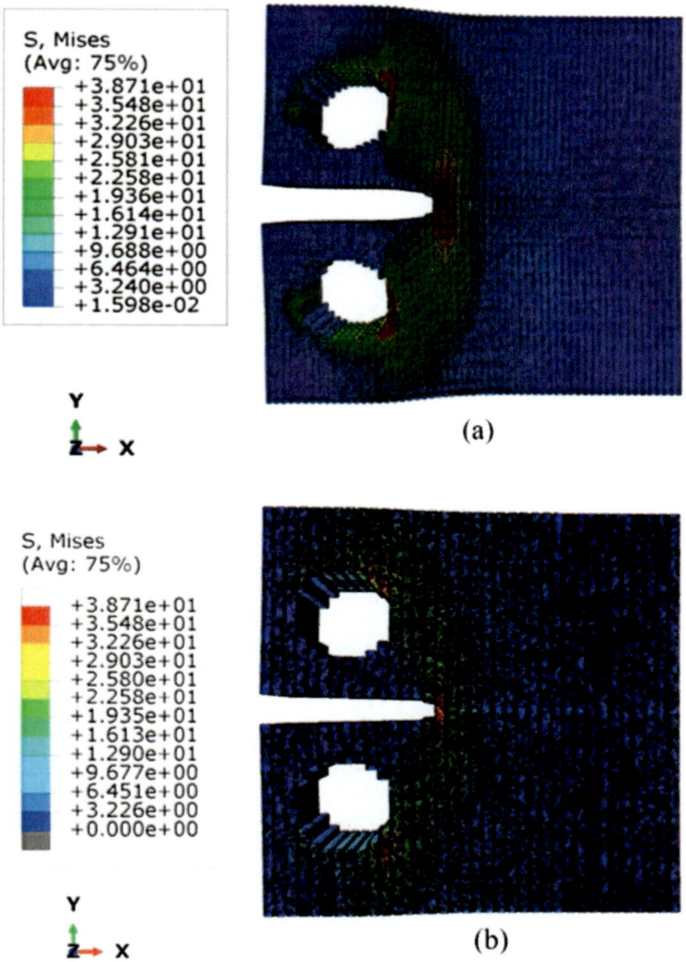

Fig. 17.6 Von Mises stress distribution for two cases of raster width: **a** 0.42 mm and **b** 0.56 mm

17.4.2 Energy Restitution Rate G

In the previous paragraph, we highlighted the influence of the raster width on the strength of the overall mechanical behaviour of the CT specimen before crack propagation failure. Once the crack begins to progress, we sought to distinguish the behaviour of the two width cases studied. To do this, we extracted from our Abaqus simulations the energy restitution rate G as a function of lip displacement (Fig. 17.7b) and as a function of crack progression (Fig. 17.7a).

From these two graphs we can clearly see the difference in behaviour between the two cases of raster width. According to Fig. 17.7b, to restore an energy of 6 N/

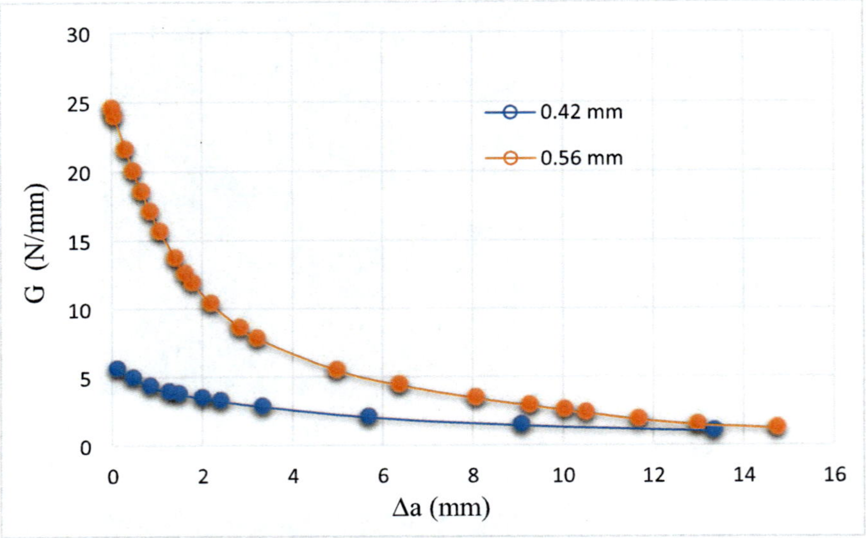

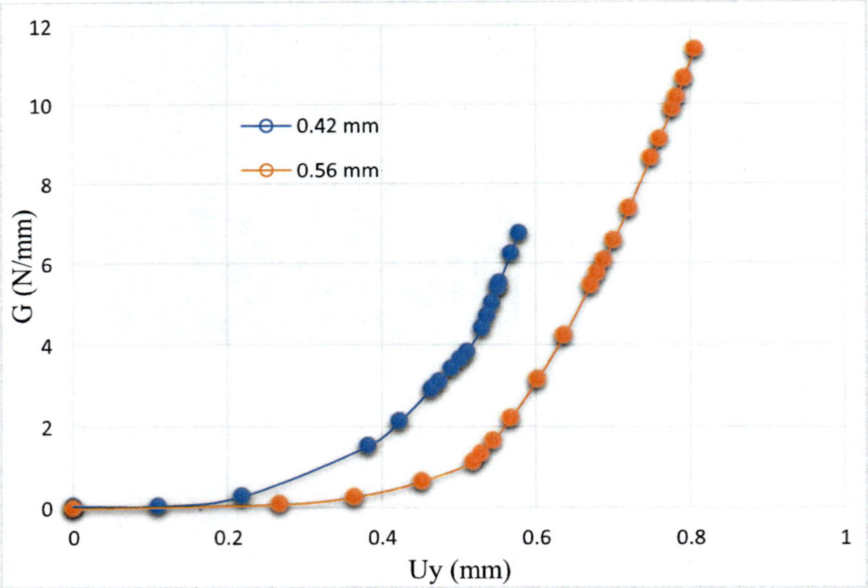

Fig. 17.7 Energy restitution rate G as a function of: **a** Crack size variation Δa and **b** Displacement Uy

mm, more displacement of the specimen lip is required in the case of l = 0.56 mm than in the case of l = 0.42 mm. This suggests that when the raster width is large, the adhesion energy is greater than when the width is small. This could be confirmed from the classical relation of the energy theory of fracture, $G = 2\gamma$ (γ is the surface energy). We recall that this relation corresponds to the propagation condition. To open a crack that corresponds to the creation of two surfaces, the energy restitution rate G must be at least twice the surface energy. Therefore, the larger the width of the raster, the larger the contact area and the higher the surface energy, which translates into more resistance to crack propagation.

This observation could be deduced from the analysis of the graph in Fig. 17.7a. Indeed, with the same displacement Δa_0 of the crack, more energy is needed in the case l = 0.56 than in the case l = 0.42.

17.4.3 Fracture Facies

As we have already developed in previous studies [10, 11], when the part is made of cross-layered filaments, crack propagation occurs by filament breakage in one layer and separation in the other layer along the larger raster angle. On the other hand, for parts made of parallel filaments between layers, crack propagation develops only by filament separation. It is the latter case that we considered to study the effect of the raster width on filament separation. In Fig. 17.8a, we can observe that the filaments are parallel and located in the same plane. This shows that there is a clear separation between filaments. For this to occur, the inter-filament adhesion energy that is generated during the FDM manufacturing process must be overcome. This inter-filament adhesion phenomenon is characterized by the cohesive forces between the filament surfaces. A preponderant effect of these cohesive forces, is material pull-out as can be seen in Fig. 17.8b. This phenomenon depends on several parameters such as the material, the temperature and the contact surface. This last parameter depends closely on the width of the raster. The larger the width, the better the resistance to crack propagation.

17.5 Conclusions

In this study, the effect of raster width on the crack propagation resistance in printed ABS CT specimens was highlighted. In particular, we focused our study on the analysis of the inter-filament cohesion. To do so, we have processed a numerical approach to simulate the mechanical behavior of a digital CT specimen formed by parallel filaments between layers. Two cases of raster width (l = 0.42 and l = 0.56 mm) were developed in our simulations.

The results obtained show that the CT specimen studied presents a better resistance with the width l = 0.56 mm compared to the one corresponding to l = 0.42 mm.

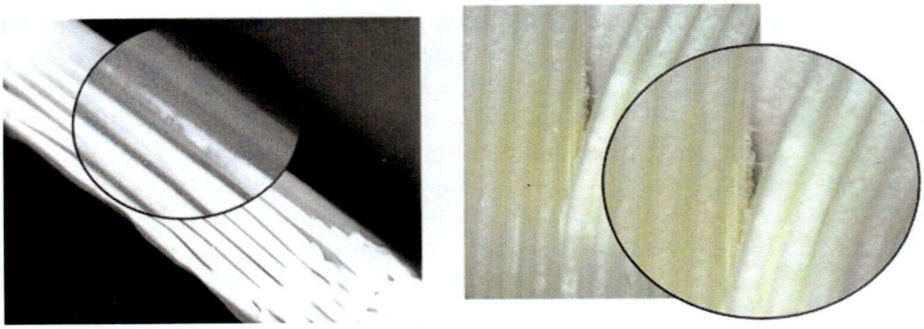

Fig. 17.8 **a** Microscopic observation of the fracture facies. **b** Rupture by filament separation

Indeed, the greater the width of the raster, the greater the contact surface and the higher the surface energy. It is this surface energy that characterizes the quality of cohesion between filaments. The results concerning the effect of the width on this cohesion are very encouraging. Therefore, we will develop our study in perspective by dealing with other widths and by taking into account the effect of temperature on the cohesion between filaments.

References

1. Winarso, R., Ismail, R., Jamari, J., Bayuseno, A.P.: Application of Fused Deposition Modeling (FDM) on Bone Scaffold Manufacturing Process: A Review
2. Li, Y., Feng, Z., Huang, L., Essa, K., Bilotti, E., Zhang, H., Peijs, T., Hao, L.: Additive manufacturing high performance graphene-based composites: a review. Compos. A Appl. Sci. Manuf. **124**, 105483 (2019)
3. Ziemian, C., Sharma, M., Ziemi, S.: Anisotropic mechanical properties of ABS parts fabricated by fused deposition modelling. In: Mechanical
4. Durgun, I., Ertan, R.: Experimental investigation of FDM process for improvement of mechanical properties and production cost. Rapid Prototyp. J. **20**, 228–235 (2014)
5. Sood, A.K., Ohdar, R.K., Mahapatra, S.S.: Experimental investigation and empirical modelling of FDM process for compressive strength improvement. J. Adv. Res. **3**, 81–90 (2012)
6. Tymrak, B.M., Kreiger, M., Pearce, J.M.: Mechanical properties of components fabricated with open-source 3-D printers under realistic environmental conditions. Mater. Des. **58**, 242–246 (2014)
7. Zhang, Y., Vassilopoulos, A.P., Keller, T.: Effects of low and high temperatures on tensile behavior of adhesively-bonded GFRP joints. Compos. Struct. **92**, 1631–1639 (2010)
8. Vicente, C.M.S., Martins, T.S., Leite, M., Ribeiro, A., Reis, L.: Influence of fused deposition modeling parameters on the mechanical properties of ABS parts. Polym. Adv. Technol. **31**, 501–507 (2020)
9. Hart, K.R., Wetzel, E.D.: Fracture behavior of additively manufactured acrylonitrile butadiene styrene (ABS) materials. Eng. Fract. Mech. **177**, 1–13 (2017)
10. Aourik, O., Othmani, M., Saadouki, B., Abouzaid, K., Chouaf, A.: Fracture toughness of ABS additively manufactured by FDM process. J. Achiev. Mater. Manuf. Eng. 109(2) (2021)
11. Aourik, O., Chouaf, A., Othmani, M.: Analysis of the resistance to crack propagation in SENT test specimens printed in ABS using parallel or crossed filaments between layers. Frat. ed Integrità Strutt. **17**(63), 246–256 (2023)

12. Li, J., Yang, S., Li, D., Chalivendra, V.: Numerical and experimental studies of additively manufactured polymers for enhanced fracture properties. Eng. Fract. Mech. **204**, 557–569 (2018)
13. Ghandriz, R., Hart, K., Li, J.: Extended finite element method (XFEM) modeling of fracture in additively manufactured polymers. Addit. Manuf. **31**, 100945 (2020)
14. ASTM D5045, Standard Test Methods for Plane-Strain Fracture Toughness and Strain Energy Release Rate of Plastic Materials. ASTM International, West Conshohocken, PA, 1999. https://doi.org/10.1520/D5045-99

Part V
AM Part Quality Control

Chapter 18
Design of a Benchmark Part with Recent Design Rules for Selective Laser Melting

Mohamed Amine Daoud, Meriem Hayani Mechkouri, Youssef Chairi, and Kamal Reklaoui

Abstract Selective laser melting (SLM) is an additive manufacturing technique for the fabrication of near-net-shaped parts directly from computer-aided design data by melting together different layers with the help of a laser source. To test the ability of the SLM process to produce consistent and defect-free parts with intended dimensions, Benchmark parts can be used to evaluate the manufacturing process and machine performance. In this paper, the selective laser melting process is presented. Design rules for SLM developed by Daniel Thomas are summarized. A benchmark part design is performed in terms of evaluating resolution, dimensional accuracy, and repeatability, in addition to investigating surface roughness.

Keywords Additive manufacturing · 3D printing · Selective laser melting · Benchmark artifact · Design rules for AM

18.1 Introduction

Additive manufacturing (AM), also known as 3D printing, is a technology of manufacturing objects by material addition in layers using data from a computer-aided 3D model, unlike conventional subtractive manufacturing processes [1]. AM has many advantages, including the potential to manufacture complex geometric shapes with minimum material wastage, wide range of part size and material option ranging [2]. In

M. A. Daoud (✉) · M. Hayani Mechkouri · K. Reklaoui
Faculty of Sciences and Techniques, University Abdelmalek Essaâdi, Tangier, Morocco
e-mail: mohamedamine.daoud@etu.uae.ac.ma

M. Hayani Mechkouri
e-mail: Mhayanimechkouri@uae.ac.ma

K. Reklaoui
e-mail: k.reklaoui@uae.ac.ma

Y. Chairi
UMR CNRS 6285 Lab-STICC, Université de Brest, Brest, France
e-mail: Youssef.chairi@univ-brest.fr

K. Zarbane and Z. Beidouri (eds.), *Proceedings of CASICAM 2022*, Springer Tracts in Additive Manufacturing, https://doi.org/10.1007/978-3-031-32927-2_18

the last decades, AM technology has known a rapid growth and has been used in many fields such as, medical, automotive, aerospace, bio-engineering and other industries. Several processes based on different principles (extrusion, melting, photopolymerization, sintering…etc.) have been developed with the benefit of offering the possibility to work with wide range of materials with high precision and for complex geometries [3].

Selective laser melting (SLM) is currently the most promising powder bed fusion technique as compared to other metal additive manufacturing processes; it became widely used in biomedical and aerospace [4]. SLM is the process through which a high power-density laser selectively melts and fuses metallic powders to produce near net-shape parts with near full density (up to 99.9% relative density) [5, 6].

In order to optimize accuracy and to quantify the ability of AM processes to produce consistent and defect-free parts with intended dimensions, many researchers proposed the use of benchmarking parts, also called "test artifact" [7]. Benchmark artefacts are representative of the manufacturing process and machine performance, they can be designed in the aim of evaluating resolution, dimensional accuracy, and repeatability, in addition of investigating the surface roughness [8]. Furthermore, the development of design rules for different AM processes could offer a better knowledge of these technologies in the aim to provide new design freedoms to its users [9].

In this paper, the selective laser melting process is presented. Design rules for SLM developed by Daniel Thomas are summarized. A benchmark part design is performed in terms of evaluation of system capacity in manufacturing different features with a good surface quality.

18.2 Selective Laser Melting

Selective Laser Melting enables the production of individual parts with complex geometries matching the mechanical properties of parts conventionally manufactured in series. Since the process deals with metallic materials like steel, aluminum, titanium, and nickel-based alloys, the build is frequently filled with nitrogen gas or argon gas to create an inert environment to prevent the hot metal pieces from oxidation.

SLM, like other 3D printing processes, includes a sequence of phases, beginning with CAD data slicing and ending with printed pieces removal from the building plate. Figure 18.1 illustrates the principle of the SLM process, which may be separated into 8 phases [10]:

1. Design the part by a CAD software;
2. Import the CAD data as a STereoLithography (STL) file into the slicing software to produce slice data and provide support structures for any overhanging elements;
3. Upload the 3D model to the Selective Laser Melting equipment.;

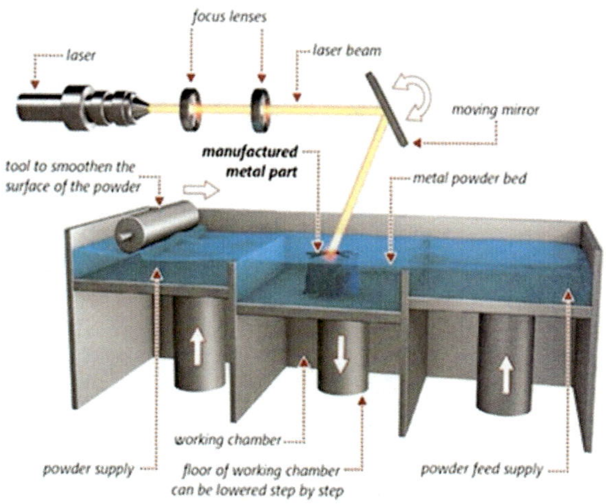

Fig. 18.1 The principle of SLM process

4. The construction procedure begins with the application of a thin coating of powder material to the substrate plate;
5. The laser beam scans the powder and fuse selected areas by following the pattern determined by the data, allowing the first layer of the part to be produced;
6. The building platform is lowered, a next layer of powder is deposited on top by a roller and the laser scans a new layer;
7. The process is repeated until the part is completely built.
8. The cooled part is removed from the powder tank to be cleaned of its supports and of the unfused powder particles.

Laser power, scanning speed, hatch spacing, and layer thickness are all regulated so that a single melt vector may entirely merge with nearby melt vectors and the previous layer. After the laser scanning process is done, loose particles are removed from the building chamber, and the component may be manually or electrically discharged and detached from the substrate plate (EDM). The whole process is automated, with the exception of data preparation and removal of produced components from the construction platform.

18.3 Case Study

The goal here is to examine the geometrical performances of an SLM system by evaluating dimensional accuracy and surface roughness in order to maximize the surface finish of produced items in 316L stainless steel. By avoiding measurement constraints, the technology will enable the production of a high-quality item that may even meet the existing tolerances of conventional machining.

18.3.1 Design Rules

Design guidelines play an important role in increasing efficiency and optimizing the quality of parts produced by the process. They have grown in tandem with the progress of manufacturing techniques, and they are now utilized as an efficient means for a designer to choose a production method and generate optimum designs. Design rules are also used to effectively convey new process capabilities to designers and engineers in order for a process to be completely explored.

Daniel Thomas [11] was among the first to develop design rules for metal additive manufacturing and more specifically for the SLM process, in order to evaluate the SLM process and identify its geometric limitations. Design rules used for the design of our benchmark test part are summarized in Table 18.1.

18.3.2 Benchmark Part Design

The benchmark artifact was designed according to the properties proposed by Byun et al. [12], in addition to the current ISO geometrical tolerance requirements [13]. It includes different sizes (small, medium, large) and forms of features (cube, hole, wall, cylinder ...etc.) set up in all three axes, in addition to both inside and outside components. The first step of our design procedure consists of an analysis of the literature and a comparison of the most developed test artifacts, then, the features type and size are selected based on the related works and design rules, and finally, the test part is designed.

The test part was designed using a CAD software with 25 mm × 22.5 mm × 20 mm dimensions.

The test part is composed of a set of features organized on a base plate, as illustrated in Fig. 18.2. The characteristics investigated by each feature are: holes of different shapes (triangle, circle, square) and different diameters (0.7 mm; 0.9 mm; 1.1 mm), three walls of different dimensions (0.4 mm; 0.6 mm; 0.8 mm), different successive slits from 0,2 mm to 1 mm and three cylinders of different diameters (0.7 mm; 1 mm; 2 mm) separated from each other by the distances 0.3 mm, 0.7 mm and 1.15 mm.

The powder in the build chamber does not provide any support to the part during its construction, so all inclined surfaces will ideally be self-supported, in this terms we have realized inclined fins by different angles (20°, 25°, 30°, 35°, 40°, 45°) to the horizontal axis. In addition, for horizontal bores, it is possible not to use any support until a diameter of about 6 mm. Beyond 6 mm, the upper surface of the bores becomes grainy, or even lifts, which can cause a blockage of the coating system and therefore requires the installation of building supports. To avoid putting supports, four holes were designed (cylindrical and circular).

Table 18.1 Design rules

Types		Design for manufacturing	Description
Minimal geometry	Gap size	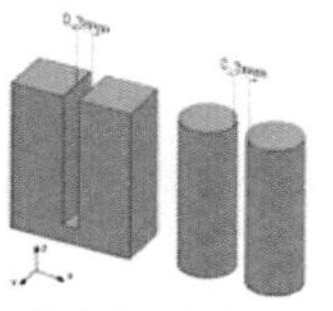	• The minimum gap size must be applied to avoid surface fusion • The minimum gap size = 0.3 mm • Supports will be required below 0.3 mm and at 45° orientation, and the gap will be filled
	Wall thickness	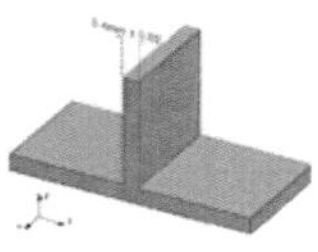	• A minimum wall thickness of 0.4 mm is required • The accuracy of the wall shall be ± 0.02 mm • Walls with a thickness of less than 0.4 mm will not be manufactured on the final metal part
	Vertical bore	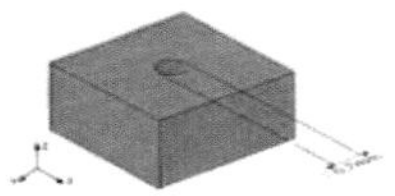	• The minimum diameter of a vertical bore = 0.7 mm • Below the limit, the bore is blocked by not fully fused powder grains
			• The greatest hole that may be produced without support is 7 mm in diameter • The smallest hole that may be produced without support is 1 mm in diameter
Holes	Horizontal bore	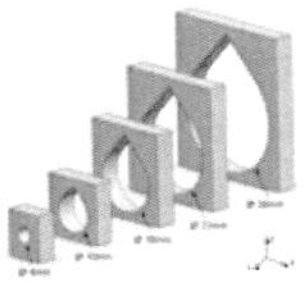	• Holes with 1 mm and 2 mm radius have a tolerance of 0.3 mm • Holes with radiuses of 3 and 4 mm will be precise to 0.2 mm • All holes with a radius of up to 15 mm measured with a 0.1 mm tolerance
Surface finish	Part orientation	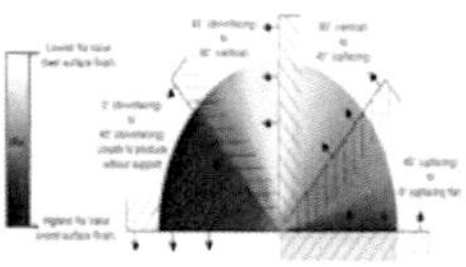	• Surfaces with an angle of 45° need a support • Surfaces with less than 45° orientation are poor surface quality • The optimal orientation is 90°

18.4 Conclusion

Artifacts may be utilized to precisely assess AM systems and possible tolerances in order to efficiently characterize and optimize them. Artifacts may also be used to evaluate different AM systems based on the same technology if the geometrical/ dimensional performance evaluation findings are paired with additional parameters like as surface roughness, mechanical qualities, production time, and prices.

In this paper, a benchmark test part for the SLM technique was designed to evaluate the dimensional accuracy, the system limitations as well as the surface quality of the

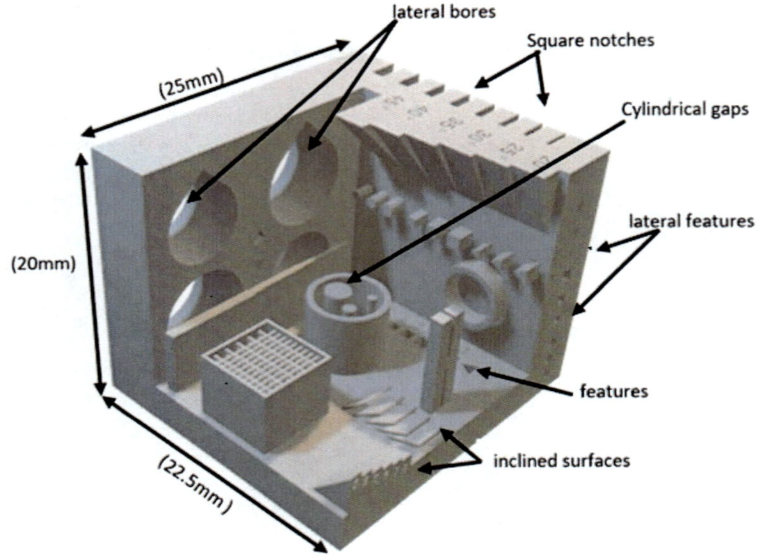

Fig. 18.2 CAD model of benchmark part

manufactured parts. The proposed test part includes different features with different forms and sizes in all three axes.

Future works will focus on the fabrication of the Benchmark test part using 316L as a material. Furthermore, we will work on optimizing the process parameters of the SLM system by printing different samples following a design of experiments to investigate the ability to produce fine details and achieve high-quality parts. Build time, cost, and material properties will also be considered in the study.

References

1. Bártolo, P., de Lemos, A., Tojeira, A., Pereira, A., Mateus, A., Mendes, A., dos Santos, C., Freitas, D., Bártolo, H., Almeida, H., dos Reis, I., Dias, J., Domingos, M., Alves, N., Pereira, R., Patrício, T., Ferreira, T.: (éds.): Direct Manufacturing Design Rules. CRC Press (2011)
2. Bremen, S., Meiners, W., Diatlov, A.: Selective laser melting: a manufacturing technology for the future? Laser Tech. J. **9**(2), 33–38 (2012). https://doi.org/10.1002/latj.201290018
3. Byun, H.-S., Lee, K.H.: Design of a new test part for benchmarking the accuracy and surface finish of rapid prototyping processes. In: V., Gavrilova, M.L., Tan, C.J.K., L'Ecuyer, P. (eds.), Computational Science and Its Applications — ICCSA 2003. Volume 2669, Lecture Notes in Computer Science Kumar, pp. 731-40. Springer, Berlin
4. Calignano, F., Lorusso, M., Pakkanen, J., Trevisan, F., Ambrosio, E.P., Manfredi, D., Fino, P.: Investigation of accuracy and dimensional limits of part produced in aluminum alloy by selective laser melting. Int. J. Adv. Manuf. Technol. **88**(1–4), 451–458 (2017). https://doi.org/10.1007/s00170-016-8788-9

5. Harun, W.S.W., Kadirgama, K., Samykano, M., Ramasamy, D., Ahmad, I., Moradi, M.: Mechanical behavior of selective laser melting-produced metallic biomaterials. In: Mechanical Behaviour of Biomaterials, pp.101–116. Elsevier (2019)
6. Kim, G.D., Oh, Y.T.: A benchmark study on rapid prototyping processes and machines: quantitative comparisons of mechanical properties, accuracy, roughness, speed, and material cost. Proc. Inst. Mech. Eng. Part B: J. Eng. Manuf. **222**(2), 201–215 (2008). https://doi.org/10.1243/09544054JEM724
7. Moylan, S., Slotwinski, J., Cooke, A., Jurrens, K., Donmez, M.A.: An additive manufacturing test artifact. J. Res. Natl. Inst. Stand. Technol. **119**, 429 (2014). https://doi.org/10.6028/jres.119.017
8. Rebaioli, L., Fassi, I.: A review on benchmark artifacts for evaluating the geometrical performance of additive manufacturing processes. Int. J. Adv. Manuf. Technol. **93**(5–8), 2571–2598 (2017). https://doi.org/10.1007/s00170-017-0570-0
9. Rupal, B.S., Ahmad, R., Qureshi, A.J.: Feature-based methodology for design of geometric benchmark test artifacts for additive manufacturing processes. Procedia CIRP **70**, 84–89 (2018). https://doi.org/10.1016/j.procir.2018.02.012
10. Seabra, M., Azevedo, J., Araújo, A., Reis, L., Pinto, E., Alves, N., Santos, R., Mortágua, J.P.: Selective laser melting (SLM) and topology optimization for lighter aerospace componentes. Procedia Struct. Integr. **1**, 289–296 (2016). https://doi.org/10.1016/j.prostr.2016.02.039
11. Thomas, D.: s. d. The development of design rules for selective laser melting 318
12. Yadroitsev, I., Thivillon, L., Bertrand, Ph., Smurov, I.: Strategy of manufacturing components with designed internal structure by selective laser melting of metallic powder. Appl. Surf. Sci. **254**(4), 980–983 (2007). https://doi.org/10.1016/j.apsusc.2007.08.046
13. ISO/ASTM standards (ISO/ASTM DIS 52904, ISO/ASTM 52911-1, ISO 12780, ISO 12781, ISO 12180, and ISO 12181)

Chapter 19
Design for Additive Manufacturing

Omar Lkadi, Mohammed Nassraoui, and Otmane Bouksour

Abstract Additive manufacturing creates objects from digital files and this layer by layer. Nowadays, additive manufacturing technologies are in full development, and are used in several industrial fields. It has a significant number of benefits, including mass reduction, design freedom, waste reduction, with a wide selection of possible material. As efforts to effectively apply additive manufacturing, design for additive manufacturing (DfAM) has increased to provide several recommendations and guidelines to designers, but knowledge, tools, rules, and methodologies related to DfAM, has not received much attention, and few works have addressed the development of this subject. This paper explores trends, issues, and challenges in design for AM, and previous Design for Additive Manufacturing (DfAM) methodologies are analysed widely and classified into distinct categories based on existing research works, however, it has been found that existing methodologies do not take full advantage of the additive manufacturing process capabilities. For that, we propose a frame for a methodology that takes advantages of additive manufacturing technologies capabilities and involve factors influencing the design, with a discussion of the additional value compared to the existent methodologies.

Keywords Additive manufacturing · Design for additive manufacturing (DfAM) · Methodology

O. Lkadi
National High School of Electricity and Mechanics (ENSEM), University Hassan II of Casablanca, Casablanca, Morocco
e-mail: omar.lkadi-etu@etu.univh2c.ma

O. Lkadi · M. Nassraoui (✉) · O. Bouksour
Laboratory of Productive Mechanics and Industrial Engineering (LMPGI), Higher School of Technology, University Hassan II (ESTC), Casablanca, Morocco
e-mail: mohammed.nassraoui@univh2c.ma

O. Bouksour
e-mail: otmane.bouksour@univh2c.ma

K. Zarbane and Z. Beidouri (eds.), *Proceedings of CASICAM 2022*, Springer Tracts in Additive Manufacturing, https://doi.org/10.1007/978-3-031-32927-2_19

19.1 Introduction

Among the key technologies that form the building blocks and stimulate research on Industry 4.0 is additive manufacturing. In additive manufacturing technology, products are manufactured layer by layer from a 3D model that allows the production of external and internal geometries that would be difficult to achieve by a traditional technique [1]. Although the advantages offered by AM technologies are several, but they present certain limits: size of parts, materials, quality of parts [2]. The evolution of additive manufacturing has significantly affected the industrial and academic world and developments continue with the purpose of improving the process, material and design. The use of additive manufacturing is often linked to the question of what advantages it offers over existing processes and technologies, and more often whether it is cheaper to manufacture [3].

In other side, it is difficult for designers to fully exploit the unique possibilities offered by AM. For that reason, many researchers are more interested in Design for Additive Manufacturing (DFAM) [4]. The term Design for Additive Manufacturing (DfAM) is far from being used consistently by researchers, [5] defines it as the operation followed by designers when seeking to make a product design that leverages the distinctive capacities of AM. While also respecting the specific limitations of the AM technology process that will be used to manufacture the product. However, a clearer hierarchy of the different types of design processes used in additive manufacturing can differentiate parameters that alter the additive manufacturing production process, change the shape of a part rather than its function to better fit the additive manufacturing process, and completely redesign the form and function of the part to make it designed for AM.

This paper is an attempt to present a methodology so designers can take advantage of AM technology to improve product performance. To achieve this goal, existing DfAM methods are discussed and categorized with different advantages and disadvantages of each DfAM method. Finally, this paper presents conclusions and opportunities for future research.

19.2 State of the Art of Design for Additive Manufacturing (DfAM)

In the literature the term "Design for Additive Manufacturing" has been usually used [6]. It is defined as a general type of design methods or tools to optimize functional performance and/or other critical aspects of the product lifecycle, depending on the characteristics of additive manufacturing technologies, such as manufacturability, reliability, and cost [5]. DfAM also considers the limitations of the AM technology that will be utilized to create the product [5].

Methods for integrating different issues into the design process are referred to as "Design for X" (DFX) [7]. For instance, "Design For Manufacturing and Assembly"

(DFMA). refers to the standards for "Design For Assembly" (DFA) and "Design For Manufacturing" (DFM), the term Design for Additive Manufacturing (DfAM) is the new name for the DFM concept in AM.

19.2.1 Design for Manufacturing and Assembly (DfMA)

DfMA is a subset of Design for X, it regroups methods for design and Assembly and design for manufacturing. Design for Manufacturing and Assembly (DfMA) can be defined according to [6] as a design approach that seeks to create a product and its manufacturing system with the goal of minimizing development costs and time while boosting performance, quality, and profitability. This is achieved by considering both design goals and production constraints, including user and market requirements, materials, processes, assembly and disassembly techniques, maintenance requirements, and more.

Three levels of cognition can be used to view DfMA. Firstly, DfMA offers designers tools, methods, and guidelines to make it easier for deigners to adapt their designs to a given set of constraints [6]. The objective of DfMA at the next level is to comprehend and measure the impact of the design process on production. At its highest level, DfMA examines the relationship between design and manufacturing and its impact on designers, design processes, and design practice.

Design For Assembly (DFA) procedures are widely used by companies in the manufacturing industries, to create products that are simple to assembly, using certain assembly techniques to save costs and assembly time. Design for assembly can take many forms. In the 1960s and 1970s, various guidelines and recommendations were proposed to help designers consider assembly issues during the design process. The most common DFA methods are the Lucas DFA Evaluation Method, Hitachi Assemblability Evaluation Method and the DFA portion of the Boothroyd-Dewhurst DFMA Method [8]. Since they aim to modify the product for a certain assembly process rather than attempting to enhance the functionalities and performance of the product, these strategies are useful in mass production [8].

The general engineering technique of designing items to be simple to fabricate is known as "Design for Manufacturing" (DFM). however, depending on the production technique, the implementation varies. The DFM approach can be direct or indirect, in the first one, the geometrical model of the part is analyzed in order to show the areas that are more or less difficult to produce according to the chosen manufacturing process. In the second one, the approach is based on the revision of an existing manufacturing process in order to reduce the cost and/or the manufacturing time [9].

19.2.2 The Necessity for Design for Additive Manufacturing

All processes, including AM processes, fall under the DfMA concept described above. However, DfAM will significantly differ from conventional DfMA in terms of the design knowledge, tools, rules, procedures, and methods at all three levels of abstraction [6]. But additive manufacturing can get over the drawbacks of conventional manufacturing processes to produce very complex parts with better functionality. Additionally, compared to traditional processes, AM techniques have distinct batch sizes, production time, and cost drivers, necessitating alternative approaches to component control [6]. For that, the characteristics of AM processes require different approaches to the design process and design practice, this includes new approaches to explore large, complex design spaces [6].

Additive manufacturing offers to designers a lot of opportunities considering the advantages of the unique capabilities of AM. To fully take advantage of any design options presented by AM technologies in a design challenge, DfAM must have enough additive manufacturing capabilities. The capabilities for additive manufacturing that have been identified are outlined below:

- **Freeform Shapes and materials:** The freedom of design complexity allowed by the AM allows designers to achieve almost any shape you desire [10]. Also, A wide range of materials are used in AM technology. Polymers, metals, and ceramic materials are among the materials used in commercial additive manufacturing equipment. In sheet lamination technology, rubber, foam, cork, wood, and paper are used [6]. Some of the additive manufacturing technologies are also capable of producing parts in colors [10]. Additionally, a variety of AM techniques have been employed to create edible products such chocolate, sugar, pasta, spreads, cheese, scallop puree, ground beef, egg whites, bug powders, and a whole pizza [6].
- **Multiple Materials**: The ability to print many materials at once is another crucial component of additive manufacturing. This feature allows composite objects with dynamically changeable topographies to be created [5, 6, 10].
- **Internal freeform geometry**: using additive manufacturing, it's possible to make Complex internal functions such as: compliant cooling channels, fluid channels, air ducts that can improve performance of a part [5, 10].
- **Embedded Components:** AM technologies can be used to produce embedded components [11], detailed a process that enabled to produce parts with embedded electronic circuits using stereolithography and direct print technologies [5].
- **Thin and small structures:** The minimal feature size is mostly governed by the 3D printer's x-y resolution, allowing AM to produce thin and small features like thin walls, tiny holes, and pins [10].
- **Textured surfaces:** Using AM technology, we can produce textured surfaces on things for the consumer goods industry [6, 10].
- **Topology optimization for additive manufacturing:** The complexity enabled by AM means that the production of parts with optimized topology is now possible. Topological optimization (TO) is a numerical methodology that optimizes the layout of materials in each design space and for given boundary conditions, so that

the resulting layout meets a prescribed set of performance objectives [4, 12] and finite element analysis (FEA) also has been frequently used in a number of research to complete topology optimization for additively manufactured parts [10].

Although AM seems to have limitless potential, but it has restricted capacities. Designers must take into account a variety of limitations, such as those related to CAD, anisotropy, porosity, support structures during production, orientation [13–17].

19.3 Design Methods for AM

There are a few design methods for additive manufacturing declared in the literature [18, 19], however the one that adopts a functional surface approach and designs a component from the bottom up is the most promising [14]. Functional surface method needs a close link between design and Finite element analysis (FEA). It is described in detail in the literature, such as the one described by [20]. This methodology (Fig. 19.1) has four main steps: (1) The definition of a design space which characterizes the space where the material constituting the final product to be designed is authorized by both the functional specifications and the manufacturing process. (2) The definition of the final theoretical geometry of the product inside the design space which respect the constraints of the problems. (3) The definition of the corresponding practical geometry which must be as close as possible to the theoretical geometry and easy as possible to fabricate. (4) Geometry estimation because for the same initial conditions, several manufacturing approaches are possible, both in terms of orientation and manufacturing path.

This methodology constitutes one of the first to open the design for AF by exploiting its opportunities and applying the constraints of the technology used and profiting of the topological optimization to establish a theoretical geometry. However, it does not consider the specificities of layered processes.

Rodrigue and Rivette [21] presented a global methodology for AM (Fig. 19.2), applicable to mechanical assemblies where the designer is guided by the possibilities and not by the constraints of the manufacturing process. The methodology's steps are displayed in Fig. 19.2 and can be defined as: (1) problem definition in which the designer determine what he wants to manufacture and why. (2) Part consolidation which consists of creating a part-level representation of the assembly and grouping together parts that can be consolidated. (3) Using The Risk in Early Design technique to identify the failure mechanisms, part optimization with reference to failure prevention. (4) Utilizing numerical optimization techniques and concept generating techniques, part-optimization with respect to user-defined goals.

Vayre [22] proposed an approach that considers the evolution of the field of possible in design and allows the generation of new shapes while respecting the manufacturing constraints of the Electron Beam Melting (EBM) technology, this approach follows the steps: (1) Generation of an initial shape, using the either the designer's expertise or the use of topological optimization. (2) choose the orientation

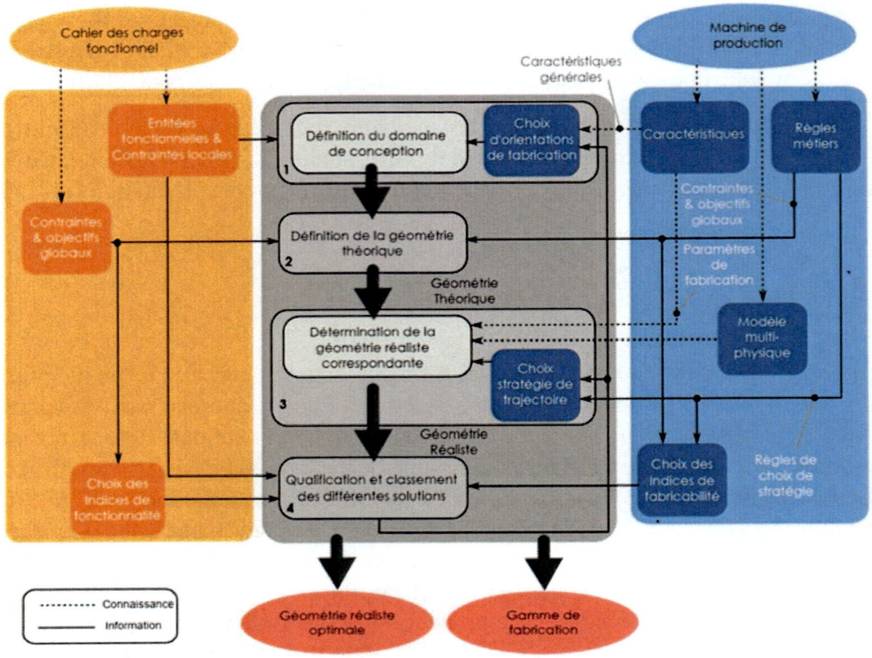

Fig. 19.1 Methodology proposed by [20]

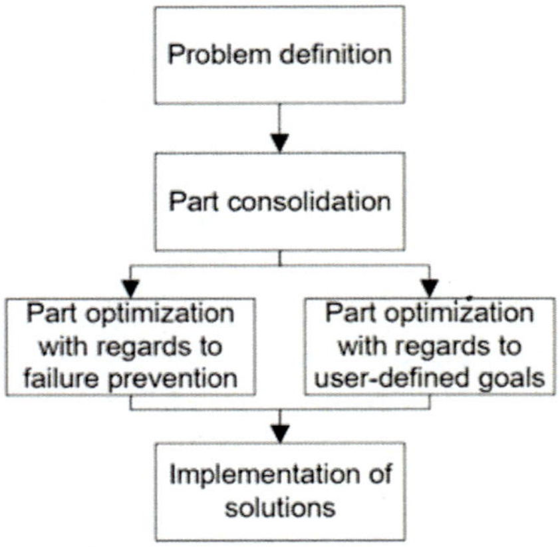

Fig. 19.2 The methodology suggested by [21]

of the future part in the space of the machine. (3) the initial geometry is adapted to consider all the requirements and parametrically optimized under constraints. (4) Validation of the part.

Boyard [8] proposed a methodology based on three steps: (1) The goal of this step is to create a basic geometry that will result in a part with physical, chemical properties, mechanical strength and functional surfaces respect the specifications. (2) It seeks to suggest an optimized geometry of the first solid to reduce the volume of material required for the manufacturing. (3) This phase aims to ensure the manufacturability of the part.

Orquéra [4] established a global methodology for a multi body mechanical systems, taking advantage of the preview methodology in order to summarize the knowledge, discovered in AM, and using different form of optimization (architectural, functional, topological).

The following steps, which are crucial for AM design, are shared by all these methodologies:

- Problem analysis
- Structural optimization
- Results interpretation
- Redesigning
- Finite Element Analysis
- Final design

However, methodologies described below are limited and don't consider the following points:

- They do not cover all the design stages
- Some of these methodologies are verified for a specific technology but not for others
- Not open to all capacities of AM
- The generation of the support, which is done by the slicing software
- Focused on a particular objective, domain, or application,
- The orientation of the part is based only on the machine space
- No information is provided on the selection of the material

19.4 Method

After examining the benefits and weaknesses of the earlier design techniques, it appears the need for a new framework or approach that integrates the benefits of AM previously announced. the technique used for designing components while considering the benefits of AM processes is introduced in this section. These benefits are described and discussed in Sect. 19.2.2.

The goal of the methodology is to provide the best optimal design that respect the client requests and profits of the unique capabilities of AM technologies. This methodology begins with the definition of the problem and end with the final validate design. It is presented as a flow chart in Fig. 19.3. Each section of the methodology is further detailed below.

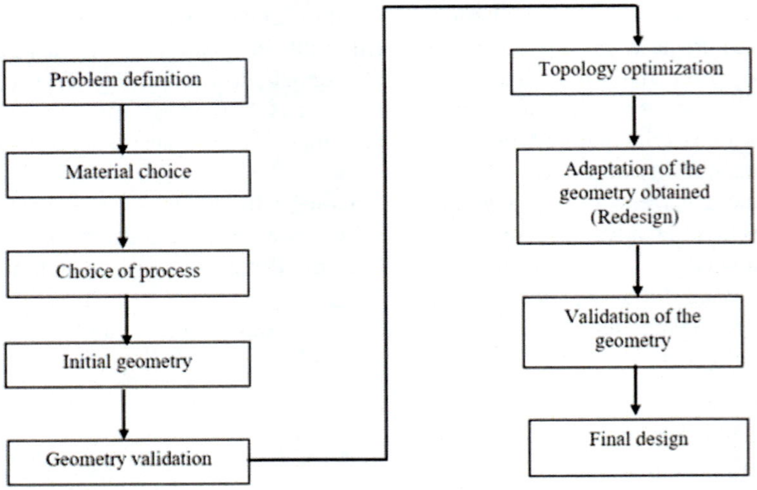

Fig. 19.3 Design methodology

19.4.1 Problem Definition

The purpose of this initial step is to define the context of the design process and comprehend the issue. For this, it's crucial to comprehend the objective and establish the design's context, by defining the design goals, for instance, by aiming to reduce the part's weight.

19.4.2 Material Choice

In this step the materials that will constitute the part is choosed. Actually, the process in the context of AM relies on the material, which affects the machine used and its workload, the possible shape, and mechanical strength of the part [8]. To make this choice, the designers take in consideration:

- Materials available
- Compatible materials
- Physical properties imposed by the specifications

19.4.3 Process Choice

Since the material has already been chosen, some process will not be used. It is enough to discard the incompatible machines among the available machines. Since the digital model of the part is not yet available, the integration of the support cannot be implemented at this time. Other constraint must be considered as time of production and speed of production.

19.4.4 Initial Geometry

At this stage, the initial geometry of the part is created using CAD software, this part is serving as a base for the topological optimization phase, with consideration of machine and material data.

19.4.5 Validation of Initial Geometry

This step is important before proceeding to topology optimization, the validation is done using finite element study. Otherwise, our methodology would lead to a dead end during the topological optimization.

19.4.6 Topology Optimization

In this step the part is topologically optimized, in order to gain to gain weight, cost and production time. For this operation some 3D software integrate this option, but there are specific software to run this mathematical methodology, to do that, the part is imported in the software, and the client specifications are entered. The results is a design often not attractive but it respect the specifications. The next step allows to correct the geometry obtained in this phase.

19.4.7 Redesign

In this phase, the geometry obtained from topology optimization, which is often unthinkable and unattractive, must be modified using 3D software in order to create a geometry adequate and attractive.

19.4.8 Validation of the Geometry

In this step, we proceed, using finite element methodology, in verification of the geometry obtained after redesigning, to make sur that it respects the specifications.

19.4.9 Final Design

The design obtained is fabricated with the technology choosed in the beginning of the methodology.

19.5 Conclusion

The DfAM approaches are intended to assist designers in making design choices to meet functional needs, guaranteeing manufacturability in AM systems, and helping manufacturers during part fabrication in AM systems. This paper describes a general methodology to design for additive manufacturing with respect to the client requirements and the machines performance, also with additional value compared to the existing DFAM methods. The main DfAM steps involved in this methodology are problem definition, material and process choice, topology optimization and redesigning. Based on the potential and constraints provided, a comprehensive collection of existing designs for AM techniques was discussed. As a perspective to this work, a prototype designed using this method is compared to other designed by the other methodology and thinking about specific methodology to each AM technology.

References

1. Omar, L., Mohammed, N., Otmane, B.: An Overview on Additive Manufacturing: Technologies, Materials and Applications. Uncertainties and Reliability of Multiphysical Systems, vol. 6 (2022)
2. Di Nicolantonio, M., Rossi, E., Alexander, T. (eds.). Advances in additive manufacturing, modeling systems and 3D prototyping. In: Proceedings of the AHFE 2019 International Conference on Additive Manufacturing, Modeling Systems and 3D Prototyping, July 24–28, 2019, Washington. Springer International Publishing, Cham (2020)
3. Salmi, M.: Additive manufacturing processes in medical applications. Materials **14**, 191 (2021)
4. Orquéra, M.: Conception pour la fabrication additive: approche méthodologique pour les systèmes mécaniques multicorps, p. 207
5. Diegel, O., Nordin, A., Motte, D.: A Practical Guide to Design for Additive Manufacturing. Springer Singapore (2019)
6. Thompson, M.K., Moroni, G., Vaneker, T., Fadel, G., Campbell, R.I., Gibson, I., et al.: Design for additive manufacturing: trends, opportunities, considerations, and constraints. CIRP Ann. **65**, 737–760 (2016)
7. Kumke, M., Watschke, H., Vietor, T.: A new methodological framework for design for additive manufacturing. Virtual Phys. Prototyp. **11**, 3–19 (2016)
8. Boyard, N.: Méthodologie de conception pour la réalisation de pièces en Fabrication Additive, p. 173
9. Rodrigue, H., Rivette, M., Calatoru, V., Richir, S.: Une méthodologie de conception pour la fabrication additive, p. 10
10. Renjith, S.C., Park, K., Okudan Kremer, G.E.: A design framework for additive manufacturing: integration of additive manufacturing capabilities in the early design process. Int. J. Precis. Eng. Manuf. **21**, 329–345 (2020)

11. Joe Lopes, A., MacDonald, E., Wicker, R.B.: Integrating stereolithography and direct print technologies for 3D structural electronics fabrication. Rapid Prototyp. J. **18**, 129–143 (2012)
12. Bahnini, I., Rivette, M., Rechia, A., Siadat, A., Elmesbahi, A.: Additive manufacturing technology: the status, applications, and prospects. Int. J. Adv. Manuf. Technol. **97**, 147–161 (2018)
13. Diegel, O., Nordin, A., Motte, D.: DfAM Strategic Design Considerations. A Practical Guide to Design for Additive Manufacturing, pp. 41–70. Springer, Singapore (2019)
14. Durakovic, B.: Design for additive manufacturing: benefits, trends and challenges. Period. Eng. Nat. Sci. (PEN). **6**, 179 (2018)
15. Zohdi, N., Yang, R. (Chunhui): Material anisotropy in additively manufactured polymers and polymer composites: a review. Polymers **13**, 3368 (2021)
16. Yang, S., Zhao, Y.F.: Additive manufacturing-enabled design theory and methodology: a critical review. Int. J. Adv. Manuf. Technol. **80**, 327–342 (2015)
17. Wiberg, A., Persson, J., Ölvander, J.: Design for additive manufacturing – a review of available design methods and software. Rapid Prototyp. J. **25**, 1080–1094 (2019)
18. Ponche, R., Kerbrat, O., Mognol, P., Hascoet, J.-Y.: A novel methodology of design for additive manufacturing applied to additive laser manufacturing process. Robot. Comput.-Integr. Manuf. **30**, 389–398 (2014)
19. Ponche, R., Hascoet, J.Y., Kerbrat, O., Mognol, P.: A new global approach to design for additive manufacturing: a method to obtain a design that meets specifications while optimizing a given additive manufacturing process is presented in this paper. Virtual Phys. Prototyp. **7**, 93–105 (2012)
20. Ponche, R.: Méthodologie de conception pour la fabrication additive, application à la projection de poudres, p. 174
21. Rodrigue, H., Rivette, M.: An Assembly-Level Design for Additive Manufacturing Methodology. Springer, Paris (2011)
22. Vayre, B.: Conception pour la fabrication additive, application à la technologie EBM. Thèse de Doctorat, pp. 7–8 (2014)

Chapter 20
Viscous Layer Formation in Electrochemical Polishing Laser-Powder Bed Fusion Parts with Different Surface Profiles

Haitao Zhu, Yingtao Tian, and Allan E. W. Rennie

Abstract The viscous layer generated on the anode surface during electrochemical polishing is essential for obtaining a smooth, mirror-like surface. However, the growth of the viscous layer for polishing Laser-Powder Bed Fusion (L-PBF) components with rough surface features of several hundred microns in height and wavelength differs from the micron/nanoscale surface finish. This study employed a Spatial Frequency Method to model rough L-PBF surfaces. The effect of height and frequency distribution parameters on the viscous layer thickness, geometry and uniformity was numerically investigated. NaCl-Ethylene Glycol-Ethanol and commercial A2 electrolytes were utilised to polish L-PBF 316L stainless steel for verification purposes. The results show that the viscous layer of the high-frequency surfaces is more homogeneous than the low-frequency surfaces, and the roughness reduction can reach 97.6% compared to 64% for the low-frequency surfaces.

Keywords Laser powder bed fusion · Electrochemical polishing · Viscous layer

20.1 Introduction

Additive manufacturing (AM; also known as 3D printing) is recognised as one of the most innovative technologies in the fourth industrial revolution (Industry4.0), where Laser-Powder Bed Fusion (L-PBF) is one of the primary manufacturing technologies for metal components. It can offer substantial benefits in manufacturing complex structures, freeing designers from geometric constraints, enhancing mechanical properties, and reducing materials waste and energy consumption [1–3]. However, as the powder particles are fused layer-by-layer, stair-stepping, powder adhesion, and

---

H. Zhu · Y. Tian · A. E. W. Rennie (✉)
School of Engineering, Lancaster University, Lancaster LA1 4YW, Lancashire, UK
e-mail: a.rennie@lancaster.ac.uk

K. Zarbane and Z. Beidouri (eds.), *Proceedings of CASICAM 2022*, Springer Tracts in Additive Manufacturing, https://doi.org/10.1007/978-3-031-32927-2_20

balling are unavoidable, resulting in poor surface quality for parts removed from L-PBF machines [4, 5]. The poor surface qualities can compromise the structural, dimensional and functional properties, obstructing applications in the aerospace, automotive and medical implantation industries [6, 7].

Electrochemical polishing (EP) is one of the post-processing technologies used for improving surface aesthetics and physical properties of metal-based AM parts by efficiently eliminating surface unevenness, contaminants, and residual stress, regardless of part geometry [8–10]. This technology was discovered and systematically studied in 1935 and 1936 by Jacquet, who brightened copper surfaces in acid solution and proposed the 'viscous layer theory', which remains one of the most reliable hypotheses today [11, 12]. However, EP is an complicated process which is difficult to be controlled and predicted. Previous investigations and findings are based on numerous experimental attempts, which waste time and resources [13]. Furthermore, the polishing parameters selected in previous studies differed due to parameters having cross-influence [14–16].

The Finite Element Method (FEM) is an effective tool to simulate the growth mechanism of the viscous layer during polishing [17]. Numerous simulation studies have investigated the effects of the total charge, concentration diffusion pattern and electrochemical reactions on the viscous layer potential, concentration distribution, thickness and uniformity [18, 19]. Additionally, simulation work on DC/pulse current machining mode, terminal effects, diluted electrochemical ionic system, rotating electrode, etc. has also been reported [20–23]. However, the effect of initial surface conditions was ignored since the models in previous studies were flat surfaces or with regular shapes [24]. The L-PBF parts typically have surface profiles with height differences exceeding 100 μm and roughness (Ra) values above 10 μm, which has a non-negligible effect on the thickness and uniformity of the viscous layer.

This work proposed a L-PBF surface profile model using a Spatial Frequency Method. The effect of surface texture on the thickness and geometry of the viscous layer was simulated by solving the Nernst-Planck equation. The surface roughness index Ra was employed to characterise the uniformity of the viscous layer for the first time. NaCl-Ethylene Glycol-Ethanol and commercial A2 electrolytes were utilised to polish 316L stainless steel test coupons, whose surfaces were separated by a Gaussian filter for validation. The results indicated that the roughness reduction could reach 97.6% for wavelengths below 150 μm while less than 64% for wavelengths exceeding 150 μm. The findings can guide the selection of welding particles, parameters, techniques, and polishing electrolytes for AM parts.

20.2 Mathematical Model Formulation

20.2.1 Problem Definition and Simulation Model

Figure 20.1a shows the schematic diagram of the EP system, including an anode and a cathode connected to a DC power supply and a beaker filled with electrolytes. To simplify the simulation model, the region of interest between the electrodes is extracted from the EP system and shown in Fig. 20.1b. The model boundaries are marked with numbers, where numbers 1 and 3 represent the anode and cathode surfaces, and numbers 2 and 4 represent the electrolyte boundary. *L* is the length of the electrodes, and *H* is the inter-electrode distance.

The L-PBF component surface texture is complicated, whereas it can be categorised into two groups: waviness profile and residual profile [1, 25]. Surface roughness can be considered as the average height over the measured length of the area enclosed by the entire profile height characteristics. This can be represented by the 2D concept of average roughness (Ra), as the model is also 2D. The cathode surface is considered to be an ideally smooth and flat boundary.

During the EP process, a viscous layer is formed on the anode surface, the composition of which is variable, including dissolved metal ions, anions, some other reaction products, etc. [26]. It is assumed that the concentration of the substances reaches its maximum at the anode surface and decreases towards the bulk electrolyte at the steady state [24]. Therefore, the substances are considered as a whole, whose concentration on the anode surface is considered as 1. The viscous layer region is considered to that with a concentration between 0.9 and 1 [27].

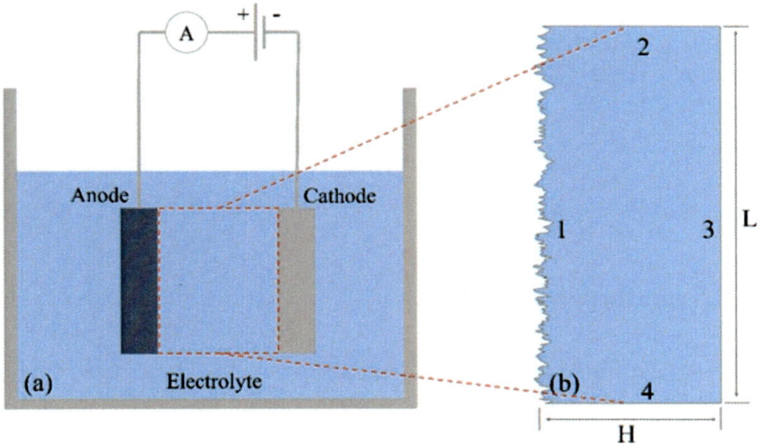

Fig. 20.1 **a** Schematic diagram and **b** simulation model of the EP system

20.2.2 Problem Definition and Simulation Model

A general rough surface can be modelled using the Spatial Frequencies Method as per Eq. (20.1) [28]. Therefore, two terms of Eq. (20.1) can be introduced to form the waviness and residual profiles for L-PBF component surfaces.

$$y = A\int_{-N}^{N}\left(n^{2\times\left(-\frac{b}{2}\right)} \times g1(n) \times \cos(\omega\pi nx + u1(n))\right). \tag{20.1}$$

where N is the spatial frequency resolution, b is the spectral exponent, A is the scale parameter in y coordinate, ω is the average spatial frequency, $g1$ is the Gaussian random function, and $u1$ is the uniform random function. Hereafter, the surface roughness of the generated profile can be calculated using Eq. (20.2).

$$Ra = \frac{\int_0^L |y - \overline{y}|dx}{L}. \tag{20.2}$$

The Nernst-Planck equation was introduced in the simulation process to describe the mass transport process, as Eq. (20.3) shows. However, convection and electromigration were ignored because no external force was applied, and hydrodynamics has a more pronounced influence on the diffusion layer formation process than migration. Therefore, the final governing equation can be expressed as Eq. (20.4).

$$\frac{\partial c_i}{\partial t} + \nabla(-D_i\nabla c_i - Z_i\mu_i c_i\nabla\Phi + \overline{u}c_i) = R_i. \tag{20.3}$$

where D_i, c_i, Z_i, μ_i and R_i are diffusion coefficient (cm^2/s), concentration (mol/m^3), charge number, ionic migration rate and electrochemical reaction rate of species i. ϕ is the electric potential (V).

$$\frac{\partial c_i}{\partial t} + \nabla(-D_i\nabla c_i) = 0. \tag{20.4}$$

20.2.3 Experiment Validation

The anode surface profile model was validated by comparing the surface roughness of the simulated surface profile with the practical surface profile of L-PBF 316L stainless steel components. The parameters for calculating the profiles and EP simulation are listed in Table 20.1. Three steel samples were selected for comparison, where the subscripts of w and R represented waviness and residual profile, and the numbers 1, 2, and 3 were the labels of the profiles. Simulated surface roughness and the height difference of the surface profile and viscous layer were calculated in Matlab.

Table 20.1 Parameters used for validating L-PBF component surface profiles and EP simulation

Anode surface	Parameters	Values
	N	2000
	b	0.5
Waviness profile	$A_{\omega 1}, A_{\omega 2}, A_{\omega 3}$	0.0012, 0.010, 0.010
	$\omega_{w1}, \omega_{w2}, \omega_{w3}$	0.08, 0.08, 0.08
Residual profile	A_{R1}, A_{R2}, A_{R3}	0.007, 0.0065, 0.008
	$\omega_{R1}, \omega_{R2}, \omega_{R3}$	0.15, 0.15, 0.15
EP simulation	Values	Other conditions
A	0.006, 0.01, 0.014, 0.018, 0.022	$\omega = 0.12$
ω	0.04, 0.08, 0.12, 0.16, 0.2	A = 0.014
D	$D = 10^{-8}/10^{-10}$ m^2/s	
H	20 mm	

The anode fixture and two-electrode system are shown in Fig. 20.2. Two types of electrolytes were introduced in the polishing process: a commercial A2 electrolyte (Struers Ltd, Rotherham, United Kingdom) and a 1 M NaCl-Ethylene Glycol-10% Ethanol electrolyte. A DC power supply (type pe86CB-20-5-25-S/GD, Plating Electronic GmbH (Sexau, Germany)) was connected to the electrodes, and net-shaped platinum was used as the cathode.

The surface morphology was characterised by an Olympus LEXT 5000 Laser Microscope, and an open-source tool, ProfilmOnline, was utilised to quantify the surface roughness. The measured surface profile was flattened by a three-order polynomial and then separated into waviness and residual surfaces by a Gaussian filter with a 150 μm cut-off value (*Psa*, *Wsa* and *Ssa*).

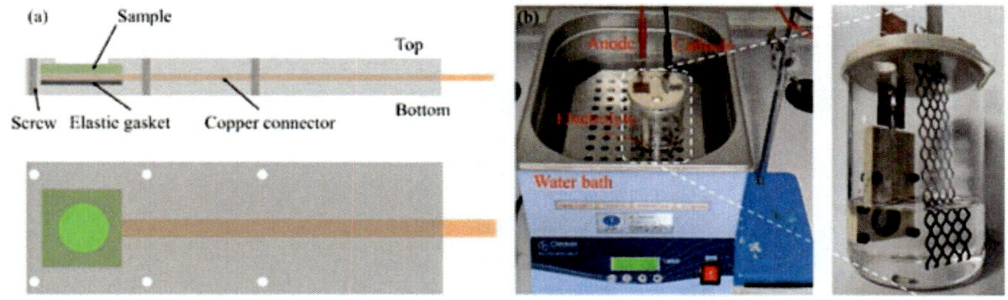

Fig. 20.2 **a** Anode fixture and **b** experimental setup for the EP process

20.3 Results and Discussion

20.3.1 Validation of the Anode Model

Figure 20.3 shows the schematic profiles and roughness values obtained in the simulation and experiment. When the roughness values of the simulated waviness and residual profiles were kept similar to the experiment results, the roughness difference of the measured profile was minimal, corresponding to 1.7 μm, 1.3 μm and 2.0 μm. The slight deviation was caused by the calculation method and accuracy, meaning that (a) the compound method was simple addition for simulation but a Gaussian filter for the experiment and (b) the waviness and residual profile can be further separated. In this study, the deviations are acceptable, whilst improved accuracy can bring high computational burdens.

20.3.2 Simulation Result

Figure 20.4a shows the surface and viscous layer profile with different height distributions. The height difference increased from 48.9 μm to 176.0 μm for the viscous layer whilst from 74.2 μm to 272.3 μm for the anode surface. The average viscous layer thickness increased from 33.64 μm to 54.92 μm with the increased anode surface height difference. The viscous layer uniformity changing with the anode surface roughness was shown in Fig. 20.4c. It is noticed that the uniformity became worse with increased anode surface roughness because the anode surface height difference increased more rapidly than the viscous layer thickness. However, the roughness difference between the anode surface and the viscous layer was more evident for rough surfaces, meaning that rough surfaces might have high roughness reductions after polishing.

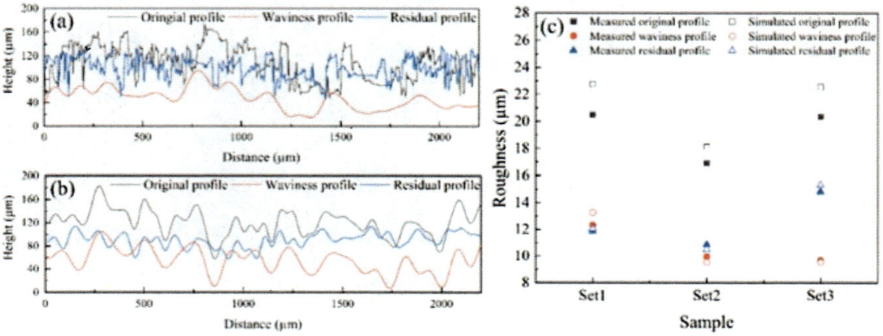

Fig. 20.3 **a** Measured L-PBF steel surface profile, **b** simulated profile and **c** roughness comparison

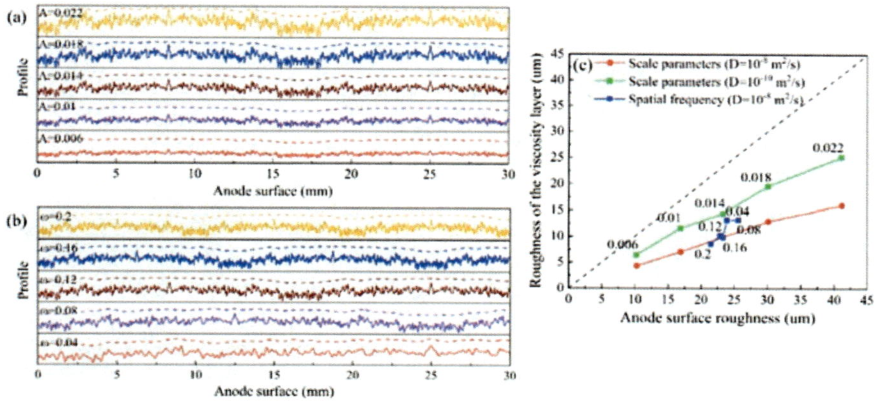

Fig. 20.4 **a**, **b** Profiles and **c** roughness of the anode surfaces and viscous layers with different height and wavelength characteristics

Figure 20.4b, c shows the surface and viscous layer profile and roughness with different frequency distribution. The height difference was not affected for the anode surface, corresponding to approximately 173 μm, while it decreased from 73.8 μm to 43.2 μm for the viscous layer. The average viscous layer thickness was not affected as well, corresponding to 45.9 μm, 41.1 μm, 43.1 μm, 45.0 μm and 47.0 μm with the increased frequency. However, the viscous layer uniformity increased with the decreased frequency distribution, and the increased rate is higher than 1 (the dashed line in Fig. 20.4c), indicating that high-frequency-dominated surfaces might generate a uniform viscous layer and then produce a good polishing effect.

20.3.3 Experiment Verification

316L stainless steel samples were polished at the transpassive region (1000 mA/cm^2) for 28 min to keep the same polishing parameters and materials removal weight. The electrode distance was 20 mm to provide sufficient space for diffusion. Figure 20.5 shows the roughness reduction and surface morphologies before and after polishing. The initial waviness and residual surface roughness ranged between 12 μm and 14 μm, while after polishing, the values reduced to approximately 12.4 μm and 0.96 μm with the NaCl-EG-Ethanol electrolyte, and 5.5 μm and 0.43 μm with the A2 electrolyte. The roughness reduction rate can reach 39.6%, 92.3%, 59.3% and 96.9%. Noticeably, the A2 electrolyte had a better polishing effect than the NaCl-EG-Ethanol electrolyte, and both electrolytes had a better polishing effect on the residual surfaces than on the waviness surfaces. This is because charges preferred to gather at the sharp region due to the tip-shape effect, meaning that the sharp regions could have a high current density, and therefore, materials removal rate. Additionally, when

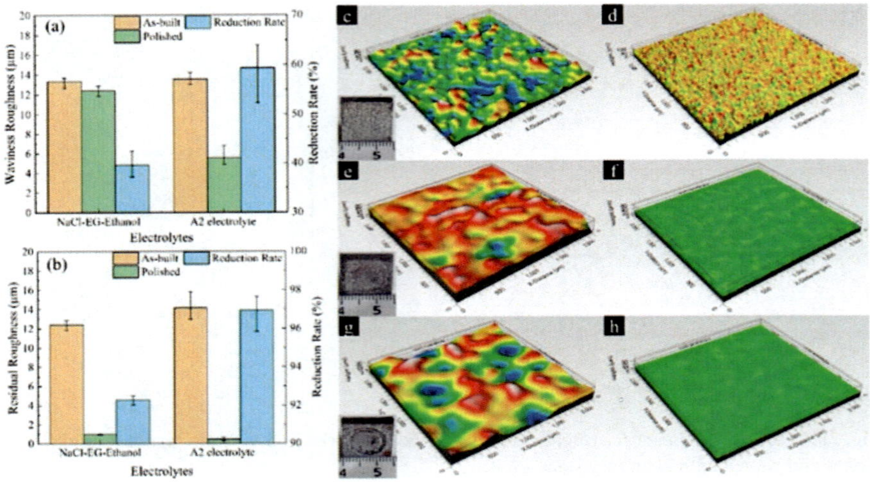

Fig. 20.5 Roughness reduction of **a** waviness and **b** residual surfaces before and after EP; **c**, **e**, **g** waviness and **d**, **f**, **h** residual surfaces before polishing and polished with the NaCl-EG-Ethanol and A2 electrolytes

removing the same thickness, more material needs to be removed for low-frequency regions than for high-frequency regions. Therefore, the electrolytes had a superior polishing effect on the residual surfaces than on the waviness surfaces.

20.4 Conclusions

The Spatial Frequencies Method has been applied to model the rough L-PBF component surface, and the viscous layer growth process with different anode surface textures was investigated. Based on the results, the following conclusions can be drawn.

(1) The Spatial Frequencies Method can be employed to model the L-PBF component surface with different height and frequency distributions by adjusting the scale and spatial frequency parameters.
(2) Surfaces with a large height distribution could have a high roughness reduction rate after polishing, and high-frequency-dominated surfaces can promote the polishing effect.
(3) Surface roughness of the 316L stainless steel components exhibited a reduction of 59.3% and 96.9% for the waviness and residual surfaces using the A2 electrolyte, while 39.6% and 92.3% using the NaCl-EG-Ethanol electrolytes.

Acknowledgements This research was financially supported by the School of Engineering, Lancaster University and The Royal Society International Exchanges 2018 Cost Share (China) Scheme (grant number IEC\NSFC\181278).

References

1. Lou, S., et al.: Characterisation methods for powder bed fusion processed surface topography. Precis. Eng. **57**, 1–15 (2019)
2. Chaghazardi, Z., Wüthrich, R.: Electropolishing of additive manufactured metal parts. J. Electrochem. Soc. **169**, 043510 (2022)
3. Shrivastava, A., Anand Kumar, S., Nagesha, B.K., Suresh, T.N.: Electropolishing of Inconel 718 manufactured by laser powder bed fusion: effect of heat treatment on hardness, 3D surfaces topography and material ratio curve. Opt. Laser Technol. **144**, 107448 (2021)
4. Strano, G., Hao, L., Everson, R.M., Evans, K.E.: Surface roughness analysis, modelling and prediction in selective laser melting. J. Mater. Process. Technol. **213**, 589–597 (2013)
5. Pyka, G., et al.: Surface modification of Ti6Al4V open porous structures produced by additive manufacturing. Adv. Eng. Mater. **14**, 363–370 (2012)
6. Gibson, I., Rosen, D., Stucker, B., Khorasani, M.: Additive Manufacturing Technologies. Springer (2021).
7. An, L., Wang, D., Zhu, D.: Combined electrochemical and mechanical polishing of interior channels in parts made by additive manufacturing. Addit. Manuf. **51**, 102638 (2022)
8. Tyagi, P., et al.: Reducing the roughness of internal surface of an additive manufacturing produced 316 steel components by chempolishing and electropolishing. Addit. Manuf. **25**, 32–38 (2019)
9. Hocheng, H., Pa, P.S.: Effective form design of electrode in electrochemical smoothing of holes. Int. J. Adv. Manuf. Technol. **21**, 995–1004 (2003)
10. Steinfeld, B., et al.: The role of lean process improvement in implementation of evidence-based practices in behavioral health care. J. Behav. Health Serv. Res. **42**, 504–518 (2015)
11. Jacquet, P.A.: Electrolytic method for obtaining bright copper surfaces. Nature **135**, 1076–1076 (1935)
12. Jacquet, P.A.: On the anodic behavior of copper in aqueous solutions of orthophosphoric acid. Trans. Electrochem. Soc. **69**, 629 (1936)
13. Walther, B., Schilm, J., Michaelis, A., Lohrengel, M.M.: Electrochemical dissolution of hard metal alloys. Electrochim. Acta **52**, 7732–7737 (2007)
14. Zhang, Y., Li, J., Che, S., Tian, Y.: Electrochemical polishing of additively manufactured Ti–6Al–4V alloy. Met. Mater. Int. **26**, 783–792 (2020)
15. Dong, G., Marleau-Finley, J., Zhao, Y.F.: Investigation of electrochemical post-processing procedure for Ti-6Al-4V lattice structure manufactured by direct metal laser sintering (DMLS). Int. J. Adv. Manuf. Technol. **104**, 3401–3417 (2019)
16. Wu, Y.-C., Kuo, C.-N., Chung, Y.-C., Ng, C.-H., Huang, J.C.: Effects of electropolishing on mechanical properties and bio-corrosion of Ti6Al4V fabricated by electron beam melting additive manufacturing. Materials **12**, 1466 (2019)
17. Dickinson, E.J.F., Ekström, H., Fontes, E.: COMSOL multiphysics®: finite element software for electrochemical analysis. A mini-review. Electrochem. Commun. **40**, 71–74 (2014)
18. Jemmely, P., Mischler, S., Landolt, D.: Electrochemical modeling of passivation phenomena in tribocorrosion. Wear **237**, 63–76 (2000)
19. Wilk, J.: A review of measurements of the mass transfer in minichannels using the limiting current technique. Exp. Therm. Fluid Sci. **57**, 242–249 (2014)
20. Tang, L., Gan, W.M.: Utilization of flow field simulations for cathode design in electrochemical machining of aerospace engine blisk channels. Int. J. Adv. Manuf. Technol. **72**, 1759–1766 (2014)

21. Marshall, S.L., Wolff, S.K.: Analysis of terminal effects in rectangular electrochemical cells. Electrochim. Acta **43**, 405–415 (1998)
22. Dan, C., Van den Bossche, B., Bortels, L., Nelissen, G., Deconinck, J.: Numerical simulation of transient current responses in diluted electrochemical ionic systems. J. Electroanal. Chem. **505**, 12–23 (2001)
23. Ma, N., Xu, W., Wang, X., Tao, B.: Pulse electrochemical finishing: modeling and experiment. J. Mater. Process. Technol. **210**, 852–857 (2010)
24. Tailor, P.B., Agrawal, A., Joshi, S.S.: Numerical modeling of passive layer formation and stabilization in electrochemical polishing process. J. Manuf. Process. **18**, 107–116 (2015)
25. Diaz, A.: Surface texture characterization and optimization of metal additive manufacturing-produced components for aerospace applications. In: Froes, F., Boyer, R. Eds.) Additive Manufacturing for the Aerospace Industry, pp. 341–374. Elsevier (2019)
26. Han, W., Fang, F.: Eco-friendly NaCl-based electrolyte for electropolishing 316L stainless steel. J. Manuf. Process. **58**, 1257–1269 (2020)
27. Lee, S.-J., Lai, J.-J., Lin, Y.-T.: Simulation of the formation mechanism of a viscous layer for the electropolishing process. WIT Trans. Eng. Sci. **48** (2005)
28. Danilov, I., et al.: Process understanding of plasma electrolytic polishing through multiphysics simulation and inline metrology. Micromachines **10**, 214 (2019)

Chapter 21
Control of a 3D Printed Carbon Fiber Reinforced Plate by Ultrasonic Guided Wave

Ismaine Zitouni, Hassan Rhimini, and Abdelkerim Chouaf

Abstract In this paper, we present a finite element numerical model of the interaction of ultrasonic guided waves with an internal defect from the fused filament additive manufacturing process. The material considered is a thermo-plastic composite reinforced with carbon fibers. The numerical control process using transducers is explained. An energy quantification of the incident, reflected and transmitted modes is presented. The influence of the defect dimensions on the reflection and transmission energy coefficients is studied. The technique based on the summation and subtraction of the signals obtained by the transducers positioned at two identical positions with respect to the wave propagation axis has been developed to determine the nature of the reflected and transmitted modes when we have a mode conversion phenomenon. On the basis of the results found, we were able to explain the mode conversion phenomenon related to the asymmetric position of the defect and we were able to determine the nature of the modes of the reflected and transmitted wave packets. The accuracy of the results is controlled by the establishment of an energy balance.

Keywords Guided wave · MEF simulation · Hanning window · Carbon fiber reinforced Thermo-plastic · Dispersion · Energy balance

I. Zitouni (✉) · H. Rhimini · A. Chouaf
Laboratory of Mechanics, Engineering and Innovation, National High School of Electricity and Mechanics, Hassan II University, Casablanca, Morocco
e-mail: Ismaine.zitouni.doc20@ensem.ac.ma

H. Rhimini
e-mail: h.rhimini@ensem.ac.ma

A. Chouaf
e-mail: a.chouaf@ensem.ac.ma

K. Zarbane and Z. Beidouri (eds.), *Proceedings of CASICAM 2022*, Springer Tracts in Additive Manufacturing, https://doi.org/10.1007/978-3-031-32927-2_21

21.1 Introduction

During these recent years, 3D printing has seen a major development in almost all the industrial sectors, especially in the automotive, aerospace and medical industries where we observe a great evolution and a variety of printing technologies. These techniques differ according to the type of material, the complexity of the piece to be printed as well as the manufacturing and maintenance costs. From the one appropriate for metallic materials (Selective Laser Melting SLM [1], Selective Laser Sintering SLS [2]) to the techniques used for printing polymers in filament (fused deposition modeling FDM [2]). However, various defects can appear in the printed pieces [3]. The origins of these defects are multiple. For example, they can be due to the materials used such as the non-adhesion between the fibers and the matrix in the case of composite materials [4, 5]. The manufacturing process can also generate defects, such as volume defects, porosities, pitting and cracks [6]. Other types of defects can be generated by the thermal phenomenon of the process, for example, the thermally affected zone TAZ [7] for the metal printing techniques (welding process). It is therefore necessary to control this type of structure and characterize their defects. Ultrasonic guided waves (UWG) [8] have a great potential for non-destructive testing of structures. They have a great ability to propagate over long distances without significant attenuation and to control the integrity of the structure. Practically speaking, in order to carry out an NDT by UGW, it is imperative to know the dispersion networks of these waves. The dispersion curves are used to select the appropriate mode for the inspection according to the frequency of the signal to be excited by the piezoelectric transducers. Normal displacements are collected on the surface of the plate using ultrasonic sensors. A post treatment in the form of signal processing by a fourier transform, or by wavelets or others, is essential to localize and characterize possible defects that may exist in the structure. Several authors have proposed numerical models for the simulation of UGW propagation in different types of structures, we cite Luca [9] who considered the propagation of UGW in a CFRP laminated plate by a finite element model via the ABAQUS software and studied the influence of the size and orientation of the defect on the damage of the structure. Chui [10] was interested in the control of the delamination of a 3D laminate in a four-layer composite material. Rhimini [8] studied the interaction of UGW with a hidden volume defect in a three-layer structure (aluminum/epoxy/aluminum) by ABAQUS software. Subsequently a post processing was done by the double Fourier transform.

In this present study, we aim to control a carbon fiber reinforced thermoplastic composite plate [11]. Due to these high mechanical properties compared to single thermoplastics [12], this composite is widely used in the filament manufacturing process. In addition, the high prototyping capability and the different printing parameters (Raster angle, filament size…) specific to the FDM process allow the fabrication of complex structures based on CFRTP. It is commonly used in the automotive and aerospace industries. However, the stiffness of the printed CFRTP composites decreases due to defects (cracks, layer delamination, notching…) that occur during the manufacturing process and are mainly due to the printing parameter [13]. The plate we consider has defects in the form of rectangular notches. To control this type of structure, we first draw the dispersion curves of this material which represent the (wave number, frequency) pairs susceptible to generate the OGU in our plate [14, 15]. We also plot the displacement and stress profiles normalized with an energy flow through the section. The spatial distribution of displacements is applied for 10 cycles and weighted by a Hanning window centered on the excitation frequency. Using the elements described above, we were able to develop a numerical finite element simulation in the ABAQUS software of the UGW generation process in the composite plate and the acquisition of the signals taken before and after the defect. By calculating the reflection and transmission coefficients, we have studied the influence of the size of the defect and its position relatively to the median plane in the choice of the ultrasonic mode for the control. The larger the defect size, the larger the reflected wave packet. Moreover, an asymmetric position of the defect regarding the median plane leads to the appearance of the mode conversion phenomenon that we have treated using a technique based on the summation and subtraction of the signals collected by the ultrasonic transducers.

21.2 Ultrasonic Guided Waves Theory

21.2.1 Dispersion Curves

We consider the propagation of ultrasonic guided waves along a composite plate. This composite is a thermoplastic reinforced with carbon fibers [11]. The latter is used in 3D printing, especially in the FDM fused filament manufacturing process.

The propagation of UGW in the structure allows to scan the whole waveguide. To do this, we first need to know the dispersion networks of these waves. In fact, these waves are strongly dependent on the frequency and a simple variation of the former can change the behavior of these waves. These dispersion networks also called dispersion curves allow to determine the couple's frequency f, wave number k susceptible to generate the OGU in the plate. Several methods have been used to determine this dispersive character. From those based on the use of analytical methods to find the zeros of a function [15], to arrive at numerical methods to draw these curves [16]. In this context, we have developed a Matlab program based on a numerical method [14] to plot the dispersion curves of composite materials. This consists in applying a spectral scheme method to the differential equations of motion and boundary conditions. Once the eigenvalues of this problem are found (frequency) for a range of chosen wavenumbers, we apply a mode separation algorithm based on the vibrational states of the modes present. We then obtain the dispersion curves in the form of frequency as a function of the wavenumber.

Figure 21.1 shows the dispersion curves of the CFRTP composite plate. The indexation of the modes is taken from their first appearance at the low frequencies.

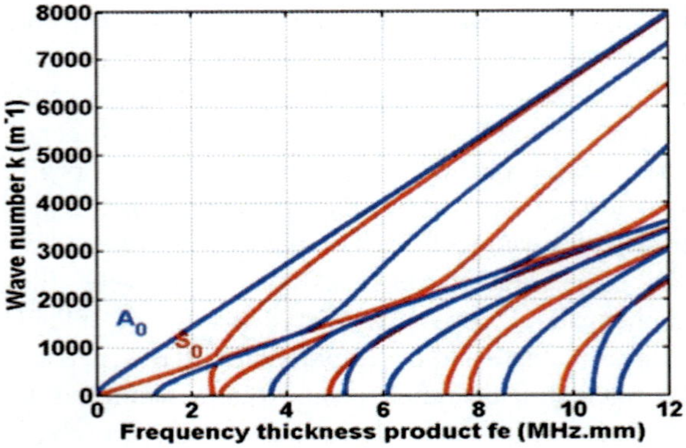

Fig. 21.1 Dispersion curves of a CFRTP plate plotted by the spectral method

There are modes with symmetric vibrations (S_n) with respect to the propagation planes and others with antisymmetric vibrations (A_n). The fundamental modes are indexed by a zero (S_0 and A_0). These modes are the most used in non-destructive testing by UGW [17].

21.2.2 *Displacement Field*

In the case of an anisotropic material, the displacement fields of the plate are of the form:

$$(u_1, u_2, u_3) = \sum_{q=1}^{6} (1, V_q, W_q) U_{1q} e^{ik(x_1 + \alpha_q x_3 - ct)} \quad (21.1)$$

where q is the summation index, x_1 is the UGW propagation direction and c is the wave phase velocity, k is the wave number, U_{1q} V_q and W_q are the displacements amplitudes and α_q is the ratio of the wave numbers along the x_1 and x_3 directions.

21.3 Modeling of Guided Wave Propagation by the Finite Element Method

The finite element method consists in discretizing the structure into finite elements. Then, interpolation functions approximate the displacement field at each node of the element, these displacements are called: nodal displacements. An assembly of the elements allows to obtain a global system of equations to solve. This system for the case of a non-damping is written:

$$[M]\{\ddot{U}\} + [K]\{U\} = \{F\} \tag{21.2}$$

where [M] represents the global mass matrix of the system, [K] the stiffness matrix, {U} is the global displacement vector, $\{\ddot{U}\}$ is the acceleration vector and {F} the force vector.

In order to solve the problem in (21.2), we need a method to approximate the spatial partial derivatives. We use here the Newmark method [18]. Using the extended mean value theorem, Newmark's method states that the first derivative with respect to time can be solved as follows:

$$\{\dot{U_{t+\Delta t}}\} = \{U_t\} + \Delta t\{\ddot{U_\gamma}\} \tag{21.3}$$

With

$$\{\ddot{U_\gamma}\} = (1-\gamma)\{\ddot{U_t}\} + \gamma\{\ddot{U_{t+\Delta t}}\} \tag{21.4}$$

Since we have partial derivatives with respect to time in the acceleration vector, then the theorem can also be written as:

$$\{U_{t+\Delta t}\} = \{U_t\} + \Delta t\{\dot{U_t}\} + \Delta t^2((1-\beta)\{\ddot{U_t}\} + \beta\{\ddot{U_{t+\Delta t}}\}) \tag{21.5}$$

where γ and β are the integration parameters of Newmark. Δt is the time step.

21.4 Numerical Simulation

21.4.1 Finite Element Method

We consider the propagation of UGW in a composite plate of thickness e = 4 mm and length L = 500 mm.

The properties of this composite are given in Table 21.1.

We consider a defect in the form of a rectangular notch inside the composite plate. The depth of the notch is p = 2 mm and its width w = 1 mm.

Table 21.1 Properties of the CFRTP composite

Densité (kg/m^3)	C11 (GPa)	C22 (GPa)	C33 (GPa)	C44 (GPa)	C55 (GPa)	C66 (GPa)	C12 (GPa)	C13 (GPa)	C23 (GPa)
1600	43.3294	43.3294	43.3294	18.4291	9.51	9.51	6.4745	6.4745	6.4712

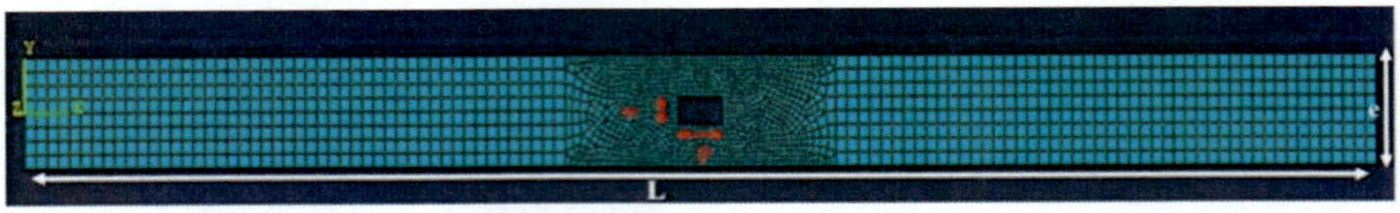

Fig. 21.2 CFRTP composite plate with a symmetrical rectangular notch in its center

We consider a spatio-temporal problem. The simulation is performed under ABAQUS. The choice of parameters for simulations time and space steps Δx and Δy (spatial and temporal discretization) will be detailed thereafter. In order to mesh the considered structure, we use an element size of 0.5 mm satisfying the following condition [19] (Fig. 21.2):

$$\max(\Delta x, \Delta y) \leq \frac{\lambda_{\min}}{10} \tag{21.6}$$

where $\lambda_{\min}$ is the minimal wavelength [19]. This condition stipulates that in order to have a computational convergence, the size of the elements meshing the waveguide must be less than one tenth of the smallest wavelength of the modes susceptible to propagate in the plate.

Concerning the temporal discretization, the temporal convergence of the ultrasonic problem requires that the smallest propagation wavelength in the ultrasonic pulse must be discretized with at least 20 nodes. We consider a frequency f = 500 kHz. On this basis, we obtain a time step of 0.1 μs, this value verifies the following condition [20]:

$$\Delta t \leq \frac{1}{20f} \tag{21.7}$$

where f is the frequency of the mode chosen to generate the UGW in the plate.

21.4.2 Generation of UGW in the CFRTP Composite Plate

To generate a propagative ultrasonic guided mode in the structure and study its interaction with the defect, we apply an analytical displacement normalized by the acoustic power through the thickness of the plate on the left face of the plate (x = 0, y). We have chosen to generate the fundamental mode A_0. According to previous studies, the most adequate modes to make a good control and which do not allow the missing of any defect are the fundamental modes A_0 and S_0 [17]. Moreover, the latter two can be generated for low frequency values.

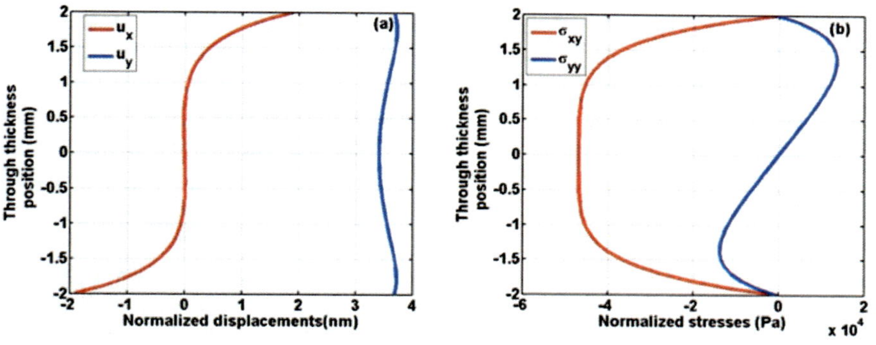

Fig. 21.3 Analytical displacements and stresses normalized by the acoustic power of the generated mode A_0 for the couple ($f = 500$ kHz, $k = 1431\ m^{-1}$)

The analytical displacements in (21.1) can be divided into two parts, a spatial and a temporal part. The spatial part demonstrates the symmetry of the displacements along the plate thickness of the generated mode. The chosen A_0 mode is characterized by antisymmetric longitudinal displacement and symmetric transverse displacement through the plate thickness (Fig. 21.3).

Concerning the temporal part, it is represented by the term $e^{ik(x1+\alpha qx3-ct)}$. Therefore, the spatial analytical displacements oscillate according to a cosinus or a sinus. In the case of isotropic materials, the displacement fields are written directly as a function of a cosinus or a sinus [21]. However, for materials with high anisotropy, as in our case, we cannot differentiate whether the displacements will take a cosinus or a sinus. To remedy this, when plotting the spatial displacements at a given position x_1 and time t (=cte), the exponential term depends only on $\alpha_q x_3$, knowing that $e^{i\varphi} = \cos(\varphi) + i\sin(\varphi)$. So if the displacement is imaginary then it follows a sinus otherwise it follows a cosinus. In our case, the displacement u_1 follows the sinus function and the displacement u_3 follows the cosinus function (mode A_0).

Now the spatial distribution of the displacement is applied for 10 cycles and weighted by a Hanning window centered on the excitation frequency [22]. Hanning windows are applied to filter out the perturbations of the incident signal.

21.5 Numerical Simulation

21.5.1 Symmetrical Defect

The UGW are generated on the left face of the plate through the thickness. Subsequently, to control their propagation along the plate, we place transducers on the $x_2 = \pm d$ faces before and after the defect. The transducers placed before the defect are intended to collect the wave packet reflected from the notch. On the other hand, the transducers placed after the defect aim to capture the quantity of energy transmitted by the notch. The position of these transducers with respect to the edges of the plate and the defect is very important. Effectively, the UGW are composed of propagative modes (which we use in the control) and evanescent modes. The evanescent modes are modes that are characterized by a strong attenuation in the material, leaving them inappropriate for non-destructive ultrasonic testing. The particularity of these waves is that when we have a reflection caused by the edges of the plate or by a defect, they appear. So if the ultrasonic sensors are positioned near the defect or the edge, we will capture the signal of the propagative mode generate and another signal issued from them. There, the interpretation of the results becomes complex especially if we do not have enough information on the behavior of these evanescent waves. In our work, we have decided to place our transducers in such a way to have a minimum distance from the defect and the edge of at least 10 mm. The transducers considered register the displacements u_1 and u_2 of 150 nodes along the face of the structure. The considered nodes are regularly spaced by 0.5 mm (mesh step). Transducers 1 and 2 control the composite plate at the same x_1 position and register the reflected wave packets at $x_2 = +d$ and $x_2 = -d$ respectively. Similarly, transducers 3 and 4 register the transmitted wave packets. Figure 21.4 shows the position of the UGW generation points, the positions of the 4 transducers as well as the captured normal displacements as a function of time. First, we will consider that the notch is positioned symmetrically with respect to the median plane of the plate, this positioning allows us not to have a mode conversion. We excite the A_0 mode at the beginning, this mode interacts with the defect, a part of the energy carried by the mode is reflected and the other is transmitted, while keeping the same nature of the mode which is A_0. Thereafter, we will consider that the notch is positioned asymmetrically with respect to the median plane of the plate. There, during the interaction of the mode with the defect, the reflected and transmitted wave packets will include an energy of the incident mode and another from the mode conversion phenomenon. Therefore, we will study the influence of the dimensions of the notch and its position on the energy balance and the mode conversion. First, we vary the defect parameters (height and length) and calculate the reflection and transmission coefficients. The error is controlled using the energy balance. The expressions of the reflection and transmission coefficients are given in (21.8) and (21.9):

$$R = \frac{\varphi_R}{\varphi_I} \tag{21.8}$$

$$T = \frac{\varphi_T}{\varphi_I} \tag{21.9}$$

Table 21.2 reports the values of the reflection and transmission coefficients as well as the error obtained. We choose a frequency lower than the cut-off frequency of the A_1 mode so that only the two modes A_0 and S_0 can coexist in the plate. The probed cases are chosen for different values of the length and height of the defect. If we increase the length of the defect (p↑), it means that the notch tends to a longitudinal

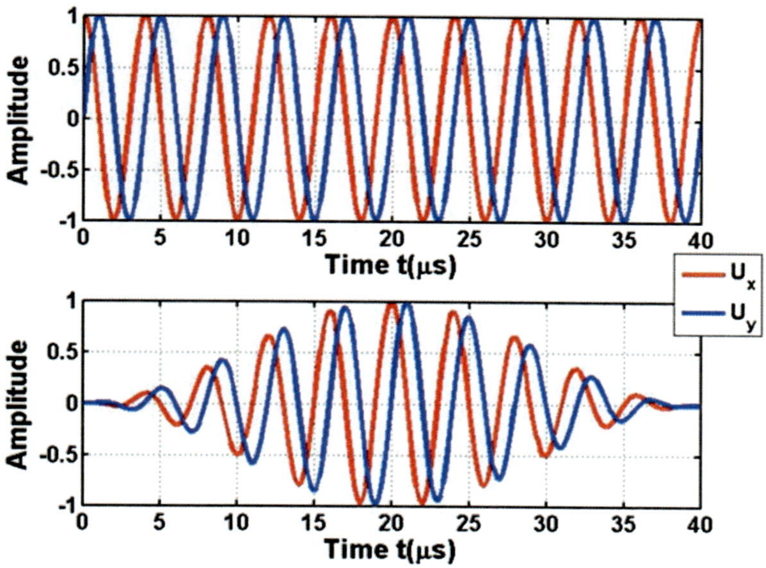

Fig. 21.4 Hanning window of the incident signal

crack, we notice that the reflected energy increases and the transmitted one decreases so that we always have an energy balance equal to 1. In the same way, an increase of the height (w↑ transverse crack) causes the increase of the reflection coefficient and the decrease of the transmission coefficient.

21.5.2 Asymmetrical Defect

Now, we study the case of a defect positioned asymmetrically with respect to the propagation axis x. Figure 21.5 illustrates the signals captured by the transducers positioned before and after the defect. Transducers 1 and 2 are used to acquire the temporal displacements at the top and bottom surfaces before the defect while transducers 3 and 4 capture the temporal displacements after the defect at the top and bottom of the plate. The signals collected by the two transducers 1 and 2 contain two wave packets, one corresponding to the incident wave and the other corresponding to the reflected wave. The signals picked up by the two transducers 3 and 4 correspond to the wave packets transmitted by the defect. The sensors 1, 2, 3 and 4 are arranged in vertical opposition two by two (1 with 3 and 2 with 4). With this arrangement, we can determine the position of the defect by determining the time of flight between the incident and reflected waves, knowing the speed of propagation of the incident wave, and we can also identify the nature of the modes contained in the reflected and transmitted wave packet, by exploiting the symmetry and antisymmetry properties of

Table 21.2 Reflection and transmission coefficients for different defect dimensions and for the two modes A_0 and S_0

Modes	Dimension of the defect	Reflection coefficient R	Transmission coefficient T	Energy balance R + T	Error (%)
A_0 mode ($k = 1431\ m^{-1}$, $f = 500$ kHz)	P = 2 mm, w = 1 mm	0.095	0.9093	1.0043	0.43
	P = 5 mm, w = 1 mm	0.7845	0.2364	1.0209	2.09
	P = 1 mm, w = 1 mm	0.0076	0.9927	1.0003	0.03
	P = 1 mm, w = 3 mm	0.8956	0.1134	1.0090	0.90
S_0 mode ($k = 621\ m^{-1}$, $f = 500$ kHz)	P = 2 mm, w = 1 mm	0.086	0.907	0.993	0.7
	P = 5 mm, w = 1 mm	0.8502	0.1672	1.0174	1.74
	P = 1 mm, w = 1 mm	0.0023	0.9966	0.9989	0.11
	P = 1 mm, w = 3 mm	0.8534	0.1542	1.0076	0.76

the displacements of the S_0 and A_0 modes. To identify the reflected and transmitted modes, several authors have used the Fourier transform of the captured signal or the wavelet transform [8, 19].

We notice in transducers 1 and 2 that we have two signals of reflections and in transducers 3 and 4 two transmitted signals. This is due to the phenomenon of mode conversion. We have generated the A_0 mode, the latter in contact with the defect is reflected with a given percentage that depends on the size of the defect. This reflected energy in our case is fragmented into two wave packets (Fig. 21.5). The same remark is seen for the transmitted part. In the present work, we propose a new method based on the symmetry of the propagative modes to characterize the mode conversion phenomenon. Effectively, a symmetric mode is characterized by a symmetric longitudinal displacement ($u_1(+d) * u_1(-d) > 0$) and an antisymmetric transverse displacement ($u_2(+d) * u_2(-d) < 0$). The antisymmetric mode has an inverse behavior ($u_1(+d) * u_1(-d) < 0$, $u_2(+d) * u_2(-d) > 0$). So, if we sum the signals picked up by transducers 1 + 2 and 3 + 4, then we can filter out the symmetric modes. If we subtract the signals captured by the transducers 1–2 and 3–4, then we can filter the antisymmetric modes. In this way, we have found that in the reflected signal, a part is proper to the A_0 signal and another part is from the A_0-S_0 conversion. The same deduction is obtained for the reflected wave packet.

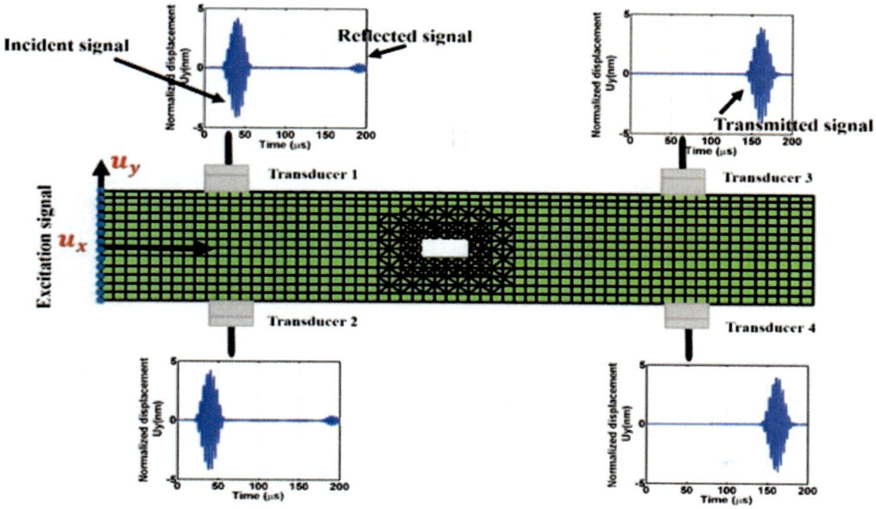

Fig. 21.5 Scan of the CFRTP composite plate by an A_0 mode (k = 1431 m^{-1}, f = 500 kHz) for a symmetric defect of dimension p = 2 mm, w = 1 mm

Table 21.3 Reflection and transmission coefficients calculated for a defect of dimensions p = 2 mm and w = 1.5 mm

Mode	Reflection coefficient R	Transmission coefficient T	Energy balance R + T	Error (%)
A_0 mode (k = 1431 m^{-1}, f = 500 kHz)	−0.8334 A_0 −0.023 S_0	−0.08 A_0 −0.048 S_0	0.9764	2.32

Then we calculate the energy coefficients of reflection and transmissions and we report our results in the Table 21.3.

In the reflected wave packet, we obtained two signals (due to the asymmetric position of the defect). The technique presented above (summation and subtraction) has allowed us to determine the reflected energy quantity that has experienced no change in the nature of the mode (A_0, R = 0.8334) and the one with change in the nature of the mode (A_0-S_0, R = 0.023). These values depend on the size of the defect. The same results are found for the transmitted wave packet (Fig. 21.6).

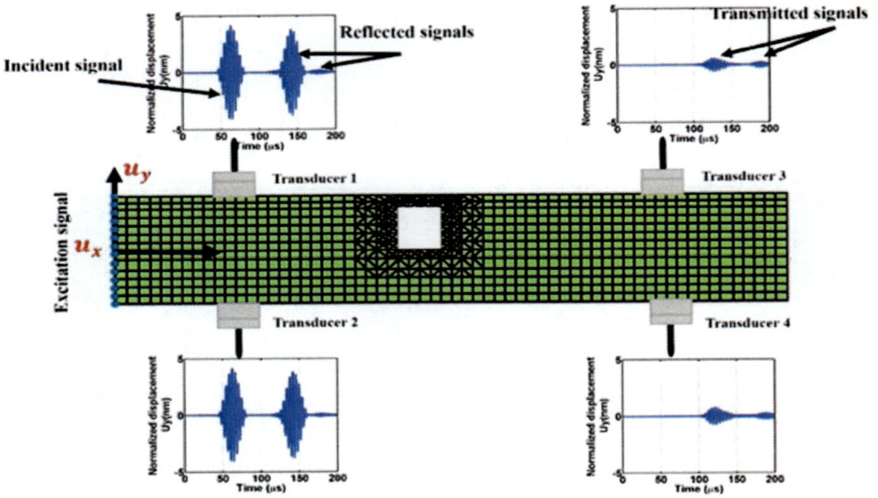

Fig. 21.6 Scan of the CFRTP composite plate by an A_0 mode ($k = 1431\ m^{-1}$, $f = 500$ kHz) for an asymmetric defect of dimension p = 2 mm, w = 1.5 mm

21.6 Conclusion

In this work, we have simulated by a finite element model the propagation of ultrasonic guide waves in a CFRTP composite plate used in the 3D printing process by FDM. This plate has an internal defect in the form of a rectangular notch. The choice of the different spatial and temporal meshing parameters was explained. The numerical model was based on the generation of analytical displacements weighted by a Hanning window on the left side of the composite plate. Transducers were positioned at $y = \pm h$ to capture the reflected and transmitted signals. The discussion of the accuracy of the obtained results was based on the establishment of an energy balance using the calculation of the reflection and transmission coefficients. The results obtained show the ability of UGW to inspect a piece with an internal defect. A study on the dimensions of the defect was implemented and we presented a new technique that can determine the nature of the modes during the mode conversion phenomenon based on the symmetry of the displacements captured at the surfaces.

References

1. Sefene, E.M.: State-of-the-art of selective laser melting process: a comprehensive review. J. Manuf. Syst. **63**, 250–274 (2022)
2. Gueche, Y.A., Sanchez-Ballester, N.M., Cailleaux, S., Bataille, B., Soulairol, I.: Selective laser sintering (SLS), a new chapter in the production of solid oral forms (SOFs) by 3D printing. Pharmaceutics **13**(8), 1212 (2021)

3. Aourik, O., Othmani, M., Saadouki, B., Abouzaid, K., Chouaf, A.: Fracture toughness of ABS additively manufactured by FDM process. J. Achievem. Mater. Manuf. Eng. **109**(2) (2021)
4. Ichihara, N., Ueda, M., Urushiyama, Y., Todoroki, A., Matsuzaki, R., Hirano, H.: Progressive damage simulation for a 3D-printed curvilinear continuous carbon fiber-reinforced thermoplastic based on continuum damage mechanics. Adv. Compos. Mater **29**(5), 459–474 (2020)
5. Blanco, I.: The use of composite materials in 3D printing. J. Compos. Sci. **4**(2), 42 (2020)
6. Tang, C., Tan, J.L., Wong, C.H.: A numerical investigation on the physical mechanisms of single track defects in selective laser melting. Int. J. Heat Mass Transf. **126**, 957–968 (2018)
7. Zhang, B., Li, Y., Bai, Q.: Defect formation mechanisms in selective laser melting: a review. Chin. J. Mech. Eng. **30**(3), 515–527 (2017)
8. Rhimini, H., El Allami, M., Sidki, M., Haddout, A., Benhadou, M.: Ultrasonic guided waves in tri-layer structure. Application to study the interaction of guided waves with hidden defect at low frequency. Техническая акустика **16**, 5 (2016)
9. De Luca, A.: Numerical simulation of the Lamb wave propagation in impacted CFRP laminate. Proc. Eng. **167**(2016), 109–115 (2016)
10. Chiua, W.K., Roseb, L.R.F., Nadarajaha, N.: Scattering of the fundamental anti-symmetric Lamb wave by a midplane edge delamination in a fiber-composite laminate. Proc. Eng. **188**(2017), 317–324 (2017)
11. Lee, J.M., Lee, C.J., Kim, B.M., Ko, D.C.: Design of prepreg compression molding for manufacturing of CFRTP B-pillar reinforcement with equivalent mechanical properties to existing steel part. Int. J. Precis. Eng. Manuf. **21**(3), 545–556 (2020)
12. Tian, X., Liu, T., Yang, C., et al.: Interface and performance of 3D printed continuous carbon fiber reinforced PLA composites. Compos. Part A Appl. Sci. Manuf. **88**, 198–205 (2016)
13. Nairn, J.A.: Matrix microcracking in composites. Polym. Matrix Compos. Elsevier Sci. **2**(13), 1–34 (2000)
14. Zitouni, I., Rhimini, H., Chouaf, A.: Modeling the propagation of ultrasonic guided waves in a composite plate by the spectral method. In: The 2nd International Conference on Mechanics, Materials and Energy (2022)
15. Zitouni, I., Rhimini, H., Chouaf, A.: Dispersion curves of composite graphite-epoxy by a hybrid analytic method. $15^{\text{éme}}$ édition du congrés marocain de mécanique (2022)
16. Quintanilla, F.H., Leckey, C.A.: Lebedev scheme for ultrasound simulation in composites. Ultrasonics **86**, 28–40 (2018)
17. Wandowski, T., Malinowski, P., Kudela, P., Ostachowicz, W.: Analysis of S0/A0 guided wave mode conversion phenomenon. In: Health Monitoring of Structural and Biological Systems XII, vol. 10600, pp. 501–515. SPIE (2018)
18. Hashamdar, H., Ibrahim, Z., Jameel, M.: Finite element analysis of nonlinear structures with Newmark method. Int. J. Phys. Sci. **6**(6), 1395–1403 (2011)
19. El Allami, M., Rhimini, H., Sidki, M.: Application of the complex mother wavelet Shan 1–1.5 processing to lamb modes signals in plates. Int. J. Sci. Res. (IJSR) **4**(1), 1849–1854 (2015). http://www.ijsr.net
20. Moser, F., Jacobs, L.J., Qu, J.: Modeling elastic wave propagation in waveguides with the finite element method. NDT and E Int. **32**(4), 225–234 (1999)
21. Nayfeh, A.H.: The general problem of elastic wave propagation in multilayered anisotropic media. J. Acoust. Soc. Am. **89**(4), 1521–1531 (1991)
22. Podder, P., Khan, T.Z., Khan, M.H., Rahman, M.M.: Comparative performance analysis of hamming, hanning and blackman window. Int. J. Comput. Appl. **96**(18) (2014)

Chapter 22
Effect of Density and Surface Quality on Fatigue Behavior of LPBF 316L Steel

Matias Jaskari, Atef Hamada, Pentti Karjalainen, and Antti Järvenpää

Abstract Metal additive manufacturing (AM) is a fabrication method to effectively produce optimized parts for various applications. Laser powder bed fusion (LPBF) is currently one of the most common AM methods. For efficiency, shorter throughput times are searched by increasing the deposited layer thickness for instance. Faster processing can, however, lead to formation of defects such as voids in the structure, which in turn may result in impaired fatigue resistance, even in the instance that the static properties are unaffected. The effect of the material density and surface quality on the fatigue behavior of an AISI 316L austenitic stainless steel fabricated by the LPBF method with two different layer thicknesses (40 and 80 μm). Some samples were tensile tested, while an extensive study was performed by axial tension-compression load-controlled fatigue testing (the loading frequency 100 Hz). Samples were tested either in the as-built or electropolished condition until failure or to 10^7 cycles. Surface topography and roughness were measured using a focal imaging laser microscope and fracture surfaces were examined by a scanning electron microscope. It was found that the layer thickness during deposition affected the material density and amount of defects in the structures. However, the tensile strength was not affected significantly by these differences. The fatigue limit of both the as-built 40 um and 80 um structures as well as electropolished 80 μm structure remained below 100 MPa at 107 cycles as a result from both near surface defects or rough surface. However, the electropolished 40 μm structure exhibited an excellent fatigue limit of 240 MPa. Therefore, both the defect density and surface roughness must be controlled in order to enhance significantly the fatigue resistance of LPBF 316L.

Keywords Additive manufacturing · Laser powder-bed fusion · 316L · Defects · Surface quality · Fatigue resistance

---

M. Jaskari (✉) · A. Hamada · A. Järvenpää
FMT-research group, Kerttu Saalasti Institute, University of Oulu, 85500 Nivala, Finland
e-mail: matias.jaskari@oulu.fi

P. Karjalainen
Centre of Advanced Steels Research, University of Oulu, 90570 Oulu, Finland

K. Zarbane and Z. Beidouri (eds.), *Proceedings of CASICAM 2022*, Springer Tracts in Additive Manufacturing, https://doi.org/10.1007/978-3-031-32927-2_22

22.1 Introduction

Additive manufacturing (AM) is currently one of the prominently emerging technologies, which is widely used in prototyping industry. Compared to conventional fabrication methods, AM allows more freedom in component design. For attaining the best efficiency, the whole fabrication process should be designed from AM point-of-view [1]. Even though the process is relatively slow, its efficiency and deposition rate can be increased for instance by increasing the deposited layer thickness. The AM fabricated products are, however, susceptible for quality variation, for even very slight changes in the process may result in increased density of defects, such as pores [2]. It is stated that there are over 100 processing parameters in AM, which can affect the outcome [3], and the risk gets higher with increasing the production rate. It is shown that static properties equal to those of fully dense material can be obtained even with some amount of porosity [4–6], but the detrimental effect of porosity appears in the dynamically loaded AM structures, as reported in numerous papers. For example, Solberg et al. [7] studied the effect of porosity and surface roughness on bending fatigue behavior of a 316L austenitic stainless steel, and they realized that the defects governed the fatigue behavior. Jaskari et al. [8] showed that the fatigue limit of AM 316L can be as low as 100 MPa in flexural bending fatigue. Kumar et al. [9] found out in their study of fatigue behavior of binder jet printed 316L that sharp lack of fusion type defects was the main cause for fatigue crack initiation and the columnar dendritic microstructure allowed fast crack propagation.

In addition to defects, also the surface quality is known to have significant influence on fatigue resistance. The surface is inherently rough in as-built AM products that tends to reduce the fatigue life, e.g., shown by Liang et al. [10] and Solber et al. [7]. Polishing and machining can improve the fatigue strength to some extent.

In this study, the effect of achieved material density, i.e., amount of defects, and surface quality on the fatigue behavior of a LPBF AISI 316L fabricated using two different layer thicknesses was studied by axial tension-compression load-controlled dynamic testing. The aim is to demonstrate that even a low amount of defects can impair the fatigue limit of LPBF 316L in the as-built condition, whereas the surface electropolishing together with low amount of defects can significantly improve the fatigue resistance.

22.2 Material and Methods

The material tested was an AISI 316L type austenitic stainless steel, the powder supplied by EOS GmbH company (Turku, Finland). The chemical composition, measured using a glow discharge optical emission spectroscope (Spectruma GDA750), was as follows (in wt.-%): 0.017C, 1.5Mn, 0.34Si, 18.0Cr, 12.8Ni, 3.0Mo, 0.24Cu, Fe Bal.

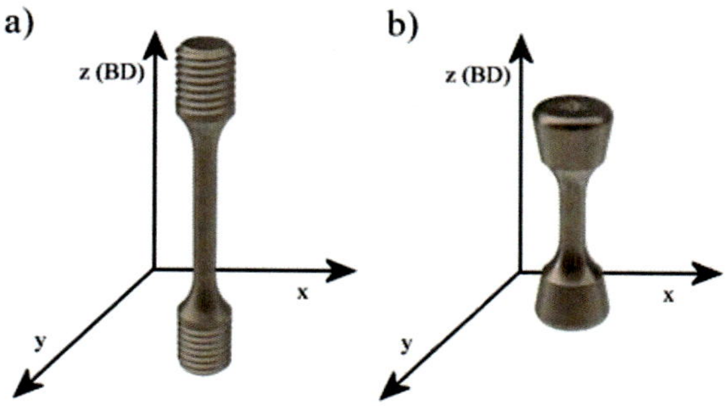

Fig. 22.1 Schematic illustration of the deposited tensile and fatigue samples and their built direction

Specimens were additively manufactured (AMed) using an EOS M 290 equipment and EOS standard parameters (parameter set 316L-040-080-Core M291). Two build layer thicknesses were chosen for the study, 40 and 80 μm, while the corresponding heat inputs were 57.7 and 40.0 J/mm^3 for the structures, respectively. Conical, round test samples were AMed "in-shapeâŁž in vertical orientation, as illustrated in Fig. 22.1.

Quasistatic tensile test was conducted using a Zwick Z100 testing equipment according to ISO 6892-1 for threaded round tensile samples at the strain rate of 0.008 s^{-1} after yielding. The test length, gauge length and diameter were 25, 20 and 5 mm, respectively. Strains during the tensile tests were recorded with a MacroXtens extensometer. At least 3 samples were tested for each condition for statistics.

For high cycle fatigue testing, some samples were tested in the as-built condition and the others were carefully machined and ground to 600 Grit surface quality, and then electropolished using a commercial electropolishing equipment (DLyte 100HF+). Polishing was continued for several hours until mirror finish was achieved. The test length and diameter were 12 and 6 mm, respectively. Experiments were carried out in total load control using a resonator, where the achieved average loading frequency 100 Hz. The load ratio was −1, and testing was continued until failure, or until the cut-off number of 10^7 cycles.

Microstructural examinations and surface roughness measurements were conducted using a Keyence VK-200 confocal imaging laser optical microscope (LOM). The fracture surfaces were examined using JEOL JSM-7900F field emission scanning electron microscope (FESEM), using 5 kV acceleration voltage. Densities were also determined from polished cross-sections (XZ-plane) with the same equipment. The total measured cross-sectional area was 15 mm^2 for both the structures.

22.3 Results and Discussion

22.3.1 Initial Microstructure and Surface Quality After Additive Manufacturing

Densities of the two material variants due to different build layer thickness were measured from the polished cross-sections of the printed samples. The pore sizes are shown as cumulative distribution plots (Fig. 22.2a) and on cross-sections (Fig. 22.2b, c). It was noticed that the 80 μm structure contains a higher number of defects (N = 355/15 mm^2) compared to that in the 40 μm (N = 25/15 mm^2) structure. The measured total densities were 99.80 and 99.98% for the 80 and 40 μm structures, respectively. As seen from Fig. 22.2b and c, most of the pores reside near the surface.

The microstructures of both the structures seen in Fig. 22.2b and c display the typical melt pool structure solidified during the fabrication process. Because of the remelting phenomenon, some epitaxial grain growth is present in both the structures. Between the used layer thickness, the main difference in the microstructure was in the melt pool depth, which was 150 and 240 μm for the 40 and 80 μm structures, respectively.

Results of surface roughness measurements are listed in Table 22.1. Measured R_a values were equal 9.6 μm for the 40 and 80 μm structures, and the R_z value was 65 and 75 μm for these structures, respectively. Topography profiles plotted in Fig. 22.3 reveal the similarity of the surface features, although the 80 μm structure shows somewhat higher peaks than the 40 μm structure does. After electropolishing the measured R_a and R_z values were 0.09 and 1.0 μm, respectively, for both the structures.

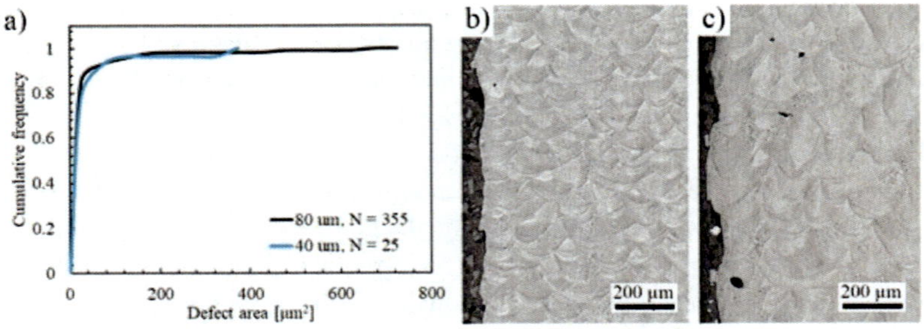

Fig. 22.2 **a** Defect distribution plot of the 40 and 80 μm structures, and cross-sectional LOM images showing the edge porosity in the **b** 40 μm and **c** 80 μm structures. N in **a** depicts the total number of defects measured

Table 22.1 Surface roughness values for the structures in as-built and electropolished conditions

As-built			Electropolished	
Layer thickness [μm]	R_a [μm]	R_z [μm]	R_a [μm]	R_z [μm]
40	9.6	65	0.09	1.0
80	9.6	75	0.09	1.0

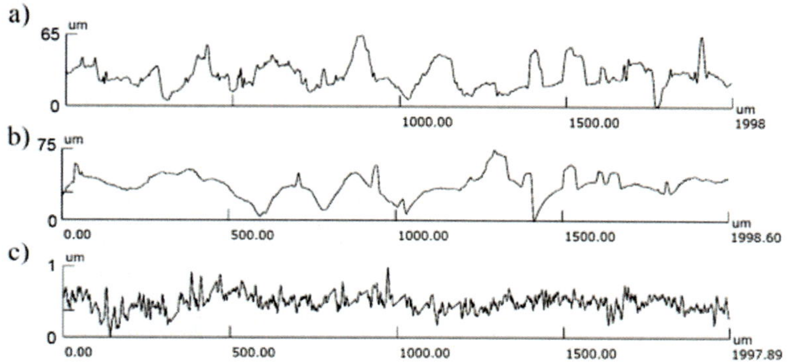

Fig. 22.3 Surface topography profiles of as-built samples fabricated with **a** 40 and **b** 80 μm layer thickness. **c** The average profile of the ground and electropolished sample

22.3.2 Tensile Properties

Tensile stress-strain curves are plotted in Fig. 22.4, showing that the tensile strength (TS) is practically equal 628 and 637 MPa for the 40 and 80 μm structures, respectively. A minor difference can be seen in the yield strength (YS) and a more distinct one in the elongation. For the 40 μm structure, the YS was measured to be 515 MPa, which is 28 MPa higher than that of the 80 μm structure. For the 40 μm structure, the elongation is good, the uniform elongation (UE) and total elongation (TE) being 24 and 47%, respectively, but they were inferior for the 80 μm structure, the UE and TE being 21 and 39%, respectively. The tensile test results evidence that the number of flaws present in these structures do not affect the static strength considerably but affects the ductility. This is agreement with observations of Li et al. [5], where the tensile ductility was shown to be sensitive to flaws despite the high material density. Further, increasing the energy density is seen to improve the ductility as decreasing the number of defects, which is consistent with the findings of Jaskari et al. [4].

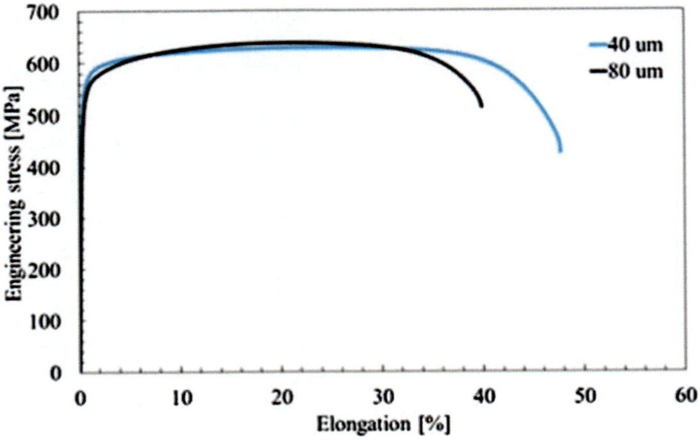

Fig. 22.4 Tensile test curves of LPBF 316L manufactured with 40 or 80 μm deposited layer thickness

22.3.3 Fatigue Behavior

The fatigue resistance of the studied structures can be assessed from Fig. 22.5, where first it is seen that the high-cycle fatigue life is practically equal among the as-built structures. In both the structures, the estimate curve for 50% fatigue limit decreases well below 150 MPa when nearing 107 cycles, and the calculated fatigue limit (FAT) was determined to be 72 and 98 MPa for the as-built 80 and 40 μm structures, respectively. Thus, the results evidence that decreasing defect density cannot improve the fatigue resistance in the instance of the as-built surface condition. In fair agreement, Liang et al. [10] reported the fatigue limit of 92.5 MPa at 2×10^6 cycles for an as-built LPBF 316L.

Secondly, we can notice that even in the electropolished condition, the fatigue life of the 80 μm structure was low, the fatigue limit 90 MPa. Thus, only the increase of 18 MPa was attained by the improved surface quality in this structure. The decrease in the fatigue limit agrees with the literature data, as the defects and surface roughness are known to influence on the cyclic behavior [7, 11, 12]. Sharp surface features and lack of fusion defects cause stress concentrations, which in turn cause local excess plastic straining and induce fatigue crack initiation. However, very drastic improvement in the fatigue resistance by the electropolished surface quality was obtained for the 40 μm structure, which only contained few defects. The fatigue limit at 10^7 cycles is 240 MPa, which value is very high, even higher than for wrought structures, e.g., 184 MPa given by Huang et al. [13]. Thus, the surface quality has a distinct beneficial contribution only in the structure with a low defect density, i.e., presently in the 40 μm structure but not in the 80 μm structure.

Concluding from the fatigue test data, it seems that only by decreasing enough both the internal defects and surface roughness, the cyclic behavior in the high-

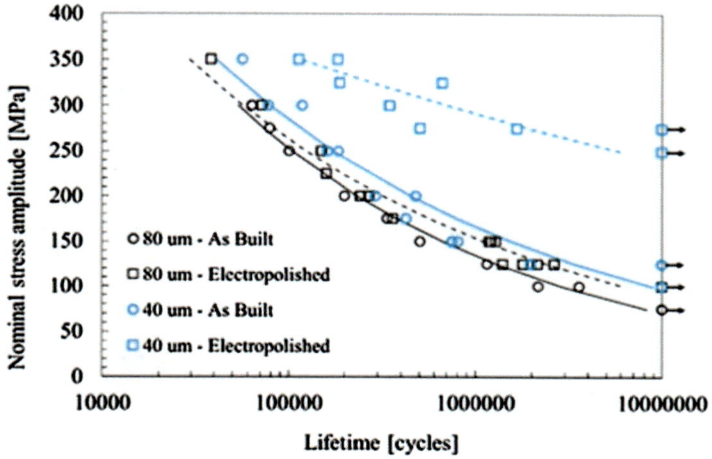

Fig. 22.5 S-N curves of LPBF 316L manufactured with 40 and 80 μm layer thicknesses. The solid lines depict the 50% fatigue limit estimates of as-built samples, whereas the dashed lines depict the estimates for the electropolished samples

cycle fatigue regime can be significantly enhanced. In the low-cycle fatigue regime, when the stress amplitudes are considerably higher, the lifetime estimate curves show convergence moving towards the YS of the structures, meaning that the static strength is the dominant feature affecting the fatigue resistance.

22.3.4 Crack Initiation

Crack initiation in the studied structures were examined from the fracture surfaces of fatigued samples. It was found out that the main reason for the initiation was either the features of the rough surface or the defects residing on or near the surface, but no fracture starting from internal defects were detected in this study. This indicates that in all instances, more severe crack initiation sites existed at surface region than inside the samples. As shown as an example in Fig. 22.6a, the rough surface and the partially melted particles on the surface of an as-built 40 μm sample, fatigued at an 150 MPa amplitude, form an effective site for crack initiation.

The insert in Fig. 22.6b reveals how the crack has initiated between a partially melted particle and surface in the as-built 40 μm structure sample, and then grown deeper into the structure, as pointed out by the blue arrows. Crack has propagated in several directions until the threshold, being shown as a white dashed line in Fig. 22.6a. In electropolished condition a typical failure site for the 40 μm structure is shown in Fig. 22.6c. There fracture has initiated from the surface, either from a slip band or a small surface flaw, such as a polishing pit.

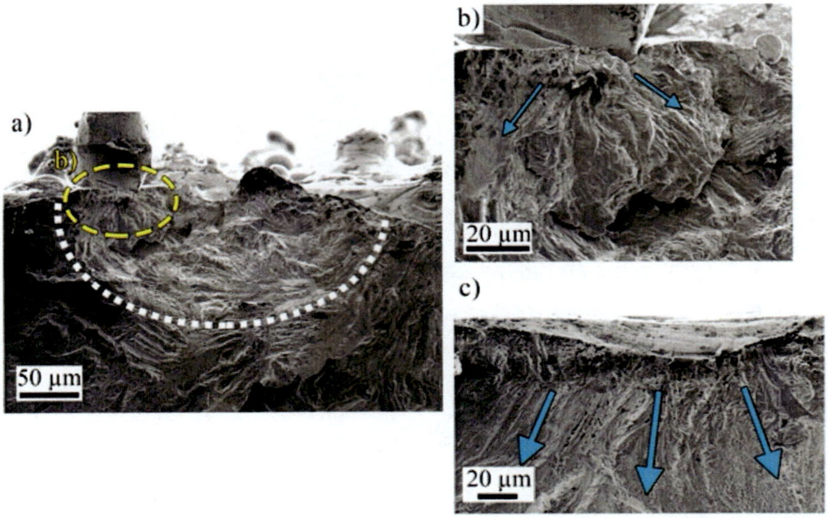

Fig. 22.6 Fatigue crack initiation site (**a**) in the as-built 40 µm structure after fatigue at 150 MPa amplitude until failure. Insert in **b** shows an initiation point at a partially melted particle adjacent to the surface. **c** Surface failure initiation in the electropolished 40 µm structure, fatigued at 275 MPa

In the as-built 80 µm structure the crack initiation occurred similarly. As shown in Fig. 22.7a, a sharp notch on the surface has initiated the fatigue failure. In the electropolished condition, however, the defects have a significant effect on crack initiation. As shown in Fig. 22.7b, a defect residing adjacent to the surface has initiated the failure, and then the crack has propagated in the same manner as in the 40 µm structure.

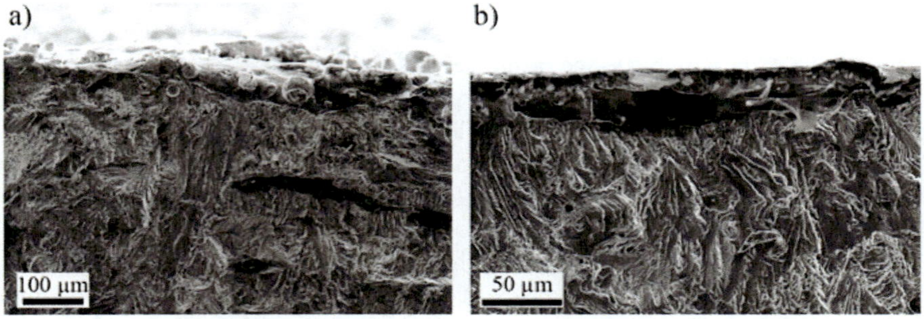

Fig. 22.7 **a** Fatigue crack initiation from crevice in the as-built 80 m sample fatigued at 175 MPa amplitude and **b** from lack of fusion defect adjacent to surface of an electropolished 80 µm sample fatigued at 200 MPa amplitude

According to this data and the related literature, it is obvious that both the surface flaws as well as the defects near the surface have their effect on the fatigue behavior of LPBF 316L, and the behavior is dictated by the stress concentrating features. Sharp edges, such as crevices, act as pre-cracks on the surface and can cause premature failure. Further, the effect of residual stresses needs to be considered, as the AM materials are known to exhibit even relatively high stresses after fabrication. Tensile residual stresses can be very detrimental from fatigue point-of-view. When combined with the effect of local stress concentrators, such as pores, the effect can be high enough to induce a crack initiation even at low stress amplitudes. Residual stresses can be removed efficiently by heat treatment, as commonly applied for AM materials. This would also enhance the fatigue properties, as proposed e.g., by Blinn et al. [14]. Also, removing surface material by polishing or machining can affect surface stress state in addition to the topography. This was not studied in the present work, however.

22.4 Conclusions

In this study, the effect of defects and surface roughness on fatigue performance of LPBF 316L was assessed by uniaxial compression-tension dynamic testing. Tests were conducted for the as-built and electropolished samples under total load control, and the main findings are as follows:

- Increasing the productivity by increasing the layer thickness led to decrease in measured density, i.e., the number of defects increased.
- The as-built surfaces were relatively rough and contained sharp crevices and partially melted particles. Increasing the layer thickness resulted in the formation of slightly rougher surface, which was seen as an increase in the R_z value.
- The tensile strength was almost equal in the studied structures despite different layer thickness used in their manufacturing.
- High-cycle fatigue test results indicated that in the instance of rough surface (as-built condition), defect density had a secondary influence only, and the fatigue limit was low, below 100 MPa.
- Electropolishing improved significantly the fatigue life of the 40 μm structure (with low defect density), and the fatigue limit achieved was very high, about 240 MPa at 10^7 cycles.
- Only by reducing both defects and rough surface features, the fatigue behavior can be enhanced considerably.

References

1. Mäntyjärvi, K., Iso-Junno, T., Niemi, H., Mäkikangas, J.: Design for additive manufacturing in extended DFMA process. Key Eng. Mater. **786**, 342–347 (2018). https://doi.org/10.4028/www.scientific.net/kem.786.342
2. Liverani, E., Toschi, S., Ceschini, L., Fortunato, A.: Effect of selective laser melting (SLM) process parameters on microstructure and mechanical properties of 316L austenitic stainless steel. J. Mater. Process. Technol. **249**, 255–263 (2017). https://doi.org/10.1016/j.jmatprotec.2017.05.042
3. Oliveira, J.P., LaLonde, A.D., Ma, J.: Processing parameters in laser powder bed fusion metal additive manufacturing. Mater. Des. **193**, 108762 (2020). https://doi.org/10.1016/J.MATDES.2020.108762
4. Jaskari, M., Ghosh, S., Miettunen, I., Karjalainen, P., Järvenpää, A.: Tensile properties and deformation of AISI 316L additively manufactured with various energy densities. Materials **14**(19), 5809 (2021). https://doi.org/10.3390/MA14195809
5. Li, Z., Voisin, T., McKeown, J.T., Ye, J., Braun, T., Kamath, C., Wang, Y.M.: Tensile properties, strain rate sensitivity, and activation volume of additively manufactured 316L stainless steels. Int. J. Plast. **120**, 395–410 (2019). https://doi.org/10.1016/j.ijplas.2019.05.009
6. Wang, X., Muñiz-Lerma, J.A., Attarian Shandiz, M., Sanchez-Mata, O., Brochu, M.: Crystallographic-orientation-dependent tensile behaviours of stainless steel 316L fabricated by laser powder bed fusion. Mater. Sci. Eng. A **766**, 138395 (2019). https://doi.org/10.1016/j.msea.2019.138395
7. Solberg, K., Guan, S., Razavi, S.M.J., Welo, T., Chan, K.C., Berto, F.: Fatigue of additively manufactured 316L stainless steel: The influence of porosity and surface roughness. Fatigue Fract. Eng. Mater. Struct. **42**(9), 2043–2052 (2019). https://doi.org/10.1111/ffe.13077
8. Jaskari, M., Mäkikangas, J., Järvenpää, A., Mäntyjärvi, K., Karjalainen, P.: Effect of high porosity on bending fatigue properties of 3D printed AISI 316L steel. Procedia Manuf. **36**, 33–41 (2019). https://doi.org/10.1016/j.promfg.2019.08.006
9. Kumar, P., Jayaraj, R., Suryawanshi, J., Satwik, U.R., McKinnell, J., Ramamurty, U.: Fatigue strength of additively manufactured 316L austenitic stainless steel. Acta Mater. **199**, 225–239 (2020). https://doi.org/10.1016/J.ACTAMAT.2020.08.033
10. Liang, X., Hor, A., Robert, C., Salem, M., Lin, F., Morel, F.: High cycle fatigue behavior of 316L steel fabricated by laser powder bed fusion: Effects of surface defect and loading mode. Int. J. Fatigue **160**, 106843 (2022). https://doi.org/10.1016/J.IJFATIGUE.2022.106843
11. Jeon, J.M., Park, J.M., Yu, J.H., Kim, J.G., Seong, Y., Park, S.H., Kim, H.S.: Effects of microstructure and internal defects on mechanical anisotropy and asymmetry of selective laser-melted 316L austenitic stainless steel. Mater. Sci. Eng.: A **763**, 138152 (2019). https://doi.org/10.1016/J.MSEA.2019.138152
12. Hatami, S., Ma, T., Vuoristo, T., Bertilsson, J., Lyckfeldt, O.: Fatigue strength of 316 L stainless steel manufactured by selective laser melting. J. Mater. Eng. Perform. **29**(5), 3183–3194 (2020). https://doi.org/10.1007/S11665-020-04859-X/FIGURES/16
13. Huang, J.Y., Yen, J.J., Jeng, S.L., Chen, C.Y., Kuo, R.C.: High-cycle fatigue behavior of Type 316L stainless steel. Mater. Trans. **47**(2), 409–417 (2006). https://doi.org/10.2320/MATERTRANS.47.409
14. Blinn, B., Krebs, F., Ley, M., Teutsch, R., Beck, T.: Determination of the influence of a stress-relief heat treatment and additively manufactured surface on the fatigue behavior of selectively laser melted AISI 316L by using efficient short-time procedures. Int. J. Fatigue 131 (2020). https://doi.org/10.1016/J.IJFATIGUE.2019.105301

Part VI
Potential of Additive Manufacturing

Chapter 23
Comparative Study of Dimensional and Surface Specification: Additive Manufacturing and Injection Molding

Mohammed Lamrhari, Ali Allouch, and Mohamed Elghadoui

Abstract In the last decade, Additive Manufacturing (AM) has taken an important part in the production of prototypes and finished parts. Given the functional aspects of parts and the new applications of components produced by additive manufacturing, it is necessary to study the dimensional specifications and surface roughness of the printed parts according to the production parameters. In this study, samples were manufactured by Injection Molding (IM) and the additive manufacturing process, to compare the impact of the two processes on the functional aspect of the part, a comparative study on the dimensional, geometrical and surface specifications was used, the samples manufactured according to the axes X, Y and Z were realized both of Stereo lithography process and injection molding, then a series of controls of the samples were taken. The comparison of the results confirmed the effects of the modification of the production processes. For the samples examined, the IM process gives a good quality of the surface compared to the AM process; but the dimensional quality is better for the AM process. The documented results are beneficial for future designs and the optimization of production processes.

Keywords Injection molding · Additive manufacturing · Stereo lithography · Surface roughness · Dimensional · Specifications

M. Lamrhari (✉)
ENSAM, MEKNES, Moulay Ismail University of Meknès, Meknès, Morocco
e-mail: lamrha@yahoo.fr

A. Allouch
ESEF, Ibn Tofail University, Kenitra, Morocco

M. Elghadoui
ENSAM, Casablanca, Hassan II University, Casablanca, Morocco

K. Zarbane and Z. Beidouri (eds.), *Proceedings of CASICAM 2022*, Springer Tracts in Additive Manufacturing, https://doi.org/10.1007/978-3-031-32927-2_23

23.1 Introduction

Due to the continued demand for more complex design and production cost optimization, Additive Manufacturing (AM), has grown exponentially in terms of application and development. It is based on the creation of parts by adding material layer by layer from a CAO model of the part. The advantages of AM over conventional manufacturing are very obvious in terms of energy saving, waste reduction, shortening of the manufacturing cycle, minimization of the costs of small production and the manufacture of geometrically complex parts. Different fields such as aeronautics, civil engineering…, [1] has been used this technique. In addition, during this decade, the community of research continuously presents new achievements on materials [2], methodologies [3], and mechanical properties of AM parts [4]. There have been significant developments in the field of AM and it has expanded in terms of new and wide range of materials, often revealing improved mechanical properties. Different AM processes have been used to print from small to large size such as home [5]. In 1986, the first functional 3D printer has Stereo lithography (SLA) to manufacture plastic 3D parts with complex geometry. Since then, several subsequent technologies have been developed, Selective Laser Sintering (SLS), Fused Deposition Material (FDM), Material Jetting (MJ) etc. Several recent articles developments and advances in additive manufacturing techniques and related materials [6]. The technology used in this work is Stereo lithography, it produces parts by scanning an ultraviolet laser on a liquid causing photo polymerization which transforms the liquid resin into a solid. The part is thus built on a platform inside a tank of liquid. Moving the platform to a position below the liquid surface equivalent to one layer thickness produces the first layer of the part. Sine as SLA parts require support structures, they often require post-processing which opens up opportunities for new part features. For similar materials, the closest processes for obtaining parts are plastic processing processes. Among them, injection molding (IM) is a process that consists of injecting molten plastic into a mold to shape the piece. They are considered less expensive than other processes, such as 3D printing. Different types of thermoplastics are widely used in the manufacture of products using an injection molding process. With recent advancements in many elements of the injection molding process, various researchers have study the mechanic and geometric characteristics of injection molded parts. For polypropylene materials, the Taguchi desirability function approach is used to optimize the injection molding parameters [7]. For high density polyethylene, the lowest shrinkage is obtained by optimizing the injection molding parameters; the most important factor is the cooling time [8]. Kitayama et al. improved plastic injection molding parameters to increase part accuracy [9]. The finite element analysis platform is used to study the mechanical characteristics of thermoplastics manufactured by injection moulding and 3D printing [10]. They reported that additive manufacturing could be an excellent alternative to injection moulding in the production of functional parts of manufactured products. In this work, a new approach is presented on the factors influencing the dimensional specifications and surface properties of the produced parts in order to obtain how the parts can be manufactured (printed, injected). The methodology is based on an experimental procedure. The objective of this study is to reduce the gap between dimensional specifications and surface roughness of the two processes, AM and IM of parts by optimizing the production parameters of the two processes.

23.2 Experimental Procedure

23.2.1 Sample Geometry

The geometry and dimensions of the samples (Fig. 23.1) were designed in accordance with standard NF EN ISO 527-5A (Table 23.1). The CAD design was created on CATIA V5.

A polycarbonate material, PC, was used to fabricate the samples, a commercial quality of thermoplastic material according to the European region LEXAN™ 101 (Table 23.2) gives the physical properties of the material.

23.2.2 Preparation of Samples by Injection Molding IM

The samples were fabricated using a horizontal plastic injection machine (Fig. 23.2).

The experiments were conducted on a horizontal injection press and the mold was made of steel. The real mold scheme is presented in Fig. 23.3.

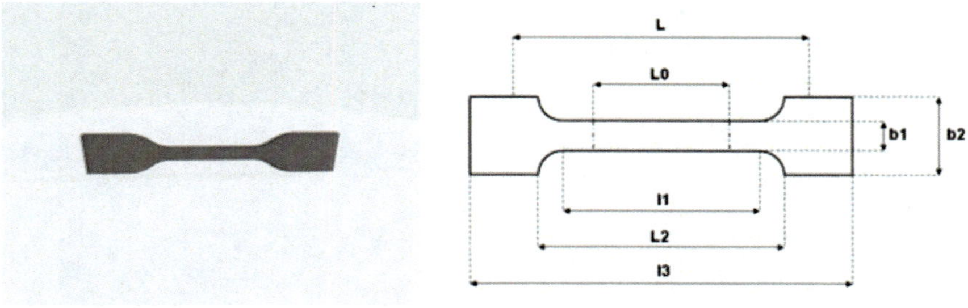

Fig. 23.1 Specimen geometry

Table 23.1 Dimensional sample geometry

Norm	Sample type	L3	L2	b1	b2	L0	L	Form
NF EN ISO 527	5A	75	25	4	12.5	20	50	Haltère

Table 23.2 Physical properties of the polycarbonate material

Properties	Values	Units	Test methods
Density	1.12	g/cm^3	ISO 1183
Water absorption (23 °C/saturated)	0.35	%	ISO 62-1
Tensile stress, yield	63	MPA	ISO 527
Tensile stress, break,	70	MPA	ISO 527
Elongation at Yield	6	%	ISO 527
Elongation at Break	12	%	ISO 527
Tensile modulus, 1 mm/min	2350	MPA	ISO 527
Thermal conductivity	0.2	W/m-°C	Iso 8302
Melt temperature	280–320	°C	

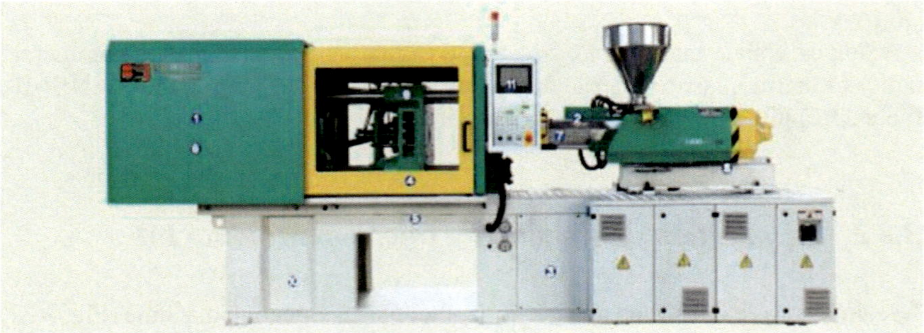

Fig. 23.2 Horizontal injection machine

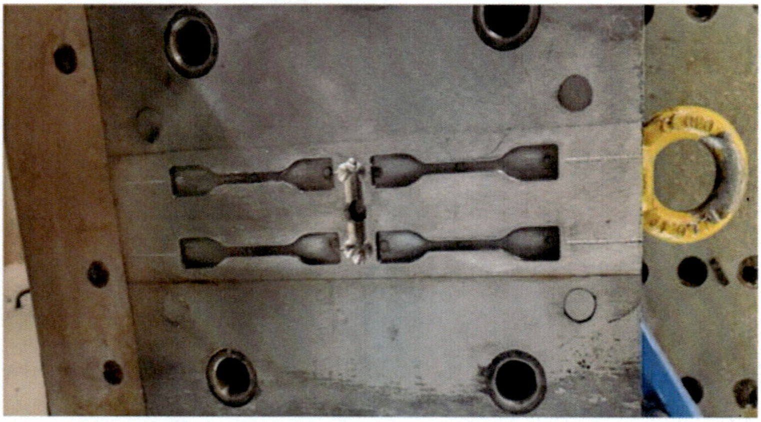

Fig. 23.3 Mold used in the fabrication of samples

23.2.3 Elaboration of Samples by Additive Manufacturing

The process used is SLA, with standard ISO/ASTM 52,900; it is classified as Vat Photo polymerization (VP) processes. SLA is a rapid prototyping technology that consists of manufacturing parts by adding material from digital files. The materials used are photosensitive liquid resins which solidify by means of ultra-violet (UV) light.

The successive layers created will reproduce the shape of your part, with a quality almost equivalent to plastic injection. The part printed in stereolithography requires post-processing. It must first be cleaned with a solvent to remove the uncured resin residue and then baked to complete the polymerization in order to increase the strength of the material as much as possible. The specification of the 3D printer used in this study is given in Table 23.3.

Indeed, the resolution and the nominal precision of the catalogs of the machines do not make it possible to achieve the characteristics with these precision values. Because the size and intensity of the laser spot, the optical properties of the resin and its hardening profile, build orientation, and hatch style vary the accuracy of parts. For these reasons, it is necessary to study the dimensional and surface roughness accuracy. The samples by Additive Manufacturing used in this study are given in Fig. 23.4.

Table 23.3 Specification of the 3D printer

Tehnology	Company	Device	Nominal XY resolution	Z-axis resolution	Accuracy
SLA	3DSystems	ProJet 6000 HD	6.35 μm	50 μm	0.025–0.05 mm

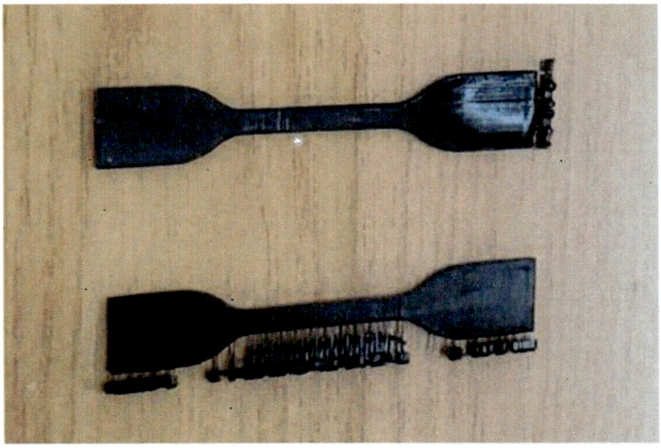

Fig. 23.4 Samples of additive manufacturing

23.3 Results and Discussion

23.3.1 *Dimensional Accuracy*

In this study, the dimensional accuracy of the 3D printed SLA and IM samples is evaluated, the concordance between the fabricated and designed dimension of the fabricated parts is evaluated. In this work, we examined the dimensional accuracy characteristics of a part made with two different technic. The accuracy of a part is evaluated by its dimensional tolerance. The dimensional variations in the current study were addressed for length (L3), width (b2) and thickness h, as shown in Fig. 23.3. The variation in dimensions considerably influences the precision of the parts. The dimensional accuracy of 3D printed parts manufactured by IM is assessed using a profile projector. Measurements are taken for five directions of five random locations for each feature on samples printed (SLA) and fabricated by IM technique Table 23.4. Then, the measurement results are compared to their designed specifications. To check the influence of the technique on the precision of the parts, the analysis of variance ANOVA technique is used; if the p value of ANOVA is less than 0.05 then the production technique has an effect on the precision of the part. In addition the value of the variance of the measured dimensions gives an idea of the tolerance of the dimension.

For the evaluation of the dimensional accuracy, the reference samples were produced with two different techniques: SLA and IM. Table 4 summarizes the measurement results. The results show that the width and height dimensions of the samples manufactured by SLA are very similar to the dimensions of the designed part; however the length is offset from the length of the designed part. In addition, the length, width and height dimensions of the samples manufactured by IM are long away to the dimensions of the designed part. The ANOVA results confirmed that the fabrication technique has a significant effect on the dimensional accuracy. As shown in Table 23.5, the p-value is less than 0.05 for all dimensions, which means that the measured dimensions vary wildly from manufacturing technique.

Table 23.4 Results of the dimensional measurements for SLA and IM

Measures	SLA			IM		
	Dimension (mm)			Dimension (mm)		
	$L_3 = 75$	$b_2 = 12.5$	$h = 2$	$L_3 = 75$	$b_2 = 12.5$	$h = 2$
Mean	74,6511	12,505	2,022	74,43,968	12,387	2,180
Standard deviation	0,03,835	0,022	0,025,184	0,018,204	0,021,023	0,061,475

Table 23.5 ANOVA results

ANOVA	Dimension (mm)			
	L3 = 75	b2 = 12.5	b1 = 4	h = 2
p-value	4,2399E-29	3,4622E-24	8,4753E-10	4,4778E-16

23.3.2 Roughness Surface Study

In order to study the roughness surfaces of the samples fabricated by the two techniques, a 3D laser microscope (Alti Surf 500) was used. A measurement of a 3 × 3 mm^2 scanning area was made at three random positions of each sample. Figures 23.5 and 23.6 show the surface topography of the samples printed by SLA and fabricated by IM. We find that the roughness of surfaces manufactured by plastic injection is much reduced than the surfaces of parts printed by SLA. Visual inspection of the samples proved that certain defects such as the positions of the ejectors and the burrs of the joint plane for the samples manufactured by IM, and the supports for the parts manufactured by AM. According to the ISO 4287 standard, the microscopic examination showed that the average arithmetic roughness of the surfaces of the samples was reduced from 6.91 μm of the surfaces printed by SLA to 0.885 μm for the surfaces elaborated by IM.

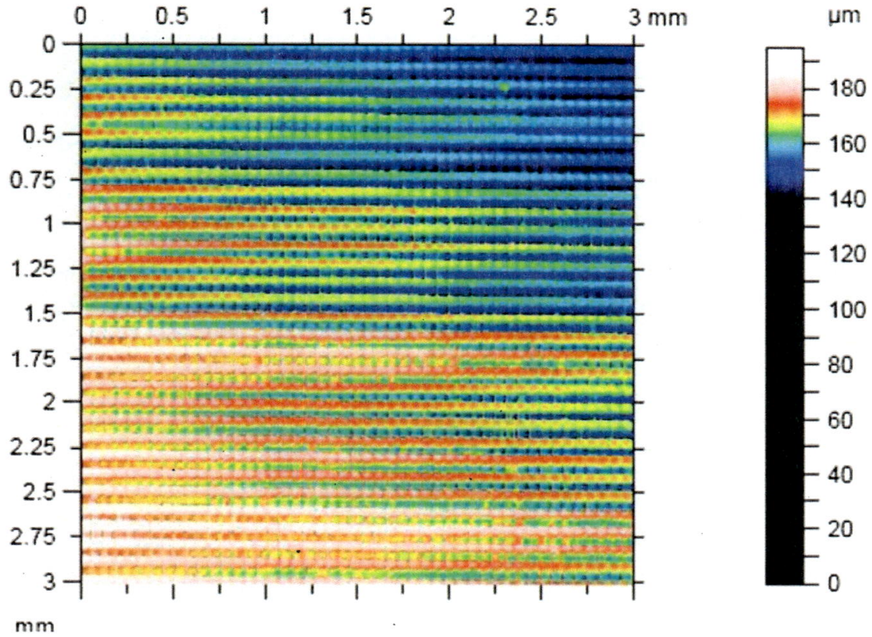

Fig. 23.5 Topology of the surface printed by SLA

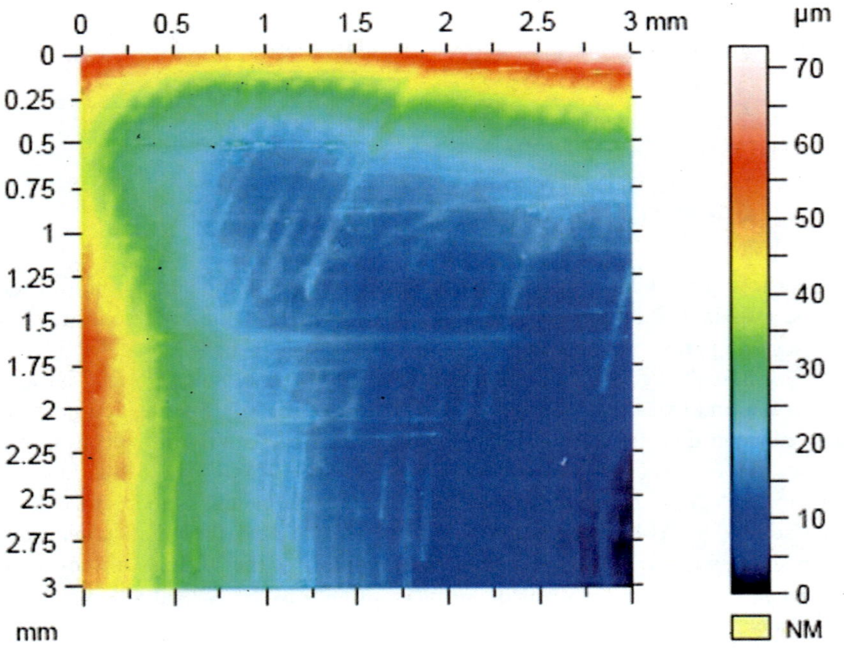

Fig. 23.6 Topology of the surface manufactured by IM

23.4 Conclusions

In this study, samples were produced with two techniques: SLA and IM. The dimensional accuracy of the fabricated samples was evaluated. The measurements were made for four different dimensional characteristics. The following characteristics were observed than the standard deviation for the measurements and the deviation from the nominal values have the lowest values for the specimen printed by the SLA technique. SLA technology is the best method which provided the accuracy to the designed dimensions. In addition, the surfaces of the specimens were studied under a 3D laser microscope in order to measure their roughness. Reported results have confirmed that the surface of parts made by IM has a better surface than printed parts. In this case, judgment is needed to balance the dimensional accuracy and surface roughness of the parts.

References

1. Wang, X., Jiang, M., Zhou, Z., Gou, J.: Hui, D3D printing of polymer matricomposites: a review and prospective. Compos. Part B. **110**, 442–458 (2017)
2. Ngo, T.D., et al.: Additive manufacturing (3D printing): a review of materials, methods, applications and challenges. Compos. B **143**, 172–196 (2018)
3. Papacharalampopoulos, A., Bikas, H., Stavropoulos, P.: Path planning for the infill of 3D printed parts utilizing Hilbert curves. Proc. Manuf. **21**, 757–764 (2018)
4. Dizon John Ryan, C., Espera Alejandro, H., Qiyi, C., Advincula Rigoberto, C.: Mechanical characterization of 3D-printed polymers. Addit. Manuf. **20**, 44–67 (2018)
5. Camacho, D.D., Clayton, P., O'Brien, W.J., Seepersad, C.: Applications of additive manufacturing in the construction industry—a forward-looking review. Autom. Constr. **89**, 110–119 (2018)
6. Calignano, F., Manfredi, D., Ambrosio, E.P., Biamino, S., Lombardi, M., Atzeni, E., et al.: Overview on additive manufacturing technologies. Proc. IEEE **105**, 593–612 (2017)
7. Singh, G., Pradhan, M.K., Verma, A.: Multi response optimization of injection moulding process parameters to reduce cycle time and warpage. Mater. Today: Proc. **5**(2), 8398–8405 (2018)
8. Osarenmwinda, J.O., Olodu, D.D.: Optimization of injection moulding process parameters in the moulding of High Density Polyethylene (HDPE). J. Appl. Sci. Environ. Manage. **22**(2), 203–206 (2018)
9. Kitayama, S., Tamada, K., Takano, M., Aiba, S.: Numerical and experimental investigation on process parameters optimization in plastic injection molding for weldlines reduction and clamping force minimization. Int. J. Adv. Manuf. Technol. **97**(5–8), 2087–2098 (2018)
10. Althammer, F., Ruf, F., Middendorf, P.: Size optimization methods to approximate equivalent mechanical behaviour in thermoplastics. J. Comput. Des. Eng. **8**(1), 170–188 (2021)

Chapter 24
Potential of Additive Manufacturing for Complex System Configurations to Improve Heat Transfer

Justin Byiringiro, Meriem Chaanaoui, and Sébastien Vaudreuil

Abstract This paper presents a review of recent research work on the use of additive manufacturing for devices with complex structures to enhance heat transfer. Different techniques have been developed for heat transfer enhancements, such as; active, passive and compound techniques. Passive techniques are commonly used since they don't need external support. It was shown that some potential configurations cannot be fabricated using conventional manufacturing methods, due to complex geometries. Because of the manufacturing restrictions many studies especially for big-sized heat transfer devices, are limited to numerical analysis without doing experiments. Additive manufacturing provides opportunities to produce these systems. The most commonly used method is powder bed fusion (PBF). Several limitations are associated with additive manufacturing technology, for instance; porosity, shrinkage, dimensional accuracy, surface quality and cost. This paper gives an overview of the advantages of AM as a benchmark for future research and development, by comparing economically a parabolic trough collector tube produced using additive manufacturing and a conventional receiver tube. It was shown that the cost of PTC electric energy can be reduced through gains in efficiency from the additive manufactured receiver that reduce the solar field cost by 8.5%.

Keywords Additive manufacturing · Additive manufacturing for metals · Heat transfer enhancement · Inserts · Turbulators

J. Byiringiro (✉) · M. Chaanaoui · S. Vaudreuil
Euromed Polytechnic School, Euro-Mediterranean University of Fes, Route de Meknes, 30000 Fes, Morocco
e-mail: j.byiringiro@ueuromed.org

M. Chaanaoui
e-mail: m.chaanaoui@ueuromed.org

S. Vaudreuil
e-mail: s.vaudreuil@ueuromed.org

K. Zarbane and Z. Beidouri (eds.), *Proceedings of CASICAM 2022*, Springer Tracts in Additive Manufacturing, https://doi.org/10.1007/978-3-031-32927-2_24

24.1 Introduction

The rising population worldwide and the advance in the standard of living due to advancements in technology as well as economic growth have motivated the attention gained by energy efficiency enhancement techniques. It was reported that energy demand is anticipated to rise to thirty percent from 2019 to 2040 [1]. Therefore, much effort is needed to ensure improving energy efficiency. Researchers have recently focused on techniques for enhancing heat transfer within fluid channels because of their importance in heat transfer devices. Nevertheless, heat transfer enhancement always goes along with pressure drop which increases the pumping power leading to additional costs. Heat transfer enhancement techniques are categorized into active, passive, and compound [2]. While passive approaches do not need external support to improve heat transfer, active procedures do, and compound approach will incorporate both passive and active methods. Passive heat transfer enhancement includes using pin-fins, insert with fins, wavy channels, tubes with internal ribs, dimples, protrusions, and channels with twisted tape insert among others. Those configurations are used in heat transfer devices such as; heat sinks, and heat exchangers used in electronic equipment, power and energy industries, and aerospace technologies. Some of these configurations are complex to be manufactured using traditional manufacturing techniques because of the limited machining possibilities. This manufacturing restriction can be eradicated through additive manufacturing. Additive manufacturing (AM) is a process where 3D solid objects are produced from digital CAD files through successive layers of material sacked and either bonded or melted together [3]. In recent years, additive manufacturing has gained popularity as a technology that enables the manufacturing of complex structures and designs that would be difficult or impossible by conventional methods. AM is already used in many fields such as aerospace, energy industry, biomedical, and micro-manufacturing, among others [4]. However, AM has some limitations such as its costs when compared to conventional methods, post-processing requirement, lack of parts repeatability, and high surface roughness and porosity [5]. This review aims to highlight the opportunities and challenges of using additive manufacturing in designing structures to be used in the heat transfer enhancement process. The techniques for heat transfer enhancement are discussed. The study also highlights the effects of some inherent features on heat transfer and the potential in manufacturing large-sized structures.

24.2 Heat Transfer Enhancement Passive Techniques

Passive methods are commonly used forms of enhancement because they can be integrated with existing technologies. They involve changing the geometry of the channel by inserting turbulators or by modifying the walls of the tubing itself (dimples, wire inserts, and tape inserts). With this technique, the boundary layers are disturbed, thus resulting in increased effective surface area and improved heat transfer coefficient.

24.2.1 Twisted Tape Inserts

Twisted tapes (TT) are among the most commonly used techniques to enhance heat transfer notably. Twisted tape improves the heat transfer by restricting the fluid flowing in the twisted pathway which leads to increasing the length of the flow. In addition, because of the insert characteristics, swirl mixing takes place leading to enhanced agitation and turbulence that enhance heat transfer. The most important parameters in the use of twisted tape are the twisting ratio and clearances between tapes and channel walls. When the twisting ratio increases, the swirl flow decreases as well as the heat transfer rate. Increasing the clearance between the tape and the tube wall leads to low heat performance [6]. In the study done by Mwesigye et al. [7], wall-detached twisted inserts were investigated to analyse the effect of TT on heat transfer. It was found that heat transfer improves at a low twist ratio. Jafar and Sivaraman [8] investigated experimentally the use of twisted tape. It was shown that TT improves heat transfer with better performance in the laminar flow. Several articles have been done on heat transfer performance improvement using different types of twisted tape inserts, and it was shown that there is an improvement in the performance. Zheng et al. [9] studied numerically, dimple twisted tape as shown in Fig. 24.1a. It was remarked that this configuration increase the heat transfer coefficient by 25.5%. Hong et al. [10] studied helical tapes with short lengths and achieved 2.64 times increase in heat transfer compared to a smooth tube. In a study done by Li et al. [11] on helical-type twisted tape, the heat transfer was intensified by up to 14.7%. Nakhchi et al. [12] used V-cut twisted tape (Fig. 24.1b) in their study, and heat transfer was improved by 117%.

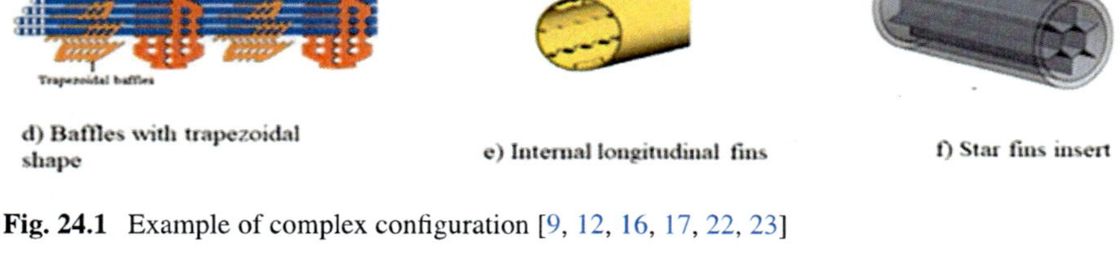

Fig. 24.1 Example of complex configuration [9, 12, 16, 17, 22, 23]

24.2.2 Wire Coil

This technique increases the rate of heat transfer by increasing the area of heat transfer, as well as the flow turbulence. Factors that affect the enhancement of heat transfer are wire pitch, the distance separating the coils and mass flow rate [13]. In a study done by Garcia et al. [14], the effect of helical steel-type wire coil on heat transfer and pressure drop was investigated. It was found that in the laminar flow region, there is a small increase in the Nusselt number which can be up to 1.4 times compared to that of a tube without coil. In the turbulent flow region, a significant increase in the Nusselt number was observed (up to 2.8 times that of a tube without coil), increasing the heat transfer rate. However, an increase in pressure drop was observed as well. In the laminar region, friction increased up to 1.8 times that of a tube without coil insert while in the turbulent region, the increase of friction factor can be up to 9 times that of tube without coil insert. Bahabadi et al. [15] conducted a study on the influence of wire coil inserts on heat transfer and pressure drop. It was found that the heat transfer increased by 85% compared to the tube without coil inserts, while the pressure drop increased by 4.75 times that of the tube without inserts. The benefit from wire coil and twisted tape can be combined to get a heat transfer performance. Promvonge [16] analyzed the effect of utilizing wire coil and twisted tape insert in the same tube using a configuration shown in Fig. 24.1c. The heat transfer was found to be significantly increased due to the combined effects of the wire coil and twisted tape inserts. A larger increase in pressure drop was observed, caused by the increased surface area and flow restriction.

24.2.3 Fins and Baffles

Baffles and fins are devices that enhance the contact surface area and cause turbulence in the fluid flow, which intensifies heat transmission. The thermal–hydraulic performance of a channel with internal longitudinal fins and sinusoidal lateral surfaces shown in Fig. 24.1e, was examined by Kursum [17]. It was observed that the Nusselt number increased by 1.66 while the pressure drop increased by 5.75. The overall performance of the channel found was poor because the Performance evaluation criterion (PEC) was found to be 0.93. This is because the change in pressure drop is higher than the Nusselt number increase. Hosseinzadeh et al. [18] analyzed the effect of fins on heat transfer performance, using a configuration with fins in a star structure as shown in Fig. 24.1f. This enhanced the heat transfer due to the increased contact surface area produced by the fins. By altering the orientation, angle, and length of fins, the level of heat transfer enhancement can be controlled. Borhani et al. [19] numerically studied spiral fins. Their results showed that heat transfer improved by 56%. Helical baffles direct the flow to improve heat transfer by reducing the pressure drop. Sahel et al. [20] investigated the effect of baffles on heat transfer performance. It was found that thermal performance improved compared to a plain tube. In a study conducted by Abbasi et al. [21], segmental baffles were studied numerically, with results showing that the heat transfer coefficient increased by 11.15%. Gu et al. [22] studied numerically and experimentally baffles with a trapezoidal shape (Fig. 24.1d.). The results showed an improved heat transfer coefficient of 10.2%.

In the review done by Muhammad et al. [24] on heat transfer enhancement in parabolic trough solar receivers, it was shown that using fillets on fins could lower pressure drop. They are not however easy to fabricate. Because of this, some configurations that have complex geometries cannot be manufactured using traditional manufacturing approaches. Because of these manufacturing restrictions, many studies are limited to a simulation analysis without doing experiments, especially for big-size heat transfer devices. Through additive manufacturing, manufacturing restrictions could be alleviated or eradicated.

24.2.4 Comparison of Heat Transfer Enhancement from Different Configuration

Wei et al. [25] compared different configurations of straight fins that can be produced using conventional manufacturing (Fig. 24.2a) and interrupted fins produced using additive manufacturing (Fig. 24.2b). It was found that the interrupted fins have an excellent enhancement of heat transfer, being 1.4 times that of straight fins and 2.6 times that of the smooth tube.

Without a hydraulic analysis, the thermal investigation of a configuration and its comparison is not complete since heat transfer enhancement always goes along with pressure drop. This is caused by the Colburn analogy which fundamentally links heat and momentum transfer [26]. Therefore, enhancement techniques are compared in terms of Nusselt number and pressure drop through Performance evaluation criteria (PEC) which combine hydraulic and thermal characteristics. Table 24.1 shows heat transfer enhancement comparison of different configurations in terms of PEC.

Mohammad et al. [24], combined geometries (rod insert, ribbed tube) to benefit from their characteristics. From Table 24.1, it can be seen that configurations with combined geometries of rib and rod insert [27], presents the highest performance as far as PEC is considered at 6.87 and 6.03 respectively. It can be noted that

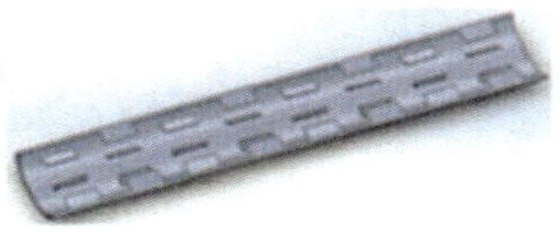

Fig. 24.2 Example of fins [25] licensed under a creative common attribution 4.0 International (CC BY 4.0) (http://creativecommons.org/licenses/by/4.0)

Table 24.1 Comparison of Different Configurations

Type of insert	(Nu/Nu_o)	(f/f_o)	PEC
Combined rod and ribbed insert	15.43	11.34	6.87
Rod insert	13	10	6.03
Perforated plate	7.12	12.75	3.05
Conical strip	6.50	14	2.7
Tape	4.17	4.98	2.44
Ring attached twisted tape	2.32	1.83	1.90
Porous	2.83	4.54	1.75
Internally finned	2.33	2.45	1.73
Star insert	2.98	7.28	1.54
Helical fins	2.35	4.6	1.41
Twisted tape	1.75	3.14	1.2
Helical screw tape	1.73	3.51	1.14
Porous rings	1.68	3.52	1.10
Wall-detached twisted tape	2.56	13.28	1.08

although some configurations (Star insert and wall-detached twisted tape) have a good enhancement in Nusselt number, they cannot overcome large friction factor values. Additionally, although the configuration with ring-attached twisted tape does not present a good Nusselt number, the low friction factor makes it efficient.

24.3 Potential of Additive Manufacturing in Heat Transfer Enhancement

24.3.1 Additive Manufacturing for Metals

Additive manufacturing can be regarded as one of the biggest innovations in manufacturing [28]. The advantage of AM is that complex geometry can be produced with good accuracy and that raw materials are saved. AM for metals is grouped into four major classes; Powder bed fusion (PBF), Directed energy deposition (DED), Binder jetting (BJ), and Sheet lamination (SL) [29]. Among the AM groups, Powder bed fusion (PBF), and Directed energy deposition (DED) are the most used. PBF is a metal powder material deposition. It involves the spread of thin layers of fine powders, which are fused by the use of a laser beam or binder. Using PBF, higher accuracy products can be fabricated. However, powder is very expensive and the size of the component that can be printed is generally limited because of the build-in envelope. DED involves the use of laser, electron beam or electric arc as the source of heat for melting the metal wire or metal powder to form molten droplets accumulating layer by layer to become the part. DED has economic advantage compared to

PBF because of its low operating cost as metal wire cost is lower compared to metal powder [30]. However, parts produced by DED have lower resolution compared to parts from PBF, and it requires post-machining process.

24.3.2 Application of Additive Manufacturing in the Fabrication of Heat Transfer Devices

The ability of additive manufacturing to produce complex structures has enabled the design of improved heat transfer devices. Different AM methods have been used with multiple materials to produce various configurations. This section focus on the analysis of heat transfer enhancement for structures produced using additive manufacturing. Kirstch [31] investigated the wavy channel produced by the direct metal laser sintering method using Inconel 718 as material. It was shown that excessive surface roughness of AM played a role in increasing the heat transfer coefficient by 15%. However, these surface features affect performance because of the resulting increase in pressure drop. Zhang et al. [32] studied experimentally the convective heat transfer of a honeycomb structure produced by additive manufacturing using 316L stainless steel powder. It was observed that this structure improved the convective heat transfer. In a study done by Tiwari et al. [33], a unique shell and tube structure produced by the filament deposition method (FDM) was studied to investigate its performance. The results revealed an enhancement in heat transfer. Kwon et al. [34] investigated the heat transfer enhancement in channels with static mixers produced using additive manufacturing. It was found that heat transfer can be increased by up to two times compared to channel without static mixers. Parts made by additive manufacturing are characterized by their surface roughness which is a direct result of the manufacturing process. While this can be beneficial for heat transfer enhancement, it can be also a disadvantage as it increases the resulting pressure drop. For instance, selective laser Sintering (SLS) and Fused Deposition Modeling (FDM) produce parts that have surface roughness respectively ranging from 5 to 35 μm and 9 to 40 μm [35]. The AM process and machine used influence the level of surface roughness. Several studies have been done to analyze the effect of the roughness of AM surfaces, specifically on heat performance and pressure drop. In a study done by Stimpson et al. [36], small channels produced using direct metal laser sintering (DMLS) were investigated experimentally to have a clear understanding of the effect of roughness on heat transfer performance. It was seen that there is a significant increase in the Nusselt number and friction factor compared to a smooth tube. Generally, the thermal performance of the channel was enhanced since the increase of the Nusselt number was proportional to the increase of friction factor, which means that the required Nusselt number was higher to overcome the pressure drop. Ventola et al. [37] carried out an experiment to investigate the effect of the rough surface of finned heat sinks manufactured using DMLS. The enhancement of convective heat transfer compared to smooth finned surface was observed. Saltzman et al. [38]

analyzed the effect of surface roughness on the heat transfer performance of the heat exchanger by using a crossflow produced using the laser-based PBF method. It was seen that the overall heat transfer improved by 10%, while pressure drop increased by up to two times. The other factor that affects the performance of heat transfer devices manufactured by AM is the porosity. The AM technique being employed affects the porosity defect. For instance, shrinkage, gas trapping during solidification, and the attachment of partially molten particles to surfaces between layers all contribute to the porosity in SLM parts [39]. Thermal performances are affected because porosity affects the thermal conductivity of the material as well as the fatigue and tensile strength. Collins et al. [5] investigated the heat transfer in a channel produced by the laser sintering method using aluminium alloy. It was found that thermal performance is lowered by 15 to 20% compared to a channel of the same size produced by conventional manufacturing. This was attributed to the small-scale voids found in the material, which cause a reduction in thermal conductivity. With PBF there is a size limit in part produced related to the build envelope. This is problematic in some heat transfer applications found in specific fields, for instance in concentrated solar power applications. In this case, the receiver tube can have a length of up to four meters [24]. There are other additive manufacturing techniques like directed energy deposition (DED), where large size parts can be manufactured and have a lower cost compared to powder bed fusion. For instance, Stainless steel 316 PBF powder cost 40 $/kg whereas DED powder cost 12 $/kg [30].

24.4 Economic Aspect of Configuration Produced Through Am

To analyze the economic aspect of using additive manufacturing to produce heat transfer devices, an example of a Parabolic Trough Collector (PTC), the most common concentrated solar power (CSP) technology, can be used. Thermal and electric energy from Parabolic trough concentrating solar technology (PTC) is expensive at about 0.2 USD/kWh [40]. This is mainly due to the high capital cost of a PTC Plant. The estimated cost of producing one meter of the receiver will include the cost of producing the absorber tube, its selective coating cost and the glass envelope. If the absorber tube is made using selective laser melting (SLM) methods, the cost of the receiver is estimated at around 691 $/meter, while the use of Directed energy deposition (DED) can reduce the costs to around 391$/meter. These results clearly show that producing one meter of the receiver using additive manufacturing is higher when compared to the existing commercial receiver which is calculated at 250 to 350$/meter [41]. As seen in the literature, some configurations of receiver tubes may enhance the heat transfer efficiency by up to 6 times what is achieved with the plain tube. An additive-manufactured receiver will thus be advantageous by the increased heat transfer and efficiency of the receiver, yielding a reduction in capital cost for a PTC plant through a lower length of receivers and collectors while decreasing thermal losses. For instance, a PTC power plant of 100 MW capacity designed by Benhadji et al. [42], has a solar field consisting of 1168 solar collectors (Eurothrough ET 150) and 42,048 receivers (Scott PTR 70). The cost of the solar field is found at 377.69 $\$/m^2$. Normally, the solar receivers account for 30% of the total cost of solar fields [43]. This means that the cost of receivers in this PTC Plant is 113.31 $\$/m^2$. When using directed energy deposition (DED) to produce a receiver, where the cost of the receiver is approximated to 391 $/meter, the cost of the receiver may roughly increase by 23%. However, as the efficiency of the receiver will be increased, the number of receivers needed will be reduced by close to 30%, reducing the number of receivers to 29,433. When comparing the cost of conventional and additive manufactured receivers, this would reduce the cost of the solar field by 8.5%, and in turn the capital cost and eventually the cost of energy.

24.5 Conclusion

In this paper, heat transfer enhancement techniques were discussed. There are configurations that can enhance heat transfer while minimizing pressure drop increase. Potential configurations cannot be produced using conventional manufacturing due to their complexity. This manufacturing restriction can be addressed by additive manufacturing. Comparison of the PTC plant having receiver tubes produced using AM and conventional receiver tubes showed that the cost of the solar field will be reduced by 8.5%, which will eventually reduce the cost of energy.

Acknowledgements The present work is part of a research funded by African Scientific and Research Innovation Council (ASRIC) with a close collaboration with the Euro-Mediterranean University of Fes (UEMF).

References

1. British, P.: BP Energy outlook 2019 edition The energy outlook explores the forces shaping the global energy transition out to 2040 and the key uncertainties surrounding that. In: BP Energy Outlook (2019)
2. Kareem, Z.S., Mohd Jaafar, M.N., Lazim, T.M., Abdullah, S., Abdulwahid, A.F.: Passive heat transfer enhancement review in corrugation. Exp. Thermal Fluid Sci. **68**, 22–38 (2015)
3. Louvis, E., Fox, P., Sutcliffe, C.J.: Selective laser melting of aluminium components. J. Mater. Process. Technol. **211**(2), 275–284 (2011)
4. Li, J., Duan, C., Zhao, M., Luo, X.: A review of metal additive manufacturing application and numerical simulation. IOP Conf. Ser. Earth Environ. Sci. **252**(2) (2019)
5. Collins, I.L., Weibel, J.A., Pan, L., Garimella, S.V.: Evaluation of additively manufactured microchannel heat sinks. IEEE Trans. Compon. Packag. Manuf. **9**(3), 446–457 (2019)
6. Hasanpour, A., Farhadi, M., Sedighi, K.: Intensification of heat exchangers performance by modified and optimized twisted tapes. Chem. Eng. Process. **120**, 276–285 (2017)

7. Mwesigye, A., Bello-Ochende, T., Meyer, J.P.: Heat transfer and entropy generation in a parabolic trough receiver with wall-detached twisted tape inserts. Int. J. Therm. Sci. **99**, 238–257 (2016)
8. Jafar, K.S., Sivaraman, B.: Performance characteristics of parabolic solar collector water heater system fitted with nail twisted tapes absorber. J. Eng. Sci. Technol. **12**(3), 608–621 (2017)
9. Zheng, L., Xie, Y., Zhang, D.: Numerical investigation on heat transfer performance and flow characteristics in circular tubes with dimpled twisted tapes using Al2O3-water nanofluid. Int. J. Heat Mass Transf. **111**, 962–981 (2017)
10. Hong, Y., Du, J., Wang, S., Huang, S.M., Ye, W.B.: Effect of decaying swirl flow on tubular turbulent heat transfer enhancement by using short length helical tapes. Chem. Eng. Res. Des. **138**, 1–12 (2018)
11. Li, Z., Sheikholeslami, M., Jafaryar, M., Shafee, A., Chamkha, A.J.: Investigation of nanofluid entropy generation in a heat exchanger with helical twisted tapes. J. Mol. Liq. **266**, 797–805 (2018)
12. Nakhchi, M.E., Esfahani, J.A.: Performance intensification of turbulent flow through heat exchanger tube using double V-cut twisted tape inserts. Chem. Eng. Process. Process Intensif. **141**, 107533 (2019)
13. Diwan, K., S Soni, M.: Heat transfer enhancement in absorber tube of parabolic trough concentrators using wire-coils inserts. Univ. J. Mech. Eng. **3**(3), 107–112. (2015)
14. García, A., Vicente, P.G., Viedma, A.: Experimental study of heat transfer enhancement with wire coil inserts in laminar-transition-turbulent regimes at different Prandtl numbers. Int. J. Heat Mass Transf. **48**(21–22), 4640–4651 (2005)
15. Akhavan-Behabadi, M.A., Shahidi, M., Aligoodarz, M.R.: An experimental study on heat transfer and pressure drop of MWCNT-water nano-fluid inside horizontal coiled wire inserted tube. Int. Commun. Heat Mass Transf. **63**, 62–72 (2015)
16. Promvonge, P.: Thermal augmentation in circular tube with twisted tape and wire coil turbulators. Energy Convers. Manage. **49**(11), 2949–2955 (2008)
17. Kurşun, B.: Thermal performance assessment of internal longitudinal fins with sinusoidal lateral surfaces in parabolic trough receiver tubes. Renew. Energy **140**, 816–827 (2019)
18. Hosseinzadeh, K., Mogharrebi, A.R., Asadi, A., Paikar, M., Ganji, D.D.: Effect of fin and hybrid nano-particles on solid process in hexagonal triplex latent heat thermal energy storage system. J. Mol. Liq. **300**, 112347 (2020)
19. Borhani, S.M., Hosseini, M.J., Ranjbar, A.A., Bahrampoury, R.: Investigation of phase change in a spiral-fin heat exchanger. Appl. Math. Model. **67**, 297–314 (2019)
20. Sahel, D., Ameur, H., Benzeguir, R., Kamla, Y.: Enhancement of heat transfer in a rectangular channel with perforated baffles. Appl. Therm. Eng. **101**, 156–164 (2016)
21. Abbasi, H.R., Sharifi Sedeh, E., Pourrahmani, H., Mohammadi, M.H.: Shape optimization of segmental porous baffles for enhanced thermo-hydraulic performance of shell-and-tube heat exchanger. Appl. Therm. Eng. **180** (2020)
22. Gu, X., Luo, Y., Xiong, X., Wang, K., Wang, Y.: Numerical and experimental investigation of the heat exchanger with trapezoidal baffle. Int. J. Heat Mass Transf. **127**, 598–606 (2018)
23. Mousa, M.H., Miljkovic, N., Nawaz, K.: Review of heat transfer enhancement techniques for single phase flows. Renew. Sustain. Energy Rev. **137**, 110566 (2021)
24. Shahzad Nazir, M., Shahsavar, A., Afrand, M., Arıcı, M., Nižetić, S., Ma, Z., Öztop, H.F.: A comprehensive review of parabolic trough solar collectors equipped with turbulators and numerical evaluation of hydrothermal performance of a novel model. Sustain. Energy Technol. Assess. **45** (2021)
25. Wei, C., Diaz, G.A.V., Wang, K., Li, P.: 3D-printed tubes with complex internal fins for heat transfer enhancement-CFD analysis and performance evaluation. AIMS Energy **8**(1), 27–47 (2020)
26. Zheng, N., Yan, F., Zhang, K., Zhou, T., Sun, Z.: A review on single-phase convective heat transfer enhancement based on multi-longitudinal vortices in heat exchanger tubes. Appl. Therm. Eng. **164**, 114475 (2020)

27. Chang, C., Sciacovelli, A., Wu, Z., Li, X., Li, Y., Zhao, M., Deng, J., Wang, Z., Ding, Y.: Enhanced heat transfer in a parabolic trough solar receiver by inserting rods and using molten salt as heat transfer fluid. Appl. Energy **220**, 337–350 (2018)
28. Guo, D., Lu, Y.: Overview of extreme manufacturing. Int. J. Extreme Manuf. **1**(2), (2019)
29. Cunningham, C.R., Flynn, J.M., Shokrani, A., Dhokia, V., Newman, S.T.: Invited review article: strategies and processes for high quality wire arc additive manufacturing. Addit. Manuf. **22**, 672–686 (2018)
30. Li, Y., Su, C., Zhu, J.: Comprehensive review of wire arc additive manufacturing: hardware system, physical process, monitoring, property characterization, application and future prospects. Results Eng. **13**, 100330 (2022)
31. Kirsch, K.L., Thole, K.A.: Experimental investigation of numerically optimized wavy microchannels created through additive manufacturing. J. Turbomach. **140**(2) (2018)
32. Zhang, Y., Kong, D., Liu, S.: Experimental study of forced convection heat transfer of graded metal honeycomb fabricated by additive manufacturing. Int. Commun. Heat Mass Transf. **98**, 67–73 (2018)
33. Tiwari, R., Andhare, R.S., Shooshtari, A., Ohadi, M.: Development of an additive manufacturing-enabled compact manifold microchannel heat exchanger. Appl. Therm. Eng. **147**, 781–788 (2019)
34. Kwon, B., Liebenberg, L., Jacobi, A.M., King, W.P.: Heat transfer enhancement of internal laminar flows using additively manufactured static mixers. Int. J. Heat Mass Transf. **137**, 292–300 (2019)
35. Kumbhar, N.N., Mulay, A.V.: Post processing methods used to improve surface finish of products which are manufactured by additive manufacturing technologies: a review. J. Inst. Eng. (India) Ser. C **99**(4), 481–487 (2018)
36. Stimpson, C.K., Snyder, J.C., Thole, K.A., Mongillo, D.: Roughness effects on flow and heat transfer for additively manufactured channels. J. Turbomach. **138**(5) (2016)
37. Ventola, L., Robotti, F., Dialameh, M., Calignano, F., Manfredi, D., Chiavazzo, E., Asinari, P.: Rough surfaces with enhanced heat transfer for electronics cooling by direct metal laser sintering. Int. J. Heat Mass Transf. **75**, 58–74 (2014)
38. Saltzman, D., Bichnevicius, M., Lynch, S., Simpson, T.W., Reutzel, E.W., Dickman, C., Martukanitz, R.: Design and evaluation of an additively manufactured aircraft heat exchanger. Appl. Therm. Eng. **138**(April), 254–263 (2018)
39. Tapia, G., Elwany, A.H., Sang, H.: Prediction of porosity in metal-based additive manufacturing using spatial Gaussian process models. Addit. Manuf. **12**, 282–290 (2016)
40. Bashir, A.A.A., Özbey, M.: Modelling and analysis of an 80-MW parabolic trough concentrated solar power plant in Sudan. Clean Energy **6**(3), 512–527 (2022)
41. Shinde, T.U., Dalvi, V.H., Patil, R.G., Mathpati, C.S., Panse, S.V., Joshi, J.B.: Thermal performance analysis of novel receiver for parabolic trough solar collector. Energy **254**, 124343 (2022)
42. Benhadji Serradj, D.E., Sebitosi, A.B., Fadlallah, S.O.: Design and performance analysis of a parabolic trough power plant under the climatological conditions of Tamanrasset, Algeria. Int. J. Environ. Sci. Technol. **19**(4), 3359–3376 (2022)
43. Wu, Z., Li, S., Yuan, G., Lei, D., Wang, Z.: Three-dimensional numerical study of heat transfer characteristics of parabolic trough receiver. Appl. Energy **113**, 902–911 (2014)

Chapter 25
Smart Materials Moisture-Responsive Use in 4D Printing

Bassam Badr Mohammed Abdo Al Nahari, Khalid Zarbane, and Zitouni Beidouri

Abstract 4D printing is a revolutionary modern manufacturing technique that will have a significant impact in many fields. This technique uses 3D printing to create an object that alternates between two controllable shapes or properties over time and under specific conditions such as exposure to moisture, air, heat, or electrical current. To this end, 4D printing uses special smart materials. These materials expand our understanding through their broad stimulation and diversity of uses. This article presents a state of the art of Smart materials moisture-responsive and their applications.

Keywords 3D printing · 4D printing · Smart materials and moisture-responsive

25.1 Introduction

In the case of conventional manufacturing, the transformation of raw materials into finished goods necessitates a great number of steps, such as planning and the execution of a great number of manufacturing processes, among other things, all of which require a significant amount of time and result in an increase in costs. We can quickly create and print a wide variety of items, including those with intricate designs, with the help of additive manufacturing and 3D printing. This can be accomplished in a shorter amount of time, using fewer production procedures, and at a lower cost. In

B. B. M. A. Al Nahari (✉) · K. Zarbane · Z. Beidouri
Laboratory of Advanced Research on Industrial and Logistics Engineering (LARILE), National Higher School of Electricity and Mechanics (ENSEM), Hassan II University of Casablanca, Casablanca, Morocco
e-mail: bassam.alnahari-etu@etu.univh2c.ma

K. Zarbane
e-mail: khalid.zarbane@univh2c.ma

Z. Beidouri
e-mail: zitouni.beidouri@univh2c.ma

K. Zarbane and Z. Beidouri (eds.), *Proceedings of CASICAM 2022*, Springer Tracts in Additive Manufacturing, https://doi.org/10.1007/978-3-031-32927-2_25

recent years, the technique known as 3D printing has received prominence by manufacturing components used in a variety of applications. These applications include the biomedical sector, the food industry, electrical components, aerospace applications, and soft robots.

At the MIT Conference, Professor Skylar Tibbits was the one who first used the phrase “4D printing.” He gave the following definition of 4D printing: “Using 3-D printer in the creation of objects that could transform their form whilst they are eliminated from the 3-D printer.” To put it another way, the term “4d printing” refers to a process that combines the methods of 3D printing with the novel usage of phase change materials (PCMs). The form of these PCMs may evolve as a result of their interactions with their environment. A direct result of this is that it undergoes changes in the properties of individual components, which in turn gives rise to the presence of a fourth dimension, the time.

The phrases “shape memory materials” (SMMs), “self-healing materials,” and “metamaterials” refer to the three most important categories of phase change materials. The term “shape memory materials” SMMs refers to materials that are able to alter form in response to some external stimuli. As its name suggests, self-healing materials may repair themselves whenever damage or defect appears, replacing the affected region with new material. Materials that are produced and have qualities that are not found in natural materials are called meta materials. To ensure optimal plant development and safe grain storage, SMPs rely on the humidity-response effect, which is especially important in greenhouses and silos. Although beneficial, commercial applicability is limited by high costs and a lack of scalable technology for humidity-sensitive materials [1–3] and in this article, we well discover the most frequent 4D printing smart materials they are moisture-sensitive.

25.2 Cellulos

Zhang et al. are made a self-standing films moisture-responsive from sustainable cellulose. The cellulose was initially changed by stearoyl moieties, producing cellulose stearoyl esters (CSEs) with varying degrees of substitution (DSs). CSE films with a DS of 0.3 (CSE 0.3) were moisture-responsive, whereas those with DSs of 1.3 or 3 were not. Due to water absorption and desorption at the film surface, CSE 0.3 films may reversibly fold and unfold within a local moisture gradient. Spray-coating CSE3 nanoparticles (NPs) over CSE 0.3 films developed films that react to moisture while having a nonwetting exterior that may undergo continual morphing and shape-shifting on water with fast bending and unbending. Dual-layer films with one CSE0.3 layer and one CSE3 layer showed combination moisture and temperature responsiveness. By altering CSE 0.3 film thickness, the minimum degree of bend could change because of varying mechanical resistances, allowing degree of bend to begin with the thinner side (Fig. 25.1) [4].

Figure 25.1 is a comparison of the properties of CSE 0.3, CSE 1.3 and CSE 3 films using a width of 21.2 ± 1.6, 20.6 ± 1.6 and 20.3 ± 2.2 μm, each in order. One

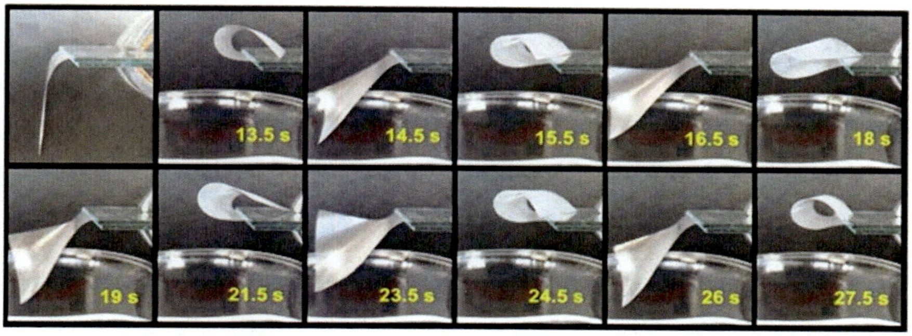

Fig. 25.1 Characteristics of CSE films

side of The film is mounted on glass plates, when 37 degrees Celsius of warm water was placed beneath the film, the moisture-responsive motions occurred instantly [4].

By integrating tunicate cellulose nanocrystals (TCNCs) into polymeric networks via host-guest liaison, Cui et al. created artificial muscles in a hydrogel with a tendril-like structure. Hydrogel muscles responded to stimuli by increasing their actuation rate, exhibiting actuation strain, and retaining their form. As shown in Fig. 25.2, the trigger properties of hydrogel muscles varied with their chirality, tension, twisting rate, and transitory form. Given their high water content and the fact that they can be made to contract like real muscles, developed hydrogels provide a lot of prospects for the biomedical industry [5].

Eco-friendly and low-cost options are essential for 4D printing to become popular. When it comes to 4D printing of responsive structures, Mulakkal et al. reported the development and physical characterisation (stability, swelling potential, and rheology) of the cellulose-hydrogel composite to prove its appropriateness. Ink with a high total cellulose content and an excellent dispersion of fibers inside the hydrogel matrix was created using the carboxymethyl cellulose (CMC) hydrocolloid with

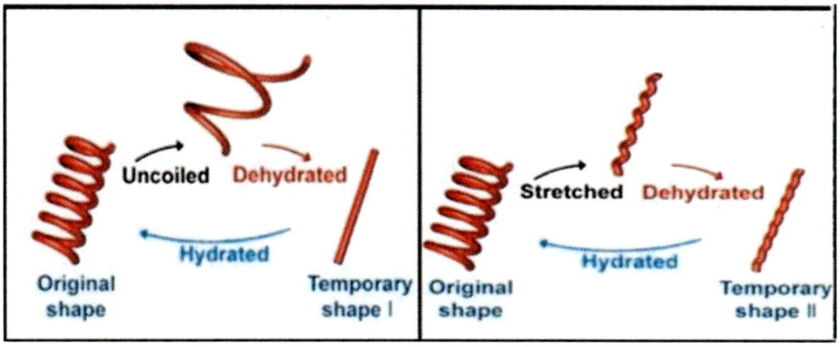

Fig. 25.2 The illustration presents that the hemochorial hydrogel's may remember their previous forms and put them to use in a variety of ways

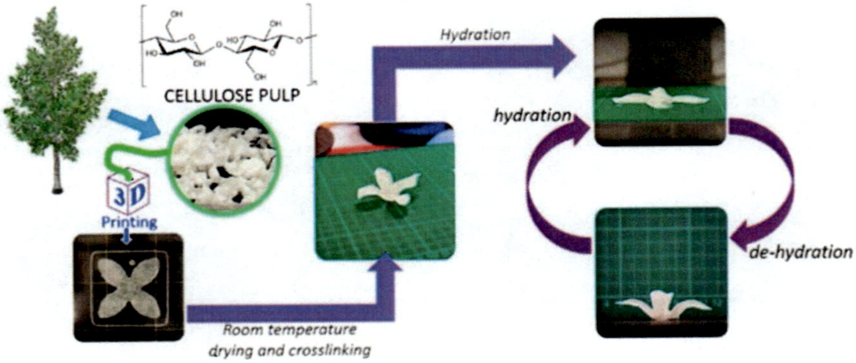

Fig. 25.3 A stimuli responsive cellulosic pulphydrogel composite ink

integrated cellulose pulp fibers, Fig. 25.3. The most utilization of cellulose is a resistive humidity sensor with high sensitivity and linearity was manufactured by coating cellulosic paper with carboxylic acid functionalized carbon nanotubes (CNTs) [6, 7].

Figure 25.3 present a stimuli responsive cellulosic pulphydrogel composite ink was developed and the petal architecture fabricated using this ink could deploy to flat configuration upon hydration and recover from drying [6].

25.3 Silk Fibroin Films

Using regenerated silk fibroin sheets, Ganesan et al. have shown that multivapor responsive actuation is fully reversible and bidirectional. A multi-vapor response may be obtained in a film of single-layer silk fibroin. In response to water and ethanol vapor, the silk fibroin films bend upwards while remaining flexible when exposed to air. Simplicity of experimentation with silica gels of varying hydration levels confirms the film's contraction due to vapor pressure. They are focused on how the amount of water present in the regenerated silk fibroin films affects the overall efficiency with which they may be used to conduct actuation. It's interesting to note that ethanol vapor's actuation effectiveness increases relative to water vapor and vice versa under situations of increased humidity. The researchers demonstrated that the modifying a vapor generator's water-to-ethanol proportion allowed them to regulate the actuation's amplitude, direction, and velocity. As an added bonus, the film's actuation capability may be completely muted by selecting the optimal water-to-ethanol ratio in a binary combination. Water's vapor pressure has a substantial effect on the amplitude of the actuation, whereas ethanol's vapor pressure has the opposite effect. Utilizing the unique actuation characteristics of multivapor-responsive homogeneous single-layer silk fibroin films, a continuous undulating wavy motion is exhibited. There are a variety of applications for silk fibroin within the domain

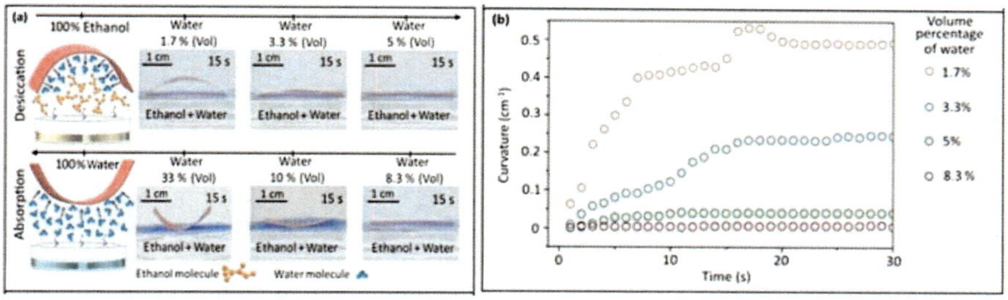

Fig. 25.4 The effect of moisture on a single-layer silk fibroin films

of biomedicine. Utilization of this technology may be utilized in the fields of bone tissue engineering, eye regeneration, nerve regeneration, skin tissue engineering, cartilage regeneration, vascular tissue engineering, spinal cord tissue engineering, gene therapy, and biological drug delivery [8, 9].

The response of a silk fibroin film to binary mixes of ethanol and water is shown in Fig. 25.4a, where the film's actuation behavior is shown to vary with the mixture's ethanol concentration. Film bowing upwards when exposed to ethanol indicates a considerable increase in water content. The Fig. 25.4b shows the silk fibroin films were exposed to vapors from a binary combination of water and ethanol at concentrations of 1.7, 3.3, 5, and 8.3%, and their curving changed with time [8].

25.4 Smart Polyurethane (PU)

PU is more environmentally friendly since it may be made from non-petroleum sources. Isocyanate functional groups and polyol groups combine to form PU. Hard segments are formed by the diisocyanate and the chain extender. On the other hand, polyol creates flexible segments that improve the PU/elasticity. Shape memory PU (SMPU) Polyols may be made from both petroleum and renewable resources and also the smart PU composites have progressed significantly in 3D and 4D printing. In spite of the fact that most printed PU or SMPU composites are still in prototype form, they show promising promise as the functional objects of the future, particularly in biomedical engineering and electronics. However, PU rely on the desorption or sorption of moisture as the source of the actuating response force [10, 11].

Baker et al. used 3D printing to make three-layer origami structures with a polyurethane hydrogel core and polyurethane elastomer skins. In order to create active hinges, discrete localized gaps had to be manufactured in the elastomeric skin. Because of changes in shape brought on by hydration, multiple complicated origami fold patterns were produced. These patterns were determined by the spatial distribution of hinges, Fig. 25.5 [12].

The Fig. 25.7 shown an origami-inspired trilayer structure folding upon hydration (core from polyurethane hydrogel and skins from polyurethane elastomer) [13].

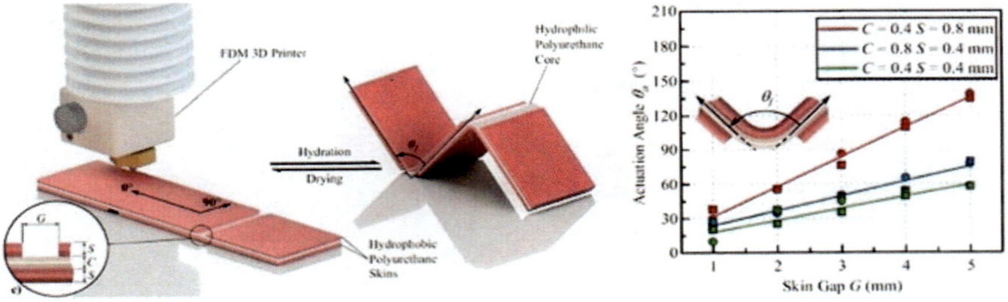

Fig. 25.5 3D printed moisture-responsive polyurethane elastomer

25.5 Polyethylene Glycol Diacrylate (PEGDA)

In the presence of UV light and benzophenone as a photo initiator, Chan et al. successfully grafted PEGDA to a microporous polypropylene membrane. The contact angle measurements were showing that the surface change made the formerly hydrophobic polypropylene membrane more hydrophilic. PEGDA membrane grafts swelled more in response to increasing humidity (Fig. 25.6). This membrane's responsiveness to moisture may increase its efficacy in stopping infections that are delivered in liquids. Grafted PEGDA membranes expanded in response to an increase in relative humidity. However, the membrane's responsiveness to water may increase its efficacy against liquid-borne infections [14].

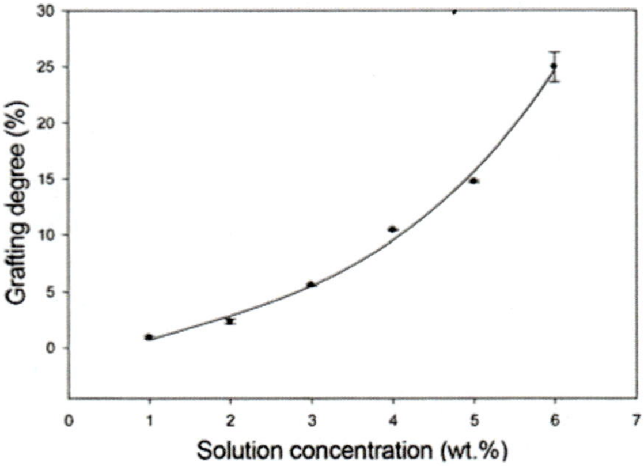

Fig. 25.6 Effect of solution concentration on grafting degree of PEGDA,

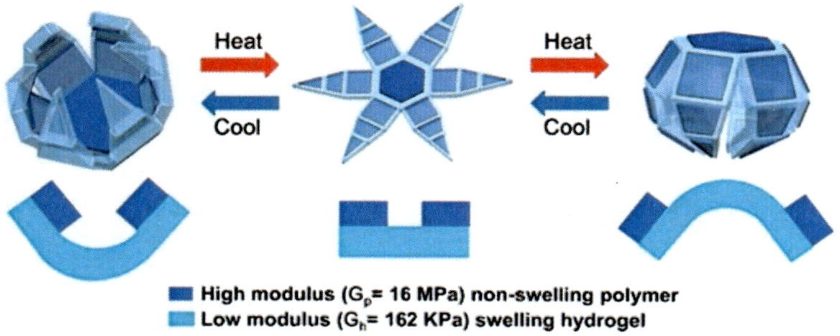

Fig. 25.7 The reversible self-folding of pNIPAM-AAc soft microgrippers

25.6 Poly (N-Isopropylacrylamide-co-Acrylic Acid) (pNIPAM-AAc)

In order to manufacture microgripper joints, Breger et al. made use of gradient crosslinked pNIPAM-AAc soft-hydrogels. The saturation point may be adjusted, which enables reversible actuation, by either heating or cooling the water in which the gripper is immersed. Changing the temperature of the water will have the opposite effect. In addition, utilizing hinges with unconventional designs is another method for preventing overswelling, besides the mechanical stability and strength of pNIPAM-AAc soft-hydrogels are found to be significantly increased after enzymatic crosslinking [15, 16].

Figure 25.7 is a simplified illustration of the temperature-induced self-folding of pNIPAM-AAc soft microgrippers. When the temperature drops below 36 degrees Celsius, water is absorbed by the pNIPAM-AAc, causing it to expand and push the microgripper open, before closing in the opposite way, exposing its pNIPAM-AAc layer to the environment [17].

25.7 Double Cross-Linked Polymer

Jiang et al. have manufactured a novel kind of humidity-responsive polymeric networks. These networks are made up of hydrogen bonds and carboxyl-Fe^{3+} coordination interactions. Because of their extreme sensitivity to variations in humidity, acid ether hydrogen bonds are able to go through rapid deformation in reaction to these shifts. In response to changes in humidity, the reversible actuation is made easier and achievable because to the stretch deformity enabled by the strong cross-linking of coordination compounds. The double-crosslinked polymer can be bent and unbent indefinitely without losing its structural integrity. Additionally, changing the layer width would be the simple way to fine-tune the trigger attitude caused by

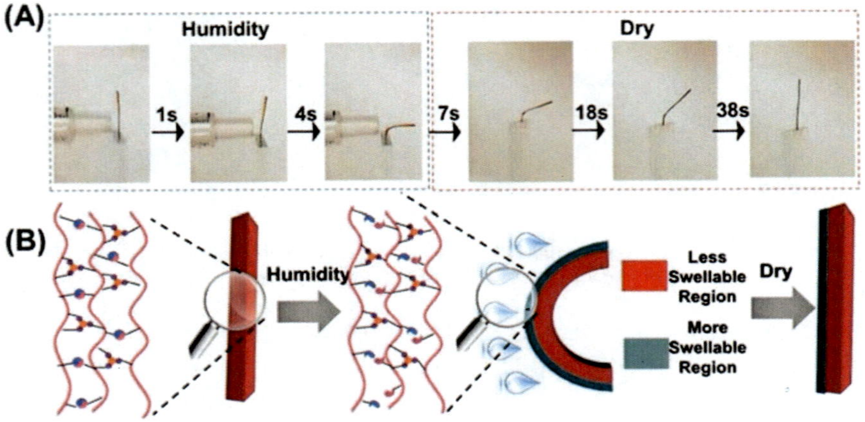

Fig. 25.8 Present the double cross-linked polymer reacts to changes in humidity

humidity. Humidity-sensitive polymeric actuators, soft robotics, and artificial muscle applications may all benefit from the usage of this double-crosslinked polymer [2].

The Fig. 25.8a pictures of double cross-linked polymer bending quickly (thickness = 50 μm). If you put film in a humid environment (relative humidity between 22 and 95%), the polymer will curve away from the moisture. The Fig. 25.8b a schematic depicting the humidity-induced development of the double-crosslinked hydrogel displaying a swelling-ratio-gradient, plus solid ion liaison and stability Cross-linkers have a potential to improve the film's elasticity and stiffness, allowing it to straighten out after being bent as it dries [2].

25.8 Conclusions

4D printing is a revolutionary technology which gives us only few examples to work with. It is mainly based on the smart materials that change their shape due to the stimulation (heat, light, moisture, PH, electricity, etc.…). In this research, we had listed the most important and well-known Smart material's; moisture-responsive used in 4D printing, which will help in future researches specialized in the technology of developing sensors, actuators and soft robots.

References

1. Mehta, P., Sahlot, P.: Application of phase change materials in 4D printing: a review. Mater. Today Proc. (2021). https://doi.org/10.1016/j.matpr.2021.05.664
2. Jiang, Z., Shen, P., Tan, M.L., Yan, Q., Viktorova, J., Cementon, C., Peng, X., Xiao, P., Connal, L.A.: 3D and 4D printable dual cross-linked polymers with high strength and humidity-triggered reversible actuation. Mater. Adv. (2021). https://doi.org/10.1039/d1ma00223f
3. Maraveas, C., Ilker, S., Bayer b, Thomas Bartzanas.: 4D printing: perspectives for the production of sustainable plastics for agriculture. Biotechnol. Adv. (2021). https://doi.org/10.1016/j.biotechadv.2021.107785
4. Zhang, K, et al.: Moisture-responsive films of cellulose stearoyl esters showing reversible shape transitions. Sci. Rep. **5**(1), 11011 (2015). https://doi.org/10.1038/srep11011
5. Cui, Y., Li, D., Gong, C., Chang, C.: Bioinspired shape memory hydrogel artificial muscles driven by solvents. ACS Nano **15**(8), 13712–13720 (2021). https://doi.org/10.1021/acsnano.1c05019
6. Mulakkal, M.C., Trask, R.S., Ting, V.P., Seddon, A.M.: Responsive cellulose-hydrogel composite ink for 4D printing. Mater. Des. (2018). https://doi.org/10.1016/j.matdes.2018.09.009
7. Li, Z., Wanga, J., Xu, Y., Shen, M., Duan, C., Dai, L., Ni, Y.: Green and sustainable cellulose-derived humidity sensors: a review. Carbohydr. Polym. (2021). https://doi.org/10.1016/j.carbpol.2021.118385
8. Ganesan, M., Kumar, R., Satapathy, D.K.: Bidirectional actuation of silk fibroin films: role of water and alcohol vapors. Langmuir **38**(19), 6066–6075 (2022). https://doi.org/10.1021/acs.langmuir.2c00315
9. Grabska-Zielinska, S., Sionkowska, A.: How to improve physico-chemical properties of silk fibroin materials for biomedical applications?—Blending and cross-linking of silk fibroin—a review. Mater. Mater. (2021). https://doi.org/10.3390/ma14061510
10. Nugroho, W.T., Dong, Y., Pramanik, A., Leng, J., Ramakrishna, S.: Smart polyurethane composites for 3D or 4D printing: general-purpose use, sustainability and shape memory effect. Compos. B Eng. 109104 (2021). https://doi.org/10.1016/j.compositesb.2021.109104
11. Arif, Z.U., Khalid, M.Y., Zolfagharian, A., Bodaghi, M.: 4D bioprinting of smart polymers for biomedical applications: recent progress, challenges, and future perspectives. React. Funct. Polym. (2022). https://doi.org/10.1016/j.reactfunctpolym.2022.105374
12. Falahati, M., Ahmadvand, P., Safaee, S., Chang, YuChung, Lyu, Z., Chen, R., Li, L., Lin, Y.: Smart polymers and nanocomposites for 3D and 4D printing. Mater. Today (2020). https://doi.org/10.1016/j.mattod.2020.06.001
13. Baker, A.B., Bates, S.R.G., Llewellyn-Jones, T.M., Valori, L.P.B., Dicker, M.P.M., Trask, R.S.: 4D printing with robust thermoplastic polyurethane hydrogel-elastomer trilayers. Mater. Des. **163**, 107544 (2022). https://doi.org/10.1016/j.matdes.2018.107544
14. Chan, M.A., Kay Obendorf, S.: Surface modification of microporous polypropylene membrane by UV-initiated grafting with poly (ethylene glycol) diacrylate. Fibers Polym. (2014). https://doi.org/10.1007/s12221-014-2032-8
15. Zhang, Z., Demir, K.G., Gu, G.X.: Developments in 4D-printing: a review on current smart materials, technologies, and applications. Int. J. Smart Nano Mater. (2019). https://doi.org/10.1080/19475411.2019.1591541
16. Ansari, M.J., Rajendran, R.R., Mohanto, S., Agarwal, U., Panda, K., Dhotre, K., Ravi Manne, A., Deepak, A.Z., Yasir, M., Praman, S.: Poly(N-isopropylacrylamide)-based hydrogels for biomedical applications: a review of the state-of-the-art. Gels (2022). https://doi.org/10.3390/gels8070454
17. Breger, J.C., Yoon, C., Xiao, R., Kwag, H.R., Wang, M.O., Fisher, J.P., Nguyen, T.D., Gracias, D.H.: Self-folding thermo-magnetically responsive soft microgrippers. ACS Appl. Mater. Interfaces (2015). https://doi.org/10.1021/am508621s

Chapter 26
Additive Manufacturing as an Enabler of Environmental Solutions to Address Food Security

Jenny Roberts, Philip Donkersley, Lisa Ashmore, and Allan Rennie

Abstract Pollinator decline is prevalent around the globe threatening food production, yields and the economic income of farmers. With reducing yields, land cultivation increases compromising natural habitats. Aligned with the UN Sustainable Development Goals of no poverty and zero hunger, the creation of artificial bumblebee nests enables a means to address habitat shortfalls. Considerations around preventing predator attacks, withstanding external environmental elements and creating a stable internal habitat were critical for success. Utilization of polymers with opacity and strength characteristics were important, whilst allowing for low volume prototype manufacturing methods (vacuum forming using additive manufactured formers). Subsequent design iterations utilised additive manufacturing as the primary production process, focusing on redistributed manufacture to enable rapid deployment in areas of crisis. Initial results from simulations and physical testing evidence the feasibility of the design in terms of strength, durability and environmental suitability. Deployment of six prototype units, with three artificially introduced bumblebee colonies, demonstrated a sustained natural reproduction cycle for a season. Subsequent deployment of empty units observed a successful wild queen habitation and sustained colony production over a season. Further field testing will ascertain how bumblebees utilise their nests long-term to drive future decisions on design and materials for environmental sustainability.

Keywords Additive manufacturing · Redistributed manufacture · Environmental solutions · Crisis response · Bio-inspired design

---

J. Roberts (✉) · P. Donkersley · L. Ashmore · A. Rennie
Lancaster University, Lancaster LA1 4YW, UK
e-mail: j.m.roberts@lancaster.ac.uk

P. Donkersley
e-mail: p.w.donkersley@lancaster.ac.uk

L. Ashmore
e-mail: l.ashmore@lancaster.ac.uk

A. Rennie
e-mail: a.rennie@lancaster.ac.uk

K. Zarbane and Z. Beidouri (eds.), *Proceedings of CASICAM 2022*, Springer Tracts in Additive Manufacturing, https://doi.org/10.1007/978-3-031-32927-2_26

26.1 Introduction

Aside from their intrinsic beauty, bumblebees are amongst the most ecologically and economically important of the many insects that pollinate our crops and wildflowers, ensuring secure food production [1, 2]. Ongoing mass declines threaten these key ecosystem service providers [3, 4] and bumblebees are suffering the most severe declines of all pollinators [5]. Although we have known for over 20 years that bumblebee populations are diminishing [3], declines are continuing despite efforts including Agri-Environment Schemes (AES) and sustainable agriculture [6]. These continued reductions indicate the methods we have employed to support population growth are not sufficient to completely address pollinator decline [7].

Contemporary pollinator conservation has been focused on feeding bees by planting wildflowers [8], but bees do not subsist on food alone. Nest sites have largely been ignored by AES resulting in two significant implications for wild bees: critically limited understanding of what makes a good nest site [9, 10] and the reduction in availability of nest sites due to intensive agricultural land management [11]. To address this shortfall and increase nest sites for bees, two distinct strategies can be employed: create wild habitats or install artificial nest boxes. The latter offers an immediate response for areas in pollinator crisis and has the potential to address the food security concerns of the future. However, despite rapid advances in materials science and manufacturing technology [12], bumblebee nest box design has not developed since the 1980s. Commercially available bee nests for horticulture are currently cardboard or ceramic-construction, which presents numerous limitations for use outdoors including lack of weather resistance and vulnerability to predators.

26.2 Materials and Methods

Limited prior studies concerning the development of artificial bumblebee nests make it challenging to definitively specify the requirements for optimum conditions. Nest site choice is predicated on many factors, most of which are poorly understood [9], yet dark and enclosed spaces are a consistent recurring theme. Most of the more common species of ground nesting bumblebee prefer dry, dark cavities and nests can turn up in a variety of unexpected places such as abandoned rodent holes, under garden sheds and in compost heaps [13, 14]. However, contemporary boxes used for commercial colonies are translucent, allowing ingress of natural light. Utilization of opaque material to limit light ingress was therefore promoted as a requirement of the nest design, limiting light ingress to reduce any light induced stress. Strength requirements are governed by deployment location and proximity to predators or curious animals. Typical deployment locations may be shared with small ruminants such as goats, therefore there was a requirement to withstand accidental force applied from a misplaced hoof. Additionally, a secure fastening and installation method was essential to prevent predators such as badgers, that have been known to destroy artificial nest boxes to access honey. The optimum internal cavity size can be approximated utilizing data on the average Bombus terrestris nest size and analysis of opportunistic nest sites such as upturned plant pots. A domed cavity with diameter 200 mm and height 125 mm was designed to best fulfil the volumetric requirement. Additional requirements included subterranean access to the nest as is typical for Bombus terrestris, management of faecal matter to reduce pathogens, environmental resistance over a prolonged deployment period ($n \geq 2$ years) and camera access port for assessing occupation status.

The design of the artificial nest box consisted of an upper dome and a lower base, sandwiching a gasket at the adjoining rim to prevent water ingress, held together with retainer clips. The base was designed to support a raised perforated platform, allowing faecal matter to collect away from the wax cells of the bumblebee nest, and with an opening for a subterranean entrance tube. A flat disc depression added to the dome component allowed the addition of a housing for data capturing electronics (Fig. 26.1). Acrylonitrile butadiene styrene (ABS) offered a balance of opacity, formability and strength, for the batch production of six prototype units via vacuum forming.

To determine the thickness of material required, the draw ratio (D) was calculated using the surface area (SA) and footprint (F) of each component (Eq. 26.1; Table 26.1). Draw ratio is defined as the proportion by which the material thickness will reduce during the forming process.

$$D = SA/F \tag{26.1}$$

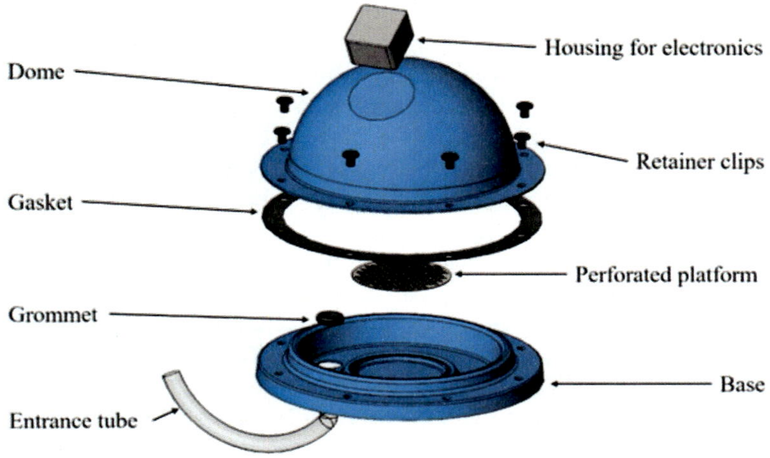

Fig. 26.1 Exploded view of artificial nest box design

Table 26.1 Component surface area, footprint and calculated draw ratio

Component	Surface area (cm^2)	Footprint (cm^2)	Draw ratio
Dome	1000	615	1.62
Base	1032	615	1.68

Therefore, if the desired thickness of the final components is 2 mm, a sheet of 3.36 mm thickness is required. 3.5 mm ABS sheet was therefore chosen as the closest standard size to ensure the final wall thickness remained greater than 2 mm.

Stress analysis was undertaken on the upper dome of the nest box, to determine whether it would withstand the accidental load from an adult male goat. Given an adult male goat typically weighs no more than 90 kg and assuming that the shape of the dome prevents the total weight of the goat to be applied, a value of 60 kg was chosen as the worst-case load scenario (half a 95-percentile male goat = 45 kg plus a safety factor (SF) of 1.33). 2 mm thickness ABS was specified and the load was applied over a circular area of 30 mm diameter at the top of the dome. The lower surface of the rim was fixed for the purpose of the analysis, given when deployed it would be attached to the base component of the nest box which in turn would be sunk into the ground.

With a load of 600 N applied, a maximum stress of 42.2 MPa and deflection of 1.74 mm (Fig. 26.2) was predicted. ABS has a yield strength of 48 MPa [15] however, given that the load scenario is possible but unlikely during the early testing phase, the value of 42.2 MPa was considered sufficient.

Traditionally, vacuum forming tools are manufactured from a variety of materials using subtractive processes for original designs and forming processes for replication of existing designs. However, additive manufacturing (AM) offers an alternative method of tool production enabling rapid manufacture and the potential for redistributed manufacture (RDM). The equipment to produce AM tooling, specifically material extrusion (MEx), is both more accessible and more economical compared with subtractive manufacturing methods, such as milling or turning on a lathe, and therefore offered an efficient means of developing the prototype nest boxes.

The formers were designed with draft angles between $5° \geq 30°$ and included equally spaced (18.6–22 mm centre to centre distance) through holes of diameter 1.5 mm at the point of gradient change from a horizontal surface. Draft and correctly placed through-holes ensure part definition and successful release from the moulded component post-forming. Male formers were designed to ensure compatibility with the vacuum former available (Formech 300XQ), with the design including compensation for shrinkage of the components post-forming.

Initial formers (Fig. 26.3) were built on a Builder Extreme 2000 Pro using white ABS filament, 40% infill, 2.5 mm wall thickness and a 0.2 mm layer height. Additional surface finishing with abrasive paper reduced the typically ridged surface that MEx produces. Subsequent designs were manufactured in two halves on an Ultimaker 3 before being bonded together, demonstrating the feasibility of using a more accessible 3d printer.

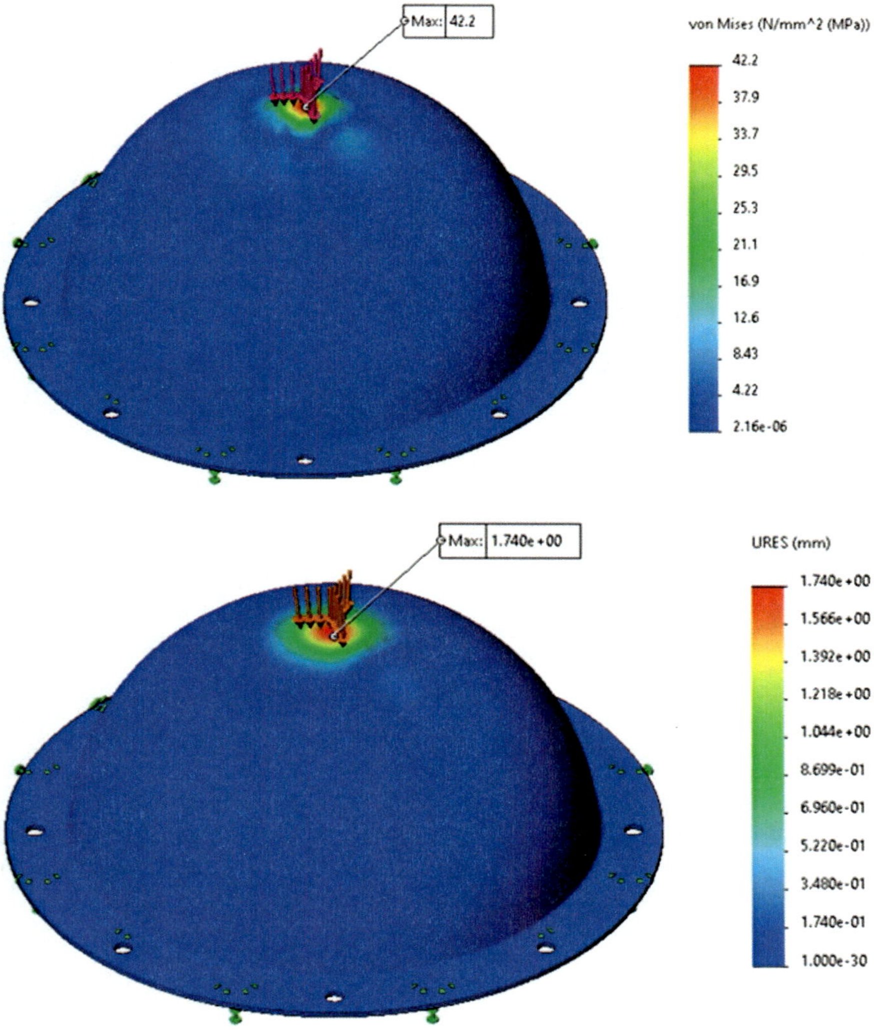

Fig. 26.2 VonMises stress analysis of dome component (top image). Displacement analysis of dome component (bottom image)

Sealing gaskets were cut from 1.5 mm thick rubber sponge sheet, perforated disks were waterjet cut from 1.5mm perforated stainless steel sheet, entrance tubes were formed into a curved tunnel and drainage holes added, and formed ABS components were manually finished using standard workshop manual/hand tools before being assembled into complete prototype units.

Six units were deployed at a field-testing site in Lancaster (UK) in May 2021, with the base and entrance tube sunk into the ground before attaching the top half

Fig. 26.3 Left to right: Base and dome being manufactured on the Builder Extreme 2000 Pro; Final base component former; Final dome component with MEx halves bonded together

Fig. 26.4 From left to right: Digging holes for nest box base and entrance hole deployment; Installed nest box base and entrance hole; Complete installation with domed top attached

of the assembly (Fig. 26.4). Captively bred Bombus terrestris colonies were placed in three of the six nest boxes, along with environmental data loggers (temperature, dew point and humidity) inside all occupied boxes and a single control data logger situated externally. The nest boxes remained deployed until September 2022.

26.3 Results and Discussion

26.3.1 Nest Box Suitability

The colonies were observed from May 2021 through to their natural senescence in September. World first video data (Fig. 26.5) captured the natural life operations of bumblebee nests, including: wax cell production, production of food stores, egg laying by the queen, hatching of imago adult bees from pupal cells, the successful lifecycle of the nest through to gyne (virgin queens) production and absconding during the senescence period.

Subsequent observation of a cleaned and empty nest box in March 2022 witnessed wild queen recruitment resulting in successful colony production, however further

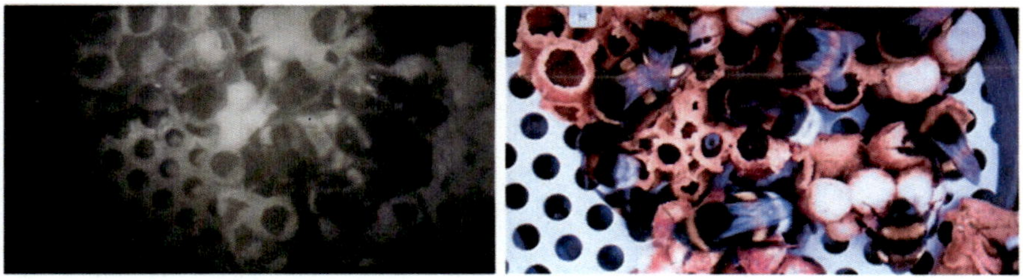

Fig. 26.5 Left: Infra-Red (IR) camera capture from inside the artificial nest box. Right: colour camera capture from inside a translucent AM iteration of the artificial nest box

research is required to ascertain whether the queen was truly wild or returning having been born in the nest box the previous year. Nevertheless, demonstration of recruitment of this type gives confidence in the suitability of the nest box for use in areas of natural habitat decline or destruction. Further field testing will ascertain how bumblebees utilise their nests long-term to drive future decisions on design and materials for environmental sustainability.

26.3.2 Environmental Data

Data loggers captured temperature, dew point and humidity data within and outside of the nest boxes during habitation. Typical data captured (Fig. 26.6) illustrated that internal temperature fluctuates in line with outside measurements, albeit with elevated peaks likely due to thermal absorption from the nest box materials. It is anticipated that this is akin to fluctuations experienced by the ground nesting bumblebee in natural habitats, although this is likely dependent on nest location above or below ground. Dew point inside the artificial nest box can be seen to fluctuate with that outside, but peaks again appear elevated, influenced by both temperature and humidity. Humidity inside the nest box fluctuates to an extent, however there are periods where it remains high despite the outside humidity dropping. Further analysis of the results suggests this correlates with a period of rainfall in combination with lower day-time temperatures and therefore less fluctuation in internal nest box temperature. Fungal and bacterial parasites will grow rapidly in high humidity conditions and are well known as a threat to bumblebee nests, leading to colony mortality [16].

The current design has a single entrance hole to allow access and ventilation. Video data captured bees fanning the entrance tube with their wings to promote air circulation, however this was not sufficient to reduce humidity effectively. Therefore, the use of AM processes which result in naturally porous structures (e.g. from scaffold production in MEx processes or the process of powder bed sintering) may well benefit this application [17]. In addition, incorporating learning from the natural world, such as the passive ventilation methods employed in termite nests, may provide a solution to providing environmentally balanced structures that remain weather resistant. The

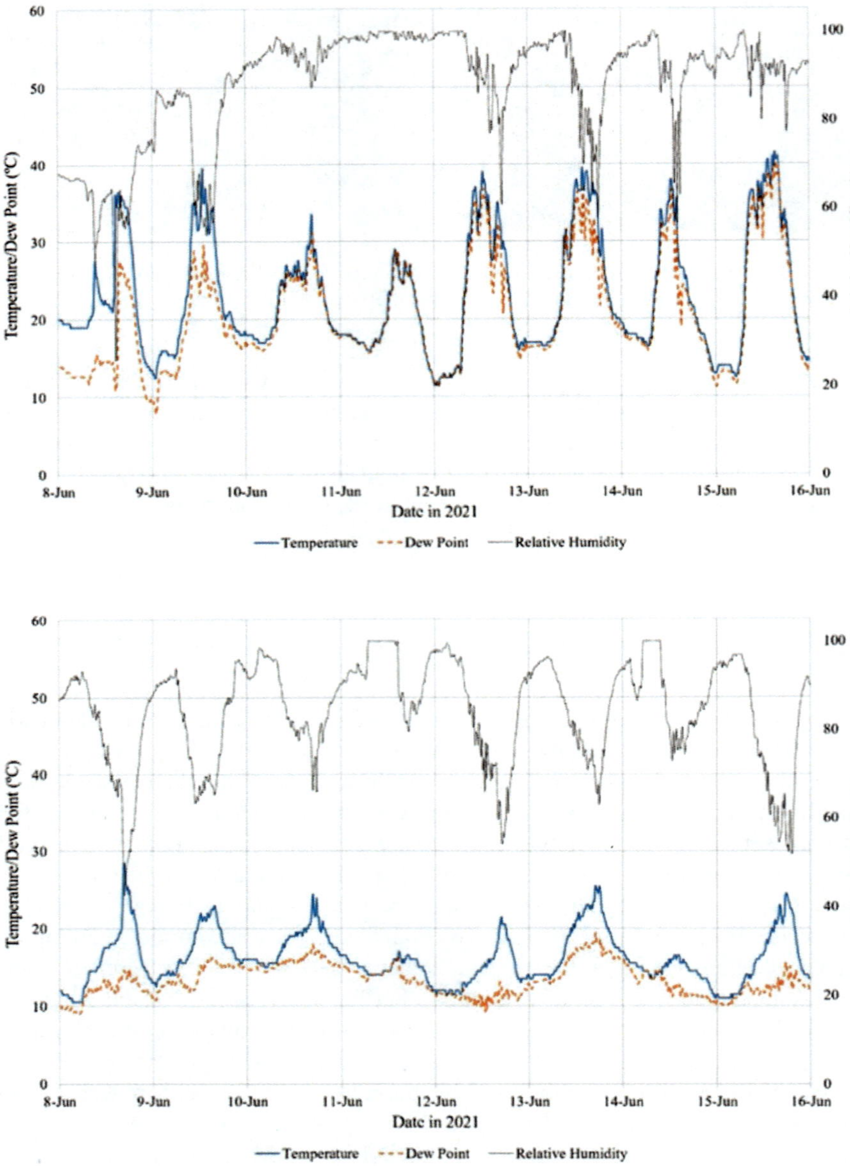

Fig. 26.6 Environmental data internal (top image) and external (bottom image) to the nest boxes

application of computational fluid dynamics (CFD) to future designs will give key insights into the effectiveness of these more organic structures in optimising air flow and internal humidity regulation. Consideration around geographical location of deployment further highlights the need for active management of humidity but also challenges how effective temperature and dew point management is currently and whether deployment in a location with a temperate climate has masked any potential shortcomings.

26.3.3 Structural Integrity

There was no evidence of tampering from predators or accidental damage from larger animals. However, the location selected for the initial field trials was a managed site with little threat from anything other than curious wild mammals. Effective design strategies to reduce the likelihood of sustained force being applied to the artificial nest box will reduce the requirement for excessive structural strength, but geographical location and associated fauna may dictate the minimum level of structural integrity required. Designing to ensure misplaced hooves glance off the nest box, rather than allow force to be applied, will reduce the force requirements substantially. RDM, where data files are sent electronically to enable production at multiple locations, will allow the freedom to select the most appropriate design for the deployment location demonstrating lean and sustainable principles [18]. For example, in areas with little or no expected damage from ruminants, strength requirements are minimal and therefore the design can be optimised to use less material. Conversely, in areas where accidental load is more likely, additional structural elements can be included to ensure the nest boxes' longevity.

26.3.4 Future Design and Manufacture

The production of six vacuum formed prototype nest boxes was labour intensive and prone to errors. With elevated production rates, employing automated manufacturing techniques becomes more feasible, enhancing quality and efficiency. However, with increased quantity, distribution logistics will become more prevalent, especially if the artificial nest box is intended for use in areas of crisis that may be a substantial distance from the place of manufacture. AM offers access to RDM to enable production at the point of need, reducing the time and cost associated with distributing products and enabling a more agile response.

Subsequent research has resulted in the replication of the existing nest box design dome utilising MEx with ABS filament. The manufactured dome was initially used solely for demonstrations of the research, in which livestream video was broadcast from within the nest box utilising a Raspberry Pi camera module. Latterly, the unit has been coated in black paint primer and deployed outdoors, where initial observational data suggests that it performs adequately. Further research is required to ascertain the comparative performance against the earlier vacuum formed version, both in terms of internal environment and strength. Whilst AM offers more design freedom and opportunity to address the current environmental limitations, strength performance is typically less favourable with polymer MEx structures [19].

Drawing on inspiration from stalactites and termite nests, exploration of internal condensing stalactite structures that deposit moisture into a reservoir is an area of imminent research. It is anticipated that these structures will also provide additional strength to the dome akin to a vaulted ceiling, thus exploiting the advantages of design for additive manufacture (DfAM) fully.

26.4 Conclusions

The research has demonstrated the success of the artificial bumblebee nest box in both the sustainment of colonies throughout the natural cycle and through wild queen recruitment. Issues around humidity control and prototype production methods have been identified as being detrimental to future success. The application of DfAM principles in conjunction with strategies to create a homeostatic environment within the nest boxes will advance the current research unit into a functional product that can be utilised around the globe. Striving to remove geographical barriers via RDM will enable use in areas where pollinator decline is hampering crop production, increasing farming yields to prevent both hunger (directly) and poverty via loss of income. The importance of this alignment to the UN Sustainable Development Goals of zero hunger and no poverty cannot be understated.

Acknowledgements The authors acknowledge the support of the Economic and Social Research Council (ESRC) Impact Acceleration Account (IAA) (Grant Reference: ESRC-IAA 2019), awarded through Lancaster University.

References

1. Brown, M.J.F., Dicks, L.V., Paxton, R.J., Baldock, K.C.R., Barron, A.B., Chauzat, M.P., Freitas, B.M., Goulson, D., Jepsen, S., Kremen, C., et al.: A horizon scan of future threats and opportunities for pollinators and pollination. PeerJ **4**, [e2249] (2016)
2. Wahengbam, J., Raut, A.M., Pal, S., Najitha Banu, A.: Role of bumble bee in pollination. Annals Biol. **35**(2), 290–295 (2019)
3. Burkle, L.A., Marlin, J.C., Knight, T.M.: Plant-pollinator interactions over 120 years: Loss of species, co-occurrence, and function. Science (New York, N. Y.) **339**(6127), 1611–1615 (2013)
4. Bartomeus, I., Stavert, J.R., Ward, D., Aguado, O.: Historical collections as a tool for assessing the global pollination crisis. Philos. Trans. R. Soc. Lond. Ser. Biol. Sci. **374**(1763), 20170389 (2018)

5. Carvalheiro, L.G., Kunin, W.E., Keil, P., Aguirre-Gutiérrez, J., Ellis, W.N., Fox, R., Groom, Q., Hennekens, S., Van Landuyt, W., Maes, D., et al.: Species richness declines and biotic homogenisation have slowed down for NW-European pollinators and plants. Ecol. Lett. **16**(7), 870–878 (2013)
6. Wagner, D.L., Grames, E.M., Forister, M.L., Berenbaum, M.R., Stopak, D.: Insect decline in the Anthropocene: death by a thousand cuts. Proc. Natl. Acad. Sci. U.S.A. **118**(2), e2023989118 (2021)
7. Donkersley, P.: Trees for bees. Agr. Ecosyst. Environ. **270–271**, 79–83 (2019)
8. Cole, L.J., Stockan, J., Helliwell, R.: Managing riparian buffer strips to optimise ecosystem services: a review. Agric. Ecosyst. Environ. **296**, [106891] (2020)
9. O'Connor, S., Park, K.J., Goulson, D.: Location of bumblebee nests is predicted by counts of nest-searching queens. Ecol. Entomol. **42**, 731–736 (2017)
10. Becher, M.A., Twiston-Davies, G., Penny, T.D., Goulson, D., Rotheray, E.L., Osborne, J.L.: Bumble-BEEHAVE: a systems model for exploring multifactorial causes of bumblebee decline at individual, colony, population and community level. J. Appl. Ecol. **55**(6), 2790–2801 (2018)
11. Wood, T.J., Holland, J.M., Goulson, D.: Pollinator-friendly management does not increase the diversity of farmland bees and wasps. Biol. Cons. **187**, 120–126 (2015)
12. Berman, B.: 3-D printing: the new industrial revolution. Bus. Horiz. **55**(2), 155–162 (2012)
13. Kells, A.R., Goulson, D.: Preferred nesting sites of bumblebee queens (Hymenoptera: Apidae) in agroecosystems in the UK. Biol. Cons. **109**(2), 165–174 (2003)
14. Osborne, J.L., Martin, A.P., Shortall, C.R., Todd, A.D., Goulson, D., Knight, M.E., Hale, R.J., Sanderson, R.A.: Quantifying and comparing bumblebee nest densities in gardens and countryside habitats. J. Appl. Ecol. **45**(3), 784–792 (2008)
15. Ebnesajjad, S.: Fluoroplastics, 2nd edn. William Andrew, Oxford (2015)
16. Goulson, D., O'Connor, S., Park, K.J.: Causes of colony mortality in bumblebees. Anim. Conserv. **21**(1), 45–53 (2018)
17. Gibson, I., Rosen, D.W., Stucker, B.: Additive Manufacturing Technologies: Rapid Prototyping to Direct Digital Manufacturing. Springer, New York (2009)
18. Ghobadian, A., Talavera, I., Bhattacharya, A., Kumar, V., Garza-Reyes, J.A., O'Regan, N.: Examining legitimatisation of additive manufacturing in the interplay between innovation, lean manufacturing and sustainability. Int. J. Prod. Econ. **219**, 457–468 (2020)
19. Ngo, T.D., Kashani, A., Imbalzano, G., Nguyen, K.T.Q., Hui, D.: Additive manufacturing (3D printing): A review of materials, methods, applications and challenges. Compos. B Eng. **143**, 172–192 (2018)

Chapter 27
Lean and Additive Manufacturing: How Can Additive Manufacturing Contribute to Lean Objectives?

Laila Driouach, Khalid Zarbane, and Zitouni Beidouri

Abstract Applying Lean Manufacturing can significantly improve a company's performance by acting on quality, productivity and costs by eliminating or reducing waste. In addition, additive manufacturing is considered a promising trend in several sectors and especially in manufacturing. However, the existing literature lacks a comprehensive and detailed conjunction of these two concepts in terms of how they complement each other and are compatible. In this paper, we propose to investigate how the use of 3D printing facilitates the achievement of Lean Manufacturing objectives. The main purpose of the study is to highlight the convergences between the two concepts and thus enable an assessment of the extent to which Lean concepts and Additive Manufacturing processes complement each other. Firstly, we build on existing work to conclude that 3D printing contributes to Lean Manufacturing objectives. Secondly, this work examines how Additive Manufacturing could concretely help avoid wastes, improve processes and achieve Lean Manufacturing.

Keywords Additive manufacturing · Lean manufacturing · Lean objectives · 3D printing · Waste

L. Driouach (✉) · Z. Beidouri
National Higher School of Electricity and Mechanics (ENSEM), Hassan II University, Casablanca, Morocco
e-mail: Laila.driouach@ensem.ac.ma

Z. Beidouri
e-mail: zitouni.beidouri@univh2c.ma

K. Zarbane
Higher School of Technology (EST), Hassan II University, Casablanca, Morocco
e-mail: khalid.zarbane@univh2c.ma

K. Zarbane and Z. Beidouri (eds.), *Proceedings of CASICAM 2022*, Springer Tracts in Additive Manufacturing, https://doi.org/10.1007/978-3-031-32927-2_27

27.1 Introduction

Nowadays, 3D printing or Additive Manufacturing (AM) is receiving considerable attention from both the academic and industrial communities [1]. AM is a material assembly process that makes objects from the data in a 3D model, typically layer by layer, as opposed to subtractive manufacturing methods [2]. It's also known as additive layer manufacturing, 3D printing, digital manufacturing, rapid prototyping, rapid manufacturing [3]. Additive Manufacturing has become an increasingly common technology for creating prototypes, complex parts and custom products. AM applications are growing in a wide variety of markets, including aerospace, automotive and medical [4].

Lean Manufacturing (LM) is a management philosophy based on the Toyota Production System. The LM aims to improve the company's overall performance by applying a range of tools, techniques and methods [5, 6]. One of the primary goals of LM is the elimination of waste of time, money and other resources. However, the relationship between 3D printing and Lean Manufacturing is rarely studied. In this paper, we will highlight the correlation between Lean Manufacturing and Additive Manufacturing. We will also explain how additive manufacturing could be a great way to achieve an effective lean manufacturing system and address the question: What can Additive Manufacturing contribute to Lean Manufacturing? To meet this objective, we use a methodology based on the analysis of some works that study the two concepts.

27.2 Methodology

This paper's methodology is based on a three-step approach (Fig. 27.1). The first step is to highlight the importance of AM in general and in particular in Morocco. The second step is to collect data from the literature on the relationship between AM and lean production. Finally, the last step is to discuss these two concepts as a tool to increase efficiency and eliminate sources of inefficiency.

27.3 General Overview

Producing a functional part requires a multitude of manufacturing processes, each with different specificities, and allowing meeting the specifications requirements. Among these processes, we can distinguish: subtractive manufacturing by removing material, forming which consists in imposing a geometric shape on a material, additive manufacturing.

Additive manufacturing encompasses all manufacturing processes using added materials. According to ISO/ASTM 52,900:2015, it is defined as a process of adding

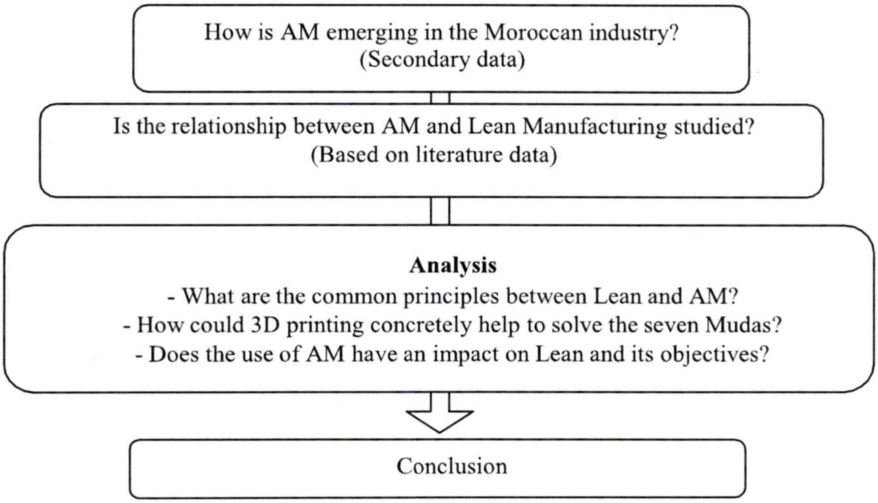

Fig. 27.1 Research methodology

materials to make parts from 3D model data, commonly layer by layer, as distinct from subtractive and shaping manufacturing methods [7]. Additive manufacturing processes appeared in the 1980s [8]. During the 1990s–2000s, AM applications involved just manufacturing of technological models and prototypes [9]. Nowadays, additive manufacturing technologies are used to produce functional parts that meet the expectations of customer specifications. Many sectors such as education, research and industry involve additive manufacturing [10].

By using 3D printing, companies can save up to 25% in cost and 40% in time compared to traditional manufacturing methods by making complex structures that are both stable and lightweight [11]. In recent years, advanced AM techniques have grown significantly, resulting in broader industrial applications including product development, finished goods production, and mass production. In fact, the 3D printing market was worth $13.7 billion in 2020, and is estimated to reach USD 63.46 billion by 2026, growing at a 29.48% compound annual growth rate CAGR over the forecast period (2021–2026) [12]. Additionally, 2.1 million units of 3D printers were shopped worldwide in 2020, this number is expected to attain 15.3 million units by 2028. Aerospace, automotive, orthodontics, tooling and production systems are the most sectors using 3D printing not only for models and prototypes, but increasingly for industrial-scale production [11].

In Morocco, some companies and start-ups have already begun the process of adopting AM. Today, they are well positioned to make more significant changes toward the democratization of this technology. A French electronics group specializing in aeronautics has created an industrial skills center in Morocco specialized in metal additive manufacturing, also known as "3D printing" in Casablanca. A Moroccan university in Fez has also created one of the largest AM platforms in

Morocco and probably in Africa; it includes about 50 3D printing machines. For years now, Moroccan industry has been using additive manufacturing, and even higher education institutions are getting involved through the opening of a network of university Fab Labs (Fabrication Laboratory) [13]. Additive manufacturing is now starting to change the world of manufacturing and is being called a new industrial revolution. However, the cost of the technology is one of the factors that sometimes discourages companies from investing in 3D printing.

Additive manufacturing is gaining popularity not only in the manufacturing industry, but also in the consumer market, as it offers a new world of opportunities, starting with the absence of geometric constraints and the reduction of waste due to material removal, typical of subtractive manufacturing. In addition, it is a great way to reinforce Lean Manufacturing objectives [14]. Lean manufacturing (LM) was initially developed within the Toyota Production System in Japan in the 1950s [15]. However, the term "lean manufacturing" was not coined until 1988 by John Krafcik in the article "Triumph of the lean production system" in the MIT Sloan Management Review. Lean production is a holistic system that includes a set of rules, guidelines, tools and techniques to eliminate significant waste in all company processes for continuous improvement [16].

The use of 3D printing could clearly contribute to setting up a lean manufacturing model. Despite the fact that LM and AM are two complementary concepts, authors rarely mention this relation. Ghobadian et al. in their paper looked at how AM can significantly reduce/eliminate waste [17]. Torres et al. in their work offer a theoretical and managerial perspective on how AM technology can support green and lean supply chain practices, thereby contributing to better supply chain performance [18]. Sini et al. attempted to further study the defects that affect 3D printed products and propose new ways to control them [14].

While the adoption and industrial use of additive manufacturing is discussed nowadays, we notice that there is little research that relates lean manufacturing and additive manufacturing, so it is necessary to address the question of the level of overlap and complementarity between AM and LM. The answer to this question is of significant importance to the future practice of LM, it is also important to know to what extent AM offers a technological solution capable of eliminating process waste, which are the core of the LM concept.

27.4 Results and Discussion

27.4.1 Contributions of 3D Printing on Lean Manufacturing

AM is in transition from manufacturing prototypes in the design phase to industrial use [2]. This transition requires anticipating many questions and, in particular, studying the expected benefits of AM and its effects on the company's performance. One of questions that arises is the impact of AM's emergence on the lean manufacturing objectives.

According to [19] additive manufacturing is an emerging tool that contributes to the sustainable manufacturing of components in several fields, based on the concept of "zero waste"; which is one of the main objective of lean manufacturing. In addition, AM technology improves supply chain efficiency by contributing to waste reduction, elimination of many assembly steps, and reduction of energy consumption, resulting in "leaner" and "greener" production processes [14]. Indeed, one of the benefits of additive manufacturing is to simplify the assembly process: some parts that had to be made from the assembly of several components are now obtained directly by additive manufacturing in a single part. This can reduce manufacturing time and simplify the components' production [7].

The layer-by-layer production methodology offers other advantages such as geometry flexibility as any geometry can be produced in a single operation without additional cost or time constraints. It also reduces resource consumption since only the precise amount of material is needed to create the final product, avoiding the usual waste of traditional manufacturing [3].

A key advantage of 3D printing that complies with the LM objectives is the small batch. With 3D printing, a product can be manufactured on demand, which eliminates the requirement to stock the product and ensures easier inventory and information system management, and thus satisfies the pulled system and zero stock concepts of lean manufacturing.

Reducing the cost of inventory obsolescence is an upside to AM deployment. Rather than searching for multiple components in an inventory, the part can simply be found in an online library and 3D printed. This eliminates the need for large storage space to quickly respond to customer needs [2].

3D printer operations are mostly automated. This property allows separating human work from the machine, which is one of the key aspects of the "Jidoka principle" of lean manufacturing, and is a key source of quality, productivity and cost reduction. Human-machine separation changes the worker's task from operating the machine to engaging in the process of continuous improvement. In addition, a 3D printer is easy to self-maintain, which facilitates the implementation of 5S actions.

Manufacturing facilities equipped with 3D printers can be isolated from other facilities, facilitating an optimal layout that helps eliminate waste, increase flexibility and spread the workload across all 3D printers, which is also in accordance with the Lean manufacturing objectives.

Just-in-time (JIT) manufacturing is a production method in which products are created to meet demand, not in excess or in advance of need. JIT manufacturing is a key Lean manufacturing objective. In the case of additive manufacturing, this objective is confirmed because it is possible to use a 3D printer to manufacture a 3D object on a just-in-time basis, taking into account the necessary printing and shipping times, the location of the customer and the order deadline.

Thanks to AM, it is possible to save up to 90% on storage and inventory costs as well as on transportation costs. Consequently, total production costs can be reduced and product availability can be improved [2]. Figure 27.2 shows how 3D printing helps to achieve the objectives of lean manufacturing.

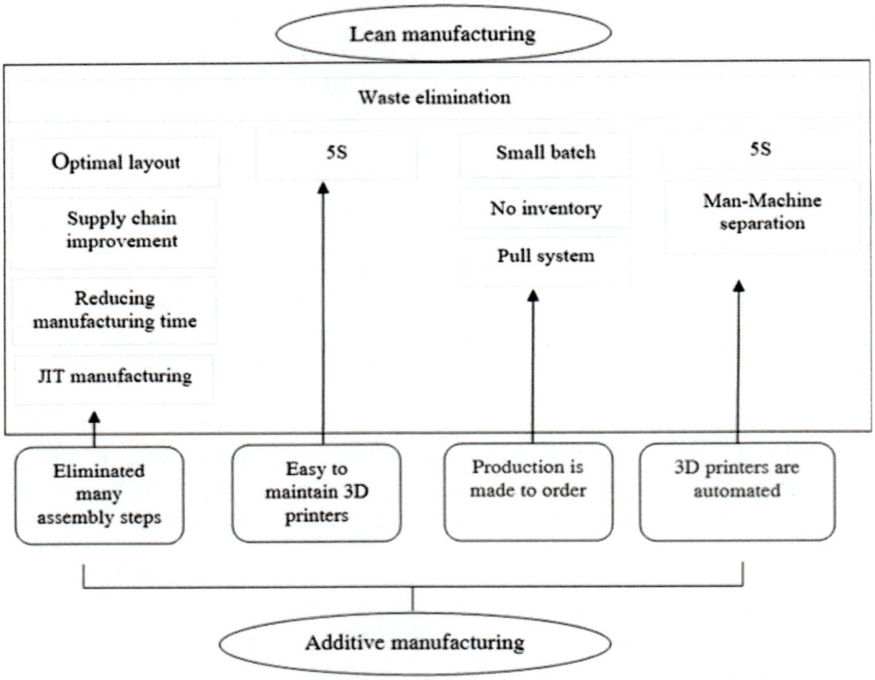

Fig. 27.2 Contribution of AM in achieving Lean Manufacturing objectives

27.4.2 3D printing facing wastes

The objective of LM is to maximize perceived value by customers by minimizing non-value- adding activities, called Muda (Japanese word for waste). Seven Mudas to avoid improving processes and achieving lean manufacturing are [20]:

- **Transportation and useless displacements**: When moving an object or materials that do not add value to the product.
- **Inventory**: This involves storing quantities greater than the quantity required for the next step in the manufacturing process, from raw material to finished product.
- **Motion**: This waste refers to any unnecessary physical motions or walking of workers that distract them from actual processing work.
- **Waiting time**: This refers to the idle time of workers or machines, i.e. the time during which the product is not transported or manufactured. Flow times within the process must be reduced, to achieve "just-in-time manufacturing" is one of the main principles of lean manufacturing.
- **Overproduction**: Producing more than the customer requires. Overproducing wastes time, energy, money and even credibility.
- **Over-processing**: Adding too much processing work compared to the customer's requirements in terms of product quality or functionality.
- **Defects**: Damaged parts require rework, which wastes time.

All of these wastes have a direct impact on costs, as they are non-value-added operations that do not improve the company's process or product. The question then is how 3D printing could concretely help to solve the seven Mudas.

- Transportation and useless displacements: Manufacturing a part using additive manufacturing does not require several production steps. Thus, 3D printing can reduce unnecessary movement of parts.
- Stock management: By using 3D printing, only what the customer has ordered is produced, thus minimizing inventory!
- Motion: When using MA, the parts are entirely manufactured in a single workstation, which is the 3D printers, so employees do not have to make unnecessary movements to create the objects or assemble the parts.
- Waiting and lead time: 3D printing reduces waiting times, as there are no different steps to follow for this type of manufacturing. The product is delivered once it is made.
- Overproduction: In this case, production is made to order, only the necessary parts are printed, and only the necessary amount of material is used.
- Over-processing: Manufacturing a part by AM does not require a lot of processing steps; to create the parts a 3D printing material and a finishing if necessary.
- Defects: The waste that cannot still be reduced by additive manufacturing is the presence of defect in the final product. Once the parts are printed, a quality control is needed [3].

Additive manufacturing promises to revolutionize the manufacturing industry by radically eliminating waste and thereby ensuring sustainability [17]. AM effectively contributes to waste elimination, which is a key component of a lean manufacturing approach. 3D printing reduces costs, avoids material waste, avoids inventory, and creates only what the customer orders. Finally, this manufacturing process improves product quality and shortens lead times.

27.5 Conclusion

Additive manufacturing appears to be a relevant solution for component manufacturing thanks to its numerous advantages such as free design, simplified assembly and others. It also allows an increasing flexibility and a better control of the production chain.

In this work, we have highlighted the importance of this technology and given an overview of its beginnings in Morocco; it has been introduced through associations, start-ups and educational institutions. A key question is highlighted: how additive manufacturing can promote lean manufacturing and achieve its objectives.

This work allowed us to conclude that 3D printing can reduce material waste, avoid inventory and create only what the customer has ordered. Indeed, a 3D printer can print a 3D object in just-in-time (JIT), by ensuring the shipment of goods when they are needed, thereby reducing inventory levels and associated costs. Just-in-time manufacturing is a key objective of lean manufacturing. In addition, the operations of a 3D printer are largely automated. This property allows for human–machine separation, which is also a key principle of lean manufacturing. In addition, the most direct advantage of 3D printing is the small batch size, which is also a goal of lean manufacturing. In addition, the additive manufacturing process actually improves product quality and shortens lead times.

There are many discussions about the effectiveness of AM as a technology contributing to the principles of sustainability, energy efficiency, and lean manufacturing in its broadest context. The paper is therefore very timely, especially with the focus on the industrial adoption and use of AM for commercial and also localized purposes for the Moroccan economy.

References

1. Jiang, J., Newman, S.T., Zhong, R.Y.: A review of multiple degrees of freedom for additive manufacturing machines. Int. J. Comput. Integr. Manuf. **34**(2), 195–211 (2021). https://doi.org/10.1080/0951192X.2020.1858510
2. Antomarchi, A.L., Durieux, S., Duc, E.: Impact de la fabrication additive sur la supply chain : état des lieux et diagnostics. Logist. Manage. **28**(1), 29–47 (2020). https://doi.org/10.1080/12507970.2019.1682950

3. Chiabert, P., Sini, F.: Lean management in Additive manufacturing : a methodological proposal for quality control. **91** (2020)
4. Zawaski, C.E., Chatham, C.A., Wilts, E.M., Long, T.E., Williams, C.B.: Using fillers to tune material properties of an ion-containing semi-crystalline polyethylene glycol) for fused filament fabrication additive manufacturing. Addit. Manuf. **39**, 101844 (2021). https://doi.org/10.1016/j.addma.2021.101844
5. Ribeiro, P., Sá, J.C., Ferreira, L.P., Silva, F.J.G., Pereira, M.T., Santos, G.: The impact of the application of lean tools for improvement of process in a plastic company: a case study. Proc. Manuf. **38**, 765–775 (2019). https://doi.org/10.1016/j.promfg.2020.01.104
6. Roriz, C., Nunes, E., Sousa, S.: Application of lean production principles and tools for quality improvement of production processes in a carton company. Proc. Manuf. **11**, 1069–1076 (2017). https://doi.org/10.1016/j.promfg.2017.07.218
7. Querard, V.: Réalisation de pièces aéronautiques de grandes dimensions par fabrication additive WAAM. L'école centrale de Nante Comue Université Bretagne Loire (2019)
8. Bourell, D.L., Beaman, J.J, Leu, M.C., Rosen, D.W.: A brief history of additive manufacturing and the 2009 roadmap for additive manufacturing: looking back and looking ahead. In: US-Turkey Workshop on Rapid Technologies (2009)
9. Cottelee, R.M.J.: 3D opportunity: additive manufacturing paths to performance, innovation, and growth. In: SIMT Additive Manufacturing Symposium, p. 23 (2014)
10. Geraedts, J., Doubrovski, E., Verlinden, J., Stellingwerff, M.: Three views on additive manufacturing: business, research and education. In: Conference TMCE 2012, no. February 2016, pp. 1–15 (2012)
11. Impression 3D : Des débouchés prometteurs pour la production (2021). http://fnh.ma. https://fnh.ma/article/-/impression-3d-des-debouches-prometteurs-pour-la-production
12. Taille, croissance, tendances du marché de l'impression 3D (2022–27) | Prévisions de l'industrie. http://www.mordorintelligence.com. Accessed 11 May 2022
13. Sophie Martin, A.: L'impression 3D se fait une place au Maroc—La Vie éco (2016). https://www.lavieeco.com/. https://www.lavieeco.com/economie/limpression-3d-se-fait-une-place-au-maroc/
14. Sini, F., Bruno, G., Chiabert, P., Segonds, F.A.: Lean quality control approach for additive manufacturing. In: Nyffenegger, F., Ríos, J., Rivest, L., Bouras, A. (Éds.) Product Lifecycle Management Enabling Smart X, Vol. 594, pp. 59–69. Springer International Publishing (2020). https://doi.org/10.1007/978-3-030-62807-9_6
15. Driouach, L., Zarbane, K., Beidouri, Z.: Literature review of lean manufacturing in small and medium-sized enterprises. Int. J. Technol. **10**(5), 930 (2019). https://doi.org/10.14716/ijtech.v10i5.2718
16. Womack, J.P., Jones, D.T., Roos, D.: The machine that changed the world. N.Y. **323**, 273–287 (1990)
17. Ghobadian, A., Talavera, I., Bhattacharya, A., Kumar, V., Garza-Reyes, J.A., O'Regan, N.: Examining legitimatisation of additive manufacturing in the interplay between innovation, lean manufacturing and sustainability. Int. J. Prod. Econ. **219**, 457–468 (2020). https://doi.org/10.1016/j.ijpe.2018.06.001
18. Torres, B., Naghshineh, B., Cardeal, G., Filipe, D., Carvalho, H., Peças, P., Ribeiro, I.: The impacts of additive manufacturing technology on lean/green supply chain management practices. In: Rossi, M., Rossini, M., Terzi, S. (Éds.) Proceedings of the 6th European Lean Educator Conference, Vol. 122, pp. 159–168. Springer International Publishing (2020). https://doi.org/10.1007/978-3-030-41429-0_16
19. Inoque, N., de Faria, L., Muñoz, R.: Additive manufacturing towards the fabrication of greener electrochemical sensors for antioxidants. J. Braz. Chem. Soc. (2022). https://doi.org/10.21577/0103-5053.20220025
20. Driouach, L., Oumami, M.E., Beidouri, Z., Zarbane, K.: Towards implementing lean manufacturing in very small businesses in Morocco : qualitative exploratory study. Int. J. Supply Chain Manage. **9**(6), 6 (2020)

Methods in Statistical Ecology

This new series in statistical ecology is designed to cover diverse topics in emerging interdisciplinary research. The emphasis is on specific statistical methodologies utilized in burgeoning areas of quantitative ecology. The series focuses primarily on monographs from leading researchers.

More information about this series at https://link.springer.com/bookseries/10235

David I. Warton

Eco-Stats: Data Analysis in Ecology

From *t*-tests to Multivariate Abundances

Springer

David I. Warton
School of Mathematics and Statistics
and the Evolution & Ecology Research Centre
The University of New South Wales
Sydney, NSW, Australia

ISSN 2199-319X ISSN 2199-3203 (electronic)
Methods in Statistical Ecology
ISBN 978-3-030-88442-0 ISBN 978-3-030-88443-7 (eBook)
https://doi.org/10.1007/978-3-030-88443-7

This Springer imprint is published by the registered company Springer Nature Switzerland AG
The registered company address is: Gewerbestrasse 11, 6330 Cham, Switzerland

Preface

The core idea in this book is to recognise that most statistical methods you use can be understood under a single framework, as special cases of (generalised) linear models—including linear regression, t-tests, ANOVA, ANCOVA, logistic regression, chi-squared tests, and much more. Learning these methods in a systematic way, instead of as a "cookbook" of different methods, enables a systematic approach to key steps in analysis (like assumption checking) and an extension to handle more complex situations (e.g. random factors, multivariate analysis, choosing between a set of competing models). A few simplifications have been made along the way to avoid "dead-end" ideas that don't add to this broader narrative.

This book won't teach readers everything they need to know about data analysis in ecology—obviously, no book could—but there is a specific focus on developing the tools needed to understand modern developments in multivariate analysis. The way multivariate analysis is approached in ecology has been undergoing a paradigm shift of late, towards the use of statistical models to answer research questions (a notion already well established in most fields, but previously difficult to apply in many areas of ecology). The first half of the book (Parts I–II) provides a general platform that should be useful to any reader wanting to learn some stats, thereafter (Part III) it focuses on the challenging problem of analysing multivariate abundances.

This book started out as notes for a couple of intensive short courses for ecologists, each taught across just one week (or less!). OK, those two courses don't quite cover all of the material in this book, and they are super intensive, but the emphasis on a few key principles makes it possible to cover a lot of ground, and importantly, modern statistics can be seen as a cohesive approach to analysis rather than being misunderstood as an eclectic cookbook of different recipes.

The book is punctuated with exercises and examples to help readers check how they are going and to motivate particular chapters or sections. *All* correspond to real people, projects, and datasets. Most come from colleagues doing interesting work or problems I encountered consulting; some are gems I found in the literature. Where project outcomes were published, a reference has been included. Solutions to exercises can be found in vignettes in the R package `ecostats`, available at CRAN (`https://cran.r-project.org`), organised by chapter.

The computing component of the book makes use of R, but the code is for the most part separated out from the main text into code boxes, the intention being that the main ideas of the book are independent of the analysis platform. However, the book will be much easier to understand, and the ideas much easier to implement, if you do follow through with it using R and work through the analysis exercises scattered throughout. Many datasets are used to demonstrate core ideas in the book, and all of these are freely available in the R package `ecostats` (unless already available elsewhere on R).

Statistics is written in the language of mathematics; anything else is a rough translation. When we communicate about statistics in English alone (or any another language), some messages can get a little distorted, and some practices can develop in applied statistics that have no mathematical basis or, worse, that are poor practice. For example, *independence* has a specific meaning in statistics that is slightly different to its meaning in English. If you corroborate a result from one dataset using what you consider to be an independent dataset (because it comes from a difference source), this may not necessarily be statistically independent if the new values are correlated with those in the original dataset (e.g. because it involved sampling at nearby locations). A bit of maths can help clarify this issue. So while this book is intended to make statistical thinking accessible to ecologists who might not be up for a lot of maths, some maths boxes have also been included to offer a deeper understanding of key results and the reasons for some of the recommendations made here. The maths boxes vary in difficulty—some don't involve much maths at all, and some might look a tad intimidating, like a proof of the central limit theorem in Chap. 1! The harder ones are flagged with  or , for maths ninjas. The maths boxes are written as stand-alone, and the remaining text is understandable without them, so while you are encouraged to give them a go, don't fret or give up if you don't understand a maths box—they are there as an opportunity, not as essential knowledge.

Sydney, NSW, Australia David I. Warton

Acknowledgements

First I'd like to thank those who permitted the use of their datasets in this text—these were invaluable in motivating and illustrating the analysis methods described in this work.

I'd like to thank those who helped me get to the point where I was writing a book on stats for ecologists, including my mother, for encouraging me to do whatever I liked for a career; my lecturers at the University of Sydney and Macquarie University for teaching me stats and ecology; my mentors who have helped me develop as a researcher—most notably Mark Westoby, William Dunsmuir, Malcolm Hudson, Matt Wand, Glenda Wardle, and Neville Weber; my workplace supervisors at UNSW, who have given me a long leash; the Australian Research Council, which provided financial support throughout much of my career to date; and my collaborators and research team, the UNSW Eco-Stats research group. This team is my "work family"; they are a pleasure to work with and I have learnt much from them and with them. There are a bunch of original ideas in this work, in the later chapters, which the Eco-Stats group have been integral to developing.

Thanks to everyone who has looked over part of this book and offered advice on different sections, including Wesley Brooks, Elliot Dovers, Daniel Falster, Rob Freckleton, Francis Hui, Michelle Lim, Mitch Lyons, Ben Maslen, Sam Mason, Maeve McGillycuddy, Robert Nguyen, Eve Slavich, Jakub Stoklosa, Sara Taskinen, Löic Thibaut, Mark Westoby, and Gordana Popovic. Gordana in particular can be credited with the idea of adding maths boxes to complement the code boxes.

Finally, this book has been something of a journey, taking over 5 years to write, which is a long enough period for anyone to go through some ups and downs. Thanks to my friends and family for helping me to find strength and confidence when it was needed and supporting me when I couldn't. And thanks to Bob for the apple crumble.

Contents

Part II Regression Analysis for Multiple Response Variables

Part III Regression Analysis for Multivariate Abundances

Part I
Regression Analysis for a Single Response Variable

Chapter 1
"Stats 101" Revision

No doubt you've done some stats before—probably in high school and at university, even though it might have been some time ago. I'm not expecting you to remember all of it, and in this Chapter you will find some important lessons to reinforce before we get cracking.

> **Key Point**
> Some definitions.
>
> Response variable y The variable you are most interested in, the one you are trying to predict. There can be more than one such variable, though (as in Chap. 11).
>
> Predictor variable x The variable you are using to predict the response variable y. There is often more than one such variable (as in Chap. 3).
>
> Regression model A model for predicting y from x. Pretty much every method in this book can be understood as a regression model of one type or another.
>
> Statistical inference The act of making general statements (most commonly about a population) based on just a sample (a smaller number of "representative" cases for which you have data.
>
> Categorical variables break subjects into categories (e.g. colour, species ID)
>
> Quantitative variables are measured on a scale (e.g. biomass, species richness)

D. I. Warton, *Eco-Stats: Data Analysis in Ecology*, Methods in Statistical Ecology,
https://doi.org/10.1007/978-3-030-88443-7_1

1.1 Regression, Predictors, and Responses

Most methods of data analysis, and all methods covered in this book, can be thought of as a type of *regression model*. A regression model is used in any situation where we are interested in understanding one or more *response* variables (which we will call y) which we believe we can describe as a function of one or more *predictor* variables (which we will call x).

The predictors might be *quantitative* variables measured on a scale, *categorical* predictors that always fall in one of a set number of predefined categories, e.g. experimental treatments, or other variables taking one of a set of possible values that might be predetermined by the sampling design.

A response variable can similarly be quantitative or categorical, but importantly, its value cannot be predetermined by the sampling design—as in its name, it is a response, not something that is fixed or constrained by the way in which sampling was done. Sometimes you might have more than one response variable, and you can't see a way around modelling these jointly; this complicates things and will be considered later (Chap. 11).

How can you tell whether a variable is a predictor (x) or a response (y)? The response is the variable you are interested in *understanding*. Usually, if one variable is thought to be causing changes in another, the variable driving the change is x and the one responding is y, but *this is not always the case*. If the goal is to predict one variable, then it is most naturally treated as the response y, irrespective of which causes which.

Key Point
The study design is critical—think carefully before you start, and get advice.

1.2 Study Design Is Critical

There is no regression method that can magically correct for a crappy design.

Garbage in $\Longrightarrow$ Garbage out

So you need to put a lot of thought into how you collect data.

1.2.1 See a Statistical Consultant Before You Start the Study

It is always worth *consulting with a statistician* in the design phase. Even if you don't get anything out of the meeting, it's better to waste an hour (and maybe $300) at the start than to find out a year later you wasted 300 h of field work (and $20,000) with a flawed design that can't effectively answer the question you are interested in.

Statistical consultants absolutely love talking to researchers during the design phase of a study. They will typically ask a bunch of questions intended to establish what the primary research question is, to check that the study is designed in such a way that it would be able to answer that question, and to understand what sort of sample sizes might be needed to get a clear answer to the question. They would also think about how to analyse data prior to their collection.

Many universities and research organisations employ statistical consultants to offer free study design advice to students, and probably to staff, or at least advice at a reduced rate. If your workplace doesn't, consider it your mission to get this situation addressed—any research-intensive organisation will want its researchers to have access to quality statistics advice!

There may be a "resident expert" in your research group or nearby who is the go-to person for study design and stats advice but who lacks formal training in statistics. Some of these people are very good at what they do, actually many of them are (otherwise why would they be considered the expert?). But as a general rule, they should not be a substitute for a qualified statistical consultant—because the consultant does this as their day job and is trained for this job, whereas the resident expert typically does statistics only incidentally and it is neither the resident expert's main job nor something they trained for. I used to be one of these resident experts as a student, in a couple of biology departments, before I retrained as a statistician. I can assure you I am a much better consultant now that I have a deeper understanding of the theory and a greater breadth of experience that comes from formal training in stats.

Having said this, the resident expert, if one exists in your research area, will be really good at relating to some of the intricacies of your research area (e.g. difficulties finding extra field sites, "hot issues" in your literature) and might raise important discipline-specific issues you need to think about. Because the resident expert might see different issues not raised by a statistical consultant, and vice versa, perhaps the best approach is to talk to both your resident expert and a statistical consultant! After all, when making any big decision it is worth consulting widely and getting a range of views. This should be no exception to that rule, because most research projects represent serious commitments that will take a lot of your time.

While you should seek advice from more qualified people, you should not pass responsibility for design and analysis decisions over to them entirely—as a researcher you have a responsibility to understand key concepts in design and analysis as they relate to your research area, to ensure design and analysis decisions are well informed. I am fortunate to know some really successful ecologists who are at the top of their field internationally, and one thing they have in common is a high level of understanding of key issues in study design and a broad understanding of data analysis methods and issues that can arise in this connection. Something else they have in common is that they know that they don't know everything, and occasionally they seek advice from me or my colleagues on trickier problems, usually in the design phase.

1.2.2 What Is the Research Question?

Before forging ahead with study design (or analysis), you need to clarify the objective of your research. A good way to do this is to summarise your research project as a question that you want to answer. Once you have framed the *research question*, you can use this as a guide in everything that follows.

Is your research question interesting, something that is important, that lots of people will be interested in finding out the answer to? If not, go find another research question! Easier said than done though— asking a good question is typically the hardest thing to do as a researcher. Finding the answer can be relatively easy, after you have found a good question.

1.2.3 Are You Trying to Demonstrate Causation?

It is common to distinguish between two main types of study: experimental and observational.

Experiment: Subjects are deliberately manipulated through the application of a treatment.

Observational study: No treatment is applied; direct measurements are taken on subjects.

You cannot demonstrate causation by observational study alone. This is the whole point of experiments—manipulate X and see if it affects Y in order to demonstrate causation.

There is a large body of literature on causal models (Shipley, 2016, for example) concerning structural equation models and related methods. These are models for testing hypothesised causal pathways, usually fitted to observational studies. However, they cannot be used to *demonstrate* causation, unless one is prepared to make the strong (and usually unrealistic) assumption that the model has been specified correctly, without any unmeasured variables that could be relevant to the response and any predictors. Causal models are better used for testing only—to test whether a causal model is consistent with the data or to see which of a few competing causal models is more consistent with the data.

So the weak point of observational studies is that they can't demonstrate causation. What is the weak point of experiments?

Experiments, by virtue of occurring in a controlled setting, do not happen in "the real world" and can be quite unrealistic. For example, it is common to try to control factors that could cause variation across replicates. In an experiment on plants, one might take steps to ensure each seedling is planted in identically sized containers with exactly the same amount of soil, fertiliser, and water, in a growth chamber with (approximately) uniform temperature and light. These sorts of practices, by reducing variation across replicates, offer a greater chance of detecting a treatment effect (greater *power*), and so are a good idea when trying to identify if a given experimental treatment has an effect. However, a side effect is that the environment of the experiment is less realistic, e.g. the average growth chamber does not look like a field site. This prompts the following question: Sure you demonstrated this in your experiment, but will it happen in the real world? Good experimental procedure either mimics the real world as well as possible or is supplemented by additional trials mimicking the real world (e.g. repeat the growth chamber experiment out in the field).

1.2.4 What Is the Target Population?

Another thing to think about early in the design of a study is what broader group of subjects your research question is intended to apply to. This is known as the *target population.*

The decision about the target population is a function both of the group of subjects you are interested in and the group of subjects you might include in the study. For example, you may be interested in insects in forests around the world, but if you only have time and funding for field trips a few hundred kilometres from home, you will need to narrow your focus, e.g. the target population could be all insects in forests within a few hundred kilometres of home.

Ideally, any narrowing of your focus should be done in a way that does not sacrifice scientific interest (or at least minimises the sacrifice). Is there a network of forested national parks within a few hundred kilometres of home? Or could you team up with others going on field trips further afield? If not, maybe you will need to switch to an interesting research question for which you can actually access a relevant target population. . .

Having decided on your target population, you will then work out how to select a smaller group of subjects (a *sample*) to actually use in your study. This sample should be *representative* of this population; one way to guarantee this is to randomly sample from the population, as discussed later. How you sample requires careful thought, and it is well worth your while seeking advice at this step.

Target populations are especially important in observational studies, but they are also important in experiments—in the absence of arm-waving, the effects of a treatment on subjects can only be generalised as far as the population from which the experimental subjects were sampled.

1.2.5 Compare, Replicate, Randomise!

If the study being undertaken is an experiment, then it will need to *compare*, *replicate* and *randomise*, as discussed in what follows.

Compare across treatments that vary only in the factor of interest. This may require the use of an experimental or "sham" control. For example, consider a study of the effect of bird exclusion on plant growth, where birds are excluded by building a cage around individual plants. A sham control would involve still caging plants, but leaving off most of the netting so that the birds can still get in—so to the extent possible, the only thing that differs across treatments is the presence/absence of birds. The cage structure itself might adversely affect plants, e.g. due to soil disturbance during its construction or as a result of shading from the cage structure, but using a sham control means that these effects would be common across treatments and so would not mess up the experiment.

Replicate the application of the treatment to subjects. This is necessary so we can generalise about the effects of treatment, which would be impossible if it were applied once to a large batch. For example, if looking at the effect of a gene through breeding knock-out fruit flies, the process of gene knock-out should be replicated, so that we know the effect is due to knock-out and not to something else particular to a given batch.

Exercise 1.1: Experimental Design Issues

Each of these experiments has a problem with one of the three essential steps (*compare*, *replicate*, and *randomise*). Which experiment had a problem with which step? And what is the specific problem?

1. Alan was interested in the effects of mite infestation on the growth of cotton plants. He randomly assigned 15 (out of 30) cotton seedlings to a mite treatment, which he applied by crumbling up mite-infested leaves and placing them near the base of each seedling. The remaining 15 no-mite seedlings received no such treatment. Alan then compared the total biomass of cotton plants in the mite and no-mite treatments to determine whether the mites had a detrimental effect on cotton plant growth. (To his surprise, plants in the mite treatment actually grew larger!)
2. Beryl was interested in the effects of oven temperature on the water content of bread. She prepared 20 loaves of bread using dough prepared in exactly the same way. Ten loaves of bread were baked simultaneously in an oven that had been heated to 200 °C, and the other ten loaves were baked in an oven at 220 °C. She compared the water content of loaves across treatments to draw conclusions about the effects of oven temperature on the water content of bread.
3. In the Lanarkshire Milk Experiment of 1930 (Student, 1931), towards the start of the Great Depression, Gerald and Peter looked at whether giving milk to school students affected growth. In total 10,000 school children received $\frac{3}{4}$pint of milk each day for 4 months, and another 10,000 received none. Teachers assigned students to treatments using their own judgement to ensure that neither group had "an undue proportion of well fed or ill nourished children". All students were weighed before and after the experiment to measure growth. Subsequent analysis revealed that not only did those receiving milk grow more, but they started off weighing significantly less, so initial size was a confounding factor that could explain the difference in growth (e.g. due to the timing of growth spurts).

Randomise the application of subjects to treatment groups using a random number generator, drawing numbers from a hat, or using some other technique to decide which subjects get which treatment. This accomplishes several things.

- It removes control from the experimenter over the selection of subjects and their assignment to treatment groups, thereby eliminating selection bias.
- It makes it possible to consider uncontrolled variation as being random, which is essential in any analysis. Even if one is of the view that everything in the world has a deterministic cause, introducing randomness in the method of sampling (or applying treatments) gives a justification for treating the response variable(s) observed, and any unmeasured predictors, as random.
- In many instances, this justifies an assumption that responses are independent of each other—a core assumption in most statistical procedures discussed in this book.

Failure to carry out any of the aforementioned steps can invalidate an experiment, but problems are not always obvious until afterwards, as happened in Exercises 1.1.

Maths Box 1.1: Probability Functions, Conditioning, and Independence
Any random variable can be characterised by its *probability distribution*, $f_Y(y)$ for random variable Y. This function tells us how to calculate probabilities of observing different values of Y.

If there is another random variable X for which we observed the value x, we can also look at the *conditional distribution* of Y—how likely different values of Y will be after having observed $X = x$. We could write this as $f_{Y|X=x}(y)$. A regression model for a response Y as a function of a predictor X is a model for the conditional distribution $f_{Y|X=x}(y)$.

If Y is related to X (if they are *dependent*), then the probability distribution $f_{Y|X=x}(y)$ will change depending on which value of x has been observed. For example, the biomass of a tree Y tends to be positively related to its basal area X. If we observe a tree with a large basal area, then the tree probably has a large biomass, so the probability distribution $f_{Y|X=x}(y)$ will be centred on larger values.

If two random variables X and Y are independent, then X does not carry information about Y, so the conditional distribution $f_{Y|X=x}(y)$ is the same as the distribution of Y ignoring X, $f_Y(y)$ (the *marginal distribution* of Y):

$$X \text{ and } Y \text{ are independent} \Longleftrightarrow f_{Y|X=x}(y) = f_Y(y) \quad \text{for all } x, y$$

This key result defines independent variables. When we assume independence of X and Y, we assume the (conditional) distribution of Y does not change as X changes.

Maths Box 1.2: Random Samples Are Identically and Independently Distributed

A *random sample* from a population is a sample taken in such a way that all possible samples have an equal chance of being selected.

In a random sample, all subjects have an equal chance of being the ith in the sample (for each i), which means that all observations Y_i have the same distribution, $f_{Y_i}(y) = f_Y(y)$ (the Y_i are *identically distributed*).

If the population sampled from is large, knowing which is the first subject in your sample gives no information about which will be the second, since all are equally likely. This means that the conditional distribution $f_{Y_2|Y_1=y_1}(y) = f_Y(y)$ is not a function of y_1, and so Y_1 and Y_2 are independent (similarly, all n observations in the sample are *independently distributed*).

Hence, we can say that a random sample from a variable, $Y_1, \ldots, Y_n$, is *independently and identically distributed*, or *iid*.

Maths Box 1.3: Mean and Variance Defined

The mean of a variable Y, μ_Y, is its long-run average. For a discrete random variable, μ_Y can be calculated from its probability function as

$$\mu_Y = \sum_{\text{all possible } Y} y f_Y(y)$$

and if Y is continuous, we integrate instead of summing over y. The mean is the most common measure of the central tendency of a variable. We can estimate μ_Y by averaging the values observed in a sample dataset to get $\bar{Y}$.

The variance of Y is the mean of the squared distance of Y from its mean:

$$\sigma_Y^2 = \mu_{(Y-\mu)^2} \tag{1.1}$$

This is the most common measure of the spread of a variable—the more spread out the values are, the larger the values $(Y - \mu)^2$ tends to take and, hence, the larger the variance. Variance can be estimated from sample data by averaging the values of $(y - \bar{y})^2$, although this is usually rescaled slightly to remove small-sample bias.

The square root of the variance σ is called the *standard deviation*. It is often used in place of the variance because it has the same units as Y, so its values make more sense.

The mean and variance, and assumptions made about them in analysis, are central to regression methods and their performance.

Very similar principles also apply to observational studies—we need to *replicate*, in the sense that we take measurements on multiple subjects, these subjects need to have different levels of X so we can *compare* subjects (if we are to demonstrate an association between Y and X), and we need to *randomise* in the sense that subjects are randomly sampled. Random sampling from a population has the important benefit of ensuring that the sample is representative of the target population (and also assists with independence assumptions—see Maths Boxes 1.1–1.2).

Randomly sampling from a population can be difficult in practice—it might not be possible for logistical reasons, e.g. there are only eight potentially suitable field sites and you will use them all. What happens then? Well, you have to wave your arms a bit and hope all your assumptions are valid, despite your not having done anything in the design phase to ensure they are valid. It is worth thinking carefully about this problem—could the sampling method have inadvertently introduced dependence? You should never be completely assured about things when randomization was not possible. It is important to avoid such a situation if possible so your results won't be vulnerable to criticisms concerning the representativeness of sites and the validity of independence assumptions.

Key Point
In data analysis, always *mind your Ps and Qs*—let your decisions on what to do be guided by the research *q*uestion and data *p*roperties (especially the properties of your response variable).

1.3 When Do You Use a Given Method?

When thinking about how to analyse data, there are two key things to think about: let's call them the Ps and Qs of data analysis:

P What are the main properties of my data?
Q What is the research question?

And whenever analysing data, we need to *mind our Ps and Qs*—make sure we check that the assumed properties align with what we see in our data and that the analysis procedure aligns with the research question we want to answer.

1.3.1 Qs—What Is the Research Question?

Consider the data in Table 1.1, which reports the number of ravens at 12 sites before and after a gun is fired. Consider the problem of visualising the data, the first step in any analysis. How should we graph the data? It depends on what we want to know.

Table 1.1: Raven counts at 12 different sites before and after the sound of a gunshot, courtesy of White (2005)

Before	0	0	0	0	0	2	1	0	0	3	5	0
After	2	1	4	1	0	5	0	1	0	3	5	2

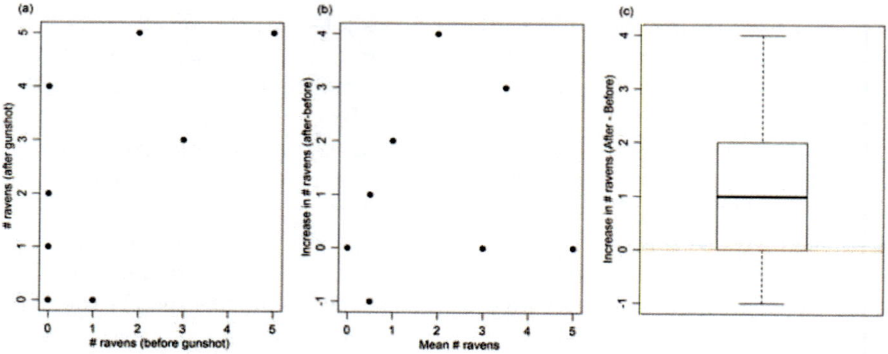

Fig. 1.1: The method of analysis depends on the research question. Consider the set of paired observations from White (2005), as in Table 1.1. Here are three example plots, all of which are valid, but they answer different research questions—see Exercise 1.2

A few options are given in Fig. 1.1, where each graph corresponds to one of the following three research questions:

1. Are counts larger after the gunshot sound than before?
2. Are counts at a site after the gunshot related to counts before the gunshot?
3. Do before and after counts measure the same thing?

The right approach to analysis depends on the question!

So this is number one on the list of things to think about when analysing your data—what question are you trying to answer?

Exercise 1.2: Which Plot for Which Research Question?
Consider the three plots in Fig. 1.1. These correspond to three different research questions:

1. Are counts larger after the gunshot sound than before?
2. Are counts after the gunshot related to counts before the gunshot?
3. Do before and after counts measure the same thing?

Can you tell which graph answers which question?

There are many different approaches to data analysis, depending on the exact question of primary interest. Broadly speaking, the main approaches fall into one of the following categories (which will be discussed later):

- Descriptive statistics (playing with data)
- Hypothesis testing (a priori hypothesis of key interest)
- Estimation/confidence intervals (CIs)(to estimate the key quantity/effect size)
- Predictive modelling (to predict some key response variable)
- Model selection (which theory or model is more consistent with my data? Which predictor variables are related to the response?)

Sometimes more than one of these is appropriate for a particular problem, e.g. I would never test a hypothesis without first doing descriptive statistics to visualise my data—or anything, actually. There are some things you can't combine, though; in particular, when doing model selection, usually you cannot do hypothesis testing or CI estimation on the same data because standard procedures don't take into account the fact that you have done model selection beforehand.

A key distinction is between descriptive and inferential statistics.

Descriptive statistics is what you would do if the goal were to *describe*—understanding the main features of the data. Common examples are graphing (histograms, scatterplots, barplots, ...), numerical summaries (mean, median, sd, ...), and "pattern-finders" or data-mining tools like cluster analysis and ordination. Graphing data is a key step in any analysis: one analysis goal should always be to find a visualisation of the data that directly answers the research question.

Inferential statistics is what you would do if the goal were to *infer*—making general statements beyond the data at hand. Common examples include hypothesis tests (t-tests, binomial tests, χ^2 test of independence, ANOVA F-test, ...), CIs (for mean difference, regression slope, some other measure of effect size), predictive modelling, and variable or model selection.

The distinction between descriptive and inferential statistics is important because inference tends to involve making stronger statements from the data, and doing so typically requires stronger assumptions about the data and study design. Thus, it is important to be aware of when you are moving beyond descriptions and making inferences and to *always check assumptions before making inferences*!

1.3.2 Ps—What Are the Main Properties of My Data?

When working out how to analyse data, we need to think not just about our research question, but also about the properties of our data. Some of these properties are implied by our study design.

A few aspects of our data of particular interest:

- Categorical or quantitative?
- If the data are quantitative, they are *discrete*, taking a "countable" number of values, like $0, 1, 2, \ldots$ (e.g. species richness). Otherwise, they are probably *continuous*, taking any value in some interval (e.g. biomass). Discrete vs continuous is an important distinction for "highly discrete" data with lots of zeros (Warton et al., 2016), and there is a separate set of regression tools specifically for this type of data (Chap. 10). You can usually treat larger counts as continuous.
- If the data are categorical, they are *ordinal*, with the categories having a natural ordering, e.g. abundance as one of {absent, present in trace amounts, a fair bit, it's pretty much everywhere}. Otherwise, they are *nominal*, e.g. colour such as red, blue, green, purple, . . .
- Is there one variable, or are we looking at an association between at least two variables?
- Design considerations: any pairing (or blocking)?

In regression problems, what really matters are the properties of the *response variable(s)*, that is, the properties of the variable(s) you are trying to predict.

Exercise 1.3: Raven Count Data—What Data Properties?
Which type of data is each of the following variables—categorical or quantitative? Discrete or continuous? Ordinal or nominal?

- # ravens
- Sampling time (before or after gunshot)

1.3.3 Putting It Together: Mind Your Ps and Qs

You have probably already taken a stats course. In your first stats course you would have come across a bunch of different techniques—histograms, clustered bar charts, linear regression, χ^2 tests of independence, and so forth. Figure 1.2 is a schematic summarising how these all fit together. Down the rows are different techniques of analysis that depend on the research question (Qs)—am I doing a hypothesis test, just making inferences about some key parameter I want to estimate, etc? Across the columns we have different data properties (Ps)—does the research question involve one or two variables, and are they categorical or quantitative?

Figure 1.2 isn't the end of matters when it comes to doing statistics, obviously, although unfortunately many researchers have no formal training that takes them beyond these methods. Clearly sometimes you have more than two variables, or one of them is discrete and takes small values (unlikely to satisfy normality assumptions), or the response variable is ordinal. Such situations will be covered later in this book.

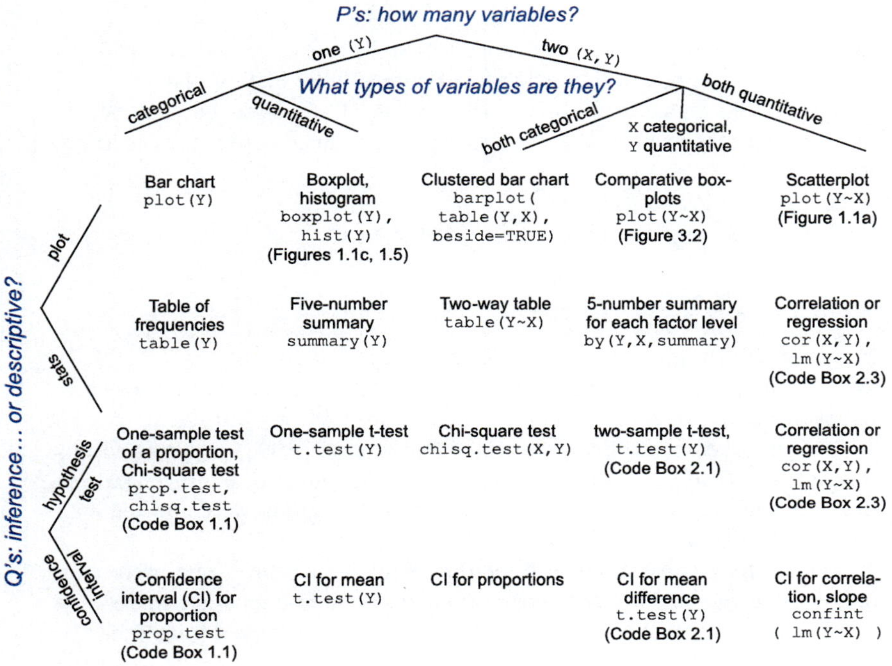

Fig. 1.2: A schematic of "Stats 101" methods—tools you likely saw in an introductory statistics course, organised according to the research question they answer (rows) and the properties the data have (columns). In this text, we will find commonalities across these methods and extend them to deal with more complex settings

Exercise 1.4: Gender Ratios in Bats
Kerry goes bat counting. She finds 65 female bats and 44 male bats in a colony. She would like to know whether there is evidence of gender bias.

What do the data tell us—one variable or two? Categorical or quantitative?

What does the question tell us—descriptive, estimation, hypothesis testing, etc?

So how would you analyse the data?

What graph would you use to visualise the data?

Exercise 1.5: Ravens and Gunshots
In the paper titled "Hunters ring dinner bell for ravens..." (White, 2005), ecologist Crow White posed the question of whether ravens actually fly towards (not away from!?) the sound of a gunshot on the chance they'll find a carcass

to scavenge. He went to 12 locations, counted the ravens he saw, then shot his gun, waited 10 min, and counted again. The results are in Table 1.1.

What do the data tell us—one variable or two? Categorical or quantitative?

What does the question tell us—descriptive, estimation, hypothesis testing, etc?

So how would you analyse the data?

What graph would you use to visualise the data?

Exercise 1.6: Pregnancy and Smoking
What is the effect of a mother's smoking during pregnancy on the resulting offspring?

A randomised controlled experiment (Johns et al., 1993) investigated this by injecting 10 pregnant guinea pigs with 0.5 mg/kg nicotine hydrogen tartrate in saline solution. The 10 guinea pigs in the control group were injected with a saline solution without nicotine.

The learning capabilities of the *offspring* of these guinea pigs were then measured using the number of errors made trying to find food in a maze.

Number of errors made by guinea pigs in the maze (derived from Table 2):

Control	11	19	15	47	35	10	26	15	36	20
Treatment	38	26	33	89	66	23	28	63	43	34

What do the data tell us—one variable or two? Categorical or quantitative?

What does the question tell us—descriptive, estimation, hypothesis testing, etc?

So how would you analyse the data?

What graph would you use to visualise the data?

When analysing your data you should always *mind your Ps and Qs*—make sure the assumptions about data properties are reasonable, and make sure that the analysis procedure aligns with the research question you want to answer. We will discuss this in more detail in Sect. 1.5.

Key Point
Statistical inference involves making general statements about population or "true" quantities (e.g. mean) based on estimates of these quantities from samples. We need to take into account uncertainty due to the fact that we only have a sample, which is commonly done using a CI or a hypothesis test.

1.4 Statistical Inference

When we use the term statistical inference, we are talking about the process of generalising from a sample—making statements about general ("population") patterns based on given data ("sample"). In an observational study, there is typically some statistic that was calculated from a sample, and we would like to use this to say something about the true (and usually unknown) value of some *parameter* in the target population, if we sampled all possible subjects of interest. In an experiment, we use our data ("sample") to calculate a treatment effect and want to know what this says about the true treatment effect if we repeat the experiment on an ever-increasing set of replicates ("population").

It is always important to be clear whether you are talking about the sample or the population—one way to flag what you mean is through notation, to distinguish between what you know (sample) and what you wish you knew (population).

Notation for some common sample estimators and their corresponding population parameters are listed in Table 1.2. Sample estimates usually use the regular alphabet, and population parameters usually use Greek letters (e.g. s vs σ). The Greek letter used is often chosen to match up with the subject (e.g. σ is the Greek letter s, s for standard deviation, μ is the Greek letter m, m for mean). An important exception to the rule of using Greek letters is proportions—the population proportion is usually denoted by p (because the Greek letter p is π, which we reserve for a special number arising in circle geometry).

Another common way to write sample estimators is by taking the population parameter and putting a hat on it (e.g. $\hat{\beta}$ vs β). So there really is a bit of a mixture of terminologies around, which doesn't make things easy. A way to be consistent with notation, instead of chopping and changing, is to go with hats on all your estimators—you can write an estimator of the mean as $\hat{\mu}$ and an estimator of the standard deviation as $\hat{\sigma}$, and most of us will know what you mean.

We usually refer to the number of observations in a sample as the sample size, n.

Table 1.2: Notation for common sample estimators and the corresponding population parameters being estimated

	Sample	Population
Mean	$\bar{x}$ (or $\bar{y}$)	μ (or μ_y)
SD	s	σ
Proportion	$\hat{p}$	p
Regression slope	$\hat{\beta}$	β

Exercise 1.7: Inference Notation—Gender Ratio in Bats
Kerry goes bat counting. She finds 65 female bats and 44 male bats in a colony. She would like to know if there is evidence of gender bias.

What is n in this case?

The proportion of female bats in this dataset is $\frac{65}{65+44} \simeq 0.596$. Do we write this as $\hat{p}$ or p?

We are interested in the true proportion of bats in the colony that are female, which is unknown because we only have a sample of data. How do we write this true proportion, as $\hat{p}$ or p?

Exercise 1.8: Inference Notation—Raven Counts
Consider again Exercise 1.5. The average number of ravens before the gunshot sound was 0.92, whereas after the sound, it was 2.

We are interested in whether there has been an increase in the true mean number of ravens after the gunshot compared to before.

Which value do we know from the data—$\bar{x}_{\text{after}} - \bar{x}_{\text{before}}$ or $\mu_{\text{after}} - \mu_{\text{before}}$?

Which value is unknown but something we want to make inferences about—$\bar{x}_{\text{after}} - \bar{x}_{\text{before}}$ or $\mu_{\text{after}} - \mu_{\text{before}}$?

When making inferences, we need to account for *sampling error*—the fact that we did not sample all subjects of potential interest, only some of them, so any statistic we calculate from the sample will likely differ from what the true answer would be if we took measurements on the whole population.

If we repeated a study, any statistic we calculated might take a different answer. So statistics vary from one sample to the next. A key idea in statistical inference is to treat a statistic as a random variable that varies across samples according to its *sampling distribution*. We can often use theory (or simulation) to study the sampling distribution of a statistic to understand its behaviour (as in Maths Boxes 1.4–1.5) and to design inference procedures, as discussed in what follows.

1.4.1 Two Common Types of Inference

Consider Exercise 1.4, where the target we want to make inferences about is p, the (true) sex ratio in the bat colony. There are two main ways to make inferences about the true value of some parameter, such as p. If you use the `prop.test` function to analyse the data, as in Code Box 1.1, the output reports both analyses.

Code Box 1.1: Analysing Kerry's Gender Ratio Data on Bats

```
> prop.test(65,109,0.5)

1-sample proportions test with continuity correction

data:  65 out of 109, null probability 0.5
X-squared = 3.6697, df = 1, p-value = 0.05541
alternative hypothesis: true p is not equal to 0.5
95 percent confidence interval:
 0.4978787 0.6879023
sample estimates:
        p
0.5963303
```

The P-value was computed here using an approximate test, but for this situation it is simple enough to use the binomial distribution directly:

```
> 2*pbinom(64,109,0.5,lower.tail=FALSE)
[1] 0.05490882
```

Hypothesis test—there is a specific hypothesis we want to test using data (the null hypothesis, often written H_0). In Exercise 1.7, we could test the hypothesis that there is no gender bias, $H_0 : p = 0.5$. We observed 65 female bats out of a colony of 109, and we could use probability to work out how likely it would be to get this many females if there were no gender bias:

$$P\text{-value} = 2 \times P\left(\hat{p} > \frac{65}{109}\right) = 0.055$$

(See Code Box 1.1 for P-value computation.) A statistic this far from 50:50 is reasonably unlikely (almost 5%) so there is reasonable evidence against H_0, i.e. reasonable evidence of gender bias.

Confidence interval—we don't know the true sex ratio, but we can construct an interval which we are pretty sure contains the true sex ratio.

e.g. from the CI in Code Box 1.1 we can say that we are 95% confident that the true p is between 0.498 and 0.688.

1.4.2 Interpreting P-values

Loosely speaking, a P-value measures how unlikely your data are under H_0. (More specifically, it is the probability of getting a test statistic as extreme as or more extreme than the observed one if the null hypothesis is true.) This can be used as a measure of how much evidence there is *against* H_0 (but not as a measure of evidence for H_0).

Small P-value $\Longrightarrow$ data unlikely under H_0 $\Longrightarrow$ evidence against H_0.
Large P-value $\Longrightarrow$ data not unlikely under H_0 $\Longrightarrow$ no evidence against H_0.
Here are a few hypothetical scenarios to illustrate this idea.

Scenario 1—Evidence Against H_0
Let's say that out of the 109 bats, Kerry found that 75 were female. We can then work out that the P-value is 0.001, which tells us that there is strong evidence (against H_0) of gender bias. We can also compute a CI and find that we are 95% confident that the true sex ratio is in the interval (0.59, 0.77). So you can see there is clear evidence that the proportion of females exceeds 0.5.

Scenario 2—No Evidence Against H_0
Now let's say that out of 109 bats, 57 were female. We could then crunch the numbers and find that the P-value is 0.70, meaning there is no evidence (against H_0) of gender bias. Further computation tells us that we are 95% confident that the true sex ratio is in the interval (0.43, 0.62). Note that 0.5 is in the interval, so we don't have evidence of a bias towards females. But we can't rule out a bias, e.g. maybe the true proportion is 0.6! *So a large P-value doesn't mean that H_0 is true.*

Often researchers use the 0.05 significance level in testing—declaring results are statistically significant at the 0.05 level if the P-value is less than 0.05. Using this hard rule, in Exercise 1.7 we would say results are not statistically significant. Others, such as myself, would interpret P-values along a scale and think of anything in the neighbourhood of 0.05 as marginally significant, whether it was larger or smaller than 0.05. The arguments against this are in terms of objectivity (if not sticking with 0.05, what exactly is your decision rule?) and convention (what is your reason for departing from 0.05 exactly?). Neither of these issues is insurmountable, but they are nonetheless issues to consider.

It always helps to look at the CI to help with interpretation; in fact, this is strongly advised (Nakagawa & Cuthill, 2007, for example). If you have done a hypothesis test and found a significant effect, the next natural question is how big of an effect it is. So you find yourself computing a CI for the parameter of interest. Conversely, if there is no significant effect, the next natural question is whether this means the effect is small. Or is it plausible that maybe there was a big effect that has gone undetected? Which leads you back to the CI.

Some argue you should look at CIs and other techniques and not bother with hypothesis tests at all (see for example the Special Issue in Ecology, introductory remarks by Ellison et al., 2014). But while hypothesis tests and CIs are different ways of presenting the same information, they put the focus on different aspects of it, and both have their place in the toolbox. If there is a specific hypothesis of primary interest (e.g. is there an effect?), that is what a hypothesis test is for. It will quantify how likely your data are under the null hypothesis and, hence, to give you a measure of how much evidence there is against this hypothesis of primary interest (e.g. evidence for an effect). On the other hand, if the main game is not to test a specific hypothesis but rather to estimate some population parameter, then hypothesis testing has no place in analysis, and a CI is your best bet.

While usually we have the luxury of moving interchangeably between hypothesis tests and CIs, depending on what is of interest, things get tricky for interval estimation when you have lots of response variables (Chap. 11). In that case it is hard to work out a single quantity that you want to estimate (instead, there are lots of separate quantities to jointly estimate), making it difficult to define a target parameter that you want to construct an interval around. There are, however, no such difficulties for hypothesis tests. Thus, you might notice in the multivariate literature a little more reliance on hypothesis testing (although arguably too much).

1.4.3 Common Errors in Hypothesis Testing

Hypothesis tests are often abused or misused, to the point where some discourage their use (Johnson, 1999, for example). The American Statistical Association released a joint statement on the issue some years ago (Wasserstein & Lazar, 2016) in an attempt to clarify the key issues. Some of the main issues are discussed in what follows.

Don't conclude that H_0 is true because P is large: This mistake can be avoided by also looking at your question in terms of estimation (with CIs)—what is a range of plausible values for the effect size?

Don't test hypotheses you didn't collect the data to test.

Don't "search for significance", testing similar hypotheses many times with slightly different data or methods: This gets you into trouble with multiple testing—every time you do a test, at the 0.05 significance level, by definition, 5% of the time you will accidently conclude there is significant evidence against the null. So if you trawl through a bunch of response variables, doing a test for each of 20 possible response variables, on average you would expect one of them to be significant at the 0.05 level by chance alone, even if there were no relation between these responses and the predictors in your model. To guard against this, tests should be used sparingly—only for problems of primary interest—and if you really do have lots of related hypotheses to test, consider adjusting for multiple testing to reduce your rate of false positives, as in Sect. 3.2.4.

Don't test claims you know aren't true in the first place: I have reviewed papers that have included P-values concerning whether bigger spiders have bigger legs, whether taller trees have thicker trunks, etc.. These are hypotheses we all know the answer to; what the researchers were really interested in was, for example, *how* leg length scales against body size or *how* basal diameter scales against tree height. So there is no need to do the test!

1.4.4 Confidence Intervals

Most sample estimates of parameters ($\bar{x}, \hat{p}, \hat{\beta}, \ldots$) are approximately normally distributed, and most software will report a standard error (standard deviation of the estimator) as well as its estimate. In such cases, an approximate 95% CI for the true ("population") value of the parameter ($\mu, p, \beta, \ldots$) is

$$(\text{estimate} - 2 \times \text{standard error},\ \text{estimate} + 2 \times \text{standard error}) \qquad (1.2)$$

About 95% of the time, such an interval will capture the true value of the parameter if the assumptions are met.

Equation (1.2) uses two times the standard error, but to three significant figures this so-called critical value is actually 1.960 when using a normal distribution. And sometimes a t distribution can be used, which can give a value slightly greater than 2. Your software will probably know what it is doing (best to check yourself if not sure), but on occasion you might need to compute a CI manually, and unless sample size is small (say, less than 10), the rule of doubling standard errors in Eq. (1.2) should do fine.

You can often compute CIs on R using the `confint` function.

Key Point
There are assumptions when making inferences—you need to know what they are and to what extent violations of them are important (to the validity and efficiency of your inferences). Independence, mean, and variance assumptions tend to be important for validity, distributional assumptions not so much, but skew and outliers can reduce efficiency.

1.5 Mind Your Ps and Qs—Assumptions

Recall that when analysing data you should always *mind our Ps and Qs*—make sure the assumptions about data properties are reasonable, and make sure that the analysis procedure aligns with the research question you want to answer. The role of assumptions and assumption checking is critical when making inferences (e.g. computing a CI or a P-value)—it is not possible to make big general statements based on samples without making a set of assumptions, they need to be interrogated to have any assurance that you are on the right track.

One key implicit assumption when making inferences is that you are *sampling from the population of interest*. If you think about the broader set of subjects or events you would like to generalise to, is your sample representative of this broader set? This could be assured in an observational study by randomly sampling from the population of interest in a survey. In an experiment, replicating the treatment of interest and randomising the allocation of subjects to these treatments ensures that we can generalise about any impacts of treatment (at least the impacts of treatment on this set of subjects). If you did not do one of the aforementioned steps (in an observational study, it may not be possible), then some amount of arm-waving and hand-wringing is needed—you will need to convince yourself and others that your sample is representative of some broader population of interest.

Exercise 1.9: Assumptions—Gender Ratio in Bats
Recall the study of bat gender ratios (Exercise 1.7). We used a one-sample test of proportions (for bat gender), which made the following assumptions:

> Each observed bat in the sample has the same probability of being female, independent of (i.e. unaffected by) the gender of any other bats in the sample.

This is also commonly referred to as assuming the observations are "*identically and independently distributed*" ("iid").

How could you guarantee that this assumption is satisfied?

Exercise 1.10: Assumptions—Raven Example
Consider the raven data of Exercise 1.5 and whether there is evidence that ravens increase in number, on average, after a gunshot sound. A natural way to answer this question is to consider after–before differences and whether there is evidence that the mean difference is greater than zero. This is commonly done using a one-sample t-test on the differences (otherwise known as a paired t-test), and it makes the following assumptions:

> Each after–before difference is independent and comes from the same normal distribution (with constant variance).

Here we assume that the observations being analysed (after–before differences) are identically and independently distributed. We also assume the differences are normally distributed, which we can check using a normal quantile plot, as in Code Box 1.2.

Have a look at the normal quantile plot. Do you think the normality assumption is reasonable in this case?

Other assumptions are specific to the analysis method, and how you check assumptions varies to some extent from one context to the next. The bat (Exercise 1.9) and raven (Exercise 1.10) analyses, and indeed most statistical models, involve an *independence assumption*, an important assumption and one that is quite easily violated. This is not something that is easy to check from the data; it is more about checking the study design—in an experiment, randomising the allocation of subjects to treatments guarantees independence is satisfied, and in a survey, random sampling goes some way towards satisfying the independence assumption. One exception to this rule is that if you sample randomly in space (or time) but include in your analysis predictor variables measured at these sites that are correlated across space (or time), this can induce spatial (temporal) autocorrelation that should be accounted for in analysis. So, for example, if we were to investigate bat gender ratio across colonies and whether it was related to temperature, then because temperature is a spatial variable, we would need to consider whether this induced spatial dependence in the response.

Often there are additional assumptions, though. The main types of assumptions a model makes can be broken down into the following categories:

Independence: Most (but not all) methods in this book require observations to be independent, conditional on the values of any predictors (x).

Mean model: Common regression methods involve assuming a model for the mean response (the true mean of y, as in Maths Box 1.3) as a function of x. e.g. We can assume that the mean of y is linearly related to x. In the raven example, our mean model under H_0 is that the true mean (of the after–before differences) is zero.

Variance model: Usually we need to make assumptions about the variance, the most typical assumption being that the variance of y under repeated sampling (at a given value of x) will be the same no matter what. We made this equal variance assumption (of the after–before differences) in the raven example.

Distributional assumption: Parametric methods of regression (which are the focus of this book) involve assuming the response y comes from a particular distribution. Most commonly we will assume a normal distribution, as in the raven example, but for something different see Chap. 10 and later chapters.

Strictly speaking, you could consider mean and variance assumptions as special cases of distributional assumptions, but it is helpful to tease these apart because mean and variance assumptions often have particular importance.

Some especially useful tools for checking assumptions are normal quantile plots and residual vs fits plots, although it is also worth thinking about the extent to which assumptions can be violated before it becomes a problem.

1.5.1 Normal Quantile Plot of Residuals

In the raven example, we have a normality assumption. How do we check it?

Normal quantile plots are the best method of checking if a variable is normally distributed—they plot the variable against values expected from a normal distribution. When checking normality you are checking that there are no systematic departures from a straight line. Some examples, good and bad, of what normal quantile plots might look like are given in Fig. 1.3. An example of how to use R to construct such a plot, for the differences in raven counts, is in Code Box 1.2.

It is hard to tell from looking at a normal quantile plot, or indeed any diagnostic plot, whether apparent departures from the expected pattern are greater than what would be expected by chance. For example, in Fig. 1.3 (left), data were simulated

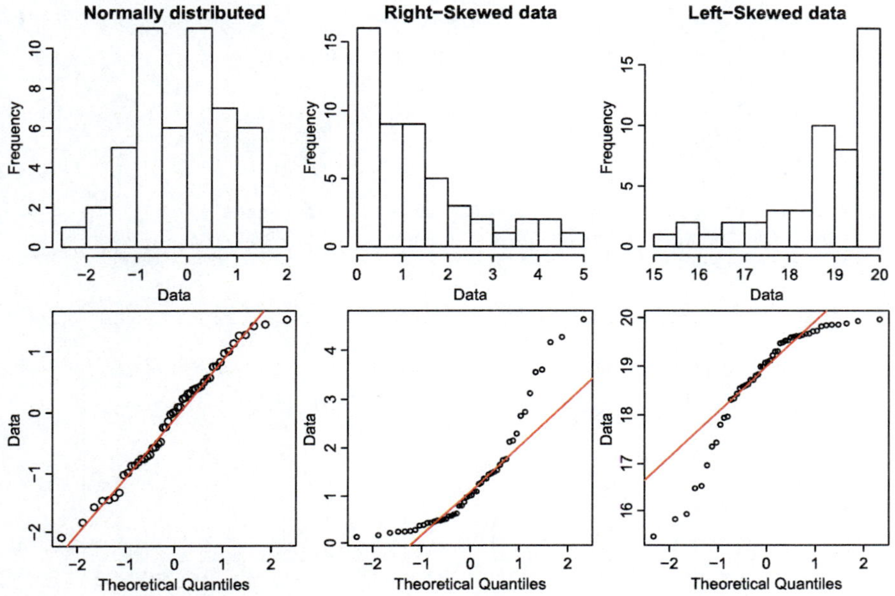

Fig. 1.3: Interpreting normal quantile plots. Example histograms (top row) and normal quantile plots (bottom row) for data simulated from a normal distribution (left), a right-skewed distribution (centre), or a left-skewed distribution (right). On the normal quantile plot, data stay fairly close to a straight line if it is close to a normal distribution, are J-shaped if right-skewed, or r-shaped if left-skewed

from a normal distribution, so we know that the normality assumption is satisfied here. However, points do not all lie exactly on the line. To help understand the extent to which departures from a line can be explained by sampling error, I wrote a function (`qqenvelope` in the `ecostats` package, as in Code Box 1.2), that will put an envelope around the region we would expect points to be in, if assumptions were satisfied. This envelope is constructed by simulating many datasets under the normality assumption, and it is a global envelope, meaning that if *any* points lie outside the envelope, then we have evidence that the normality assumption is not satisfied.

One might think a normal quantile plot was only useful for checking the normality assumption, but it can be used to check other distributional assumptions too—if you use a special type of residual that maps values onto the normal (Dunn & Smyth, 1996). This can be a really handy trick we will discuss in Chap. 10.

Residual vs fits plots are also often useful, especially for checking the mean and variance models. We will look at this in Sect. 2.2.3.

Code Box 1.2: Normal Quantile Plot for Raven Data

```
# Enter the data
Before = c(0, 0, 0, 0, 0, 2, 1, 0, 0, 3, 5, 0)
After  = c(2, 1, 4, 1, 0, 5, 0, 1, 0, 3, 5, 2)

# Construct a normal quantile plot of the differences
qqnorm(After-Before, main="")
qqline(After-Before,col="red")
```

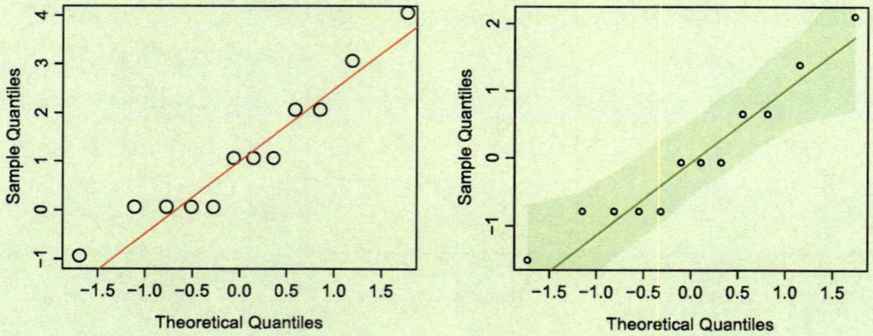

This produces the plot on the left. There are lots of tied values on this plot (several dots fall along the same horizontal line), and a small sample size makes it hard to see a pattern (but maybe a slight suggestion of right skew).

The plot on the right was constructed using the `qqenvelope` function as follows:

```
library(ecostats)
qqenvelope(After-Before)
```

This puts a simulation envelope around where data should lie if they were normal. All points lie inside the envelope, so we have no evidence of a violation of the normality assumption.

Don't be too fussy about the normality assumption, though—even when some points lie outside their simulation envelope, this is not necessarily a problem. Violations of the normality assumption tend not to matter unless one of the following applies:

- Sample size is small (e.g. $n < 10$),
- Data are strongly skewed or have big outliers, or
- You are using a really, really small significance level (e.g. only declaring significance at 0.001), which might happen for example if doing multiple testing.

1.5.2 Robustness to Assumption Violations

What happens if your data aren't normally distributed? Or the variances aren't equal? There is error in any model fitted to data, and one of the main points of statistics is to try to deal with this error appropriately. If assumptions are not satisfied, this might not be done successfully. There are two things to consider:

Validity A valid method deals with error appropriately. Although your model doesn't give exactly the true answer, does it give estimates centred around it (*unbiased*)? Is the estimate of the amount of error reasonable? Is a 95% CI really going to capture the true parameter 95% of the time? Or, equivalently, is the 0.05 significance level really going to be exceeded only 5% of the time when the null is true?

Efficiency An efficient method has relatively small error and so is relatively good at answering your research question. Is the standard error small and CI narrow? Or, equivalently, is a test going to have good power at detecting an effect?

What happens depends on which assumption we are talking about. But here are some general rules (we will elaborate on them when we have a specific example to talk about in Sect. 2.1.2).

Violations of independence assumption: If your data are not independent, but you have assumed your data are independent, then you are stuffed.[1] When using methods that assume independence, it is really important to make sure your observations in each sample are independent of each other—and sometimes it is quite easy to guarantee this assumption is satisfied through randomisation. For a mathematical definition of independence and how randomisation can help satisfy this assumption see Maths Box 1.1.

The most common violation of this assumption is when everything is positively correlated (e.g. observations close together in space are more similar), and the effect this has is to make estimates of standard errors smaller than they should be (as in Maths Box 1.5). If standard errors are too small, you tend to get false confidence—CIs are too narrow (missing the true value more often than they claim to) and test statistics are too large (more likely to declare false significance than they ought to). This is really bad news.

False confidence is really bad news because one of the main points of inferential statistics is to guard against this (or, more specifically, to control the rate of false discoveries). If our data violate the independence assumption, and this is not accounted for in our inferential procedure, then it can't do the job it was designed to. People have a habit of finding an explanation for anything, even really random things that make no sense (there is a lot of literature behind this, for example Shermer, 2012, and a word for it, "apophenia"). An important benefit of inferential statistics is providing a reality check to scientists, so that when they think they see a pattern, the analysis procedure can tell them if it should be taken seriously. If the independence assumption is violated, the stats procedure can't do this job and there is little point using it, you might as well just eyeball the data...

Another common example of where the independence assumption is often violated is commonly referred to in ecology as *pseudo-replication* (Hurlbert, 1984), where there is structure from the sampling design that is not accounted for in modelling, such that some groups of observations are related to each other and cannot

[1] That is, you are in a world of pain!

be assumed to be independent. For example, if a treatment were applied in batches to subjects, we might expect observations within batches to be more closely related to each other than to observations across different batches. In the raven example, imagine if there were not actually 12 sites and 12 gunshots, but instead there were 4 sites and 4 gunshots but 3 observers (they could be standing 100 m apart, say). Then we would expect the groups of three measurements to be similar to each other and the after–before differences to be similar across observers at a site, introducing dependence. For many years pseudo-replication was seen as a bit of a no-no in ecology, because failing to account for it is really bad news (leading to false confidence, as previously). But if the pseudo-replication is reflected in analysis, then this can provide useful additional information. For example, with the raven example, we could add an observer term to the analysis model to account for multiple observers and, thus, estimate how reliable our estimates of raven abundance are using the methods of Chap. 6. So one lesson here is to make sure the analysis model reflects the study design.

Maths Box 1.4: Effects of Assumption Violations: Bias of $\bar{Y}$

When we are primarily interested in the true mean of Y (μ_Y), we typically estimate it using the mean of a sample ($\bar{Y}$) and use inference procedures that assume observations are independent and normally distributed with equal variance. If these assumptions are violated, does $\bar{Y}$ estimate the wrong thing?

What we want to do here is check $\bar{Y}$ for *bias*, that is, to see whether sample means tend to be centred around the wrong value when assumptions are violated. We will check here only for *estimation bias*, i.e. if there is bias because something is wrong with the analysis procedure. Other important sources of bias to consider are whether something is wrong with the way data are collected (*sampling bias*) or recorded (*response bias*).

To look at estimation bias we will need the following rules for means:

Rescaling If $Y = aX$, $\mu_Y = a\mu_X$ (e.g. doubling all values doubles the mean).

Adding If $Y = X_1 + X_2$, $\mu_Y = \mu_{X_1} + \mu_{X_2}$ (sum two variables, sum their means).

The adding rule generalises to more than two variables, $\mu_{\sum_i X_i} = \sum_i \mu_{X_i}$.

We can work out what value $\bar{Y}$ is centred around directly from these rules:

$$\begin{aligned} \mu_{\bar{Y}} = \mu_{\frac{1}{n}\sum_i Y_i} &= \frac{1}{n}\mu_{\sum_i Y_i} \quad \text{using the rescaling rule} \\ &= \frac{1}{n}\sum_{i=1}^{n}\mu_{Y_i} \quad \text{using the adding rule} \\ &= \frac{1}{n}\sum_{i=1}^{n}\mu_Y \quad \text{if all } Y_i \text{ have mean } \mu_Y \\ &= \mu_Y \end{aligned}$$

So $\bar{Y}$ is unbiased, being centred around the true mean μ_Y, if all Y_i have mean μ_Y.

Note we did not use assumptions of independence, equal variance, or normality in the preceding problem—all those assumptions could be violated, and we could still get an unbiased estimate of the mean. We did assume a simple type of *mean model*—that all observations share the same mean μ_Y.

For similar reasons, most regression models in this book give unbiased estimates of the mean if the mean model is satisfied, irrespective of violations of independence, equal variance, and distributional assumptions. Violations of assumptions will, however, often lead to other problems (Maths Box 1.5).

Maths Box 1.5: Effects of Assumption Violations: Standard Error of $\bar{Y}$

Consider again using the sample mean of a quantitative variable ($\bar{Y}$) to make inferences about the true mean (μ_Y), as in Maths Box 1.4. How much error does $\bar{Y}$ have, and how does this change when assumptions are violated?

In answering this question we will use the following rules for the standard deviation (σ, square root of the variance) of a variable:

Rescaling If $Y = aX$, $\sigma_Y = a\sigma_X$.

Adding If $Y = X_1 + X_2$, $\sigma_Y = \sqrt{\sigma^2_{X_1} + \sigma^2_{X_2} + 2\sigma_{X_1,X_2}}$ where σ_{X_1,X_2} is the covariance of X_1 and X_2.

As before, the adding rule generalises to n random variables in a natural way.

Inference using $\bar{Y}$ typically requires its standard error, i.e. its standard deviation across replicate surveys/experiments. The standard error of $\bar{Y}$ can be derived as

$$\begin{aligned}
\sigma_{\bar{Y}} = \sigma_{\frac{1}{n}\sum_i Y_i} &= \frac{1}{n}\sigma_{\sum_i Y_i} \text{ using the rescaling rule} \\
&= \frac{1}{n}\sqrt{\sum_{i=1}^{n}\sigma^2_{Y_i}} \text{ using the adding rule, if all } \sigma_{Y_i,Y_j} = 0 \\
&= \frac{1}{n}\sqrt{\sum_{i=1}^{n}\sigma^2_{Y}} \text{ if all values have standard deviation } \sigma_Y \\
&= \frac{1}{n}\sqrt{n\sigma^2} = \frac{\sigma_Y}{\sqrt{n}}
\end{aligned}$$

A critical assumption here was that all covariances σ_{Y_i,Y_j} are zero, which is satisfied *if the observations are independent* (e.g. by taking a random sample; see Maths Box 1.2). If there were pseudo-replication, for example, covariances would be positive and the true standard error could potentially be much larger than $\sigma_Y/\sqrt{n}$. In this situation, assuming independence and using $\sigma_Y/\sqrt{n}$ in analyses would be *invalid*, using too small a value for the standard error, giving us false confidence in $\bar{Y}$. We also assumed in the preceding problem that all standard deviations were equal (our *variance model*). Note that normality was not used; it tends to be less important than other assumptions.

For similar reasons, all regression models in this book that assume independence rely heavily on this assumption and risk false confidence when it is not satisfied. Similarly, beyond the mean and variance models, distributional assumptions tend to be less critical for validity throughout this book.

Violations of mean and variance assumptions—The consequences of violations of mean and variance assumptions vary slightly depending on the test being conducted, so they will only be discussed here briefly. Both the mean and variance models are very important, for different reasons.

Usually, the mean is the target of interest in regression models. If the model for the mean is wrong, there isn't much hope in terms of getting anything sensible out of your analysis—you will probably get biased answers, as explained in Maths Box 1.4.

The variance model is important because it has a key role in determining the variance (hence the standard error) of the statistics calculated from the data. If the variance model is wrong, your standard errors will likely be wrong, and any subsequent inferences that are made from the statistic will not be valid (unless adjusted for, e.g. using resampling, as in Chap. 9). Violations of your variance model tend also to make your procedure less efficient, so your estimates of the mean aren't as good as they could have been.

The simplest and arguably the best way to diagnose the mean and variance models is using residual plots—violations of both can be seen on a *residual vs fits plot*, as in Chap. 2.

Violations of distributional assumptions: Most regression methods (including all the ones in this book) are actually pretty robust to violations of their distributional assumptions. This is somewhat ironic because this is the thing most people tend to focus on in their assumption checks. Our obsession with checking distributional assumptions is often misplaced.

The main reason for robustness to assumption violations is the central limit theorem, which ensures that most statistics we use in practice become approximately normally distributed as sample size increases, *irrespective of the distribution of the original data*. This approximation gets better and better as sample size increases, such that you can be pretty confident that your method is valid even in cases where your distributional assumptions are very wrong—that is, if independence is reasonable and the mean and variance models are correct. The central limit theorem applies very generally, indeed to every analysis technique in this book, affording *all methods in this book* some level of robustness to violations of distributional assumptions. See Maths Box 1.6 if you are game to see the mathematics of the central limit theorem and Maths Box 1.7 for a discussion of its sensitivity to skewed and long-tailed data. The maths of the proof are surprisingly straightforward, but it does use moment generating functions (MGFs) and Taylor series (typically taught in first-year mathematics courses), so it is not for everyone.

Maths Box 1.6: Proof of Central Limit Theorem

Consider a random sample $Y_1, \ldots, Y_n$ with mean μ and standard deviation σ. As n increases, the limiting distribution of $\sqrt{n}(\bar{Y} - \mu)$ is $\mathcal{N}(0, \sigma^2)$, *irrespective of the distribution of* Y.

We will prove the central limit theorem using the *moment generating function* (mgf) of Y, written as $m_Y(t)$ and defined as the mean of e^{tY}. The mgf of a variable, when it exists, uniquely defines its distribution. For $\mathcal{N}(0, \sigma^2)$, the MGF is $e^{\sigma^2 t^2/2}$, so our goal here will be to show that as $n \to \infty$, $\log(m_{\sqrt{n}(\bar{Y}-\mu)}) \to \sigma^2 t^2/2$. But to do this, we will use a few tricks for MGFs. We will use this Taylor series expansion of $m_Y(t)$:

$$m_Y(t) = 1 + \mu_Y t + \frac{\mu_{Y^2}}{2}t^2 + \frac{\mu_{Y^3}}{6}t^3 + \frac{\mu_{Y^4}}{24}t^4 + \ldots \tag{1.3}$$

where μ_{Y^3} is a function of skewness, and μ_{Y^4} is a function of long-tailedness.

Rescaling If $Y = aX$, $m_Y(t) = \mu_{\exp(Yt)} = \mu_{\exp(aXt)} = \mu_{\exp(Xat)} = m_X(at)$.

Adding If $Y = X_1 + X_2$ where X_1 and X_2 are independent, $m_Y(t) = \mu_{\exp(X_1+X_2)t} = \mu_{\exp(X_1 t)\exp(X_2 t)} = \mu_{\exp(X_1 t)}\mu_{\exp(X_2 t)} = m_{X_1}(t)m_{X_2}(t)$

The adding rule generalises to n independent variables as $m_{\sum_i X_i}(t) = \prod_i m_{X_i}(t)$. Independence is critical for the adding rule! Now we prove the central limit theorem.

$$\begin{aligned}
\log m_{\sqrt{n}(\bar{Y}-\mu)}(t) &= \log m_{\sum_i (Y_i-\mu)/\sqrt{n}}(t) \\
&= \sum_{i=1}^{n} \log m_{(Y_i-\mu)/\sqrt{n}}(t) \text{ by the adding rule, if } Y_i \text{ are independent} \\
&= \sum_{i=1}^{n} \log m_{(Y_i-\mu)}(t/\sqrt{n}) \text{ by the rescaling rule} \\
&= n \log m_{(Y_i-\mu)}(t/\sqrt{n}) \text{ since the } Y_i \text{ are identically distributed} \\
&= n \log\left(1 + \frac{\sigma^2}{2}\left(\frac{t}{\sqrt{n}}\right)^2 + \ldots\right) = n \log\left(1 + \frac{\sigma^2 t^2}{2n} + \ldots\right)
\end{aligned} \tag{1.4}$$

Equation (1.4) was obtained by applying Eq. (1.3), noting that the mean of $Y_i - \mu$ is 0 and the mean of $(Y_i - \mu)^2$ is σ^2 (Eq. (1.1)), and ignoring third- and higher-order terms because for large n they are all multiplied by terms $\left(\frac{t}{\sqrt{n}}\right)^3$ or larger, and this higher power of n sends these terms to zero as n gets large.

Now as $x \longrightarrow \infty$, it can be shown that $x \log(1 + ax^{-1}) \longrightarrow a$, so as $n \longrightarrow \infty$,

$$\log m_{\sqrt{n}(\bar{X}-\mu)}(t) \rightarrow \frac{\sigma^2 t^2}{2}$$

which completes the proof.

Maths Box 1.7: How Fast Does the Central Limit Theorem Work?
So sample means are approximately normal when n is large, irrespective of the distribution of the response variable Y, but how large does n need to be?

There is no simple answer; it depends on the distribution of Y, in particular, it depends on *skewness*, which we define as $\kappa_1 = \mu_{(Y_i-\mu)^3}$, and *kurtosis* (or *long-tailedness*), defined as $\kappa_2 = \mu_{(Y_i-\mu)^4}$.

The main approximation in the proof was when the third and fourth moments of $Y_i - \mu$ (and higher-order terms) in Eq. (1.4) were ignored. If we go back to this equation and stick these terms back in, we get

$$\begin{aligned}
\log m_{\sqrt{n}(\bar{Y}-\mu)}(t) &= n \log m_{(Y_i-\mu)}(t/\sqrt{n}) \\
&= n \log \left(1 + \frac{\sigma^2}{2}\left(\frac{t}{\sqrt{n}}\right)^2 + \frac{\kappa_1}{6}\left(\frac{t}{\sqrt{n}}\right)^3 + \frac{\kappa_2}{24}\left(\frac{t}{\sqrt{n}}\right)^4 + \ldots\right) \\
&= n \log \left(1 + \frac{\sigma^2 t^2}{2n} + \frac{\kappa_1 t^3}{6n\sqrt{n}} + \frac{\kappa_2 t^4}{24n^2} + \ldots\right) \\
&\simeq \frac{\sigma^2 t^2}{2} + \frac{\kappa_1 t^3}{6\sqrt{n}} + \frac{\kappa_2 t^4}{24n} + \ldots
\end{aligned}$$

The higher-order terms, having n in the denominator, will always become ignorable when n is large enough, but the n that is "large enough" is bigger when data come from a distribution with a larger (in absolute terms) value of κ_1 or κ_2. The third moment κ_1 is zero for symmetric distributions, non-zero for skewed, and larger (in absolute terms) for more strongly skewed distributions. The fourth moment κ_2 is larger when Y has a more long-tailed distribution.

So if averaging a variable that is skewed or that has the occasional large outlier, the sample size needed for the central limit theorem to take effect needs to be larger. Simulations suggest that $n = 10$ is plenty for fairly symmetrical, short-tailed distributions, but $n = 100$ might not be enough for more pathological long-tailed cases. If data are strongly skewed or if they come with large outliers, methods that model the mean should not really be used anyway, without data transformation. Keeping this in mind, $n = 30$ is usually plenty for appropriately transformed data.

The central limit theorem applies surprisingly quickly—if taking a simple average of a symmetric variable without long tails, five observations are usually plenty for it to kick in and ensure that the sample mean will be close to being normally distributed. It works more slowly for data from skewed distributions, or when a statistic is highly influenced by just a few data points (long-tailed distributions and outliers being a particular issue). Some useful insights can be found in Miller Jr. (1997), and there are plenty of online applets as well to discover the robustness of the central limit theorem yourself, a good one is at `http://onlinestatbook.com/stat_sim/sampling_dist`.

The convergence to normality of most statistics, in most situations, explains why the mean model and variance model are the most important things to focus on. The mean and the variance are the only parameters that matter in a normal distribution. Roughly speaking, if you get your mean model right, then the mean of any statistic capturing (mean) trends in the data will be right, and if you get the variance model right, plus your independence assumption, then the variance (standard error) of your statistic will be right too. Throw in some central limit theorem and you're done, you know the distribution of your test statistic (in large samples) irrespective of whether your distributional assumptions are right. In which case, if your sample size is large enough for the central limit theorem to apply (usually, $n > 30$ is plenty), then any inferences you make based on this statistic will be *valid*.

That doesn't mean that your inferences will be *efficient*. If you have significant departures from your distributional assumptions, in particular, ones that add additional skewness to the distribution, or long-tailedness (propensity for outliers), this can have substantial effects on the efficiency of your inferences, i.e. how likely you are to pick up patterns that are there. So if you know your distributional assumptions are quite wrong (especially in terms of skewness and long-tailedness), you should try to do something about it, so that your inferences (tests, CIs) are more likely to pick up any signal in your data. One thing you could do is transform your data, as in the following section; another option is to use a different analysis technique designed for data with your properties. A more flexible but more sophisticated option is to try *non-parametric statistics* (Corder & Foreman, 2009) or *robust statistics* (Maronna et al., 2006)—branches of statistics that have developed a suite of techniques intended to maintain reasonable efficiency even for long-tailed or skewed data.

Assessing skewness is easy enough—is there a J shape (right-skewed) or an r shape (left-skewed) on a normal quantile plot of residuals. You can also look at a histogram of residuals, but this tends to be less sensitive as a diagnostic check. Assessing long-tailedness is a little problematic because by definition you don't see outliers often, so from a small sample you don't know if you are likely to get any. It will show up in a larger sample, though, and, if present, can be seen on a normal quantile plot as departures from a straight line above the line for large values and below the line for small values (since observed values are more extreme than expected in this case).

You can also compute sample estimates of skewness and kurtosis (long-tailedness) on most stats packages, but I wouldn't worry about these—they are unreliable unless you have a big enough sample size that the answer is probably screaming at you from a graph anyway.

So in summary, what you need to worry about depends on sample size, along the lines of the key point stated below. The preceding rules represent a slight relaxation of those in Moore et al. (2014); many use tougher criteria than suggested here. I arrived at the rules stated in this key point using a combination of theory (Maths Box 1.7) and practice, simulating data (and corresponding normal quantile plots) and seeing what happened to the validity of the methods.

An important and difficult situation arises as a result of small sample sizes ($n < 10$)—in this case, distributional assumptions matter but are hard to check. They matter because the central limit theorem doesn't help us too much, and they are hard to check because the small sample size doesn't give us a lot of information to detect violations of our assumptions. This situation is a good case for trying *non-parametric statistics* (Corder & Foreman, 2009) or design-based inference (Chap. 9) to ensure the inference procedure is valid.

Key Point

You don't need to be too fussy when checking distributional assumptions on your model fit. Usually, we can use a model fitted to our data to construct a set of residuals that are supposed to be normally distributed if the model is correct. How carefully you check these residuals for normality depends on your sample size:

$n < 10$ Be a bit fussy about the normality of residuals. They should look symmetric and not be long-tailed (i.e. on a normal quantile plot, points aren't far above the line for large values and aren't far below the line for small values).

$10 < n < 30$ Don't worry too much; just check you don't have strong skew or big outliers/long-tailedness.

$n > 30$ You are pretty safe unless there is really strong skew or some quite large outliers.

1.5.3 Hypothesis Tests of Assumptions Are Not a Great Idea

There are plenty of formal tests around for checking assumptions, including tests of normality (Anderson-Darling, Shapiro-Wilk, ...), or tests for equal variance (Levene's test, F-test, ...). These tests are not such a good idea; in fact, I actively advise *against* using them. Checking a graph is fine (using the rules in the previously stated key point).

There are two main reasons not to use hypothesis tests to check assumptions. First, it is a misuse of the technique, and second, it doesn't work, as explained below.

Hypothesis tests on assumptions are a misuse of the approach because you should only test hypotheses you collected the data to test! Was the primary purpose of your study really to test the assumption of normality?

Hypothesis tests on assumptions don't work well because this approach doesn't answer the question we want answered. A hypothesis test would give you a sense of whether there is evidence in your data against the hypothesis of normality, but recall we want to know whether violations of normality are sufficiently strong that we should worry about them (in particular, skewness and long-tailedness). These are quite different things. Further, hypothesis tests of assumptions behave in the wrong way as the sample size increases. A larger sample size gives us more information about the distribution of our data, so when n is large, we are more likely to detect violations of normality, even if they are small. But we know, from the central limit theorem, that statistics are more robust to violations of normality when sample size is larger, so tests of normality are going to make us worry more in situations where we should be worrying less!

Key Point
Transformation can be useful; in particular, when data are "pushed" up against a boundary, transformation can remove the boundary and spread the data out better. The log transformation is a good one because it is easier to interpret.

1.6 Transformations

Consider y_{new}, formed as some function of a variable y. Examples:

$$y_{\text{new}} = \sqrt{y} \quad y_{\text{new}} = 1000y \quad y_{\text{new}} = \log(y)$$

If y_{new} is a function of y, we say y_{new} is a transformation of y. The act of calculating y_{new} is referred to as *transforming* y.

Transforming is super easy using R, as in Code Box 1.3.

Code Box 1.3: $\log(y+1)$**-transformation of the Raven Data**

```
# Enter the data
Before = c(0, 0, 0, 0, 0, 2, 1, 0, 0, 3, 5, 0)
After  = c(2, 1, 4, 1, 0, 5, 0, 1, 0, 3, 5, 2)

# Transform the data using y_new = log(y+1):
logBefore = log(Before+1)
logAfter  = log(After+1)
```

```
# Construct a normal quantile plot of the transformed data
qqenvelope(logAfter-logBefore)
```

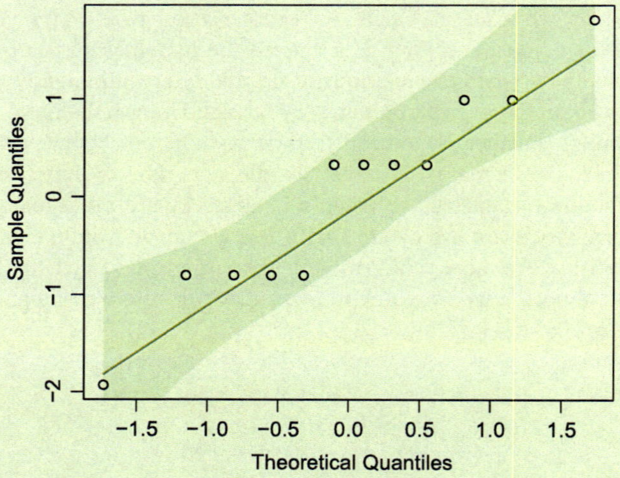

This plot looks okay, but the previous one (Code Box 1.2) looked reasonably good too. Another option to consider is to model the data as discrete, as in Chap. 10.

Why transform data? Usually, *to change its shape*, in particular, to get rid of strong skew and outliers. (As before, if residuals have strong skew or outliers, most analysis methods don't work well.) Figure 1.4 is a schematic to gain some intuition for why transformation can change the shape of data.

A special type of transformation is the linear transformation, which is used to change the scale (e.g. from grams to kilograms) rather than the shape of a distribution. Linear transformations have the form $y_{new} = a + by$ and are so called because if you plot y_{new} against y, you get a straight line. The most common linear transformation in statistics is when you want to standardise a variable, subtracting the mean and dividing by the standard deviation. Note that a linear transformation doesn't change the shape at all; it only changes the scale. Only non-linear transformations are shape changers.

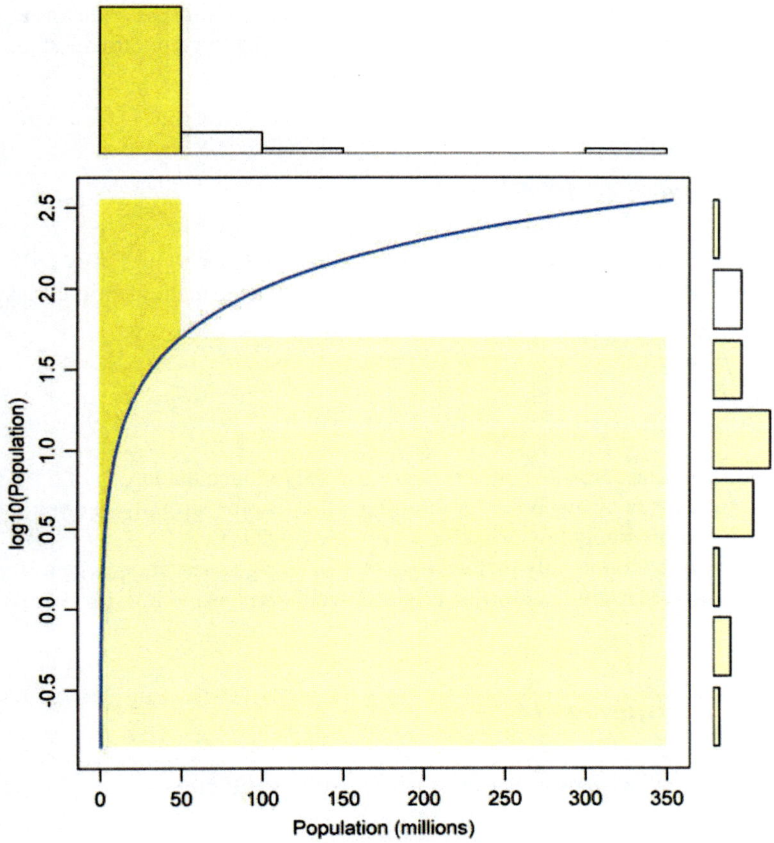

Fig. 1.4: How transformation changes the shape of data. The $\log_{10}$ function (blue curve) has a steeper slope at smaller values, so if most observations are small (yellow vertical band), then after log transformation, these will get spread out over a much broader range of values (light yellow horizontal band). The histograms in the margins show the effect this has on the shape of a right-skewed variable. In this case, population size of first-world countries becomes much more symmetric after log transformation, and the apparent outlier on untransformed data (USA) looks fine after transformation

A cost of transformation is loss of interpretability, because the scale being used for analysis is different from the one on which values were measured. Predictions from a model to transformed data can be back-transformed to the original scale, but doing this introduces *retransformation bias* (Duan, 1983)—the mean of a function of data is not the same as applying that same function to the mean (except for linear transformations). Thus, where possible, it is advisable to transform data to scales that themselves are meaningful, e.g. the log transformation, and to use transformation with caution if predictions are required on the original scale (e.g. to estimate standing biomass of a forest).

1.6.1 Common Transformations

The most common situation is where your data take positive values only ($y > 0$) and are right-skewed. In this situation, the following transformations might make it more symmetric:

- $y_{\text{new}} = \sqrt{y}$
- $y_{\text{new}} = y^{1/4}$
- $y_{\text{new}} = \log y$

These transformations are all "concave down", so they reduce the length of the right tail. They are in increasing order of strength—that is, for strongly skewed data, $y_{\text{new}} = \log y$ is more likely to work than $y_{\text{new}} = \sqrt{y}$ (Fig. 1.5).

They are also monotonically increasing—that is, as y gets larger, y_{new} gets larger. (A transformation that didn't have this property would be hard to interpret!)

1.6.2 Log Transformation

The logarithmic or log transformation is particularly important:

$$y_{\text{new}} = \log_a y$$

where a is the "base", commonly $\log_{10} y$, $\log_e y$, $\log_2 y$.
(The base doesn't matter—it only affects the scale, not the shape.)

Logs have the following key property:

$$\log(ab) = \log a + \log b$$

and more generally,

$$\log(y_1 \times y_2 \times \ldots y_n) = \log y_1 + \log y_2 + \ldots + \log y_n$$

In words, *the logarithm transforms multiplicative to additive.*

Often, many variables can be understood as the outcome of a series of multiplicative processes, and there is a case for the use of log transformations in such situations (Kerkhoff & Enquist, 2009). It often seems to work really well, as if by magic, perhaps because many processes are best understood as multiplicative (this times this times that, not this plus this plus that). For example:

- Wealth
- Size

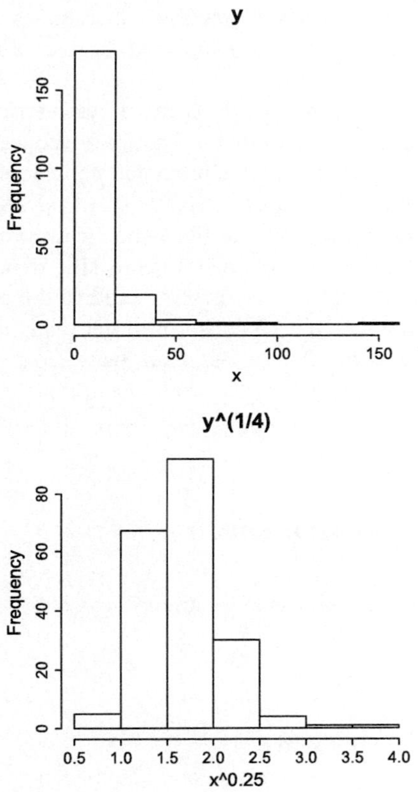

Fig. 1.5: Examples of the effect of transformation of some strongly right-skewed data. The transformations are in increasing order of strength—they progressively reduce the extent of right skew, until the log transformation, which does a good job of symmetrising the data in this particular case

- Profit
- Population

By log-transforming such processes, they change from multiplicative to additive. See, for example, the data on population sizes of first-world countries in Fig. 1.4.

1.6.3 Another Situation Needing Transformation—Boundaries

If your data are "pushed up" against a boundary, you should think about transforming the data to remove the boundary.

For example, a population can be small, but it can't be negative, so values are pushed up against zero. A log transformation removes this boundary (because as y approaches 0, $\log(y)$ approaches $-\infty$).

Removing boundaries in this way often makes the data look more symmetric, but it also prevents some awkward moments. Boundaries that are not removed can lead to nonsensical predictions from regression models, e.g. a predicted population of −2 million! A rather infamous example of this is a paper in Nature that analysed Olympic 100 m sprint times for males and females, without transformation to remove the boundary at zero seconds. Both male and female plots had a decreasing trend over time, but the fitted line was steeper for females, so the authors predicted that by the mid-22nd century, a woman would complete the 100 m sprint faster than the winning time for the men's event (Stephens et al., 2004). But there was also the rather embarrassing problem that by the 27th century the race would be completed in less than zero seconds (Rice, 2004)!

1.6.4 Situations Needing a Different Transformation

Proportions: If data are between 0 and 1, try the logit transformation:

$$y_{\text{new}} = \log\left(\frac{y}{1-y}\right)$$

(This stretches data over the whole real line, from $-\infty$ to ∞.)
The arcsine transform was often used historically—it is not a good idea because it is not monotonic and hard to interpret (Warton & Hui, 2011). If your proportions arise from counts, methods specially developed for discrete data might be worth a go (Chap. 10). Otherwise, beta regression is also worth considering (Ferrari & Cribari-Neto, 2004, an extension of methods of Chap. 10).

Right-skewed with zeros: This often happens when data are counts. The problem is that you can't take logs because $\log 0$ is undefined. Try

$$y_{\text{new}} = \log(y+1)$$

This might work for you, unless you have lots of small counts. If there are lots of zeros, you will need to use generalised linear models, described in Chap. 10.

Left-skewed data: This is less common. But if data are left-skewed and negative, then $-y$ is right-skewed and positive, in which case the transformations previously discussed can be applied to $-y$. Maybe take the negative of them afterwards so they remain monotonic increasing (if a value is bigger on the untransformed scale, it is bigger after transformation too).

For example, $y_{\text{new}} = -\log(-y)$ takes negative, left-skewed values of y and tries to make them more symmetric.

Exercise 1.11: Height and Latitude
Angela and her friends (Moles et al., 2009) collected some data on how tall plants are in lots of different places around the world. She wants to know how plant height is related to latitude.

What does the question tell us—descriptive, interval estimation, hypothesis testing?

What do the data tell us—one variable or more? What type of response?

Exercise 1.12: Transform Plant Height?
You can access Angela's height data (Exercise 1.11) using `R`; it is the `height` variable from the `globalPlants` dataset in the `ecostats` package:

```
library(ecostats)
data(globalPlants)
globalPlants$height
```

Construct a histogram. Do you see any boundaries that could cause problems for analysis? What sort of transformation would you suggest? Try out this transformation on the data and see if you think it has fixed the problem.

Exercise 1.13: Snails on Seaweed
David and Alistair looked at the density (per gram of seaweed) of invertebrate epifauna (e.g. snails) settling on algal beds (also known as seaweed) with different levels of isolation (0, 2, or 10 m buffer) from each other (Roberts & Poore, 2006), to study potential impacts of habitat fragmentation. They want to know: *Does invertebrate density change with isolation?*

What does the research question ask us—descriptive, estimation, hypothesis testing, etc?

What do the data tell us—one variable or two? What type of response variable?

What graph would you use to visualise the data?

Exercise 1.14: Transform Snails?
You can access David and Alistair's invertebrate data (Exercise 1.13) in `R`; it is the `Total` variable from the `seaweed` dataset in the `ecostats` package.

Load the data and construct a histogram. Do you see any boundaries that could cause problems for analysis? What sort of transformation would you suggest? Try out this transformation on the data and see if you think it has fixed the problem.

Chapter 2
An Important Equivalence Result

In this chapter we will revise two of the most commonly used statistical tools—the two-sample t-test and simple linear regression. Then we will see a remarkable equivalence—that these are actually exactly the same thing! This is a very important result; it will give us some intuition for how we can write most of the statistical techniques you have previously learnt as special cases of the same linear model.

2.1 The Two-Sample t-Test

The two-sample t-test is perhaps the most widely used method of making statistical inferences. I reckon it has had a hand in maybe half of the significant scientific advances of the twentieth century, maybe more. So it will be quite instructive for us to revise some of its key properties and some issues that arise in its application.

Exercise 2.1: Two-Sample t-Test for Guinea Pig Experiment
Is the number of errors made by guinea pigs related to nicotine treatment?

Recall the experiment looking at the effect of nicotine on the development of guinea pig offspring. The number of errors made by guinea pigs in a maze is given below:

Control	11	19	15	47	35	10	26	15	36	20
Nicotine	38	26	33	89	66	23	28	63	43	34

We can use a *two-sample t-test* to answer the research question.

Rather than thinking of this as comparing # errors in two samples (Control and Nicotine) think of it *as testing for an association between # errors and treatment group*. That is, we think of this problem as having two variables—# errors is quantitative, treatment group is categorical.

So the two-sample t-test is *a test for association* between # errors and treatment. We test H_0 : no association between # errors and treatment using the test statistic:

D. I. Warton, *Eco-Stats: Data Analysis in Ecology*, Methods in Statistical Ecology,
https://doi.org/10.1007/978-3-030-88443-7_2

$$t = \frac{\bar{y}_{\text{Control}} - \bar{y}_{\text{Nicotine}}}{\text{standard error of } (\bar{y}_{\text{Control}} - \bar{y}_{\text{Nicotine}})}$$

which (if H_0 is true) comes from a t distribution with degrees of freedom $n-2$, where n is the total sample size ($n = n_{\text{Control}} + n_{\text{Nicotine}}$).

For the guinea pig data, this test statistic works out to be −2.67. We would like to know whether this statistic is unusually large compared to the sorts of values you would expect if there were actually no effect of treatment. What do you think?

The strategy for constructing any t-statistic, to test for evidence against the claim that a parameter is zero, is to divide the quantity of interest by its standard error and to see whether this standardised or t-statistic is unusually far from zero. (As a rough rule of thumb, values beyond 2 or −2 start to get suspiciously far from zero.) This type of statistic is commonly used and is known as a Wald test. (The test is named after Abraham Wald, who in math genealogy terms `http://genealogy.math.ndsu.nodak.edu/` happens to be my great-great-great-grandfather!)

In a two-sample t-test we are looking for evidence that the true (population) means differ across two groups. The quantity of primary interest here is the mean difference, and we want to know if it is non-zero. So as in Exercise 2.1, we construct a two-sample t-statistic by dividing the sample mean difference by its standard error and considering whether this statistic is unusually far from zero.

A confidence interval for the true mean difference is returned automatically in a t-test using R, and it is calculated in the usual way (but typically using a critical value from the t-distribution, not using 2 as a rough approximation, which would get things a bit wrong if sample size is small).

Code Box 2.1: A Two-Sample t-Test of the Data from the Guinea Pig Experiment

```
> library(ecostats)
> data(guineapig)
> t.test(errors~treatment, data=guineapig, var.equal=TRUE,
     alternative="less")

Two Sample t-test

data:  errors by treatment
t = -2.671, df = 18, p-value = 0.007791
alternative hypothesis: true difference in means is less than 0
95 percent confidence interval:
      -Inf -7.331268
sample estimates:
mean in group C mean in group N
           23.4            44.3
```

2.1.1 Mind Your Ps and Qs When Doing a Two-Sample t-Test

In a two-sample t test of data y we make the following assumptions:

1. The y-values are *independent* in each sample, and samples are also independent. This can be guaranteed—*how?* (Hint: see Sect. 1.2.5)
2. The y-values are *normally distributed* with *constant variance*

$$y_{ij} \sim \mathcal{N}(\mu_i, \sigma^2)$$

and *normality* can be checked with a *normal quantile plot*
Constant variance can be checked by comparing standard deviations or using a residuals vs fits plot (as is done later for linear regression).

Code Box 2.2: Smoking and Pregnancy—Checking Assumptions

The normal quantile plots of Fig. 2.1 were generated using the following code:

```
> qqenvelope(guineapig$errors[guineapig$treatment=="N"])
> qqenvelope(guineapig$errors[guineapig$treatment=="C"])
> by(guineapig$errors,guineapig$treatment,sd)

guineapig$treatment: C
[1] 12.30357
------------------------------------------------------
guineapig$treatment: N
[1] 21.46858
```

Do you think assumptions are reasonable?

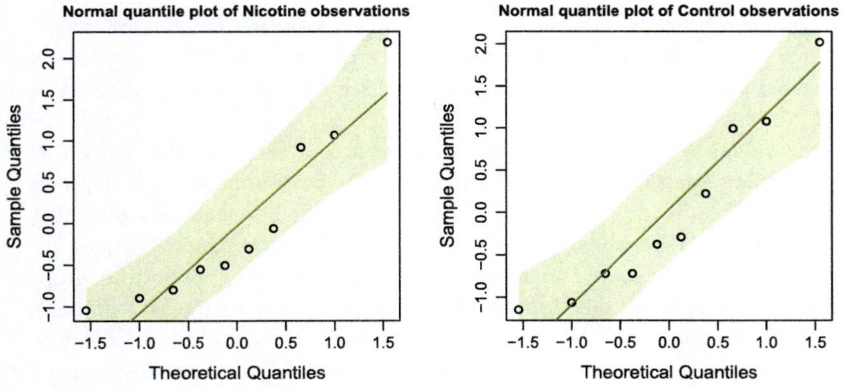

Fig. 2.1: Normal quantile plots of the treatment and control data from the smoking-during-pregnancy study of Exercise 2.1

2.1.2 Robustness of t-Tests

What happens if your data aren't normally distributed? Or the variances aren't equal? Recall that there are two main things to consider:

Validity Am I actually going to exceed my 0.05 significance level only 5% of the time when the null is true? (Or, equivalently, is a 95% confidence interval really going to capture the true parameter 95% of the time?)

Efficiency Is my test going to have good power at detecting a change in mean? Or, equivalently, is a confidence interval going to be narrow, to give me a precise estimate of the true mean difference?

Recall also that most analyses can be broken down into four different types of assumption: independence, mean model, variance model, and distributional assumptions. What does the t-test say about each of these components? And how much do the different assumptions really matter?

Independence: Observations are assumed independent within and across samples. As previously, if your data are not independent but you have assumed your data are independent, then you are stuffed. Your methods will probably not be valid, most typically with standard errors being too small and false declarations of significance. In an experiment, randomly assigning subjects to treatment groups guarantees this assumption will be satisfied.

Distribution assumption: We assume data are normally distributed. As previously, distributional assumptions (in this case, the assumption that the data are normally distributed) rarely matter for the validity of our method, thanks to the central limit theorem. Only if sample size is very small do we need to take this assumption seriously. However, if the data are strongly skewed or have outliers, then the t-test will not be very efficient, so we need to check for skew and outliers—see the key point from Sect. 1.6 for rough guidelines.

Mean model: The only assumption here is that the mean is the same across observations in each sample. This is satisfied if we randomly sample observations from the same population, or in an experiment, if we randomly assign subjects to treatment groups.

Variance model: We assume constant variance across samples, also known as *homoscedasticity*. Apart from independence, this is the most important assumption. How robust the t-test is to violations of this assumption will depend on the relative sample sizes in the two groups. If your sample sizes in the two treatment groups are equal, $n_1 = n_2$ or nearly equal (on a proportional scale, e.g. they only differ from each other by 10% or less), then the *validity* of the two-sample t-test will be quite robust to violations of the equal variance assumption. If your sample sizes are quite different, e.g. one is four times the size of the other ($n_1 = 4n_2$), then violations of the equal variance assumption will spell trouble. What happens is that the estimate of the standard error of the mean difference (used on the denominator of a t-statistic and in the margin of error of a confidence interval for mean difference) is either over-

or underestimated, depending on whether the group with the larger sample size has a larger (or smaller) variance. This leads to an overly conservative (or overly liberal) test and confidence intervals that are too wide (or too short).

In practice, you rarely get substantial differences in variance across samples unless they are also accompanied by skewed data or outliers. Transformation is the first thing to consider to fix the problem.

2.2 Simple Linear Regression

Linear regression, when there is only one x variable, is often referred to as simple linear regression. Let's quickly review what it's about.

Exercise 2.2: Water Quality
One way to study whether river health deteriorates downstream is to study the association between water quality and catchment area. Thierry and Robert used fish composition to construct an index of biotic integrity (IBI) and studied how it related to catchment areas in the Siene River basin in France (Oberdorff & Hughes, 1992). The results are in the `waterQuality` dataset in the `R` package `ecostats` and look like this:

log Catchment area (km^2)	2	2.9	3.15	3.2	3.55	...	4.85
Water quality (IBI)	50	40	34	4.05	40.5	...	20

What does the research question ask us—descriptive, estimation, hypothesis testing, etc?

What do the data tell us, i.e. data properties—one variable or two? Categorical or quantitative?

What graph would you use to visualise the data?

So how would you analyse the data?

For the water quality example of Exercise 2.2, we study the relationship between two quantitative variables, so the obvious graphical summary is a scatterplot, as in Fig. 2.2.

A regression line was added to the plot in Fig. 2.2. This regression line was fitted using *least squares*. That is, we found the line that gave the least value of the sum of squared errors from the line, as measured in the vertical direction. (Why just the vertical direction? Because this is the best strategy if your goal is to *predict* y using the observed x.)

A nice applet illustrating how this works can be found (at the time of writing) at https://phet.colorado.edu/sims/html/least-squares-regression/latest/least-squares-regression_en.html

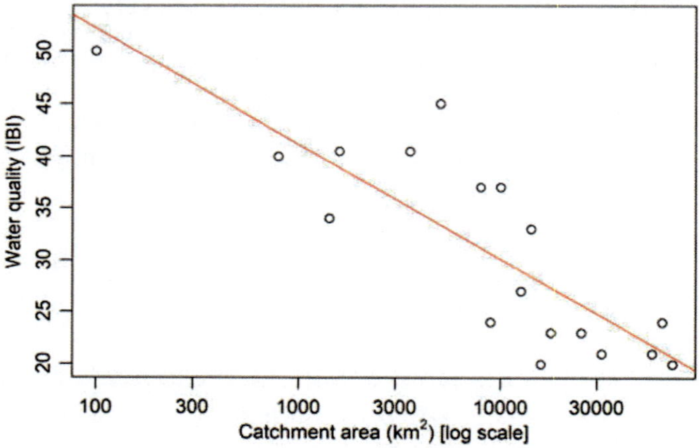

Fig. 2.2: Scatterplot of water quality (IBI) against catchment area (km^2) for data of Exercise 2.2. A simple linear regression line was added to the plot. This line was fitted using least squares, i.e. minimising the sum of squared vertical distance between each point and the line (as indicated by arrows)

But if you Google "least squares regression applet", something useful should come up that you can interact with to get a sense for how regression works.

Simple linear regression is a straight line, taking the algebraic form

$$\mu_y = \beta_0 + \beta_1 x$$

where

$$\beta_0 = \text{intercept on y-axis (x=0)}$$
$$\beta_1 = \text{slope of the line.}$$

β_1 represents the *magnitude of the effect of x on y* (Fig. 2.3).

When we fit this model to a sample of data, we don't get the true values of the slope and intercept; rather, we get sample estimates of them. Which we can write like this:

$$\hat{\mu}_y = \hat{\beta}_0 + \hat{\beta}_1 x$$

When looking at computer output from a fit, the key information about the line that has been fitted can usually be found in a table, as in Code Box 2.3. Information about the estimate of the y-intercept is usually referred to as "intercept" or something similar, and information about the slope estimate is usually referred to by the name of the x variable being used to predict y (because this is the variable that the slope is multiplied by in the regression equation). A column labelled "Estimate" or some-

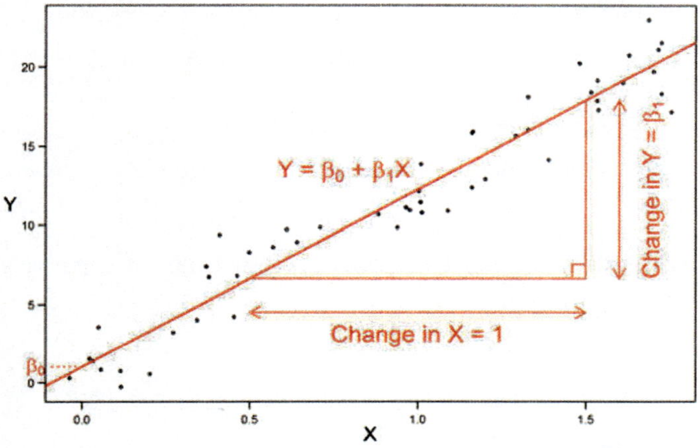

Fig. 2.3: β_1 represents the magnitude of the effect of x on y: if x changes by one, we predict that the mean of y will change by β_1. For this reason, the slope is the important parameter describing how x and y are related

thing similar gives the estimated values of these parameters (or "coefficients"), and estimated standard errors will be given in a column of the table labelled something like "Standard error" or "SE".

Code Box 2.3: Fitting a Linear Regression to the Water Quality Data

```
> data(waterQuality)
> fit_qual=lm(quality~logCatchment, data = waterQuality)
> summary(fit_qual)

Call:
lm(formula = quality ~ logCatchment, data = waterQuality)

Residuals:
   Min     1Q Median     3Q    Max
-7.891 -3.354 -1.406  4.102 11.588

Coefficients:
             Estimate Std. Error t value Pr(>|t|)
(Intercept)    74.266      7.071  10.502 1.38e-08 ***
logCatchment  -11.042      1.780  -6.204 1.26e-05 ***
---
Signif. codes:  0 '***' 0.001 '**' 0.01 '*' 0.05 '.' 0.1 ' ' 1

Residual standard error: 5.397 on 16 degrees of freedom
Multiple R-squared:  0.7064,Adjusted R-squared:  0.688
F-statistic: 38.49 on 1 and 16 DF,  p-value: 1.263e-05
```

Exercise 2.3: Water Quality—Interpreting R Output
The model is

$$\mu_{\text{quality}} = \beta_0 + \beta_1 \times \text{logCatchment}$$

Use the preceding output to compute an approximate 95% confidence interval for β_1.

Explain what β_1 means in words.

NB: Recall that if you want to get more accurate confidence intervals in R, you can always use `confint`.

2.2.1 The Importance of Testing Slope = 0

Notice there is a t-statistic and P-value for each coefficient in the foregoing output. These test the null hypothesis that the true value of the parameter is 0.

Why 0?

For the y-intercept—no reason; usually it would be a stupid idea!
For the slope—this tests for an *association between* y *and* x. If you are interested in testing for an association between y and x, this is the P-value to pay attention to.

Recall that the regression line is for predicting y from x. If there were no association between y and x, then the (true) predicted value of y would be the same irrespective of the value of x. This makes a straight horizontal line, as in Fig. 2.4b.

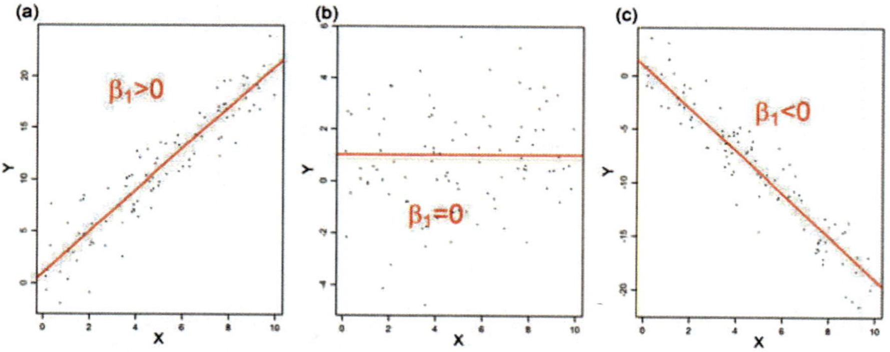

Fig. 2.4: The slope of the regression line, β_1, is the most important parameter; it tells us how y relates to x: (**a**) if β_1 is positive, y is expected to increase as x increases; (**b**) if β_1 is zero, there is no relationship between y and x; (**c**) if β_1 is negative, y is expected to decrease as x increases

2.2.2 Mind Your Ps and Qs When Using Linear Regression

For inference about simple linear regression, we make the following assumptions:

1. The observed y-values are *independent*, after accounting for x.
   This assumption can often be guaranteed to be satisfied—*how???*
2. The y-values are *normally distributed* with *constant variance*.

$$y_i \sim \mathcal{N}(\mu_i, \sigma^2)$$

   Normality usually doesn't matter (due to the central limit theorem) except for small samples/strongly skewed data/outliers. The discussion in Sect. 2.1.2 applies here. You can check normality on a *normal quantile plot of residuals* (as in Code Box 2.4).
   Constant variance can be checked using a *residuals vs fits plot* to see if there is any fan-shape pattern—as in the following section.
3. There is a *straight-line relationship* between mean of y and x

$$\mu_i = \beta_0 + \beta_1 x_i$$

   A straight-line relationship is crucial—no point fitting a straight line to non-linear data! Check for no obvious pattern e.g. a U-shape, on a *residuals vs fits plot*.

NB: This looks a lot like two-sample t-test assumptions. . . .

2.2.3 Residuals vs Fits Plot

A *residuals vs fits plot* ($y - \hat{\mu}_y$ vs $\hat{\mu}_y$ as in Fig. 2.5b) can be used to check whether data are linearly related and whether the variance is constant for different values of x. The original data in Fig. 2.5a are squished (linearly transformed and, sometimes, flipped around) in a residual plot so that the linear trend is removed, to focus our attention on the errors around the line. A residuals vs fits plot typically has a horizontal line at a residual of zero, which represents the original regression line, since any point that fell exactly on the regression line would have had a residual of zero. Residuals with positive values correspond to points above the regression line; residuals with negative values correspond to points below the regression line.

There should be *no pattern* on a residual plot—if there is, then the assumptions made previously are violated, and we cannot make inferences about the true regression line using simple linear regression. Which assumption is violated depends on what sort of pattern you see.

The two most common diagnostic plots used in regression are a normal quantile plot of residuals and a residuals vs fits plot (check for no pattern), which can be constructed in R using the `plot` function on a fitted regression object (Code Box 2.4). Do you think the regression assumptions seem reasonable for this dataset?

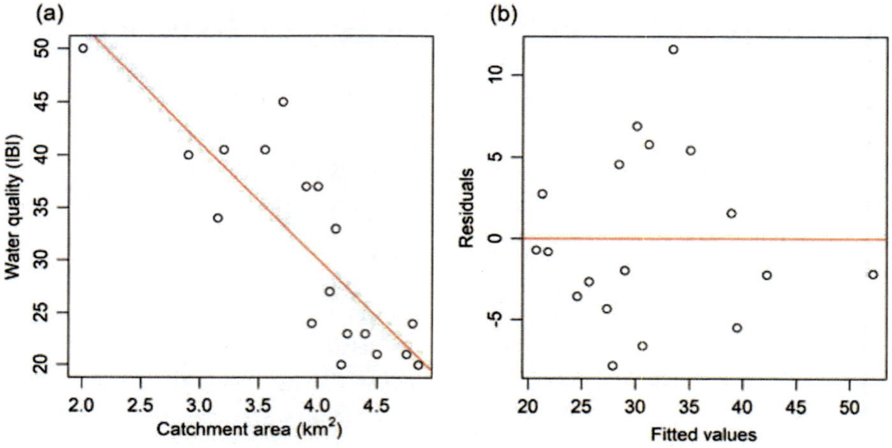

Fig. 2.5: A residuals vs fits plot for water quality data. (**a**) Original data, with linear regression line added in red. (**b**) Residuals ($y - \hat{\mu}_y$) plotted against fitted values ($\hat{\mu}_y$), which can be understood as a transformation of the data so that the regression line is now horizontal and passes through zero. Because fitted values decreased when catchment area increased, the plot against fitted values is flipped horizontally, with the leftmost observation from (**a**) appearing as the rightmost observation in (**b**). We can check regression assumptions using this plot by checking for a pattern

As with a normal quantile plot, it can be hard to determine visually whether patterns on a residuals vs fits plot are large compared to what we would expect under normal sampling variation, when assumptions are satisfied. The `plotenvelope` function in the `ecostats` package can help with this by simulating lots of datasets that satisfy assumptions and constructing an envelope around what we expect to see. For a residual vs fits plot, the envelope covers the smoother rather than the data points (see Code Box 2.4).

Key Point
A simple linear regression assumes independence, normality, constant variance, and a straight-line relationship.

There are two scary patterns to watch out for in a residuals vs fits plot:

- A U-shaped pattern, suggesting a problem with your straight-line assumption.
- A fan-shaped pattern, suggesting a problem with the constant variance assumption.

Code Box 2.4: Diagnostic Plots for Water Quality Data

To produce a residuals vs fits plot and a normal quantile plot of residuals, you just take a fitted regression object (like `fit_qual`, produced in Code Box 2.3) and apply the `plot` function:

```
> plot(fit_qual, which=1:2)
```

The `which` argument lets you choose which plot to construct (1 = residuals vs fits, 2 = normal quantile plot).

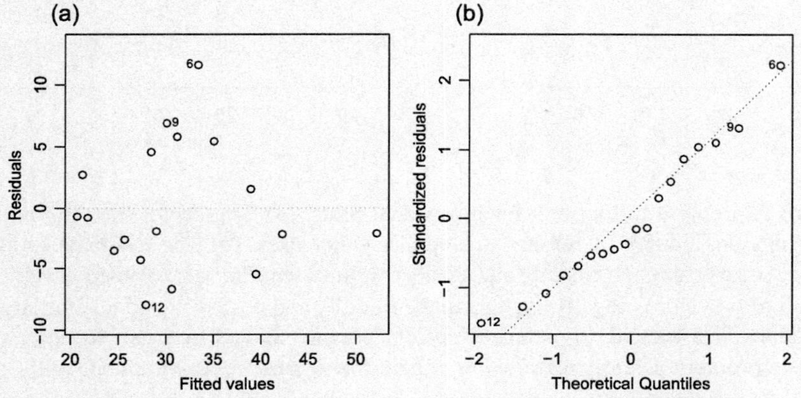

Alternatively, we can use `plotenvelope` to add simulation envelopes around points on these plots, to check whether any deviations from expected patterns are large compared to what we might expect for datasets that satisfy model assumptions:

```
> library(ecostats)
> plotenvelope(fit_qual, which=1:2)
```

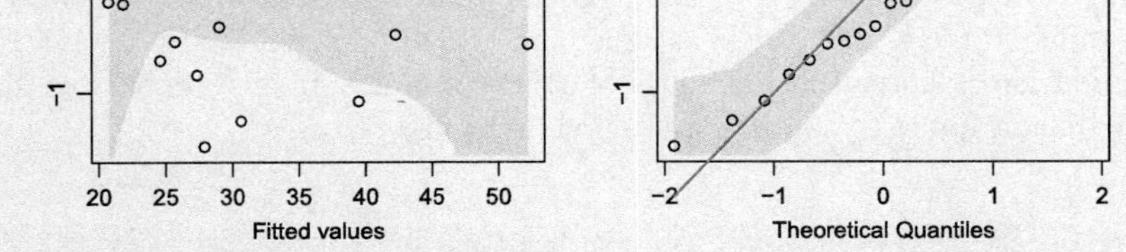

Do the assumptions look reasonable?

If there is a U-shaped pattern in your residuals vs fits plot, *you should not be fitting a straight line* to the data, as in Fig. 2.6. A U shape in the residual plot means that points tend to start off above the regression line, then they mostly move below

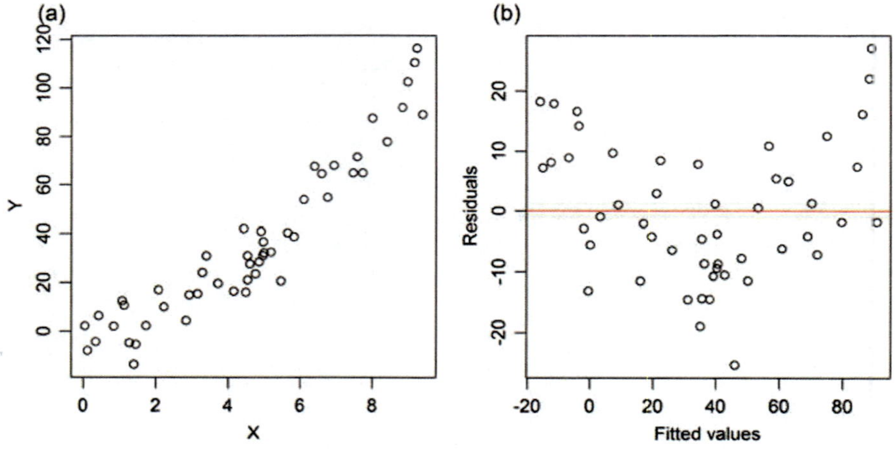

Fig. 2.6: A problem to look out for in residual plots. (**a**) Scatterplot showing non-linear function, a curve increasing in slope as x increases. (**b**) The residuals vs fits plot suggests a problem by having a U shape, residuals tending to take positive values for small fitted values, negative values in the middle, and positive residuals for large fitted values. We wanted no pattern in residuals as we moved from left to right, so we have a problem. Clearly *we should not be fitting a straight line* to the data

the line, then they finish off above it again. That is, the relationship is non-linear, and you should not assume it is linear. Your next step is either to consider transforming your data or to look into how to fit a non-linear model. You could also see this type of pattern but upside-down (n-shape).

A common example of where we would expect a non-linear relationship is when we are looking at some measure of habitat suitability as a function of an environmental variable (Austin, 2002, for example), e.g. species abundance vs temperature. All species have a range of thermal tolerance and an "optimal" range that they are most likely to be found in, with suitability dropping (sometimes quite abruptly) as you move away from the optimum (when it becomes too warm or too cold). Hence, linear regression is unlikely to do much good when modelling some measure of species suitability (e.g. abundance) as a function of environmental variables, at least if measured across a broad enough range of the environment. In this sort of situation we should know not to try a linear regression in the first place.

Exercise 2.4: Water Quality—Assumption Checks
Code Box 2.4 presents diagnostic plots for the water quality data.

Which regression assumptions can be checked using which of these plots?

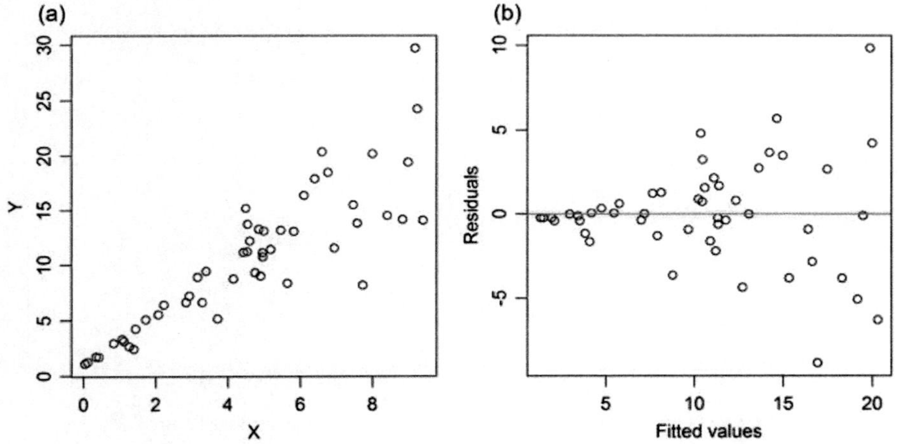

Fig. 2.7: Another problem to look out for in residual plots. (**a**) Scatterplot of data with variance increasing as predicted value increases. (**b**) The residuals vs fits plot with fan shape, residuals tending to take values closer to zero for small fitted values (left), and residuals tending to take values farther from zero for large fitted values (right). We wanted no pattern in residuals as we moved from left-to-right, i.e. similar spread in residuals at all fitted values, so we have a problem. Clearly *we should not be assuming equal variance* for these data

The other main type of pattern to watch out for is a fan-shaped pattern, meaning that the variance changes with x, as in Fig. 2.7. The equal variance assumption means that we should expect similar amounts of spread around the fitted line at each point along the line. So you have a problem with the equal variance assumption if your residuals vs fits plot suggests a smaller amount of spread around the line at some points compared to others. Typically, this shows up as residuals fanning out as fitted values increase, from left to right, as in Fig. 2.7. A common example of this is when studying abundance—the variance of abundance tends to increase as the mean increases (because for small means, data are squished up against the boundary of zero). In principle, residuals could fan inwards, but this is less common.

2.2.4 Influential Observations

An influential observation is one that has an *unusual x-value*. These are often easily seen in residuals vs fits plots as outlying values on the x-axis (outlying fitted values), although it gets more complicated for multiple linear regression (coming in Chap. 3).

Influential observations are dangerous because they have *undue influence* on the fitted line—pretty much the whole fit can come down to the location of one point. Once detected, a simple way to determine whether the whole story comes down to

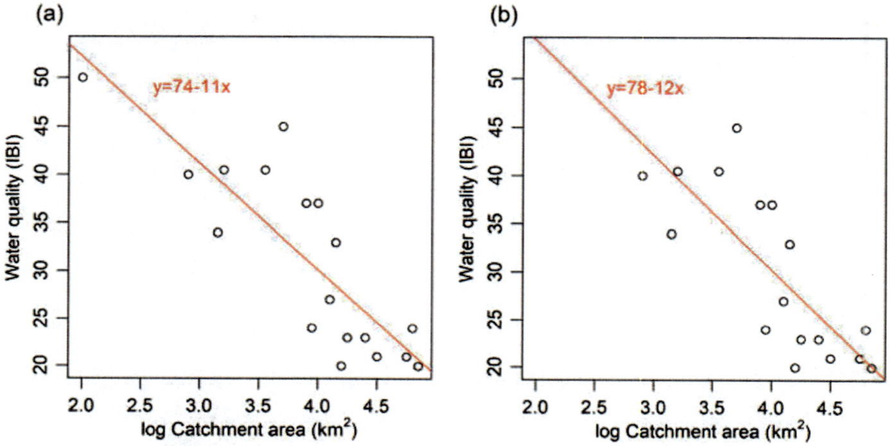

Fig. 2.8: Water quality example—testing the effect of the most influential value, the value with the largest catchment area. Compare (**a**) the full dataset to (**b**) the dataset with the most influential value removed. This had little effect on the regression fit

these influential values is to remove them and see if this changes anything. If little changes, there is little to worry about.

To check for high-influence points, always plot your x variables before you use them in analyses—you could use a histogram or boxplot, but if you have many x variables, try a `pairs` plot (a scatterplot matrix). You are looking for outliers and strongly skewed x variables.

High-influence points can often be avoided by transformation of the x variable, just as outliers can often be avoided by transformation. This doesn't always work, though.

Notice in Fig. 2.8 that removal of the most extreme x-value (smallest catchment area) had little influence on the fitted line. This is largely because, although that point was quite influential, it lay close to the main trend, so it didn't really change things compared to the information gained from the other data points.

2.2.5 R^2 as Proportion of Variance Explained

A nice way to summarise the strength of regression is the R^2 value, *the proportion of variance* in the y variable that has been *explained by regression* against x. In the water quality example output of Code Box 2.3, there was a line that read

```
Multiple R-squared:  0.118, Adjusted R-squared:  0.06898
```

so 11.8% of variance in water quality can be explained by catchment area.

But R^2 is a function of the sampling design as well as the strength of association, so it is difficult to generalise. For example, sampling a broader range of catchment sizes would increase R^2 without changing the underlying water quality-catchment area relationship. Thus, R^2 values shouldn't really be compared across different datasets where the x variable has been sampled in different ways.

2.3 Equivalence of t-Test and Linear Regression

This section describes one of the most important results in this book, so it is worth reading through it carefully. It is also one of the more difficult results. . . .

Consider again the output for the two-sample t-test of the smoking-during-pregnancy data. The output is repeated in Code Box 2.5, but this time using a two-sided test.

Key Point
A two-sample t-test can be thought of as a special case of linear regression, where the predictor variable only takes two values. Thus, you can fit a t-test as a regression and check the same assumptions in the same way as you would for linear regression.

Code Box 2.5: Two-Sample t-Test Output for Smoking-During-Pregnancy Data, Again

```
> t.test(errors~treatment, data=guineapig, var.equal=TRUE)

Two Sample t-test

data:  errors by treatment
t = -2.671, df = 18, p-value = 0.01558
alternative hypothesis: true difference in means is not equal to 0
95 percent confidence interval:
 -37.339333  -4.460667
sample estimates:
mean in group C mean in group N
           23.4            44.3
```

We previously established that a two-sample t-test can be understood as a type of test for association—testing for an association between number of errors and treatment. We could actually have used linear regression to analyse the data instead—constructing an "indicator variable" that takes a value of one for treatment guinea pigs and zero for controls, then testing for an association. Such a variable was already constructed in the `guineapig` dataset in the `ecostats` package. The results are in Code Box 2.6. *Can you see any similarities with Code Box 2.5?*

Code Box 2.6: Linear Regression Analysis of Smoking-During-Pregnancy Data. Compare to Code Box 2.5

```
> ft_guineapig = lm(errors~treatment,data=guineapig)
> summary(ft_guineapig)
Call:
lm(formula = errors ~ treatment, data = guineapig)

Residuals:
   Min     1Q Median     3Q    Max
-21.30 -11.57  -5.35  11.85  44.70

Coefficients:
            Estimate Std. Error t value Pr(>|t|)
(Intercept)   23.400      5.533   4.229 0.000504 ***
treatmentN    20.900      7.825   2.671 0.015581 *
---
Signif. codes:  0 '***' 0.001 '**' 0.01 '*' 0.05 '.' 0.1 ' ' 1

Residual standard error: 17.5 on 18 degrees of freedom
Multiple R-squared:  0.2838,Adjusted R-squared:  0.2441
F-statistic: 7.134 on 1 and 18 DF,  p-value: 0.01558
```

Two important things to notice:

- The regression intercept equals the sample mean of the control group.
- The t-statistic and P-value from the t-test match up with output for the regression slope (well, the t-statistic has a different sign, but exactly the same magnitude).

But *why???*

The regression line goes through the sample means: Recall that the regression line is estimated by least squares. For a single sample, it turns out that *the sample mean is the least-squares estimate* of the centre of the sample. If you remember differentiation and summation notation, the proof of this result is pretty easy (Maths Box 2.1). For two samples (lined up along the x-axis), a line will need to go through each sample mean if it wants to be the least-squares estimate. So simple linear regression just joins the two sample means (Fig. 2.9).

Maths Box 2.1: The Sample Mean Is a Least-Squares Estimator

We have a sample $y_1, y_2, \ldots, y_n$, and we want to find the least-squares estimate, i.e. the value a that minimises the sum of squared errors:

$$SS(a) = \sum_{i=1}^{n} (y_i - a)^2$$

It turns out that this least-squares estimate is the sample mean, $\bar{y} = \frac{1}{n}\sum_{i=1}^{n} y_i$, as shown below.

Notice that as a gets very large or very small, the sum of squared errors $SS(a)$ goes to infinity. Thus, the least-squares estimator must be a stationary point on this function. Differentiating $SS(a)$ with respect to a yields

$$\frac{d}{da}SS(a) = -2\sum_{i=1}^{n}(y_i - a)$$

and setting to zero at the stationary point $\hat{a}$:

$$0 = -2\sum_{i=1}^{n}(y_i - \hat{a})$$

$$n\hat{a} = \sum_{i=1}^{n} y_i$$

$$\hat{a} = \frac{1}{n}\sum_{i=1}^{n} y_i = \bar{y}$$

Since $\hat{a} = \bar{y}$ is the only stationary point on $SS(a)$, it must be the minimiser, i.e. the least-squares estimate.

What does this mean for the slope and elevation? In our regression analysis, # of errors is the y-variable and treatment is the x variable. Now there are only two values on the x-axis (Control and Nicotine). R by default will put Control values at $x = 0$ and Nicotine values at $x = 1$. As in Fig. 2.9, this means the following:

- The y-intercept is the Control mean.
- The slope is the mean difference between Nicotine and Control.

Further, because the parameter estimator of the slope is a quantity we saw earlier, the sample mean difference, the standard errors will be the same also, hence t- and P-values too. This is why key parts of the output match up between Code Boxes 2.5 and 2.6.

2.3.1 The t-Test as a Linear Regression

We can write any two-sample t-test as a linear regression. Recall that a simple linear regression model for the mean has the form

$$\mu_y = \beta_0 + \beta_1 x$$

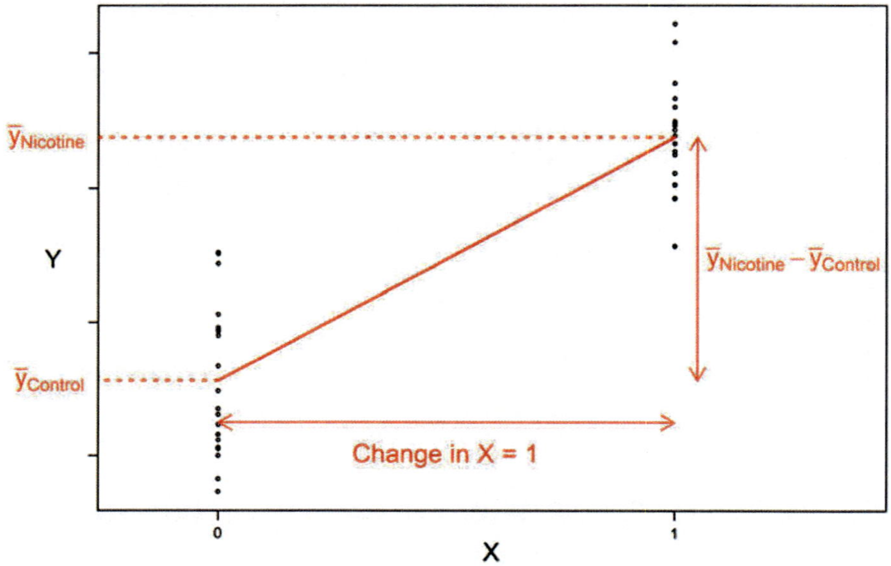

Fig. 2.9: Equivalence of a two-sample t-test (as in Exercise 2.1) and linear regression: a least-squares line will join each mean, so if the difference in x between the two groups is one, the slope of the least-squares line will be the difference between sample means (and if one group, such as the Control, has $x = 0$, the sample mean of that group becomes the estimated y-intercept)

whereas the two-sample t test models the mean as

$$\mu_y = \begin{cases} \mu_{\text{Control}} & \text{for subjects in the Control group} \\ \mu_{\text{Nicotine}} & \text{for subjects in the Nicotine group} \end{cases}$$

If we let $x = 0$ for Control and $x = 1$ for Nicotine, then the two-sample t-test is exactly the regression model with

$$\beta_0 = \mu_{\text{Control}}$$
$$\beta_1 = \mu_{\text{Nicotine}} - \mu_{\text{Control}}$$

So testing H_0: "slope is zero" can test H_0: "means are equal"!

Recall the two-sample t-test assumptions:

1. The y-values are *independent* in each sample.
2. The y-values are *normally distributed* with *constant variance*.

$$y_{ij} \sim \mathcal{N}(\mu_i, \sigma^2)$$

And the simple linear regression assumptions:

1. The observed y-values are *independent*, after accounting for x.
2. The y-values are *normally distributed* with *constant variance*.

$$y_i \sim \mathcal{N}(\mu_i, \sigma^2)$$

3. There is a *straight-line relationship* between the mean of y and x:

$$\mu_i = \beta_0 + \beta_1 x_i$$

These are the *same set of assumptions*. They can be *checked the same way*. Well OK, regression had a third assumption about linearity, but this turns out not to be important for the two-sample t-test—we only have two points on the x-axis and can join them however we want, straight line, curve, ...

2.3.2 So What?

So now you don't need to worry about two-sample t-tests anymore—just think of it as linear regression with a binary predictor variable. One less thing to remember.

Distribution of the response is what matters: The equivalence between t-tests and linear regression helps us realise that the distribution of our predictor variable x doesn't really matter when choosing an analysis method—it is really the response variable y we have to focus our attention on. Recall that in Fig. 1.2 we had the following rules for when to use two-sample t vs regression:
Two-sample t: when y is quantitative and x is binary
Linear regression: when y is quantitative and x is quantitative

The fact that two-sample t = linear regression means that *it doesn't matter what type of variable x is*. We can use the following rule instead:
If y is quantitative, try a linear regression ("linear model").

Why don't we have to worry about the distribution of the predictor when choosing a method of analysis? Because regression *conditions* on x. A regression model says: "If x was this value, what would we expect y to be?" The thing we are treating as random is y. The x-value has been given in the question; we don't need to treat it as random (even if it was when sampling).

In the next chapter we will learn about multiple regression, i.e. regression with multiple x variables. After that you can think of pretty much any (fixed effects) sampling design under a single framework—they are all just linear regressions. This is a major simplification for us because then we can fit most of the analysis methods we need to use in experimental work using the same function, checking the same assumptions in the same way, and interpreting results in the same way. Really, once you understand that t-tests (and ANOVA-type problems) can be understood as types of linear regression, then you can do almost anything you want as a *linear model*, i.e. as a linear regression with multiple x variables (which may be categorical or quantitative predictors).

Exercise 2.5: Global Plant Height Against Latitude
Angela collected some data on how tall plants were in lots of different places around the world (Moles et al., 2009). She wants to know: *How does plant height change as latitude changes?*

The data can be found as the `globalPlants` dataset in the `ecostats` package. What sort of analysis method are you thinking of using? Apply this method to estimate the relationship between height and latitude. Don't forget to mind your Ps and Qs. . . .

Exercise 2.6: Transform Guinea Pigs?
In analysing the data from the guinea pig experiment of Exercise 1.6, there was a slight suggestion of right skew (e.g. Fig. 2.1). Load the `guineapig` data (available in the `ecostats` package), transform, and reanalyse.

Did this change your results?

Exercise 2.7: Influential Value in Water Quality Data
Return to Thierry and Robert's `waterQuality` data, and consider again the influence of the observation with the smallest catchment area (as in Fig. 2.8). Reanalyse this dataset, excluding this influential value (by adding `subset=logCatchment>2` to your `lm` call), and compare results.

Did the R^2 increase or decrease on reanalysis? Is that what you expected? What about the P-value?

Chapter 3
Regression with Multiple Predictor Variables

Key Point

- Multiple regression is pretty much the same as simple linear regression, except you have more than one predictor variable. But effects should be interpreted as conditional not marginal, and multi-collinearity should be checked (if important).
- ANOVA can be understood as a special case of multiple regression.

In this chapter we will discuss what changes in a linear regression when you have more than one predictor variable, and we will consider the special case where you have one predictor, which is categorical and takes more than two values—commonly known as *analysis of variance* (ANOVA).

3.1 Multiple Regression

Exercise 3.1: Global Plant Height
Angela collected some data on how tall plants are in a variety of places around the world (Moles et al., 2009). She wants to know: *Can latitudinal variation in plant height be explained by climate?*

What does the question tell us—descriptive, interval estimation, hypothesis testing?

What do the data tell us—one variable or more? What type of response?

What sort of analysis method are you thinking of using?

D. I. Warton, *Eco-Stats: Data Analysis in Ecology*, Methods in Statistical Ecology,
https://doi.org/10.1007/978-3-030-88443-7_3

Multiple linear regression is a special name for linear models that have *more than one x variable* that we want to use to predict y.

It's an extension of simple linear regression to two or more x variables.

The multiple regression model for two predictor variables x_1 and x_2 is

$$y \sim \mathcal{N}(\mu_y, \sigma^2)$$
$$\mu_y = \beta_0 + \beta_1 x_1 + \beta_2 x_2$$

Geometrically, this fits a plane in three dimensions, not a line (in two dimensions).

The model for the mean can be written in vector notation as

$$\mu_y = \beta_0 + \mathbf{x}^T \boldsymbol{\beta}$$

where $\boldsymbol{\beta}$ and $\mathbf{x}$ are vectors. This notation is a simple way to write a linear model for any number of predictor variables; it just means "multiply the corresponding values in $\mathbf{x}$ and $\boldsymbol{\beta}$, then add them up".

3.1.1 Mind Your Ps and Qs When Using Multiple Regression

When making inferences about a multiple regression, we make the following assumptions:

1. The observed y-values are *independent*, after conditioning on x (that is, all information contained in the data about what y you might observe is found in x, and y-values for other observations wouldn't help predict the current y-value at all).
2. The y-values after conditioning on x are *normally distributed* with *constant variance*
$$y \sim \mathcal{N}(\mu_y, \sigma^2)$$
(this is the assumed distribution of y under repeated sampling if all predictors are kept at a fixed value).
3. There is a *straight-line relationship* between the mean of y and each x variable
$$\mu_y = \beta_0 + \mathbf{x}^T \boldsymbol{\beta}$$

Look familiar? Same assumptions as before, checked the same way as before (residuals vs fits plots, normal quantile plots).

It can also be useful to plot residuals against each x variable to check that *each predictor* is linearly related to μ_y.

Key Point
Multiple regression makes the same assumptions we saw earlier—independence, normality, constant variance (all these assumptions are conditional on x), and linearity.

So we can mind our Ps and Qs the same way as before—normal quantile plot of residuals (but don't be fussy about normality, especially if you have a large sample) and a residuals vs fits plot. Best to also check residuals against each predictor.

3.1.2 Multiple Regression in R—Same Same

You fit multiple regression models the same way as simple linear regression. The output looks pretty much the same too. For example, compare the code and output from simple linear regression (Code Box 3.1) and multiple regression (Code Box 3.2). If you are interested in seeing the mathematics of how least-squares regressions are fitted, it isn't too hard to follow; try out Maths Box 3.1.

Code Box 3.1: Simple Linear Regression of Global Plant Height Data—Predicting Height as a Function of Latitude Only

```
> library(ecostats)
> data(globalPlants)
> ft_heightLat=lm(height~lat, data=globalPlants)
> summary(ft_heightLat)

Coefficients:
            Estimate Std. Error t value Pr(>|t|)
(Intercept) 17.00815    2.61957   6.493 1.66e-09 ***
lat         -0.20759    0.06818  -3.045  0.00282 **
---
Signif. codes:  0 '***' 0.001 '**' 0.01 '*' 0.05 '.' 0.1 ' ' 1

Residual standard error: 13.59 on 129 degrees of freedom
  (47 observations deleted due to missingness)
Multiple R-squared:  0.06705,Adjusted R-squared:  0.05982
F-statistic: 9.271 on 1 and 129 DF,  p-value: 0.002823
```

Code Box 3.2: Multiple Linear Regression of Global Plant Height Data on R—Predicting Plant Height as a Function of Annual Precipitation and Latitude

Note the code is almost the same as for simple linear regression—just add an extra predictor variable!

```
> ft_heightRainLat=lm(height~rain+lat, data=globalPlants)
> summary(ft_heightRainLat)

Coefficients:
             Estimate Std. Error t value Pr(>|t|)
(Intercept)  5.223135   4.317148   1.210  0.22856
rain         0.005503   0.001637   3.363  0.00102 **
lat         -0.052507   0.080197  -0.655  0.51381
---
Signif. codes:  0 '***' 0.001 '**' 0.01 '*' 0.05 '.' 0.1 ' ' 1

Residual standard error: 13.08 on 128 degrees of freedom
  (47 observations deleted due to missingness)
Multiple R-squared:  0.1428,Adjusted R-squared:  0.1294
F-statistic: 10.66 on 2 and 128 DF,  p-value: 5.226e-05
```

Maths Box 3.1: Deriving the Least-Squares Regression Estimator

If we have responses $\mathbf{y} = (y_1, \ldots, y_n)$ and predictors $\mathbf{x}_1, \ldots, \mathbf{x}_n$, the multiple regression model can be written $\mathbf{y} = \mathbf{X}\boldsymbol{\beta} + \boldsymbol{\epsilon}$, where predictors are stored in the matrix

$$X = \begin{pmatrix} 1 & \mathbf{x}'_1 \\ 1 & \mathbf{x}'_2 \\ \vdots & \vdots \\ 1 & \mathbf{x}'_n \end{pmatrix}$$

The least-squares estimate finds regression parameters $\boldsymbol{\beta}$ to minimise

$$SS(\boldsymbol{\beta}) = (\mathbf{y} - \mathbf{X}\boldsymbol{\beta})'(\mathbf{y} - \mathbf{X}\boldsymbol{\beta}) = \mathbf{y}'\mathbf{y} - 2\boldsymbol{\beta}'\mathbf{X}'\mathbf{y} + \boldsymbol{\beta}'\mathbf{X}'\mathbf{X}\boldsymbol{\beta}$$

Differentiating to find its stationary point yields

$$\frac{\partial SS(\boldsymbol{\beta})}{\partial \boldsymbol{\beta}} = -2\mathbf{X}'\mathbf{y} + 2\mathbf{X}'\mathbf{X}\boldsymbol{\beta}$$

where differentiation has been applied element-wise. Differentiating vectors has its subtleties, and there is a bit of thinking behind the above line. From here however it is quite straightforward to find the value $\widehat{\boldsymbol{\beta}}$ for which $\frac{\partial SS(\boldsymbol{\beta})}{\partial \boldsymbol{\beta}} = \mathbf{0}$:

$$\mathbf{0} = -2\mathbf{X}'\mathbf{y} + 2\mathbf{X}'\mathbf{X}\widehat{\beta}$$
$$\widehat{\beta} = (\mathbf{X}'\mathbf{X})^{-1}\mathbf{X}'\mathbf{y}$$

There is only one stationary point, and it will be a global maximum or minimum of $SS(\boldsymbol{\beta})$. A little more work (e.g. looking at the second derivative) shows that it is a minimum, hence $\widehat{\boldsymbol{\beta}} = (\mathbf{X}'\mathbf{X})^{-1}\mathbf{X}'\mathbf{y}$ is the least-squares estimate of $\boldsymbol{\beta}$.

3.1.3 New Ideas in Multiple Regression

The maths and the computation are pretty much the same as for simple linear regression. But there are a few new ideas to look out for when you have multiple x-values, as below.

Key Point
Regression with multiple predictors is pretty much the same as regression with one predictor, except for the following new ideas that you should keep in mind:

1. Interpret coefficients as conditional effects (not marginal).
2. You can plot partial residuals to visualise conditional effects.
3. You can test hypotheses about multiple slope parameters at once.
4. Multi-collinearity makes estimates of coefficients inefficient.

1. Interpret as conditional effects: Multiple regression coefficients should be interpreted *conditionally on all other predictors* in the model. For example, in Code Box 3.2, the coefficient of `lat` tells us the effect of `lat` after controlling for the effect of `rain`. That is, what is the association of latitude with height *after controlling for the effect of rain*?

When ignoring all other predictors, as in simple linear regression, we are studying the *marginal* effect of a predictor.

Recall that in simple linear regression (Code Box 3.1), the slope of the marginal effect of latitude was $\hat{\beta}_{\text{lat}} = -0.17$ (and significant). After controlling for the effect of rain, the slope was much flatter $\hat{\beta}_{\text{lat}} = -0.004$ (and not significant). This means that (if assumptions are satisfied), most of the association between latitude and height is explained by precipitation.

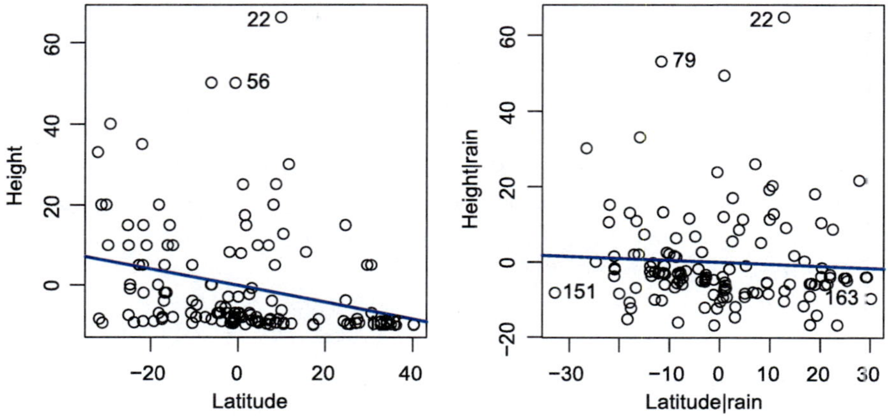

Fig. 3.1: (**a**) Scatterplot of height against latitude, and (**b**) a partial residual plot, after controlling for the effect of rain, for the global plant height data of Exercise 3.1. The partial residual plot helps us visualise the effects of latitude after controlling for the effects of rainfall. *Any problems with assumptions?*

2. *Partial residual plots*: Partial residual plots let us look at the effect of a variable after controlling for another. ("Added variable plot", `avPlots` from `car` package.) This is a good way of *visualising the conditional effect of one variable given another.*

To get a partial residual plot of height against latitude controlling for rain (Fig. 3.1b), we take the residuals from a regression of height against rain and plot them against the residuals from a regression of latitude against rain, to remove any linear trend with rain from either variable. If latitude had an association with height not explained by rain, then there would still be an association on this partial residual plot, and the slope of the (least-squares) trend on the partial residual plot would equal the multiple regression slope.

Partial residual plots can also be handy as an assumption check, just like scatterplots. Looking at Fig. 3.1b it is clear that there is a problem, with lots of points packed in close together below the line and spread out above the line. Height is right-skewed (because it is being "pushed up" against the boundary of zero). *Can you think of a transformation that might help here?*

Code Box 3.3: R Code to Produce the Plots of Fig. 3.1

```
library(car)
avPlots(ft_heightLat, terms = ~lat, xlab="Latitude", ylab="Height",
    grid=FALSE) ##left plot
avPlots(ft_heightRainLat, terms = ~lat, xlab="Latitude|rain",
    ylab="Height|rain", grid=F) ## right plot
```

3. Testing of multiple slope parameters: Notice the bottom line of the output in Code Box 3.2:

```
F-statistic: 14.43 on 2 and 175 DF,  p-value: 1.579e-06
```

This tests the null hypothesis that y is unrelated to *any* x variables.
What do you conclude?
What if we wanted to know the answer to the following question:

Does latitude explain the effect of climate on plant height?

We could use annual precipitation and mean temperature to represent "climate". Then, to answer this question, we would want to compare two models:

$$\mu_{\text{height}} = \beta_0 + \text{lat} \times \boldsymbol{\beta}_{\text{lat}} + \text{temp} \times \boldsymbol{\beta}_{\text{temp}} + \text{rain} \times \boldsymbol{\beta}_{\text{rain}}$$
$$\mu_{\text{height}} = \beta_0 + \text{lat} \times \boldsymbol{\beta}_{\text{lat}}$$

We would want a *single test* of the hypothesis that there was no effect of `rain` or `temp` (i.e. $H_0 : \boldsymbol{\beta}_{\text{temp}} = \boldsymbol{\beta}_{\text{rain}} = 0$), while accounting for the fact that there was a relationship with latitude. To do this, we would fit both models and compare their fits. We could do this in R using the **anova** function (Code Box 3.4). *How would you interpret these results?*

Code Box 3.4: Tests of Multiple Parameters on R Using anova Function

```
> ft_Lat=lm(height~lat,data=globalPlants)
> ft_LatClim=lm(height~lat+rain+temp,data=globalPlants)
> anova(ft_Lat,ft_LatClim)
ANOVA Table

Model 1: height ~ lat
Model 2: height ~ lat + rain + temp
  Res.Df   RSS Df Sum of Sq     F   Pr(>F)
1    129 23818
2    127 21794  2    2023.9 5.897 0.003556 **
---
Signif. codes:  0 '***' 0.001 '**' 0.01 '*' 0.05 '.' 0.1 ' ' 1
```

The foregoing approach only works for *nested models*—when one model includes all the terms in the other model plus some extra ones.

For instance, you can't use the **anova** function to compare

$$\mu_{\text{height}} = \beta_0 + \text{temp} \times \boldsymbol{\beta}_{\text{temp}} + \text{rain} \times \boldsymbol{\beta}_{\text{rain}} \tag{3.1}$$
$$\mu_{\text{height}} = \beta_0 + \text{lat} \times \boldsymbol{\beta}_{\text{lat}} \tag{3.2}$$

because latitude only appears in the second model, so it would need to appear in the first model as well for the second model to be nested in the first model.

4. Multi-collinearity: Multi-collinearity is where some of the predictor variables are highly correlated. It is a problem when making inferences about coefficients as the *standard errors become inflated.* (So confidence intervals for slope parameters are wider, and *P*-values are larger.) An example is in Code Box 3.5.

A point that often goes unnoticed is that multi-collinearity is not always a problem. In particular, it is not a problem for prediction, or for global tests, so whether or not you need to worry about multi-collinearity depends on what your question is about. For example, imagine a situation where you have one predictor in the model, and you add a second predictor that contains almost exactly the same information (its correlation with the first predictor might be 0.999, say). Because this second predictor contains hardly any new information, the model that is fitted hardly changes—it will be hardly any different in terms of predictive performance (and there is no reason to think it would do any worse).

Code Box 3.5: Multi-Collinearity Example—Another Rainfall Variable?

Let's consider adding rainfall in the wettest month (rain.wetm) to a model that already has annual precipitation (rain) in it:

```
> ft_climproblems=lm(height~rain+rain.wetm+lat,data=globalPlants)
> summary(ft_climproblems)

Coefficients:
             Estimate Std. Error t value Pr(>|t|)
(Intercept)  4.907914   4.607746   1.065    0.289
rain         0.004911   0.003370   1.457    0.148
rain.wetm    0.004686   0.023283   0.201    0.841
lat         -0.046927   0.085140  -0.551    0.582
...
```

Note that standard errors are larger and suddenly everything is non-significant

Maths Box 3.2: Multi-Collinearity and Variances

The variance matrix of the least-squares estimator is

$$\text{var}(\widehat{\boldsymbol{\beta}}) = \text{var}\left\{(\mathbf{X}'\mathbf{X})^{-1}\mathbf{X}'\mathbf{y}\right\}$$

In regression, we (usually) condition on $\mathbf{X}$, meaning that we treat it as a constant. Using a vector generalisation of the rescaling rule for variances:

$$= (\mathbf{X}'\mathbf{X})^{-1}\mathbf{X}'\text{var}(\mathbf{y})\mathbf{X}(\mathbf{X}'\mathbf{X})^{-1}$$
$$= \sigma^2(\mathbf{X}'\mathbf{X})^{-1}\mathbf{X}'\mathbf{X}(\mathbf{X}'\mathbf{X})^{-1} = \sigma^2(\mathbf{X}'\mathbf{X})^{-1}$$

assuming that responses are conditionally independent with variances σ^2.

If predictors are highly collinear, $\mathbf{X}'\mathbf{X}$ is nearly singular, meaning that its inverse contains large values. This makes the variance (and hence standard errors) of slope coefficients large.

Consider, however, the variance of predictions for new values, at the same set of $\mathbf{x}$ values for which we have data:

$$\begin{aligned}\text{var}(\mathbf{X}\widehat{\boldsymbol{\beta}}) &= \text{var}\left\{\mathbf{X}(\mathbf{X}'\mathbf{X})^{-1}\mathbf{X}'\mathbf{y}\right\}\\ &= \mathbf{X}(\mathbf{X}'\mathbf{X})^{-1}\mathbf{X}'\text{var}(\mathbf{y})\mathbf{X}(\mathbf{X}'\mathbf{X})^{-1}\mathbf{X}'\\ &= \sigma^2\mathbf{X}(\mathbf{X}'\mathbf{X})^{-1}\mathbf{X}'\end{aligned}$$

The sum of the variances of these predictions can be calculated using the *trace* function (tr), which has the special property that $\text{tr}(\mathbf{AB}) = \text{tr}(\mathbf{BA})$:

$$\begin{aligned}\text{tr}(\text{var}(\mathbf{X}\widehat{\boldsymbol{\beta}})) &= \text{tr}\left(\sigma^2\mathbf{X}(\mathbf{X}'\mathbf{X})^{-1}\mathbf{X}'\right)\\ &= \sigma^2\text{tr}\left(\mathbf{X}'\mathbf{X}(\mathbf{X}'\mathbf{X})^{-1}\right) = \sigma^2\text{tr}(\mathbf{I})\end{aligned}$$

where $\mathbf{I}$ is known as the identity matrix. If there are p parameters in the model, then $\text{tr}(\mathbf{I}) = p$. The main point to notice is that the sum of the variances of predictions is not a function of $\mathbf{X}$ and, hence, is not affected by multi-collinearity. More generally, we can conclude (by similar methods) that multi-collinearity affects inferences about individual coefficients but has little effect on predictions (unless you extrapolate!).

A common way to check for multi-collinearity is to compute *variance inflation factors* (VIFs) for each explanatory variable x, as in Code Box 3.6. These tell us the factor by which standard errors are larger than they would have been if explanatory variables had been uncorrelated. VIFs near 1 are good, values in the 5–10 range ring alarm bells.

Code Box 3.6: Computing Variance Inflation Factors to Check for Multi-Collinearity

```
> library(car)
> vif(ft_heightRainLat)
    rain      lat
1.494158 1.494158
> vif(ft_climproblems)
     rain rain.wetm       lat
 6.287835  7.031092  1.671396
```

Clearly adding `rain.wetm` to the model has done some damage (but to `rain` only, not so much to `lat`).

Another way to see what is going on is to simply *look at the correlation* between predictor variables, as in Code Box 3.7.

Code Box 3.7: Correlations and Pairwise Scatterplots to Look for Multi-Collinearity

```
> X = data.frame(globalPlants$lat,globalPlants$rain,globalPlants$rain.wetm)
> cor(X)
                       globalPlants.lat globalPlants.rain globalPlants.rain.wetm
globalPlants.lat              1.0000000        -0.5750884              -0.633621
globalPlants.rain            -0.5750884         1.0000000               0.917008
globalPlants.rain.wetm       -0.6336210         0.9170080               1.000000
> pairs(X)
```

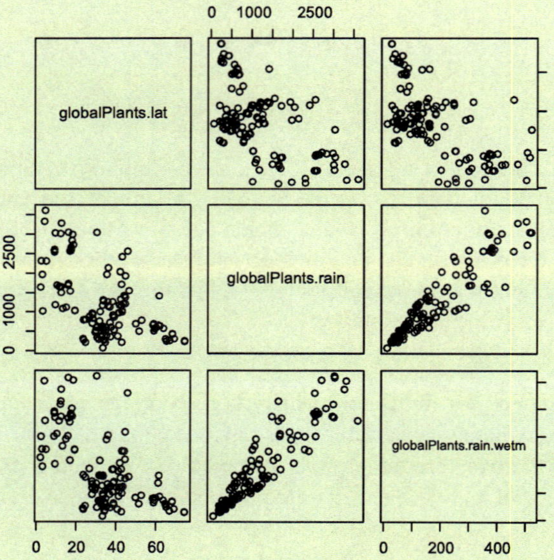

The rainfall variables are strongly correlated.

Exercise 3.2: Plant Height Data—Transform Response?
Consider again the global plant height data of Exercise 3.1:

Can latitudinal variation in plant height be explained by rainfall?

We already looked at this question in Code Box 3.2, but recall that when we looked at the residual plots (Fig. 3.1), it was clear that the height variable was strongly right-skewed.

Transform height and rerun your analyses. Note that this changes the results. Which set of results do you think is more correct, and why do you think results changed in connection with the data transformation?

Exercise 3.3: Plant Height—Skewed Rainfall Data?
Consider again the global plant height data of Exercise 3.1. The plots in Code Box 3.7 suggest that maybe rainfall is right-skewed.

Construct histograms of rainfall and see if you agree. Does log transformation assist? Do results change qualitatively based on whether or not you transform rainfall? (For example, do any new variables become significant? Is the R^2 appreciably different?)

3.2 ANOVA

Exercise 3.4: Snails on Seaweed
Recall how David and Alistair looked at the density (per gram of seaweed) of invertebrate epifauna (e.g. snails) settling on algal beds (also known as seaweed) with different levels of isolation (0, 2, or 10 m buffer) from each other (Roberts & Poore, 2006) to study the potential impacts of habitat fragmentation. They wanted to answer this question: *Does invertebrate density change with isolation?*

What does the research question ask us—descriptive, estimation, hypothesis testing, etc?

What do the data tell us—one variable or two? What type of response variable?

What sort of analysis method are you thinking of using?

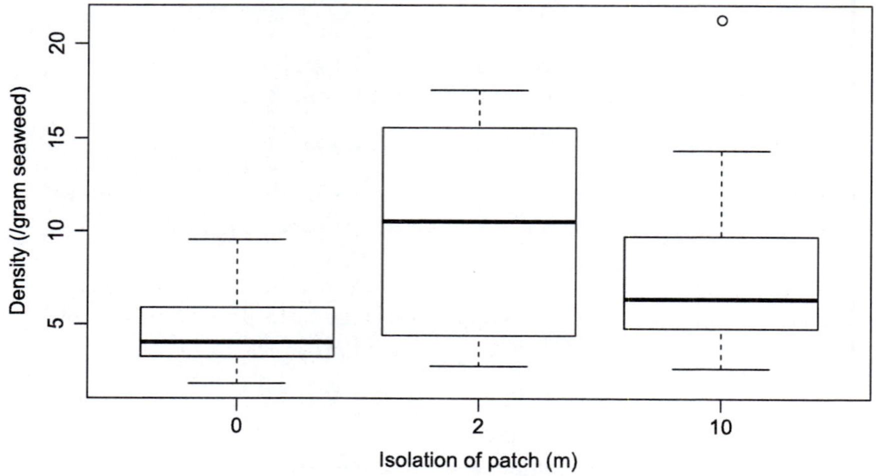

Fig. 3.2: Epifauna counts in algal beds with different levels of isolation

Consider Exercise 3.4. The response variable in this case is the density of epifauna—a quantitative variable, meaning we should be thinking of some variation on linear regression as a method of analysis. The predictor variable, distance of isolation of an algal bed, takes one of three different values (0, 2, or 10 m). This can be understood as a *categorical* variable with three levels. The analysis method for this situation, which you have probably already seen, is conventionally referred to as *analysis of variance*. But we won't go into details because it is just another example of a linear model. . . .

3.2.1 Equivalence of ANOVA and Multiple Regression

Recall that a two-sample t-test was just linear regression with a binary predictor variable.

Multiple regression is an extension of simple linear regression, and ANOVA is a special case of multiple regression (linear models).

Recall that a multiple regression equation with two predictors has the form

$$\mu_y = \beta_0 + x_1\beta_1 + x_2\beta_2$$

ANOVA, for Exercise 3.4, uses the model

$$\mu_y = \begin{cases} \mu_0 & \text{for subjects in the Isolation} = 0 \text{ group} \\ \mu_2 & \text{for subjects in the Isolation} = 2 \text{ group} \\ \mu_{10} & \text{for subjects in the Isolation} = 10 \text{ group} \end{cases}$$

If we let $x_1 = x_2 = 0$ for Isolation = 0, $x_1 = 1, x_2 = 0$ for Isolation = 2, and we let $x_1 = 0, x_2 = 1$ for Isolation = 10, then the ANOVA model follows the regression model with

$$\begin{aligned}\mu_0 &= \beta_0 \\ \mu_2 &= \beta_0 + \beta_1 \\ \mu_{10} &= \beta_0 + \beta_2\end{aligned}$$

So we can fit the ANOVA model as a multiple regression and interpret the parameters in the model as follows:

β_0 is the mean of the control group (distance of isolation is 0 m).
β_1 is the change in mean when moving from the control to the group at 2 m isolation, i.e. it tells us if there is an effect of 2 m (as compared to zero).
β_2 is the change in mean when moving from the control to the group at 10 m isolation, i.e. it tells us if there is an effect of 10 m (as compared to zero).

The foregoing approach is the most common way to code the predictors x_1 and x_2 in order to do an ANOVA as a regression, but there are other ways, too, which will give equivalent results for the ANOVA test. The predictors x_1 and x_2 are known as dummy variables or *indicator variables*. Two dummy variables were needed to fit this ANOVA model, which had a factor with three levels. More generally, to use multiple regression to fit an ANOVA model for a factor that has K levels, we would need $K - 1$ dummy variables (with one "baseline" level in which all dummy variables are set to zero).

3.2.2 Mind Your Ps and Qs When Using ANOVA

The assumptions of ANOVA are similar to those of multiple regression:

1. The observed y-values are *independent*, conditional on x.
2. The y-values are *normally distributed* with *constant variance*

$$y \sim \mathcal{N}(\mu_y, \sigma^2)$$

The only difference from multiple regression assumptions is that we don't have to worry about any *linearity* assumption. (Because we only observe data at two points on each x-axis, you can join them however you want.) That is, unless we have multiple categorical variables and we want to assume additivity (as in Chap. 4).

How do we check the foregoing assumptions?

3.2.3 ANOVA Output in R

You can fit ANOVA the same way as linear regression in R, just *make sure your x variable is a factor* before fitting the model, as in Code Box 3.8. *Factor* is just another word for a predictor variable that is categorical. Its categories are often referred to as *factor levels*.

Code Box 3.8: Analysis of Variance in R for the Seaweed Data of Exercise 3.4 Using `lm`

```
> data(seaweed)
> seaweed$Dist = factor(seaweed$Dist)
> ft_seaweed=lm(Total~Dist,data=seaweed)
> anova(ft_seaweed)
 ANOVA Table

Response: Total
          Df Sum Sq Mean Sq F value    Pr(>F)
Dist       2 300.25 150.123  8.5596 0.0005902 ***
Residuals 54 947.08  17.539
```

Any evidence that density is related to distance of isolation?

One thing that has been done differently here, compared to previous multiple regression examples, is that we have used the `anova` function to test for an effect of `Dist` instead of using the `summary` function. The reason for this is that the `Dist` term has more than one slope parameter in it (β_1 and β_2 as described previously), and when we want to test for a `Dist` effect, we want to test simultaneously across these two slope parameters. As discussed previously, the `anova` function is the way to test across multiple parameters at once (for nested models).

Key Point

When you have lots of related hypotheses that you want to test, you should correct for multiple testing, so that you can control the chance of falsely declaring significance. A common example of this is when doing pairwise comparisons in ANOVA. (Another example, coming in Chap. 11, is when there are multiple response variables and you want to do separate tests of each response variable.)

3.2.4 Multiple Comparisons

After using ANOVA to establish that there is an effect of treatment, the next question is: What is the effect? In particular, which pairs of treatments are different from which? If we try to answer this question using `confint` on the fitted linear model, we don't get what we want. Firstly, this will not give all pairwise comparisons: it compares Isolation = 2 with Isolation = 0, and it compares 10 with 0, but it doesn't compare 10 with 2. Secondly, it doesn't correct for the fact that we did more than one test.

Code Box 3.9: Running `confint` on the Seaweed Data Doesn't Give Us What We Want

```
> confint(ft_seaweed)
                2.5 %   97.5 %
(Intercept) 2.7785049 6.533423
Dist2       2.9211057 8.460686
Dist10      0.4107071 5.720963
```

This compares each of the "2" and "10" groups to "0".
But:

- What about "2" vs "10"?
- When doing multiple tests we should correct for this in assessing significance.

Every time you do a hypothesis test and conclude you have significant evidence against H_0 when $P < 0.05$, you have a 5% chance of accidently rejecting the null hypothesis (assuming the null hypothesis is in fact true). This is called a Type I error. This happens by definition, because the meaning of a P-value of 0.05 is that there is a 5% chance of observing a test statistic this large by chance alone. If doing three pairwise comparisons (2–0, 10–0, 10–2), then this would give about a 15% chance of accidently declaring significance.[1] And the more groups being compared, the greater the chance—with 10 different treatment groups you are almost guaranteed a false positive!

Tukey's "Honestly Significant Differences" are more conservative to account for this, so that across all comparisons, there is a 5% chance of accidently declaring significance. We will use the `multcomp` package in R because of its flexibility, but another option is to use the `TukeyHSD` function.

[1] Not exactly 15%, for a couple of reasons, one being that the probability of no false significance from three independent tests is $1 - 0.95^3 \simeq 14.3\%$. Another reason is that these tests aren't independent, so we shouldn't be multiplying probabilities.

Code Box 3.10: Analysis of Variance of the Seaweed Data of Exercise 3.4 with Tukey's Multiple Comparisons via the `multcomp` Package

```
> library(multcomp)
> contDist = mcp(Dist="Tukey") # telling R to compare on the Dist factor
> compDist = glht(ft_seaweed, linfct=contDist) # run multiple comparisons
> summary(compDist) # present a summary of the results
 Simultaneous Tests for General Linear Hypotheses

Multiple Comparisons of Means: Tukey Contrasts

Fit: lm(formula = Total ~ Dist, data = seaweed)

Linear Hypotheses:
            Estimate Std. Error t value Pr(>|t|)
2 - 0 == 0     5.691      1.382   4.119   <0.001 ***
10 - 0 == 0    3.066      1.324   2.315   0.0623 .
10 - 2 == 0   -2.625      1.382  -1.900   0.1483
---
Signif. codes:  0 '***' 0.001 '**' 0.01 '*' 0.05 '.' 0.1 ' ' 1
(Adjusted p values reported -- single-step method)

> plot(compDist)
```

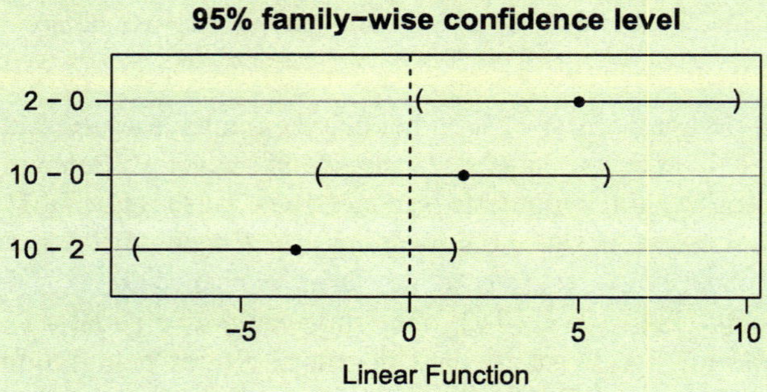

Also try using `confint` on `compDist`.

3.2.5 ANOVA with a Factor That Has Just Two Levels

Say we took the two-sample data of Sect. 2.1 and tried an ANOVA on it. After all, we have a quantitative response variable (number of errors) and a categorical predictor (treatment); it's just that the predictor has only two possible levels (control and treatment) rather than three or more. What would happen?

We would get exactly the same P-value as we get from a two-sample t-test (for a two-sided test). The methods are mathematically equivalent. So this is yet another reason why you don't need to worry about two-sample t-tests any more—alternative methods are available that are more general and, hence, able to be used for this problem as well as a bunch of other ones you will come across. There's nothing wrong with using the t-test; you just don't need to use it given the equivalent alternatives that are available.

Exercise 3.5: Plant Height—Climate Explains Patterns?
Consider again the global plant height data of Exercise 3.1, but now Angela would like to answer the following question:

Can latitudinal variation in plant height be explained by climate?

What sort of model(s) should be fitted to answer this question? Fit the models and answer this question. Remember to mind your Ps and Qs!

Exercise 3.6: Habitat Configuration Study—Mind Your Ps and Qs
For Exercise 3.4, we have undertaken an ANOVA in Code Box 3.8 and multiple comparisons in Code Box 3.10. But we never checked our assumptions!

Check assumptions now, using the data in the `seaweed` dataset in the `ecostats` package, paying particular attention to the equal variance assumption. Notice in Fig. 3.2 that the spread of values increases as the mean increases, suggesting a violation of equal variance that might be addressed by transformation.

Suggest a transformation that might address this, and reanalyse the data. Did the results become more or less significant after transformation? Why do you think this happened?

Exercise 3.7: Habitat Configuration Study—Small Plots
For Exercise 3.4, we used all the plots from the study, but they were set out in two different sizes.

Subset the data to just the small plots (`Size=="SMALL"`) and rerun the analyses. Did the results become more or less significant? Why do you think this happened?

Chapter 4
Linear Models—Anything Goes

In this chapter we will look at some other common fixed effects designs, all of which can be understood as special cases of the linear model.

Key Point
There are lots of different methods designed for analysing data with a continuous response that can be assumed to be normally distributed with constant variance, e.g. factorial ANOVA, multiple regression, analysis of covariance (ANCOVA), variations for paired data, and blocked designs. All these methods are special cases of the linear model.

4.1 Paired and Blocked Designs

Consider again the raven data, as in Exercise 4.1—measurements of the number of ravens at 12 sites, before and after a gunshot sound. This type of data looks like it could be analysed using a two-sample t-test, but there is a problem—the independence assumption won't be satisfied. Independence of observations is violated by any sampling design with *paired data*—in the case of the raven data, if the count was high before the gunshot sound, then it is likely to be high after the gunshot sound also.

Exercise 4.1: Ravens and Gunshots
In a paper titled "Hunters ring dinner bell for ravens. . . " (White, 2005), ecologist Crow White asked the question of whether ravens actually fly towards (not away from!?) the sound of gunshots in hopes of scavenging a carcass.

D. I. Warton, *Eco-Stats: Data Analysis in Ecology*, Methods in Statistical Ecology,
https://doi.org/10.1007/978-3-030-88443-7_4

White visited 12 locations, counted the number of ravens he saw, shot his gun, waited 10 min, and recounted the ravens. The results follow.

Before	0	0	0	0	0	2	1	0	0	3	5	0
After	2	1	4	1	0	5	0	1	0	3	5	2
After-Before	2	1	4	1	0	3	-1	1	0	0	0	2

We cannot use a two-sample *t*-test on before and after measurements because we cannot assume observations are independent across samples. But we could analyse the paired differences and check whether their mean is significantly different from zero.

What assumptions are made in analysing the paired differences? Do they seem like reasonable assumptions?

A common approach to the analysis of paired data is to *calculate paired differences and analyse those*, as described in Exercise 4.1. If there is no difference between the means, then the mean of the differences will be zero (if $\mu_{\text{before}} = \mu_{\text{after}}$, then $\mu_{\text{before-after}} = 0$). So we can test for an effect of gunshot sounds by testing if there is evidence that the before–after differences have a non-zero mean. We no longer assume independence across sites and sampling times; instead, we assume *differences* are independent across sites (and normally distributed with constant variance). In R, we could use the `t.test` function, as in Code Box 4.1.

Code Box 4.1: Paired *t*-Test for Raven Data

```
> library(ecostats)
> data(ravens)
> crowGun = ravens[ravens$treatment == 1,]
> t.test(crowGun$Before, crowGun$After, paired=TRUE, alternative=
        "less")

Paired t-test

data:  crowGun$Before and crowGun$After
t = -2.6, df = 11, p-value = 0.01235
alternative hypothesis: true difference in means is less than 0
95 percent confidence interval:
      -Inf -0.335048
sample estimates:
mean of the differences
              -1.083333
```

Is there evidence that ravens fly towards gunshot sounds?

4.1.1 Paired t-Test as a Main Effects ANOVA

Another way to think of what is going on with paired data is that there is a third variable—a blocking factor that pairs up the data. In the raven example, the blocking factor is site. The problem with a two-sample *t*-test would be that it would ignore this third variable.

Another way to analyse paired data is via a linear model, where you include the blocking factor in the model to control for the pairing structure in the data. For the raven example, we could fit a model to predict the number of ravens as a function of time (before–after) and site, using a site categorical variable to control for site-to-site variation in abundance. This is done in Code Box 4.2.

Code Box 4.2: Paired *t*-Test for Raven Data via Linear Model

```
> library(reshape2)
> crowLong = melt(crowGun,measure.vars = c("Before","After"),
    variable.name="time",value.name="ravens")
> head(crowLong)
  delta    site treatment few.0..or.many..1..trees   time ravens
1     2 pilgrim         1                        1 Before      0
2     1 pacific         1                        1 Before      0
3     4 uhl hil         1                        1 Before      0
4     1 wolff r         1                        1 Before      0
5     0 teton p         1                        1 Before      0
6     3 glacier         1                        1 Before      2
> ravenlm = lm(ravens~site+time,data=crowLong)
> anova(ravenlm)
Analysis of Variance Table

Response: ravens
          Df Sum Sq Mean Sq F value   Pr(>F)
site      11 55.458  5.0417    4.84 0.007294 **
time       1  7.042  7.0417    6.76 0.024694 *
Residuals 11 11.458  1.0417
---
Signif. codes:  0 '***' 0.001 '**' 0.01 '*' 0.05 '.' 0.1 ' ' 1
```

How do these results compare to those of Code Box 4.1?

Comparing results with the paired *t*-test output of Code Box 4.1, you might notice that the *P*-value from the test of significance of the `time` term (Code Box 4.2) is exactly double the value for the paired *t*-test (Code Box 4.1). It turns out that these methods are *mathematically equivalent*, but the ANOVA (Code Box 4.2) did a two-tailed test while the paired *t*-test (Code Box 4.1) used a one-tailed test, hence the *P*-value was twice as large. But aside from this, these are two alternative ways of doing exactly the same analysis.

The foregoing result is handy for a few reasons. For example, it gives us a way forward when we start thinking about more complicated data types for which we can't just analyse paired differences (Chaps. 10 and 11). It also gives us a way to handle more complicated study designs, such as Exercise 4.2.

Exercise 4.2: Ravens, Guns, and Air Horns
Crow White didn't just fire a gun off; he also tried other noises to see if it was specifically the gun that was attracting ravens. Like an air horn and a whistle:

site	pilgrim	pacific	uhl hil	wolff r	teton p	glacier	ant fla	btb n	btb s	hay e	hay s	grove
After gun	2	1	4	1	0	5	0	1	0	3	5	2
After horn	0	0	1	0	1	1	3	2	0	0	0	0
After whistle	1	0	1	0	6	1	1	1	0	1	4	0

We want to know: *Is there evidence that the response to a gunshot different to the response to other treatments?*

How would you analyse this dataset?

4.1.2 Mind Your Ps and Qs for Paired/Blocked Designs

1. The observed y-values are *independent* given x—in other words, for Exercise 4.2, after accounting for site and treatment.
2. The y-values are *normally distributed* with *constant variance*

$$y_{ij} \sim \mathcal{N}(\mu_{ij}, \sigma^2)$$

3. The treatment effect is *additive* across blocks. For raven counts

$$\mu_{ij} = \beta_{\text{site}i} + \beta_{\text{treatment}j}$$

("Additive effects"=Linearity)

Same as before, right. Check assumptions the same way.

Note that the independence of y is *conditional on x*: We *don't* assume abundances at each site and treatment are independent. We do assume that *beyond* site and treatment; there are no further sources of dependence between observations. As before, this would be satisfied in an experiment that randomly allocated subjects to treatment groups. For Crow's data, randomly choosing locations to sample at would help a lot with this assumption.

Code Box 4.3: A Linear Model for Blocked Design Given by Raven Counts in Exercise 4.2

To analyse, we first subset to the three treatments of interest (1=gunshot, 2=air horn, 3=whistle):

```
> crowAfter = ravens[ravens$treatment <=3,]
> ft_crowAfter = lm(After~site+treatment,data=crowAfter)
```

```
> anova(ft_crowAfter)
Analysis of Variance Table

Response: After
          Df Sum Sq Mean Sq F value Pr(>F)
site      11 28.667  2.6061  0.9269 0.5327
treatment  1  2.667  2.6667  0.9485 0.3402
Residuals 23 64.667  2.8116
```

Is there evidence of a treatment effect? What assumptions were made here, and how can they be checked?

Note that when using `anova` in R, *the order matters*—to test for an effect of `treatment`, after accounting for a blocking effect of `site`, the formula for the linear model has to be written `ravens~site+treatment` rather than `ravens~treatment+site`. This can be understood as using Type I Sums of Squares. For Type II Sums of Squares, i.e. for output that looks at the conditional effect of each term in the model after adding all other terms, you can use the `drop1` function.

Key Point
Order matters when doing an ANOVA in R using `anova`—make sure you specify model terms in the correct order to answer the research question you are interested in. Alternatively, use `drop1` to look at what happens when each term is left out of the model while keeping all remaining terms.

4.1.3 Randomised Block Designs

You can think of Exercise 4.2 as a *blocked* design, where for each site ("block") we get three measurements—one for each treatment. Paired designs can also be understood as blocked, with a block size of two. Code Box 4.3 shows how to fit such a model using R—it is pretty straightforward once you know about multiple regression!

Another term for this sort of design is a *randomised block design*, where the randomised part comes about when treatments are randomly allocated to observations within each block (e.g. the order in which gunshot and air horn treatments are applied could be randomised). This randomisation would give us a leg up when it comes to satisfying independence assumptions. For a more conventional example of a randomised blocks design see

http://www.r-tutor.com/elementary-statistics/analysis-variance/randomized-block-design

The point of blocking is to *control for known major sources of variability* that are not of direct interest to the research question (such as species-to-species variation).

By controlling for these extra sources of variation and having less unaccounted-for sampling error, it is easier to see patterns in the data. Statistically speaking, we can more efficiently estimate the quantity of interest, e.g. in Exercise 4.2 we can have more power when testing for an effect of treatment.

The blocking factor is not of interest—so ignore its P-value in output!

4.2 Analysis of Covariance

Analysis of covariance (ANCOVA) is a common term for situations where we are primarily interested in the effect of some categorical predictor on the response, but we measure another predictor that is quantitative and that we want to control for. This is desirable if we know that the response is related to some known quantitative predictor because, as with blocking, it makes our estimates of the effect of interest more efficient to remove known sources of variation.

Exercise 4.3: Seaweed, Snails, and Seaweed Mass
Consider again David and Alistair's seaweed data of Exercise 1.13. They measured the wet mass of algal beds as well, because it is expected to be an important predictor of epifauna density.

They want to know: *Is there an effect of distance of isolation after controlling for wet mass?*

What does the research question tell us—descriptive, estimation, hypothesis testing. . . ?

What do the data tell us—how many variables? Distribution of the response variable?

What type of graph might you start with?

This problem is similar to the randomised block design—we have a variable (in the case of Exercise 4.3, `Wmass`) that we know is important but not of primary interest, i.e. we only want to control for this "covariate". Like `site` in the raven example, but the difference is that the covariate is *quantitative* rather than being categorical. But that's no big deal for us—it can still be handled using a linear model. The code for analysis doesn't need to change from what was used in a randomised block design (see for example Code Box 4.5).

Code Box 4.4: Scatterplot of Data from Exercise 4.3

```
data(seaweed)
seaweed$Dist = factor(seaweed$Dist)
plot(Total~Wmass, data=seaweed, col=Dist,
     xlab="Wet Mass [log scale]",ylab="Density (per gram) [log scale]")
legend("topright",levels(seaweed$Dist),col=1:3,pch=1)
```

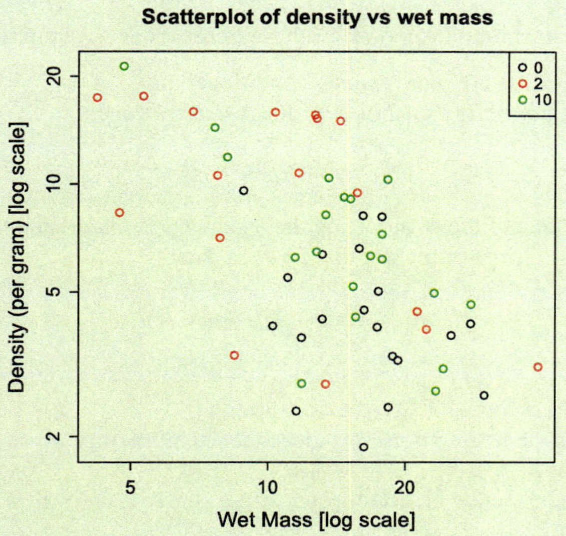

Code Box 4.5: Analysis of Covariance for Seaweed Data of Exercise 4.3

```
> seaweed$logTot = log(seaweed$Total)
> seaweed$logWmass = log(seaweed$Wmass)
> lmMassDist=lm(logTot~logWmass+Dist,data=seaweed)
> anova(lmMassDist)

Analysis of Variance Table

Response: log(Total)
          Df  Sum Sq Mean Sq F value    Pr(>F)
logWmass   1  7.7216  7.7216 35.7165 1.975e-07 ***
Dist       2  2.1415  1.0708  4.9528   0.01067 *
Residuals 53 11.4582  0.2162
---
Signif. codes:  0 '***' 0.001 '**' 0.01 '*' 0.05 '.' 0.1 ' ' 1
```

(The log-transform on `Total` was chosen based on exploratory plots along the lines of Code Box 4.4. It was also applied to `Wmass` because total abundance may be proportional to wet mass, so it should be included in the model with the transformation.)

4.2.1 Mind Your Ps and Qs When Using ANCOVA

We are still using a linear model, so we still have linear model assumptions:

1. The observed y-values are *independent*, conditional on x.
2. The y-values are *normally distributed* with *constant variance*

$$y_{ij} \sim \mathcal{N}(\mu_{ij}, \sigma^2)$$

3. *Linearity*—the effect of the covariate on the mean of y is *linear*, and the effect of factors is *additive*. For example, for the seaweed data:

$$\mu_{ij} = \beta_{\text{Dist}i} + x_j \beta_{\text{Wmass}}$$

Exercise 4.4: Checking ANCOVA Assumptions
ANCOVA makes the assumptions of independence, normality, constant variance, and linearity.

How can independence be established in the study design? How can we check the remaining assumptions?

As always, the independence assumption is critical to valid inference; the equal variance assumption is also important, and linearity of response to quantitative predictors is needed so that when we include these terms in the model, we are in fact controlling for the effect of these variables.

The assumption of additivity of the treatment effect (`Dist` in Exercise 4.3) can be checked for by adding interaction terms to the model, as in Sect. 4.4.1.

Sometimes it is said that ANCOVA requires an assumption that the mean value for the quantitative predictor is equal across groups. This is not in fact a formal assumption of the method; however, when different groups have quite different means for the quantitative predictor, the method can become quite sensitive to violations of the linearity assumption (Warton, 2007). So if your means along the quantitative variable differ appreciably, your inferences will rely heavily on the linearity assumption, so you will need to check it especially carefully. If quite different treatment means for the quantitative variable are expected and sample sizes are too small to reliably check the linearity assumption, you are taking something of a "leap of faith" when using ANCOVA (or indeed any method attempting to correct for a covariate).

A similar argument applies to multiple regression—if predictors in multiple regression are correlated, some inferences are quite sensitive to violations of linearity.

4.2.2 *Order Matters*

Recall again that *order matters* when using the anova function in R (except for perfectly balanced designs). If the categorical predictor from an ANCOVA were listed first in the model, rather than the quantitative predictor, different results would be obtained (Code Box 4.6) with a different interpretation.

Code Box 4.6: ANCOVA with Order of Terms Switched

Notice that switching the order changes the results in the ANOVA table and their interpretation

```
> lmDistMass=lm(logTot~Dist+logWmass,data=seaweed)
> anova(lmDistMass)
Analysis of Variance Table

Response: logTot
          Df  Sum Sq Mean Sq F value    Pr(>F)    
Dist       2  4.8786  2.4393  11.283 8.273e-05 ***
logWmass   1  4.9845  4.9845  23.056 1.329e-05 ***
Residuals 53 11.4582  0.2162                      
---
Signif. codes:  0 '***' 0.001 '**' 0.01 '*' 0.05 '.' 0.1 ' ' 1
```

We have a very significant effect of wet mass, after controlling for the effects of distance of isolation.

The anova function in R uses Type I sums of squares—they *sequentially add terms* to the model and test if each added term explains additional variation compared to those already in the model. So for example an anova call of a model with formula `lm(Total~Dist+log(Wmass),data=seaweed)`, as in Code Box 4.6, will fit the following sequence of models:

1. "Intercept model", no terms for `Dist` or `log(Wmass)`.
2. `Dist` only.
3. `Dist` and `log(Wmass)`.

And in the anova call:

- The first row tests the first term in the model (model 2 vs 1); is there any effect of `Dist` (ignoring `log(Wmass)`)?
- The second row tests the second term in the model (model 3 vs 2); is there any additional effect of `log(Wmass)` after controlling for `Dist`?

The results are different from Code Box 4.5, and they mean something different—these tests would answer questions different to what would be answered if the terms were entered into the model the other way around. Which way we should enter the terms depends on the research question.

Exercise 4.5: Order of Terms in Writing Out a Model for Snails and Seaweed
Recall from Exercise 4.3 that David and Alistair measured the wet mass of algal beds as well, because it is expected to be an important predictor of epifauna density. Recall that they want to answer the following question:

Is there an effect of distance of isolation after controlling for wet mass?

Which way should we specify the linear model:
`...logWmass+Dist` or `...Dist+logWmass`?

You can beat this "problem" of order dependence in R using the `drop1` function, as in Code Box 4.7. I say "problem" because it is not usually a problem if you are clear about exactly which hypothesis your study was designed to test and you order your model accordingly. The `drop1` function tests for an effect of each term in the model after including all other terms in the model. Hence, the order in which the model was specified no longer matters—in Code Box 4.7, we test for an effect of `Dist` after controlling for `log(Wmass)`, and we test for an effect of `log(Wmass)` after controlling for the effect of `Dist`.

Code Box 4.7: Type II Sums of Squares for the ANCOVA of Snails and Seaweed

```
> drop1(lmMassDist,test="F")
Single term deletions

Model:
log(Total) ~ log(Wmass) + Dist
           Df Sum of Sq    RSS     AIC F value    Pr(>F)
<none>                  11.458 -83.448
log(Wmass)  1    4.9845 16.443 -64.861 23.0561 1.329e-05 ***
Dist        2    2.1415 13.600 -77.681  4.9528   0.01067 *
---
Signif. codes:  0 '***' 0.001 '**' 0.01 '*' 0.05 '.' 0.1 ' ' 1
```

4.3 Factorial Experiments

Exercise 4.6: Snails, Isolation, and Time
David and Alistair also looked at the density (per gram of habitat) of invertebrate epifauna settling on habitat patches over time periods of two different

lengths (5 and 10 weeks), as well as different isolation (0, 2, or 10 m buffer). They want to answer the question: *Does invertebrate density change with isolation? Does the isolation effect vary with time period?*

What does the research question tell us—descriptive, estimation, hypothesis testing. . . ?

What do the data tell us—one variable or two, distribution of the response variable?

What type of graph would you start with?

Code Box 4.8: Comparative Boxplot of Snail Density at Each of Six Possible Combinations of Sampling Time and Distance of Isolation

```
plot(Total ~ interaction(Dist,Time), data=seaweed,
   log="y")  ## and as usual use xlabel, ylabel to name axes
```

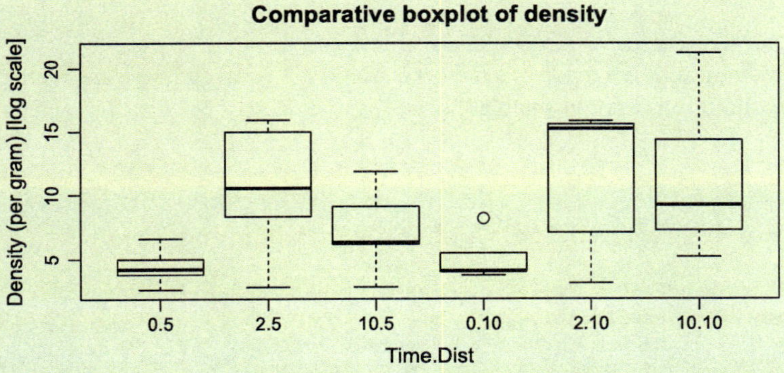

Consider Exercise 4.6. There are now two explanatory variables:

- `Dist` (of isolation): 0, 2 or 10 m
- `Time`: 5 or 10 weeks

David and Alistair took 10 replicate measurements at each combination of `Dist` and `Time`, across a total of 60 plots. This is commonly referred to as a *factorial design*—we take measurements at each possible combination of the two factors in the model.

To analyse the data in this case, we could use a linear model. One option would be to add `Time` as a blocking factor:

```
> ft_seaweed=lm(log(Total)~Time+Dist,data=seaweed)
```

But this assumes the isolation (`Dist`) effect is the same at each sampling time. This doesn't answer the question:

Does the isolation effect vary with time period?

To answer this question, we need an *interaction* term.

4.3.1 Interactions

An interaction between two variables tells us whether the nature of the effect of one variable *changes as the other variable changes*.

To answer the question of whether the isolation effect varies with time period, we need to test for an *interaction* between `Dist` and `Time`.

This is often written `Dist*Time`, but in R, it is written `Dist:Time`. In R, `Dist*Time` is useful as shorthand for "a factorial design with main effects and interactions", i.e. `Dist + Time + Dist:Time`, as in Code Box 4.9. This fits a factorial ANOVA, a special case of linear models (hence, it is fitted using the same function; it makes the same assumptions, checked in the same way).

Exercise 4.7: Factorial ANOVA Assumptions
What assumptions are made in a factorial ANOVA? What can be done to check these assumptions are reasonable?

Code Box 4.9: Factorial ANOVA of Snails, Isolation, and Time

```
> ft_seaweedFact = lm(logTot~Time*Dist,data=seaweed)
> anova(ft_seaweedFact)
            Df Sum Sq Mean Sq F value  Pr(>F)
Time         1  0.243  0.2433   0.851 0.36055
Dist         2  5.032  2.5161   8.802 0.00052 ***
Time:Dist    2  1.467  0.7337   2.567 0.08668 .
Residuals   51 14.578  0.2859
---
Signif. codes:  0 '***' 0.001 '**' 0.01 '*' 0.05 '.' 0.1 ' ' 1
```

How would you interpret these results?

Hey, why use `Time*Dist`, as opposed to `Dist*Time`?

A helpful way to visualise interactions between factors is by using an *interaction plot*, as in Fig. 4.1.

You can create basic interaction plots using the `interaction.plot` function in R, but Fig. 4.1 was created using the `ggplot2` package (Code Box 4.10), a powerful graphical tool that enables the creation of complex plotting objects relatively quickly. I also used the `dplyr` package to compute the treatment means, because this package has some intuitive functions for doing these sorts of calculations.

Interactions are cool—a lot of interesting ideas can be expressed as interactions.

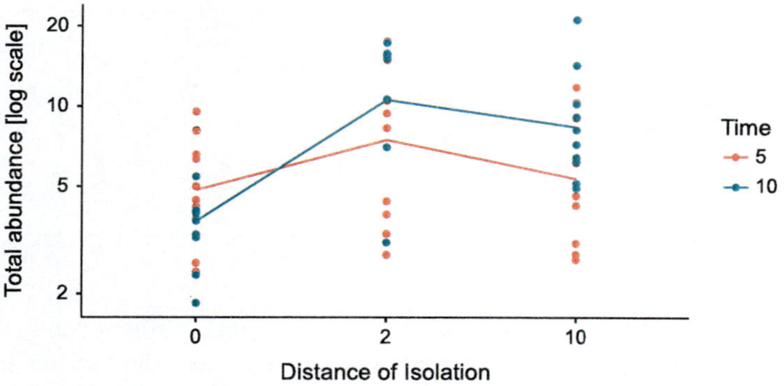

Fig. 4.1: Interaction plot of effects of distance of isolation and sampling time on density of epifauna on seaweed, as in Exercise 4.6. This type of display can be useful for factorial ANOVA

For example, when monitoring for an environmental impact using a so-called BACI design—if we have monitoring data before and after an impact, with control and impacted treatments, the treatment×time interaction tells us whether there was an environmental impact (Fig. 4.2, left).

Another example is so-called fourth corner models—across several co-occurring species, if we model species abundance as a function of environment, species traits, and the environment×trait interaction, the interaction tells us how traits mediate different environmental responses of different species (Fig. 4.2, right).

Code Box 4.10: R Code for Interaction Plot in Fig. 4.1

```
> library(dplyr)
> seaweed$Time = as.factor(seaweed$Time)
> by_DistTime = group_by(seaweed,Dist,Time)
> distTimeMeans = summarise(by_DistTime, logTotal=mean(log(Total)))
> distTimeMeans
> library(ggplot2)
> library(ggthemes) #loads special themes
> ggplot(seaweed, aes(x = factor(Dist), y = Total, colour = Time)) +
   geom_point() + geom_line(data = distTimeMeans, aes(y = exp(logTotal),
   group = Time)) + theme_few() + xlab("Distance of Isolation") +
   ylab("Total abundance [log scale]") +
   scale_y_log10(breaks=c(1, 5, 10, 50, 100, 500))
  Dist  Time  logTotal
  <fct> <fct>    <dbl>
1 0     5         1.58
2 0     10        1.31
3 2     5         2.01
4 2     10        2.36
5 10    5         1.68
```

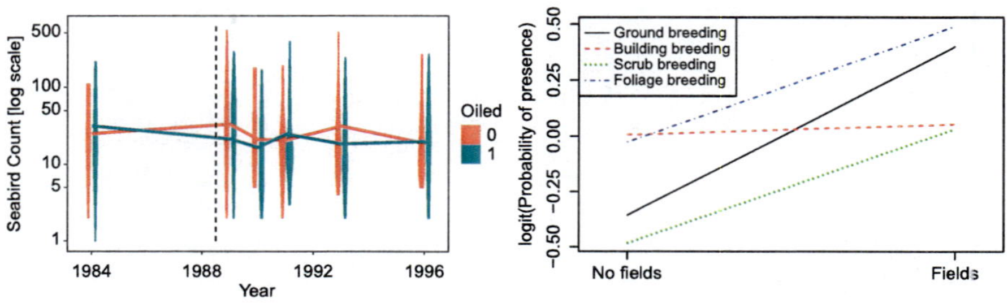

Fig. 4.2: Interactions are cool. (left) Seabird counts in Prince William Sound following the Exxon-Valdez spill of 1989 (dotted vertical line), data from McDonald et al. (2000). In BACI designs like this, we can test for an environmental impact by testing for an interaction between time (before–after) and treatment (control–impact). An interaction is seen here with a rather dramatic relative change in mean abundance (solid lines) immediately following the spill. (right) A fourth corner model for how bird response to environment varies with species traits along a rural–urban gradient in France (Brown et al., 2014). This plot focuses on how bird presence at sites near fields is associated with breeding habit. There is an environment–trait interaction, with the likelihood of encountering most bird species increasing as you move to sites near fields, except for species that breed in buildings

```
6 10     10          2.13
```

Alternatively, for a simpler plot without the data points on it, try

```
> interaction.plot(seaweed$Dist, seaweed$Time, ft_seaweedFact$fitted,
  xlab="Isolation of patch", ylab="Total density [log]",
  trace.label="Time")
```

Interactions are complicated—if David and Alistair have an interaction between `Dist` and `Time`, this means that there is no universal explanation for what is going on as `Dist` varies. So interactions complicate things.

The next step (if significant interaction) is to "break it down"—look at the effects of `Dist` separately for each sampling `Time`.

4.3.2 What Df?

Recall from Chap. 3 that ANOVA can be understood as multiple regression, where the factor in the model is represented as a set of *indicator variables* (ones and zeros) in a regression model. The number of indicator variables that are needed to include a factor in a model, known as its *degrees of freedom* (df), is # levels – 1. For example, for David and Alistair's factorial seaweed experiment (Exercise 4.6), `Time` has 1 df (two sampling times, 5 and 10) and `Dist` has 2 df (three distances, 0, 2, and 10).

For an interaction, the rule is to multiply the df for the corresponding main effects, e.g. `Dist:Time` has $2 \times 1 = 2$ df.

For a quantitative variable x, only one predictor (x) is included in the linear model so it adds one df.

Why are the df important? Well, they aren't really! They used to be, back in the day, when they were used to manually compute stuff (variance estimates). But this is not really relevant to modern statistics, where the computer can do most of the work for us, and the computations it does rely less on df than they used to.

Degrees of freedom are mostly useful just as a *check* to make sure nothing is wrong with the way you specified a model. In particular, if you accidently treat a factor as quantitative, it will have one df no matter how many levels the factor has. If we forget to turn `Dist` into a factor, the table will look like Code Box 4.11, and we will reach the wrong conclusion!

Code Box 4.11: Uh oh... anova Gone Wrong

This is what your output might look like if your computer thinks your factors are quantitative variables—each term has only one df, when `Dist` should have two (because it is a factor with three levels). The biggest problem with this is that we don't necessarily want to assume the effect of `Dist` is linear, but that is what is done when `Dist` is entered into the model as a quantitative variable rather than as a factor.

```
> data(seaweed)
> ft_nofactor=lm(log(Total)~Time*Dist,data=seaweed)
> anova(ft_nofactor)
            Df Sum Sq Mean Sq F value Pr(>F)
Time         1  0.243  0.2433   0.667 0.4177
Dist         1  0.716  0.7164   1.964 0.1669
Time:Dist    1  1.030  1.0303   2.825 0.0987 .
Residuals   53 19.331  0.3647
---
Signif. codes:  0 '***' 0.001 '**' 0.01 '*' 0.05 '.' 0.1 ' ' 1
```

4.3.3 Multiple Comparisons in Factorial Designs

You can use the `multcomp` package, as before, for multiple comparisons. Things get more complicated now, though—if you try the same approach as previously, the `multcomp` package will give us a warning (Code Box 4.12).

Code Box 4.12: Tukey's Comparisons Don't Work for Main Effects in an Orthogonal Design

Using the data from Exercise 4.9:

```
> seaweed$Dist = factor(seaweed$Dist)
> seaweed$Time = factor(seaweed$Time)
> seaweed$logTot = log(seaweed$Total)
> ft_seaweedFact = lm(logTot~Time*Dist, data=seaweed)
> library(multcomp)
> contFact = mcp(Dist="Tukey") # telling R to compare on the Dist
  factor
> compFact = glht(ft_seaweedFact, linfct=contFact)
Warning message:
In mcp2matrix(model, linfct = linfct) :
  covariate interactions found---default contrast might be
  inappropriate
```

The problem is that you can't make inferences about the main effects in a fixed effects model where there are interactions. We need to either remove the interactions (if they are non-significant) or do multiple comparisons on the interaction terms.

The reason we get a warning is that when you fit a model with interactions such as `Time*Dist`, you are saying the effect of `Dist` varies with `Time`. So it no longer makes sense to look at the main effects for `Dist`; we have to look within each level of `Time` to see what the effect is.

We have a few options:

1. Fit a main effects model and then do multiple comparisons on the main effect of interest (Code Box 4.13). You should *not* do this if you have a significant interaction!!
2. Compare all treatment combinations with each other (Code Box 4.14). But this is wasteful if some treatment combinations are not of interest. For example, David and Alistair do not want to know if isolation = 10 at time = 5 is significantly different from isolation = 0 and time = 10!
3. Manually specify the contrasts you are interested in testing (Code Box 4.15). For example, David and Alistair were interested in the effect of `Dist`, so they wanted to specify all pairwise comparisons of levels of `Dist` at each sampling time. (This got messy to code because the names for the sampling levels were numbers, which `multcomp` gets upset about, so the first step in Code Box 4.15 was to change the factor levels from numbers to words.)

Code Box 4.13: Tukey's Comparisons for a Main Effect of `Dist` for Exercise 4.6, Assuming No Interaction

```
> ft_seaweedMain=lm(log(Total)~Time+Dist,data=seaweed) # "+" not "*"
> contrast = mcp(Dist="Tukey") # telling R to compare on the Dist
  factor
```

```
> compDistMain = glht(ft_seaweedMain, linfct=contrast)
> confint(compDistMain)

 Simultaneous Confidence Intervals

Multiple Comparisons of Means: Tukey Contrasts


Fit: lm(formula = logTot ~ Time + Dist, data = seaweed)

Quantile = 2.4119
95% family-wise confidence level


Linear Hypotheses:
             Estimate lwr      upr
2 - 0 == 0    0.72650  0.28760  1.16539
10 - 0 == 0   0.45838  0.03872  0.87805
10 - 2 == 0  -0.26812 -0.70701  0.17078
```

Or you could use **summary** on the multiple testing object `compDistMain`. Note that the intervals don't cover zero when comparing isolation = 2 to isolation = 0, and when comparing 10 to 0, meaning there is significant evidence of an effect in these instances.

Code Box 4.14: Tukey's Comparisons for All Possible Treatment Combinations for Exercise 4.6

This approach is wasteful as it compares some pairs we are not interested in (e.g. 2.10 vs 0.5).

```
> td = interaction(seaweed$Dist,seaweed$Time)
> ft_seaweedInt=lm(logTot~td,data=seaweed) # Time*Dist as a single term
> contInt = mcp(td="Tukey") # so R compares on all Time*Dist levels
> compDistInt = glht(ft_seaweedInt, linfct=contInt)
> summary(compDistInt)
 Simultaneous Tests for General Linear Hypotheses

Multiple Comparisons of Means: Tukey Contrasts

Fit: lm(formula = log(Total) ~ td, data = seaweed)

Linear Hypotheses:
                  Estimate Std. Error t value Pr(>|t|)
2.5 - 0.5 == 0      0.4356     0.2391   1.822  0.46046
10.5 - 0.5 == 0     0.1013     0.2391   0.424  0.99815
0.10 - 0.5 == 0    -0.2643     0.2391  -1.105  0.87659
2.10 - 0.5 == 0     0.7852     0.2635   2.980  0.04761 *
10.10 - 0.5 == 0    0.5512     0.2391   2.305  0.21028
10.5 - 2.5 == 0    -0.3343     0.2391  -1.398  0.72720
0.10 - 2.5 == 0    -0.6999     0.2391  -2.927  0.05417 .
2.10 - 2.5 == 0     0.3496     0.2635   1.327  0.76842
```

```
10.10 - 2.5 == 0     0.1156     0.2391   0.483  0.99654
0.10 - 10.5 == 0    -0.3656     0.2391  -1.529  0.64688
2.10 - 10.5 == 0     0.6839     0.2635   2.596  0.11654
10.10 - 10.5 == 0    0.4499     0.2391   1.882  0.42421
2.10 - 0.10 == 0     1.0495     0.2635   3.983  0.00277 **
10.10 - 0.10 == 0    0.8155     0.2391   3.411  0.01522 *
10.10 - 2.10 == 0   -0.2340     0.2635  -0.888  0.94750
---
Signif. codes:  0 '***' 0.001 '**' 0.01 '*' 0.05 '.' 0.1 ' ' 1
(Adjusted p values reported---single-step method)
```

Code Box 4.15: Tukey's Comparisons for `Dist` Within Each Sampling Time, for Exercise 4.6

This is the best approach to use if you think there is an interaction and are primarily interested in `Dist`.

```
> levels(seaweed$Time) = c("five","ten") #mcp needs non-numeric levels
> levels(seaweed$Dist) = c("Zero","Two","Ten")
> td = interaction(seaweed$Dist,seaweed$Time)
> ft_seaweedInt=lm(log(Total)~td,data=seaweed) # Time*Dist as one term
> contDistinTime = mcp(td = c("Two.five - Zero.five = 0",
                              "Ten.five - Zero.five = 0",
                              "Ten.five - Two.five = 0",
                              "Two.ten - Zero.ten = 0",
                              "Ten.ten - Zero.ten = 0",
                              "Ten.ten - Two.ten = 0"))
> compDistinTime = glht(ft_seaweedInt, linfct=contDistinTime)
> summary(compDistinTime)
 Simultaneous Tests for General Linear Hypotheses

Multiple Comparisons of Means: User-defined Contrasts

Fit: lm(formula = log(Total) ~ td, data = seaweed)

Linear Hypotheses:
                          Estimate Std. Error t value Pr(>|t|)
Two.five - Zero.five == 0   0.4356     0.2391   1.822  0.31172
Ten.five - Zero.five == 0   0.1013     0.2391   0.424  0.99084
Ten.five - Two.five == 0   -0.3343     0.2391  -1.398  0.57117
Two.ten - Zero.ten == 0     1.0495     0.2635   3.983  0.00124 **
Ten.ten - Zero.ten == 0     0.8155     0.2391   3.411  0.00717 **
Ten.ten - Two.ten == 0     -0.2340     0.2635  -0.888  0.87445
---
Signif. codes:  0 '***' 0.001 '**' 0.01 '*' 0.05 '.' 0.1 ' ' 1
(Adjusted p values reported---single-step method)
```

At a sampling time of 10 weeks, there are significant differences between isolation = 2 and isolation = 0, and between 10 and 0, after correcting for multiple testing.

4.4 Interactions in Regression

4.4.1 Analysis of Covariance

Recall that David and Alistair also measured the wet mass of algal beds (also known as seaweed) as well, because that is expected to be important to epifauna density.

We assumed additivity—that distance of isolation had an additive effect on log(density). Could there be an interaction between isolation and wet mass?

Graphically, an ANCOVA interaction would mean that the slope of the relationship between log(density) and `Wmass` changes as `Dist` changes, as in Fig. 4.3. The slope of the line at a distance of isolation of 10 seems steeper than for the other treatment levels. But is it significantly steeper, or could this be explained away by sampling variation? We can test this using an interaction term in an ANCOVA, in just the same way as we test for interactions in factorial ANOVA, as in Code Box 4.16.

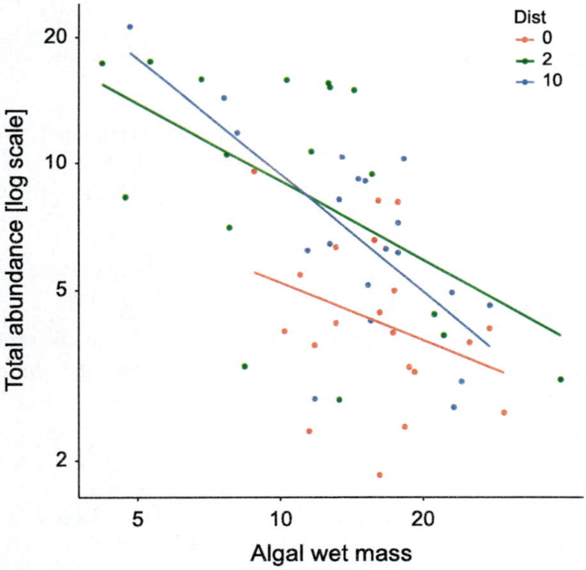

Fig. 4.3: An interaction in an ANCOVAmeans that the slopes of the regression lines differ across levels of the treatment variable. For David and Alistair's seaweed data (Exercise 4.3), the slope of the line at a distance of isolation of 10 seems steeper than for the other treatment levels. But is it significantly steeper, or could this be explained away by sampling variation?

Code Box 4.16: Testing for an Interaction in an ANCOVA for Density of Epifauna as a Function of Dist and Algal Wet Mass

```
> lmMassDistInter=lm(logTot~log(Wmass)*Dist,data=seaweed)
> anova(lmMassDistInter)
Analysis of Variance Table

Response: log(Total)
                Df  Sum Sq Mean Sq F value    Pr(>F)
log(Wmass)       1  7.7216  7.7216 35.3587 2.489e-07 ***
Dist             2  2.1415  1.0708  4.9032   0.01128 *
log(Wmass):Dist  2  0.3208  0.1604  0.7345   0.48475
Residuals       51 11.1374  0.2184
---
Signif. codes:  0 '***' 0.001 '**' 0.01 '*' 0.05 '.' 0.1 ' ' 1
```

Is there evidence of an interaction between the effects on density of wet mass and distance of isolation?

4.4.2 Interactions in Multiple Regression

In multiple regression models, you can also have interactions between two continuous covariates, too, but it's a little tricky. Consider for example Exercise 4.8, where Angela would like to know if the associations of height with precipitation and latitude interact.

Exercise 4.8: Global Plant Height
Angela collected some data on how tall plants are in lots of different places around the world. She wants to know: *Do the effects on plant height of latitude and rainfall interact?*

What sort of linear model would you fit here?

The interaction between two quantitative variables (`rain:lat` in this case) is a *quadratic* term. Other quadratic terms are `rain^2` and `lat^2`. It doesn't make sense to include interactions without including other quadratic terms also. This could lead to a really weird looking response surface, as in Fig. 4.4. (Such "saddlepoint" surfaces are still possible when all quadratic terms are included, but they are less likely.)

The simplest way to fit quadratics on R is to use the `poly` function as in Code Box 4.17.

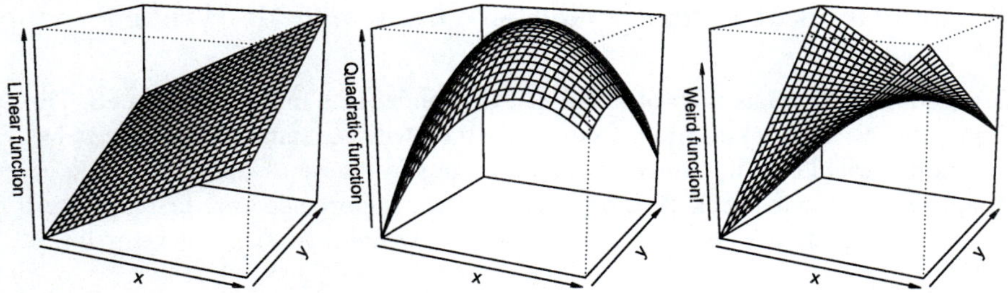

Fig. 4.4: Interactions without other quadratic terms look weird: from left to right we have a linear model, a model with all quadratic terms, and a model with just a quadratic interaction (`x:y`) without quadratic main effects (`x^2` and `y^2`). Notice the last one is a weird saddle-shaped function that doesn't make a whole lot of sense, so it shouldn't really be our default fit!

Code Box 4.17: Using R to Fit a Quadratic Model to the Plant Height Data of Exercise 3.1

```
> ft_latRain2 = lm(log(height)~poly(rain,lat,degree=2),
        data=globalPlants)
> summary(ft_latRain2)

Call:
lm(formula = log(height) ~ poly(rain, lat, degree = 2),
        data = globalPlants)

Residuals:
    Min      1Q  Median      3Q     Max
-3.3656 -0.9546 -0.0749  0.9775  3.1311

Coefficients:
                              Estimate Std. Error t value Pr(>|t|)
(Intercept)                     1.3314     0.2298   5.794 5.25e-08 ***
poly(rain, lat, degree = 2)1.0  7.2939     2.8268   2.580   0.0110 *
poly(rain, lat, degree = 2)2.0 -1.4744     2.6221  -0.562   0.5749
poly(rain, lat, degree = 2)0.1 -5.6766     2.2757  -2.494   0.0139 *
poly(rain, lat, degree = 2)1.1 -9.1362    56.1632  -0.163   0.8710
poly(rain, lat, degree = 2)0.2 -2.5617     2.7153  -0.943   0.3473
---
Signif. codes:  0 '***' 0.001 '**' 0.01 '*' 0.05 '.' 0.1 ' ' 1

Residual standard error: 1.441 on 125 degrees of freedom
  (47 observations deleted due to missingness)
Multiple R-squared:  0.2706,Adjusted R-squared:  0.2415
F-statistic: 9.277 on 5 and 125 DF,  p-value: 1.558e-07
```

Is there a significant interaction? How would you check if this model is a better choice than one with just linear terms for precipitation and latitude?

4.5 Robustness of Linear Models—What Could Go Wrong?

As we add more variables to our analysis, the model gets more complicated. The assumptions we are making are the same as before, with the same consequences, but with more variables in the model there are more ways these assumptions could go wrong, hence more to check. Assumptions and their importance were first discussed in general in Sect. 1.5, in what follows we review these in the context of linear models.

4.5.1 Violations of Independence Assumption

Linear models assume the response observations are independent, conditional on all the predictors. For example, in a block design, we assume independence across blocks, and we only assume independence within a block after controlling for the effect of a block (i.e. we know some sites have more ravens than others, so we assume there is no correlation in repeat raven counts beyond that explained by site-to-site variation in abundance).

If this independent assumption is not satisfied, you are stuffed—standard errors will be wrong (usually too small) and you will be more likely to make incorrect decisions (most critically, you can get false declarations of significance).

The main ways to guard against this are to use *randomisation* in your study design and to include in your analysis any variables used in the study design (e.g. blocks).

Some study designs will inevitably introduce dependence between observations in a special way, e.g. in an observational study sampling across different places, we might expect the response to be more similar between locations that are closer together spatially (this example is especially common in ecology). In such cases we can modify linear models to replace the assumption of independence with an assumption that observations are correlated spatially (or temporally, or phylogenetically, *etc.*), as considered in Chap. 7.

4.5.2 Violations of Mean and Variance Assumptions

The mean assumption in a linear model is that the mean of y is linearly related to our predictor variables. If any predictors are quantitative, it is assumed that the mean of y is a linear function of each of these, and if any of them are categorical, then the only real assumption imposed here is additivity (the effect of the categorical variable is the same at all values of other x variables). If you include interactions between categorical predictors and others, then you no longer make the additivity assumption.

Our variance assumption in a linear model, as previously, is that the error variance is constant irrespective of the value of predictors.

As previously, next to the independence assumption, the mean and variance assumptions are the most important assumptions we have. If linearity is violated, then our predictions will be biased (the straight line will miss the main trend, at some places) and our inferences unreliable. If the equal variance assumption is violated, then this usually doesn't introduce bias, but it does mess with standard errors and the inferences that are made from the model.

Mean and variance assumptions are best diagnosed using residual plots, as previously.

4.5.3 Violations of Distributional Assumptions

Linear models assume a response is normally distributed, but as previously, they are pretty robust to violations of this assumption, because parameter estimates and fitted values are approximately normally distributed irrespective of whether or not the data are normally distributed, thanks to the central limit theorem, often with surprisingly small sample sizes (Miller Jr., 1997). But as previously, just because a statistical procedure is valid doesn't mean that it is efficient. We should be on the lookout for strongly skewed distributions and outliers, because these make our inferences less *efficient*. The simplest course of action in this case is to transform the data, if it makes sense to. Sometimes that doesn't work, especially for rare counts or binary data, and alternative methods designed specially for this type of data have been developed (Chap. 10).

An alternative regression approach is to use *quantile regression* (Koenker & Hallock, 2001)—instead of modelling the mean of y as a function of x, model a quantile (such as the median, or the 90th percentile). Occasionally you may see it argued that quantile regression is an alternative to linear modelling that is suitable for data that don't satisfy the normality assumption. Strictly speaking this is true—quantile regression makes no normality assumption—but a better reason to consider using it would be if your research question were actually about quantiles. For example, Angela could hypothesise that plants were height-limited by climate, so short plants would be found everywhere, but tall plants were only found in high-rainfall areas. Quantile regression would be useful here—she could estimate how the 90th percentile (say) of height varied with rainfall and look at how much steeper this line was than that relating the 10th percentile of height to rainfall.

4.5.4 What If I Haven't Measured All the Important Predictors?

Few if any models are perfect, and often factors that we don't know about or haven't measured affect our response. What effect do they have on results?

The first thing to worry about is *confounding*. If you leave out a potentially important predictor that is correlated with other predictors, then it can change the results. If an omitted variable is independent of those included in the analysis, then there is no confounding. In a well-designed experiment, confounding variables are controlled for by randomising the application of treatments to subjects, which makes the treatments independent of confounders, so they can't bias the results in any way.

The second effect of missing predictors is *increased error*—because an important predictor is missing, the fitted values are not as close as they could be to the observed values, so there is more error in a linear model. In the absence of confounding, our parameters might not be all that different to how they were previously; the main change will be in the error variance (σ^2), which gets larger to absorb the effects of the missing predictors. In practical terms, if there is more error, it is harder to see the signal, so our statistical procedure becomes less *efficient*. When designing a study, the simplest way to account for unmeasured predictors is to use blocking.

The foregoing ideas can be illustrated mathematically relatively easily, as in Maths Box 4.1. It is worth thinking about these ideas, and their impact on your research, carefully—no model is perfect, so we can reasonably expect missing predictors most of the time. How they impact your next project depends on how you design the study and on the variables you measure to include in analysis.

Maths Box 4.1: Missing Predictors in Linear Models

What is the effect of leaving an important predictor (z) out of a linear model? Let's consider what happens if we model y as a function of a single predictor x when the true regression model is

$$y_i = \beta_0 + x_i\beta_x + z_i\beta_z + \epsilon_i$$

where $\epsilon_i \sim \mathcal{N}(0, \sigma^2)$, and z_i is correlated with x_i, satisfying the linear model

$$z_i = \gamma x_i + \delta_i$$

where $\delta_i \sim \mathcal{N}(0, \sigma_z^2)$.

We can write the linear model for $\mathbf{y}$ as follows:

$$\begin{aligned} y_i &= \beta_0 + x_i\beta_x + (\gamma x_i + \delta_i)\beta_z + \epsilon_i \\ &= \beta_0 + x_i(\beta_x + \gamma\beta_z) + \epsilon_i + \delta_i\beta_z \\ &= \beta_0 + x_i\beta^* + \epsilon_i^* \end{aligned}$$

where

$$\beta^* = \beta_x + \gamma\beta_z \tag{4.1}$$

$$\epsilon_i^* = \epsilon_i + \delta_i\beta_z \sim \mathcal{N}(0, \sigma^2 + \sigma_z^2\beta_z^2) \tag{4.2}$$

Equations 4.1 and 4.2 show the two key consequences of missing predictors.

Equation 4.1 shows the idea of *confounding*—if a missing predictor is correlated with those in the model, the value of the slope is no longer centred around its true value β_x; instead, it is shifted (*biased*) by $\gamma\beta_z$—an amount that gets larger as the relationship with the included predictor (γ) gets stronger and as the relationship with the response (β_z) gets stronger. Note that if the missing and included predictors are uncorrelated ($\gamma = 0$), there is no confounding.

Equation 4.2 illustrates the idea of *increased error*—if a predictor is missing from the model, regression error increases from σ^2 by the amount $\sigma_z^2\beta_z^2$, which is larger if the missing predictor is more variable (σ_z^2) or more important to the response (β_z^2).

Obviously, if a missing predictor is not important to the response, then its exclusion has no effect on the model. So if $\beta_z = 0$, there is no bias or increased error.

Exercise 4.9: Snowmelt and Time to Flowering
Snowmelt dates have been getting earlier in Europe in response to climate change, and Julia was interested in the effects this might be having on shrub development (Wheeler et al., 2016). She established 120 plots on 3 mountains in the Swiss Alps, varying in elevation and in expected timing of snowmelt, and measured time from snowmelt to flowering of creeping willow plants (*Salix herbacea*). She wants to know: *Does time from snowmelt to flowering vary with snowmelt date, beyond that explained by elevation? If so, how does it vary?*

The data are available as `snowmelt` in the `ecostats` package, averaged over measurements taken at three sampling times. Answer the research questions using the appropriate linear model, being sure to check your assumptions.

Exercise 4.10: Bird Exclusion and Biological Control
Invertebrates are often used as biological controls, but their effectiveness may be reduced by interactions with other species in the food web. Ingo looked at the effect of bird exclusion on the effectiveness of biological control of aphids in an oat field in Germany (Grass et al., 2017).

He established eight plots, four of which were randomly allocated a plastic netting treatment to exclude birds. He measured the number of aphids in each plot a few days after the treatment was applied, and again six weeks later. He wants to know: *Does the netting treatment affect trends in aphid numbers across the sampling times?*

The data are available as `aphidsBACI` in the `ecostats` package. Note that the `Plot` variable contains the plot ID for each of the eight plots.

Answer the research question using the appropriate linear model, being sure to check your assumptions.

Exercise 4.11: Seaweed, Snails, and Three Factors
In David and Alistair's seaweed experiment of Exercise 4.6, in addition to varying the distance of isolation and sampling times (5 and 10 days), they also had plots of two difference sizes. Hence, they have a three-factor design and want to answer the following questions:

> How does epifaunal density vary with distance of isolation? Does this vary across plots of different size or across different sampling times?

Answer these questions using the appropriate linear model. Be sure to mind your Ps and Qs!

Chapter 5
Model Selection

Often it is not clear which model you should use for the data at hand—maybe because it is not known ahead of time which combination of variables should be used to predict the response, or maybe it is not obvious how the response should be modelled. In this chapter we will take a look at a few strategies for comparing different models and choosing between them.

Key Point

How do you choose between competing models? A natural approach to this problem is to choose the model that has the *best predictive performance* on new, independent data, whether directly (using training data to fit the model and separate test data to evaluate it) or indirectly (using information criteria).

A key issue to consider is the level of *model complexity* the data can support—not too simple and not too complex! If the model is too simple, there will be bias because of important features missing from the model. If the model is too complex, there will be too much variance in predictions, because the extra parameters will allow the model to chase the data too much. (Occasionally, it is better to leave out a term even when it is thought to affect the response if there are insufficient data to do a good job of estimating its effect.)

Exercise 5.1: Plant Height and Climate

Which climate variables best explain plant height?

Angela collects data on how tall plants are in lots of different places around the globe. She also has data on 19 different climate variables (precipitation and temperature summarised in many different ways). She is interested in how plant height relates to climate and which climate variables height relates to most closely.

D. I. Warton, *Eco-Stats: Data Analysis in Ecology*, Methods in Statistical Ecology, https://doi.org/10.1007/978-3-030-88443-7_5

What does the question tell us—descriptive, hypothesis test, interval estimation, . . . ?

What do the data tell us—one variable/more, what type of variable is the response?

So what sort of analysis method are you thinking of using?

Consider Exercise 5.1. The key difference in this example, compared to those of previous chapters, is that we are primarily interested in choosing the best x variables (which climate variables height relates to most closely). This is a *variable selection* or *model selection* problem—the goal is to select the best (or a set of best) models for predicting plant height.

The paradigm we will use for model selection is to maximise predictive capability—if presented with new data, which model would do the best job of predicting the values of the new responses?

5.1 Understanding Model Selection

Model selection is a new way of thinking about things compared to what we have seen before and introduces some new issues we have to consider.

5.1.1 The Bias–Variance Trade-Off

In model selection, as well as trying to choose the right predictors, you are trying to choose the right number of them, i.e. the right level of model complexity. When making a decision about model complexity, you are making a decision about how to trade off bias against variance (Geman et al., 1992). If you make a model too simple, leaving out important terms, predictions will be systematically wrong (they will be *biased*). If you make a model too complex, adding terms that don't need to be there, it will "overfit" the data, chasing it too much and moving away from the main trend. This will increase the *variance* of predictions, essentially absorbing some of the error variance into predictions. The outcome of this is usually a J curve in predictive error, with a steep decrease in predictive error as bias is removed, a gradual increase in variance with overfitting, and a minimum somewhere in between that optimally manages this *bias–variance trade-off* (Maths Box 5.1).

The idea of the bias–variance trade-off applies any time you are choosing between models of differing complexity—most commonly, when deciding which predictors and *how many* predictors to add to a model, but in many other contexts, too.

One example useful for illustrating the idea is when predicting a response as a non-linear function of a single predictor. Many responses in ecology are thought

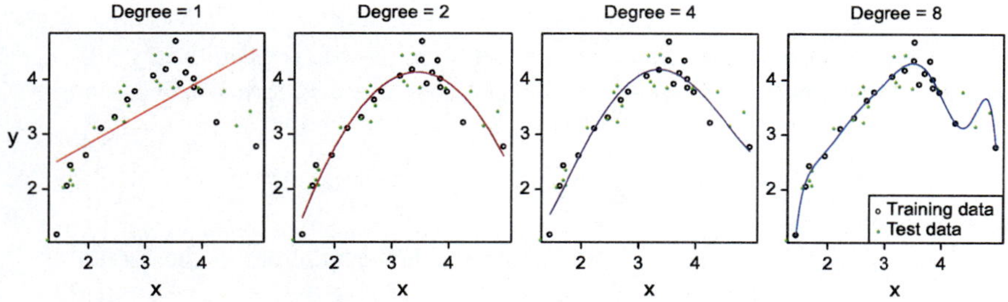

Fig. 5.1: Overfitting a model increases the variance in predictions. Data were generated using a quadratic model (degree = 2), and training data (black circles) were fitted with a model that is linear (degree = 1), quadratic (degree = 2), quartic (degree = 4) or a polynomial with degree 8. Test data (green stars) were used to assess model fit but were not used in model fitting. The straight line is biased; all other models can capture the true trend. But as the degree increased beyond 2, the extra model parameters enabled better tracking of the training data, at the cost of pulling the fitted model away from the true trend and, hence, away from test data. This is especially apparent for x-values between 3 and 4, where several large y-values in the training data "dragged" the fit above the test data for the quartic and 8-degree polynomial models

to follow such non-linear functions (e.g. growth rate as a function of temperature, as in Cooper et al., 2001). In Fig. 5.1, the true model is a quadratic trend, and we consider modelling it using a linear model (a polynomial with degree 1), quadratic (a polynomial with degree 2), quartic (degree 4), or a polynomial with degree 8. As the degree of the polynomial increases, so does the number of parameters needed to fit the model. Note that the linear model misses the curve in the trend (Fig. 5.1) and so is biased. The quartic and 8-degree polynomials are too complex for the data and seem to chase them. This reduces the error variance of the data to which the model was fitted (training data, black curve in Fig. 5.2), but the real test is how well the model would perform on new test data (green curve). Note the J shape of the green curve in Fig. 5.2 (well, a mirror image of a J).

Maths Box 5.1: Bias–Variance Trade-Off

Consider a linear model fitted to n observations, giving predictions $\hat{\boldsymbol{\mu}} = (\hat{\mu}_1, \ldots, \hat{\mu}_n)$, where $\hat{\mu}_i = \hat{\beta}_0 + \mathbf{x}_i'\hat{\boldsymbol{\beta}}$ for $i = 1, \ldots, n$. The predictive performance of the model can be measured using mean squared error (MSE) estimating the true means μ_i:

$$\text{MSE}(\widehat{\boldsymbol{\mu}}) = \frac{1}{n}\sum_{i=1}^{n}(\hat{\mu}_i - \mu_i)^2$$

In practice, we don't know the μ_i, so we don't know MSE($\widehat{\mu}$). But theory tells us a bit about how it will behave. First, we can write it in terms of bias and variance:

$$\text{MSE}(\widehat{\mu}) = \frac{1}{n}\sum_{i=1}^{n} \text{bias}(\hat{\mu}_i)^2 + \frac{1}{n}\sum_{i=1}^{n} \text{var}(\hat{\mu}_i)$$

Bias—one way to get bias is to include too few predictors in the model, missing some important ones. For example, in Maths Box 4.1, we studied the situation where there was one predictor x_i in the model and one missing predictor z_i, where $\mu_i = \beta_0 + x_i\beta_x + z_i\beta_z$, and we fit $\mu_i = \beta_0 + x_i\beta^*$. In this situation the bias is

$$\text{bias}(\hat{\mu}_i) = \beta_0 + x_i\beta^* - (\beta_0 + x_i\beta_x + z_i\beta_z) = (x_i\gamma - z_i)\beta_z = -\delta_i\beta_z$$

and $\frac{1}{n}\sum_i \text{bias}(\hat{\mu}_i)^2 \simeq \sigma_z^2\beta_z^2$, where σ_z^2 is the error variance when predicting $\mathbf{z}$ from $\mathbf{x}$. Bias gets larger if the missing predictor is more important (larger β_z) and more different from other predictors in the model (larger σ_z^2).

Variance—variance increases when there are more predictors in the model, because estimating extra terms introduces extra uncertainty into predictions. In Maths Box 3.2, for a linear model with p terms in it, the sum of variances of $\hat{\mu}_i$ is $p\sigma^2$, when predicting on the same set of predictors the model was fitted to:

$$\frac{1}{n}\sum_{i=1}^{n} \text{var}(\hat{\mu}_i) = \frac{1}{n}p\sigma^2 \tag{5.1}$$

If we use too many predictors, p will be too large, so var($\hat{\mu}_i$) will be too large.

Our goal in analysis is to build a model that is big enough to capture the main trends (not too much bias), but not excessively large (not too much variance). For models with an increasing number of terms, MSE typically follows a J curve—there is a steep initial decrease in MSE as bias is reduced, because there are fewer missing predictors, then a gradual increase as too many predictors are added. The increase is gradual because the $1/n$ in (5.1) keeps the variance small relative to bias, except in the analysis of a small, noisy (large σ) dataset, in which case a small model is often best.

The aim is to find the optimal point in the bias–variance trade-off. This point will be different for different datasets because it depends not just on how much data you have and the relative complexity of the models being compared but also on how well each model captures the true underlying process. In Fig. 5.2, the optimum was at degree = 2, which was the correct answer for this simulation, since a quadratic model was the true underlying process from which these data were simulated. (Sometimes, the optimum can be much smaller than the true model, if the true model is too complex for our data to fit it well.)

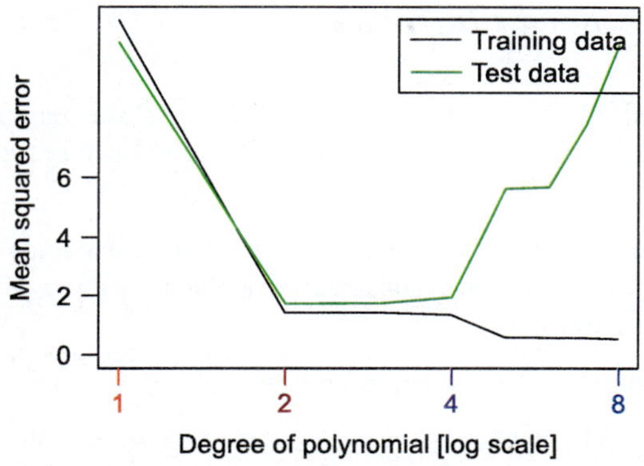

Fig. 5.2: The bias–variance trade-off for polynomial models in Fig. 5.1. Note that for the test data (green curve), the biased model (degree = 1) has a high predictive error, and as the model gets overfitted (degree > 2), the predictive error increases due to an increased variance of predictions. The predictive error for the training data does not account for overfitting, so it always decreases as more parameters are added to the model (black curve). Predictive error was measured here using MSE, defined later

5.1.2 The Problem with R^2 and P-Values for Model Selection

Many use R^2 and P-values to decide how well a model fits, but these aren't good tools to use for model selection.

R^2 makes *no attempt* to account for the costs of model complexity—it keeps going up as you add more terms, even useless ones! If you used R^2 as a basis for including potential predictors in a model, you would end up putting all of them in because that would maximise R^2, irrespective of whether or not each predictor was useful. The same is true of estimated error variance for the data the model was fitted to ($\hat{\sigma}^2$), except this (usually) decreases as you add more terms to the model, as in Fig. 5.2 (black line).

OK, well why not use hypothesis tests? Why not add terms to the model if they are significant and remove terms if they are not significant? This is commonly done and for many years was the main strategy for model selection. Much software still uses this approach as the default, which encourages its use. But there are a few problems with the technique. From a philosophical perspective, it is not what hypothesis testing was designed for, there is not really an *a priori* hypothesis being tested. So it is not really the right way to think about the problem. From a pragmatic perspective, using hypothesis testing for model selection has some undesirable properties. In particular, it is not variable selection consistent, i.e. is not guaranteed to pick the right model even when given as much data as it wants in order to do so—it *overfits*, choosing too complex a model, especially when considering a large number of potential predictors.

5.1.3 Model Selection as Inference

Recall that statistical inference is the process of making some general claim about a population based on a sample. So far we have talked about two types of statistical inference:

1. Hypothesis testing—to see if data are consistent with a particular hypothesis
2. Confidence interval estimation—constructing a plausible range of values for some parameter of key interest.

Now we have a third type of statistical inference:

3. Model selection—which model (or which set of predictor variables) best captures the true underlying process from which the data were generated.

This can be understood as statistical inference because again we are using a sample to make general claims—this time about models (or combinations of predictors), and how well they predict, instead of about parameters.

Note that model selection should *never* be used in combination with hypothesis testing or confidence interval estimation to look at related questions on the same dataset – these methods of inference are not compatible. The process of model selection will only include a term in the model if it is considered important – hence it is doing something kind of like significance testing already. If you were to perform model selection to choose key predictors, then do a hypothesis test on one of these predictors, this is known to lead to high rates of false significance, and similarly, performing model selection then constructing a confidence interval is known to lead to intervals that are biased away from zero. It is best to think of model selection and hypothesis testing/CIs as mutually exclusive: you either use one of these approaches or the other. Although having said this, there is a growing literature on post-selection inference (Kuchibhotla et al., 2022) which offers approaches to address this, the simplest of which is *data splitting* – splitting your data into two independent sets, and applying model selection to one part, and inference to the other.

Key Point
Model selection can be thought of as a method of inference, alongside hypothesis testing and confidence interval estimation. However, it should *not* be applied to the same data you plan to use for a hypothesis test or confidence interval to answer a related question, because these methods won't work correctly in this situation, unless using methods specifically designed for post-selection inference.

5.1.4 It Gets Ugly Quickly

Consider a situation where you have a set of predictor variables and you want to fit all possible models ("all subsets"). If there are p predictor variables, there are 2^p possible models—this gets unmanageable very quickly, as in Table 5.1. If you have 200 observations and 10 variables, the use of all subsets means trying to choose from 1000+ models using just 200 observations. Good luck!

Table 5.1: Variable selection gets ugly quickly—the number of candidate models increases exponentially with the number of predictor variables, such that it is no longer feasible to explore all possible models with just 20 or 30 predictors

# variables	# models to fit
2	4
3	8
5	32
10	1024
20	1,048,576
100	1.27×10^{30}
300	More than the number of electrons in the known universe!

This means two things:

- When there are a lot of possible or *candidate models* being compared, what the data say is the "best model" should be taken with a grain of salt. When there are lots of possible models, it is very hard for the data to make the right call.
- Simplify! The fewer candidate models you are comparing, the better—don't bother with anything you think is unrealistic, and if you know something is important, then include it in all candidate models. Try to refine your question. Do you really need all those variables?

5.1.5 A Cautionary Tale—Building a Spam Filter

In a third-year class several years ago, I asked students, as a group assignment, to construct an e-mail spam filter. The idea was that they would study regular e-mails and spam e-mails from their inbox, find predictors to distinguish between them, and build a model that could predict which e-mails in their inbox were spam.

Some groups put a huge amount of effort into this assignment, using complex models—usually via logistic regression or some variation thereof (as in Chap. 10), with a dozen or so carefully constructed predictors that did a near perfect job of distinguishing spam from real e-mails in the data they built the model on (the training data). One group even wrote a program for their spam filter and put it on a website, so if you copy-pasted text into it, it would return the predicted probability that your e-mail was spam.

At the other extreme, one group put hardly any effort into the assignment—it seemed like they had forgotten about the assignment and slapped something together at the last minute, with a handwritten report and a simple linear model with just a few terms in (for which model assumptions looked like they wouldn't be satisfied, if they had thought to check them).

As part of the assessment, I brought five e-mails to class and asked students to classify them using their filter (a "test" dataset). Most groups did poorly, getting one or two out of five correct, but one group managed four out of five correct—the last-minute group with the simple linear model.

The problem was that students weren't thinking about the costs of model complexity and were assuming that if they could do a really good job of modelling their training data, their method would work well on new data, too. So they built overly complex models that overfitted their data, chasing them too much and ending up with highly variable predictions, a long way past the optimum on the bias–variance trade-off. The group with the last-minute, simple linear model had the best predictive performance because their model did not overfit the data, so they ended up a lot closer to the optimum choice for model complexity.

The way to beat this problem is to use model selection tools, as described in the following sections, to make sure the level of model complexity is appropriate for the data at hand—not too complex and not too simple. Directly or indirectly, all such methods work by thinking about how well a model can predict using new data.

5.2 Validation

The simplest way to compare predictive models is to see how well they predict using new data, *validation*. In the absence of new data, you can take a *test* or *hold-out* sample from the original data that is kept aside for model evaluation. The remaining *training* data are used to fit each candidate model. Overfitted models may look good on the training data, but they will tend to perform worse on test data, as in Figs. 5.1 and 5.2 or as in the spam filter story of the previous section.

It is critical that the test sample be *independent* of the training sample; otherwise this won't work (see Maths Box 5.2). If all observations are independent (given x), then a random allocation of observations to test/training will be fine. If you have spatial data that are not independent of each other (Chap. 7), a common approach is to break it into coarse spatial blocks and assign these to training and test datasets (Roberts et al., 2017, for example).

Maths Box 5.2: Validation Data Can Be Used to Estimate Mean Squared Error

Predictive performance can be measured using MSE:

$$\text{MSE}(\widehat{\mu}) = \frac{1}{n}\sum_{i=1}^{n}(\hat{\mu}_i - \mu_i)^2$$

But how can we calculate this when we don't know the true mean, μ_i? We can use new observations, since $y_i = \mu_i + \epsilon_i$. We can compare the new responses to their predicted values by estimating the variance of $y_i - \hat{\mu}_i$. Using the adding rule for standard deviations (from Maths Box 1.5) yields

$$\begin{aligned}\sigma^2_{y_i-\hat{\mu}_i} &= \sigma^2_{\mu_i+\epsilon_i-\hat{\mu}_i} \\ &= \sigma^2_{\hat{\mu}_i-\mu_i} + \sigma^2_{\epsilon_i} + 2\sigma_{\epsilon_i,\hat{\mu}_i-\mu_i}\end{aligned}$$

The first term $\sigma^2_{\hat{\mu}_i-\mu_i}$ is another way of writing MSE. The second term is a constant. The third term, the covariance of ϵ_i and $\hat{\mu}_i$, is zero if ϵ_i is independent of $\hat{\mu}_i$. This independence condition is satisfied if y_i is a new observation that is independent of those used in fitting the model.

So when using a set of new "test" observations that are independent of those used to fit the model, estimating $\sigma^2_{y_i-\hat{\mu}_i}$ will estimate $\text{MSE}(\widehat{\mu})$, up to a constant.

How should we choose the size of the test sample? Dunno! (There is no single best answer.)

One well-known argument (Shao, 1993) is that as sample size n increases, the size of the training sample should increase, but as a proportion of n it should decrease towards zero. This ensures "variable selection consistency", guaranteeing the correct model is chosen for very large n.

An example strategy Shao (1993) suggested (which hence became a bit of a thing) is to use $n^{3/4}$ observations in the training sample. This can be quite harsh, though, as in Table 5.2. This rule tends not to be used so much in ecology, but the general strategy is certainly worth keeping in mind—using a smaller proportion of data in the training set when analysing a larger dataset, rather than sticking with the same proportion irrespective of sample size.

Table 5.2: How the suggested number of training observations changes with sample size if using the $n^{3/4}$ rule mentioned in Shao (1993)

n	20	50	100	200	1000
$n^{3/4}$	9	19	32	53	178

How should we measure predictive performance? For linear regression, the obvious answer is MSE:

$$\frac{1}{n_{\text{test}}} \sum_{i=1}^{n_{\text{test}}} (y_i - \hat{\mu}_i)^2$$

where the summation is over test observations, for each of which we compare the observed y-value, y_i, to the value predicted by the model fitted to the training sample, $\hat{\mu}_i$. This criterion was used in Fig. 5.2. Maths Box 5.2 explains how this quantity estimates the MSE of predictions. It makes sense to use this criterion for models where we assume equal variance—if not assuming equal variance, then it would make sense to use a criterion that weighted observations differently according to their variance. In later chapters, we will learn about models fitted by maximum likelihood, and in such a situation, it would make sense to maximise the likelihood on test data rather than minimising MSE.

An example using validation via MSE for model selection is in Code Box 5.1.

Code Box 5.1: Using Validation for Model Selection Using Angela's Plant Height Data

Comparing MSE for test data, for models with `rain` and considering inclusion of `rain.seas` (seasonal variation in rainfall)

```
> library(ecostats)
> data(globalPlants)
> n = dim(globalPlants)[1]
> indTrain = sample(n,n^0.75) #select a training sample of size n^0.75:
> datTrain = globalPlants[indTrain,]
> datTest = globalPlants[-indTrain,]
> ft_r = lm(log(height)~rain,dat=datTrain)
> ft_rs = lm(log(height)~rain+rain.seas,dat=datTrain)
> pr_r = predict(ft_r,newdata=datTest)
> pr_rs = predict(ft_rs,newdata=datTest)
> rss_r = mean( (log(datTest$height)-pr_r)^2 )
> rss_rs = mean( (log(datTest$height)-pr_rs)^2 )
> print( c(rss_r,rss_rs) )
[1] 2.145927 2.154608
```

So it seems from this training/test split that the smaller model with just `rain` is slightly better.

Try this yourself—do you get the same answer? What if you repeat this again multiple times? Here are my next three sets of results:

```
> print( c(rss_r,rss_rs) )
[1] 2.244812 2.116212
> print( c(rss_r,rss_rs) )
[1] 2.102593 2.143109
> print( c(rss_r,rss_rs) )
[1] 2.575069 2.471916
```

The third run supported the initial results, but the second and fourth runs (and most) gave a different answer—suggesting that including `rain.seas` as well gave the smaller MSE. But when the answer switches, it suggests that it is a close run thing and the models are actually quite similar in performance.

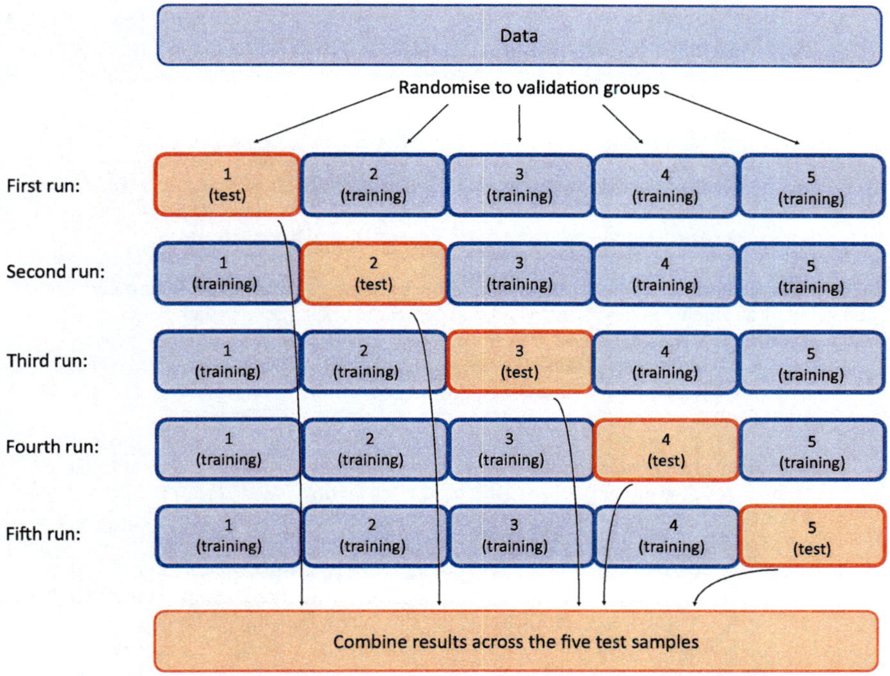

Fig. 5.3: A schematic diagram of five-fold CV. Each observation in the original dataset is allocated to one of five validation groups, and the model is fitted five times, leaving each group out (as the test dataset) once. Estimates of predictive performance are computed for each run by comparing predictions to test observations, then pooled across runs, for a measure of predictive performance that uses each observation exactly once

When using a test dataset to estimate predictive performance on new data, clearly the *test/training split matters*—it is a random split that introduces randomness to results. In Code Box 5.1, four different sets of results were obtained, leading to different conclusions about which model was better. This issue could be handled by repeating the process many (e.g. 50) times and averaging results (and reporting standard errors, too). The process of repeating for different test/training splits is known as *cross-validation* (CV).

5.3 *K*-fold Cross-Validation

A special case of CV is when you split data into K groups (usually $K = 5$, 5-fold CV, or $K = 10$) and fit K models—using each group as the test data once, as in Fig. 5.3.

Results tend to be less noisy than just using one training/test split, because each observation is used as a test observation once, so one source of randomness (choice of test observation) has been removed.

Code Box 5.2: 5-Fold Cross-Validation for Data of Exercise 5.1

```
> library(DAAG)
> ft_r = lm(log(height)~rain,dat=datTrain)
> ft_rs = lm(log(height)~rain+rain.seas,dat=datTrain)
> cv_r = cv.lm(data=globalPlants, ft_r, m=5, printit=FALSE) # 5 fold CV
> cv_rs = cv.lm(data=globalPlants, ft_rs, m=5, printit=FALSE) # 5 fold CV
> print( c( attr(cv_r,"ms"),attr(cv_rs,"ms") ), digits=6 )
[1] 2.22541 2.15883
```

suggesting that the models are very similar, the model without `rain.seas` predicting slightly better, but by an amount that is likely to be small compared to sample variation. For example, repeating analyses with different random splits (controlled through the `seed` argument):

```
> cv_r = cv.lm(data=globalPlants, ft_r, m=5, seed=1, printit=FALSE)
> cv_rs = cv.lm(data=globalPlants, ft_rs, m=5, seed=1, printit=FALSE)
> print( c( attr(cv_r,"ms"),attr(cv_rs,"ms") ), digits=6 )
[1] 2.21103 2.16553
> cv_r = cv.lm(data=globalPlants, ft_r, m=5, seed=2, printit=FALSE)
> cv_rs = cv.lm(data=globalPlants, ft_rs, m=5, seed=2, printit=FALSE)
> print( c( attr(cv_r,"ms"),attr(cv_rs,"ms") ), digits=6 )
[1] 2.22425 2.14762
> cv_r = cv.lm(data=globalPlants, ft_r, m=5, seed=3, printit=FALSE)
> cv_rs = cv.lm(data=globalPlants, ft_rs, m=5, seed=3, printit=FALSE)
> print( c( attr(cv_r,"ms"),attr(cv_rs,"ms") ), digits=6 )
[1] 2.2783 2.2373
```

we are now getting consistent results on different runs, unlike in Code Box 5.1, suggesting that adding rain. seas to the model improves predictive performance. Also note the answers are looking much more consistent now than before, with predictive errors within 2–3% of each other across runs.

How do you choose K? Dunno! (There is no single correct answer to this.)

The most common choices are N-fold or "leave-one-out" CV, 10-fold, or 5-fold CV. For no particular reason.

You could use the $n^{3/4}$ rule again; no one ever does, though, in K-fold CV. You could try some compromise between this and current K-fold conventions; I use something like the following:

- N-fold or "leave-one-out" cross-validation for small datasets ($n < 20$, say)
- 10-fold CV for medium-sized datasets ($20 < n < 100$, say)
- 5-fold CV for large datasets ($n > 100$, say).

For larger datasets you could go further, in the spirit of Shao (1993), and use two-fold or start to use increasing values of K but just use one of the folds for training and the rest for testing (rather than the other way around, as is done for small samples). This tends not to be done in practice, but there are good theoretical arguments for such an approach.

5.4 Information Criteria

Another way to do model selection is to use the whole dataset (no training/test split) and to penalise more complex models in some way to try to account for the additional variance they introduce. Such approaches are referred to as information criteria, largely for historical reasons (specifically, the first such criterion was derived to minimise an expected Kullback-Leibler information). The two most common criteria are AIC and BIC, which for linear models can be written

$$AIC = n \log \hat{\sigma}^2 + 2p$$
$$BIC = n \log \hat{\sigma}^2 + p \log(n)$$

(sometimes plus a constant), where p is the number of parameters in the model, n is sample size, and $\hat{\sigma}^2$ is the estimated error variance from the linear model.

The aim of the game is to

choose the model that minimises the information criterion

If we were to try to minimise $\hat{\sigma}^2$ alone, this would tend to favour complex models, as if we tried to maximise R^2. By adding $2p$ or $p \log(n)$ to the criterion, there is a larger penalty on models with more terms in them (larger p), so larger models are only chosen if they appreciably improve the fit of the model (by reducing $\hat{\sigma}^2$ appreciably). This penalty is intended to approximate the effects on the variance of predictions when adding unneeded terms to a model.

AIC stands for Akaike information criterion, named after Akaike (1972), although he originally intended the acronym to stand for "an information criterion" (Akaike, 1974). The $2p$ is motivated by thinking about predictive performance on test data and is an approximation to the amount of so-called optimism bias (Efron, 2004) that comes from estimating predictive performance on the same dataset used to fit the model in the first place.

BIC stands for Bayesian information criterion, and while the criterion looks similar, it has quite a different motivation. It was derived as an approximation to the posterior probability of a model being correct, integrating over all its parameters (Schwarz, 1978). Despite this Bayesian motivation, it tends not to be used in Bayesian statistics (Clark, 2007, Chapter 4), but is common elsewhere.

Both criteria can be understood as taking a measure of predictive error and adding a penalty for model complexity (as measured by the number of parameters in the model). For linear models, the measure of predictive error is a function of the error variance, but it will take other forms for other models. The better a model fits the sample data, the smaller the predictive error. While bigger models will always fit the sample data better than smaller models, the penalty for model complexity aims to correct for this.

Both criteria, despite quite different derivations, end up being different only in the value that is multiplied by the number of parameters in the model. AIC uses a 2, whereas BIC uses $\log n$, which will (as long as $n > 7$) take a larger value and, hence, penalise larger models more harshly. Hence when AIC and BIC differ, it is BIC that is choosing the smaller model. AIC is known to overfit, in the sense that even if the sample size is very large, it will often favour a model that is larger than the best-fitting one. BIC, in contrast, is known to be model selection consistent in a relatively broad range of conditions, i.e. as sample size increases, it will tend towards selecting the best model all the time.

The similarities in the form of the criteria motivate a more general approach, aptly named the generalised information criterion (GIC) (Nishii, 1984), which takes the form

$$GIC = n \log \hat{\sigma}^2 + \lambda p$$

where λ is an unknown value to be estimated by some method (preferably from the data). Usually, we would want to estimate λ in such a way that if sample size (n) were to increase, λ would get larger and larger (going to infinity) but at a slower rate than n. One way to ensure this would be to use CV, but with an increasing proportion of the data in the test sample in larger datasets, as in Table 5.2. This is something of a hybrid approach between those of the two previous sections, using both information criteria and CV. The main advantage of this idea is that predictive error is better estimated, because it is estimated by fitting the model to the whole dataset all at once (using information criteria), while at the same time, the appropriate level of model complexity is chosen using CV, to ensure independent data are used in making this decision.

Code Box 5.3: Computing Information Criteria on R for Exercise 5.1

The AIC or BIC function can be used to compute information criteria for many models:

```
> ft_r = lm(log(height)~rain,dat=globalPlants)
> ft_rs = lm(log(height)~rain+rain.seas,dat=globalPlants)
> c( AIC(ft_r), AIC(ft_rs) )
[1] 479.6605 475.4343
> c( BIC(ft_r), BIC(ft_rs) )
[1] 488.2861 486.9351
```

These results favour the larger model, although not by a lot, so results should be interpreted tentatively. Note that there is an advantage to the larger model when measured using the BIC score, because this criterion penalises complexity more harshly.

5.4.1 Pros and Cons of Information Criteria

Information criteria have the advantage that there are no random splits in the data—you get the same answer every time. This makes them simpler to interpret. (An exception is when using GIC with λ estimated by CV—in that case, the choice of λ can vary depending how the data are split into validation groups.)

The disadvantages are that they are slightly less intuitive than CV, derived indirectly as measures of predictive performance on new data, and in the case of AIC and BIC, their validity relies on model assumptions (essentially, the fitted models need to be close to the correct model). CV requires only the assumption that the test/training data are independent—so it can still be used validly when you aren't sure about the fitted model.

This is the first example we shall see of the distinction between *model-based* and *design-based* inference; we will see more about this in Chap. 9.

5.5 Ways to Do Subset Selection

It's all well and good if you only have a few candidate models to compare, but what if you have a whole bunch of predictor variables and you just want to find the subset that is best for predicting y? Common approaches:

- Forward selection—add one variable at a time, adding the best-fitting variable at each step
- Backward selection—add all variables, then delete one variable at a time, deleting the worst-fitting variable at each step
- All subsets—search all possible combinations. For p predictors there are 2^p possible combinations, which is not easy unless there are only a few variables (p small), as in Code Box 5.4.

There are also hybrid approaches that do a bit of everything, such as the `step` function in R, as in Code Box 5.5.

Code Box 5.4: All Subsets Selection for the Plant Height Data of Exercise 5.1

```
> library(leaps)
> fit_heightallsub<-regsubsets(log(height)~temp+rain+rain.wetm+
      temp.seas, data=globalPlants,nbest=2)
```

The results are most easily accessed using `summary`, but we will look at two parts of the summary output side by side (the variables included in models, stored in `outmat`, and the BICs of each model, stored in `bics`):

```
> cbind(summary(fit_heightallsub)$outmat,summary(fit_heightallsub)$bic)
         temp rain rain.wetm temp.seas
1  ( 1 ) " "  " "  "*"       " "        "-21.06175277099"
1  ( 2 ) " "  "*"  " "       " "        "-19.2868231448677"
2  ( 1 ) "*"  "*"  " "       " "        "-24.8920679441895"
2  ( 2 ) "*"  " "  "*"       " "        "-23.9315826810965"
3  ( 1 ) "*"  "*"  " "       "*"        "-20.9786934545272"
3  ( 2 ) "*"  "*"  "*"       " "        "-20.3405400349995"
4  ( 1 ) "*"  "*"  "*"       "*"        "-16.4229239023018"
```

The best single-predictor model has just `rain.wetm` and has a BIC of about −21; the next best single-predictor model has just `rain`. But including both `temp` and `rain` does the best among the models considered here, with a BIC of about −25.

Code Box 5.5: Stepwise Subset Selection for Plant Height Data of Exercise 5.1

```
> ft_clim = lm(log(height)~temp+rain+rain.wetm+temp.seas,
   data=globalPlants)
> stepClim=step(ft_clim,trace=0)
> stepClim$anova
         Step Df  Deviance Resid. Df Resid. Dev       AIC
1             NA        NA       126   260.6727 100.13694
2 - rain.wetm  1 0.6363946       127   261.3091  98.45637
3 - temp.seas  1 1.9256333       128   263.2347  97.41819
```

This table lists the three steps that were taken along the variable selection path, which ended up being backward selection (as indicated by the - signs at Steps 2 and 3). In Step 1, all four predictors are in the model, and the AIC is about 100. At Step 2, `rain.wetm` is removed from the model, which has little impact on the log-likelihood (residual deviance is changed only a little) and reduces the AIC by almost 2. At Step 3, `temp.seas` is removed from the model, reducing the AIC slightly further and leaving a final model with `temp` and `rain` as the remaining predictors.

To do forward selection, you need to start with a model with no terms in it and specify a `scope` argument with a formula including all the terms to be considered:

```
> ft_int = lm(log(height)~1,data=globalPlants)
> stepForward <- step(ft_int,scope=formula(ft_clim),
     direction="forward",trace=0)
> stepForward$anova
         Step Df Deviance Resid. Df Resid. Dev       AIC
1             NA       NA       130   355.9206 132.93585
2 + rain.wetm -1 74.59845       129   281.3221 104.12370
3      + temp -1 16.15030       128   265.1718  98.37867
```

Again the path has three steps in it, but this time they involve the addition of variables to the model (as indicated by the + sign). Step 1 starts with an intercept model, but the AIC is much improved upon the addition of `rain.wetm` at Step 2. Step 3 adds `temp`, which also decreases the AIC.

Notice that we ended up with a slightly different model! But recall that `rain` and `rain.wetm` are highly correlated, so the two final models are actually quite similar.

So which method is best? Dunno! (There is no simple answer.) You can explore this yourself by simulating data and seeing how different methods go—no method is universally best. Results of a small simulation looking at this question are given in Fig. 5.4, but the simulation settings could readily be varied so that any of the three methods was the best performer.

All-subsets selection is more comprehensive but not necessarily better—because it considers so many possible models, it is more likely that some quirky model will jump out and beat the model with the important predictors, i.e. it is arguably less robust. In Fig. 5.4, all-subsets selection seemed to perform best when looking at AIC on training data (left), but when considering how close predicted values were to their true means, forward selection performed better (right). This will not necessarily be true in all simulations. Recall also that all subsets is simply not an option if you have lots of x variables.

Backward selection is not a good idea when you have many x variables—because you don't really want to use all of them, and the model with all predictors (the "full model") is probably quite unstable, with some parameters that are poorly estimated. In this situation, the full model is not the best place to start from and you would be better off doing forward selection. In Fig. 5.4 (right), backward selection was the worst performing measure, probably because the sample size (32) was not large compared to the number of predictors (8), so the full model was not a good starting point.

Forward selection doesn't work as well in situations where the final model has lots of terms in it—because the starting point is a long way from the final answer, there are more points along the way where things can go wrong. In Fig. 5.4 (right), it was the best performing method, probably in part because in this simulation only two of the predictors were associated with the response, so this method started relatively close to the true answer.

Multi-collinearity (Page 70) can muck up stepwise methods—well it will cause trouble for any variable selection method, but especially for stepwise methods where the likelihood of a term entering the model is dramatically reduced by correlation with a term already in the model, the end result being that the process is a lot more noisy. In Fig. 5.4, all predictors had a correlation of 0.5, and the true model was correctly selected about 20% of the time. When the correlation was increased to 0.8, the correct model was only selected about 5% of the time.

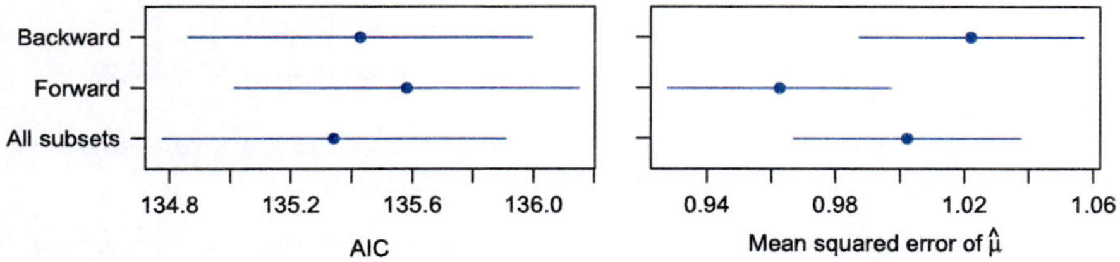

Fig. 5.4: Results of simulation looking at performance of different subset selection routines as measured using (left) AIC on training data and (right) MSE predicting the true mean response, for multiple linear regression with 8 predictors and 32 observations, of which 2 were related to the response. The mean AIC or MSE is reported across 1000 simulated datasets, together with a 95% confidence interval for the mean. All subset selection routines used AIC minimisation as their selection criterion. *All* pairwise contrasts were significant at the 0.05 level (using paired t-tests). Note that while all-subsets selection produced the smallest AIC values (indeed it always chose the model with the smallest possible AIC), it did not have the best predictive performance, as MSE was about 4% smaller for forward selection

5.6 Penalised Estimation

A modern and pretty clever way to do subset selection is to use penalised estimation. Instead of estimating model parameters (β) to minimise least squares

$$\min\left\{\sum_{i=1}^{n}(y_i - \mu_i)^2\right\}$$

we add a penalty as well, which encourages estimates towards zero, such as this one:

$$\min\left\{\sum_{i=1}^{n}(y_i - \mu_i)^2 + \lambda\sum_{j}|\beta_j|\right\}$$

This approach is known as the LASSO (Tibshirani, 1996, least absolute shrinkage and subset selection operator). This is implemented in a lot of recently developed statistical tools, so many ecologists will have used the LASSO without realising it, e.g. in MAXENT software under default settings (Phillips et al., 2006).

This looks a bit like GIC, but the penalty term is a function of the size of the parameters in the model, rather than just the number of parameters. The effect is to push parameter estimates towards zero (so their penalty is smaller), especially for coefficients of variables that aren't very useful in predicting the response. This biases parameter estimates in order to reduce their sampling variance.

Penalised estimation is a good thing when

- the main goal is prediction—it tends to improve predictive performance (by reducing variance);
- you have lots of parameters in your model (or not a large sample size)—in such cases, reducing the sampling variance is an important issue.

The λ parameter is a *nuisance parameter* that we need to estimate to fit a LASSO model. The value of this parameter determines how hard we push the slope parameters towards zero, i.e. how much we bias estimates, in an effort to reduce variance. So this parameter is what manages the bias–variance trade-off.

λ is large $\Longrightarrow$ most $\beta_j = 0$

λ is small $\Longrightarrow$ few $\beta_j = 0$

The parameter λ controls model complexity, determining how many predictors are included in the model. The full range of model sizes is possible, from having no predictors included (if λ is large enough) to including all of them (as λ approaches zero and we approach the least-squares fit). We can choose λ using the same methods we used to choose model complexity previously—CV is particularly common, but BIC is known to work well also.

The LASSO can equivalently be thought of as constrained minimisation:

$$\min\left\{\sum_{i=1}^{n}(y_i - \mu_i)^2\right\} \text{ such that } \sum_{j}|\beta_j| \le t$$

In other words, it can be understood as a least-squares estimator that insists that the sum of the absolute values of all parameters is no larger than some nuisance parameter t (which is a known function of λ and the data).

Code Box 5.6: LASSO for Plant Height Data of Exercise 5.1

```
> library(glmnet)
> X = cbind(globalPlants$temp, globalPlants$rain,
  globalPlants$rain.wetm, globalPlants$temp.seas)
> ft_heightcv=cv.glmnet(X,log(globalPlants$height))
> plot(ft_heightcv)
> ft_lasso=glmnet(X,log(globalPlants$height),
  lambda=ft_heightcv$lambda.min)
> ft_lasso$beta
```

Some good news about the LASSO:

- It reduces the sampling variability in parameters, and in predictions, by shrinking them towards zero.
- It does model selection as part of the estimation process. This happens because some (or many, depending on the data) parameter estimates are forced to zero by the LASSO penalty term, and if a parameter estimate is zero, that term has been excluded from the model.
- It's fast compared to other model selection methods (Friedman et al., 2010).
- It predicts well (by reducing variance). In fact, use of the LASSO is probably the main reason why MAXENT software has done well in comparisons of different methods of species distribution modelling (Elith et al., 2006; Renner & Warton, 2013).
- It simplifies the problem of model selection to one of estimating a single parameter (λ). Pretty cool that a problem involving choosing between 2^p candidate models (which could be in the millions or larger, see Table 5.1) can be simplified to estimating just one nuisance parameter.

And the bad news:

- It biases parameter estimates—relationships are flatter than they should be. There are variations on the LASSO to address this (such as the *adaptive LASSO*, Zou, 2006).
- Obtaining standard errors is complicated. (And how useful are standard errors when we know our estimates are biased anyway?) There are approximate methods for getting standard errors (Fan & Li, 2001), but they are rarely implemented in software and hard to interpret without concurrent information about the bias in estimates.

The LASSO, and related developments in fitting sparse models to data (i.e. models forcing lots of parameters to zero), is one of the more exciting developments in statistics over the last couple of decades, and there is a huge and rapidly growing literature on it in statistics journals.

5.7 Variable Importance

Exercise 5.2: Relative Importance of Climate Variables
Consider again Angela's height data (Exercise 5.1) and four variables—average annual temperature (`temp`), total precipitation (`rain`), rainfall in the wettest month (`rain.wetm`), and variation in mean monthly temperature (`temp.seas`).

How important are the different climate variables in explaining plant height?

Sometimes we are interested not just in which variables best predict a response, but how important they are relative to each other, as in Exercise 5.2. There are a few options here in terms of how to approach this sort of problem.

Recall the difference between *marginal* and *conditional* effects (Sect. 3.1.2)—the estimated effect of a predictor will change depending on what other terms are included in the model, because linear models estimate conditional effects. So when measuring the relative importance of predictors, we can expect to get different answers depending on what other terms are included in the model, as indeed happens in Code Boxes 5.7 and 5.8.

One option is to use forward selection to sequentially enter the predictor that most reduces the sum of squares at each step (Code Box 5.7). This is one way to order variables from most important to least, but not the only way (e.g. backward selection, which often leads to a different ordering). The table in Code Box 5.7 is intuitive, breaking down the overall model R^2 into components due to each predictor, but it does so in a misleading way. By adding terms sequentially, the R^2 for the first predictor `temp` estimates the *marginal* effect of temperature, because it was added to the model before any other predictors. But by the time `temp.seas` was added to the model, all other predictors had been included, so its *conditional* effect was being estimated in Code Box 5.7. So we are comparing "apples with oranges"—it would be better to either include all other predictors in the model or none of them when quantifying the relative effects of predictors.

Code Box 5.7: Sequential R^2 for Variable Importance
Proportion of variance explained R^2 will be used to measure importance of each predictor; it can be calculated by dividing the sum of squares explained by the sum of squares from a model with no predictors in it (`ft_int`).

Let's enter the variables sequentially:

```
> ft_clim = lm(log(height)~temp+rain+rain.wetm+temp.seas,
    data=globalPlants)
> ft_int = lm(log(height)~1,data=globalPlants)
> stepAnova = step(ft_int, scope=formula(ft_clim),
    direction="forward", trace=0, k=0)$anova
> stepAnova$R2=stepAnova$Deviance/deviance(ft_int)
```

```
> stepAnova
          Step Df  Deviance Resid. Df Resid. Dev       AIC          R2
1                NA        NA       130   355.9206 130.93585          NA
2 + rain.wetm -1 74.598450       129   281.3221 100.12370 0.209592965
3      + temp -1 16.150298       128   265.1718  92.37867 0.045376129
4      + rain -1  2.586703       127   262.5851  91.09452 0.007267641
5 + temp.seas -1  1.912441       126   260.6727  90.13694 0.005373225
```

`Deviance` means the same thing as sum of squares for a linear model, and `deviance(ft1)` gets the total sum of squares needed to construct R^2. In the line calling the `step` function, setting `k=0` ensures no penalty when adding terms, so that all four variables get added, in decreasing order of importance.

We can see that `rain.wetm` explains about 21% of variation in plant height, `temp` adds another 5%, and the other two variables do very little.

But there are different ways we could add these terms to the model! See Code Box 5.8 for alternatives.

Code Box 5.8: Marginal and Conditional R^2 for Variable Importance

In Code Box 5.7, we sequentially calculated R^2 on adding each term to the model in a way that maximised the explained variation at each step. Alternatively, we could look at the effect of the predictors one at a time, their *marginal* effect:

```
> stepMargin=add1(ft_int,scope=formula(ft_clim))
> stepMargin$R2=stepMargin$`Sum of Sq`/deviance(ft_int)
> stepMargin
           Df Sum of Sq    RSS    AIC      R2
<none>                  355.92 132.94
temp        1    66.224 289.70 107.97 0.18607
rain        1    70.761 285.16 105.90 0.19881
rain.wetm   1    74.598 281.32 104.12 0.20959
temp.seas   1    46.401 309.52 116.64 0.13037
```

It seems that either `temp` or `rain` explains about as much variation in plant height as `rain.wetm` did, some information we were missing from Code Box 5.7.

Alternatively, we could measure the *conditional* effect of each predictor by sequentially leaving each predictor out of the model while keeping all others in:

```
> leave1out=drop1(ft_clim)
> leave1out$R2=leave1out$`Sum of Sq`/deviance(ft_int)
> leave1out
           Df Sum of Sq    RSS     AIC       R2
<none>                  260.67 100.137
temp        1   16.0581 276.73 105.968 0.045117
rain        1    3.4438 264.12  99.856 0.009676
rain.wetm   1    0.6364 261.31  98.456 0.001788
temp.seas   1    1.9124 262.58  99.095 0.005373
```

These values are much smaller because they only capture variation explained by a predictor that is not explained by anything else. Leaving out `temp` increases the sum of squares by 4.5%; leaving out other terms increases error by 1% or less.

Code Box 5.8 presents two alternatives: first, estimating the *marginal* effect of a predictor, with the variation in height explained by this predictor when included in the model by itself; or, second, estimating the *conditional* effect of a predictor, with the variation in height it explains not being captured by other predictors. Fitting the full model and looking at standardised coefficients is another and more or less equivalent way to look at conditional effects (Code Box 5.9). The advantage of using standardised coefficients is that this method can be readily applied to other types of models (e.g. using a LASSO).

Looking at marginal vs conditional effects can give quite different answers (as in Code Box 5.8), especially when predictors are correlated. Neither of these is a perfect way to measure what is happening, and both seem to miss some details.

Code Box 5.9: Standardised Coefficients for Angela's Height Data

Looking at standardised coefficients:

```
> # first create a dataset with standardised predictors:
> globalPlantStand=globalPlants
> whichVars=c("temp","rain","rain.wetm","temp.seas")
> globalPlantStand[,whichVars]=scale(globalPlantStand[,whichVars])
> # then fit the model:
> ft_climStand = lm(log(height)~temp+rain+rain.wetm+temp.seas,
                 data=globalPlantStand)
> summary(ft_climStand)
             Estimate Std. Error t value Pr(>|t|)
(Intercept)    1.1947     0.1257   9.507  < 2e-16 ***
temp           0.5715     0.2051   2.786  0.00616 **
rain           0.4185     0.3244   1.290  0.19934
rain.wetm      0.1860     0.3353   0.555  0.58013
temp.seas      0.2090     0.2174   0.961  0.33816
---
Signif. codes:  0 `***' 0.001 `**' 0.01 `*' 0.05 `.' 0.1 ` ' 1
```

Again we see that `temp` is the most important predictor, followed by `rain`.

The problem with looking at marginal effects only is that if two predictors are highly correlated, measuring very similar things, then both can have large marginal effects, even if one predictor is not directly related to the response. For example, the marginal effect of temperature seasonality is $R^2 = 13\%$ (Code Box 5.8, `temp.seas`), but there seems to be little effect of `temp.seas` on plant height after annual temperature (`temp`) has been added to the model (Code Box 5.7, $R^2 < 1\%$ for `temp.seas`). It seems that the marginal R^2 for `temp.seas` was as high as 13% simply because it is correlated with `temp`. The `temp` predictor on the other hand does seem to be important, because even after including other predictors in the model, it still explains about 5% of variation in plant height (Code Box 5.8).

The problem with looking at conditional effects only is that if two predictors are highly correlated, measuring very similar things, then the conditional effect of each is small. For example, total precipitation (`rain`) and rainfall in the wettest month (`rain.wetm`) are very highly correlated (Code Box 3.7). Hence the conditional effect of each, after the other has already been included in the model, is small (Code Box 5.8) because most of the information captured in `rain` has already entered the model via `rain.wetm`, and vice versa. However, rainfall is clearly important to the distribution of plant height—it was in all models produced by best-subsets selection

(Code Box 5.4) and actually explains about 8% of variation after temperature has been added to the model; a leave-one-out approach misses this part of the story because rainfall enters the model via two variables. We would need to leave both rainfall variables out to see this—as in Code Box 5.10.

Exercise 5.3: Variable Importance Output
Compare the R^2 results of Code Boxes 5.7 and 5.8. Which table(s) do you think Angela should report when describing variable importance?

Now look at the standardised coefficients in Code Box 5.9. Do these coefficients measure marginal or conditional effects? Which of the R^2 tables in Code Box 5.8 are they most similar to in relative size (e.g. ranking from largest to smallest)? Is this what you expected?

Code Box 5.10: Importance of Temperature vs Rainfall
Yet another approach is to classify predictors as either temperature or rainfall predictors and study their relative effect. Looking at the effect of rainfall after temperature:

```
> ft_onlyTemp = lm(log(height)~temp+temp.seas,data=globalPlants)
> tempAn=anova(ft_int,ft_onlyTemp,ft_clim)
> tempAn$R2=tempAn$`Sum of Sq`/deviance(ft_int)
> tempAn
Analysis of Variance Table

Model 1: log(height) ~ 1
Model 2: log(height) ~ temp + temp.seas
Model 3: log(height) ~ temp + rain + rain.wetm + temp.seas
  Res.Df    RSS Df Sum of Sq      F     Pr(>F)       R2
1    130 355.92
2    128 289.00  2    66.917 16.173 0.00000056 0.188011
3    126 260.67  2    28.331  6.847 0.00150359 0.079599
```

Looking at the effect of temperature after rainfall:

```
> ft_onlyRain = lm(log(height)~rain+rain.wetm,data=globalPlants)
> rainAn=anova(ft_int,ft_onlyRain,ft_clim)
> rainAn$R2=rainAn$`Sum of Sq`/deviance(ft_int)
> rainAn
Model 1: log(height) ~ 1
Model 2: log(height) ~ rain + rain.wetm
Model 3: log(height) ~ temp + rain + rain.wetm + temp.seas
  Res.Df    RSS Df Sum of Sq       F    Pr(>F)       R2
1    130 355.92
2    128 279.80  2    76.118 18.3964 0.0000001 0.213863
3    126 260.67  2    19.130  4.6233 0.0115445 0.053747
```

temperature seems able to explain about 19% of global variation in plant height; then rainfall can explain about 8% more, whereas over 21% of variation can be explained by rainfall alone. This idea is visualised in Fig. 5.5.

For Angela's height data, one solution is to aggregate variables into types (temperature vs rainfall) and look at the importance of these variable types as a unit,

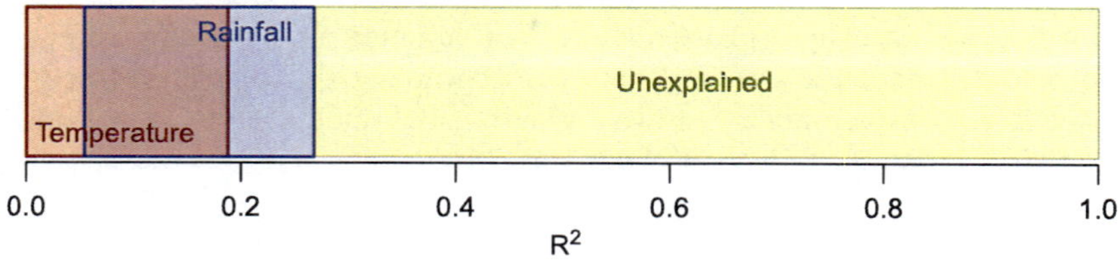

Fig. 5.5: Schematic diagram of relative importance of temperature and rainfall for Angela's height data, based on results in Code Box 5.10. Temperature and rainfall variables jointly explain about 27% of global variation in plant height for Angela's data, but temperature and rainfall each on their own explain closer to 20% of variation. About 5% of variation can be attributed to temperature, about 8% to rainfall, and the remaining 14% could be explained by either (it is *confounded*). This sort of plot, while conceptually helpful, is difficult to generalise to several predictors

as in Code Box 5.10. Each of temperature and rainfall, on its own, seems able to explain about 20% of variation in plant height, but adding the other variable type as well explains an additional 5–8% of variation, as visualised in Fig. 5.5. So we may conclude that temperature and rainfall are both important, separately, rainfall perhaps slightly more so, but we can do a better job (about 5% better) at explaining global variation in plant height by looking at temperature as well.

Plenty of alternative approaches could be used here. The simplest is to reduce multi-collinearity by removing highly correlated responses—this reduces the overlap, so conditional and marginal effects become more comparable. For Angela's data, we could have reduced the dataset to one temperature and one rainfall variable—a model with just `temp` and `rain`, for example, which ended up being suggested by `step` anyway (Code Box 5.5). Another option is to use structural equation modelling (Grace, 2006) to explicitly build into the model the idea that while temperature and rainfall are important, each is measured using multiple predictors. A more controversial option is to use a technique that averages measures of variable importance across different choices of model, which has been very popular in some parts of ecology under the name *hierarchical partitioning* (Chevan & Sutherland, 1991). The issue with that type of approach is that coefficients have different meanings depending on what other terms are included in the model—recall linear models estimate conditional effects, so changing what terms are in the model changes what we are conditioning on. So it makes little sense to average measures of variable importance across different models, which condition on different things, meaning we are measuring different things.

5.8 Summary

Say you have a model selection problem, like Angela's (Exercise 5.1). We have seen a suite of different tools that can be used for this purpose. So what should she actually do? Well, she could try a number of these methods; the important thing is to abide by a few key principles:

- Model selection is difficult and will be most successful when there are only a few models to choose between—so it is worth putting careful thought into what you actually want to compare, and deciding whether you can shortlist just a few candidate models.
- A key step is choosing model complexity—how many terms should be in the model? Too few means your model will be biased, too many means its predictions will be too variable, and we are looking to choose a model somewhere in the middle. A good way to choose the model complexity for your data is to consider how well different models predict to new, *independent data*—directly, typically using some type of CV, or indirectly, using information criteria.
- If you do have a whole heap of predictors, penalised estimation using methods like the LASSO is a nice solution that returns an answer quickly, meaning it is applicable to big data problems. Stepwise methods can also be useful, but it is worth starting them from a model near where you think the right model will be. For example, if you have many predictors but only want a few in the final model, it would be much better to use forward selection starting with a model that has no predictors in it than to use backward selection from a model with all predictors in it.

In the analyses for Angela's paper (Moles et al., 2009), there were 22 potential predictors, which we shortlisted to 10; then I specially wrote some code to use all-subsets selection and CV to choose the best-fitting model. With the benefit of hindsight I don't think the added complexity implementing an all-subsets algorithm justified the effort, as compared to, say, forward selection. A decade on, if dealing with a similar problem, I would definitely still shortlist variables, but then I would probably recommend a LASSO approach.

Model selection is an active area of research, and the methods used for this problem have changed a lot over the last couple of decades, so it is entirely possible that things will change again in the decades to come!

Exercise 5.4: Head Bobs in Lizards—Do Their Displays Change with the Environment?
Terry recorded displays of 14 male *Anolis* lizards in the wild (Ord et al., 2016). These lizards bob their head up and down (and do push-ups) in attempts to attract the attention of females. Terry measured how fast they bobbed their heads and wanted to know which environmental features (out of temperature,

light, and noisiness) were related to head bobbing speed. The data, with one observation for each lizard, can be found in the `headbobLizards` dataset.

What type of inference method is appropriate here?

What sort of model would you fit?

Load the data and take a look. Would it make sense to transform any of the variables in the data prior to analysis?

Which environmental variables best predict head bob speed?

Exercise 5.5: Plant Height Data and Precipitation
Consider Angela's global plant height data of Exercise 5.1. Angela collected data on how tall plants are in lots of different places around the globe. She also has data on eight different precipitation variables. She is interested in how plant height relates to precipitation and *which precipitation variables height relates to most closely*.

Find a subset of precipitation variables that optimally predicts plant height. Try a couple of different methods of model selection.

Any issues with multi-collinearity among the precipitation variables? Try to address any multi-collinearity by culling one or two of the main culprits. Does this affect your previous model selection results?

Chapter 6
Mixed Effects Models

Key Point
A random effect is a set of terms in a model that are assumed to come from a common distribution (usually a normal distribution). This technique is most commonly used to capture the effects of random factors in a design, i.e. factors whose levels are sampled at random from a larger population of potential levels.

If you have in your design

- a factor (categorical predictor) that takes a large number of potential levels
- only a random sample of these levels has actually been included in your study

then you have yourself a random factor. You can incorporate it into your model using *random effects*. Mathematically, this puts a distribution on the coefficients for that factor.

Any factor that is not treated as random is referred to as *fixed*. To this point, we have treated everything as fixed (fixed effects models). A model with both fixed and random effects in it is called a *mixed effects model*.

Exercise 6.1: Effects of Water Pollution on Subtidal Marine Micro-Invertebrates
Graeme is interested in the effects of water pollution on subtidal marine micro-invertebrates—in particular, the effect on invertebrate abundance (Clark et al., 2015). He samples in seven estuaries along the New South Wales coast (three of which are *pristine*, four are *modified*), and in each estuary, he placed settle-

D. I. Warton, *Eco-Stats: Data Analysis in Ecology*, Methods in Statistical Ecology,
https://doi.org/10.1007/978-3-030-88443-7_6

ment plates at four to seven sites, and counted the invertebrates that established on those plates (Clark et al., 2015).

What factors are there? Fixed or random?

Exercise 6.1 is an example of the most common scenario when a random effect pops up, *nested factors*. A factor B is nested in A if each level of B only occurs in one level of A. In Exercise 6.1, the estuary factor is nested within modification, because each estuary is classified as either pristine or modified (Fig. 6.1). Nested factors are not necessarily random, but they probably should be—otherwise you would have a tough job making inferences about at the higher sampling level ("factor A").

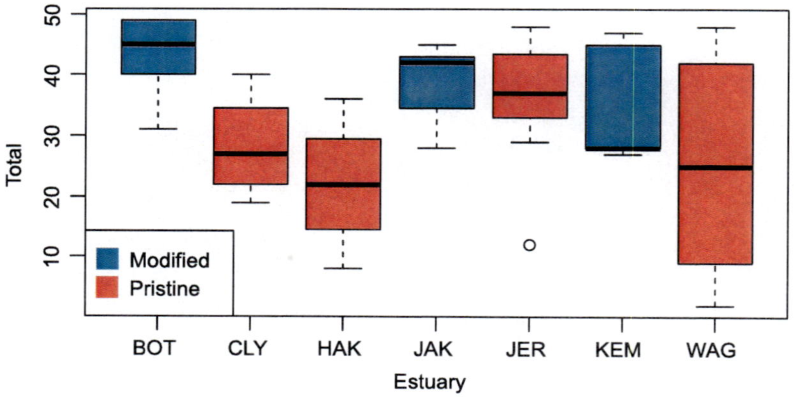

Fig. 6.1: Total invertebrate counts from Graeme's estuary data (Example 6.1), with replicate measurements at each of seven estuaries that were classified as either modified or pristine. Because of the multiple levels of sampling (sampling estuaries, then sites within estuaries), Graeme will need a mixed effects model to make inferences about the effect of estuary modification on invertebrates

A multi-level or hierarchical sampling design, where one randomly samples at different levels (e.g. site then transect or species), commonly generates random factors. Exercise 6.1 can also be understood as a multi-level design, with sampling of estuaries then sites within estuaries.

Pseudo-replication (Hurlbert, 1984), where one takes replicates at a level not of primary interest, can be considered as a multi-level design. Hurlbert (1984) originally described this as a big no-no, but provided that the multi-level nature of the sampling design is reflected in analysis, this can be quite a good thing to do. When using such a design, one needs to be particularly careful to make sure that there is sufficient replication at the level required to answer the research question of interest (e.g. do we need more sites or more measurements at each site?). A useful starting point is that you need to focus replication at the level where the factor of interest operates.

For example, if one is studying an environmental impact on sites, in the first instance look to increase replication by sampling more sites (rather than by taking more measurements within sites). But one is studying the effects of fine-scale (within site) environmental variation, replication within sites clearly has a more important role.

Key Point
Factor B is nested within A if each level of B only occurs within one level of A
In Exercise 6.1, Graeme's estuaries were nested within modification—each of the seven estuaries (B) was classified as either modified or pristine (A).
A nested factor, like estuary, is typically sampled as a random factor.

Why Not Average Within Estuaries?

If you have multiple sampling levels in your design, and you are interested in effects at the higher sampling level, why not average within the lower sampling units? For example, in Exercise 6.1, Graeme sampled estuaries and then sites within estuaries, but he was interested in cross-estuary differences. So why not average across the sites within each estuary? This would considerably simplify the design.

Well in many instances you can actually do this. A few things you will miss out on, though:

- Estimates of where the main sources of variance are (across or within estuaries);
- If there are different sampling intensities (e.g. some estuaries had four samples, some had as many as seven), then there would be different sampling errors across different samples. This would not be reflected in a naïve analysis of sample means;
- In some instances, this will be less powerful, because there is less information on sampling error in group averages, i.e. this approach might be less likely to detect patterns in the data that you are looking for.

But if you have balanced sampling and expect similar within-group variation across the different samples, and if you don't have an interest in partitioning variation across the different hierarchies in your design, the simpler option of averaging across samples is definitely worth considering. A passionate case for this was made by Murtaugh (2007), arguing the benefits of simplifying the analyses and their communication.

6.1 Fitting Models with Random Effects

A common tool for fitting mixed models is the `R` package `lme4` (Bates et al., 2015) as in Code Box 6.1. A brief outline is given here, but a full text is available online that gets into the gory details (Bates, 2010). Models are fitted using the `lmer` function, with random effects specified in brackets.

Code Box 6.1: Fitting a Linear Mixed Model to the Estuary Data of Exercise 6.1

```
> library(ecostats)
> data(estuaries)
> library(lme4)
> ft_estu = lmer(Total~Mod+(1|Estuary),data=estuaries)
> summary(ft_estu)
Linear mixed model fit by REML
Formula: Total ~ Mod + (1 | Estuary)
   Data: estuaries
   AIC   BIC logLik deviance REMLdev
 322.4 329.3 -157.2    322.4   314.4
Random effects:
 Groups   Name        Variance Std.Dev.
 Estuary  (Intercept)  10.68    3.268
 Residual             123.72   11.123
Number of obs: 42, groups: Estuary, 7
Fixed effects:
            Estimate Std. Error t value
(Intercept)   39.053      3.237  12.066
ModPristine  -11.243      4.287  -2.623
```

Any effect of modification on invertebrate abundance? What effect?

6.1.1 (Huh | What)?

Note in Code Box 6.1 the part of the formula `(1|Estuary)`. In `R` formulas, 1 means fit a y-intercept (because the intercept can be understood as a coefficient of 1, $\beta_0 = 1 \times \beta_0$). Using `(1|Estuary)` means "shift the y-intercept to a different value for each level of the factor `Estuary`". The vertical bar (`|`) means "given" or "conditional on"—so we are saying that the value of the y-intercept depends on `Estuary`, and it needs to be different for different levels of `Estuary`. In short, this introduces a main effect for `Estuary` that is random.

You can also use this notation to introduce random slopes—`(pH|Estuary)` would fit a different slope against pH (as well as intercept) in each estuary.

6.2 Linear Mixed Effects Model

The linear mixed effects model for the response of observation i, y_i, in its general form, is as follows:

$$\begin{aligned} y_i &\sim \mathcal{N}(m_i, \sigma^2) \\ m_i &= \beta_0 + \mathbf{x}_i^T \boldsymbol{\beta} + \mathbf{z}_i^T \mathbf{b} \\ \mathbf{b} &\sim \mathcal{N}(0, \boldsymbol{\Sigma}) \text{ independently of } y_i \end{aligned}$$

Basically, it looks just like the standard (fixed effects) linear model, except now there is a third line—this line says that some of the coefficients in the model are random (and normal and independent of y_i). Vector notation is used here, and it is worth noting that $\mathbf{x}$ is a list of any fixed number of predictor variables, as is $\mathbf{z}$. These lists do not need to have the same length—you can have different numbers of fixed vs random effects in a model. Every predictor should, however, have a value for every observation (every i); otherwise, you may need to use special methods to deal with the missing data.

Graeme (Exercise 6.1) had a model with random intercepts for `Estuary`, i.e. the $\mathbf{z}_i$ were indicator variables for each estuary.

The $\boldsymbol{\Sigma}$ in the foregoing equation is a variance–covariance matrix, an idea that will be discussed in more detail in Chap. 7. Basically this allows us to introduce random effects that are correlated with each other instead of being independent. In this chapter we will assume all random effects are independent with constant variance (except for Exercise 6.2), so for the jth element of $\mathbf{b}$ we could rewrite the assumption as

$$b_j \sim \mathcal{N}(0, \sigma_b^2)$$

The $\boldsymbol{\Sigma}$ notation allows for situations where there is more than one type of random effect in the model, and they should be related to each other, such as when including random slopes as well as random intercepts (as needed in Exercise 6.4).

6.2.1 Mind Your Ps and Qs—Mixed Effects Models

1. The observed y-values are *independent* (after conditioning on x).
   This can often be guaranteed to be satisfied by sampling in a particular way *(how???)* (Hint: See Sect. 1.2.5.)
2. The y-values are *normally distributed* with *constant variance*

   $$y \sim \mathcal{N}(\mu_y, \sigma^2)$$

   As before, the *normality* of y doesn't really matter (due to the central limit theorem), except for small samples/strongly skewed data/outliers. You can check this on a normal quantile plot.

   The *constant variance* assumption, as before, is important and can be checked using a *residual plot* for no fan shape
3. *Straight-line relationship* between the mean of y and each x (and z)

   $$\mu_i = \beta_0 + \mathbf{x}_i^T \boldsymbol{\beta} + \mathbf{z}_i^T \mathbf{b}$$

(As before, when x variables are factors, this assumption doesn't really matter.) This can be checked by looking for no pattern on a *residual plot*; a U shape is particularly bad news.

4. The random effects **b** are *independent of* y.
   This assumption can be guaranteed by sampling in a particular way *(how??)*.
5. The random effects **b** are *normally distributed*, sometimes with *constant variance*.
   These assumptions don't seem to matter much (but not due to CLT, for a different reason. See Schielzeth et al., 2020, for example).

Note these are the same old linear model assumptions, checked in the same way, but with a couple of assumptions on the additional random terms in the model, the random effects **b**.

Code Box 6.2: Residual Plots from a Mixed Model for Exercise 6.1

```
ft_estu = lmer(Total~Mod+(1|Estuary),data=estuaries)
scatter.smooth(residuals(ft_estu)~fitted(ft_estu),
    xlab="Fitted values",ylab="Residuals")
abline(h=0,col="red")
```

Or to plot residuals against "unconditional" predicted values (using the fixed effects term only):

```
scatter.smooth(residuals(ft_estu)~predict(ft_estu,re.form=NA),
    xlab="Fitted values (no random effects)",ylab="Residuals")
abline(h=0,col="red")
```

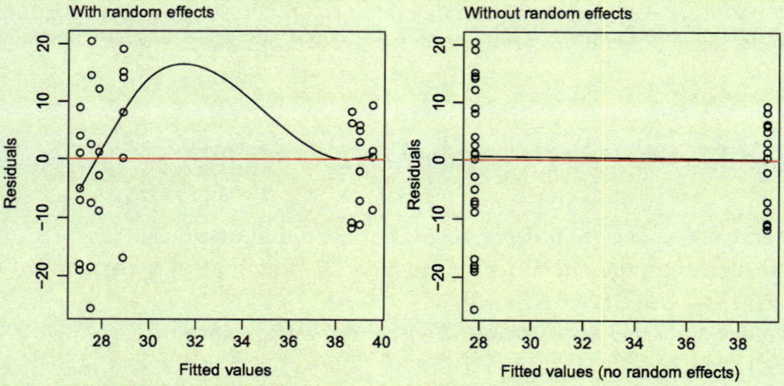

What do these plots tell you about model assumptions?

There are two types of residual plot that you could produce, depending on whether or not you include random effects as predictors in the model. Including them sometimes creates artefacts (inducing a correlation between residuals and predicted values), so it is often a good idea to consider both types of plot, as in Code Box 6.2. Studying these residuals, there is a suggestion that residuals are actually less variable in one group (modified) than the other. Perhaps modified estuaries are more homogeneous? Although having said that, we are talking about quite small sample sizes here (seven estuaries sampled in total) and making generalisations about changes in variability from this sample size is a tough ask. We will consider this in Exercise 6.2.

6.3 Likelihood Functions

Recall that (fixed effects) linear models are fitted by *least squares*—minimise the sum of squared errors in predicting y from x. Mixed effects models are fitted by *maximum likelihood* or *restricted maximum likelihood*.

Maths Box 6.1: Maximum Likelihood Estimation

The *likelihood function* of some data $\mathbf{y}$ is defined as the probability function of the data, treated as a function of its parameters $\boldsymbol{\theta}$. If $\mathbf{y}$ are independent observations, the likelihood function $\mathcal{L}(\mathbf{y})$ is the product of probability functions of the Y_i:

$$\mathcal{L}(\boldsymbol{\theta}, \mathbf{y}) = \prod_{i=1}^{n} f_{Y_i}(y_i)$$

We usually take the log of this, which tends to be easier to work with.

For example, in a linear model, we assume (conditionally on $\mathbf{x}$) that observations are independent and come from a normal distribution with mean $\mu_i = \beta_0 + \mathbf{x}_i'\boldsymbol{\beta}$ (for observation i) and standard deviation σ. The probability function of observation i is

$$f_{Y_i}(y_i) = \frac{1}{\sqrt{2\pi}\sigma} e^{-\frac{1}{2\sigma^2}(y_i-\mu_i)^2}$$

and so the log-likelihood of the linear model with parameters $\beta_0, \boldsymbol{\beta}, \sigma$ is

$$\ell(\beta_0, \boldsymbol{\beta}, \sigma; \mathbf{y}) = \sum_{i=1}^{n} \log f_{Y_i}(y_i) = -\frac{n}{2}\log 2\pi - n\log\sigma - \frac{1}{2\sigma^2}\sum_{i=1}^{n}(y_i - \mu_i)^2 \tag{6.1}$$

In practice, we don't know the values of these parameters and want to estimate them. We can estimate the parameters by *maximum likelihood estimation*—finding the parameter values that make $\ell(\beta_0, \boldsymbol{\beta}, \sigma; \mathbf{y})$, hence $\mathcal{L}(\beta_0, \boldsymbol{\beta}, \sigma; \mathbf{y})$, as big as possible.

Notice that to maximise $\ell(\beta_0, \boldsymbol{\beta}, \sigma; \mathbf{y})$ with respect to β_0 and $\boldsymbol{\beta}$, we need to minimise $\sum_{i=1}^{n}(y_i - \mu_i)^2$, which is what least-squares regression does. That is, least-squares regression gives the maximum likelihood estimate of a linear model.

To test whether model $\mathcal{M}_1$ (with parameter estimates $\widehat{\boldsymbol{\theta}}_1$) has a significantly better fit than a simpler model $\mathcal{M}_0$ (with estimates $\widehat{\boldsymbol{\theta}}_0$), we often use a (log-)likelihood ratio statistic:

$$-2\log\Lambda(\mathcal{M}_0,\mathcal{M}_1) = 2\ell_{\mathcal{M}_1}(\widehat{\boldsymbol{\theta}}_1;\mathbf{y}) - 2\ell_{\mathcal{M}_0}(\widehat{\boldsymbol{\theta}}_0;\mathbf{y}) \tag{6.2}$$

For a linear model, this is a function of the usual F statistic. The multiplier of -2 looks weird but ends up being convenient theoretically.

We often measure how well a model $\mathcal{M}$ fits data, as compared to some "perfect" model $\mathcal{S}$ (which makes predictions $\widehat{\boldsymbol{\theta}}_{\mathcal{S}}$ as close to data as possible), using the *deviance*:

$$\mathcal{D}_{\mathcal{M}}(\mathbf{y}) = 2\ell_{\mathcal{S}}(\widehat{\boldsymbol{\theta}}_{\mathcal{S}};\mathbf{y}) - 2\ell_{\mathcal{M}}(\widehat{\boldsymbol{\theta}};\mathbf{y}) \tag{6.3}$$

which for a linear model is a function of the sum of squares.

In a mixed model, the **b**, as well as the **y**, are random and so also contribute to the likelihood. This makes the maths trickier and will be discussed later.

The likelihood function, for a given value of model parameters, is the joint probability function of your data. The higher the value, the more likely your data. The key idea of *maximum likelihood* estimation is to choose as your parameter estimates the values that make your data most likely, i.e. the values for parameters that would have given the highest probability of observing the values of the response variables you actually observed. Maths Box 6.1 gives the likelihood function for linear models (which ends up simplifying to least-squares estimation), and Maths Box 6.2 discusses challenges extending this to mixed models.

Under a broad set of conditions that most models satisfy, maximum likelihood estimators

- are consistent (in large samples, they go to the right answer),
- are asymptotically normal (which makes inference a lot easier),
- are efficient (they have minimum variance among consistent, asymptotically normal estimators).

Basically, they are awesome, and these properties give most statisticians license to base their whole world on maximum likelihood[1]—provided that you can specify a plausible statistical model for your data (you need to know the right model so you are maximising the right likelihood).

Many familiar "classical" statistical procedures can be understood as maximum likelihood or some related likelihood-based procedure. The sample mean can be understood as a maximum likelihood estimator, as can sample proportions; ANOVA is a likelihood-based procedure, as are χ^2 tests for contingency tables. . . .

Restricted maximum likelihood (REML) is a slight variation on the theme in the special case of linear mixed models—it is more or less a cheat fix to ensure all parameter estimates are exactly unbiased when sampling is balanced (they are only approximately unbiased for maximum likelihood, not exactly unbiased). This is worth doing if your sample size is small relative to the number of terms in the model. In a

[1] Well, Bayesians multiply the likelihood by a prior and base their whole world around that.

linear model, i.e. a fixed effects model with a normally distributed response variable, least-squares regression is (exactly) restricted maximum likelihood. REML is the default method in the `lme4` package; for maximum likelihood you use the argument `REML=FALSE` as in Code Box 6.3.

Maths Box 6.2: Random Effects Make Everything Harder
Random effects complicate estimation and inference in a couple of ways.

As in Maths Box 6.1, for independent observations $\mathbf{y}$, the likelihood is a product of the probability function for the observed data, $\mathcal{L}(\boldsymbol{\theta}, \mathbf{y}) = \prod_{i=1}^{n} f_{Y_i}(y_i)$. But this is more difficult to calculate in a mixed model because the values that $\mathbf{y}$ might take depend on the random effects $\mathbf{b}$. We know the conditional distribution of data $f_{Y_i|\mathbf{B}=\mathbf{b}}(y_i)$, given random effects, and we know the probability function of random effects $f_{\mathbf{B}}(\mathbf{b})$. But we don't directly observe the values of the random effects, so we have to *marginalise* over all possible values of $\mathbf{b}$:

$$f_{Y_i}(y_i) = \int_{\mathbf{b}} f_{Y_i|\mathbf{B}}(y_i) f_{\mathbf{B}}(\mathbf{b}) d\mathbf{b}$$

This integral is our first difficulty—it often makes it hard to calculate the likelihood, and often we can't write down a simple expression for the maximum likelihood estimator (in contrast to Maths Box 6.1, for example). Sometimes it is possible to solve integrals exactly, but sometimes it is not, and they have to be approximated. The linear mixed model log-likelihood works out OK (it has a *closed form*), but extensions of it may not (e.g. Sects. 10.6.3 and 12.3).

A second difficulty for mixed models is that tools we conventionally use for inference don't always work. For example, the likelihood ratio statistic $-2\log\Lambda$ of Maths Box 6.1 can be shown to often approximately follow a known distribution (a χ^2 distribution) when the null hypothesis is true, a result that we often use to make inferences about key model terms (in R, using the anova function). However, this result does not hold if the null hypothesis puts a parameter on a boundary of its range of possible values. When testing whether a fixed effect should be included in the model, we test to see whether a slope coefficient (β) is zero, which is not on a boundary because slopes can take negative or positive values. But when testing to determine whether a random effect should be included in the model, we are testing to see whether its variance (σ_b^2) is zero, which is as small as a variance can get. Hence we have a boundary problem, and the theory that says $-2\log\Lambda$ has a χ^2 distribution does not hold when testing random effects.

6.4 Inference from Mixed Effects Models

Inference from mixed effects models is a little complicated, because the likelihood theory that usually holds sometimes doesn't when you have random effects (Maths Box 6.2).

Note in Code Box 6.1 that there are no P-values for the random effects or the fixed effects—these were deliberately left out because the package authors are a little apologetic about them; they would only be approximately correct. The t-values give you a rough idea, though anything larger than 2 is probably significant at the 0.05 level. (You could also try the `nlme` package for slightly more friendly output.)

You can use the `anova` function as usual to compare models, as in Code Box 6.3. This uses a *likelihood ratio test* (comparing the maximised likelihood under the null and alternative models, as in Maths Box 6.1), and curiously, this often does return P-values in the output. When comparing models that differ only in fixed effects terms, these P-values are fairly reliable (although a little dicey for small sample sizes). It is advised that you use `REML=FALSE` when fitting the models to be compared (models should be fitted by maximum likelihood to do a likelihood ratio test).

Code Box 6.3: Using anova to Compare Mixed Effects Models for Estuary Data

```
> ft_estu = lmer(Total~Mod+(1|Estuary),data=estuaries,REML=F)
> ft_estuInt = lmer(Total~(1|Estuary),data=estuaries,REML=F)
> anova(ft_estuInt,ft_estu)
Data: estuaries
Models:
ft_estuInt: Total ~ (1 | Estuary)
ft_estu: Total ~ Mod + (1 | Estuary)
           npar    AIC    BIC  logLik deviance  Chisq Df Pr(>Chisq)
ft_estuInt    3 334.18 339.39 -164.09   328.18
ft_estu       4 330.14 337.09 -161.07   322.14 6.0396  1    0.01399 *
---
Signif. codes:  0 `***' 0.001 `**' 0.01 `*' 0.05 `.' 0.1 ` ' 1
```

Is there evidence of an effect of modification?

You can use `anova` to similarly test for random effects, but this gets a little complicated for a couple of reasons:

- It doesn't work when you have a single random effect (as Graeme does); Zuur et al. (2009) proposes a workaround using the `nlme` package and `gls`.
- P-values are very approximate, and the theory used to derive them breaks down when testing for random effects (Maths Box 6.2). The P-values tend to be roughly double what they should be, but even then are still very approximate.

6.4.1 *Confidence Intervals for Parameters*

Use the `confint` function (just like for `lm`), as in Code Box 6.4.

Code Box 6.4: Confidence Intervals for Parameters from a Mixed Effects Model for Estuary Data

```
> confint(ft_estu)
Computing profile confidence intervals ...
                 2.5 %     97.5 %
.sig01              NA   7.613337
.sigma        8.944066  14.084450
(Intercept)  32.815440  45.246159
ModPristine -19.461059  -2.994392
```

For particular values of a random effect (e.g. the size of the effect for a given estuary in Exercise 6.1) the mechanics of inference are messier, but you can get intervals that can be interpreted more or less as approximate 95% confidence intervals for the random effects (true values of shift in each estuary), as in Code Box 6.5.

Code Box 6.5: Prediction Intervals for Random Effects Terms in a Mixed Effects Model

```
> rft=ranef(ft_estu,condVar=T)
> library(lattice)
> dotplot(rft)
```

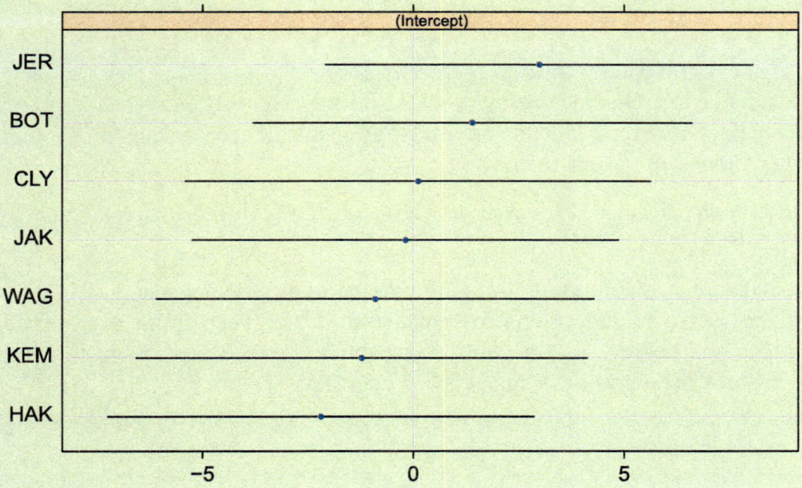

This is a dotplot with 95% prediction intervals for random effects from the mixed effects model to the estuary data of Code Box 6.5. Any evidence of an effect of estuary?

Exercise 6.2: Fitting Random Effects with Different Variances
Notice from Code Box 6.2 that the two different levels of modification seem to lead to different levels of variability in total abundance. Fit a mixed effects model to the estuary data that allows modified and pristine estuaries to have different variances by constructing the following variables:

```
estuaries$isMod = as.numeric(estuaries$Mod=="Modified")
estuaries$isPri = as.numeric(estuaries$Mod!="Modified")
```

and including in your model formula a random effect for each of these: `(0+isMod|Estuary)+(0+isPri|Estuary)`.

Compare this to your previous model, using a model selection criterion of your choice. Which better fits the data?

Exercise 6.3: Bird Exclusion and Biological Control
Recall Exercise 4.10, a bird exclusion experiment studying the effects on biological control of aphids. Ingo wants to know: *Is there an effect of the netting treatment on trends in aphid numbers across the sampling times?* The data are available as `aphidsBACI` in the `ecostats` package.

Note that two measurements were taken at each of eight plots, so we want to include `Plot` in the model. This variable is better understood as a random rather than as a fixed factor.

Reanalyse the data to answer the research question, using the appropriate mixed model, being sure to check your assumptions.

Exercise 6.4: Estuary Data in Different Zones
Consider again Graeme's estuary data of Exercise 6.1. For each estuary, Graeme actually took four to seven measurements at each of the "inner" and "outer" zones. He wants to know:

> Is there an effect of modification on total abundance (`Total`)? Is this effect different in different estuary zones?

The data are available as `estuaryZone` in the `ecostats` package.

Construct a suitable graph to visualise the data. Then fit an appropriate mixed model, being sure to mind your Ps and Qs, then use `anova` to get a quick-and-dirty answer to Graeme's key questions.

6.5 What If I Want More Accurate Inferences?

Key Point
Inferences from standard software for mixed models are only approximate and are quite arm-wavy when testing for a random effect. If you want a better answer, you should use a simulation-based approach like the parametric bootstrap—this is advisable if testing for a random effect.

So we know that inferences are a bit arm-wavy in mixed models, especially if testing for a random effect, such as if we wanted to test for an effect of estuary in Exercise 6.1. There are alternative and more accurate methods of inference to use, based on simulation—generating the distribution of a test statistic by simulation, rather than estimating it from theory.

To make inferences from a sample, we need to know how much the statistic of interest might vary from one sample to the next, so we can take into account uncertainty estimating the statistic from our data (one of the many possible samples we could have ended up with). The idea of using simulation for inference is that we randomly generate multiple datasets, recompute the statistic for each, and see how it varies. The problematic part is working out how to simulate data. One of the better simulation approaches for mixed models is the *parametric bootstrap*, the general idea being to simulate new responses directly from our fitted model (because that is our best estimate of the model for the response variable). Like all simulation methods, this requires a bit of coding and computation time, especially for large datasets. Code Box 6.6 uses this technique to estimate the standard error of a fixed effect for Graeme's data. Note that in this code only the response variable is being re-estimated—we are conditioning on the observed values of predictors, as is commonly done in regression problems. Inference about fixed effects tends to work pretty well for mixed models, so the standard error we get by simulation is very close to what we saw in `summary` output of Code Box 6.1. A more important use of the parametric bootstrap is in testing whether a random effect needs to be included in the model. The `anovaPB` package in the `ecostats` package was specially written to make it easy to use a parametric bootstrap to perform an `anova` to compare two fitted objects, as in Code Box 6.7. This function works not just for mixed models but for most of the models seen in this book.

Code Box 6.6: Using the Parametric Bootstrap to Compute the Standard Error of the `Mod` Fixed Effect in Exercise 6.1

We will estimate the standard error of the `Mod` effect using a parametric bootstrap—by simulating a large number (`nBoot`) of datasets from the fitted model, re-estimating the `Mod` effect for each, then taking the standard deviation.

```
> nBoot=500
> bStat=rep(NA,nBoot)
> ft_estu = lmer(Total~Mod+(1|Estuary),data=estuaries)
> for(iBoot in 1:nBoot)
+ {
+   estuaries$TotalSim=unlist(simulate(ft_estu))
+   ft_i = lmer(TotalSim~Mod+(1|Estuary),data=estuaries)
+   bStat[iBoot] = fixef(ft_i)[2]
+ }
> sd(bStat) #standard error of Mod effect
[1] 4.294042
```

How does this compare to the standard error from the original model fit?

Code Box 6.7: A Parametric Bootstrap to Test for an Effect of Estuary in Exercise 6.1

We will test for an effect of `Estuary` using the anovaPB function in the `ecostats` package. This computes a likelihood ratio statistic (LRT) to compare a model with an `Estuary` effect (`ft_estu`) to a model without (`ft_noestu`), then repeats this process a large number (`n.sim`) of times on datasets simulated from our fitted model under the null hypothesis (`ft_noestu`). The P-value reports the proportion of test statistics for simulated data exceeding the observed value.

```
> ft_noestu = lm(Total~Mod,data=estuaries)
> library(ecostats)
> anovaPB(ft_noestu,ft_estu,n.sim=99)
Analysis of Deviance Table

ft_noestu: Total ~ Mod
ft_estu: Total ~ Mod + (1 | Estuary)

          df deviance   LRT Pr(>LRT)
ft_noestu 3     322.2
ft_estu   4     314.4 7.832     0.27

P-value calculated by simulating 99 samples from ft_noestu.
```

If the null hypothesis were true, we would expect a larger test statistic to arise by chance about 27% of the time.

Any evidence of an effect of estuary?

6.6 Design Considerations

When you have random effects, there are now multiple sample sizes to worry about:

Total sample size—The larger it is, the better your estimates of lower-level fixed effects (and residual variance).

The number of levels of the random factor—The larger it is, the better your estimate of the random effect variance and any higher-level effects depending on it.

You could have a million observations, but if you only have a few levels of your random effect, you have little information about the random effect variance and, hence, about anything it is nested in. For example, if Graeme had taken 100 samples at each of 3 estuaries, he would probably have less of an idea about the effect of modification than he does now, with his 4–7 samples at each of 7 estuaries.

There are broad principles but no hard guidelines on precisely how large your various sample sizes need to be. Some would argue you need to sample at least five levels of a factor in order to call it random, but this is a fairly arbitrary number; the idea is that the more levels you sample, the better you can estimate the variance of the random effect. A good way to decide on your design in any setting is to think about how accurately you want target quantities to be estimated and to study how accurately you can estimate these target quantities under different sampling scenarios. Usually, this requires some pilot data, to get a rough estimate of parameters and their uncertainty, and some power analyses or margin of error calculations. For mixed models this may need to be done by simulation, e.g. using the `simR` package.

Another issue to consider is what sample size to use within each level of the random factor, e.g. how many samples Graeme should take within each estuary. It is usually a good idea to aim for balanced sampling (i.e. the same number of samples in each treatment combination), but mixed effects models do *not* require balanced sampling. Some texts have said otherwise, largely because old methods of fitting mixed effects models (via sums of squares decompositions) did require balanced sampling for random effects estimation. But that was last century. (Restricted) maximum likelihood estimation has no such constraint. It is usually a great idea to aim for balanced sampling—it is usually better for power and for robustness to assumption violations (especially the equal variance assumption). But it is not necessary.

Exercise 6.5: Accurate Inferences About Estuary Data
Consider again Graeme's estuary data, as in Exercise 6.4. We previously did a quick-and-dirty test to see if the effect of modification was different in inner and outer sampling zones.

Use the parametric bootstrap to get a formal test for a `zone:mod` interaction. How do results compare to those from when you were using the `anova` function?

This would all have been so much easier if there wasn't a random effect in the model.... Do we really need `Estuary` in there?

6.7 Situations Where Random Effects Are and Aren't Used

So you have a random factor in your study design. Does it really need to be treated as a random effect in modelling? Maybe not. Use random effects if *both* the following conditions are satisfied:

- If you have a random factor (i.e. large number of levels, from which you have a random sample)
- You want to make general inferences across the random factor itself (across all its possible levels, not just those that were sampled).

Usually, if the first condition is satisfied, the second will be too, but it need not be. If you are happy making inferences conditional on your observed set of levels of the random factor, then there is no harm in treating the effect as fixed and saving yourself some pain and suffering. (But this is not always possible!)

Using fixed effects models has the advantage that estimation and inference are much simpler and better understood—indeed some more advanced models (discussed in later chapters) can handle fixed effects models only. Using random effects has the advantage that inferences at higher levels in the hierarchy are still permissible even when there is significant variation at lower levels. For example, in Exercise 6.1, if `Estuary` were treated as a fixed effect, then if different estuaries had different invertebrate abundances, we could not have made any inferences about the effect of modification. If we knew that different estuaries had different abundances, then by default `mod` would have had a different abundance (because different estuaries had different levels of `mod`). But treating `Estuary` as a random effect, we can estimate the variation in abundance due to estuary and ask if there is a `mod` effect above and beyond that due to variation with `Estuary`.

What if I need to treat my factor as random (to make inferences about higher-level effects) but I didn't actually sample levels of this factor at random? Well that's a bit naughty. Recall that for a factor to be random, the levels of it used in your study need to be a *random sample* from the population of possible values. Wherever possible, if you want to treat a factor as random in analysis, you should sample the levels of the factor randomly. (Among other things, this ensures independence assumptions are satisfied and that you can make valid inferences about higher-level terms in the model.)

That said, the interpretation of random effects as randomly chosen levels is often stretched a bit; random effects are sometimes used as a mathematical device that can

- induce correlation between groups of correlated observations (invertebrate abundance across samples from the same estuary are correlated). This idea can be extended to handle spatial or temporal correlation, as in the following chapter;
- stabilise parameter estimates when there are lots of parameters (fixed effects models want the number of parameters to be small compared to the sample size n).

In these instances we are using random effects as a mathematical device rather than using them to reflect (and generalise from) study design. A cost of doing this is that inference becomes more difficult.

Recall the LASSO—a shrinkage method for improving predictive performance by reducing the variance in parameter estimates. The LASSO minimises

$$\min \left\{ \sum_{i=1}^{n} (y_i - \mu_i)^2 + \lambda \sum_j |\beta_j| \right\}$$

The LASSO can be thought of as putting random effects on regression parameters—but not assuming they are normal, instead assuming they come from a longer-tailed distribution with a big peak at zero (the so-called double exponential or Laplace distribution). Hence, the LASSO is an example of using random effects as a mathematical device—to stabilise parameter estimates. A cost of doing this, when the parameters of interest may not actually be random, is that inference becomes more difficult (e.g. recall that we don't have good techniques for estimating the standard error of LASSO coefficients).

Chapter 7
Correlated Samples in Time, Space, Phylogeny...

Recall from Chap. 1 that a critical assumption we often make in statistics is independence—when this assumption is violated, you are stuffed. Well, unless you have a good idea of how this assumption is violated. This chapter describes a few such situations where we expect data to be dependent and can specify a model to describe this dependence, so that we can make valid conclusions from our data.

The previous chapter introduced one situation where we have an idea how the independence assumption is violated—in a multi-level or hierarchical sampling design, we have "pseudo-replication", with replicate samples being more similar within a cluster than across clusters, which can be addressed using a mixed model (with a random effect in the model for cluster). This chapter considers some other common examples where we have some idea how independence may be violated:

- Repeated measures in time—if you go back to the same places or subjects multiple times, often it is reasonable to assume the correlation across these repeat measures decreases as a (possibly known) function of time between measurements.
- Spatial autocorrelation—when sampling in different places, often observations would be expected to be more similar in response if they were located closer together. (This is called *auto*correlation because the response variable is correlated with itself.)
- Phylogeny—if sampling multiple taxa, responses might be expected to be more similar in response if the taxa are more closely related.

All of these examples involve *structured correlation* in the data—there is dependence but we know the main variables imposing dependence and so can try to include them in our model. An example correlation structure is given in Maths Box 7.1.

D. I. Warton, *Eco-Stats: Data Analysis in Ecology*, Methods in Statistical Ecology,
https://doi.org/10.1007/978-3-030-88443-7_7

Maths Box 7.1: An Example Correlation Structure

Consider a sequence of k observations equally spaced in time, $\mathbf{y} = y_1, \ldots, y_k$, where you believe that the next value you observe is the current value plus noise (but rescaled to have constant variance). If the y_i are also normally distributed with constant variance, then it can be shown that $\mathbf{y}$ has a *multivariate normal* distribution with a *variance–covariance matrix*:

$$\sigma^2 \begin{pmatrix} 1 & \rho & \rho^2 & \rho^3 & \ldots & \rho^{k-1} \\ \rho & 1 & \rho & \rho^2 & \ldots & \rho^{k-2} \\ \rho^2 & \rho & 1 & \rho & \ldots & \rho^{k-3} \\ \rho^3 & \rho^2 & \rho & 1 & \ldots & \rho^{k-4} \\ \vdots & & & & & \\ \rho^{k-1} & \rho^{k-2} & \rho^{k-3} & \rho^{k-4} & \ldots & 1 \end{pmatrix}$$

where the value in row i and column j tells us the correlation between y_i and y_j. This is known as an *autoregressive process* with lag one, or AR(1).

Data from such a process can be described as having *autocorrelation* because values of the response are correlated with themselves. One way to estimate ρ would be to look at neighbouring values. Since $\text{cor}(y_i, y_{i-1}) = \rho$, we could estimate ρ using the sample correlation coefficient of the values $y_2, y_3, \ldots, y_k$ against their "lag 1" values $y_1, y_2, \ldots y_{k-1}$. A better estimator, e.g. a maximum likelihood estimator, would consider other lags too, because these also contain some information about ρ.

A useful summary of the autocorrelation structure is the *sample autocorrelation function*, the sample autocorrelation $\widehat{\text{cor}}(y_i, y_{i-h})$ plotted against lag h. For an AR(1) process we would expect the sample autocorrelation to (approximately) decay exponentially towards zero as the lag k increased.

For spatial data, we might instead assume that correlation is a function of how far apart two points are. For phylogenetic data, we might assume correlation is a function of the phylogenetic distance between the relevant nodes. The sample autocorrelation functions are then slightly more difficult to compute and would be plotted against distance rather than lag. We could similarly use sample autocorrelations to estimate parameters in our assumed correlation structure for spatially or phylogenetically structured data.

Key Point

Sometimes data are collected in a way that (potentially) introduces dependence that is a function of some known variable such as time, space, or phylogeny, sometimes called a *structured correlation*. In such cases we can fit a statistical model that estimates and accounts for this dependence—usually done using random effects that have a structured correlation.

A structured correlation can appear in a response variable for a couple of reasons. Obviously, the response variable could have inherent dependence in it (*endogenous*), e.g. repeated measures of abundance over time may be temporally correlated because the number of organisms at a location is a function of how many organisms were there previously. But another important source of dependence is model misspecification (*exogenous*)—if you fit a model to repeated measures data but omit an important term that varies over time (weather, for example), then your response will appear to have a temporally structured correlation in it. This type of effect can also be observed when a predictor is added to a model incorrectly, e.g. as a linear term when its effect is non-linear. This is an important source of dependence because in ecology we rarely believe that we have the correct model—there are often predictors we can't measure (or can't measure correctly) or predictors that may not have been entered into the model in the right way.

Failure to model the aforementioned sources of dependence can have important consequences—most notably, false confidence in our results. A structured correlation like that mentioned previously tends to lead to a positive correlation in data, which makes the standard errors of coefficients larger than we think they are if the dependence is ignored. This affects all inferences about this parameter—not accounting for positive dependence will make confidence intervals too narrow, P-values from hypothesis tests too small and model selection techniques too confident in a particular model. Trying to avoid false confidence is one of the main reasons for doing stats in the first place—we want a reality check against the apophenia we are known to be guilty of (Shermer, 2012) to distinguish between science and a scientist's opinion. So a situation where a statistical procedure gives false confidence, by underestimating uncertainty, undermines the procedure's validity and its ability to do the job it was originally intended for.

Just specifying a model for correlation isn't good enough; we need a (close to) correct model—just as valid inferences from linear models rely critically on the independence assumption, valid inferences from dependent data models rely critically on the assumed dependence structure being (close to) correct. The additional difficulty here, however, is that while we can often guarantee that the independence assumption will be satisfied (e.g. in an experiment, by randomising the allocation of subjects to treatment groups), we cannot guarantee by design that any particular structured dependence assumption will be satisfied. Instead we propose a model for dependence, just as we propose a model for other aspects of a response, and we need to use diagnostic checks to see if the model seems reasonable. Because of the importance of the dependence assumption to the validity of inferences, it is also often a good idea to try a few different models for dependence to see how sensitive results are to switching between competing, plausible models.

One way to understand linear models with a dependence structure is to view them as linear models fitted to contrasts that have been constructed so that they are independent of each other (Maths Box 7.2). If the assumed dependence structure is wrong, the contrasts that are assumed to be independent will not be (Maths Box 7.3).

Maths Box 7.2: Contrasts and Structured Correlation
Assume that we have a linear model for a set of observations **y**, and the model for the *i*th observation is

$$y_i = \beta_0 + \mathbf{x}_i'\boldsymbol{\beta} + \epsilon_i$$

as usual, except that we will not assume the errors ϵ_i are independent. Instead we assume the vector of errors $\boldsymbol{\epsilon}$ satisfies

$$\boldsymbol{\epsilon} \sim \mathcal{N}(\mathbf{0}, \boldsymbol{\Sigma})$$

where $\mathcal{N}$ now refers to the *multivariate normal distribution* and $\boldsymbol{\Sigma}$ is the *variance-covariance matrix*, a matrix that captures the dependence structure across observations. Both will be explored in more detail in Chap. 11. We will assume the standard deviation of each error is σ, as usual.

Linear combinations of a multivariate normal distribution are normal. Specifically, for any full rank, square matrix **A** (which represents a linear transformation), $\mathbf{A}\boldsymbol{\epsilon} \sim \mathcal{N}(\mathbf{0}, \mathbf{A}\boldsymbol{\Sigma}\mathbf{A}')$. We will consider what happens when we choose the transformation matrix $\sigma\boldsymbol{\Sigma}^{-1/2}$. Applying this transformation to our linear model:

$$\begin{aligned}\sigma\boldsymbol{\Sigma}^{-1/2}y_i &= \sigma\boldsymbol{\Sigma}^{-1/2}\beta_0 + \sigma\boldsymbol{\Sigma}^{-1/2}\mathbf{x}_i'\boldsymbol{\beta} + \boldsymbol{\Sigma}^{-1/2}\epsilon_i \\ y_i^* &= \beta_0^* + x_i^*\boldsymbol{\beta} + \epsilon_i^*\end{aligned}$$

where y_i^* and x_i^* can be understood as contrasts to **y** and **x**, calculated using the contrast matrix $\sigma\boldsymbol{\Sigma}^{-1/2}$, and the vector of errors $\boldsymbol{\epsilon}^*$ is multivariate normal with variance–covariance matrix

$$\sigma^2\boldsymbol{\Sigma}^{-1/2}\boldsymbol{\Sigma}\boldsymbol{\Sigma}^{-1/2} = \sigma^2\mathbf{I}$$

where **I** is the *identity matrix*, with all diagonal elements equal to one and all off-diagonal elements (covariances) zero. This means that the transformed errors σ_i^* are independent with variance σ^2, and our dependent data linear model is equivalent to fitting an ordinary (independent errors) linear model to contrasts constructed using the assumed correlation matrix of errors.

Maths Box 7.3: Consequences of Misspecifying the Correlation Structure in a Linear Model
What if you assume the wrong dependence structure, $\boldsymbol{\Sigma}_{\text{wrong}}$? From Maths Box 7.2, this is equivalent to fitting a linear model using contrasts $\sigma^2\boldsymbol{\Sigma}_{\text{wrong}}^{-1/2}$ and assuming independence of contrasts. But the variance–covariance matrix of contrast errors is

$$\sigma^2\boldsymbol{\Sigma}_{\text{wrong}}^{-1/2}\boldsymbol{\Sigma}\boldsymbol{\Sigma}_{\text{wrong}}^{-1/2}$$

which still contains a correlation. So what we are doing here is equivalent to fitting a linear model on contrasts, assuming contrasts are independent, when they violate the independence assumption. We know such violations lead to incorrect standard errors (as in Maths Box 1.5) and possibly false confidence. The extent to which this is an issue is a function of how far the assumed correlation structure ($\boldsymbol{\Sigma}_{\text{wrong}}$) is from the true structure ($\boldsymbol{\Sigma}$).

Accounting for violations of independence is one reason for fitting a model with a structured correlation; another reason is to model heterogeneity—if your treatment effect varies across time/space/phylogeny, you will need to incorporate this into your model in order to quantify it.

In this chapter, the main strategy for introducing a structured correlation into models will be to use a random effect or an error term that contains the structured correlation that is desired. The `nlme` package in `R` can do this and is good for observations that are correlated in space or time (although it is more difficult to use for discrete data, as in Chap. 10).

Another common strategy, not considered here, is autoregressive models, i.e. models that assume the response at a given location (in space, time, or phylogeny) is a function of nearby responses. Autoregressive models are typically the best option for *time series data*, where you have a single chain of many measurements over time. Time series analysis will not be considered here, as it requires a suite of additional tools (Hyndman & Athanasopoulos, 2014) and is arguably less common in ecology than the repeated measures setting considered below.

7.1 Longitudinal Analysis of Repeated Measures Data

Consider Exercise 7.1. In eight plots, Ingo took measurements of the number of aphids present on seven sampling occasions following application of a bird exclusion treatment. The issue that arises in this design is that repeated measures will be more similar to each other within plots than they will be across plots, and the autocorrelation may be higher for times that are closer together. Analysis of such repeated measures data, in a way that can account for these features, is often referred to as *longitudinal data analysis*.

Exercise 7.1: Biological Control of Aphids over Time
Recall the bird exclusion experiment of Exercise 6.3, which studied the effects on biological control of aphids in an oat field. Ingo wants to know: *Is there an effect of the netting treatment on trends in aphid numbers across the sampling times?* While counts from only two sampling times were discussed previ-

ously, Ingo actually took measurements in the oat field seven times following application of the bird exclusion treatment.

This larger dataset is available as the `oat` object in the `aphids` dataset in the `ecostats` package.

What sort of model is appropriate here?

7.1.1 Four Approaches for Longitudinal Data

There are four main approaches for modelling longitudinal data, which we will refer to as

- ignoring it,
- the random intercept model,
- the random slope model, and
- use of a temporally structured random effect.

Ignoring it: While this approach needs to be used with caution, it is sometimes OK to ignore correlation across repeated measures and use standard linear modelling approaches for analysis, assuming repeated measures are independent of each other. Fixed effects terms could be added to the model to try to account for changes over time. This approach is appropriate if repeated measures are far enough apart in time that residuals do not show evidence of autocorrelation, or if the autocorrelation is effectively captured by predictors that vary over time—this *should be checked*, either by constructing a sample autocorrelation plot of residuals or by fitting some of the following models for comparison (Code Box 7.2). Remember the importance of the independence assumption in linear models—if data are dependent, but you make inferences using a linear model (assuming independence), you are stuffed!

Random intercept model: In this approach we include a random effect for the subject that repeated measures are taken on (in the case of Exercise 7.1, this is the plot). For example, for the aphid data of Exercise 7.1, we could model the mean of plot i at time j using

$$\mu_{ij} = \beta_0 + x_{ij}\beta + u_i$$

where $u_i \sim \mathcal{N}(0, \sigma^2)$, i.e. we have a random effect in the model that takes different values for different plots. An example of this is in Code Box 7.2, stored as `aphid_int`. This is fine for situations where you have measurements at two time points (as in Exercise 1.5 or Exercise 6.3), but when there are several, it is less commonly used. The random intercept term induces correlation across repeated measures of the subject, but a potential issue is that the correlation is the same strength irrespective of how close together two sampling points are in time, which doesn't make much sense. This type of model is usually better suited to clustered or multi-level data (such as Graeme's estuary data, Exercise 6.1) but is simpler than the following model and so is worth considering.

Random slope model: A random effect can also be included for the slope of the relationship between response and time, for each subject. This is sometimes referred to as a random slope model. For example, for the aphid data:

$$\mu_{ij} = \beta_0 + x_{ij}\beta + u_{1i} + t_{ij}u_{2i}$$

where t_{ij} is the time of the jth sample from plot i, and (u_{1i}, u_{2i}) is a bivariate normal random effect. (That is, each of u_{1i} and u_{2i} is normal, and they are correlated. The variance of each random effect and the correlation between them are estimated from the data.) An example of this is in Code Box 7.2, stored as `aphid_slope`. The random slope induces a correlation that is stronger between repeated measures that are closer together in time.

Temporally structured random effect: A temporally structured random effect could be added to the model that is more highly correlated for points closer together in time. For example, for the aphid data, we could model the mean of plot i at time j using

$$\mu_{ij} = \beta_0 + x_{ij}\beta + u_i + u_{ij}$$

where (as previously) u_i is a random intercept term across plots (this term is optional but often helpful), and now we have a multivariate normal random effect u_{ij} that has the same variance at every sampling time but a correlation that changes as a function of time between repeat samples. For example, we could assume that the correlation between sampling times t_{ij} and $t_{ij'}$ is

$$\rho^{|t_{ij}-t_{ij'}|}$$

which will decay to zero as the difference between sampling times increases. This function is plotted in Fig. 7.1, and it defines what is known as an autoregressive process with order one (well, it is a continuous time extension of this process). An example of this is in Code Box 7.2, stored as `aphid_CAR1`. There are many different ways of introducing temporally structured random effects into a model; this is one of the simplest and most common.

7.1.2 Which Longitudinal Model Should I Use?

Recall that the introduction to this chapter emphasised that not just any model for correlation would do; we need to work towards specifying a correct model—because inferences from a model with structured random effects rely critically on their dependence assumptions, just as inferences from ordinary linear models rely critically on their independence assumptions. So the goal, if we wish to make inferences from a model, is to find the best-fitting model for the dependence in the data. You could treat this as a model selection problem (Chap. 5), fitting each longitudinal model and for example selecting the model with the smallest Bayesian

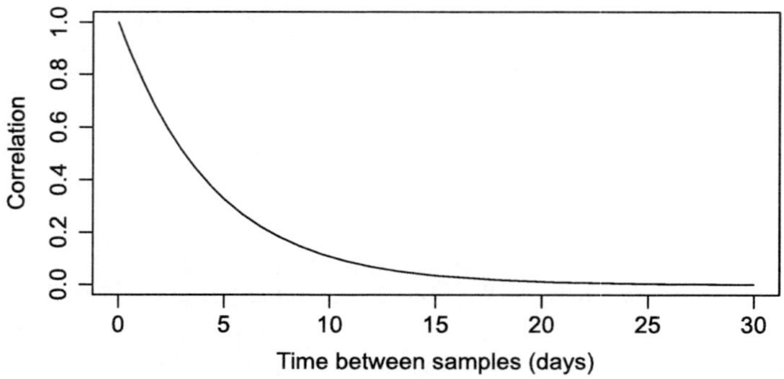

Fig. 7.1: Example of how autocorrelation might decay as time between samples increases. Correlation between sampling times t_{ij} and $t_{ij'}$ is assumed here to be $\rho^{|t_{ij}-t_{ij'}|}$, with $\rho = 0.8$. Note in this case that the correlation is quite high between points sampled at similar times (separated by a few days) but is almost zero when points are sampled at least a couple of weeks apart (> 15 days)

information criterion (BIC). In the case of Exercise 7.1, we ended up with a random intercept model, as indicated in Code Box 7.2.

If you don't wish to make inferences from the model and are just after predictions, then there is an argument that you could ignore dependence and fit a standard linear model, to keep things simple. The main problems from ignoring dependence—confidence intervals too narrow and P-values too small—aren't relevant if you are interested in neither confidence intervals nor P-values. However, you can expect to lose some predictive power if ignoring dependence, especially if predicting the value of future repeated measures for existing subjects. I would only ignore dependence if predicting to new, independent observations, and only if dependence was relatively weak.

Of the remaining options for longitudinal analysis, the random intercept and random slope models have the advantage that they can be implemented on pretty much any mixed modelling software. The last option, including a temporally structured random effect, needs more specialised software. It can be fitted in `nlme` in `R` for normally distributed data.

As such, if there are a couple of candidate models for dependence that seem reasonable, it might be advisable to try each of them to see if the results are robust to the choice of longitudinal model. In the case of the aphid data of Exercise 7.1, we found that the marginally significant effect of treatment seen in the random intercept model (Code Box 7.3) did not remain when fitting a random slope model (Code Box 7.4).

7.1.3 Minding Your Ps and Qs

All of the aforementioned options involve fitting a linear (mixed) model and so making the usual linear modelling assumptions, which are checked in the usual way. Additionally, we now consider adding a correlation structure to repeated measures of a subject. We can use model selection to choose between competing models for the correlation structure, but it would be helpful to visualise it in some way also.

One common graphical technique for longitudinal data is a *spaghetti plot*, joining the dots across repeated measures for each observation in the dataset, which can be constructed using the `interaction.plot` function in R as in Code Box 7.1. Note that the `interaction.plot` function treats time as a factor, so the sampling times are distributed evenly along the x-axis, rather than having points closer together if they are closer in time. This could be addressed by constructing the plot manually (e.g. using `lines` to build it up subject by subject).

Sometimes a spaghetti plot reveals a lot of structure in data—for example, if the lines cross less often than you would expect by chance, this suggests a positive correlation across subjects. In Fig. 7.2a, there is a suggestion of a treatment effect, with the reduction in aphid numbers being steeper in the bird exclusion treatment. Beyond this pattern (looking within the red lines or within the black lines) there is a suggestion of correlation in the data, with the lines not crossing each other very often, considering the number of lines on the plot.

For a large number of subjects, a spaghetti plot can quickly become quite dense and hard to interpret. One approach that helps with this is using transparent colours (which can be done in R using the `rgb` function and `alpha` argument; see the second line of Code Box 7.1) to reduce ink density; another option is to plot one (or several) random samples of observations, at some risk of loss of information.

Code Box 7.1: R Code to Produce Fig. 7.2

```
data(aphids)
cols=c(rgb(1,0,0,alpha=0.5),rgb(0,0,1,alpha=0.5)) #transparent colours
par(mfrow=c(2,1),mar=c(3,3,1.5,1),mgp=c(2,0.5,0),oma=c(0,0,0.5,0))
with(aphids$oat, interaction.plot(Time,Plot,logcount,legend=FALSE,
  col=cols[Treatment], lty=1, ylab="Counts [log(y+1) scale]",
  xlab="Time (days since treatment)") )
legend("bottomleft",c("Excluded","Present"),col=cols,lty=1)
mtext("(a)",3,adj=0,line=0.5,cex=1.4)
with(aphids$oat, interaction.plot(Time,Treatment,logcount, col=cols,
  lty=1, legend=FALSE, ylab="Counts [log(y+1) scale]",
  xlab="Time (days since treatment)"))
legend("topright",c("Excluded","Present"),col=cols,lty=1)
mtext("(b)",3,adj=0,line=0.5,cex=1.4)
```

Also useful as a diagnostic tool is a plot of the sample autocorrelation, graphing the strength of correlation between pairs of points, as a function of the difference between sampling times. Whereas usually correlation is computed between one

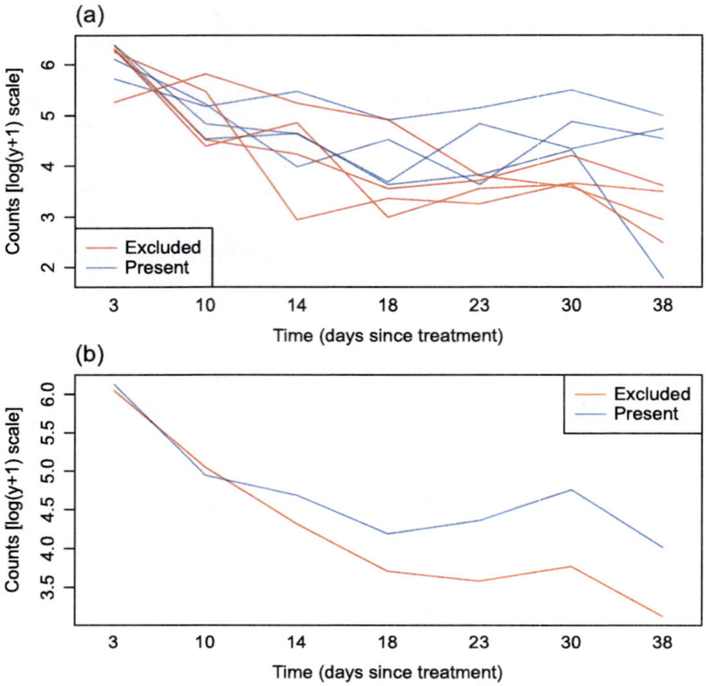

Fig. 7.2: Exploratory plots looking at the effect of bird exclusion on aphid abundance. (**a**) A "spaghetti" plot of how abundance changes over time within each plot. (**b**) An interaction plot looking at how mean (transformed) abundance changes over time across the two treatments. Notice in (**a**) that there is relatively little crossing of lines, suggesting some correlation, specifically a "plot" effect, and notice in (**b**) that under bird exclusion, aphid numbers tend to be lower (increasingly so over time since exclusion)

variable and another, here we compute a correlation between a variable and itself at different time lags, an *auto*correlation. This can be plotted using the `nlme` package in R for evenly spaced sampling times, and while we used this technique to produce the figure in Code Box 7.3, the uneven intervals between samples was not accounted for (as with the interaction plot). This could be addressed by manually constructing such a function, binning pairs of residuals based on differences in sampling time, as is commonly done for spatial data.

For temporally structured data, we would expect the autocorrelation to decrease smoothly towards zero, approximating something like Fig. 7.1. In Code Box 7.3, there was no such smooth decay to zero (the correlation quickly jumped to weak levels), suggesting little need for a temporally structured term in addition to the random intercept already in the model. This verifies our findings in Code Box 7.2, where the preferred model appeared to be the random intercept model.

Code Box 7.2: Choosing a Longitudinal Model for Aphid Data

A quadratic term for time will be included in the model to approximate the curve seen in Fig. 7.2b. The question then becomes: How do we model the correlation across repeated measures of each plot?

```
> library(lme4)
> aphid_int = lmer(logcount~Treatment*Time+Treatment*I(Time^2)+
          (1|Plot),data=aphids$oat,REML=FALSE) # random
          intercept model
> aphid_slope = lmer(logcount~Treatment*Time+Treatment*I(Time^2)+
   (Time|Plot), data=aphids$oat, REML=FALSE) # random slope model
> library(nlme) # refit random intercept model in nlme to get a ACF:
> aphid_int2 = lme(logcount~Treatment*Time+Treatment*I(Time^2),
   random=~1|Plot, data=aphids$oat, method="ML")
> plot(ACF(aphid_int2),alpha=0.05) # only works for nlme-fitted mixed
  models
> # now try a model with a temporally structured random effect:
> aphid_CAR1 = update(aphid_int2,correlation=corCAR1(,form=~Time|Plot))
> BIC(aphid_int,aphid_int2,aphid_slope,aphid_CAR1)
            df      BIC
aphid_int    8 144.5429
aphid_int2   8 144.5429
aphid_slope 10 147.4243
aphid_CAR1   9 148.1673
```

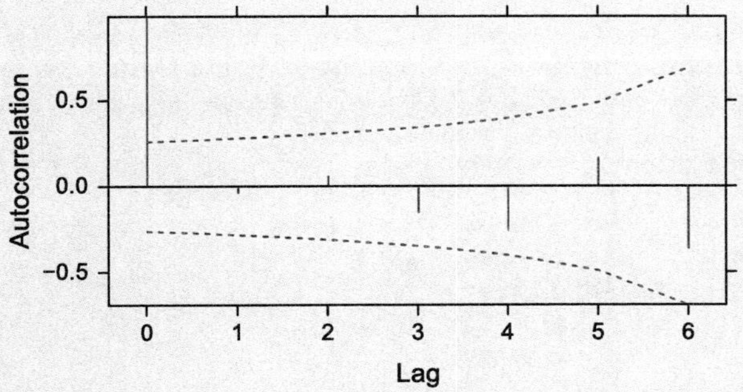

The autocorrelation plot suggests a negligible residual autocorrelation, after including a random intercept in the model, and the BIC results further suggest that additional terms are not needed in the model.

Code Box 7.3: Exploring Random Intercept Fit to Aphid Data

```
> print(aphid_int)
...
Random effects:
 Groups   Name        Std.Dev.
 Plot     (Intercept) 0.202
```

```
 Residual             0.635
...
```

The random effect for plot is relatively small compared to residual variation, i.e. compared to within-plot variation not explained by the temporal trend.

```
> anova(aphid_int)
Analysis of Variance Table
                    Df  Sum Sq Mean Sq F value
Treatment            1  2.0410  2.0410  5.0619
Time                 1 24.7533 24.7533 61.3917
I(Time^2)            1  5.7141  5.7141 14.1717
Treatment:Time       1  1.6408  1.6408  4.0693
Treatment:I(Time^2)  1  0.0413  0.0413  0.1024
```

Any treatment effect is relatively small compared to the broader temporal variation in aphid abundance; judging from mean squares, there is almost 10 times more variation across time compared to across treatments.

```
> aphid_noTreat = lmer(logcount~Time+I(Time^2)+(1|Plot),
    data=aphids$oat, REML=FALSE)
> anova(aphid_noTreat,aphid_int)
              Df    AIC    BIC  logLik deviance  Chisq Df Pr(>Chisq)
aphid_noTreat  5 130.26 140.39 -60.131   120.26
aphid_int      8 128.34 144.54 -56.170   112.34 7.9224  3    0.04764 *
```

The treatment effect is marginally significant.

Code Box 7.4: Exploring Random Slope Fit to Aphid Data

We will repeat the analyses of Code Box 7.3, but using a random slope model, to study the robustness of results to the choice of longitudinal model.

```
> print(aphid_slope)
...
Random effects:
 Groups   Name        Std.Dev. Corr
 Plot     (Intercept) 0.11725
          Time        0.01909  -1.00
 Residual             0.57875
...
```

Note the residual error is smaller than it was in the previous model, suggesting that the random slope term has explained more of the variation in trajectories of aphid numbers across plots. This is reflected in the smaller mean squares in the following table.

```
> anova(aphid_slope)
Analysis of Variance Table
                    Df  Sum Sq Mean Sq F value
Treatment            1  0.7472  0.7472  2.2309
Time                 1 12.7915 12.7915 38.1889
I(Time^2)            1  5.7141  5.7141 17.0593
Treatment:Time       1  0.8479  0.8479  2.5313
Treatment:I(Time^2)  1  0.0413  0.0413  0.1232
> aphid_noTreatS = lmer(logcount~Time+I(Time^2)+(Time|Plot),
    data=aphids$oat, REML=FALSE)
> anova(aphid_noTreatS,aphid_slope)
```

```
                Df    AIC    BIC  logLik deviance  Chisq Chi Df Pr
                (>Chisq)
aphid_noTreatS  7 125.04 139.22 -55.519   111.04
aphid_slope    10 127.17 147.42 -53.585   107.17 3.8666      3
                 0.2762
```

Now the treatment effect is not significant, so the results do not seem robust to the choice of longitudinal model! This suggests that any evidence for an effect of treatment, in this dataset, is weak (note this is a subset of the full data of Grass et al., 2017).

Exercise 7.2: Biological Control of Aphids in a Wheat Field
Ingo repeated his experiment in a wheat field, and data from this field are available as the `wheat` object in the `aphids` dataset in the `ecostats` package.

Repeat the preceding longitudinal analyses for data from the wheat field. Which longitudinal model better handles repeated measures in this case? Is there evidence that bird exclusion improves biological control of aphids?

Exercise 7.3: Biological Control of Aphids Across Both Fields!
Ideally, we should use a single model to answer Ingo's question of whether there is an effect of bird exclusion on the biological control of aphids. We could also use such a model to test whether the effect differs across fields.

Combine Ingo's data from the wheat and oat field. Construct a single, larger model to test for an effect of biological exclusion and to check whether this effect differs across fields.

7.2 Spatially Structured Data

Ecological data are typically collected across different spatial locations, and it is often the case that the data are spatially structured, with observations from closer locations more likely to be similar in response. As previously, this autocorrelation could have exogenous (predictors are spatial) or endogenous (response is spatial) sources, or both. An example is found in Exercise 7.4, where species richness appears to have spatial clusters of low and high richness.

Exercise 7.4: Eucalypt Richness as a Function of the Environment
Ian is interested in the species richness of eucalypts and related plants (the *Myrtaceae* family) in the Blue Mountains area west of Sydney, Australia. This area was declared a World Heritage Site in part because of its diversity of

eucalypts, so Ian wanted to know the answer to the following question: How does *Myrtaceae* species richness vary from one area to the next, and what are the main environmental correlates of richness? Ian obtained data on the number of *Myrtaceae* species observed in 1000 plots and obtained estimates of soil type and climate at these sites.

Plotting richness against spatial location, he found spatial clusters of high or low species richness (Fig. 7.3).

What sort of analysis method should Ian consider using?

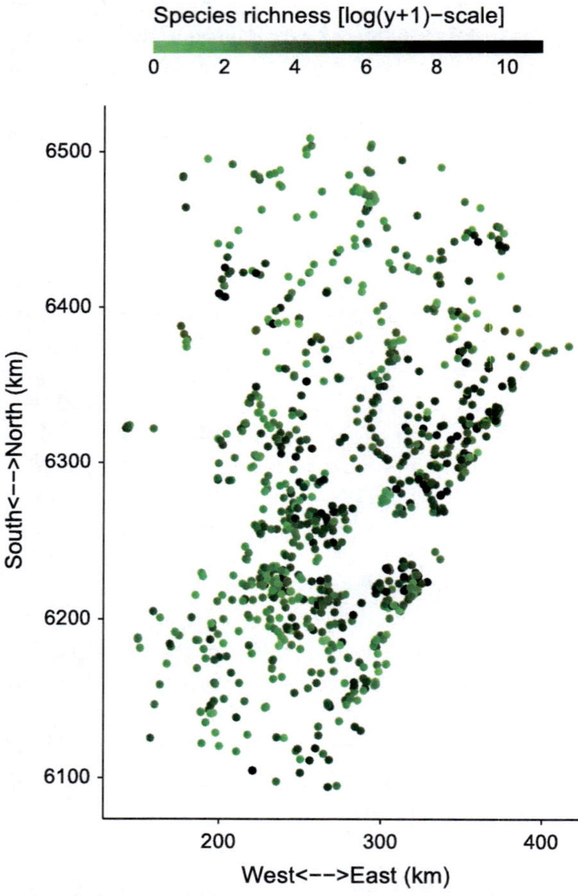

Fig. 7.3: Observed *Myrtaceae* species richness in Blue Mountains World Heritage Site (with a 100-km buffer) near Sydney, Australia. Note there are patches of high and low species richness, suggestive of the need for a spatial model

Strategies similar to those used for spatial analysis can be used for longitudinal data analysis, with one important difference. For longitudinal data, there are *clusters* of (independent) subjects on which repeated measures are taken. In Ingo's case (Exercise 7.1), repeated measures were taken in each of eight plots, so while there was temporal dependence among repeated measures within plots, responses were still assumed independent across plots. There is no such clustering in the spatial context—we typically assume all observations are spatially correlated with each other (which is like what happens in a time series model). In effect, we have one big, fat cluster of correlated observations.

Four main strategies are used for spatial modelling: ignore it, use spatial smoothers, use autoregressive models, or use spatially structured random effects.

Ignore it: The biggest risk from ignoring a correlation is that uncertainty in predictions will be underestimated. You can also make estimates that are a lot poorer if there is a spatial structure and it is not used to improve predictions. This is OK as a strategy, though, if you cannot see evidence of spatial structure in residuals. One scenario where this might happen is if samples are sufficiently far apart that spatial autocorrelation is negligible. But you should use your data to check this!

Spatial smoothers: We could include a bivariate smoother (Chap. 8) for latitude and longitude in the model in order to approximately account for spatial structure. This does not always work, though, e.g. in the next chapter we will see an example of where a temporal structure is still evident in a residual plot, even though a smoother for time is in the model (Code Box 8.8). The same thing could happen in a spatial context. The main advantage of using smoothers is that if it works, it is the simplest method of handling spatial structure.

Autoregressive models: This is a commonly used class of models for spatial data, where we predict the response at a location as a function of the response at nearby locations (as well as predictor variables). This method won't be discussed in detail here because it is a bit of a departure from the techniques we have used elsewhere; for a quick description see Dormann et al. (2007).

Spatially structured correlation: This is closely analogous to the use of temporally structured correlation for longitudinal data in Sect. 7.1; it simply involves adding a random effect (or changing the error term) so that it is spatially structured. This can be implemented in `R` using the `lme` or `gls` function in the `nlme` package. Generalised least squares can be understood as doing a linear model on contrasts that have been calculated in such a way that they are independent of each other (as in Maths Box 7.2).

There are a few options for correlation structures that could be assumed, and standard model selection tools could be used to choose between them. In Code Box 7.5, an exponential autocorrelation is assumed, presupposing the autocorrelation decays exponentially towards zero as in Fig. 7.4. Sometimes, a response does not vary smoothly in space but seems to have some random ("white") noise on top of the spatial signal. In this situation, we do not want to assume that observations at a distance of zero from each other have a correlation of one, which can be done by including what is known as a "nugget" in the model—a white noise component. This was done in Code Box 7.5 using `corExpo(~x+y,nugget=TRUE)`.

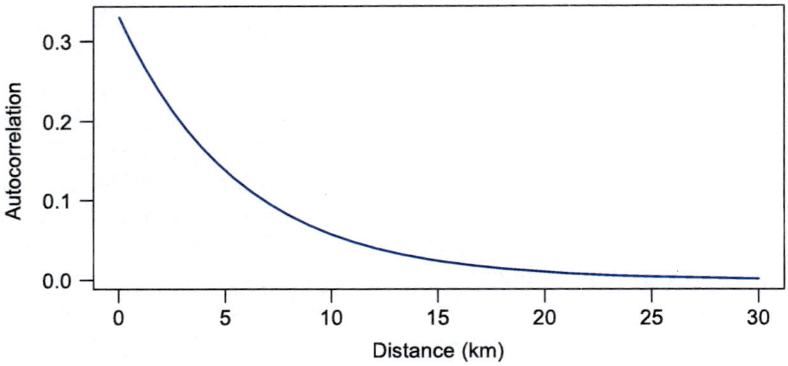

Fig. 7.4: An example of how spatial autocorrelation might decay as distance between samples increases. A correlation between samples at a distance d from each other is assumed here to follow an exponential model, $(1-n)e^{-d/r}$, where n is the nugget effect and r the range parameter. The curve plotted here uses $n = 0.67$ and $r = 5.7$, as was the case for Ian's richness data in Code Box 7.6. The correlation becomes negligible at a distance of about 15 km

Code Box 7.5: Model Selection to Choose Predictors, and a Spatial Model, for Ian's Richness Data

Quantitative climatic predictors should be added as quadratic rather than linear terms, to enable an "optimum" climate rather than a function that is always increasing or decreasing as temperature or rainfall increases. But should this include interactions between climate predictors or not?

```
> data(Myrtaceae)
> Myrtaceae$logrich=log(Myrtaceae$richness+1)
> ft_rich = lm(logrich~soil+poly(TMP_MAX,TMP_MIN,RAIN_ANN,degree=2),
     data=Myrtaceae)
> ft_richAdd = lm(logrich~soil+poly(TMP_MAX,degree=2)+
     poly(TMP_MIN,degree=2)+poly(RAIN_ANN,degree=2), data=Myrtaceae)
> BIC(ft_rich,ft_richAdd)
           df      BIC
ft_rich    19 1014.686
ft_richAdd 16 1002.806
```

This suggests that we don't need interactions between environmental predictors. But further study suggested an apparent weak spatial signal in the residuals, as indicated later in Code Box 7.7. So consider a spatial model with exponential spatial correlation (potentially including a nugget also):

```
> library(nlme)
> richForm = logrich~soil+poly(TMP_MAX,degree=2)+poly(TMP_MIN,degree
     =2)+
     poly(RAIN_ANN,degree=2)
> ft_richExp = gls(richForm,data=Myrtaceae,correlation=corExp(form
  =~X+Y))
> ft_richNugg = gls(richForm,data=Myrtaceae,
```

```
          correlation=corExp(form=~X+Y,nugget=TRUE))
> BIC(ft_richExp,ft_richNugg)
            df       BIC
ft_richExp  17 1036.2154
ft_richNugg 18  979.5212
```

These models take several minutes to run!

The model with a nugget in it has a much smaller BIC, suggesting that species richness does not vary smoothly over space. Note the BIC of this model is slightly smaller than for the non-spatial model, suggesting that it is worth including spatial terms.

Spatial models of this form can take a long time to fit! Code Box 7.5 contains 1000 observations, meaning that there are a lot (half a million) of pairwise correlations to compute, and the model takes several minutes to run. The computational complexity of these models increases rapidly with sample size, such that it becomes infeasible to fit this sort of model when there are hundreds of thousands of observations. There are, however, some nice tricks to handle larger models—some autoregressive models have fewer parameters to estimate so can be fitted for larger datasets, and there are also tricks like fixed rank kriging (Cressie & Johannesson, 2008) or using predictive processes (Banerjee et al., 2008; Finley et al., 2009) that make some simplifying approximations so the model can be fitted using a smaller number of random terms.

The impact of including spatial terms in a model tends to be to increase the size of standard errors and reduce the significance of effects in the model. This is seen in Code Box 7.6, where one of the terms in the model changed from having a P-value of 0.0003 to being marginally significant!

Code Box 7.6: Inferences from Spatial and Non-Spatial Models for Ian's Richness Data

```
> ft_richNugg
...
Correlation Structure: Exponential spatial correlation
 Formula: ~X + Y
 Parameter estimate(s):
   range   nugget
5.691041 0.649786
Degrees of freedom: 1000 total; 985 residual
Residual standard error: 0.3829306
```

The model includes a nugget effect of 0.65 (meaning that the correlation increases to a maximum value of $1 - 0.65 = 0.35$ at a distance of zero) and has a range parameter of 5.7 (meaning that as distance increases by 5.7 km, the correlation between points decreases by a factor of $e^{-1} \approx 37\%$). This is plotted in Fig. 7.4.

```
> anova(ft_richAdd)
                          Df  Sum Sq Mean Sq F value    Pr(>F)
soil                       8  15.075 1.88438 12.9887 < 2.2e-16 ***
poly(TMP_MAX, degree = 2)  2   5.551 2.77533 19.1299 7.068e-09 ***
poly(TMP_MIN, degree = 2)  2   5.162 2.58082 17.7892 2.573e-08 ***
poly(RAIN_ANN, degree = 2) 2   2.382 1.19081  8.2081 0.0002915 ***
```

```
Residuals                        985 142.902 0.14508

> anova(ft_richNugg)
                                 numDF  F-value p-value
(Intercept)                          1 5687.895  <.0001
soil                                 8    5.926  <.0001
poly(TMP_MAX, degree = 2)            2   10.528  <.0001
poly(TMP_MIN, degree = 2)            2    7.672  0.0005
poly(RAIN_ANN, degree = 2)           2    2.719  0.0664
```

Notice that the F statistics are a factor of two or three smaller in the spatial model. It is worth re-running model selection algorithms to see if all these predictors are still worth including, e.g. the effect of precipitation is now only marginally significant, so inclusion of this term may no longer lower the BIC.

7.2.1 Minding Your Ps and Qs—The Correlogram

The spaghetti plot, which we considered as a diagnostic tool for longitudinal data, has no equivalent in the spatial world, for a couple of reasons. Most fundamentally, we do not have independent clusters of observations we can plot as separate lines. The idea behind a temporal autocorrelation function does, however, translate to the spatial context.

The spatial correlogram is a tool that works much like the temporal autocorrelation function—it estimates the correlation between pairs of observations different distances apart and plots this autocorrelation as a function of distance between observations. The spatial autocorrelation might be close to one for neighbouring observations (but then again it might not be), and correlation typically decays towards zero as distance increases. There are a few techniques for computing spatial autocorrelation; one of the more common is known as Moran's I statistic. For a correlogram, pairs of observations are "binned" into groups depending on how far apart they are, then Moran's I is computed within each distance class.

There are a lot of software options for computing correlograms, the `correlog` function in the `pgirmess` R package being one example (Code Box 7.7), that allow the user to control how the data are "binned" into groups of distances for computation of the statistic.

When fitting a regression model, note the correlogram should be constructed using the *residuals* after modelling the response of interest as a function of its predictors, as in (b) from Code Box 7.7. The response is strongly spatially correlated, as in (a) from Code Box 7.7, but a lot of this spatial structure is due to the relationship between richness and climatic predictors, which are themselves strongly spatially structured and capable of explaining much of the spatial signal. Hence the residuals seem to have a much weaker spatial autocorrelation, which seems to operate over shorter distances. In some situations, the predictors may seem to explain all of the spatial autocorrelation, such that the autocorrelation curve of residuals would be flat and near zero, in which case there would be no need to fit a spatial model.

A related tool is the variogram, which is like a correlogram but flipped around to compute variances of differences between pairs of points. This means that as distance increases, the variance increases. Variograms will not be considered further here in order to keep the development analogous to what is usually done for longitudinal data, where the correlogram is more common.

Code Box 7.7: Spatial Correlogram for Ian's Species Richness Data

Correlograms will be computed for the richness data (log(y + 1)-transformed) and for residuals from a linear model as a function of soil type and additive, quadratic climatic predictors:

```
library(pgirmess)
corRich = with(Myrtaceae,correlog(cbind(X,Y),logrich))
plot(corRich,xlim=c(0,150),ylim=c(-0.05,0.2))
abline(h=0,col="grey90")

Myrtaceae$resid = residuals(ft_richAdd)
corRichResid = with(Myrtaceae,correlog(cbind(X,Y),resid))
plot(corRichResid,xlim=c(0,150),ylim=c(-0.05,0.2))
abline(h=0,col="grey90")
```

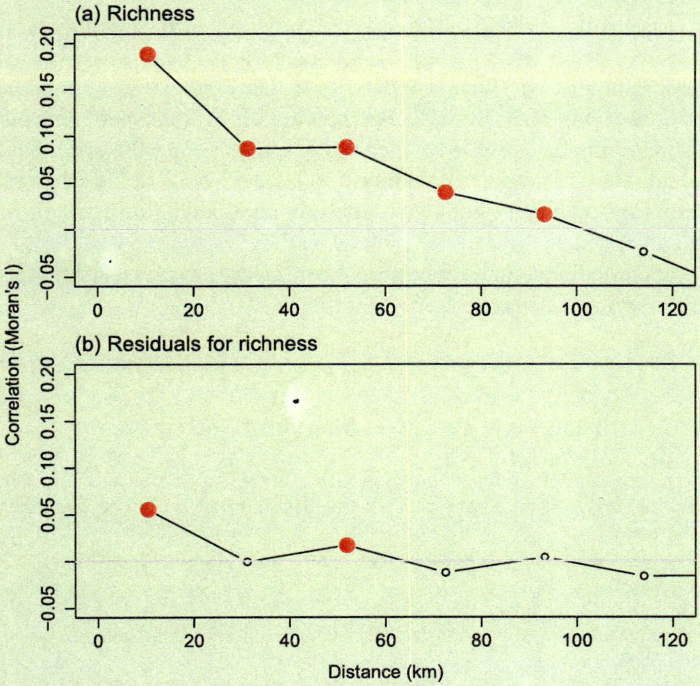

There is still spatial autocorrelation after including climatic predictors, but less so.

7.3 Phylogenetically Structured Data

Another way for data to have a structured correlation is if the subjects are species (or some other taxonomic group), with some more closely related than others due to shared ancestry. The evolutionary relationships between any given set of species can be mapped out on a phylogenetic tree (Bear et al., 2017), which reconstructs as best we can the evolutionary path from a common ancestor to all of the species used in a study (usually based on molecular analysis of DNA sequences). Some pairs of species will be closer together on the tree than others, which is usually measured in terms of the number and length of shared branches until you get to a common ancestor. If a response is measured that is expected to take similar values for closely related species, there is *phylogenetically structured* correlation in the data, with autocorrelation that is a function of phylogenetic distance.

As an example, consider Exercise 7.5. Data were collected across 71 species of bird, some of which were more closely related than others. More closely related species tended to be closer in body or egg size (Code Box 7.8). Hence, we expect phylogenetically structured correlation in the data, and it is unlikely we would be able to satisfy the independence assumption that we usually need in linear modelling.

Exercise 7.5: Egg Size When Dads Incubate
Egg size is typically limited by the size of the adult bird, probably in part because a parent needs to sit on eggs to incubate them. For species where the male does the incubating, Terje wondered whether egg size was specifically limited by male body size. So he collected data on 71 species of shorebird, where the male incubates the egg, measuring egg size and the size of adult males and females (Lislevand & Thomas, 2006). He wants to know whether, after accounting for female body size, there is an association between male body size and egg size.

What sort of model might be appropriate here? Can see you see any potential problems satisfying independence assumptions?

Code Box 7.8: Phylogenetic Tree of 71 Shorebird Species
We will use the phylosignal library to plot a phylogenetic tree and map variables onto it. The shorebird data of Exercise 7.5 are available in the caper package. The species trait data (shorebird.data) and phylogenetic tree data (shorebird.tree) are stored as separate data frames.

To load the data and plot a phylogenetic tree against size variables for each species:

```
library(caper)
data(shorebird)
shore4d=phylobase::phylo4d(shorebird.tree,shorebird.data)
library(phylosignal)
barplot.phylo4d(shore4d,c("Egg.Mass","F.Mass","M.Mass"))
```

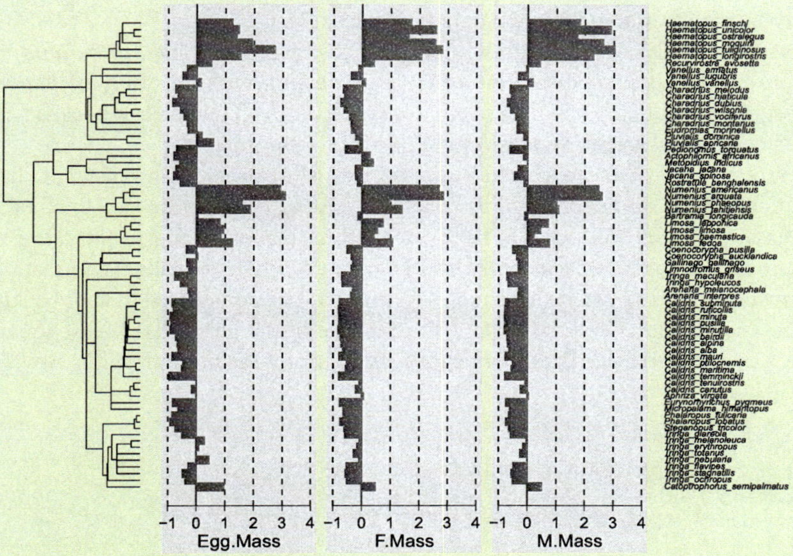

The phylogenetic tree is on the left of this plot, and species connected by a common ancestor closer to the tips are more closely related. Note that closely related species seem to have similar sizes (e.g. *Haematopus* species, oystercatchers, are all large). Note also that branches have different lengths—because different pairs of species are thought to have diverged from a common ancestor at different times during their evolutionary history.

When measuring a response across multiple species it is commonly (but not always) the case that there is a phylogenetic correlation, just as when measuring a response across space it is commonly (but not always) the case that there is a spatial correlation. As previously, this can be endogenous or exogenous autocorrelation. Many variables that can be measured across species, or *species traits*, have some phylogenetic signal, with more similar values for more closely related species (endogenous). This is more likely to be the case for responses that vary over long (evolutionary) timescales rather than short (ecological) timescales. But some responses can have a phylogenetic signal simply because they respond to phylogenetically structured predictors (exogenous). For example, abundance can vary considerably from one season to the next (sometimes from one day to the next! Morrisey et al., 1992), and so we might not expect to see trends in abundance across evolutionary timescales (at least not due to an endogenous process). But if abundance were associated with species traits that do carry a phylogenetic signal, we might see a phylogenetic correlation in abundance when these traits are not included in the model (Li & Ives, 2017).

A phylogenetically structured correlation can be accounted for much as temporal or spatial autocorrelation is—by fitting a generalised least-squares (GLS) model or by fitting a mixed model with phylogenetically structured random effects. This is often called *phylogenetic regression*, after Grafen (1989). A linear model with structured dependence in the response can be fitted by a GLS approach using the `caper` package, as in Code Box 7.10. Egg size data were log-transformed data given that this seemed to symmetrise the data (Code Box 7.9).

This type of analysis is sometimes referred to in ecology as a *comparative analysis*, and more broadly, the subdiscipline of ecology dealing with the study of patterns across species is often known as *comparative ecology* (presumably because this involves comparing across lineages).

There are many other approaches to comparative analysis, many of which have connections to the approach described earlier. An early and influential method of phylogenetic analysis (Felsenstein, 1985) is to assume that branching events along the tree are independent, so we can construct phylogenetically independent contrasts between species and between species means. This won't be considered further because it can be understood as equivalent to a special case of the generalised least-squares approach considered here (Blomberg et al., 2012), along the lines of Maths Box 7.2.

Code Box 7.9: Exploratory Analysis of Egg Size Data

For a fancy scatterplot matrix of log-transformed data:

```
library(GGally)
ggpairs(log(shorebird.data[,2:4]))
```

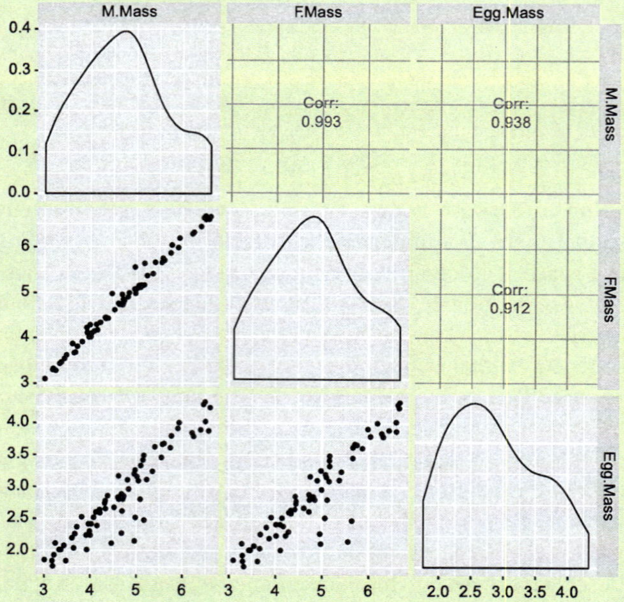

Note all correlations are quite high, in particular, there is multi-collinearity between male and female bird size. This will make it harder to detect an effect of one of these variables when the other is already in the model. Note also that the correlation of egg size with male body size is slightly larger than with female body size, which is in keeping with Nerje's hypothesis (Exercise 7.5).

Code Box 7.10: Comparative Analysis of Egg Size Data

```
> library(caper)
> shorebird = comparative.data(shorebird.tree, shorebird.data,
                    Species, vcv=TRUE)
> pgls_egg = pgls(log(Egg.Mass) ~ log(F.Mass)+log(M.Mass),
                 data=shorebird)
> summary(pgls_egg)
Branch length transformations:

kappa  [Fix]  : 1.000
lambda [Fix]  : 1.000
delta  [Fix]  : 1.000

Coefficients:
            Estimate Std. Error t value Pr(>|t|)
(Intercept) -0.37902    0.23172 -1.6357 0.106520
log(F.Mass) -0.22255    0.22081 -1.0079 0.317077
log(M.Mass)  0.89708    0.22246  4.0325 0.000142 ***
---
Signif. codes:  0 `***' 0.001 `**' 0.01 `*' 0.05 `.' 0.1 ` ' 1

Residual standard error: 0.03343 on 68 degrees of freedom
Multiple R-squared: 0.8476,Adjusted R-squared: 0.8431
F-statistic: 189.1 on 2 and 68 DF,  p-value: < 2.2e-16
```

What would you conclude from these analyses? (Provided assumptions are satisfied. . .)

7.3.1 Mind Your Ps and Qs in Phylogenetic Regression

The phylogenetic regression model used here is a type of linear model that (under a simplifying assumption that the tree is ultrametric, with constant total branch lengths from the root of the tree to each species) can be written as follows:

$$y_i \sim \mathcal{N}(\mu_i, \sigma^2)$$
$$\mu_i = \beta_0 + \mathbf{x}_i^T \boldsymbol{\beta}$$
$$\text{cov}(y_i, y_j) = \sigma^2 \left(\sum_{k=1}^{K} \left(\lambda l_{ij}^{(k)} \right)^{\kappa} \right)^{\delta}$$

where cov means covariance (basically, correlation times standard deviations). We assume a pair of species (i and j) share K branches on the tree, and the lengths of these branches are $l_{ij}^{(1)}, \ldots, l_{ij}^{(K)}$. The covariance between this pair of species is a function of these branch lengths and three scaling parameters λ, δ, κ, which can be estimated from the data. The λ parameter, attributed to Pagel (1997, 1999), measures the strength of phylogenetic signal (and is related to nugget effects in spatial statistics)—values near zero suggest no phylogenetic signal, values near one suggest most (co)variation across species is explained by phylogeny. The other parameters determine the extent to which correlation across species is determined by recent divergences vs ancestral ones. By default, the `pgls` function of Code Box 7.10 assumes all parameters are one, meaning that covariances are proportional to the total shared branch length (assuming traits evolve entirely via so-called Brownian motion).

So in summary, we have the following assumptions:

1. The observed y-values are *phylogenetically dependent* in such a way that, after accounting for predictors, the correlation between values is a fixed function of shared branch lengths (as earlier).
2. The y-values are *normally distributed* with *constant variance*

$$y_i \sim \mathcal{N}(\mu_i, \sigma^2)$$

   (Strictly speaking, variance is a function of the total branch length of a species from the root of the tree, so σ sometimes is assumed to vary across species.)
3. *linearity*—the effect of the covariate on the mean of y is *linear*, and the effect of any factor is *additive*. For example, for the egg size data

$$\mu_i = \beta_0 + \text{female}_i \beta_{\text{female}} + \text{male}_i \beta_{\text{male}}$$

As usual, we can use residuals to diagnose these assumptions. We can plot residuals against fitted values and check for no pattern, or construct a correlogram of residuals, as in Code Box 7.11. Given the small sample sizes often involved, Keck et al. (2016) suggest using a smoother (Chap. 8) when constructing a phylogenetic correlogram. Note that residuals should be used when constructing a correlogram, as opposed to the response variable (egg size), because the correlation assumptions in our model are made after accounting for predictors, i.e. they are on residuals after taking out the effects of predictors. Even though we had a strong phylogenetic signal in egg size, we should check our residuals also because the use of body size as a predictor could potentially remove much of this phylogenetic signal.

Under the foregoing assumptions, if $\kappa = 1$, and if the total path length from the root of the tree to each species were a constant value L, we would expect a phylogenetic correlogram to decay towards zero as a predictable function of phylogenetic distance (usually defined as total branch length shared by a pair of species):

$$(1 - \lambda) + \lambda \left(1 - \frac{d}{L}\right)^{\delta}$$

for a pair of species with phylogenetic distance d. This function is plotted in Fig. 7.5 for a few different values of δ. If $\delta = 1$, we get a straight line (Brownian motion). The correlogram in Code Box 7.11 does not seem far from a straight line, so the Brownian motion assumption ($\delta = 1$) seems reasonable here.

Code Box 7.11: Residual Diagnostics for Egg Size Data

```
par(mfrow=c(2,2))
plot(pgls_egg)
```

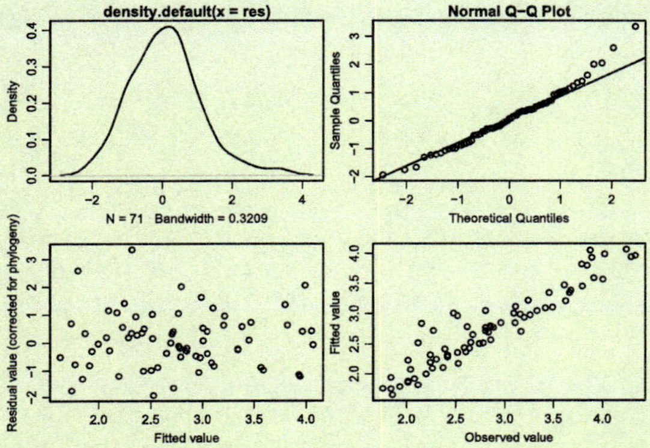

The first two of these plots check normality (which, as previously, is not critical!) and the first in the second row is our residual vs fits plot, suggesting no pattern.

For a phylogenetic correlogram of residuals

```
res.df = data.frame(Species = shorebird.data$Species,
                    res = residuals(pgls_egg))
res4d = phylobase::phylo4d(shorebird.tree,res.df)
res.pg = phyloCorrelogram(res4d,trait="res")
plot(res.pg)
```

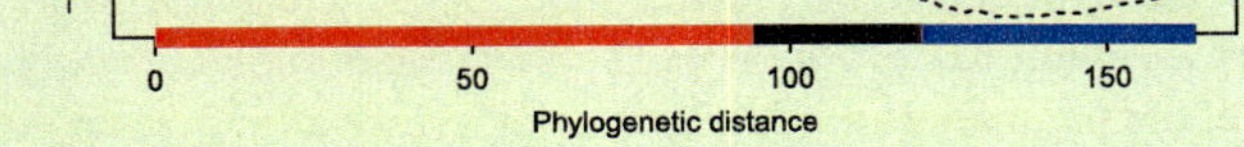

There is a significant amount of autocorrelation at short phylogenetic distances (less than 100), rising to high correlation values (0.7).

Another way to get at the question of whether the phylogenetic dependence model is adequate is to refit the model using different assumptions:

- The `pgls` model can easily be refitted estimating some of its phylogenetic parameters from the data and returning approximate confidence intervals for these parameters. Of particular interest is whether confidence intervals for λ contain zero (no phylogenetic signal) or one (all error is phylogenetic). Note, however, that

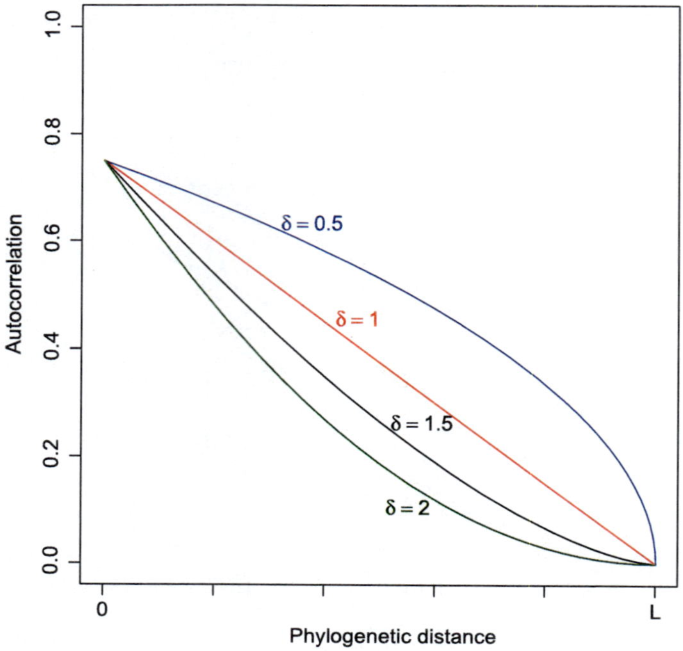

Fig. 7.5: Assumed phylogenetic correlograms used in `pgls` for different values of δ, assuming $\kappa = 1$. In this situation there is a direct relationship between phylogenetic distance (total shared branch length) and phylogenetic autocorrelation. The value of λ determines the y-intercept, and we used $\lambda = 0.75$ here

δ and κ are known to be biased, as non-linear parameters in covariance models often are (Freckleton et al., 2002), so if their values are of particular interest they should be estimated in a different way.

- Refit the model assuming independence and use model selection techniques to see which model is preferred by the data (and maybe check results to see if they are different). `AIC` and `BIC` work on `pgls` objects, although currently they don't seem to count the parameters used in estimating error variances and correlations.

Both of these options are considered further in Exercise 7.6.

Exercise 7.6: Comparative Analysis of Egg Size Data Revisited
Return to the egg size data analysed in Code Box 7.10. As previously, there are a few parameters used in modelling the phylogenetic correlation structure (λ, δ, κ) and choosing different values for these parameters changes the value of the assumed correlation between species. Let's look at how robust results are to different choices of assumed phylogenetic structure.

Refit the model allowing λ to be estimated from the data (using `lambda="ML"`) or allowing δ to be estimated from the data. Does this change the results and their interpretation?

Now fit a linear model ignoring phylogeny, via `lm`. What happens here? Is this what you would expect? Look at the log-likelihood (using the `logLik` function) to help decide which of these models is a better fit to the data.

7.4 Confounding—Where Is the Fixed Effect You Love?

Relatively recently there has been greater appreciation of the issue of confounding when using structured random effects in models—as Hodges and Reich (2010) put it, a structured random effect has the potential to "mess with the fixed effect you love". Mixed models work like multiple regression, in the sense that any terms that enter into the model have their effects estimated conditionally on everything else that is in the model. So we can in effect get multi-collinearity if some of our predictors share a similarly structured correlation to that assumed in the model. This will often happen—in our spatial example (Exercise 7.4), we used temperature as a predictor, which will be highly spatial, and in our phylogenetic example (Exercise 7.5), we used body mass as a predictor, which had a strong phylogenetic signal. In these situations, inclusion of both the predictor and the structured dependence term with which it is correlated will weaken each other's estimated effects, much as collinearity weakens signals in multiple regression. This is a type of confounding—there are two possible explanations for a pattern in the data, and it is hard to tell which is the cause (without experimentation). In Exercise 7.4, is richness varying spatially largely because of its association with temperature, or is the association with temperature largely because of spatial autocorrelation? In Exercise 7.5, is the strong phylogenetic signal in egg size largely because of an association with body mass, or is the association with body mass largely because of the strong phylogenetic signal?

Maths Box 7.4: Confounding in Linear Models with Structured Correlation

Confounding is easier to explain if we split the error into two terms: δ_i, which captures structured dependence, and ϵ_i, which captures independent noise (sometimes called measurement error or a nugget effect). A linear model for the ith observation is

$$y_i = \beta_0 + \mathbf{x}_i'\boldsymbol{\beta} + \delta_i + \epsilon_i$$

where we assume ϵ_i are independent normal variables with variance σ^2, and we assume the vector of correlated errors $\boldsymbol{\delta}$ has a multivariate normal distribution with mean $\mathbf{0}$ and variance–covariance matrix $\boldsymbol{\Sigma}$. (This representation is actually equivalent to that used in Maths Box 7.2 since sums of normal variables are normal.) We can use contrasts $\mathbf{\Sigma}^{1/2}$ to rewrite this in the form of a linear mixed model (Chap. 6):

$$y_i = \beta_0 + \mathbf{x}_i'\boldsymbol{\beta} + \mathbf{z}_i\mathbf{b} + \epsilon_i$$

where $\mathbf{b} = \mathbf{\Sigma}^{-1/2}\boldsymbol{\delta}$ are independent random effects, and $\mathbf{z}_i$ is the ith row of the matrix of contrasts $\mathbf{\Sigma}^{1/2}$ that captures structured dependence in the data. Now that the random effects are independent, we can treat this model as a standard linear mixed model and apply ideas from earlier chapters.

If the predictors $\mathbf{x}$ have a similar correlation structure to that assumed for the response, then $\mathbf{x}$ will be correlated with the contrast matrix $\mathbf{z}$ that was designed to capture the structured dependence.

But recall that in a linear model, coefficients estimate the effects of predictors *conditional* on all other terms in the model. So, for example, in a spatial model, the coefficient of $\mathbf{x}$ will tell us only about the effects of $\mathbf{x}$ on responses that are "non-spatial", not able to be explained by $\mathbf{z}$. This is a different quantity to the *marginal* effect of $\mathbf{x}$, if we had not conditioned on $\mathbf{z}$. The mathematics of missing predictors (Maths Box 4.1, Eq. 4.1) can tell us the extent to which fixed effects coefficients change when you include structured correlation in the model.

The models considered so far in this chapter will behave like multiple regression and effectively condition on the assumed correlation structure when estimating fixed effects. That is, a model for species richness will estimate the effect of temperature after removing the variation in temperature that is spatially structured. A model for egg size will estimate the effect of body mass after removing the variation in body mass that is phylogenetically structured. Often dealing with confounding in this way is appropriate if the question of interest requires studying the effect of a predictor conditional on other terms in the model (just as multiple regression is often appropriate, in attempts to remove the effects of confounding variables).

Sometimes this sort of effect is not desirable. We might wish to describe how richness is associated with climatic variables, estimating the full (marginal) climate effect rather than attributing part of it to a spatial error term. But in a spatial model, the climate term represents only the portion of the climate effect that is uncorrelated with the spatial random effect, so it would be smaller (sometimes surprisingly so) and it would be harder to detect associations with climate. Hodges and Reich (2010) proposed an approach that can adjust for spatial dependence but not attribute variation in response to the structured random effect, unless it is uncorrelated with predictors in the model.

A well-known debate in the comparative ecology literature (Westoby et al., 1995; Harvey et al., 1995) can be understood as arising around this same issue but in the context of phylogenetically structured correlations. By the early 1990s, comparative methods had been developed for the analysis of data across multiple species (or other taxa) to account for differing levels of correlation across species due to shared phylogeny. Westoby et al. (1995) argued that these methods should not necessarily be used by default in comparative studies, essentially because these methods look for effects on a species trait conditional on phylogeny, e.g. estimating a cross-species association between seed size and seed number beyond that which can be explained by phylogeny. Their argument was based around the idea that ecological determinants of a trait would have been present over the evolution of the species, rather than just in the present day, so some of the trait effects that have been attributed to phylogeny are confounded with ecologically relevant effects. This is visualised nicely in Figure 1 of Westoby et al. (1995). How this confounded information should be treated is subject to interpretation, and Westoby et al. (1995) argue it is related to the type of question one is trying to answer—very much as Hodges and Reich (2010) argued in the context of spatial models. In response, Harvey et al. (1995) emphasised the importance of satisfying the independence assumption in analyses (or accounting for dependence when it is there), but they did not propose a way forward to estimate or address this "phylogenetic confounding". Perhaps the methods of Hodges and Reich (2010) could be adapted to the phylogenetic context for this purpose. In the meantime, many comparative ecologists analyse their data with and without phylogenetic terms in the model, to understand the extent to which cross-species patterns they observe are confounded with phylogeny.

7.5 Further Reading

This chapter provided a very brief introduction to dependent data, which is a very big area. We have focused on the use of GLS and random effects to account for structured correlation and the use of correlograms to diagnose the type of correlation we are dealing with. However, other techniques and other data types require different approaches. For more details, there are entire separate texts on each of spatial, longitudinal, and comparative analyses. A well-known text on repeated measures in time is Diggle et al. (2002), and for spatial data a comprehensive but quite technical text is that by Cressie (2015). For a relatively quick tour of spatial tools for ecology see Dormann et al. (2007), who fitted a range of different methods to simulated data, with a quick description of each. For the phylogenetic case, an accessible introduction is Symonds and Blomberg (2014). For time series data, a good introductory-level text is Hyndman and Athanasopoulos (2014).

Chapter 8
Wiggly Models

Recall from Sect. 4.4.2 that a "linear model" does not need to be linear. Mathematically, we say a model is linear if it is a linear function of its coefficients ("something times β_1 plus something times β_2 ..."). But if we include non-linear functions of x as predictors, we can use this framework to fit a non-linear function of x to data. The simplest example of this is including quadratic terms in the model, as in Fig. 8.1, or cubic terms. This approach has a theoretical basis as a Taylor series approximation (Maths Box 8.1).

Maths Box 8.1: Polynomial predictors as a Taylor series approximation
Let's say we want to model the mean of a response as some unknown function of a predictor $f(x)$. One way to do this is to use a *Taylor series*—any smooth function $f(x)$ can be written

$$f(x) = b_0 + b_1(x-a) + b_2(x-a)^2 + b_3(x-a)^3 + \dots$$

for any a and suitably chosen values of b_i. If we drop some of the higher-order terms (e.g. just taking the first three terms from this expansion), we get a *local approximation* to the function $f(x)$ near a. The approximation works better when we use more terms (hence are ignoring less higher-order terms). It is a local approximation around a because it works better for values of x that are close to a since $x - a$ is small, so the higher-order terms we ignore are also small.

Taylor approximations are linear in the coefficients, so they can be fitted using linear model techniques. When we use just $\mathbf{x}$ as a predictor in a linear model, we get a linear approximation, which we can think of as using just the first two terms from a Taylor series expansion to approximate $f(x)$ (estimating

D. I. Warton, *Eco-Stats: Data Analysis in Ecology*, Methods in Statistical Ecology,
https://doi.org/10.1007/978-3-030-88443-7_8

b_0 and b_1 from data). If we add x^2 to the linear model as a second predictor, we use the first three terms from a Taylor approximation (estimating b_0, b_1, and b_2 from data) and get a slightly better approximation. Adding x^3 too is better still, and so on.

Note that while Taylor expansions work well locally, they can be quite bad at predicting extreme values of x. This is especially the case for higher-order polynomials (powers higher than 3), which tend to be avoided in data analysis for this reason, especially if there is a need to extrapolate beyond the range of observed x-values.

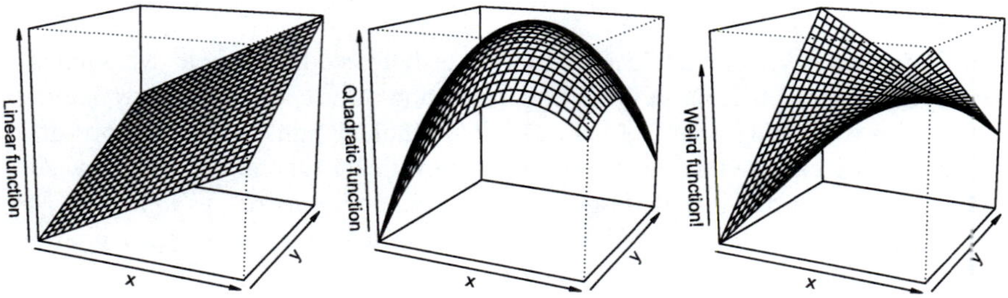

Fig. 8.1: Linear models are not necessarily linear functions of explanatory variables (but they are linear in terms of the parameters). All of the above response surfaces can be fitted using linear modelling software: from left-to-right we have we have a linear model, a model with all quadratic terms, and a saddle-shaped surface

Lots of other cool functions of x can be fitted in a linear modelling framework. We will look at two important examples—using spline smoothers to fit non-linear functions in a flexible way and trigonometric functions to handle cyclical variables. Cyclical variables are something that most ecologists have to deal with because a lot of predictors return to where they started from (seasonality, time of day, aspect, ...), and this issue needs to be thought about carefully when analysing data.

8.1 Spline Smoothers

A good way to relax the linearity assumption is often to break down the explanatory variable into pieces (at *knots*) and fit a multiple regression against all of these pieces. A piecewise approach usually works better than using polynomials (e.g. x, x^2), because we are essentially doing a local approximation (as in Maths Box 8.1) for each piece, instead of trying to apply a single approximation across a broad range of values for x.

Key Point
Do you have a regression with a potentially non-linear response and data at lots of different values of your predictor (x)? In that case, you can get fancy and try fitting a smoother to your data, e.g. using software for generalised additive models (GAMs).

A simple example of this is piecewise linear model fits (as used in the well-known MAXENT software, Phillips et al., 2006). A specific example of a piecewise linear fit is in Fig. 8.2. Piecewise linear fits are a bit old school, at least for functions of one variable. They don't look smooth, and in most problems a "kinky" function is not realistic, e.g. why should the red line in Fig. 8.2 suddenly change slope in the year 2000?

Exercise 8.1: Annual carbon dioxide measurements at Mauna Loa observatory
Mauna Loa is a volcano in Hawaii that has an observatory near its peak (over 4000 m above sea level) which has been taking atmospheric measurements since the late 1950's. Because of its remoteness (being in the middle of the Pacific) and the length of time data have been collected for it, this is a famous dataset presenting compelling instrumental evidence of climate change over the last half-century. We will study carbon dioxide measurements presented as monthly averages in January every year over a 60-year period and how they change over time. Data are available in the `ecostats` package as `maunaloa`, thanks to the Global Monitoring Laboratory, within the US National Oceanic and Atmospheric Administration (NOAA). Updated data can be obtained from the NOAA website (`https://www.esrl.noaa.gov/gmd/ccgg/trends/data.html`).

How many response variables? What type of problem—hypothesis testing, model selection, estimation, predictive modelling?

A more common approach is to keep the function smooth by fitting a piecewise cubic[1] function, a function where only the cubic term is broken into pieces, the linear and quadratic terms are not. This gives the curve a smooth appearance, as in the blue curve of Fig. 8.2, because the slope and curvature of the function (as defined mathematically by the first and second derivatives) change continuously and don't have any jumps in them (Maths Box 8.2). These functions are known as *spline smoothers*. We can ensure these curves won't be very wiggly by fitting the curve using penalised likelihood, as in Sect. 5.6 (but usually using a quadratic penalty on parameters, not a LASSO penalty). This idea was introduced in the early 1990s by

[1] A cubic function is a third-order polynomial, e.g. $y = \beta_0 + \beta_1 x + \beta_2 x^2 + \beta_3 x^3$.

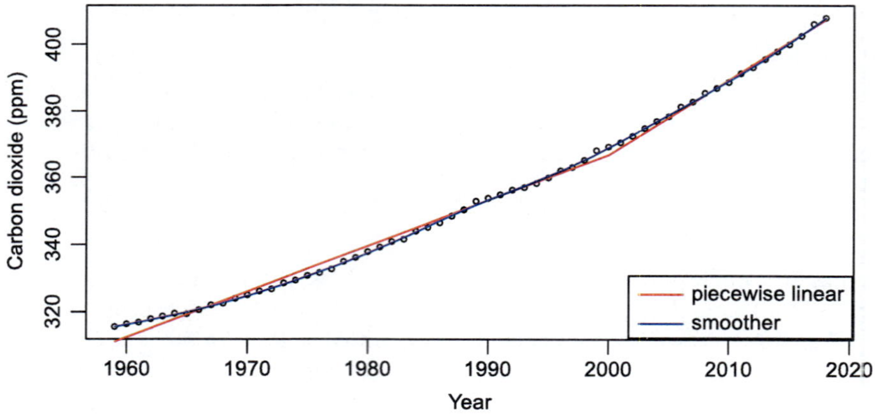

Fig. 8.2: Non-linear models fitted to the annual Mauna Loa data of Exercise 8.1. A piecewise linear model is fitted in red with a change in slope at the year 2000. Note the fit is a bit weird, with a sudden change in slope in the year 2000. A spline smoother is fitted in blue, which smooths out the kink

Hastie and Tibshirani (1990) and quickly became a mainstay in applied statistics. It was popularised in ecology, especially for modelling species' distributions, by papers like Yee and Mitchell (1991) and, in particular, Guisan and Zimmerman (2000). An excellent modern text on the subject is Wood (2017).

Maths Box 8.2: Spline smoothers

A piecewise linear model can be fitted by choosing K knots, $t_1, \ldots, t_K$, and using predictors of the form $b_k(x) = \max(x - t_k, 0)$ in the model. These predictors take the value zero if $x < t_k$; then they "kick in" at the knot t_k as a linear trend. They are sometimes called "hockey-stick functions", given their appearance, as shown below.

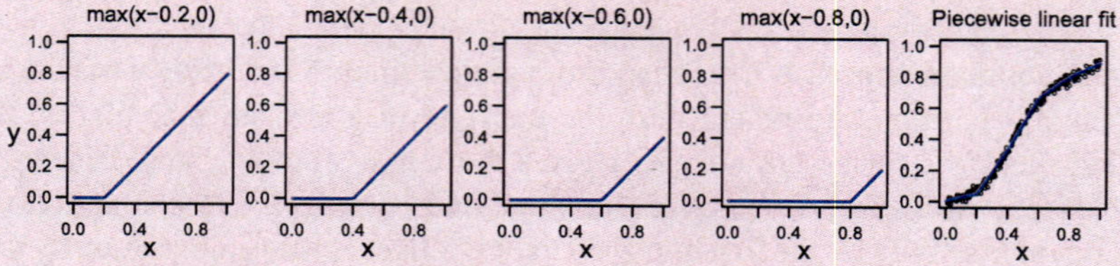

We fit the linear model

$$\mu_y = \beta_0 + x\beta + \sum_{k=1}^{K} b_k(x)\gamma_k$$

The prediction function from this model has "kinks" in it at the knots. This is illustrated above (right) for a model with knots at 0.2, 0.4, 0.6, 0.8.

The function telling us the slope of this model (the *gradient function*) is $\beta + \sum_{t_k \leq x} \gamma_k$, which is not continuous. The gradient starts with a slope of β, then at $x = t_1$ the slope suddenly jumps to $\beta + \gamma_1$, then at $x = t_2$ it jumps to $\beta + \gamma_1 + \gamma_2$, and so forth.

We can get a smooth, kink-free fit using piecewise polynomials of order two or higher; piecewise cubics are the most commonly used. The model is as follows:

$$\mu_y = \beta_0 + x\beta_1 + x^2\beta_2 + x^3\beta_3 + \sum_{k=1}^{K} b_k(x)^3\gamma_k$$

Below are the *basis functions* $b_k(x)^3$ (with knots at 0.2, 0.4, 0.6, 0.8), and a model fit:

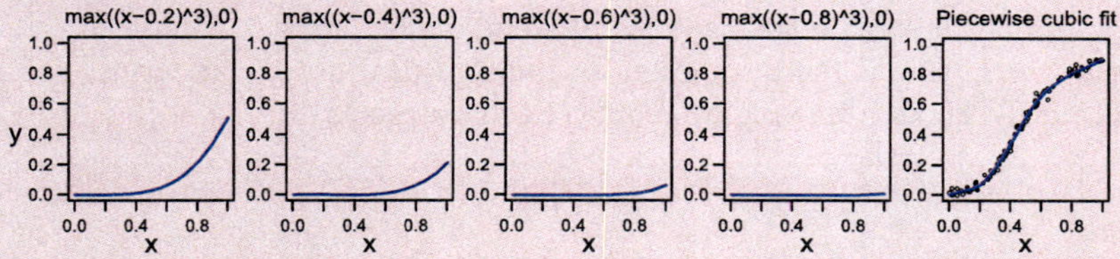

The fit from this model is smooth because its gradient function is continuous (it has no sudden jumps at any value of x). Specifically, the gradient is

$$\beta_1 + 2x\beta_2 + 3x^2\beta_3 + \sum_{t_k \leq x} 3b_k(x)^2 b_k'(x)\gamma_k$$

which is continuous, thanks to the presence of $b_k(x)$ in the summation term.

Why "splines"? The term spline comes from woodworking, especially shipbuilding, and the problem of bending straight sections of wood to form curves, e.g. a ship hull. This is done by fixing the wood to control points or "knots" as in Fig. 8.3. The statistician who came up with the term "spline smoothing" clearly spent a lot of time in his/her garage. . . .

The most common way to fit a model with a spline smoother in R is to use the `mgcv` package (using the methods of Wood, 2011). This package is well written and is quite fast considering what it does. A model with a spline smoother is often called a *generalised additive model (GAM)*, so the relevant function for fitting a smoother is called `gam`. You can add a spline smoother in the formula argument as `s(year)` (for a spline for `year`) as in Code Box 8.1. The `s` stands for `spline`.

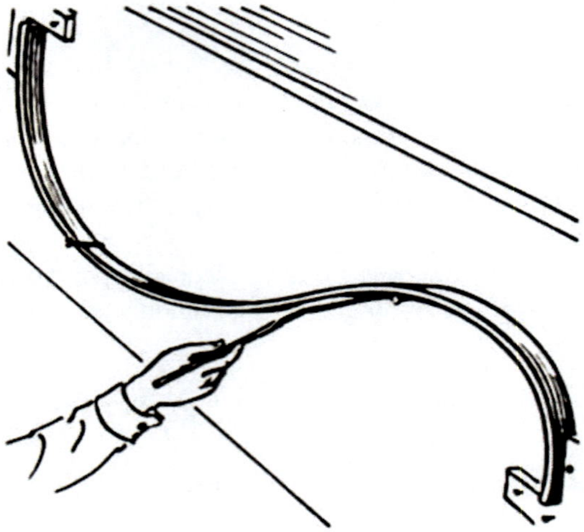

Fig. 8.3: Splines in woodwork—a piece of wood is bent around "knots" to create a smooth curve. This is the inspiration for the statistical use of the terms "spline smoother" and "knots" (drawing by Pearson Scott Foresman)

Code Box 8.1: Fitting a spline smoother to the Mauna Loa annual data of Exercise 8.1 on R.

```
> library(mgcv)
> data(maunaloa)
> maunaJan = maunaloa[maunaloa$month==1,]
> ft_maunagam=gam(co2~s(year), data=maunaJan)
> summary(ft_maunagam)
Parametric coefficients:
            Estimate Std. Error t value Pr(>|t|)
(Intercept) 356.68048    0.05378    6632   <2e-16 ***
---
Signif. codes:  0 `***' 0.001 `**' 0.01 `*' 0.05 `.' 0.1 ` ' 1

Approximate significance of smooth terms:
          edf Ref.df     F p-value
s(year) 8.427  8.908 33847  <2e-16 ***
---
Signif. codes:  0 `***' 0.001 `**' 0.01 `*' 0.05 `.' 0.1 ` ' 1

R-sq.(adj) =      1   Deviance explained =  100%
GCV = 0.2143  Scale est. = 0.18223   n = 63
```

The fit to the data is strikingly good (deviance explained 100%!), which does make sense if you look at the data (Fig. 8.2), which has little variation around a smooth trend, and the smoother fits the data almost exactly. The effective degrees of freedom are about 8.5, suggesting that about nine knots were used in making this smoother. (Digging deeper, you can confirm that exactly nine knots were used here.)

8.1.1 Choosing the Number of Knots for a GAM

When fitting a GAM, a decision has to be made as to how many knots to include in the model. This controls how wiggly the curve is—the more knots, the more wiggles. Choosing the number of knots is a form of bias–variance trade-off, like choosing the number of predictors in regression, or choosing λ in a LASSO regression. In fact, mathematically this is exactly like changing the number of predictors in a linear model. Too few knots might not fit the data well and might introduce bias; too many knots might overfit the data and increase the variance of predictions to new values (because the fit chases the data too much instead of following the broader trends).

In the `mgcv` package, you can specify an upper limit to the number of knots as `k`, an argument to the spline, e.g. using `s(lat,k=5)` in the formula argument. By default `k` is usually somewhere near 10 (but it depends). If you think your fit might need to be extra wiggly, it is a good idea to try changing `k` to a larger value (20? 50?) to see if it changes much, as in Code Box 8.3. The `gam` function by default will actually estimate the number of knots to use (treating the input k as the upper limit), and it will often end up using many fewer knots than this. The effective degrees of freedom (`edf` in Code Box 8.1) tell you roughly how many knots were used. This is not always a whole number, because it tries to adjust for the fact that we are using penalised likelihood (hence shrinking parameters a bit), and it usually reports a value slightly less than the actual number of knots used. You can also get a P-value out of some `gam` software, which is only approximate, because of the use of penalised likelihood methods in estimation.

8.1.2 Mind Your Ps and Qs

The assumptions of a GAM as in Code Box 8.1 are:

1. The observed y-values are *independent* given x—in other words, for Code Box 8.1, after accounting for year.
2. The y-values are *normally distributed* with *constant variance*:

$$y_i \sim \mathcal{N}(\mu_i, \sigma^2)$$

3. The mean of y varies as a smooth (and usually additive) function of x.

The third assumption has been relaxed from what we saw in linear models—rather than assuming linearity, we now assume the mean of y is some smooth function of x. That is, this model does not allow for sudden jumps in the mean response (which are rare but possible in some contexts). If there are multiple x variables, we usually assume the mean of y is an additive function of these (meaning no interaction between predictors), although this assumption can be relaxed (Sect. 8.2).

As previously (e.g. Page 26), independence is a critical assumption, and we can often do something about it in study design. However, the Mauna Loa data of

Fig. 8.2 are a time series, collected sequentially over time in the same place in the same way. There are concerns about the independence assumption here, because CO_2 measurements in a given year might be similar to those in the previous year for reasons not explained by the smoother. We can use the techniques of Chap. 7 to check for autocorrelation and, if needed, to account for it in modelling.

As previously, normality is rarely important, but we should check that residuals are not strongly skewed and that there aren't outliers that substantially affect the fitted model. The constant variance assumption remains as important as it was previously in terms of making valid inferences about predictions and about the nature of the association between y and x.

Residual plots, as usual, are a good idea for assumption checking; unfortunately, these aren't produced by the `plot` function, but they are produced by the `plotenvelope` function in the `ecostats` package, as in Code Box 8.2.

Code Box 8.2: Residual plot from a GAM of the annual Mauna Loa data of Exercise 8.1

The `plot` function doesn't give us diagnostic tools for a `gam` object; it plots the smoothers instead. But `plotenvelope` from the `ecostats` package works fine (although it will take a while to run on large datasets):

```
plotenvelope(ft_maunagam)
```

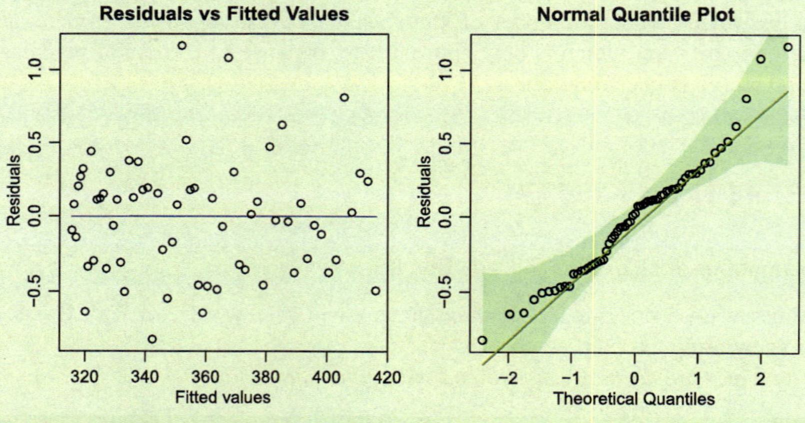

Do you think the assumptions are satisfied?

Was a smoother worthwhile, or could we have gotten away with a simpler fit (e.g. the piecewise linear fit of Fig. 8.2)? Is the default smoother OK, or should there be extra knots? To answer these questions, we would want to compare the various models, and one way to do this (if the models are nested) is to use the `anova` function. This would only give an approximate answer for models with smoothers (`anovaPB` could be used for a better answer, as in Sect. 6.5), but a bigger problem is that this approach is not a good match for the question being asked, which is a *model selection* problem. Better approaches, using some model selection techniques from Chap. 5, are illustrated in Code Box 8.3.

Code Box 8.3: Comparing curves for Mauna Loa data

The piecewise linear model of Fig. 8.2 and a GAM with extra knots also explain most (> 99.6%) of the variation in the Mauna Loa data. They can be fitted as follows:

```
> maunaJan$year00 = pmax(2000,maunaJan$year)
> ft_maunaPiece = lm(co2~year+year00,data=maunaJan)
> ft_maunagam20 = gam(co2~s(year,k=20), data=maunaJan)
> summary(ft_maunagam20)$edf # this gam added about 6 extra knots:
[1] 14.80805
```

Does either curve fit better than ft_maunagam, the original smoother? Try BIC:

```
> BIC(ft_maunaPiece,ft_maunagam,ft_maunagam20)
                    df      BIC
ft_maunaPiece  4.00000 252.2672
ft_maunagam   10.42712 104.5190
ft_maunagam20 17.01983 104.7728
```

The piecewise model does not fit the data as well as a smoother (as in Fig. 8.2). The two smoothers fit similarly well, suggesting that the extra knots don't really add value.

Alternatively, we could validate models by treating the most recent 15 years of data as test data—essentially predicting the "future", as things stood in 2006:

```
> isTrain = which(maunaJan$year<=2006)
> datTrain = maunaJan[isTrain,]
> datTest = maunaJan[-isTrain,]
> ft_piece = lm(co2~year+year00,dat=datTrain)
> ft_gam = gam(co2~s(year),dat=datTrain)
> ft_gam20 = gam(co2~s(year,k=20),dat=datTrain)
> pr_piece = predict(ft_piece,newdata=datTest)
> pr_gam = predict(ft_gam,newdata=datTest)
> pr_gam20 = predict(ft_gam20,newdata=datTest)
> preds = cbind( predict(ft_piece,newdata=datTest),
  predict(ft_gam,newdata=datTest), predict(ft_gam20,newdata=datTest))
> print( apply((datTest$co2-preds)^2,2,sum)) # getting SS by column
[1] 233.36905  99.43189  15.72148
```

The smoother is clearly a better idea than a piecewise linear fit, but now we see that a more complex smoother would be a better option for predicting to more recent years.

Note we might get a different answer if we chose a different time interval to forecast to—in particular, piecewise linear gets better as the time interval we predict to gets shorter (because there are fewer changes in the rate of increase over shorter time periods).

8.2 Smoothers with Interactions

Models with spline smoothers are often called additive models or GAMs, where *additive* means that if there are multiple predictors, they are assumed not to interact, and so only main effects for each have been included. However, while additivity is a default assumption in most GAMs, these models don't actually need to be additive—you can include interactions, too.

A bivariate spline for x1 and x2 on the same scale can be fitted using

```
s(x1,x2)
```

but if not on the same scale

`te(x1,x2)`

where `te` stands for "`tensor` product smoother".

Exercise 8.2: Eucalypt richness as a function of the environment
Recall from Exercise 7.4 that Ian is interested in species richness of eucalypts and related plants (the *Myrtaceae* family) in the Blue Mountains area west of Sydney, Australia. Ian wanted to know: How does *Myrtaceae* species richness vary from one area to the next, and what are the main environmental correlates of richness? Ian obtained data on the number of *Myrtaceae* species observed in 1000 plots and climate variables. He thinks species richness could respond to temperature and rainfall in a non-linear fashion and may interact.

What sort of model should Ian consider using?

This technique is not always suitable because it is quite data hungry (i.e. you need a big dataset) and can be computationally intensive—the number of required knots may be much larger for a bivariate smoother. For example, the Mauna Loa dataset with annual measurements is too small (only 60 observations, but also only one predictor so no need!). However Ian (Exercise 8.2) has 1000 observations and could certainly consider predicting richness using a bivariate smoother for temperature and rainfall, as in Code Box 8.4. You can extend this idea when you have more than two predictors, but it becomes much harder computationally and quickly becomes infeasible when using spline smoothers.

A simpler alternative is to try handling interactions in the usual way, with a quadratic interaction term. Keeping additive smoothers in the model preserves some capacity to account for violations of the quadratic assumption, as also illustrated in Code Box 8.4. This approach has the advantage of ensuring only one model coefficient is devoted to each pairwise interaction between predictors, so it can handle situations where the dataset isn't huge or where you have more than two predictors (but not too many).

Code Box 8.4: Handling interactions in a GAM for Ian's richness data of Exercise 8.2.
You could use a bivariate smoother:

```
data(Myrtaceae)
ft_tmprain=gam(log(richness+1)~te(TMP_MIN,RAIN_ANN),data=Myrtaceae)
vis.gam(ft_tmprain,theta=-135) #rotating the plot to find a nice view
summary(ft_tmprain)$edf
[1] 14.89171
```

The `vis.gam` command produced Fig. 8.4 and used about 15 effective degrees of freedom (but actually required 24 knots).

We could try a quadratic interaction term combined with smoothers, for something less data hungry:

```
> ft_tmprain2=gam(log(richness+1)~s(TMP_MIN)+s(RAIN_ANN)+
```

```
    TMP_MIN*RAIN_ANN,data=Myrtaceae)
> summary(ft_tmprain2)
Parametric coefficients:
                  Estimate Std. Error t value Pr(>|t|)
(Intercept)      6.970e-02  3.268e-02   2.133   0.0332 *
TMP_MIN          2.045e-02  7.029e-02   0.291   0.7712
RAIN_ANN         1.333e-03  1.501e-04   8.883   <2e-16 ***
TMP_MIN:RAIN_ANN 1.415e-05  4.060e-05   0.348   0.7276
---
Signif. codes:  0 `***' 0.001 `**' 0.01 `*' 0.05 `.' 0.1 ` ' 1

Approximate significance of smooth terms:
              edf Ref.df     F  p-value
s(TMP_MIN)  3.945  5.054 2.838   0.0145 *
s(RAIN_ANN) 5.295  6.459 9.695 1.01e-10 ***
---
Signif. codes:  0 `***' 0.001 `**' 0.01 `*' 0.05 `.' 0.1 ` ' 1
```

Note that this used fewer degrees of freedom than the bivariate smoother. Any evidence of a `rain`×`temp` interaction?

Exercise 8.3: Smoothers for climate effects on plant height

Consider Angela's height data and the question of how height is associated with climate. She would like to know *how plant height relates to climate—in particular, whether there are interactions between predictors and whether the relationship is non-linear.* The data can be found in `globalPlants` on the `ecostats` package.

Choose a temperature and rainfall variable from the dataset to use to answer these questions.

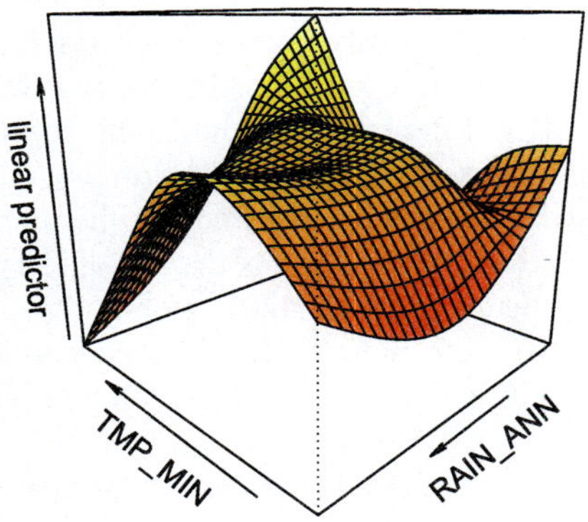

Fig. 8.4: A plot of a bivariate smoother for height as a function of temperature and rainfall, produced as in Code Box 8.4

8.3 A Smoother as a Diagnostic Tool in Residual Plots

A smoother can be used to help diagnose lack of fit in residual plots by drawing a line through the main trend in the data. In a residuals vs fits plot, we want no relationship between the residuals and fitted values, i.e. the trend curve for residuals should track zero pretty closely. Some residual plotting functions, such as the default `plot` function as applied to `lm` in R, will include a smoother by default. This smoother is not actually fitted using spline terms; it uses a different technique. The `plotenvelope` function, on the other hand, uses the `gam` function from the `mgcv` package to fit a smoother. Both of these are illustrated in Code Box 8.5.

Code Box 8.5: Residual plot with a smoother to diagnose a model.
Most residual plot functions on R include a smoother by default, e.g.

```
data(globalPlants)
globalPlants$logHt = log(globalPlants$height)
ft_heightlm = lm(logHt~lat,dat=globalPlants)
plot(ft_heightlm,which=1)
```

which produces Fig. 8.5a.

The `plotenvelope` function also includes a smoother by default:

```
ft_temp = gam(logHt~s(temp), dat=globalPlants)
ecostats::plotenvelope(ft_temp, which=1, main="")
```

which produces Fig. 8.5b.

Do these plots suggest any evidence of lack of fit?

The method used to fit a smoother in the residual plot of Fig. 8.5a uses the `scatter.smooth` function, which (by default) uses, not splines, but a different method (local or kernel fitting) to achieve a similar outcome. There are actually heaps of different ways to fit smooth curves to data. Splines are the most common in the regression setting because they allow us to stay in the linear modelling framework, with all its benefits (especially diagnostic and inferential tools).

Exercise 8.4: Non-linear predictors of species richness?
Consider Ian's species richness data stored in `Myrtaceae`, with measurements of species richness at 1000 different spatial locations, together with a set of environmental variables to use as predictors. Previously (Code Box 7.5) we found that interactions did not appreciably improve the model, so we fitted a model with additive quadratic terms. But is the non-linearity in response to environmental variables better explained by a smoother?

Refit the model using smoothers for each environmental variable and compare it to a model with additive quadratic terms (assume conditional independence for the moment).

Recall that these data were spatially structured. We could try refitting the model to account for this, using the `gamm` function in the `mgcv` package. Do you think this would make any difference to our conclusion? Justify your argument.

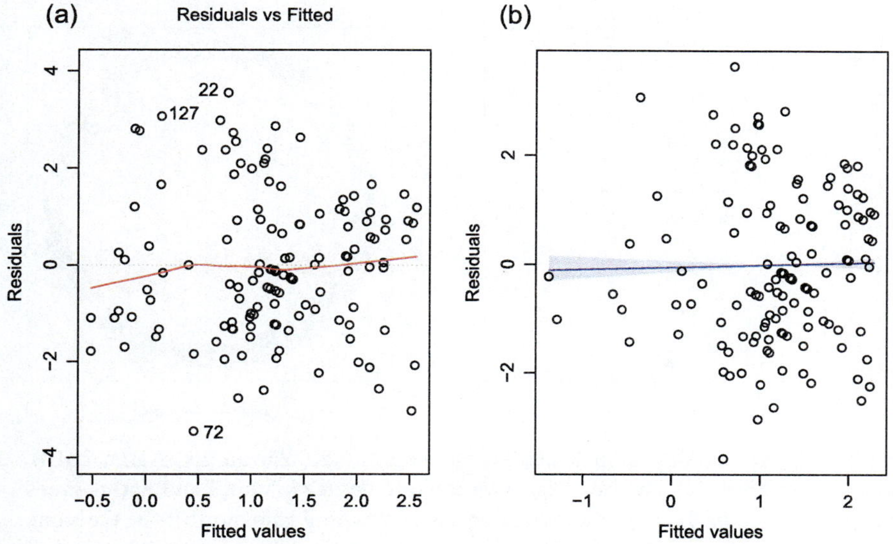

Fig. 8.5: Residual plots for global height data of Exercise 8.3, with smoothers, produced in Code Box 8.5

8.4 Cyclical Variables

Many things in nature have a cyclical pattern to them—most often in time (e.g. season, time of day) but also space (e.g. aspect). These are often called *circular variables* or "circular statistics" because they are more naturally understood by mapping them onto a circle rather a straight line. For example, consider the aspect of slopes on which sheep were found (in degrees, 0 = 360 = due north) in Fig. 8.6.

> **Key Point**
> Do you have a cyclical variable, one that finishes the same place it starts from (aspect, time of day, season, . . .)? These should enter the model in a cyclical way, which can be done using sin and cos functions (which can map a variable onto the unit circle). It is critical, however, to get the period right (so that your cyclical variable gets back to the start at the right value).

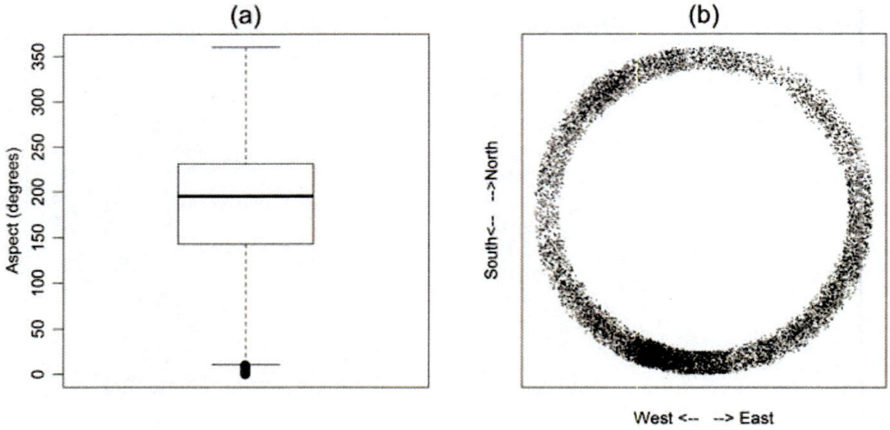

Fig. 8.6: Aspect of slopes on which sheep were found (Warton and Aarts, 2013), plotted as follows: (**a**) The raw aspect variable in degrees, with 0 and 360 degrees at opposite ends of the scale but meaning the same thing (due north); (**b**) the same data mapped onto a circle (and jittered), with each point representing the aspect of the slope where a sheep was observed (with both 0 and 360 at the top). You can see from (**b**), more so than from (**a**), that there is a higher density of points in southerly aspects (especially SSW)

8.4.1 How Do You Map Variables onto a Circle?

If you cos- *and* sin-transform a variable, it gets mapped onto a circle (cos does the horizontal axis and sin does the vertical axis), as in Maths Box 8.3. It is *critical* to get the period of the transformation right, though—you have to time it so that a full cycle has length 2π, the angle of a full rotation (in radians).

Maths Box 8.3: The cos and sin functions map a variable onto a circle
Consider a cyclical variable x with period t—that is, something that ends up where it started at t. For example, time of year has period $t = 12$ months. We want to map this variable onto a circle, so that the cyclical variable ends off where it starts. We will use a circle of radius one.

We will start on the positive horizontal axis (at January on the following figure) and work our way around in an anti-clockwise direction, getting back to the start at the value $x = t$ (the end of the year). The angle of a full rotation of the circle is 2π radians, and for this to correspond to one cycle of x we compute angles using

$$\theta = \frac{2\pi x}{t}$$

For example, the start of April is $x = 3$ months into the year, a quarter of the way through the year. Its angle is $\theta = \frac{2\pi \times 3}{12} = \frac{\pi}{2}$ radians, a quarter of the way around the circle.

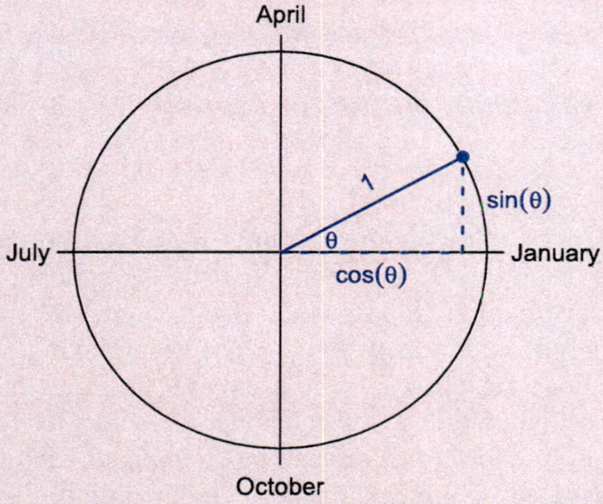

To work out how to map from a value x to a point on the circle, we first form a right-angled triangle with angle $\theta = \frac{2\pi x}{t}$ by dropping from a point on the circle to the horizontal axis. The hypothenuse is the radius of the circle, which has length one. Simple trigonometric rules then tell us that the horizontal coordinate is $\cos\theta$ ("adjacent over hypotenuse") and the vertical coordinate is $\sin\theta$ ("opposite over hypotenuse"). So we can map from a cyclical variable x to points on a circle by transforming to $\left(\cos\frac{2\pi x}{t}, \sin\frac{2\pi x}{t}\right)$. When plotting a cyclical variable x or using it in a model, it is more natural to work with $\left(\cos\frac{2\pi x}{t}, \sin\frac{2\pi x}{t}\right)$, rather than x, to reflect its cyclical nature.

Example: transforming aspect, measured in degrees (which goes from 0 to 360 degrees):

$$\cos\left(\frac{2\pi\,\texttt{aspect}}{360}\right), \sin\left(\frac{2\pi\,\texttt{aspect}}{360}\right)$$

If you plot these two variables against each other, you will get points on a circle, with a point for the aspect at each sampling location, as in Fig. 8.6. (Jittering was used in Fig. 8.6 because of the large sample size, and the axes were swapped over so that co-ordinates match compass directions.)

Transforming time of day, measured in hours (which goes from 0 to 24 h):

$$\cos\left(\frac{2\pi\,\texttt{time}}{24}\right), \sin\left(\frac{2\pi\,\texttt{time}}{24}\right)$$

If time were measured in minutes, you would divide by $24 \times 60 = 1440$.

Transforming time of year, measured in months (which goes from 0 to 12 months):

$$\cos\left(\frac{2\pi\,\texttt{month}}{12}\right), \sin\left(\frac{2\pi\,\texttt{month}}{12}\right)$$

8.4.2 Cyclical Predictors in Linear Models

Exercise 8.5: Carbon dioxide measurements at Mauna Loa observatory
Consider again the Mauna Loa carbon dioxide data, which are actually available monthly starting in 1958. We would like to model the dataset to characterise the key trends within as well as across years, taking into account that carbon dioxide does not follow a linear trend over time and that there is a periodic seasonal oscillation (due to the behaviour of deciduous trees, which occur predominantly in the Northern Hemisphere). See Fig. 8.7 for a time series plot with the points joined (to indicate the sequential nature of the data).

What sort of model would you use?

Consider Exercise 8.5. We would like to find a statistical model that characterises the main features in this dataset. Two things stand out in Fig. 8.7: the increasing trend and the periodic wiggles (seasonal variation).

For the long-term trend we could try `poly` or `gam`.

For the periodic wiggles, we should include `cos` and `sin` terms for the time variable in the model, which ends up being `month`. This adds a *sine curve* to the

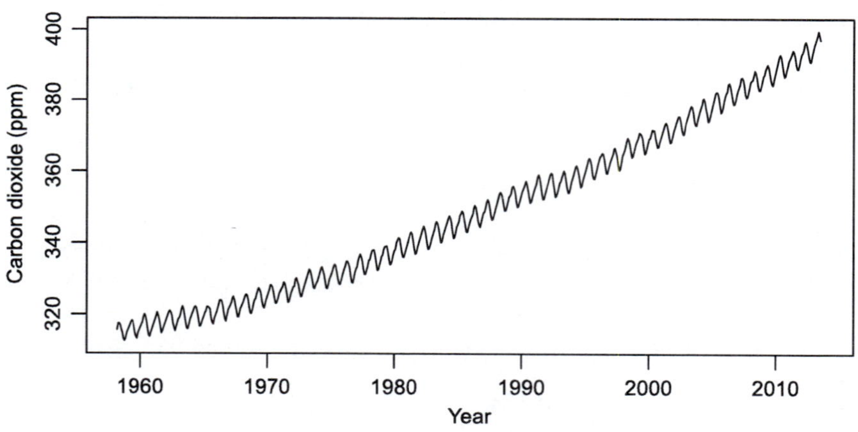

Fig. 8.7: Monthly carbon dioxide measurements at Mauna Loa observatory. The data points have been joined to reflect that this is a time series where each value is always close to the value in the previous month

model. It is important to include *both* `sin` and `cos` of month as predictors, to allow the data to choose both the amplitude of the sine curve (how big the wiggles are) and the phase (the time of year that it reaches the top of the curve), as in Maths Box 8.4. As previously, it is also important to multiply `month` by the right quantity before applying `sin` and `cos`, to make sure the sine curve completes a full cycle in exactly 1 year. Because month has been coded from 1 to 12, we will use `sin(month*2*pi/12)` and `cos(month*2*pi/12)`.

Such a model is fitted as a first step in Code Box 8.6.

Maths Box 8.4: Adding a cyclical predictor to a linear model

A cyclical predictor $\mathbf{x}$, with period t (e.g. time of year, which has period $t = 12$ months), should be added to the model in such a way that the effect of the predictor is the same at t as it is at 0—it comes "full circle". A simple way to do this is to use a sinusoidal curve with period t, which has the form $A\sin(\frac{2\pi\mathbf{x}}{t} + \psi)$, where A is the *amplitude*, controlling how big the wiggles are, and ψ is the *phase*, controlling when the curve passes through zero, as in the following figure:

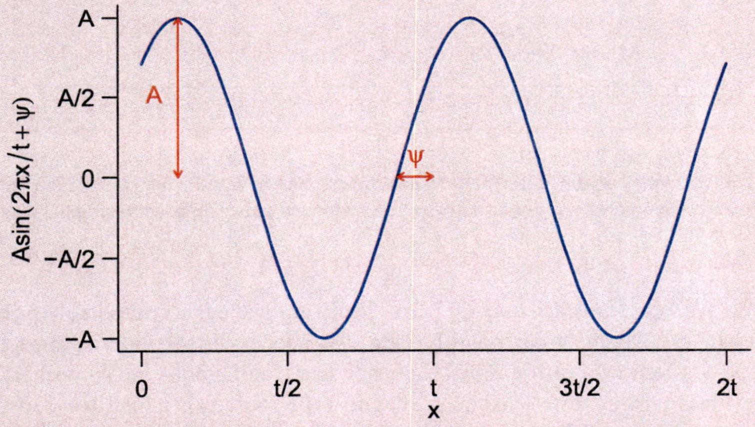

But the addition rule of trigonometry tells us

$$A\sin\left(\frac{2\pi\mathbf{x}}{t} + \psi\right) = A\cos\left(\frac{2\pi\mathbf{x}}{t}\right)\sin(\psi) + A\sin\left(\frac{2\pi\mathbf{x}}{t}\right)\cos(\psi)$$
$$= \beta_1\cos\left(\frac{2\pi\mathbf{x}}{t}\right) + \beta_2\sin\left(\frac{2\pi\mathbf{x}}{t}\right)$$

for some constants β_1 and β_2. So to fit a sinusoidal curve to data, we just add $\cos\frac{2\pi\mathbf{x}}{t}$ and $\sin\frac{2\pi\mathbf{x}}{t}$ as predictors to a linear model. When we use data to estimate these coefficients, we are estimating both the phase and amplitude.

Code Box 8.6: A simple model for the Mauna Loa monthly data with a cyclical predictor

A smoother (via a GAM) is used to handle non-linearity in the annual trend—since the rate of increase in the concentration of carbon dioxide is not constant over time, it seems to get steeper over time. A sine curve is added to handle the seasonality effect.

```
data(maunaloa)
library(mgcv)
ft_cyclic=gam(co2~s(DateNum)+sin(month/12*2*pi)+cos(month/12*2*pi),
  data=maunaloa)
plot(maunaloa$co2~maunaloa$Date,type="l",
  ylab=expression(CO[2]),xlab="Time")
points(predict(ft_cyclic)~maunaloa$Date,type="l",col="red",lwd=0.5)
```

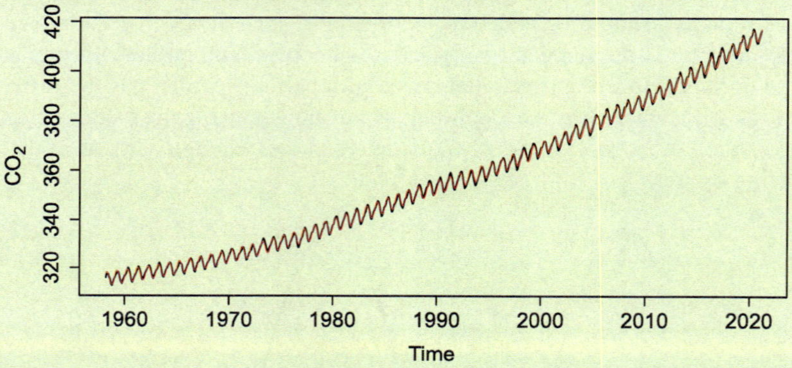

There are two curves on the plot—the raw data (black) and the fitted model (red). Note that at first glance the model seems to fit fairly well, with the red curve almost covering the black one.

How do we *mind our Ps and Qs*? To check we got each component right in the model of the Mauna Loa monthly data, we need residual plots against each component. As well as a residual plot against `time` (to diagnose the smoother), we need a residual plot against season (to diagnose periodicity), which we construct manually in Code Box 8.7. Notice in Code Box 8.7 that the lines have been joined across time points, so we can see how residuals change over time and better diagnose any problems with the model as a function of time.

Code Box 8.7: Residual plots across time and season for the Mauna Loa monthly data

In what follows we construct residual plots joining the dots across time to see if there are any patterns over time and, hence, any problems with the model as a function of time. Two plots will be constructed—one across years, another within years (using `sin` of month) to diagnose the smoother and the periodic component of the model, respectively. Using `sin` of month means that December and January will appear in the same part of the `Season` predictor (the middle, near zero). `cos` could equally well be used.

```
par(mfrow=c(1,2))
```

```
plot(residuals(ft_cyclic)~maunaloa$Date,type="l", xlab="Time")
plot(residuals(ft_cyclic)~sin(maunaloa$month/12*2*pi),
     type="l",xlab="Season")
```

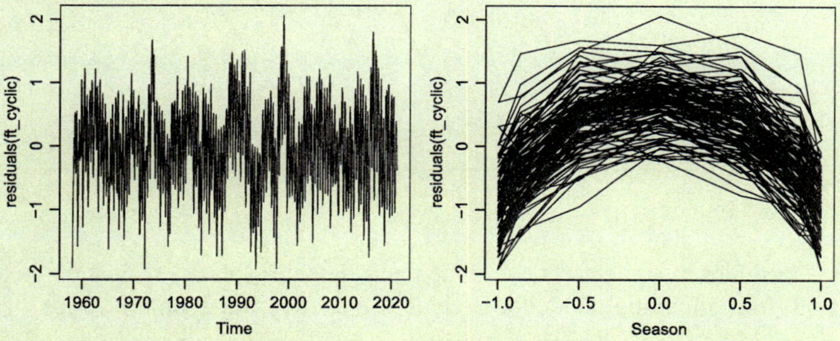

For comparison, here are some plots for simulated data to show you what they might look like if there were no violation of model assumptions:

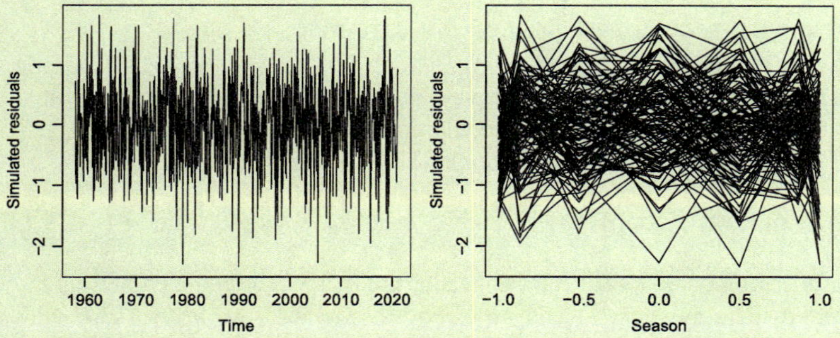

Do you think assumptions are satisfied?

As with any residual plot, we are looking for no pattern, but in Code Box 8.7, note the hump shape on the residual plot against season. This is because a sine curve did not adequately capture the seasonal trend. You can see this more clearly by zooming in on the 2000s as in Fig. 8.8.

For quantitative predictors, if a linear trend doesn't work, a simple thing you can try is quadratic terms. For circular variables, if a sine curve doesn't work, a simple thing you can try is adding sin and cos terms with half the period by multiplying the variable by two before cos- and sin-transforming, as in Code Box 8.8. This gives the circular world's equivalent of quadratic terms (Maths Box 8.5). Or if that doesn't work, try multiplying by three as well (like cubic terms)! This idea is related to Fourier analysis (Maths Box 8.5), and the terms are sometimes called *harmonics*.

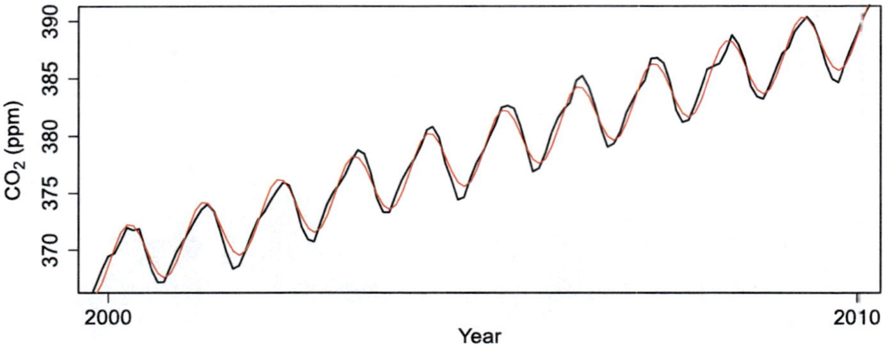

Fig. 8.8: Trace plot of some carbon dioxide measurements (black) against predicted values (red) from the initial model of Code Box 8.6, zooming in on the 2000s

Maths Box 8.5: Fancier periodic terms and Fourier series
If a linear model doesn't fit data, a simple thing you can try is to add a quadratic term to the model. In much the same way, if data aren't well fitted by a sinusoidal curve, using $\cos(\theta)$ and $\sin(\theta)$, you could try using the quadratic terms $\cos^2(\theta)$, $\sin^2(\theta)$, and $\sin(\theta)\cos(\theta)$. But the *double-angle rules* of trigonometry are

$$\sin(2\theta) = 2\sin(\theta)\cos(\theta) \quad \text{and} \quad \cos(2\theta) = \cos^2(\theta) - \sin^2(\theta)$$

so information on quadratic terms of a sine curve are captured by doubling the angle, i.e. including $\cos\frac{4x\pi}{t}$ and $\sin\frac{4x\pi}{t}$ in a model for a cyclical variable with period t, as well as $\cos\frac{2x\pi}{t}$ and $\sin\frac{2x\pi}{t}$. This adds predictors to the model with half the period, for extra wiggles.

This idea extends to higher-order terms. The analogue of a Taylor series (Maths Box 8.1) in a "circular world" is a *Fourier series*, which allows us to write any smooth periodic function $s(x)$ with period t as

$$s(x) = a_0 + a_1\cos\left(\frac{2\pi x}{t}\right) + b_1\sin\left(\frac{2\pi x}{t}\right) + a_2\cos\left(\frac{4\pi x}{t}\right) + b_2\sin\left(\frac{4\pi x}{t}\right) + a_3\cos\left(\frac{6\pi x}{t}\right) + b_3\sin\left(\frac{6\pi x}{t}\right) + \ldots$$

so if the mean of $\mathbf{y}$ is some unknown function of a circular variable $\mathbf{x}$, using $\cos\frac{2x\pi}{t}$ and $\sin\frac{2x\pi}{t}$ as predictors gives a sinusoidal approximation, estimating just the first pair of terms from the Fourier expansion. Adding $\cos\frac{4x\pi}{t}$ and $\sin\frac{4x\pi}{t}$ as extra predictors will give a slightly better approximation, adding $\cos\frac{6x\pi}{t}$ and $\sin\frac{6x\pi}{t}$ is better still, and so on. Whereas higher-order Taylor approximations have problems extrapolating, this is not really an issue for cyclical data, provided the data have been observed all the way around the circle (e.g. throughout the whole year).

Code Box 8.8: Another model for the Mauna Loa monthly data, with an extra sine curve in there to better handle irregularities in the seasonal effect

```
ft_cyclic2=gam(co2~s(DateNum)+sin(month/12*2*pi)+cos(month/12*2*pi)+
        sin(month/12*4*pi)+cos(month/12*4*pi),data=maunaloa)
par(mfrow=c(1,2))
plot(residuals(ft_cyclic2)~maunaloa$Date,type="l", xlab="Time")
plot(residuals(ft_cyclic2)~sin(maunaloa$month/12*2*pi),
        type="l",xlab="Season")
```

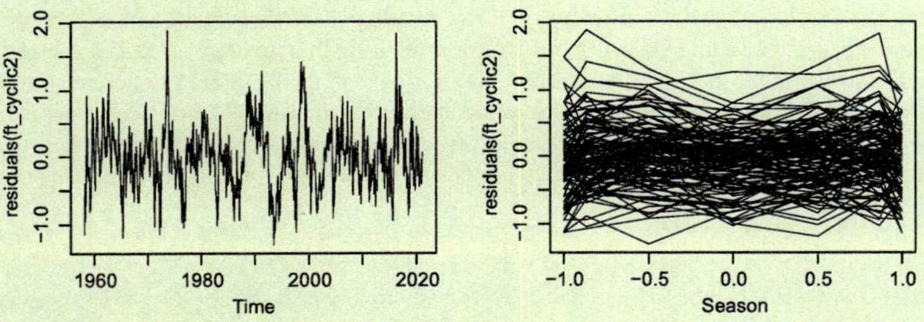

Do you think the assumptions are satisfied?

The mean trend seems to have been mostly dealt with in the residual plots of Code Box 8.8, but there is some residual correlation between variables (since CO_2 this month depends on CO_2 last month). This is evident in the left-hand plot with residuals tending to stay near previous values (e.g. a cluster of residuals around 0.5 in 1990, then a drop for several negative residuals around 1993). Residual correlation is also evident in the right-hand plot, with residuals often running across the page (left to right), remaining large if the previous value was large or remaining small if the previous value was small. This correlation (or "time lag") becomes a problem for inference—standard errors and P-values will be too small, differences in BIC too big. . . , and one thing we could do here is try adding temporal autocorrelation via the `gamm` function, as in Code Box 8.9. This function combines ideas for modelling dependent data from Chap. 7 with smoothers as used earlier.

Code Box 8.9: Mauna Loa model with autocorrelation

We will try using the `gamm` function to add temporal autocorrelation to the model, assuming errors follow an AR(1) structure:

```
> ft_gamm = gamm(co2~s(DateNum)+sin(month/12*2*pi)+cos(month/12*2*pi)+
                    sin(month/12*4*pi)+cos(month/12*4*pi),
                    correlation=corAR1(), data=maunaloa)
> acf(residuals(ft_gamm$gam))
> acf(residuals(ft_gamm$lme,type="normalized"))
```

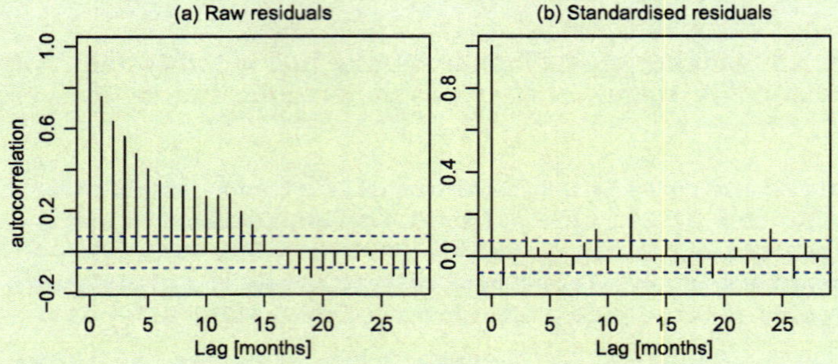

Note that the first autocorrelation function, of raw residuals, shows a strong autocorrelation signal, lasting about a year (12 months). But after modelling this (using an AR(1) structure), standardised residuals no longer have this pattern, suggesting the AR(1) structure does a much better job of capturing the temporal autocorrelation structure. There is, however, still a pretty large correlation at a lag of 12 months, and again at 24 months, suggesting that the model for longer-term trends is not quite adequate, with some year-to-year variation not being picked up.

Exercise 8.6: Mauna Loa monthly data—an extra term in seasonal trend?
Consider the Mauna Loa monthly data again. We tried modelling the trend by fitting sine curves with two frequencies—using `(sin(month/12*2*pi),cos(month/12*2*pi))`, which gives a sine curve that completes one full cycle per year, and `(sin(month/12*4*pi),cos(month/12*4*pi))`, for two cycles per year. Is this sufficient, or should we add another frequency, too, like `month/12*6*pi` (for curves with three cycles a year)?

Use `gam` and the model selection technique of your choice to consider these options.

Exercise 8.7: Mauna Loa annual data—temporal autocorrelation?
Consider the *annual* Mauna Loa monthly data of Code Box 8.1, which uses the measurements from January only.

Check for autocorrelation by fitting a `gamm` and estimating the autocorrelation function.

Is there evidence of temporal autocorrelation in this dataset?

Exercise 8.8: Aspect as a predictor of species richness
Consider Ian's species richness data stored in `Myrtaceae`, with measurements of species richness at 1000 different spatial locations, together with a set of environmental variables to use as predictors. One predictor of interest is `aspect`, the direction in which a transect is facing, measured in degrees (from 0 to 360).

How would you enter `aspect` into the model?

Add aspect to a linear model and see if it is significantly associated with richness for this dataset.

Recall that species richness had some spatial structure to it. Refit the model using the `gls` function from the `nlme` package to account for the spatially structured correlation. Did this change your inferences about `aspect` in any appreciable way?

Chapter 9
Design-Based Inference

Recall that we have met three main types of inferential procedure:

- Confidence interval (CI) estimation
- Hypothesis testing
- Model selection

Now in each case we can take one of two general strategies for inference: a model-based or a design-based approach.

Model-based inference—assume your model is correct (or nearly correct) and use theory, or sometimes simulation from your model, to make inferences. Most methods of inference we have discussed so far have been model-based—the `confint` function, `summary` and `anova` as applied to `lm` or `glm` objects, Akaike information criterion, Bayesian information criterion, the parametric bootstrap.

Design-based inference—exploit independent units in your study design as a basis for inference (usually via simulation). The only example of this we have seen to date is cross-validation—exploiting the independence of sampling units to construct an independent test dataset by subsetting the original data. These test data were then used to make inferences about which model is better (in terms of which would better predict to new data).

Methods of design-based inference typically have the following properties:

- Less restrictive assumptions—the model does not have to be (nearly) correct in order for inferences to be valid.
- Almost always takes longer—it is a simulation-based approach. If your model was hard to fit in the first place, it will be very hard now.
- Less efficient (when assumptions are reasonable)—simulation introduces noise to results, which sometimes makes patterns harder to see.

So in summary, we are relaxing some of our assumptions, at costs in computation time and possibly in efficiency.

This chapter will focus specifically on design-based methods of hypothesis testing. These methods can be used for pretty much any hypothesis test and will be illustrated

D. I. Warton, *Eco-Stats: Data Analysis in Ecology*, Methods in Statistical Ecology,
https://doi.org/10.1007/978-3-030-88443-7_9

here for linear models, but note they are most commonly used in more complex situations where it is difficult to do model-based inference (so they will feature strongly in Chap. 14—a more complex situation!).

Key Point
Methods of design-based inference enable inference even when the model used for analysis is not correct—by exploiting independent units in the study design. This can be handy in situations where you trust your independence assumptions but not much else. Common methods of design-based inference include the permutation test, cross-validation, and the bootstrap.

9.1 Permutation Tests

A permutation test is applicable whenever the null hypothesis being tested is "no effect of anything" (sometimes called an intercept model).

Exercise 9.1: Smoking during pregnancy
Consider again the guinea pig experiment of Exercise 1.6. The number of errors made by guinea pigs in the maze is as follows:

C	C	C	C	C	C	C	C	C	C	N	N	N	N	N	N	N	N	N	N
11	19	15	47	35	10	26	15	36	20	38	26	33	89	66	23	28	63	43	34

C = control, N = nicotine treatment

We would like to know if there is evidence that the guinea pigs are making more errors under the nicotine treatment, i.e. is there evidence that "smoking mums" can inhibit development of offspring.

The observed test statistic (using a two-sample t-test) is 2.67. Normality assumptions are not satisfied for these data. How can we make inferences about the treatment effect without assuming normality?

Consider Exercise 9.1. In this experiment, guinea pigs were randomly allocated to treatment and control groups. If the nicotine treatment had no effect on response, we would expect this random allocation of subjects to treatment groups to give results no more unusual than under any other random allocation (where unusualness will be measured here as size of the test statistic). That is, under the null hypothesis of no treatment effect, randomly permuting the Control/Treatment labels as in Exercise 9.2 will not affect the distribution of the test statistic.

Exercise 9.2: Three example permutations of treatment labels in the guinea pig data
If nicotine treatment has no effect on the number of maze errors, we can permute treatment labels without affecting the distribution of the test statistic. A few examples are given below.

C	C	C	N	N	N	N	C	C	N	C	C	N	N	C	C	C	N	N	N
11	19	15	47	35	10	26	15	36	20	38	26	33	89	66	23	28	63	43	34

The test statistic for this permutation is 0.76.

C	N	N	C	C	C	C	C	C	N	N	C	N	N	N	N	N	N	C	C
11	19	15	47	35	10	26	15	36	20	38	26	33	89	66	23	28	63	43	34

The test statistic for this permutation is −2.09.

N	N	N	C	N	N	C	C	C	C	C	N	N	C	N	C	C	N	C	N
11	19	15	47	35	10	26	15	36	20	38	26	33	89	66	23	28	63	43	34

The test statistic for this permutation is −1.08.

Does the observed statistic of 2.67 seem large compared to these values? What does this tell you?

In a permutation test, this process is repeated a large number of times (e.g. 999) to assess whether the the observed statistic is unusually large. This will produce an estimate of the null distribution of the test statistic, as in Fig. 9.1.

In an ordinary two-sample t-test, we compute the test statistic, as in Exercise 9.1, and compare it to a t distribution that is derived (under the null hypothesis) assuming our model assumptions are true. In a permutation test, we instead compare the test statistic to a distribution derived by simulation, permuting treatment labels and recomputing the test statistic a large number (e.g. 999) of times, as in Exercise 9.2. This permutation-based null distribution tells us the sort of test statistics we would expect to see if in fact treatment had no effect on response, in which case assignment of treatments to subjects would be arbitrary. We can then use this null distribution directly to get a P-value—the proportion of permuted samples (or *resamples*) that beat the observed test statistic, as in Fig. 9.1. Typically the observed dataset is included as one of the permutations considered (so adding this to 999 permuted datasets would give us 1000).

9.1.1 Permutation Testing Using `mvabund` Package

There are plenty of examples of permutation testing software available for simple study designs like this one. Code Box 9.1 uses the `mvabund` package, developed by myself and collaborators at UNSW Sydney, which reduces the problem to three familiar-looking lines, or two if you have already loaded the package. The package name `mvabund` stands for "multivariate abundance" data—the main type of data it was designed for (Chap. 14). But it can be used for plenty of other stuff, too. To use `mvabund` for a permutation test of all terms in a linear model, you just change the function from `lm` to `manylm` (this helps R work out what to do when you call generic functions like `anova`).

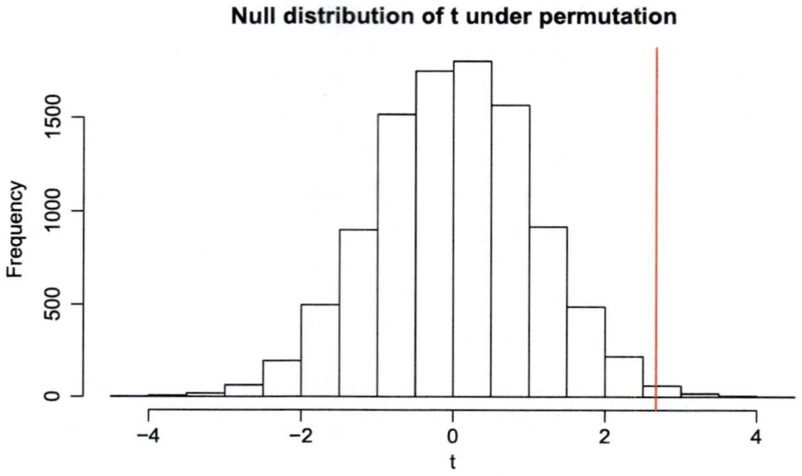

Fig. 9.1: Null distribution of two-sample t statistic, estimated by permuting treatment labels 999 times, along the lines of Exercise 9.2. Notice that the observed test statistic is in the right tail of this distribution, suggesting it is unusually large (under the hypothesis of no treatment effect). Specifically, we only see a test statistic as large as or larger than 2.67 four times in the resampled data, counting our observed statistic too makes it five, so our estimated P-value is $\frac{5}{1000} = 0.005$

Technically, what `mvabund` does here is to permute residuals (residual resampling will be explained in more detail in Sect. 9.5), but for tests of the hypothesis of no effect, this is equivalent to permuting treatment labels as in Exercise 9.2.

Code Box 9.1: A permutation test for the guinea pig data using `mvabund`

```
> library(mvabund)
> ft_guinea = manylm(errors~treatment,data=guineapig)
> anova(ft_guinea)
Analysis of Variance Table

Model: manylm(formula = errors ~ treatment, data = guineapig)

Overall test for all response variables
Test statistics:
            Res.Df Df.diff val(F) Pr(>F)
(Intercept)     19
treatment       18       1  7.134  0.015 *
```

```
---
Signif. codes:  0 `***' 0.001 `**' 0.01 `*' 0.05 `.' 0.1 ` ' 1
Arguments: P-value calculated using 999 iterations via residual
(without replacement) resampling.
```

9.1.2 Different Permutation Tests, Different P-Values

Comparing Fig. 9.1 with Code Box 9.1, you will notice the *P*-values are quite different. The *P*-value for the *t*-test in Fig. 9.1 was 0.005, whereas mvabund gave a much larger value (0.015). Why? There are two reasons.

The first reason for the difference is that we are using a *random* set of permutations of treatment labels, so results are random, too. If you repeat Code Box 9.1 yourself, you will typically get different answers on different runs (similarly for the code behind Fig. 9.1). We can actually use probability to quantify how much random error (*Monte Carlo error*) there is in a resampled *P*-value. It depends on the number of permutations used, controlled by `nBoot` in the mvabund package, and is also a function of the size of the true *P*-value (specifically, it is a binomial proportion with parameters `nBoot` and p where p is the true *P*-value). The more permutations, the smaller the error; `nBoot=999` (default) is usually good enough, in the sense that it usually gives an answer within about 0.015 of the correct *P*-value when it is marginally significant.

The second reason for the difference is that the mvabund package uses a two-sided test (is there evidence of a *difference* with treatment, not just an *increase*?), so it usually gives about double the one-sided *P*-value.

The test statistics are also slightly different—`manylm`, like `lm`, reports an (ANOVA) F-statistic in an anova call. This is the square of the t stat ($7.13 \simeq 2.67^2$). The F-test and two-sided t-test are equivalent, meaning they return the same *P*-value, so this difference in test statistics has no implications when assessing significance.

9.1.3 Permutation Tests Work for Regression Too

You can use a permutation test for any situation where the null hypothesis says that all observations come from exactly the same distribution (same mean, same variance, ...; if there is any correlation across observations, this has to be the same too). So, for example, you could use a permutation test in regression, to test whether there is an association between predictor(s) and response, as in Code Box 9.2.

Code Box 9.2: Permutation test for a relationship between latitude and plant height

```
> data(globalPlants)
> ft_height = manylm(height~lat, data=globalPlants)
> anova(ft_height)
Analysis of Variance Table

Model: manylm(formula = height ~ lat, data = globalPlants)

Overall test for all response variables
Test statistics:
            Res.Df Df.diff val(F) Pr(>F)
(Intercept)    130
lat            129       1  9.271  0.003 **
---
Signif. codes:  0 `***' 0.001 `**' 0.01 `*' 0.05 `.' 0.1 ` ' 1
Arguments: P-value calculated using 999 iterations via residual
(without replacement) resampling.
```

9.1.4 Randomisation Tests and Small Datasets

Some refer to this procedure as a randomisation test (Edgington, 1995), reserving the term permutation test for situations were all possible permutations of treatment levels are considered in computing the P-value, rather than a random sample of them (which is what `anova.manylm` does by default). Sometimes, for small datasets, the total number of possible permutations is quite small, e.g. if there were 5 replicates of each treatment group in Exercise 9.1, there would have been only 126 possible permutations of treatment labels. In this situation, sampling 999 permutations at random is wasteful computationally, but more importantly, it introduces unnecessary Monte Carlo error. You can compute all possible permutations using software, e.g. the `permute` package (Simpson, 2016) using R.

It is only for very small datasets that it is feasible to compute all possible permutations. While 5 replicates in each treatment only gives 126 possible permutations, the number of possible permutations increases rapidly with the number of replicates. For 10 replicates of each treatment, as in Exercise 9.1, there are over 90,000 possible permutations. With 20 replicates of each treatment there would be about 69 billion possible permutations!

9.1.5 How Do Permutation Tests Compare to Model-Based Tests?

We mentioned previously that standard output from `lm` uses model-based inference, and P-values from standard output are *exact* as long as your assumptions are satisfied, i.e. the P-value is exactly the probability it claims to be if the model is correct. However, if assumptions are violated, especially the independence or equal variance assumptions, then P-values can be very approximate.

Permutation tests, on the other hand, are *always exact* (for the no-effect hypothesis), even when assumptions of the model being fitted aren't satisfied, provided that the independence assumption is satisfied. Well strictly speaking we don't even need independence, just exchangeability (which roughly speaking means all correlations are equal). Thus some have argued permutation tests should be used as a matter of routine (Ludbrook and Dudley, 1998). But they can take longer to compute (especially for large datasets) and usually give similar answers.

In the examples we have done so far (Code Boxes 9.1–9.2), the P-values were not very different from what they would have been using model-based inference, as in Chaps. 1–4. This is probably because violations of model assumptions were not serious enough to cause problems for the model-based procedures in these cases; recall that the central limit theorem (CLT) gives model-based approaches a fair bit of robustness to violations of normality, so this should not really be too surprising. CLT gives no protection against violations of the equal variance assumption, but balanced sampling designs do, which was probably the saving grace for the guinea pig dataset (Exercise 9.1).

9.2 Bootstrapping

Another method of design-based inference is the bootstrap. The motivation for the bootstrap is slightly different, but the end result is very similar. Bootstrapping can be used not just for hypothesis testing, but also for estimating standard errors (SEs) and CIs, it can even be tweaked for model selection (Efron and Tibshirani, 1997).

If we knew the true distribution of our data, we could compute a P-value by simulating values directly from the true distribution (under the *null* hypothesis). But all we have is our observed data. The idea of the bootstrap is to use our observed data to estimate the true distribution. We then *resample* some data—generating a new sample from our best estimate of the true distribution.

Consider the guinea pig data of Exercise 9.1. If treatment had no effect, all 20 observations could be understood as coming from the same distribution. Without any further assumptions, our best estimate of the true distribution from the observed data would be to say it takes the values $11, 19, 15, \ldots, 34$ with equal probability ($\frac{1}{20}$). So in each resample, each observation has a $\frac{1}{20}$ probability of taking each value from the observed dataset. Four example bootstrap datasets, constructed in this way, are in Table 9.1.

Maths Box 9.1: Bootstrapping as sampling from an empirical distribution function

If we knew the correct model for our data (and the values of all its parameters), we could work out the sampling distribution of any statistic we wished to calculate from our data by simulation. We could take repeated samples from our data-generating model, calculate the statistic of interest for each sample, and then look at the sampling distribution of the statistic (e.g. in a histogram). Random samples from Y can be simulated using its *distribution function* $F(y) = P(Y \leq y)$. The problem is that we do not know the distribution function of our data.

A parametric bootstrap (Sect. 6.5) assumes the fitted model is the correct model for our data and uses the sample estimates of parameters as if they were the true values. But what if we don't want to assume the fitted model is the correct one?

A (non-parametric) bootstrap resample of an observation, Y^*, has an equal chance $\frac{1}{n}$ of being each value in the observed sample. The distribution function of Y^* is

$$F^*(y) = P(Y^* \leq y) = \frac{\#\text{ values in dataset no larger than } y}{n}$$

This function is known as the *empirical distribution function* of the data. A few example empirical distribution functions are plotted below, for different sample sizes. Note that as the sample size increases, they tend towards the true distribution function $F(y)$, in fact $\lim_{n\to\infty} F^*(y) = F(y)$.

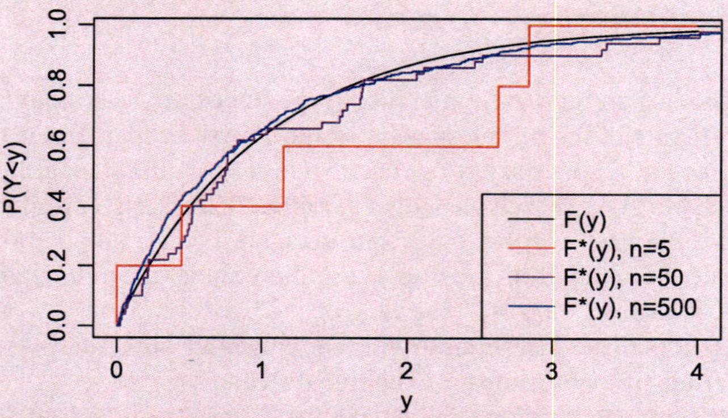

This idea, that bootstrapping can be understood as taking a random sample from the empirical distribution function $F^*(y)$, is fundamental to studying the properties of the bootstrap. For example, because $\lim_{n\to\infty} F^*(y) = F(y)$, bootstrapping a statistic approximates its true sampling distribution (under the random sampling assumption) and does so increasingly well as sample size increases.

The mvabund package can be used to perform bootstrap hypothesis tests, as in Code Box 9.3. Note that in Code Box 9.3 there is an argument `resamp="residual"` that tells the `anova.manylm` function to use the bootstrap rather than a permutation test. As previously, this bootstraps residuals (Sect. 9.5) rather than y-values, but again, this is equivalent to bootstrapping y values if the null hypothesis is that all means are equal.

Table 9.1: Guinea pig data and four bootstrap resamples, obtained by resampling the observed values with replacement

Treatment	C	C	C	C	C	C	C	C	C	C	N	N	N	N	N	N	N	N	N	N
Obs. data	11	19	15	47	35	10	26	15	36	20	38	26	33	89	66	23	28	63	43	34
Bootstrap 1	38	63	26	43	26	43	19	43	89	38	28	38	15	15	33	38	35	38	15	26
Bootstrap 2	43	26	26	26	26	23	34	43	63	11	19	35	34	89	43	28	19	28	10	47
Bootstrap 3	19	66	28	20	35	38	33	36	26	33	15	43	33	47	23	47	66	89	38	28
Bootstrap 4	15	66	28	89	47	10	28	11	19	11	66	89	36	36	47	15	28	47	63	89

Code Box 9.3: Using the mvabund package for a bootstrap test of guinea pig data

```
> library(mvabund)
> ft_guinea = manylm(errors~treatment, data=guineapig)
> anova(ft_guinea, resamp="residual")
Analysis of Variance Table

Model: manylm(formula = errors ~ treatment, data = guineapig)

Overall test for all response variables
Test statistics:
            Res.Df Df.diff val(F) Pr(>F)
(Intercept)     19
treatment       18       1  7.134  0.016 *
---
Signif. codes:  0 `***' 0.001 `**' 0.01 `*' 0.05 `.' 0.1 ` ' 1
Arguments: P-value calculated using 999 iterations via residual
resampling.
```

There are many ways to bootstrap data. The approach in Table 9.1 resampled the response variable while keeping the explanatory variables fixed—this is a natural approach to use when testing an "all means equal" hypothesis. This approach keeps the number of treatment and control replicates fixed, which is desirable in a planned experiment where these values were in fact fixed. Sometimes (especially in observational studies) the explanatory variables do not really take fixed values but are most naturally considered random. In such cases we could bootstrap whole cases (most commonly the rows) of our dataset, not just the response variable, commonly known as "case resampling" (Davison and Hinkley, 1997). This idea is illustrated for the guinea pig data in Exercise 9.3. Case resampling doesn't make a whole lot of sense for the guinea pig data, a designed experiment with 10 replicates in each treatment group, because it doesn't guarantee 10 replicates in resampled data. It would make more sense on Angela's height data or Ian's species richness data, these being observation studies where sites (hence rows of data) were sampled at random.

Case resampling is fine for estimating SEs or CIs but needs to be used with care in hypothesis testing (Hall and Wilson, 1991). Case resampling can be implemented using `mvabund`, with corrections along the lines of Hall and Wilson (1991), by specifying `resamp="case"`.

Exercise 9.3: Case resampling in guinea pig data.
We can construct new datasets that should have similar properties to the original data, by resample cases (columns) of data from Table 9.1. Below are three examples.

C	N	N	N	C	N	N	C	N	N	C	C	N	N	C	C	N	N	N	N
10	33	28	33	47	66	33	26	63	66	36	20	38	28	35	15	66	33	43	26

N	N	N	N	C	N	N	N	N	N	C	C	C	C	N	C	C	C	N	C
33	89	34	89	35	43	38	33	23	63	10	26	11	10	43	47	10	19	66	35

N	N	C	C	N	N	N	C	N	C	N	C	N	N	N	C	C	C	C	C
66	66	19	47	63	43	43	20	33	15	28	26	89	38	43	47	20	20	11	20

Count the number of controls in each case resampled dataset. Did you get the number you expected to?

9.3 Do I Use the Bootstrap or a Permutation Test?

There are two main differences in resampling procedures between bootstrapping and permutation testing, which imply different models for the underlying data-generating mechanism:

- A bootstrap treats the response variable y as random and resamples it, whereas a permutation test treats the treatment labels as random and permutes them.
- A bootstrap resamples *with replacement*, whereas a permutation test uses each response variable exactly once.

These differences may be important conceptually, but in practice, they actually have little by way of implications. In a test of a "no-effect" hypothesis, results should not be expected to differ in any practically meaningful way between a bootstrap and permutation test. So, from a practical perspective, really you could equally well use either method in that context. One important practical difference, however, is when moving beyond tests of "no-effect" null hypotheses—the bootstrap can readily be extended to such contexts, e.g. it can be used to estimate SEs or CIs. The notion of permutation testing does not extend as naturally to these settings (although it is possible). Another more subtle difference is that whereas permutation tests are exact for "no-effect" nulls, no such claim appears to have been proven for the bootstrap. But experience suggests that if bootstrap tests for no effect are not exact, they are at least very close. . . .

So how do you decide what to do in practice? Well, as previously, it doesn't really matter! But it would be prudent to (where possible) use a method of resampling that reflects the underlying design—this is, after all, the intention of design-based inference. So, for example, in an experiment testing for an effect of a treatment applied to a randomised set of subjects (e.g. the guinea pigs), the permutation test is the most intuitive approach to use because it tests explicitly whether the observed randomisation led to significantly different results compared to other possible random assignments of subjects to treatment groups. In other settings, where the response variable is treated as the random quantity (such as the global plant height data, where the goal is to model changes in plant height), bootstrapping the response tends to make more sense in order to resample in such a way that the response is the quantity that is treated as random.

When fitting a regression model to data collected in a fixed design (where the values taken by explanatory variables were for the most part under the control of the researcher), I tend to bootstrap, keeping the design fixed. In an observational study where the explanatory variables and responses were random, I would case resample.

9.4 Mind Your Ps and Qs!

Bootstrapping assumes the units can be resampled at random, that is, it assumes these units are *independently* and identically distributed (iid). This is a familiar assumption from linear models, where we assume errors are iid. In the special case of a "no-effect" null hypothesis, the iid assumption in linear models applies to the response variable, which coincides with the iid assumption of the response we made when bootstrapping the response in Code Box 9.3.

Permutation tests assume observations are *exchangeable* (can be swapped around without changing the joint distribution). Technically this is slightly more general than the iid assumption, where the independence part of the assumption is relaxed to an assumption of equal correlation. In practice it is quite rare to be able to argue that the exchangeability assumption is valid but the iid assumption is unreasonable, so the distinction being made here is of limited utility.

There are *no* further assumptions when testing no-effect null hypotheses. That's the value of design-based inference—we can make inferences using only independence assumptions that we can often guarantee are satisfied via our study design. *(How can you guarantee observations are independent?)* This sounds fantastic, but recall that when we did model-based inference on the guinea pig dataset, we weren't really worried about the additional assumptions we needed to make (because of robustness to assumption violations). So in this case, design-based inference added little value, only freeing us from making assumptions we weren't worried about making. The real power of design-based inference is in more difficult statistical problems, where model-based inference does not apply, is difficult to derive, or is known not to work well.

9.4.1 Pulling Yourself up by the Bootstraps

The term "bootstrap" comes from the expression "to pull yourself up by the bootstraps" from a Baron von Munchausen story, in which the protagonist did precisely this to get himself out of a sticky situation. The implication is that the bootstrap kind of lets the data get themselves out of a sticky situation, without external help (without model-based assumptions). It allows us to make valid, distribution-free inferences from very small samples where previously it was thought impossible.

The bootstrap is one of the most exciting developments in statistics over the last half-century; it generated a lot of chatter in the late 1970s and 1980s and is now so pervasive it is even in some high school syllabuses.

9.4.2 The Parametric Bootstrap Is Not *Design-Based*

In Chap. 6 we met the parametric bootstrap. The idea there was that we added some assumptions—we assumed we knew the form of the distribution of our data and that all we were missing were the true values of parameters. We used our sample estimates of parameters as if they were the true values, to simulate observations to use to estimate our null distribution (or SE, for example). On `R`, you can construct a parametric bootstrap for many types of model (including `lm`, `lmer`, and others we will meet later) using the `simulate` function.

The parametric bootstrap looks and sounds like a design-based method—its name mentions the bootstrap, it is a simulation-based method, and for complex models it takes a very long time to run. While these features all sound quite like those of design-based methods, the parametric bootstrap is in fact a model-based approach—because it is *assuming that the fitted model is correct* and exploiting that information to make general inferences from the sample. A design-based approach, in contrast, exploits independent units implied by the study design to make inferences, whether permuting across randomised treatment allocations, resampling responses, or some other technique.

The practical implication of this distinction is that the parametric bootstrap generates valid P-values when we are reasonably happy with our model, but may not necessarily work so well under violations of assumptions (it depends which assumptions are violated, and how badly). Design-based methods tend to make fewer assumptions and thus tend to be more robust to violations. Because the parametric bootstrap adds extra assumptions, it tends to work a bit better when those extra assumptions are satisfied, but it can be misleading when the assumptions are not satisfied. (This rule about assumptions is a pretty universal one—the "no free lunch" principle of statistics.)

A good situation for using the parametric bootstrap is when we just don't know how to use theory to get good P-values directly from the model, but we are not excessively worried by violations of model assumptions. The parametric bootstrap is often used for linear mixed models because we often find ourselves in precisely this situation.

9.5 Resampling Residuals

These resampling methods are all well and good if the goal is to test a hypothesis of no effect—when we can freely resample observations. But what if we don't want to do that? For example, in Exercise 9.4, Angela wants to test for an effect of latitude *after controlling for* the effect of rainfall. In this case, Angela doesn't have an "all means equal" null hypothesis, so she can't use the standard permutation and bootstrap tests.

Exercise 9.4: Global plant height—does rainfall explain latitude effect?
Angela collects data on plant height and climate characteristics from sites around the world. She wants to answer the following question:

Does latitude explain any variation in plant height beyond that due to rainfall?

What model should be fitted under the null hypothesis? Does it include any predictor variables?

The null hypothesis in Exercise 9.4 is that plant height is explained by rainfall. We can't freely permute or bootstrap observed data—this would break up the data structure under the null (removing the relationships between plant height and rainfall, or between latitude and rainfall). We want to resample under the null hypothesis that there is a relationship between rainfall and plant height (and possibly between rainfall and latitude). We can do this by *resampling residuals* using the fitted model under the null hypothesis (as in Edgington, 1995) (Fig. 9.2).

(a)

y	28.0	26.6	0.3	1.6	0.2	1.7	0.5	10.0	40.0	0.5
=										
$\hat{\mu}$	7.9	16.3	4.8	5.8	7.1	9.4	8.6	14.7	8.4	11.1
+										
$\hat{\epsilon}$	20.1	10.3	-4.5	-4.2	-6.9	-7.7	-8.1	-4.7	31.6	-10.6

resample residuals

(b)

$\hat{\epsilon}^*$	20.1	-6.9	-4.5	-4.5	-4.7	20.1	-8.1	-7.7	31.6	20.1
+										
$\hat{\mu}$	7.9	16.3	4.8	5.8	7.1	9.4	8.6	14.7	8.4	11.1
=										
y^*	28.0	9.5	0.3	1.3	2.4	29.5	0.5	7.0	40.0	31.2

Fig. 9.2: Residual resampling of the first 10 observations from the plant height data. (**a**) The observed responses (plant heights, y) can be split into fitted values ($\hat{\mu}$) and residuals ($\hat{\epsilon}$); (**b**) the residuals are resampled ($\hat{\epsilon}^*$), then added back onto the fitted values ($\hat{\mu}$), to construct resampled responses (y^*)

Residual resampling is the default in `mvabund`, as in Code Box 9.4 for the plant data of Exercise 9.4. Residual resampling works for *any* linear model—you could rerun any of the analyses of Chaps. 2–4 using the same lines of code as in Code Box 9.4 but with the relevant changes to the linear model specification in the first line, i.e. you could use residual resampling in factorial ANOVA, in paired or blocked designs, in ANCOVA, ... But a limitation of `mvabund`, and of residual resampling more generally, is that while it works for any fixed designs, it can't handle random effects, at least not in a natural way. The main difficulty is that there are multiple sources of randomness in such a model (specifically, some of the parameters are then treated as random, as well as the response), so resampling would need to happen at multiple levels. A few resampling approaches have been proposed for mixed effects models, but the issue is not completely settled in the literature, and a parametric bootstrap is usually a better option.

Code Box 9.4: Residual resampling using `mvabund` for Exercise 9.4.

```
> ft_heightRL=manylm(height~rain+lat, data=globalPlants)
> anova(ft_heightRL, resamp="perm.resid")
ANOVA Table

Model: manylm(formula = height ~ rain + lat, data = globalPlants)

Overall test for all response variables
Test statistics:
```

```
                Res.Df Df.diff val(F) Pr(>F)
(Intercept)       177
rain              176       1 28.648  0.002 **
lat               175       1  0.326  0.571
---
Signif. codes:  0 `***' 0.001 `**' 0.01 `*' 0.05 `.' 0.1 ` ' 1
Arguments: P-value calculated using 999 iterations via residual
(without replacement) resampling.
```

Can the latitude effect be explained by rainfall?

Does the order in which the terms are entered into the model matter? Why not enter the formula as `height~lat+rain`?

9.5.1 Residual Resampling Is Not Exact

When permuting data under the no-effect hypothesis, permutation testing is exact.

Under residual resampling, permuting and bootstrapping are *only approximate*. Two tips for ensuring this approximation is good:

- Make sure you resample residuals from a plausible null model. If your null model is wrong (e.g. forgot to transform data), your P-values can be wrong and the whole thing can be invalid. So check assumptions.
- Only estimate the resampling distribution of a standardised (or "pivotal") statistic, e.g. t-stat, Z-stat, likelihood ratio stat. Do *not* estimate the resampling distribution of an unstandardised statistic (e.g. $\hat{\beta}$). In some settings, bootstrap resampling is known *not to help* improve validity for unstandardised statistics (as compared to using standardised statistics, e.g. Z, t), but it will help if the statistic is standardised (Hall and Wilson, 1991).

9.5.2 Assumptions of Residual Resampling

Residual resampling makes a few extra assumptions compared to resampling the original observations.

In residual resampling, we assume either

- that residuals are exchangeable (if permuting them) or
- that residuals are independently and identically distributed (if bootstrapping them).

We do not make any assumptions about the shape of the distribution of residuals, but in all other respects, we need our model to be correct for inferences to be valid.

To see what this means in practice, recall, for example, the assumptions of linear models from Chap. 4:

1. The y-values are *independent* (after conditioning on x).
2. The y-values are *normally distributed* with *constant variance*

$$y \sim \mathcal{N}(\mu_y, \sigma^2)$$

3. There is a *straight-line relationship* between the mean of y and each x

$$\mu = \beta_0 + \mathbf{x}^T \boldsymbol{\beta}$$

These are the assumptions we make when using model-based approaches to inference, as for example when making an `anova` call to a `lm` object.

Key Point
Residual resampling extends the ideas of permutation tests and bootstrap tests to general fixed effects designs, but at some cost—we need to assume residuals are exchangeable or iid, which for linear models implies that we still require the linearity and equal variance assumptions.

Now compare that to the assumptions we make when bootstrapping residuals from a linear model (for example when making an `anova` call to a `manylm` object):

1. The y-values are *independent* (after conditioning on x).
2. The y-values have *constant variance*

$$y \sim \mathcal{F}(\mu_y, \sigma^2)$$

   for some distribution $\mathcal{F}$ with additive errors.
3. There is a *straight-line relationship* between the mean of y and each x

$$\mu_y = \beta_0 + \mathbf{x}^T \boldsymbol{\beta}$$

Residual resampling relaxes the normality assumption—*and that's it*! And you may recall normality was the least important assumption in the first place! (The CLT gives us a lot of robustness to violations of this assumption.) So residual resampling isn't a solution to all the world's problems.

It is also a bit of a stretch to call residual resampling design-based inference—a model is needed to compute residuals! So really this approach sits in a grey area between model-based and design-based inference—we are relaxing some of the assumptions of the model but still require key model assumptions to be satisfied, and we can't assume our inferences are valid by pointing to independent units in our study design.

9.5.3 Results Are Similar to `lm` Output

Often you will find that the results you get from design-based inference are similar to what you might get from model-based inference. For example, the P-value for the guinea pig data of Code Box 2.5 is 0.016, and a resampling-based P-value using a large number of resamples (like 9999) settles around 0.012. For the plant height data of Code Box 3.1, most P-values are similar up to two significant figures.

The reason most P-values work out similarly is that, as previously, the only assumption that is being relaxed when fitting these models is normality, which the CLT gave us a lot of protection against anyway. So unless you have quite non-normal data and quite a small sample, you should not expect appreciable differences in results. As such, unless you have a small sample ($n \leq 10$, say), you don't really need to bother with design-based inference at all when analysing a single quantitative response variable. The main case for using design-based inference, considered in this text, is when analysing highly multivariate data (Chap. 14).

9.6 Limitations of Resampling: Still Mind Your Ps and Qs!

We have seen that under resampling, the validity of P-values (or SEs or CIs or other inference procedures) rests primarily on the independence assumption—there is no longer an assumption about the actual distribution of y, although under residual resampling we also require some model assumptions (for linear models, we still need linearity and constant variance).

But we still should check linear model assumptions even when they aren't important for the test to be valid. Why check linear model assumptions that are not needed when resampling?

$$\textit{valid} \neq \textit{efficient}$$

This is a really important idea to understand—many procedures have been proposed for ecological data, which use resampling for inference and which claim to be generally applicable because they make no assumptions beyond independence. But valid procedures can still work very badly—*no* procedure should be applied blindly to data! A procedure should only be applied if it is valid *and* it can be expected to work well for data like what you have (i.e. should lead to relatively small SEs, or relatively high power).

Key Point

$$\textit{valid} \neq \textit{efficient}$$

Just because your inferential procedure is valid doesn't mean it is efficient. That is, just because you use design-based inference, and so don't require all your model assumptions to be satisfied for valid inference, doesn't mean you can ignore model assumptions. Design-based methods will work better (in terms of having better power, shorter CIs, and so forth) when model assumptions are closer to being satisfied.

Just because a method is valid doesn't mean it works well. For example, linear models are more efficient (more power) when our assumptions are closer to being satisfied—so try to satisfy them as closely as you can to get the best answer you can!

Code Box 9.5: Plant height data—checking assumptions

Consider the plant height data, where we were testing for an effect of latitude:

```
ft_heightRLlm=lm(height~rain+lat, data=globalPlants)
plotenvelope(ft_heightRLlm)
```

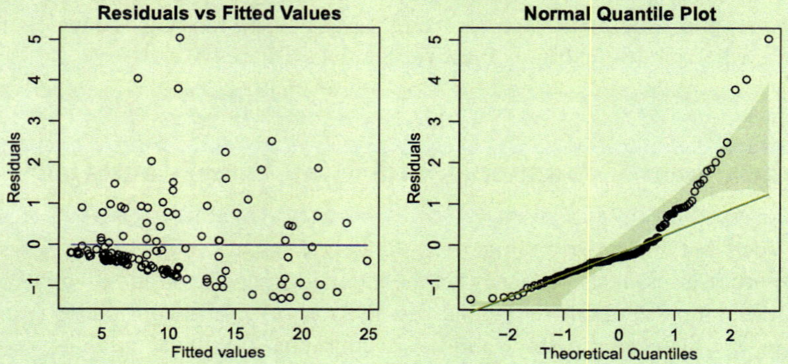

What do you reckon?

Exercise 9.5: Plant height data—log transformation

Considering the results in Code Box 9.5, we probably should have log-transformed our data.

Refit the linear model to the plant height data, available as the `globalPlants` dataset in the `ecostats` package, using a log transformation of the response. Use residual resampling to test for an effect of latitude after controlling for the effect of rainfall.

How do results compare to the analysis without a log transformation?

How do results compare to what you would have got if you used model-based inference by applying `anova` to the `lm` function? Is this what you expected?

Exercise 9.6: Guinea pig data—log transformation

Diagnostic checks of the guinea pig dataset suggest that number of errors is a right-skewed variable.

Log-transform the number of errors and check assumptions. Does this better satisfy assumptions than the model on untransformed data?

Repeat the permutation test of Code Box 9.1 on log-transformed data.

How do the results compare to the analysis without a log transformation? Is this what you expected to happen?

Consider, for example, the height data of Code Box 9.4. If we construct a residual plot, we see that the linear modelling assumptions are not satisfied (Code Box 9.5)—in particular, there is something of a fan shape, suggesting residuals do not in fact have equal variance, and hence they should not be permuted. The main cause of the problem is that values are "pushing up" against the boundary of zero—a plant can't have negative height. The solution in Exercise 9.5 is to log-transform height, in effect removing the lower boundary and putting height onto a proportional scale. This ends up doing a much better job of satisfying assumptions, but it also changes our answer—there ends up being a significant effect of latitude, after controlling for rainfall. The most likely reason for this change is that the original analysis was not efficient—by analysing a strongly skewed variable, with heteroscedastic errors, the linear model test statistic was not able to detect the effect of latitude that was in fact present.

So in summary, resampling can make your method valid even when assumptions fail, but to get a method that works well—has good power, small SEs, for example—you need to use a good model for the data at hand. Only with a good model for your data can you ensure that you have a good statistic for answering the research question of interest. You can get completely different results from different analyses even with resampling (e.g. log-transformed vs untransformed) because of big differences in how reasonable your fitted model is, hence big differences in how good your test statistic is.

Exercise 9.7: Revisiting linear models past

Go back to a couple of linear models (with fixed effects terms only) you have previously fitted, e.g. in the exercises of Chap. 4, and reanalyse using (residual) resampling for inference.

Did the results work out differently? Is this what you expected? (Think about sample size and the normality assumption.)

9.7 Design-Based Inference for Dependent Data

The design-based procedures we have considered so far (in particular, `anova.manylm`) resample residuals independently across observations. *But what if observations are not independent?*

There are a few types of dependence we often encounter and can deal with in a design-based way. In particular:

- Clustered data due to pseudo-replication
- Autocorrelation from temporal, spatial, or phylogenetic sources.

Any of these sources of dependence lead to positively correlated observations, and the impact of ignoring them is to make SEs and, hence, CIs too small and *P*-values too small. That is, *too many false positives*, which as we have discussed, is really bad news.

9.7.1 Block Resampling

Sometimes we can assume blocks of sites are independent (or approximately so), in which case we can exploit independence across blocks and resample those as a basis for inference about terms in the model that vary across blocks.

In `mvabund`, there is a `block` argument that can be used in `anova.manylm` in order to resample blocks of observations and keep all observations within a block together in resampling. This is appropriate in a multi-level (hierarchical) design, where there is an interest in making inferences at the higher sampling level, as in Graeme's estuary data. Note, though, that using this function in combination with residual resampling requires balanced sampling within blocks and assumes that if there is an order to observations within blocks, values are entered in the same order within each block. (This is needed so that each resampled residual matches up with the appropriate fitted value from the same block.) With unbalanced blocks, the `block` argument can only be used in combination with case resampling (i.e. when jointly resampling rows of *X* and *Y*). Note also that because `manylm` only works with fixed effects designs, there is no opportunity to include random effects in a fitted model (you would need to code your own function to do this). An example using `manylm` with the `block` argument to analyse Graeme's estuary data is at Code Box 9.6.

Code Box 9.6: Block resampling using mvabund for estuary data

To test for an effect of modification, using block resampling of estuaries:

```
> data(estuaries)
> ft_estLM = manylm(Total~Mod,data=estuaries)
> anova(ft_estLM,resamp="case",block=estuaries$Estuary)
Using block resampling...
Analysis of Variance Table

Model: manylm(formula = Total ~ Mod, data = estuaries)

Overall test for all response variables
Test statistics:
            Res.Df Df.diff val(F) Pr(>F)
(Intercept)     41
```

```
Mod                40        1  9.916   0.48
Arguments: P-value calculated using 999 iterations via case block
resampling.
```
Case resampling was used because the design is unbalanced.

Sometimes a more sophisticated resampling scheme is desired—not resampling blocks, but resampling treatment levels *within* blocks. This is needed if the treatment of interest is applied at a lower sampling level, as for example in the raven data—12 locations were sampled, then each treatment was applied within each location. So the resampling scheme we desire is to permute the treatment levels *within* each location as in Table 9.2. This can be achieved by first using the `permute` package to construct a resampling scheme, then including this as input to `anova.manylm` to ensure this resampling scheme is used for inference, as in Code Box 9.7. This is sometimes referred to as restricted permutation (since Brown and Maritz, 1982) or restricted resampling,[1] and it can be used to ensure an exact test when the null hypothesis is that observations come from the same distribution (same mean or variance, for example) in each of several different blocks of observations, provided that we permute observations only within blocks.

Table 9.2: Three examples of restricted permutation of the raven counts within sites compared to the observed data. In a paired design, this involves switching the ordering of the pairs at random

Observed Data	
Before	0 0 0 0 0 2 1 0 0 3 5 0
After	2 1 4 1 0 5 0 1 0 3 5 2
Bootstrap 1	
Before	2 0 0 1 0 2 0 1 0 3 5 2
After	0 1 4 0 0 5 1 0 0 3 5 0
Bootstrap 2	
Before	2 1 0 1 0 5 1 0 0 3 5 2
After	0 0 4 0 0 2 0 1 0 3 5 0
Bootstrap 3	
Before	2 1 4 1 0 2 1 1 0 3 5 0
After	0 0 0 0 0 5 0 0 0 3 5 2

[1] Not to be confused with restricted randomisation, a technique often used to allocate patients to treatment groups in clinical trials.

Code Box 9.7: Block resampling using `permute` for raven data

We will start by taking the `ravens` data for the gunshot treatment only and arranging in long format:

```
data(ravens)
crowGun = ravens[ravens$treatment == 1,]
library(reshape2)
crowLong = melt(crowGun,measure.vars = c("Before","After"),
    variable.name="time",value.name="ravens")
```

as constructed in Code Box 4.2. To construct a matrix to use for the permutation of the raven data, permuting treatment labels within each sampling location:

```
library(permute)
CTRL = how(blocks=crowLong$site)
permIDs = shuffleSet(24,nset=999,control=CTRL)
```

Now to use this in mvabund to test for a treatment effect using restricted resampling:

```
> ravenlm = manylm(ravens~site+time,data=crowLong)
> anova(ravenlm,bootID=permIDs)
Using <int> bootID matrix from input.
Analysis of Variance Table

Model: manylm(formula = ravens ~ site + time, data = crowLong)

Overall test for all response variables
Test statistics:
            Res.Df Df.diff val(F) Pr(>F)
(Intercept)     23
site            12      11   3.27  0.995
time            11       1   6.76  0.025 *
---
Signif. codes:  0 `***' 0.001 `**' 0.01 `*' 0.05 `.' 0.1 ` ' 1
Arguments: P-value calculated using 999 iterations via residual
(without replacement) resampling.
```

How do the results compare to what we got previously (Code Box 4.2) using model-based inference?

9.7.2 Block Resampling in Space

If observations are correlated in space, a design-based inference option to account for this is the *moving block bootstrap*. The general idea is that while observations are spatially dependent over short distances, there may be little dependence over larger distances, so we could use resampling of blocks of observations if the blocks are far enough apart from each other. A moving block bootstrap resamples blocks of observations by choosing random points and including in the block anything within a certain distance of that point. This idea is illustrated in Fig. 9.3.

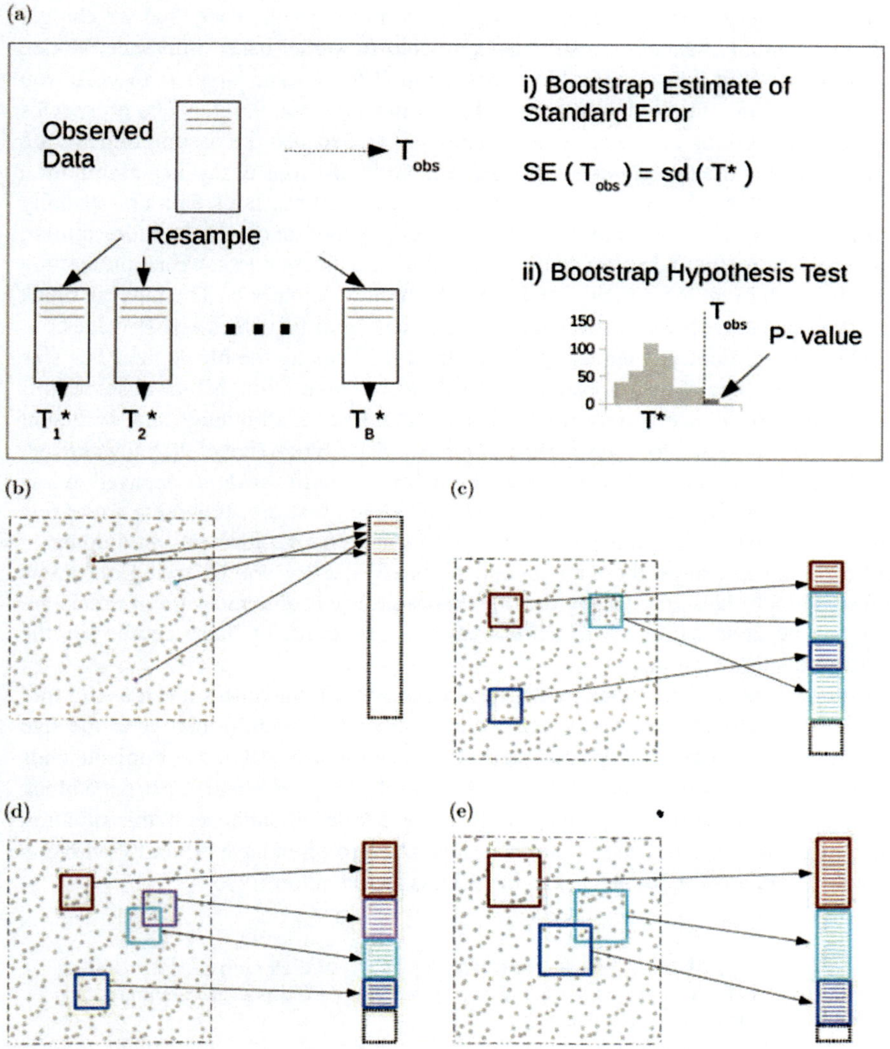

Fig. 9.3: Moving block bootstrap idea (from Slavich & Warton, *in review*). (**a**) Bootstrap procedure to construct distribution of a statistic, T. (**b**) Sampling units are single sites. (**c**) Sampling units are non overlapping blocks of sites. (**d**) Sampling units are overlapping blocks of sites, with a small block length. (**e**) Sampling units are overlapping blocks of sites, with a larger block length

Some code to do a spatial moving block bootstrap is included in the mvabund package, and is applied to Ian's species richness data in Code Box 9.8 and 9.9.

Previously we saw that resampling linear models had little impact on inferences, because the only assumption being relaxed was the least important assumption, that of normality. For block resampling, this is no longer the case, and we can get quite substantial changes in results as compared to model-based inference. Notice that in Code Box 9.8, moving block bootstrap P-values are larger than what you would get using a model-based approach, and in Code Box 9.9, the SEs on species richness predictions also tend to be larger. This is because the assumption we are relaxing here is the *independence assumption*, and violations of this key assumption can have substantial effects on the validity of our method. Ian's data are spatially autocorrelated, with observations that are closer together tending to be more similar, and if we ignore this and pretend our observations are independent, we are pretending there is much more information in the data than there actually is. The moving block bootstrap can correct for this effect, leading to less optimistic SEs and P-values.

The main difficulty when using this method is choosing the block size. The size of the spatial blocks to be resampled needs to be large enough that observations have little dependence across blocks, but it should be small enough that sufficient replication is possible for inference. In Code Box 9.8 a block size of 20 km was used, an educated guess given that the spatial autocorrelation in *residuals* decayed to low values by about this distance in Code Box 7.7. Given that the study area was a few hundred kilometres in diameter, this was also small enough for a decent amount of replication. You can let the data choose the block size for you by trying out a few different block sizes and finding the one that minimises some measure of precision. This can be done using the `blockBootApply` function; for more details see the software documentation.

Note that the moving block bootstrap is not always applicable—it is a useful tool when the spatial scale over which dependence operates is small relative to the size of the region. If there is large-scale dependence, with observations at opposite ends of the region still being dependent, then it is simply not possible to construct a block size such that observations in different blocks have little dependence. In this situation one is stuck with model-based approaches, which also often have difficulty with this situation (specifically, robustness to model misspecification).

Code Box 9.8: Moving block bootstrap test for species richness modelling

First we will fit Ian's model, with quadratic terms for average maximum and minimum daily temperature and annual precipitation:

```
> data(Myrtaceae)
> Myrtaceae$logrich=log(Myrtaceae$richness+1)
> mft_richAdd = manylm(logrich~soil+poly(TMP_MAX,degree=2)+
                   poly(TMP_MIN,degree=2)+poly(RAIN_ANN,degree=2),
                                   data=Myrtaceae)
```

Now we will use a spatial moving block bootstrap for the species richness data, with a block size of 20 km, generating 199 bootstrap resamples computed along on a 5 km grid, via the `BlockBootID` function in the `ecostats` package. This takes a while!

```
> BootID = BlockBootID(x = Myrtaceae$X, y = Myrtaceae$Y, block_L = 20,
             nBoot = 199, Grid_space = 5)
```

Now using this to test the significance of the various terms in Ian's model:

```
> anova(mft_richAdd, resamp="case", bootID=BootID)
```

```
using <int> bootID matrix from input.
Overall test for all response variables
Test statistics:
                          Res.Df Df.diff val(F) Pr(>F)
(Intercept)                  999
soil                         991       8 11.971  0.005 **
poly(TMP_MAX, degree = 2)    989       2 18.244  0.005 **
poly(TMP_MIN, degree = 2)    987       2 17.533  0.530
poly(RAIN_ANN, degree = 2)   985       2  8.208  0.005 **
---
Signif. codes:  0 `***' 0.001 `**' 0.01 `*' 0.05 `.' 0.1 ` ' 1
Arguments: with 199 resampling iterations using case resampling and
           response assumed to be uncorrelated
```

How does this compare to what you would get if you just used model-based inference using the `lm` function?

Code Box 9.9: Moving block bootstrap SEs for species richness predictions

Ian was originally interested in species richness predictions, and we can use the moving block bootstrap to estimate these, adjusting for spatial autocorrelation (using BootID, constructed in Code Box 9.8):

```
> ft_richAdd = lm(logrich~soil+poly(TMP_MAX,degree=2)+
          poly(TMP_MIN,degree=2)+poly(RAIN_ANN,degree=2),
          data=Myrtaceae)
> nBoot=199
> predMat = matrix(NA,length(Myrtaceae$logrich),nBoot)
> for(iBoot in 1:nBoot)
  {
    ids = BootID[iBoot,]
    ft_i = update(ft_richAdd,data=Myrtaceae[ids,])
    predMat[ids,iBoot] = predict(ft_i)
  }
> bootSEs = apply(predMat,1,sd,na.rm=TRUE)
```

Compare these to the SEs on predicted values from the `lm` function:

```
> lmSEs   = predict(ft_richAdd,se.fit=TRUE)$se.fit
> cbind(bootSEs,lmSEs)[1:10,]
         sePreds
  [1,] 0.03756633 0.02982530
  [2,] 0.03166692 0.02396074
  [3,] 0.04107874 0.02853756
  [4,] 0.04217585 0.03348368
  [5,] 0.06295291 0.06279115
  [6,] 0.03707599 0.03413518
  [7,] 0.05155717 0.02808283
  [8,] 0.09053780 0.08272950
  [9,] 0.03058480 0.02759009
 [10,] 0.05739949 0.06622908
```

Is this what you expected to see?

Exercise 9.8: Does block length matter?
For Ian's species richness data, compute SEs using a few different block sizes (say 5, 10, 20, and 40 km) and compare.

Did the results change as you changed the block size? Compute the mean of the SEs at each block size. Is there a trend?

9.7.3 Block Resampling in Time or Phylogeny

The moving block bootstrap was illustrated above for the spatial context, and the software used was developed specifically for spatial data. However, the idea is quite general and could be applied to other types of autocorrelated data, such as phylogenetically structured data. Resampling is commonly used in phylogenetic analysis, but, at the time of writing, the moving block bootstrap has not been. The method could be useful in a phylogenetic context if the traits being measured have "local" dependence, i.e. phylogenetic structuring is seen over small phylogenetic distances, but observations across more distant branches of the tree can be considered to be largely independent of each other. Some thought would be required to adapt block bootstrap methods to the phylogenetic context, and existing software could not be applied to the problem directly.

Chapter 10
Analysing Discrete Data: The Generalised Linear Model

Exercise 10.1: Crabs on seaweed
In Exercise 1.13, David and Alistair looked at invertebrate epifauna settling on algal beds (seaweed) with different levels of isolation (0, 2, or 10 m buffer) from each other, at two sampling times (5 and 10 weeks). They observed the following *presence (+)/absence (−)* patterns for crabs (across 10 replicates):

Time	5 5 5 5 5 5 5 5 5 5 ... 10
Distance (m)	0 0 0 0 0 2 2 2 2 2 ... 10
Crabs	− + + − − − − + − − ... +

They would like to know if there is any evidence of a difference in crab presence patterns with distance of isolation. *How should they analyse the data?*

Exercise 10.2: Do offshore wind farms affect fish communities?
Lena (and colleagues, Bergström et al., 2013) studied the effect of an offshore wind farm on eel abundance by collecting paired data before and after wind farm construction, at 36 stations in each of three zones (Wind Farm, North, or South):

Zone	Impact	Station	Abundance
Wind Farm	Before	WF1	0
Wind Farm	After	WF1	0
South	Before	S1	5
South	After	S1	0
North	Before	N1	0
North	After	N1	0
Wind Farm	Before	WF2	1
Wind Farm	After	WF2	1
⋮	⋮	⋮	⋮
North	After	N36	0

D. I. Warton, *Eco-Stats: Data Analysis in Ecology*, Methods in Statistical Ecology,
https://doi.org/10.1007/978-3-030-88443-7_10

Lena wants to know if there is any evidence of a change in eel abundance at wind farm stations, compared to others, following construction of the wind farm. *How should she analyse the data?*

Exercise 10.3: Invertebrate response to bush regeneration
Anthony wants to evaluate how well invertebrate communities are re-establishing following bush regeneration efforts. Here are some worm counts from pitfall traps across each of 10 sites (where C = control, R = bushregeneration):

Treatment	C	R	R	R	C	R	R	R	R	R
Count	0	3	1	3	1	2	12	1	18	0

He wants to know if there is any evidence that bush regeneration is working. *How should he analyse the data?*

Consider Exercises 10.1–10.3, in particular, look at the response variable. Something that is different about these datasets compared to what we saw previously is that the response variable is *discrete*. Well, in Exercises 10.2–10.3 the data are clearly discrete since they are counts $(0, 1, 2, \ldots)$. In Exercise 10.1 the data can be understood as discrete if we code 0 for absent (−) and 1 for present (+). Presence–absence data are commonly coded this way because when we model mean response, it then has an interpretation as probability of presence.

We learnt in Chaps. 2–4 that when analysing continuous responses, the place to start is linear models (LMs). Well, for discrete responses, the place to start is *generalised linear models* (GLMs). This is especially the case if you have small counts and zeros; in this scenario it is very important to use GLMs rather than LMs. But if you have counts that are all fairly large, you can ignore the discreteness and use linear models anyway. One way to think about this is that larger counts are "close enough" to being continuous, in the sense that they are not pushed up against a boundary, and the gaps between values are not large compared to the range of values observed.

Key Point
Quantitative data can be continuous or discrete, and if your response is quantitative, this has important implications for how the data are analysed.
Continuous: Can take any value in some interval (e.g. any positive value)
$\Longrightarrow$ *linear models (LMs)*
Discrete: Takes a "countable" number of values (e.g. $0, 1, 2, \ldots$)
$\Longrightarrow$ *generalised linear models (GLMs)*

If your data are discrete but never take small values (e.g. always larger than 5), you could think of it as kind of close to continuous and try LMs, but if you have small counts and zeros, GLMs are needed to handle the mean–variance relationship.

Why does discreteness matter? The main reason is that it tends to induce a *mean–variance relationship*—as the mean changes, the variance changes. This is illustrated in Figs. 10.1, 10.2, and 10.3 for Exercises 10.1–10.3, respectively.

When you have zeros and small counts, you will have trouble getting rid of the mean–variance relationship via transformation—in fact it can be shown mathematically that it is generally impossible to remove a mean-variance relationship for small counts (Warton, 2018), which leaves us violating the linear model assumption of constant variance. This happens because small counts are "pushed up" against the boundary of zero. Transformation can't fix the problem for small counts because if many values are zero or one, then transformation can't spread these value out well—the zeros will always stay together under transformation, as will all the ones. Consider the eel data of Fig. 10.2—no eels were found in northern stations after the wind farm was constructed, so the variance in this group will always be zero irrespective of data transformation, which is problematic for the equal-variance assumption of linear models.

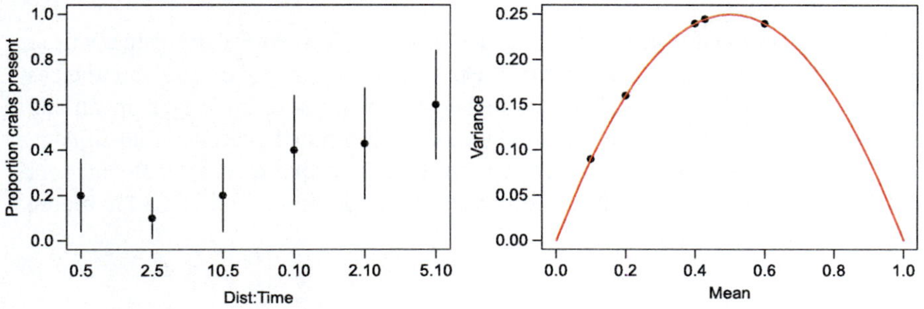

Fig. 10.1: Mean–variance relationships for crab presence–absence data of Exercise 10.1. Left: Sample proportions for the six treatment combinations, with lines for standard errors. Right: Sample means plotted against sample variances for the six treatments. Notice the treatments with the smallest proportions have smaller variances (left), and when we plot these variances against the proportions (right), they exactly fit a quadratic mean–variance trend. It is important to use a GLM to model presence–absence data to account for this mean–variance relationship

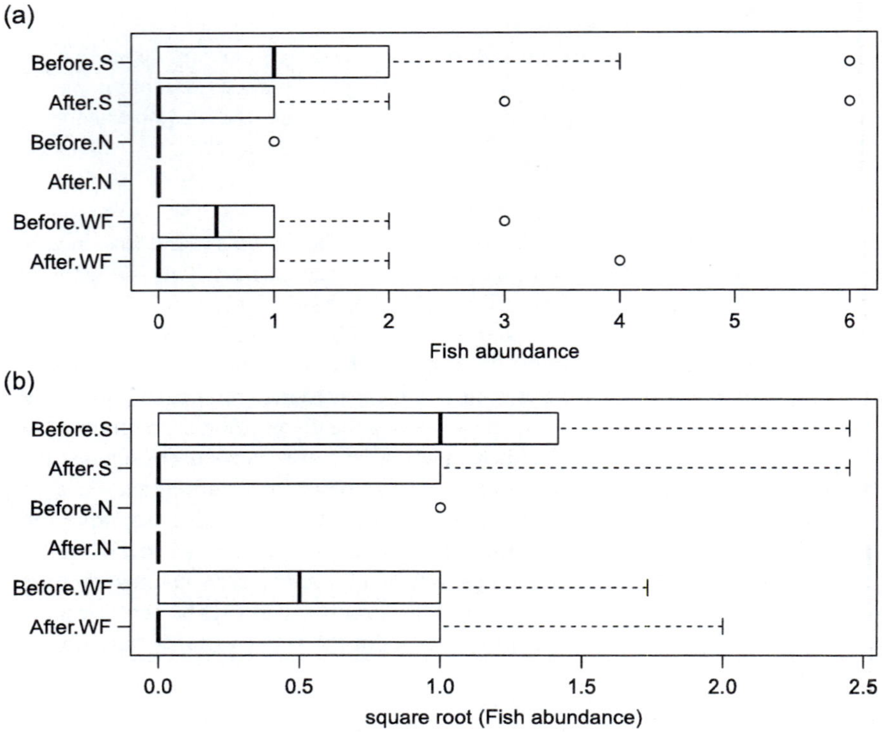

Fig. 10.2: (**a**) Untransformed and (**b**) square-root-transformed eel abundances from Exercise 10.2. Notice in (**a**) that when mean abundance is higher, the abundances are also more variable, e.g. south stations before impact have the largest mean and variance. At the other extreme, no eels were caught at north stations after impact, giving a mean and variance of zero. Note in (**b**) that after square-root transformation, these patterns are still evident. In this situation, it is important to use a GLM to model abundance

In ecology, this sort of issue arises especially often when studying the distribution or abundance of a particular species across a range of environments. Most species have some habitats in which they are rarely found, so when we sample in such habitats, we get lots of zeros and small counts, as with the eels (Fig. 10.2).

On the other hand, if the response variable is a count across different species (e.g. total abundance or species richness), counts tend to be larger, and transformation is sometimes a viable option. The reason for this is that it is not common to sample in such a way that you don't see anything at all, hence zeros (and ones) are rarely encountered when summing across species. Consider again the wind farm data,

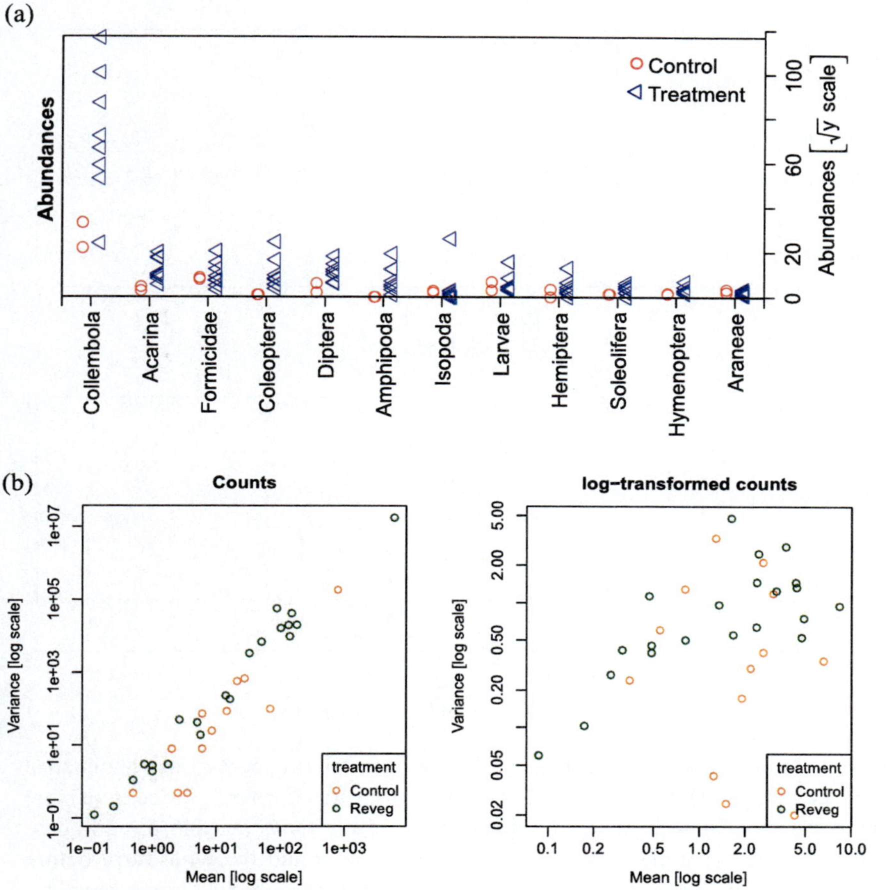

Fig. 10.3: Mean–variance relationships for bush regeneration monitoring data of Exercise 10.3, but across multiple taxonomic groups (not just worms). (**a**) Plots of the 12 most abundant taxa, $\sqrt{y}$-transformed; notice the spread of values is larger for the more abundant taxa (even after $\sqrt{y}$ transformation). (**b**) Sample mean-variance relationships across all taxa, untransformed (left) and $\log(y+1)$-transformed (right). Notice in both cases that when the mean is low, the variance is always low (and increasing with the mean), which we can account for using a GLM

but looking at all fish that were caught rather than just the eels (Fig. 10.4). While not every species was caught at every sampling station, at least one species was always caught (and commonly, 5–20 individuals). So while total fish abundance has a mean–variance relationship (Fig. 10.4a), transformation does a pretty good job of removing it in this case (Fig. 10.4b).

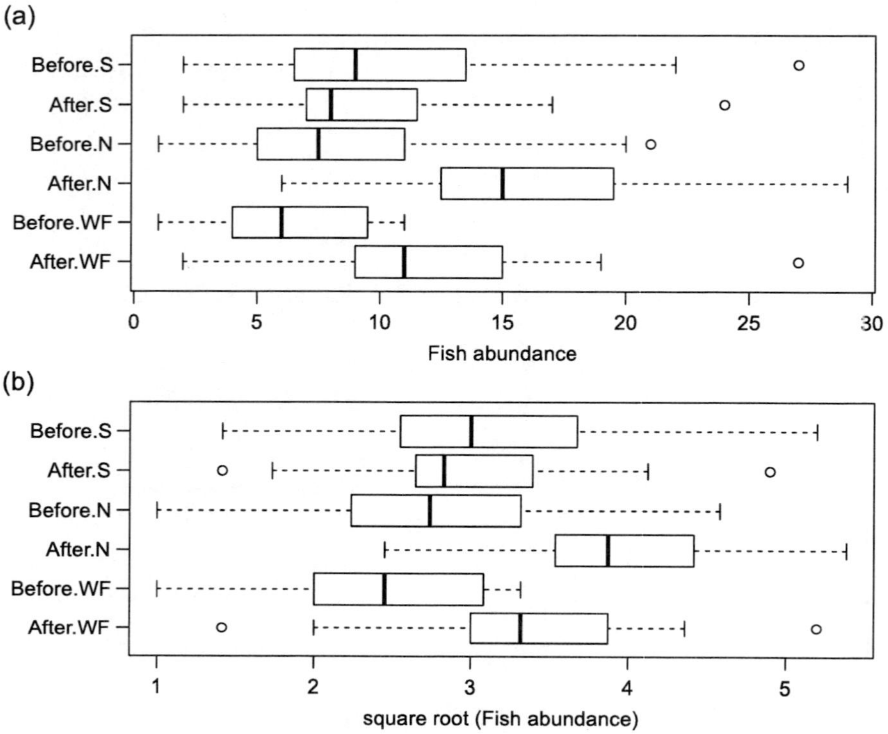

Fig. 10.4: Wind farm data of Exercise 10.2, but now looking at total fish abundance across all species (**a**) untransformed, (**b**) square-root-transformed. Notice in (**a**) that when mean abundance is higher, the abundances are also more variable, e.g. compare north stations after impact (high mean, high variance) and the wind farm before impact (low mean, low variance). Notice in (**b**) that after square-root transformation, this pattern is largely gone. This is because a look at total abundance across all species reveals there aren't many small counts, so transformation to stabilise variances becomes an option. In this situation we might be able to get away with using a LM instead of a GLM

10.1 GLMs: Relaxing Linear Modelling Assumptions

A generalised linear model (Nelder and Wedderburn, 1972, GLM) can be understood as taking a linear model and doing two things to it—changing the equal variance assumption and changing the linearity assumption.

The most important thing that is done is to change the assumptions on the variance—rather than assuming equal variance, we assume a *mean–variance relationship*, which we will write as $V(\mu)$ for a mean μ. If the variance is known to change in a predictable way as a function of the mean, a GLM will incorporate this information into the analysis model. For presence–absence data, for example, it is known that the variance is exactly a quadratic function of the mean: $V(\mu) = \mu(1-\mu)$, which is the curve drawn on the mean–variance plot in Fig. 10.1. Using a GLM there is no need to try to transform the data in advance of analysis to remove the mean–variance relationship—we just include it in the model. This is an important advance and quite a different way of approaching analysis if you are used to linear models and data transformation—by avoiding data transformation, interpretation of the model is simpler, and you keep statisticians happy, who tend to have a philosophical preference for trying to model the mechanism behind the observed data (rather than messing around with the data first to try and match them to a particular type of model).

The other change that is made in a GLM, as compared to a LM, is to the linearity assumption—rather than assuming that the mean response varies linearly with predictors, we now assume that some *known* function of the mean varies linearly with predictors. This function is known as the *link function* and will be written $g(\mu)$. The main reason the link function is introduced is to ensure that predicted values will stay within the range of possible values for the response, e.g. presence–absence data take the values zero (absence) or one (presence), so we use a link function that ensures that the predicted values will always remain between zero and one (such as the logit function, introduced later). In contrast, if you were to analyse discrete data using a linear model, it would be possible to get nonsensical predictions like 120% survival or a mean abundance of −2!

A common misconception is that the link function $g(\mu)$ transforms the data—but it is not doing anything to the data. The link function is a way of introducing special types of non-linearity to a model for the mean, e.g. assuming the mean is an exponential function of predictors or a sigmoidal ("logistic") function, for example, without messing with the data at all.

10.1.1 A More Mathematical Specification

A more mathematical specification for a generalised linear model for a response y_i as a set of predictor variables ($\boldsymbol{x}_i$) is

$$y_i \sim F(\mu_i, \phi) \text{ such that } Var(y) = V(\mu) \tag{10.1}$$

$$g(\mu_i) = \beta_0 + \boldsymbol{x}_i'\boldsymbol{\beta} \tag{10.2}$$

where, as for linear models, $\boldsymbol{x}'\boldsymbol{\beta}$ is vector notation for the linear sum $x_1\beta_1 + x_2\beta_2 + \ldots + x_p\beta_p$, and the y_i are assumed to be *independent* across observations (conditional on x).

The distribution F is assumed to come from what is known as the exponential family of distributions (Maths Box 10.1), which contains as special cases the binomial, Poisson, and (given the overdispersion parameter ϕ) the negative binomial. These are the most natural distributions to use to model discrete data.

Maths Box 10.1: Exponential family of distributions
Any one-parameter probability function $f(y; \mu)$ can be written in terms of the value of the variable y and its mean parameter μ. A distribution is said to be a member of the one-parameter exponential family if the log of its probability function has the form

$$\log f(y; \mu) = y\theta - A(\theta) + B(y) \tag{10.3}$$

where $\theta = h(\mu)$ for some function h. The key idea in this definition is that the value y and parameter μ only interact, on the log scale, through a product of the value y and some function of the parameter $\theta = h(\mu)$. The forms of the functions $h(\mu)$ and $A(\theta)$ end up being important—$h(\mu)$ suggests a "default" or *canonical link* function to use in a GLM, that has desirable properties, and $A(\theta)$ captures information about the mean–variance relationship in the data. Specifically, it turns out that the mean is the derivative of this function, $\mu = A'(\theta)$, and the *mean–variance function* is the second derivative, $V(\mu) = A''(\theta)$ (McCullagh and Nelder, 1989, Section 2.2.2). It also turns out that $h'(\mu_i) = V(\mu_i)^{-1}$ (McCullagh and Nelder, 1989, Section 2.5).

As an example, for the Poisson distribution with mean μ, $f(y; \mu) = e^{-\mu}\frac{\mu^y}{y!}$, so

$$\log f(y; \mu) = y \log \mu - \mu + \log y!$$

So the obvious or "canonical" link function to use in a GLM of Poisson counts is $\theta = \log(\mu)$. $A(\theta) = \mu = e^\theta$, and differentiating twice we see that the variance of Poisson counts is $V(\mu) = A''(\theta) = e^\theta = \mu$. Or, alternatively, $h'(\mu) = \frac{1}{\mu}$, so $V(\mu) = \left(\frac{1}{\mu}\right)^{-1} = \mu$.

Some distributions have two parameters but an exponential form, most notably, the normal, binomial, and gamma distributions. The second "scale" parameter σ affects the variance, which becomes $\sigma^2 A''(\theta)$. Some other two-parameter distributions are related to the exponential family. In particular, negative binomial and beta distributions have a second "shape" parameter and will only be members of the exponential family if the value of this shape parameter is known, making them slightly messier to work with. The binomial also has a second shape parameter, the number of trials, which typically is known.

Recall that in linear models, the normality assumption usually ends up not being that important (thanks to the central limit theorem), the more important part is the equal variance assumption, which can be understood as being implied by the normality assumption. In much the same way, the actual distribution F ends up not being that important in a GLM (thanks to the central limit theorem); the more important part is the mean–variance assumption that is implied by the chosen distribution. Thus, the key parts of a GLM to pay attention to are the independence assumption, the mean–variance relationship function $V(\mu)$, and the link function $g(\mu)$ (Maths Box 10.2).

Maths Box 10.2: Maximum likelihood and the mean–variance function
Consider a GLM for independent observations $\{y_1, \ldots, y_n\}$ with mean model

$$g(\mu_i) = x_i\beta$$

We only consider one predictor, and no intercept, to keep it simple.

Using Eq. (10.3), we can write the log-likelihood as

$$\ell(\beta) = \sum_{i=1}^{n} \log f(y_i; \theta_i) = \sum_{i=1}^{n} \{y_i\,\theta_i - A(\theta_i) + B(y_i)\}$$

where $\theta_i = h(\mu_i)$, as before.

When fitting a model by maximum likelihood, we need to find its stationary point by setting the derivative (or *score function*) to zero. Differentiating with respect to β (via the chain rule):

$$\frac{d}{d\beta}\ell(\beta) = \sum_{i=1}^{n} \frac{d\theta_i}{d\beta}\frac{\partial}{\partial\theta_i}\ell(\beta) = \sum_{i=1}^{n} x_i \frac{h'(\mu_i)}{g'(\mu_i)}(y_i - A'(\theta_i)) = \sum_{i=1}^{n} \frac{x_i}{g'(\mu_i)}\frac{y_i - \mu_i}{V(\mu_i)} \tag{10.4}$$

the last step follows since $\mu_i = A'(\theta_i)$ and $h'(\mu_i) = V(\mu_i)^{-1}$ (as in Maths Box 10.1). To show that $\frac{d\theta_i}{d\beta} = x_i\frac{h'(\mu_i)}{g'(\mu_i)}$ requires a little work—using the chain rule again we can say $\frac{d\theta_i}{d\beta} = \left(\frac{d\beta}{d\mu_i}\right)^{-1}\frac{d\theta_i}{d\mu_i}$, and recall that $\theta_i = h(\mu_i)$.

Notice that the distribution of the data only affects the score equation via its mean–variance function $V(\mu_i)$, so this is the key distributional quantity that matters. Notice also that if the link function used in fitting is the canonical link, $g(\mu_i) = h(\mu_i)$, then $g'(\mu_i) = h'(\mu_i)$, so the score equation simplifies to $0 = \sum_{i=1}^{n} x_i(y_i - \mu_i)$.

10.1.2 Generalised Linear Model Assumptions

Recall that *linear models* make the following assumptions:

1. The observed y-values are *independent*, conditional on x.
2. The y-values are *normally distributed* with *constant variance*

$$y \sim \mathcal{N}(\mu_y, \sigma^2)$$

3. There is a *straight-line relationship* between the mean of y and each x:

$$\mu_y = \beta_0 + \mathbf{x}^T\boldsymbol{\beta}$$

Generalised linear models (GLMs) extend linear models to non-normal data by making the following assumptions:

1. The observed y-values are *independent*, conditional on x.
2. The y-values come from *a known distribution* (from the exponential family) with known *mean–variance relationship* $V(\mu)$.
3. There is a *straight-line relationship* between *some known function of the mean* of y and each x

$$g(\mu_y) = \beta_0 + \mathbf{x}^T \boldsymbol{\beta}$$

The function $g(\cdot)$ is known as *the link function*.

As previously, the key changes here, as compared to linear models, are the introduction of a mean–variance relationship $V(\mu)$ and a link function $g(\mu)$.

Key Point
A GLM extends linear models by adding two features:

- A *mean–variance relationship* $V(\mu)$ (in place of constant variance)
- A *link function* $g(\cdot)$ used to transform the mean before assuming linearity.

10.2 Fitting a GLM

The key function in `R` is `glm`, and you choose which type of GLM to fit using the `family` argument, with common examples of how this is used in Table 10.1. Some simple examples using this function are in Code Box 10.1.

Code Box 10.1: Example GLM fits for Exercises 10.1–10.3.
For Alistair and David's crab data in Exercise 10.1, first we have to convert it to presence–absence, then we use the `glm` function:

```
library(ecostats)
data(seaweed)
seaweed$CrabPres = seaweed$Crab>0
ft_crab = glm(CrabPres~Time*Dist, family=binomial("cloglog"),
              data=seaweed)
```

We chose a complementary log-log link, because this is the link function theory says to use if crab counts were to follow a Poisson regression model.

For the wind farm eel counts of Exercise 10.2, assuming counts are Poisson:

```
data(windFarms)
eels = windFarms$abund[,16]
ft_eels = glm(eels~Station+Year*Zone,family="poisson",
          data=windFarms$X)
```

`Station` was included in the model to account for the pairing in the data.

To fit a negative binomial regression to Anthony's worm counts from Exercise 10.3, using the `manyglm` function from the `mvabund` package:

```
data(reveg)
Haplotaxida=reveg$abund[,12]
library(mvabund)
worms = reveg$abund$Haplotaxida
ft_worms = manyglm(worms~treatment,family="negative.binomial",
data=reveg)
```

A GLM is fitted using *maximum likelihood*, as discussed in Sect. 6.3. The main practical consequence of no longer using least squares to fit models is that when measuring goodness of fit, we no longer talk about residual sums of squares. Instead we talk about *deviance*—basically, twice the difference in (log-)likelihood between the fitted model and a perfect model (in which predicted values are exactly equal to observed values).

Remember to set the family argument—if you forget it, `glm` defaults to fitting a linear model (`family=gaussian`).

10.2.1 What Distributions Can I Use?

Not all distributions can be used with generalised linear models, but a few important ones can, as listed in Table 10.1. A few important things to know about these follow.

The **binomial** can be used for any response that has two possible outcomes (a *binary* response), also for "x-out-of-n" counts across n independent events. Three link functions are commonly used—the logit link, $\log\left(\frac{\mu}{1-\mu}\right)$, otherwise referred to as *logistic regression*, is the most common. The probit is sometimes used for theoretical reasons, $\Phi^{-1}(\mu)$, where Φ is the probability function of the standard normal. Sometimes the *complementary log-log link* is used, $\log(-\log(1-\mu))$; it is a good option for presence–absence data and should be used much more often than it currently is. This link function is derived by assuming you had counts following a Poisson log-linear model (see below) and converted them to presence–absence data (following an argument first made in Fisher, 1922). The Poisson assumption may be questionable, but this is the only common link function that was motivated by thinking of data as arising from some underlying counting process (as they often are, e.g. in David and Alistair's crab data of Exercise 10.1).

Table 10.1: Common choices of distribution and suggested link functions $g(\mu)$ in generalised linear models. Each distribution implies a particular mean–variance assumption $V(\mu)$. The required `family` argument to use each of these in R is also given

Distribution	$V(\mu)$	Good for...	Link, $g(\mu)$	family=...

Binomial	$n\mu(1-\mu)$	Binary responses (e.g. presence–absence)	$\log\left(\frac{\mu}{1-\mu}\right)$
			Probit
			$\log(-\log(1-\mu))$
				`binomial`
				`binomial("probit")`
				`binomial("cloglog")`

Poisson	μ	Counts[a]	$\log(\mu)$	`poisson`
Negative binomial	$\mu + \phi\mu^2$	Counts	$\log(\mu)$	`"negative.binomial"` (in `mvabund`)
Tweedie	$a\mu^p$	Biomass	$\log(\mu)$	`tweedie(p,link=0)` (in `statmod`)
Normal	σ^2	Continuous responses	μ	`gaussian`

[a] But does not account for overdispersion (i.e. all individuals being counted need to be independent, no missing predictors in the model.)

A **Poisson** model with a log link is often called a *log-linear model*, or sometimes *Poisson regression*. The mean–variance assumption of the Poisson, $V(\mu) = \mu$, is quite restrictive and will be violated if the individuals being counted are not independent (e.g. they cluster) or if there are potentially important covariates missing from your model. Note this assumption does not seem to work for Anthony's revegetation data (Fig. 10.5, left).

A **negative binomial** model with a log link is often called *negative binomial regression*. It is a safer option for count data than Poisson regression because it includes an overdispersion parameter (ϕ) to deal with unaccounted-for variation. Strictly speaking, this is only a GLM if the overdispersion parameter (ϕ) is known in advance, and for that reason it is not included as an option in the standard `glm` function. It can, however, be fitted using `manyglm` in the `mvabund` package or using `glm.nb` from the `MASS` package. Note this assumption seems to do quite a good job for Anthony's revegetation data (Fig. 10.5, right).

A **Tweedie** model is only a GLM when the power parameter p is known, so p needs to be specified first to use `glm` to fit a model. Usually, you would try a few values of p and check diagnostics or use model selection tools. This has two cool features for ecologists. Firstly, it has a mass at zero while being continuous for positive values, so it can do a good job when modelling biomass data (which will be zero if nothing is observed, otherwise continuous over positive values). However, it doesn't always do a good job of modelling the zeros, so assumptions need to be checked carefully. The second cool feature is that its mean–variance relationship is $V(\mu) = a\mu^p$, known as Taylor's power law, and much ecological data seem to (approximately) follow it (Taylor, 1961).

In the special case where you assume y is **normal** and use the identity link function (i.e. no transformation on μ), GLMs reduce to linear models, as discussed previously (Chaps. 2–4).

There are more distributions that you could use. I hardly ever do.

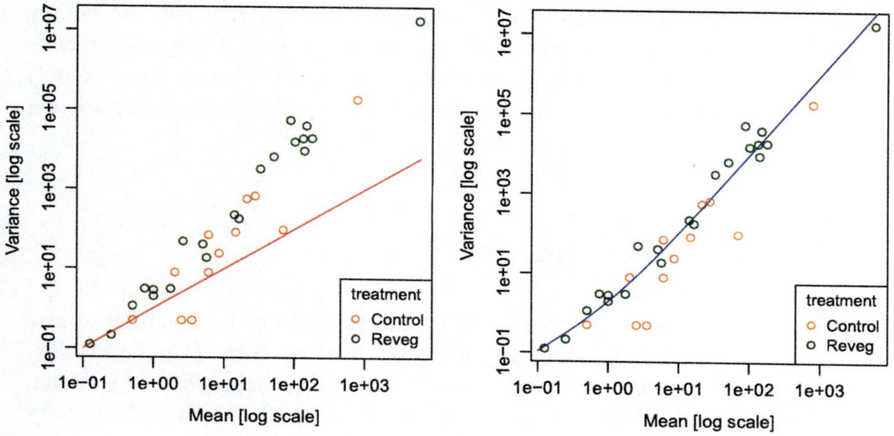

Fig. 10.5: Sample mean–variance relationship for invertebrate counts as in Exercise 10.3 (but across all orders sampled, not just worms). The assumed mean–variance relationship has been added to the plot under a Poisson (left) or negative binomial (right) model. *Which mean–variance assumption looks more plausible?*

In principle you can use any link function with any distribution. But in practice, you almost always should use a function that matches the range of allowable values (or the "domain") of the mean, e.g. Poisson means are always positive, so use a link function that works for positive values only and maps them onto the whole number line, e.g. the log function.

Note that for counts and biomass, we typically use a log link, which makes our model multiplicative rather than additive, such that we talk about effects of treatments as having a c-fold effect rather than an effect of increasing the mean by c. This is usually a good idea because counts tend to make more sense (as a first approximation) when thought of as an outcome of multiplicative rather than additive processes. For example, we talk a lot about *rates* when studying abundance (e.g. survival rates, fecundity rates), which suggests that we are thinking about multiplying things together not adding them.

10.2.2 *Notation Warning: A General Linear Model is Not a GLM!*

Some refer to the linear model as a "general linear model", GLM for short. This is a model where we assume the response is normally distributed, with no mean–variance relationship and no link function. This terminology is bad (and confusing) and we can blame software packages like SAS for popularising it.

When people talk about GLMs, make sure you are clear what they are talking about—do they mean a generalised linear model (non-normal response, mean–variance relationship, and so forth), or are they just using an ordinary linear model? If they don't mention a family or link function, they are probably just using a linear model.

10.3 Checking GLM Assumptions

It is sometimes said that a good model will have residual deviance similar in size to the residual degrees of freedom. For example, in Code Box 10.2, the residual deviance is not far from the residual degrees of freedom, so we might conclude that at this stage there does not seem to be a problem with the model fit. But this is a *very* rough rule (see section 4.4.3 of McCullagh and Nelder, 1989, for example), and it won't work if you have lots of zeros or small counts. We really should dig deeper and look at some other diagnostic tools.

Code Box 10.2: Summary of a GLM fit to the crab presence–absence data of Exercise 10.1.

```
> data(seaweed)
> seaweed$CrabPres = seaweed$Crab>0
> seaweed$Dist = as.factor(seaweed$Dist)
> ft_crab = glm(CrabPres~Time*Dist, family=binomial("cloglog"),
             data=seaweed)
> summary(ft_crab)

Coefficients:
             Estimate Std. Error z value Pr(>|z|)
(Intercept)   -2.3282     1.5046  -1.547    0.122
Time           0.1656     0.1741   0.952    0.341
Dist2         -1.5921     2.5709  -0.619    0.536
Dist10        -0.5843     2.1097  -0.277    0.782
Time:Dist2     0.1683     0.2899   0.581    0.561
Time:Dist10    0.1169     0.2399   0.487    0.626

(Dispersion parameter for binomial family taken to be 1)

    Null deviance: 71.097  on 56  degrees of freedom
Residual deviance: 62.999  on 51  degrees of freedom
AIC: 74.999
```

Is there anything here to indicate whether the model fits the data OK?

Don't just look at numerical measures (residual deviance, Akaike information criterion (AIC), and so forth)—plot your data! As with linear models (Chap. 4), we can use residual plots to check for no pattern. But if you try the default plotting function for `glm` objects using `R` or most packages, you won't have much joy for discrete data, as in Fig. 10.6.

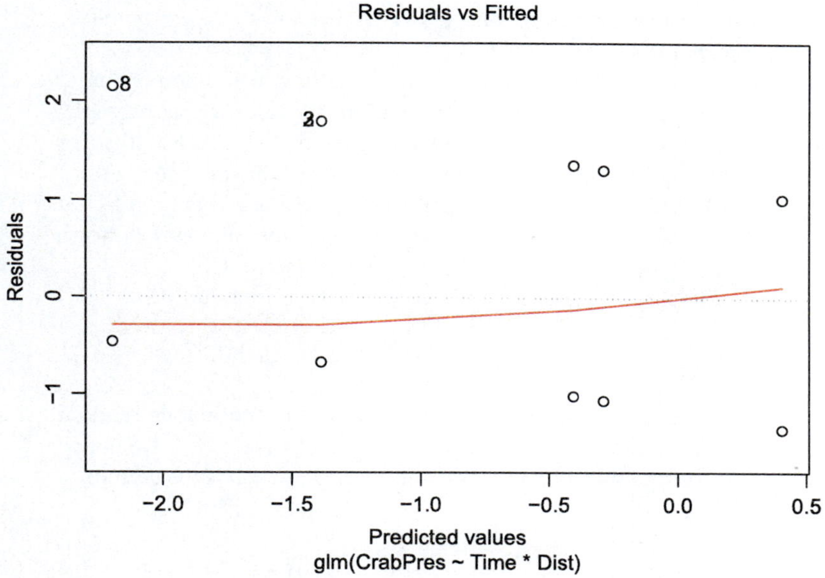

Fig. 10.6: Residual plots aren't as straightforward for a GLM. Here is the default residual plot from a `glm` fit in `R` to David and Alistair's crab data using `plot(ft_crab)`. It looks super weird and is uninformative

Figure 10.6 looks really weird for a couple of reasons. The main problem is the discreteness—there are two lines of points on the plot (for zeros and for ones), which is a pattern that distracts us from the main game. The problem is with the idea of a residual—it is not obvious how to define residuals appropriately for GLMs. Most software (including `R`) chooses badly.

The right choice of residuals is to define them via the probability integral transform (Maths Box 10.4), which we will call *Dunn-Smyth residuals* after Dunn and Smyth (1996). These residuals have been around for some time, but it was only relatively recently that their awesomeness was widely recognised.

Maths Box 10.3: The probability integral transform
Let the distribution function of Y be $F(y) = P(Y \leq y)$, and define $Q = F(Y)$. If Y is continuous,

$$Q \sim \mathcal{U}[0, 1] \tag{10.5}$$

where $\mathcal{U}[0, 1]$ is a standard uniform variable, that is, a variable equally likely to take any value between zero and one. $Q = F(Y)$ is known as the *probability integral transform (PIT)*.

Conceptually, $q = F(y)$ is a quantile function, telling us the proportion of values we would expect to be less than or equal to any value y. For example, if $F(y) = 0.2$, we expect 20% of values drawn from Y to be less than or equal to the value y. PIT notices that this means that 20% of values drawn from $Q = F(Y)$ will be less than or equal to 0.2. This argument works for any proportion q and is a special property defining the standard uniform distribution—if $P(Q \leq q) = q$ for $0 \leq q \leq 1$, then $Q \sim \mathcal{U}[0, 1]$.

The proof of this result is not so hard. Let $Q = F(Y)$. Then for $0 \leq q \leq 1$:

$$P(Q \leq q) = P(F(Y) \leq q) = P(Y \leq F^{-1}(q))$$

where this last step, inverting $F(\cdot)$, is only allowed for continuous random variables (since inversion requires functions to be one-to-one). Now we notice that this has the form of a cumulative probability for Y, which is the definition of $F(\cdot)$:

$$= F(F^{-1}(q)) = q$$

and as previously this is the distribution function of the standard uniform distribution, so the proof is complete.

Conversely, if $Q \sim \mathcal{U}[0, 1]$ and $F(\cdot)$ is an increasing function mapping to values in the interval [0, 1], then

$$Y = F^{-1}(Q) \sim F \tag{10.6}$$

meaning that Y has distribution function $F(y)$. So PIT works both ways—we can use it to transform any variable to standard uniform and to transform from the standard uniform to any distribution. We use both of these results to construct Dunn-Smyth residuals (Fig. 10.7).

Maths Box 10.4: Dunn-Smyth residuals
Dunn and Smyth (1996) proposed constructing residuals r that solve

$$\Phi(r) = F(y)$$

where $\Phi(r)$ is the distribution function of the standard normal distribution (a normal variable with mean zero, variance one). If the true distribution function of the data y is $F(y)$ and it is continuous, then by the PIT (Eq. (10.5) in Maths Box 10.3), $q = F(y)$ come from a standard uniform distribution. But since $q = \Phi(r)$, this would in turn mean that residuals r then come from a standard normal (applying the PIT inversion result, Eq. (10.6)) if $F(y)$ is the true distribution of data y. So standard diagnostic tools for linear models can be used to check if these "Dunn-Smyth" residuals are normal, to check assumptions for *any* parametric model $F(y)$ that assumes the response is continuous.

The problem for discrete Y is that $F(Y)$ will be discrete rather than standard uniform. It is still the case that if $F(y) = q$, we expect a proportion q of values less than or equal to y. The problem is that this equation only holds at special values of q corresponding to observable values of y (e.g. $y = 0, 1, \ldots$ for counts). The solution is to "jitter", solving

$$\Phi(r) = uF(y) + (1 - u)F(y^-)$$

for some standard uniform u, where y^- is the previous value of $F(y)$. The right-hand side of this equation chooses a value at random between $F(y)$ and its previous value $F(y^-)$ (Fig. 10.7). This makes the right-hand side standard uniform when $F(y)$ is the true distribution function of y, hence r is standard normal if the model is correct (Fig. 10.7).

These residuals are not in most standard packages (yet), but you can use them via the `mvabund` package by refitting your model using `manyglm`. See for example Code Box 10.3, which makes use of the `plotenvelope` function to add simulation envelopes for easier interpretation. You can also construct these residuals for many common distributions using the `qresid` function in the `statmod` package in `R` (Dunn and Smyth, 1996) or the `DHARMa` package (Hartig, 2020), which computes residuals via simulation (but note it doesn't transform to the standard normal by default).

Code Box 10.3: Dunn-Smyth residual plots for the crab data, using the mvabund package.

```
> library(mvabund)
> ftMany_crab = manyglm(CrabPres~Time*Dist,family=binomial("cloglog"),
              data=seaweed)
> plotenvelope(ftMany_crab, which=1)
```

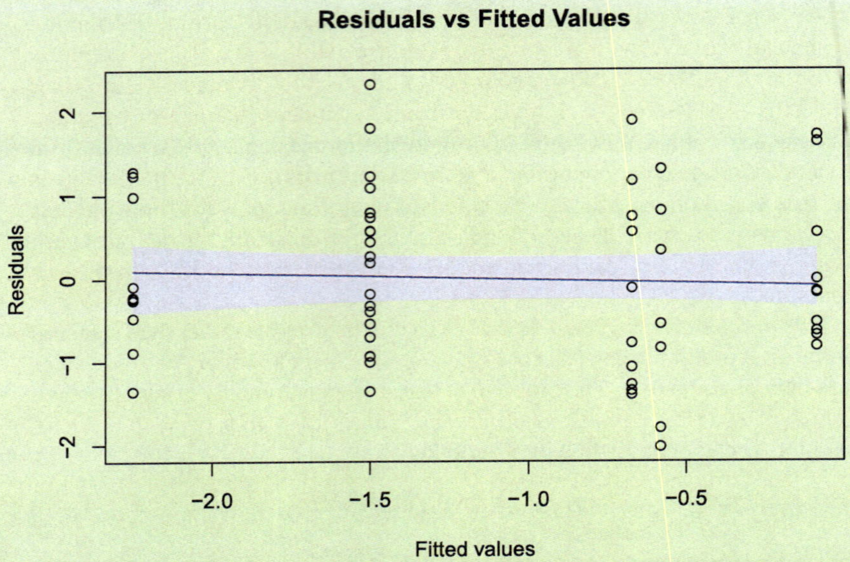

This plot has a bit of a curve on it, but notice that it stays well within the simulation envelope, so it is not doing anything unusual compared to what would happen if model assumptions were satisfied. So we don't have evidence that there is anything to worry about here.

You can interpret a plot of Dunn-Smyth residuals pretty much like a residual plot for linear models. Recall that for linear regression

- U shape $\Longrightarrow$ violation of straight-line assumption
- Fan shape $\Longrightarrow$ violation of variance assumption

Well, Dunn-Smyth plots work in much the same way:

- U shape $\Longrightarrow$ violation of linearity assumption
- Fan shape $\Longrightarrow$ violation of mean–variance assumption

(although you can get a bit of interaction between these two). As previously, the best place to see a fan shape is often a scale-location plot, where it shows up as an increasing trend (as in Code Box 10.4).

Dunn-Smyth residuals deal with the problem of discreteness through random number generation—basically, jittering points. This means different plots will give you slightly different residuals, as in Fig. 10.8. *Plot Dunn-Smyth residuals more than once* to check if any pattern is "real".

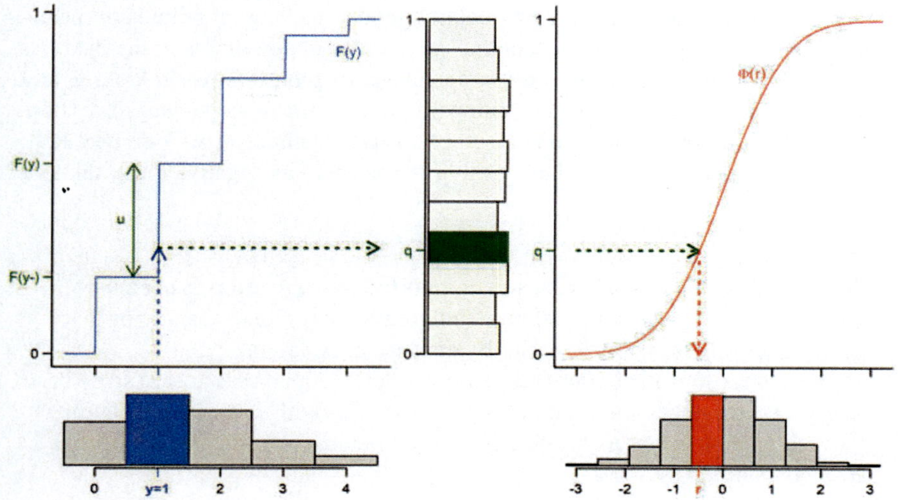

Fig. 10.7: Diagram illustrating how Dunn-Smyth residuals are calculated, for a variable that can take values $0, 1, 2, \ldots$; hence its distribution function $F(y)$ (blue, *left*) has steps at these values. Dunn-Smyth residuals find values r that satisfy $\Phi(r) = uF(y) + (1-u)F(y^-)$ $(= q)$ (Maths Box 10.4). For example, if $y = 1$, we take a value at random between $F(1)$ and its previous value, $F(0)$, and use $\Phi(r)$ to map across to a value for r. By the PIT (Maths Box 10.3), if $F(y)$ were the true distribution function of the data, quantile values q would be standard uniform (green, *centre*), and r would be standard normal (red, *right*)

If data are highly discrete, the jittering introduces a lot of noise, and assumptions can show up as relatively subtle patterns in these plots. Fitting a smoother to the data is a good thing to do in this situation; also include a simulation envelope on where you expect it to be when assumptions are satisfied, as is done by `plotenvelope`.

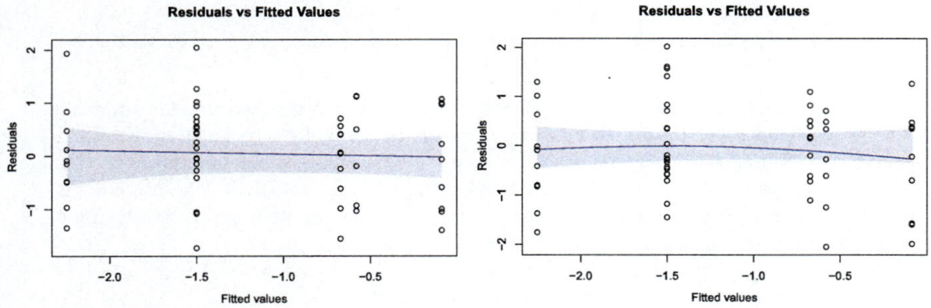

Fig. 10.8: Two Dunn-Smyth residual plots of David and Alistair's crab data. Note that the residual plot changes slightly due to jittering, as do the smoothers and their simulation envelopes. If we see a pattern in a plot, it is worth plotting more than once to check it is signal from the data rather than noise from the jittering!

Dunn-Smyth residual plots are particularly good at picking up failures of mean–variance assumptions, like not accounting for overdispersion that is in the data. As well as looking for a fan shape in the residuals vs fits plot, it is worth looking at a normal quantile plot of residuals and comparing it to a line of slope *one*, since these residuals are *standard* normal when assumptions are satisfied, as in Code Box 10.4. If Dunn-Smyth residuals get as large as four (or as small as negative four), this is a bad sign.

Exercise 10.4: Counts of ostracods in habitat configuration experiment. Recall David and Alistair's habitat configuration experiment, studying how epifauna settling on seaweed varies by distance of isolation.

They count how many ostracods (a type of small, round crustacean) settle in their seaweed patches and want to know if this varies with distance of isolation.

What sort of model would you use for ostracod counts? How would you check assumptions?

Code Box 10.4: Assumption checking for ostracod counts of Exercise 10.4.
We will start with a Poisson model, including algal wet mass as a covariate. The mvabund package will be used for model fitting to facilitate residual plots:

```
seaweed$logWmass = log(seaweed$Wmass)
ft_countOst=manyglm(Ost~logWmass+Time*Dist,data=seaweed,
        family="poisson")
plotenvelope(ft_countOst,which=1:3) # for a scale-location plot as well
```

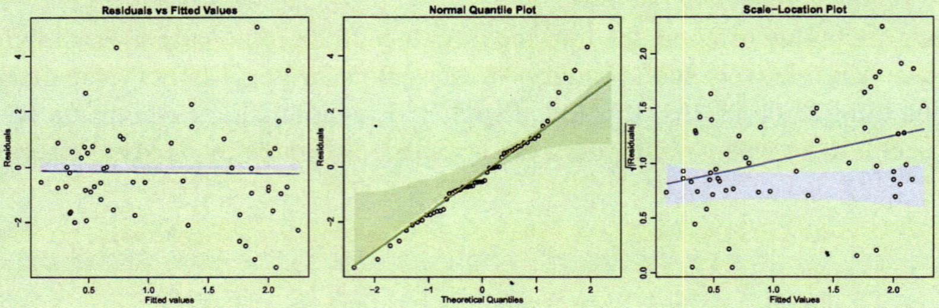

Notice that there is a bit of a fan shape, and the quantile plot suggests that the residuals are clearly steeper than the one-to-one line. The scale-location plot shows a clear increasing trend in scale of residuals as fitted values increase beyond the simulation envelope (i.e. beyond what would be expected if model assumptions were satisfied). These are signs of overdispersion. Plotting again to double-check will show the same problems. Trying a negative binomial regression to deal with the overdispersion:

```
ft.countOstNB=manyglm(Ost~logWmass+Time*Dist,data=seaweed,
        family="negative.binomial")
plot(ft_countOstNB,which=1:2)
```

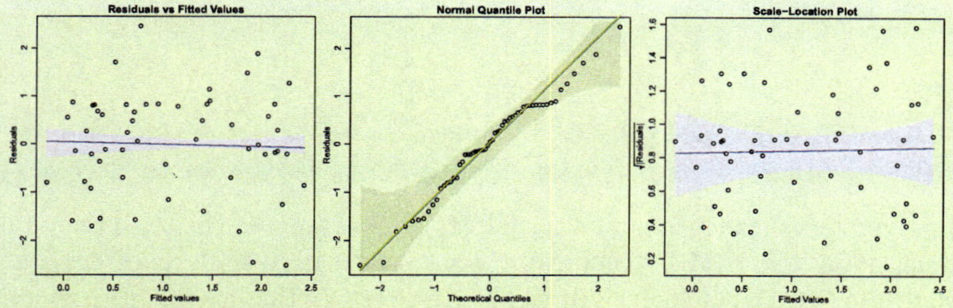

And we see the fan shape and the large residuals have largely been dealt with.

Exercise 10.5: Checking the Poisson assumption on the wind farm data. Recall the eel abundances from Lena's wind farm survey. We used Poisson regression as our initial model for eel abundance in Code Box 10.1.

Refit the model using the `manyglm` *function, and hence construct a residual plot.*

Does the Poisson assumption look reasonable?

Exercise 10.6: Checking the Poisson assumption for the worm counts. Recall the worm counts from Anthony's revegetation survey. We would like to know if we could use Poisson regression for worm abundance.

Refit the model using the `manyglm` *function under the Poisson assumption, and hence construct a residual plot.*

Also fit a negative binomial to the data and construct a residual plot.

Can you see any differences between plots?

Note it is hard to see differences because there are only two replicates in the control group. *Compare BIC for the two models using the* `BIC` *function.*

Which model has the better fit to the worm counts?

10.4 Inference from Generalised Linear Models

The same techniques can be used to make inferences from GLMs as were used for LMs, with minor changes to the underlying mathematics. In R all the same functions as for linear models work:

Confidence intervals use the `confint` function

Hypothesis testing use the `summary` or `anova` function. The latter is generally better for hypothesis testing.

Model selection use `stepAIC`, `AIC`, `BIC`, or `predict` to predict to test data in cross-validation code.

10.4.1 Test Statistics for GLMs

The main hypothesis testing function for GLMs on R is `anova`. The term `anova` is a little misleading for GLMs—technically, what we get is an analysis of deviance table (Maths Box 10.5), not analysis of variance (ANOVA). But in R the main testing function for GLMs is called `anova` anyway. For GLMs in R we have to tell `anova` what test statistic to use via the `test` argument (Code Box 10.5); otherwise it won't use any! There are a few different options for GLM test statistics.

The most common and usually the best way to test hypotheses concerning GLMs is to use a likelihood ratio test. A likelihood ratio statistic fits models via maximum likelihood under the null and alternative hypotheses and sees if the difference in likelihoods is unusually large (usually by comparing to a chi-squared distribution to get a P-value). This essentially compares deviances of models under the null and alternative hypotheses, and so is where the name "analysis of deviance" comes from, given that the procedure is closely analogous with how sums of squares are compared in order to construct an ANOVA table. The likelihood ratio statistic has some nice properties, including not being overly bothered by data with lots of zeros. In R, add the argument `test="Chisq"`, as in Code Box 10.5. This works just as well as F-tests do for linear models at larger sample sizes, but for small samples it is only approximate, as will be discussed shortly. An F-statistic from an ANOVA, incidentally, is equivalent to a type of likelihood ratio statistic.

Code Box 10.5: R code using the anova function to test the key hypotheses of interest to David and Alistair in Exercise 10.1.

```
> anova(ft_crab, test="Chisq")
          Df Deviance Resid. Df Resid. Dev Pr(>Chi)
NULL                         56     71.097
Time       1   6.6701        55     64.427 0.009804 **
Dist       2   1.0265        53     63.400 0.598553
Time:Dist  2   0.4012        51     62.999 0.818257
```

Any evidence of an effect of isolation distance on crabs?

Note the `Time` effect is significant here but not in Code Box 10.2. *Why do you think that might be?*

Another type of test for GLMs is a Wald test, which is based on studying parameter estimates and seeing if they are far from what is expected under the null hypothesis. Specifically, we would usually compare $\frac{\hat{\beta}}{\text{se}(\hat{\beta})}$ to a standard normal distribution. This looks a lot like what a t-test does—a t-test is in fact a type of Wald test. The `summary` function by default uses a Wald test. For GLMs this is less accurate than likelihood ratio tests, and occasionally, especially for logistic regression, it gives wacky answers, ignoring a really strong effect when it is there (especially for "separable models" with a "perfect" fit, for details see Væth, 1985).

Maths Box 10.5: Analysis of deviance

Consider two models $\mathcal{M}_0$ and $\mathcal{M}_1$, which are nested, in the sense that $\mathcal{M}_0$ is a special case of $\mathcal{M}_1$ (e.g. it removes a term from $\mathcal{M}_1$, equivalent to setting a parameter of $\mathcal{M}_1$ to zero). Starting from Eq. (6.3) of Maths Box 6.1, the deviance of model $\mathcal{M}_0$ can be written

$$\begin{aligned}\mathcal{D}_{\mathcal{M}_0}(\mathbf{y}) &= 2\ell_{\mathcal{S}}(\hat{\boldsymbol\theta}_{\mathcal{S}};\mathbf{y}) - 2\ell_{\mathcal{M}_0}(\hat{\boldsymbol\theta}_0;\mathbf{y})\\ &= \left(2\ell_{\mathcal{S}}(\hat{\boldsymbol\theta}_{\mathcal{S}};\mathbf{y}) - 2\ell_{\mathcal{M}_1}(\hat{\boldsymbol\theta}_1;\mathbf{y})\right) + \left(2\ell_{\mathcal{M}_1}(\hat{\boldsymbol\theta}_1;\mathbf{y}) - 2\ell_{\mathcal{M}_0}(\hat{\boldsymbol\theta}_0;\mathbf{y})\right)\\ &= \mathcal{D}_{\mathcal{M}_1}(\mathbf{y}) + -2\log\Lambda(\mathcal{M}_0, \mathcal{M}_1)\end{aligned}$$

where the log-likelihood ratio statistic $-2\log\Lambda(\mathcal{M}_0, \mathcal{M}_1)$ is defined in Eq. (6.2) of Maths Box 6.1. That is, the deviance under a null model can be partitioned into the deviance under an alternative model plus a log-likelihood ratio statistic. For a sequence of several nested models, we can further partition the deviance into several log-likelihood ratio statistics. This is often called an *analysis of deviance* and works in an analogous way to ANOVA (which can actually be understood as a special case of analysis of deviance for normally distributed responses).

The `anova` function on R, by default, uses likelihood tests, reported in an *analysis of deviance* table, which breaks down the deviance into contributions from different model terms.

Maths Box 10.6: Problems Wald statistics have with zero means

In Lena's study, recall that no eels were caught in northern sites after wind farms were constructed (Fig. 10.2), giving an estimated mean abundance of zero. This is a problem for a Poisson or negative binomial regression, because the mean model uses a log link, and the log of zero is undefined ($-\infty$). R usually returns a warning in this situation but didn't for the following fit:

```
> data(windFarms)
> eels = windFarms$abund[,16]
> ft_eels=glm(eels~Year*Zone,family="poisson",
                      data=windFarms$X)
> summary(ft_eels)
Coefficients:
              Estimate Std. Error z value Pr(>|z|)
(Intercept)    -0.4700     0.2582  -1.820   0.0687 .
Year2010       -0.2803     0.3542  -0.791   0.4288
```

```
ZoneN              -1.3218     0.5627  -2.349   0.0188 *
ZoneS               0.8002     0.3129   2.557   0.0105 *
Year2010:ZoneN    -16.2305   952.9222  -0.017   0.9864
Year2010:ZoneS     -0.2323     0.4360  -0.533   0.5942
```

The term for `Year2010:ZoneN` is -16, which is about as close to $-\infty$ as R will let it get. This doesn't sound small, but remember parameters are on the log scale: the predicted mean abundance is less than e^{-16}, which is about 1 in 10 million!

The standard error of a maximum likelihood estimator is calculated based on the inverse of the curvature (second derivative) of the likelihood function, and if the likelihood is maximised at $-\infty$, it will be pretty flat out there, meaning zero curvature, and a standard error of ∞ (estimated as 953 in the above output). The Wald statistic, the ratio of estimate over standard error, is undefined, and numerical estimates of it from software (`z value`) will be nonsensical in this case.

Likelihood ratio statistics don't have this problem, because they work on probability functions ($\log f(y; \mu)$, as in Eq. (10.3)) not on parameters (β), and probabilities still make sense in the zero-mean setting (for a count variable with mean zero, the probability of getting a zero is one!). So if a `summary` table has some very negative coefficients, with large standard errors, you have zero-mean issues and are better off with likelihood ratios.

Exercise 10.7: Testing if there is a wind farm effect
Consider Lena's wind farm study and the question of whether eel abundance varies across the three zones (north, south, and wind farm). Note that the same sites were sampled twice, so we have paired data. The model fitted in Maths Box 10.6 does not account for this.

What model should be fitted to handle the pairing structure?
Fit the appropriate model and test for an effect of wind farm.

10.4.2 Inference for Small Samples

The `summary` and `anova` tests are both approximate—they work well on large samples (well, `summary` can occasionally go a bit weird even on large samples), but both can be quite approximate when sample size is small (say, less than 30). This is quite different to what happens for linear models—when assumptions are satisfied, F-tests and t-tests based on linear models are exact for any sample size, meaning that if a significance level of 0.05 is used, the actual chance of accidently rejecting the null hypothesis when it is true ("Type I error rate") is exactly 0.05. For a GLM, however, even when all model assumptions are correct, in very small samples you might accidently reject the null hypothesis twice as often as you think you will, i.e. you can have a Type I error of 0.1 when using a significance level of 0.05 (Ives, 2015). We can beat this problem exactly the same way that we beat it for mixed models—generating P-values by simulation (Warton et al., 2016).

One of the simplest ways to do design-based inference for GLMs is to use the `mvabund` package, which uses resampling by default whenever you call `summary` or `anova` for a `manyglm` object. This sort of approach is really needed for Anthony's data, where he has a very small and unbalanced sample (Code Box 10.6), but it isn't really needed for David and Alistair's crabs (Code Box 10.7), where the sample size is already moderately large, so it is no surprise that their results come back almost identical to those from the `anova` call to `glm` (Code Box 10.5).

Code Box 10.6: Model-based inference for Anthony's worm counts from Exercise 10.3.

```
> ftmany_Hap=manyglm(Haplotaxida~treatment,family="negative.binomial",
            data=reveg)
> anova(ftmany_Hap)
Time elapsed: 0 hr 0 min 0 sec
Analysis of Deviance Table

Model: Haplotaxida ~ treatment

Multivariate test:
            Res.Df Df.diff   Dev Pr(>Dev)
(Intercept)      9
treatment        8       1 2.811    0.173
Arguments: P-value calculated using 999 resampling iterations via
           PIT-trap resampling.
```

Any evidence of an effect of revegetation on worms?

The default resampling method in the `mvabund` package is a new method called the *PIT-trap*—a residual resampling method that bootstraps residuals computed via the probability integral transform (Warton et al., 2017). Basically, it bootstraps Dunn-Smyth residuals. Previously, the only serious method available for resampling GLMs was the parametric bootstrap. Recall that we have already met the parametric bootstrap; we used it in Sect. 6.5 as a technique for making accurate inferences about linear mixed models. You can use the parametric bootstrap in `mvabund` by setting `resamp="monte.carlo"` in the `anova` or `summary` call. Recall that the parametric bootstrap uses model-based inference, so one might expect it to be more sensitive to violations of model assumptions than using the PIT trap (which, for instance, is close to exact in an example like Code Box 10.6 irrespective of model assumptions).

Code Box 10.7: Design-based inference for David and Alistair's crab data using mvabund.

```
> ftMany_crab = manyglm(CrabPres~Time*Dist,family=binomial("cloglog"),
                 data=seaweed)
> anova(ftMany_crab)
Time elapsed: 0 hr 0 min 2 sec
Analysis of Deviance Table

Model: CrabPres ~ Time * Dist

Multivariate test:
            Res.Df Df.diff   Dev Pr(>Dev)
(Intercept)     56
Time            55       1 6.670    0.011 *
Dist            53       2 1.026    0.615
Time:Dist       51       2 0.401    0.869
---
Signif. codes:  0 `***' 0.001 `**' 0.01 `*' 0.05 `.' 0.1 ` ' 1
Arguments: P-value calculated using 999 resampling iterations via
           PIT-trap resampling.
```

Any important differences from results David and Alistair previously obtained in Code Box 10.5?

10.4.3 Warning: Consider Overdispersion!

It is critical to consider overdispersion when fitting a GLM—if your data are overdispersed, and you don't account for this, you can get things spectacularly wrong. Consider for example an analysis of Anthony's worm data using the Poisson distribution (Code Box 10.8). Suddenly there is strongly significant evidence of a treatment effect, which we didn't get with a negative binomial model (Code Box 10.6). A plot of the raw data is not consistent with such a small P-value either—there is a suggestion of a trend, but nothing so strongly significant. The main problem is that the data are *overdispersed* compared to the Poisson distribution, meaning that the variance assumed by the Poisson model underestimates how variable replicates actually are. Hence standard errors are underestimated and P-values are too small, and we have false confidence in our results—the key thing we are trying to guard against when making inferences from data.

Code Box 10.8: Getting the wrong answer by ignoring overdispersion in Anthony's worm counts from Exercise 10.3.

```
> ft_wormPois = glm(Haplotaxida~treatment, family="poisson",
       data = reveg)
> anova(ft_wormPois,test="Chisq")
```

```
Analysis of Deviance Table
          Df Deviance Resid. Df Resid. Dev  Pr(>Chi)
NULL                          9     63.946
treatment  1   11.668         8     52.278 0.0006359 ***
---
Signif. codes:  0 `***' 0.001 `**' 0.01 `*' 0.05 `.' 0.1 ` ' 1
```

The treatment effect seems to be strongly significant, which is not what we saw in Code Box 10.6. If we used resampling, we would get a very different answer:

```
> ft_wormmanyPois = manyglm(Haplotaxida~treatment,family="poisson",
                            data=reveg)
> anova(ft_wormmanyPois)
Time elapsed: 0 hr 0 min 0 sec
Analysis of Deviance Table
            Res.Df Df.diff   Dev Pr(>Dev)
(Intercept)      9
treatment        8       1 11.67    0.213
Arguments: P-value calculated using 999 resampling iterations via
           resampling.
```

The reason for the difference in results is that (as previously) the data are overdispersed and so do not fit a Poisson distribution. Resampling gives some protection against this issue, but a better option is to fit a model that can account for the overdispersion, such as the negative binomial.

Resampling can offer some protection against the effects of overdispersion missing from our model. For example, if we had used resampling, we would not have seen a significant effect in Anthony's worm data (Code Box 10.8, bottom). However, the best protection is to mind your Ps and Qs and, hence, try to *use the right model for your data*! In the case of Anthony's worm data, a better model was a negative binomial regression.

A common cause of overdispersion is *missing predictors*. Recall that in linear models, if an important predictor is left out of the model, this increases the error variance. Well, the same happens in a GLM, with missing predictors increasing the variance compared to what might otherwise be expected, which has implications for the mean–variance relationship. Any time you are modelling counts, and your model for them is imperfect, you can expect overdispersion. This will actually happen most of the time. The `mvabund` package was written initially with count data in mind, and the default family for the `manyglm` function was chosen to be the negative binomial precisely because we should expect overdispersion in most models for counts.

The issue of overdispersion can also arise in models assuming a binomial distribution, but only when measurements are taken as counts across clusters of observations, e.g. if counting the number of seeds in a Petri dish that germinate. Overdispersion arises here because of missing predictors across clusters, e.g. sources of variation in germination rate across petri dishes that have not been accounted for. This is most easily handled in a slightly different way—using a mixed model with an observation-level random effect, i.e. adding a random effect to the model that takes different values for different observations, as illustrated later in Code Box 10.9. This random effect plays the role of a missing predictor. The same technique could also be used for counts, instead of using a negative binomial model. For binary responses, such as presence–absence data, observations fall exactly along a quadratic mean–variance relationship (as in Fig. 10.1b), and there is no possibility of overdispersion. This means that you cannot include an observation-level random effect in a model for presence–absence; there is not enough information in the data to estimate it.

Code Box 10.9: Using an observation-level random effect for a binomial response

Below are germination data from 29 trials of *Abutilon angulatum* seeds (taken from Sentinella et al., 2020), studying germination rate (measured as the number of seeds that germinated NumGerm, out of NumSown) at different temperatures. First we will plot the data:

```
> data(seedsTemp)
> seedsTemp$propGerm = seedsTemp$NumGerm / seedsTemp$NumSown
> plot(propGerm/(1-propGerm)~Test.Temp,data=seedsTemp,log="y")
```

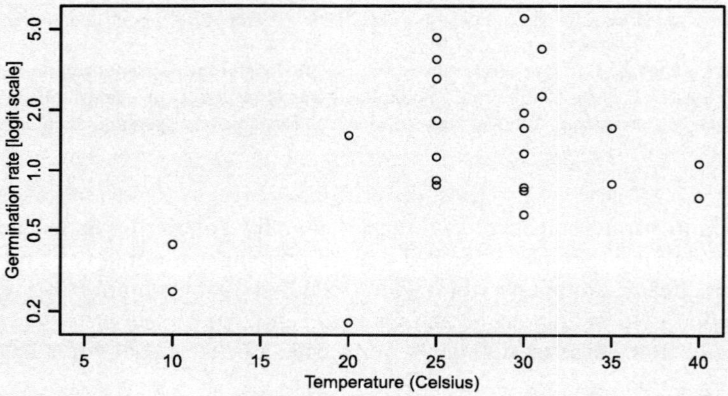

Note that repeat experiments at the same temperature have different germination rates (probably due to factors not controlled for by the experimenter). We can account for this using an observation-level random effect:

```
> library(lme4)
> seedsTemp$ID = 1:length(seedsTemp$NumGerm)
> ft_temp = glmer(cbind(NumGerm,NumSown-NumGerm)~poly(Test.Temp,2)+
              (1|ID),data=seedsTemp,family="binomial")
> summary(ft_temp)
Random effects:
 Groups Name        Variance Std.Dev.
 ID     (Intercept) 2.961    1.721
Number of obs: 29, groups:  ID, 29

Fixed effects:
                    Estimate Std. Error z value Pr(>|z|)
(Intercept)          -0.5896     0.3428  -1.720   0.0854 .
poly(Test.Temp, 2)1   2.8770     1.9395   1.483   0.1380
poly(Test.Temp, 2)2  -3.6910     1.9227  -1.920   0.0549 .
```

Note that the random effect had a standard deviation of 1.7, which is relatively large (on the logit scale), indicating a lot of trial-to-trial variation in germination rate.

Exercise 10.8: Anthony's ant data
In his revegetation study (Exercise 10.3), Anthony recorded abundances for many families of invertebrates, not just worms. We will now consider his ant counts, which you can access as the `Formicidae` object in the `reveg` dataset.

Which model do you think is more appropriate for this dataset—a Poisson or negative binomial?

Plot the data. Does there seem to be evidence of a treatment effect?

Use the `glm` *function to fit a Poisson log-linear model, and use this to test for a treatment effect. Did you get what you expected?*

Now use negative binomial regression to test for a treatment effect. Notice you got a very different answer. *Which answer is more likely to be correct in this case?*

10.5 Don't Standardise Counts, Use Offsets!

Sometimes we might be tempted to standardise or average counts across replicates before analysing them to account for differences in sampling intensity (different sized plots, different numbers of replicates). Please *don't*! This changes the distribution of your data and stuffs up the mean–variance relationship. Instead, you should use an *offset*, a term specially designed for this situation.

An offset can be used whenever there is some variable known not just to be important to the response, but its precise relationship with the mean of the response is known. This most commonly happens with sampling intensity (sample twice as hard as you expect to count twice as many things). This sort of situation is fairly common; in fact, it has already arisen in two of the examples in this chapter—Anthony had different numbers of pitfall traps at different sites (one was dug out by an animal) as in Exercise 10.9, and David and Alistair (Exercise 10.1) had different sized patches of seaweed in different replicates, and they measured seaweed wet mass so they could account for this in their analysis.

Exercise 10.9: Worm counts with different numbers of pitfall traps
Anthony actually sampled five pitfall traps in nine sites, but only four pitfall traps in one site, as follows:

Treatment	C	R	R	R	C	R	R	R	R	R
Count	0	3	1	3	1	2	12	1	18	0
# pitfalls	5	5	5	5	5	5	5	4	5	5

How can we account for the different sampling effort at different sites in our model?

Offsets are most commonly used in a log-linear model, where an offset is a predictor known to have an exactly proportional effect on the response. Our offset in a model for $\log(\mu)$ is the log of the sampling intensity variable, e.g. in Exercise 10.9 we would use log(# pitfalls) as the offset. The effect of including this term is to change our model from predicting mean abundance to modelling mean abundance *per pitfall trap*. This is what you would be trying to do if you averaged the counts across pitfalls before analysing them, but the offset achieves this without messing with the data and their mean–variance relationship. When using the `glm` function, you can add an offset using the `offset` argument, or you can add a term to your formula along the lines of `offset(log(pitfalls))`, as was done in Code Box 10.10.

Offsets are most commonly used for counts, but they can also be used for binary responses, too, but you have to be careful about your choice of link function and transformation of the offset term. Specifically, you can use a complementary log-log link, with the log of your sampling intensity variable as an offset, following the argument that a complementary log-log on a binary response implies a (Poisson) log-linear model on some underlying count variable. For example, for David and Alistair's data, we could add log(`Wmass`) as an offset term to account for different amounts of seaweed in the different plots. This assumes that doubling the wet mass of seaweed doubles invertebrate mean abundance. If we were worried about the validity of this proportionality assumption, we could also add log(`Wmass`) as a predictor and see if this term made a difference.

Code Box 10.10: Adding an offset to the model for worm counts

Because we are modelling the log of the mean, we add an offset for log(# pitfalls):

```
> ftmany_hapoffset = manyglm(Haplotaxida~treatment+offset(log(pitfalls)),
                              family="negative.binomial", data=reveg)
> anova(ftmany_hapoffset)
Time elapsed: 0 hr 0 min 0 sec
Analysis of Deviance Table

Model: Haplotaxida ~ treatment + offset(log(pitfalls))

Multivariate test:
            Res.Df Df.diff   Dev Pr(>Dev)
(Intercept)      9
treatment        8       1 2.889     0.15
Arguments: P-value calculated using 999 resampling iterations via
           PIT-trap resampling.
```

Why didn't we hear about offsets for linear models? Because linear models are additive, so if you had an offset, you could just subtract it from y before fitting the model. That is, in a linear model for `log(Haplotaxida)`, to include an offset for `log(pitfalls)` you would model `log(Haplotaxida)-log(pitfalls)` as a function of `treatment`. But that trick wouldn't work for a GLM.

10.6 Extensions

The problem with something called "generalised" is what do you call generalisations of it? There are a few important additional features and extensions of GLMs worth knowing about.

10.6.1 Zero-Inflated Models

Ecological counts often have many zeros. Consider Anthony's cockroach counts (Exercise 10.10). One option is to use a *zero-inflated* model—a model that expects more zeros than a standard model, e.g. more than expected under a Poisson or negative binomial regression. These models tend to be fitted by separately studying the question of when you observe a non-zero count and the question of how large the count is. For details on how to fit such models, see the `VGAM` package (Yee, 2010) or the `glmmTMB` package (Brooks et al., 2017), both in `R`.

Zero-inflated models are quite widely used, which may in part be because of a misunderstanding about when they are actually needed (Warton, 2005). Do you have reason to believe that the process behind presence–absence patterns is distinct from the process behind abundance? If so, a zero-inflated regression model is a good way to approach the problem. If not, and you just have lots of zeros in your dataset, then that in itself is not a good reason to use a zero-inflated regression model.

The confusion here comes from the fact that we are communicating in English, not in maths—the term "zero-inflated" sounds like it applies to any dataset with lots of zeros. But a more common reason for getting lots of zeros is that you are counting something that is rare! (Its mean is small.) For example, a Poisson distribution with $\mu = 0.1$ expects 90% of values to be zero. Also recall that regression models make assumptions about the *conditional* distribution of the response variable, after including predictors. These predictors might be able to explain many of the zeros in the data (as in Exercise 10.10). So observing many zeros is not on its own a good reason for using a zero-inflated model.

Exercise 10.10: Anthony's cockroaches
Anthony wants to evaluate how well invertebrate communities are re-establishing following bush regeneration efforts. Here is the number of cockroaches in pitfall traps across each of his 10 sites:

Treatment	C	R	R	R	C	R	R	R	R	R
Count	3	0	0	0	4	1	0	0	0	0
# pitfalls	5	5	5	5	5	5	5	4	5	5

He wants to know if there is any evidence that bush regeneration is working, in terms of having an effect on cockroach counts. *How should he analyse the data?*

Zero-inflated models can be thought of as a particular way of handling overdispersion, where the overdispersion arises via a missing predictor that affects likelihood of observing a zero, but not the size of non-zero counts. An alternative way to introduce overdispersion, which we have previously met, is the negative binomial distribution. The negative binomial also has extra zeros, relative to the Poisson distribution. When in doubt, as always, standard model selection tools are a good way to study which of these methods is best suited to a given count dataset.

10.6.2 Generalised Additive Models

Generalised additive models (GAMs) (introduced in Chap. 8) are commonly fitted using the `mgcv` package and readily handle non-normal responses via the `family` argument, which behaves just like for `glm`. There are two minor differences in implementation in `mgcv` compared to `glm`:

- No nice residual plots. But using the `statmod` package, you can apply the `qresiduals` function to a GAM fit from `mgcv` to construct Dunn-Smyth residuals.
- To access the negative binomial family, use `family=nb()`.

10.6.3 Generalised Linear Mixed Models

Recall that sometimes a design includes a random factor (e.g. nested design), so what is needed is a mixed model (Chap. 6). GLMs (the `glm` and `manyglm` functions) only handle fixed effects. If you have random effects and a non-constant assumed mean–variance relationship, then you want to fit a *generalised linear mixed model* (GLMM). A good package for this, as in the linear mixed effects case, is the `lme4` package (Bates et al., 2015). The `glmmTMB` package (Brooks et al., 2017) is also a good choice—it was written to behave similarly to `lme4` but tends to be more stable for complex models and has more options for the `family` argument, including some zero-inflated distributions.

There aren't really any new tricks when fitting a GLMM—just use the `glmer` argument as you would normally use `lmer`, but be sure to add a `family` argument. For the negative binomial family, try `nbinom2` in `glmmTMB`.

Current limitations:

- No nice residual plots. This time, the `qresiduals` function won't help us out either. However, the `DHARMa` package (Hartig, 2020) can approximate these residuals via simulation. (Note `DHARMa` returns residuals on the standard uniform scale; they need to be mapped to the standard normal.)

- GLMMs can take much longer to fit, and even then they only give approximate answers. The mathematics of GLMMs are way harder than anything else we've seen so far and get computationally challenging deceptively quickly (in particular, when the number of random effects gets large).

Maximum likelihood estimation is commonly used to fit GLMMs, but a difficulty is that the likelihood can't be computed directly in most cases and needs to be approximated (by default, this is done using a Laplace approximation in `lme4` and `glmmTMB`). In `lme4`, there is an optional argument `nAGQ` that you can try to get a better approximation for GLMMs for simple models (e.g. `nAGQ=4`) using adaptive Gauss-Hermite quadrature. This requires more computation time and gives slightly better approximations to the likelihood and standard errors. Feel free to experiment with this to get a sense of the situations where this might make a difference (typically, small sample sizes or many random effects).

10.6.4 Summary

Discrete data commonly arise in ecology, and when the data are highly discrete (e.g. a lot of zeros), there will be a mean–variance relationship that needs to be accounted for in modelling. The GLM is the simplest way to analyse such data, but there are a number of other options, some of which were discussed earlier in this chapter. Most of the remainder of this text will focus on various extensions of GLMs to handle important problems in ecology, such as multivariate abundances (Chap. 14).

Part II
Regression Analysis for Multiple Response Variables

Chapter 11
More Than One Response Variable: Multivariate Analysis

Recall that the type of regression model you use is determined mostly by the properties of the *response variable*. Well what if you have more than one response variable?

A multivariate analysis is appropriate when your research question implies more than one response variable—usually, because the quantity you are interested in is characterised by more than one variable. For example, in Exercise 11.1, Ian is interested in "leaf economics", which he quantifies jointly using leaf longevity and leaf mass per area. In Exercise 11.2, Edgar is interested in flower size/shape, which he characterised using four length measurements. In Exercises 11.3, Petrus is interested in a hunting spider community, which he quantified using abundance of organisms in three different taxonomic groups. These last two examples are quite common places where multivariate analysis is used in ecology—when studying size and shape or when studying communities or assemblages as a whole.

Exercise 11.1: Leaf economics and environment
Ian likes studying leaves. He is especially interested in so-called leaf economics—things like construction costs per unit area (`lma`), how long-lived leaves are (`longev`), and how they vary with environment. In particular:

> Is there evidence that leaves vary (in their "economics" traits) across sites with different levels of rainfall and soil nutrients?

What are the response variables? What sort of analysis is appropriate here?

D. I. Warton, *Eco-Stats: Data Analysis in Ecology*, Methods in Statistical Ecology,
https://doi.org/10.1007/978-3-030-88443-7_11

Exercise 11.2: Flower size and shape
Edgar (Anderson, 1935) was studying *Iris* flowers. For 50 flowers on each of three *Iris* species, he took four measurements—length and width of petals and length and width of sepals. He wants to know:

> How do these three species of Iris differ in terms of flower size and shape?

What are the response variables? What sort of analysis is appropriate here?

Exercise 11.3: Hunting spiders and their environment
Petrus set up 28 pitfall traps on sand dunes in different environmental conditions (in the open, under shrubs, and so forth) and classified all hunting spiders that fell into the traps into species. Consider the three most abundant genera.

> Is abundance related to environmental conditions? How?

What are the response variables? What type of analysis is appropriate here?

Key Point
multivariate = many response variables
Multivariate analysis involves simultaneously treating multiple variables as the response in order to characterise some idea that cannot be captured effectively using one response variable. Everything is a lot harder in multivariate analysis—data visualisation, interpretations, checking assumptions, analysis—so it is worth thinking carefully if you really need to do a multivariate analysis in the first place!

11.1 Do You Really Need to Go Multivariate? Really?

The first thing you should know is that multivariate analysis is much more challenging than univariate analysis, for a number of reasons:

Data visualisation Our graphs are limited to using just two (or maybe three) dimensions to visualise data. This means we can only look at responses one at a time, two at a time, or maybe three, and if there are any higher-order patterns in correlation, we will have a hard time visualising them in a graph.

Interpretation More variables means there is a lot going on, and it is hard work to get a simple answer, if there is one.

Assumption violations More response variables means more ways model assumptions can be violated.

High-dimensional data If the number of response variables is comparable to the number of observations (as we will see from Chap. 14 onwards), model-based inference is a real challenge.

Everything is a lot harder in a multivariate world. Do you really need to go there? If you can get a satisfactory answer to your key research question without going multivariate, then by all means do so!

The main potential way to avoid multivariate analysis is to find some univariate statistic to characterise what you are primarily interested in, e.g. instead of studying abundances in a community, study total biomass, species richness, or some other measure of species diversity. If this works, in the sense that it can be defended scientifically (it answers the question you are interested in, with a reasonable level of precision), then you are done. There are some ifs and buts here, but if you can take a complex situation and break it down into some simple metric that captures most of what you are interested in, then you are doing great science—it is always worth the effort to think about if and how this could be done, rather than jumping straight into a multivariate analysis. An example from my own experience involved 20 measurements of length and width of marsupial fingers, compared across species that were land-dwelling or tree-dwelling (Weisbecker and Warton, 2006). Ultimately we were able to characterise the key patterns in terms of a slenderness ratio, a ratio of length to width, where arboreal species tended to have more slender fingers. Thus, we reduced 20 measurements to just one response variable, although admittedly there was a fair bit of multivariate analysis in there as well, some of which helped get to the point where we could simplify the problem to a single "slenderness" response.

Another way you might consider avoiding multivariate analysis is to analyse each of your response variables separately, but there are two things that can go wrong with that strategy. Firstly, if the variables are correlated, structure in the data can be missed by looking at them one variable at a time, as in Fig. 11.1. Secondly, multiple comparison issues arise. Specifically, the more 95% confidence intervals (CIs) that are constructed (or the more tests that are carried out at the 0.05 significance level), the greater the chance of missing the true value of a parameter (or falsely declaring significance). For example, it turns out that constructing thirteen 95% CIs for independent parameters, when there is no effect of interest in any of them, leads to a 50:50 chance that one or more CIs will miss the true value of the parameter they are estimating. For 100 CIs, it is almost guaranteed that some CIs will miss the parameters they are trying to estimate. This can be corrected for (e.g. Bonferroni corrections), but then CIs will be very wide, and tests will have quite low power. The use of a multivariate analysis instead means power will usually be much better (unless an effect is only actually seen in one or two response variables).

If you only have a few variables and they are not strongly correlated, then it can be OK to use a one-at-a-time approach, but if you have several correlated variables and no simple way to summarise them as a metric, then multivariate analysis might be the way to go.

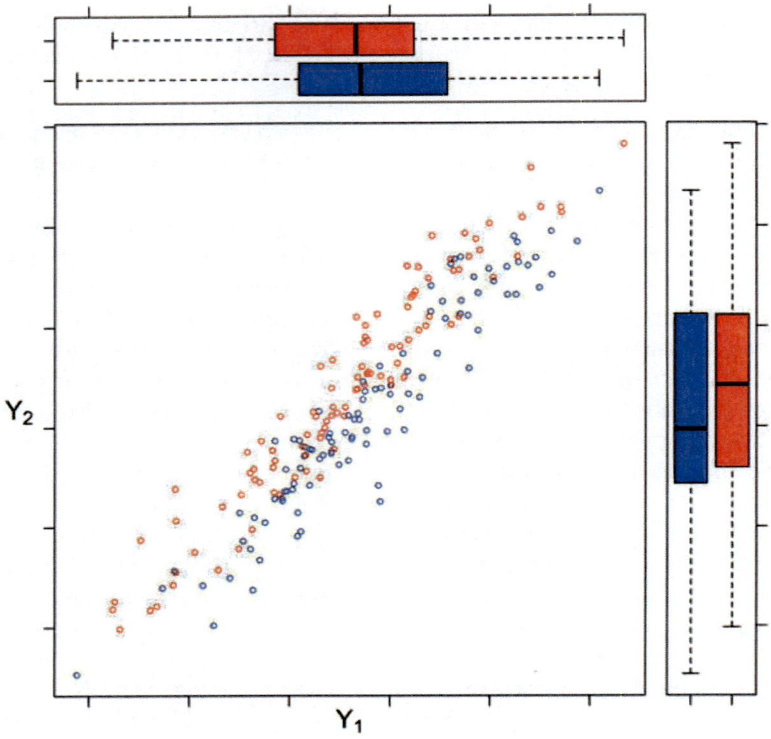

Fig. 11.1: A simulated dataset illustrating a situation where you could miss the structure in the data by analysing correlated variables one at a time. The scatterplot shows two samples (in different colours) of two response variables, Y_1 and Y_2. A multivariate analysis would readily find a difference between these two samples, because at any given value of Y_1, the red points tend to have larger values of Y_2. However, if analysing the variables one at a time (as in the boxplots in the margins), it is harder to see a difference between samples, with shifts in each mean response being small relative to within-sample variation. The specific issue here that makes it hard to see the treatment effect marginally is that it occurs in a direction *perpendicular* to the main axis of variation for the Y variables

11.2 MANOVA and Multivariate Linear Models

The multivariate generalisation of analysis of variance is known as MANOVA (Multivariate ANalysis Of VAriance). It is a special case of multivariate linear models.

A multivariate linear model, written mathematically, looks like this:

$$\mathbf{y} \sim \mathcal{N}(\boldsymbol{\mu}, \boldsymbol{\Sigma})$$
$$\mu_j = \beta_{0j} + \mathbf{x}^T \boldsymbol{\beta}_j$$

where $\mathbf{y}$ is a column vector of responses and $\boldsymbol{\mu}$ is the vector of its means, whose jth entry is μ_j.

The multivariate linear model can be broken down into the following assumptions:

1. The observed $\mathbf{y}$-values are *independent* across observations (after conditioning on x).
2. The $\mathbf{y}$-values are *multivariate normally distributed* with *constant variance–covariance matrix*
$$\mathbf{y} \sim \mathcal{N}(\boldsymbol{\mu}, \boldsymbol{\Sigma})$$
3. There is a *straight-line relationship* between the mean of each y_j and each x
$$\mu_j = \beta_{0j} + \mathbf{x}^T \boldsymbol{\beta}_j$$

Fitting a multivariate linear model using R is pretty easy—the only thing you have to do differently to the univariate case is to make sure your response variables are stored together in a *matrix* (which can be created using `cbind`). See for example Code Box 11.2.

As usual, I recommend checking residual plots for no pattern. In R, the `plot` function doesn't work for multivariate linear models, but I have made sure `plotenvelope` works for this case, as in Code Box 11.3.

11.2.1 Variance-co-What?

The variance–covariance matrix $\boldsymbol{\Sigma}$ is an important descriptor of multivariate data—it is a multivariate generalisation of the idea of variance, which captures the idea of correlation at the same time (Maths Box 11.1). Specifically, a variance–covariance matrix is a square grid of values storing the variance of each variable (along the *diagonal* that starts in the top left) and the covariances. Covariances are basically the correlations between each pair of variables, rescaled to the original units of measurement. You can estimate the variance–covariance matrix of a multivariate dataset in R using the `var` function, as in Code Box 11.1.

Maths Box 11.1: Mean vector and variance–covariance matrix
Recall the definitions of the mean and variance of a variable y from Maths Box 1.3. How can we apply these ideas to a multivariate response?

Consider a multivariate response $\mathbf{y}$, a column vector of p responses whose jth row contains y_j. A multivariate mean $\boldsymbol{\mu}_\mathbf{y}$ is a vector whose jth value is μ_j, the mean of the jth response. A multivariate variance is defined as

$$\boldsymbol{\Sigma}_\mathbf{y} = \mu_{(\mathbf{y}-\boldsymbol{\mu})(\mathbf{y}-\boldsymbol{\mu})'}$$

which gives us a matrix with p rows and p columns, whose (i, j)th value is

$$\sigma_{ij} = \mu_{(y_i-\mu_i)(y_j-\mu_j)}$$

$\mathbf{\Sigma_y}$ is often called the *variance–covariance matrix* because it stores variances as its diagonal elements (when $i = j$) and *covariances* as the off-diagonal elements. Covariances are related to correlations via $\sigma_{ij} = \rho_{ij}\sqrt{\sigma_{ii}\sigma_{jj}}$, where ρ_{ij} is the correlation between the ith and jth responses. $\mathbf{\Sigma_y}$ is symmetric about its diagonal since $\sigma_{ij} = \sigma_{ji}$.

The rescaling rules for means and variances (Maths Boxes 1.4–1.5) generalise to the multivariate case:

$$\mu_{\mathbf{Ay}} = \mathbf{A}\mu_Y, \qquad \mathbf{\Sigma_{Ay}} = \mathbf{A}\mathbf{\Sigma}_Y\mathbf{A}' \tag{11.1}$$

but note that $\mathbf{A}$ is a matrix now rather than a constant, and it describes linear functions of responses $\mathbf{y}$. So this rule tells us what happens when responses are summed as well as rescaled (unless $\mathbf{A}$ is a diagonal matrix, which just rescales).

Code Box 11.1: Sample variance–covariance matrices on R

Example sample variance–covariance matrices for Ian's leaf economics data:

```
> library(smatr)
> data(leaflife)
> Yleaf=cbind(leaflife$lma,leaflife$longev)
> colnames(Yleaf)=c("lma","longev")
> var(Yleaf)
             lma      longev
lma    4186.0264 36.7905011
longev   36.7905  0.8232901
```

Variances are along the diagonal that starts at the top left, covariances appear everywhere else. Note the covariances are symmetric, i.e. `cov(lma,longev)=cov(longev,lma)`. Note that the variance is much larger for Leaf Mass per Area (LMA) than for leaf longevity (it is on a different scale, so this doesn't really mean anything biologically!) and that the two responses are positively correlated.

Example sample variance–covariance matrices for Edgar's *Iris* data:

```
> data("iris")
> var(iris[,1:4])
             Sepal.Length Sepal.Width Petal.Length Petal.Width
Sepal.Length    0.6856935  -0.0424340    1.2743154   0.5162707
Sepal.Width    -0.0424340   0.1899794   -0.3296564  -0.1216394
Petal.Length    1.2743154  -0.3296564    3.1162779   1.2956094
Petal.Width     0.5162707  -0.1216394    1.2956094   0.5810063
```

Note that petal length is the most variable response (and sepal width the least) and that while most responses are positively correlated with each other, sepal width is negatively correlated with all other responses.

In the univariate case, we made an assumption of constant variance. In the multivariate case, we make an assumption of constant variance–covariance matrix, which means we assume:

- Every response variable has constant variance.
- The correlations between all pairs of response variables are constant.

So there are now a lot more ways our variance assumptions can be violated!

Code Box 11.2: Fitting a multivariate linear model to the leaf economics data

```
> library(smatr)
> data(leaflife)
> Yleaf = cbind(leaflife$lma,leaflife$longev)
> ft_leaf = lm(Yleaf~rain*soilp, data=leaflife)
> anova(ft_leaf, test="Wilks")
Analysis of Variance Table
            Df   Wilks approx F num Df den Df    Pr(>F)
(Intercept)  1 0.11107  248.096      2     62 < 2.2e-16 ***
rain         1 0.68723   14.108      2     62 8.917e-06 ***
soilp        1 0.93478    2.163      2     62    0.1236
rain:soilp   1 0.95093    1.600      2     62    0.2102
Residuals   63
---
Signif. codes:  0 `***' 0.001 `**' 0.01 `*' 0.05 `.' 0.1 ` ' 1
```

Any evidence of an effect of rainfall or soil nutrients?

With just two response variables, we can use a scatterplot to visualise the data:

```
plot(leaflife$lma~leaflife$longev, xlab="Leaf longevity (years)",
                    ylab="Leaf mass per area (mg/mm^2)",
                    col=interaction(leaflife$rain,leaflife$soilp))
legend("bottomright",legend=c("high rain, high soilp",
        "low rain, high soilp", "high rain, low soilp",
        "low rain, low soilp"), col=1:4, pch=1)
```

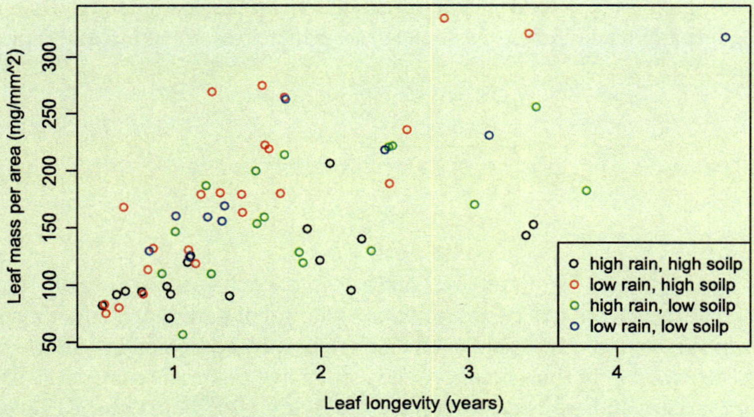

What is the nature of the rainfall effect?

Code Box 11.3: Checking multivariate linear model assumptions for leaf economics data

Multivariate normality implies assumptions will be satisfied for a linear model of each response against all other variables (against x and other y), which is a good way to check assumptions. The `plotenvelope` function constructs residuals and fitted values from this *full conditional* model.

```
par(mfrow=c(1,3),mar=c(3,3,1.5,0.5),mgp=c(1.75,0.75,0))
library (ecostats)
plotenvelope(ft_leaf,which=1:3)
```

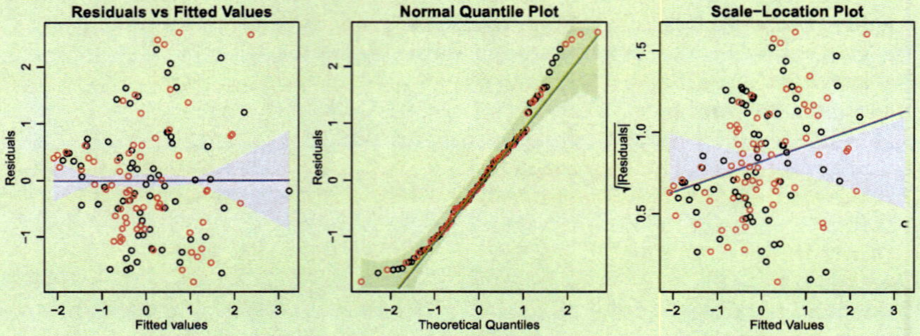

What do you think about the model assumptions made here?

Maths Box 11.2: Having many responses creates many problems for the variance–covariance matrix

There is a covariance parameter for every possible pairwise combination of responses, giving $\binom{p}{2}$ in total (in addition to the p variance parameters). So the number of parameters in $\mathbf{\Sigma}$ grows quickly as p increases, quadratically rather than linearly:

parameters in:

p	$\boldsymbol{\mu}$	$\mathbf{\Sigma}$
2	2	3
4	4	10
10	10	55
20	20	210
100	100	5050

When p gets beyond about 10, there are an awful lot of parameters to estimate in $\mathbf{\Sigma}$, which is an important challenge that needs to be addressed in building a multivariate model. In such situations the methods of this chapter are typically not appropriate, and we need to make extra assumptions on $\mathbf{\Sigma}$ so it can be estimated using fewer parameters.

A variance–covariance matrix $\mathbf{\Sigma}$ potentially stores a lot of parameters in it (Maths Box 11.2), in which case, we would need a lot of observations to estimate it. Edgar's *Iris* data have four responses, so $\mathbf{\Sigma}$ ends up having 10 parameters in it. If, however, he took 20 different types of flower measurements, there would be 210 parameters in $\mathbf{\Sigma}$! Because of the potentially large number of parameters in $\mathbf{\Sigma}$, most methods of multivariate analysis need to have far less responses than there are observations, in order to obtain a stable and reliable estimate of the variance–covariance matrix. How many responses is too many? There is no simple answer to this question, but as a rough rule of thumb, maybe aim to keep the number of responses less than $\sqrt{n/2}$, so that there are at least four(ish) observations in the dataset for every covariance parameter. Otherwise the methods of this chapter are not applicable, although there are other options (e.g. using *reduced rank* methods related to those of Chap. 12).

11.2.2 Multivariate Normality

The distributional assumption for a multivariate linear model, as stated previously, is that errors are multivariate normal. As the name suggests, this is a multivariate generalisation of normality, which means that each response when viewed separately (marginally) is normally distributed. However, marginal normality does not necessarily imply multivariate normality, so we should not use plots of marginal responses to check multivariate normality.

It has been shown, however, that pretty much any joint (multivariate) distribution can be defined as a function of its full conditional distributions (Robert and Casella, 2013), and the conditionals for the multivariate normal are linear models. So data are multivariate normal if each response variable, conditional on all other responses, satisfies a linear model (that is, its conditional mean is a linear function of all other responses, and errors are normally distributed). A good way to check multivariate normality is thus to check assumptions for linear models that predict each response variable as a function of all other variables in the model (including all other responses). The `plotenvelope` function does this by default for multivariate linear models (Code Box 11.3). Remarkably, this idea of using full conditional models to diagnose multivariate normality was not appreciated until recently, but it is easy to implement on any stats package, given the diagnostic tools we have available for linear models.

The multivariate normal is a type of elliptical distribution, i.e. if you plot the contours of its probability density, they are elliptical (in two dimensions), as in Fig. 11.2a.

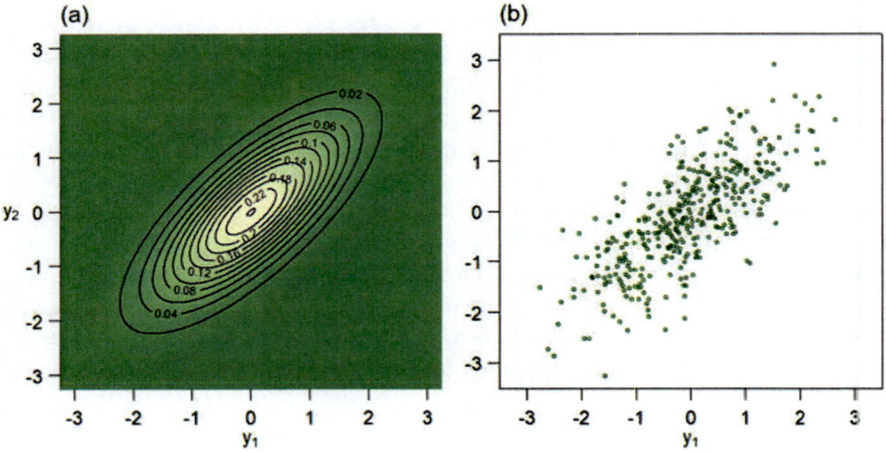

Fig. 11.2: (**a**) A contour map of the probability density for an example multivariate normal distribution. Values of y_1 and y_2 are more likely to be observed when the probability density is higher (lighter regions). Note that contours of equal probability density are ellipses. (**b**) A sample of size 500 from this distribution. Note that there tend to be more points where the probability density is higher, as expected. Note also that multivariate normality implies that each response, conditional on all others, satisfies the assumptions of a linear model

11.2.3 Different Test Statistics

There are different types of test statistics for multivariate linear modelling. The four most common, all available via the `anova` function, are Wilk's lambda (the likelihood ratio statistic), the Hotelling-Lawley statistic, the Pillai-Bartlett trace, and Roy's greatest root (using `method="Wilks"`, `"Hotelling-Lawley"`, `"Pillai"`, or `"Roy"`, respectively). For details on how these are constructed, see Anderson (2003). Some would argue for using more than one of these statistics to confirm a null result, because each is good for slightly different things and will detect slightly different types of violations of the null hypothesis; however, one must be careful with this approach to avoid "searching for significance" (Sect. 1.4.3). The most commonly used of these statistics is probably Wilk's lambda, which is a likelihood ratio statistic, an approach we saw earlier that is commonly used in other contexts (e.g. mixed models, generalised linear models). Roy's greatest root is the statistic that behaves most differently from the others (looking for the axis along which there is the largest effect) so would be another one to check if going for multiple test statistics.

In one way or another, all of these statistics can be thought of as generalisations of the classical F-statistic from univariate linear modelling.

11.2.4 Always Try to Visualise Your Results

Code Box 11.2 fits a multivariate linear model to Ian's leaf economics data and tests for an effect of environment on leaf economics traits, finding some evidence of an association with rainfall. We cannot tell, however, from the multivariate output which responses are associated with rainfall, or how.

It is always important to visualise your data and, in particular, to try to use plots of raw data to visualise effects seen in a statistical analysis. In Ian's case, because there are only two responses and categorical predictors, it is straightforward to visualise the data in a scatterplot (Code Box 11.2). We can see on the plot that the leaf variables are positively correlated, and points at high rainfall sites tend to fall slightly below the main axis along which these variables covary. Thus, at high rainfall sites, leaves of a given leaf mass per area tend to live longer (or, equivalently, leaves with a given lifespan tend to have lower leaf mass per area).

11.2.5 Multivariate Linear Models on `mvabund`

The `mvabund` package can also fit multivariate linear models, via the `manylm` function. The main differences:

- It uses (residual) resampling for design-based inference. *Rows* of residuals are resampled, which ensures correlation across responses is accounted for in testing (by preserving it in resampled data).
- It uses a statistic that takes into account correlation across responses, as in the usual multivariate linear model, if you use the argument `cor.type="R"`.
- The test statistics are called different things—Hotelling-Lawley is `test="F"`, and Wilk's lambda is `test="LR"` (Likelihood Ratio).

Using `cor.type="R",test="LR"` is equivalent to Wilk's test (even though the value of the test statistic is different). An example of this is in Code Box 11.4.

Code Box 11.4: A multivariate linear model for the leaf economics data using `mvabund`

```
> ftmany_leaf = manylm(Yleaf~rain*soilp, data=leaflife)
> anova(ftmany_leaf,cor.type="R",test="LR")
Analysis of Variance Table

            Res.Df Df.diff val(LR) Pr(>LR)
(Intercept)     66
rain            65       1  23.941   0.001 ***
soilp           64       1   4.475   0.125
rain:soilp      63       1   3.371   0.220
```

How do the results compare to those previously in Code Box 11.2?

Note that irrespective of whether the `lm` or `manylm` function was used, the results worked out pretty much the same. This is not unexpected; results should be similar unless something goes wrong with model assumptions. The `anova` function for multivariate linear models fitted using `lm` is model-based and uses large-sample approximations to get the *P*-value. This sometimes goes wrong and can be fixed using design-based inference (as in `manylm`), in two situations:

- If there are problems with the normality assumption (as usual, this is only really an issue when sample size is small).
- If you have a decent number of response variables compared to the number of observations.

The second of these points is particularly important. Testing via `lm` works badly (if at all) when there are many response variables, even when data are multivariate normal, but `manylm` holds up better in this situation. (But if there are many variables, try `cor.type="shrink"` or `cor.type="I"` for a statistic that has better properties in this difficult situation, as in Warton (2008).) In the data for Exercise 11.1, analysed in Code Box 11.2, Ian had 67 observations and only 2 response variables, so there were no issues with the number of response variables. Any lack of normality also seemed to make no difference, which is unsurprising since the sample size was reasonably large (hence the central limit theorem provided robustness to violations of distributional assumptions).

Exercise 11.4: Transforming Ian's leaf economics data
The residual plots for Ian's leaf economics data, as in Code Box 11.3, suggest transformation might be appropriate (with a fan shape, especially for leaf longevity). It might make sense to consider a log transformation, especially for leaf mass per area, which is a ratio—because it is a proportion, it probably should be analysed on a proportional scale!

Repeat the analyses and assumption checks of Code Boxes 11.2–11.3 on log-transformed data. Do the assumptions look more reasonable here? Are the results any different? Is this what you expected to happen?

Exercise 11.5: Transforming Edgar's data?
Recall that Edgar's *Iris* data are size variables, and as such, a log transformation might be appropriate (as in Sect. 1.6).

Fit a linear model to predict petal length from the remaining flower variables, and check assumptions. Is there any evidence of lack of fit? (This would imply a problem with the multivariate normality assumptions.) Does a log transformation help at all?

11.3 Hierarchical Generalised Linear Models

Consider again Petrus's spider counts (Exercise 11.3). Notice that the response variables are counts, so we should be thinking of using some multivariate extension of generalised linear models (GLMs) rather than of linear models (LMs). It turns out that multivariate extensions of GLMs are much harder than for LMs. The main difficulty is that, unlike the normal distribution, discrete distributions (like the Poisson, binomial, or negative binomial) do not have natural multivariate generalisations. Multivariate discrete distributions certainly do exist (Johnson et al., 1997), but they are often quite inflexible, e.g. some can only handle positive correlations (by assuming a common underlying counting process).

The most common solution is to use a type of *hierarchical GLM*, specifically, a GLM that includes in the linear predictor a multivariate normal random effect to induce correlation across responses. This is similar to the strategy used in Chap. 7 to introduce structured dependence in discrete data. However, this approach will work only if you have *loads of observations* and *few responses*.

The hierarchical GLM will make the following assumptions:

1. The observed y_{ij}-values (i for observation/replicate, j for variable) are *independent*, conditional on their mean m_{ij}.[1]
2. ("*Data model*") Conditional on their mean m_{ij}, the y_{ij}-values come from *a known distribution* (from the exponential family) with a known *mean–variance relationship* $V(m_{ij})$.
3. ("*Process model*") Straight-line relationship between *some known function of the mean* of y_{ij} and each x_i, with an error term ϵ_{ij}:

$$g(m_{ij}) = \beta_{0j} + \mathbf{x}_i^T \boldsymbol{\beta}_j + \epsilon_{ij} \tag{11.2}$$

4. The process model errors ϵ_{ij} are *multivariate normal* in distribution, that is, for $\boldsymbol{\epsilon}_i = (\epsilon_{i1}, \ldots, \epsilon_{ip})$:

$$\boldsymbol{\epsilon}_i \sim MVN(\mathbf{0}, \boldsymbol{\Sigma})$$

This type of model has been referred to in some parts of ecology as a *joint species distribution model* (Clark et al., 2014; Ovaskainen and Abrego, 2020). Statisticians often refer to this as a hierarchical model (Cressie et al., 2009) because it can be thought of as a hierarchy with two levels (or sometimes more)—a *process model* for $g(m_{ij})$, and a *data model* for y_{ij} conditional on m_{ij}. Random variables are present in both levels of the model.

[1] The mean is written m_{ij} here, not μ_{ij}, because in a hierarchical model, it is a random quantity, not a fixed parameter.

11.3.1 Consequences of Correlated Errors in a Hierarchical Model

The process model errors ϵ_{ij} in a hierarchical GLM have two main consequences—changing the marginal distribution of data (in particular, introducing overdispersion) and inducing correlation, as in Fig. 11.3 or Maths Box 11.4.

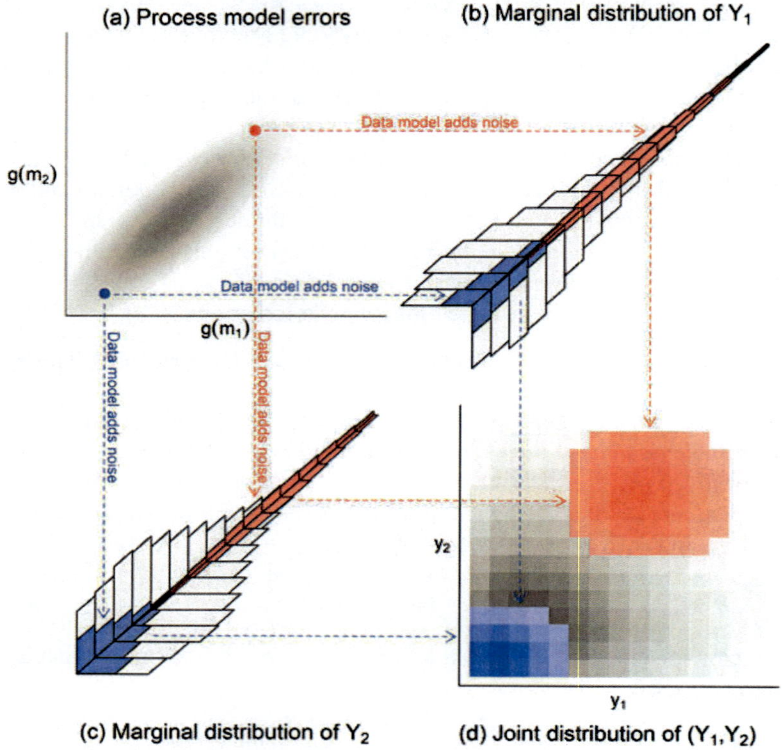

Fig. 11.3: In a hierarchical model, correlated terms m_{ij} in the process model (**a**) have consequences for marginal (**b**) and (**c**) and joint (**d**) distributions of observed data y_{ij}. The data model takes values of m_{ij}, such as the blue and red points in (**a**), and adds noise to them to form the blue and red histograms in (**b**) and (**c**) and the blue and red patches in (**d**). The marginal distribution of y_{ij} (grey histograms in b and c) is more variable ("overdispersed") than the distribution specified by the data model (blue or red histograms in b and c), with a potentially different shape, because the m_{ij} in the process model vary (**a**). The joint distribution of y_{ij} (**d**) is correlated y_{ij} because the m_{ij} in the process model are correlated (**a**). Note that the correlation in observed data (y_{ij}) is weaker because of the extra noise introduced by the data model, which maps points in m_{ij} to patches in y_{ij}.

Maths Box 11.3: The marginal distribution of a hierarchical model is messy

Our hierarchical GLM has two random components to it—the *data model*, which assumes y_{ij} comes from an exponential family with (conditional) probability distribution $f_{Y_{ij}|M_{ij}}(y_{ij}|M_{ij} = m_{ij})$, and the *process model* M_{ij} (Eq. (11.2)). These two components combine to determine the probability function of the data:

$$f_{\mathbf{Y}_i}(\mathbf{y}_i) = \int_{\mathbf{m}_i} f_{\mathbf{M}_i}(\mathbf{m}_i) \prod_{j=1}^{p} f_{Y_{ij}|M_{ij}}(y_{ij}|M_{ij} = m_{ij})$$

This is a marginal distribution for $\mathbf{Y}_i$ and can be quite different to the conditional distribution specified in the data model. For example, except for a few special cases, Y_{ij} is not a member of the exponential family.

In fact, $f_{\mathbf{Y}_i}(\mathbf{y}_i)$ often does not even have a closed form, in which case it can't be written any more simply than it was above. We would then need to use numerical integration (e.g. the trapezoidal rule, Davis and Rabinowitz, 2007, Chapter 2) to calculate values of it. This makes model fitting more difficult, because it is suddenly a lot harder to calculate the likelihood function, let alone maximise it, or (in a Bayesian approach, Clark, 2007, Chapter 4) use it to estimate a posterior distribution. We typically proceed by approximating $f_{\mathbf{Y}_i}(\mathbf{y}_i)$ in some way, using techniques like Laplace approximation or Monte Carlo integration (Evans and Swartz, 2000, Chapters 4 and 7, respectively).

Maths Box 11.4: Marginal mean and covariances of a hierarchical model

The marginal mean of Y_{ij} for our hierarchical GLM can be shown to satisfy

$$\mu_{Y_{ij}} = \mu_{M_{ij}}(\mu_{Y_{ij}|M_{ij}=m_{ij}})$$

where $\mu_Y(h(Y))$ denotes "take the mean of $h(Y)$ with respect to Y". Sometimes this can be simplified, e.g. if the data model is a Poisson log-linear model, then

$$\mu_{Y_{ij}} = \mu_{M_{ij}}\left(e^{M_{ij}}\right) = e^{\beta_{0j} + \mathbf{x}_i^T \boldsymbol{\beta}_j + \sigma_j^2/2}$$

Notice the $\sigma_j^2/2$ bit, which means that if we ignored the random effect and made predictions using fixed effects $e^{\beta_{0j} + \mathbf{x}_i^T \boldsymbol{\beta}_j}$, we would underestimate the mean of data by a factor of $e^{\sigma_j^2/2}$. This issue is somewhat analogous to *transformation bias* when transforming data to satisfy model assumptions (Section 1.6, see also Diggle et al., 2002, Section 7.4). The marginal variance of Y_{ij} can be shown to satisfy

$$\sigma^2_{Y_{ij}} = \mu_{M_{ij}}(\sigma^2_{Y_{ij}|M_{ij}=m_{ij}}) + \sigma^2_{M_{ij}}(\mu_{Y_{ij}|M_{ij}=m_{ij}})$$

The second of these terms captures *overdispersion* in the marginal model relative to the conditional model, due to variation in the random effect. This term can also sometimes be simplified, e.g. if the data model is Poisson log-linear:

$$\sigma^2_{Y_{ij}} = \sigma^2_{M_{ij}}(e^{M_{ij}}) + \mu_{M_{ij}}(e^{M_{ij}}) = \ldots = \mu_{Y_{ij}} + \left(e^{\sigma^2_j} - 1\right)\mu^2_{Y_{ij}}$$

OK, I skipped a few steps. But note this ends up having the same form as the mean–variance relationship of the negative binomial distribution, where the overdispersion parameter is now $e^{\sigma^2_j} - 1$. So when using an observation-level random effect in a model for overdispersed counts, you can often stick with the Poisson and let the process model deal with overdispersion.

The marginal covariance of Y_{ij} and $Y_{ij'}$ can be calculated similarly, and for a Poisson log-linear data model it ends up being

$$\sigma_{Y_{ij},Y_{ij'}} = (e^{\sigma_{ij}} - 1)\,\mu_{Y_{ij}}\mu_{Y_{ij'}}$$

Compared to the variance expression, this is missing the first term, which can make a big difference in the strength of the correlation induced in data. For example, if the mean counts are one and the M_{ij} are perfectly correlated with variance 0.25 ($\sigma^2_j = \sigma^2_{j'} = \sigma_{jj'} = 0.25$), the correlation between data values Y_{ij} and $Y_{ij'}$ works out to be approximately 0.22, even though the correlation between M_{ij} and $M_{ij'}$ is one!

11.3.1.1 The Marginal Distribution Has Gone Weird

The first consequence of process model errors ϵ_{ij} in a hierarchical model is that they change the marginal distribution of the responses (Fig. 11.3b–c) from what you might expect based on the data model. For example, if the responses are counts assumed to be conditionally Poisson in the data model, they will not be marginally Poisson; they will have what is known as a compound Poisson-lognormal distribution. As well as changing the marginal distribution, the interpretation of parameters also changes, making hierarchical models harder to interpret (unless you are happy conditioning on the errors in the process model).

The most striking change in the marginal distribution is that it is more overdispersed than the conditional distribution that was specified in the data model. Note for example in Fig. 11.3b–c that the grey histograms, representing the marginal distribution, have much wider variances than either of the two conditional distributions (red and blue histograms) generated by the data model around particular values of the process model error (corresponding to the red and blue points in Fig. 11.3a). Mathematically, a hierarchical model with Poisson conditional counts (for example) no longer has a mean–variance relationship of $V(y) = \mu$. Instead, the marginal mean–variance relationship becomes $V(y) = \mu + \phi\mu^2$, where ϕ is a function of the variances of the ϵ_{ij} (specifically, $\phi_j = e^{\sigma_j^2} - 1$, where σ_j^2 is the jth variance parameter in $\mathbf{\Sigma}$). The Poisson-lognormal distribution behaves a lot like the negative binomial (e.g. it has the same mean–variance relationship), so in a hierarchical model for overdispersed counts you might not need to use the negative binomial; you can often get away with a Poisson and let the ϵ_{ij} absorb any overdispersion.

This overdispersion actually becomes a problem when modelling binary responses (e.g. presence–absence), because there is no information in a binary response that can be used to estimate overdispersion. In this situation, we need to fix the variances of random effects in the process models to constants. Typically they are fixed to the value one, but the precise value doesn't actually matter, as long as it is larger than the covariances in the data.

The interpretation of marginal effects is more difficult in hierarchical models because parameters in the process model quantify effects on the conditional mean, not effects on the marginal mean, which can be affected in surprising ways. For example, in a hierarchical Poisson log-linear model, the predicted values of m_{ij} consistently underestimate the actual (marginal) mean count (by a factor of $e^{\sigma_{jj}^2/2}$). Fortunately, slope coefficients can still be interpreted in the usual way in the Poisson log-linear case, e.g. if the slope is two, then the marginal mean changes by a factor of e^2 when the x variable changes by one unit. However, the effects on marginal means of slope coefficients are not so easily understood in other types of hierarchical model, e.g. logistic regression, where a given change in the linear predictor can have quite different implications for the marginal mean (for more details see Diggle et al., 2002, Section 7.4).

11.3.1.2 Correlation in Process Model Errors Means Correlation in Responses

The second and perhaps most important consequence of ϵ_{ij} is that they induce correlation between responses. Note, however, that the correlation between responses is much weaker than the correlation between the ϵ_{ij}, because the data model introduces random noise that weakens the signal from the process model. So, for example, the ϵ_{ij} in Fig. 11.3a are highly correlated, but the observed data in Fig. 11.3d have a weaker correlation, because the data model (conditional on ϵ_{ij}) assumes counts are independent for each response variable. Thus, points in the process model (Fig. 11.3a) map to patches of potential points in the data (Fig. 11.3d) because of noise from the data model.

11.3.2 Fitting a Hierarchical GLM

Hierarchical GLMs are much more difficult to fit than multivariate linear models (Maths Box 11.3), with a lot more things that can go wrong. They can be fitted using conventional mixed modelling software like `lme4` or `glmmTMB` (Brooks et al., 2017, Code Box 11.6), but not so easily if responses are binary, where the variances in the process model need to be fixed to constants. As previously, the `glmmTMB` package is recommended in preference to `lme4`; it is noticeably faster and more stable for these sorts of models. Both require data in "long format" as in Code Box 11.5.

Alternative software, which is capable of fitting hierarchical models to binary data, is the `MCMCglmm` package (Hadfield et al., 2010, Code Box 11.7). This package is so named because it uses a Bayesian approach to model fitting via Markov chain Monte Carlo (MCMC). It accepts data in so-called short format.

Code Box 11.5: Preparing spider data for analysis on `lme4` or `glmmTMB`

Petrus's data from Exercise 11.3 are available in the `mvabund` package, but with abundances for 12 different species. First we will calculate the abundance of the three most abundant genera:

```
> library(mvabund)
> library(reshape2)
> data(spider)
> Alop=apply(spider$abund[,1:3],1,sum)
> Pard=apply(spider$abund[,7:10],1,sum)
> Troc = spider$abund[,11]
> spidGeneraWide = data.frame(rows=1:28,scale(spider$x[,c(1,4)]),
  Alop,Pard,Troc)
> head(spidGeneraWide)
  rows   soil.dry       moss Alop Pard Troc
1    1 -0.1720862  0.6289186   35  117   57
2    2  0.7146218 -0.6870394    2   54   65
3    3  0.1062154  0.1916410   37   93   66
4    4  0.2507444  0.1916410    8  131   86
5    5  0.6728333 -1.4299919   21  214   91
```

The data are in short format, with observations in rows and different responses in different columns; we need to rearrange them into long format:

```
> spiderGeneraLong = melt(spidGeneraWide,id=c("rows","soil.dry","moss"))
> names(spiderGeneraLong)[4:5] = c("genus","abundance")
> head(spiderGeneraLong)
  rows   soil.dry       moss genus abundance
1    1 -0.1720862  0.6289186  Alop        35
2    2  0.7146218 -0.6870394  Alop         2
3    3  0.1062154  0.1916410  Alop        37
4    4  0.2507444  0.1916410  Alop         8
5    5  0.6728333 -1.4299919  Alop        21
6    6  1.1247181  0.1916410  Alop         6
```

`spiderGeneraLong` is ready for model fitting using `lme4` or `glmmTMB`.

We need to have many more rows than columns when fitting the hierarchical GLM described previously. One reason, as before, is because the size of the variance–

covariance matrix Σ is related to the square of the number of response variables, so it can be very large. But in addition, discrete data are not very information rich, especially in the case of presence–absence data or discrete data with many zeros. Petrus's data originally had 12 species in it, meaning there would be 78 parameters to estimate in Σ, from just 28 observations, almost half of which were absences. The reason the data were aggregated to three genera in Exercise 11.3 was so that we would be able to reliably fit a hierarchical GLM to it. Even with just three responses, an `lme4` fit does not always converge, under default settings.

Hierarchical GLMs are technically difficult to fit because the likelihood function doesn't have a closed form and needs to be approximated. The `lme4` and `glmmTMB` packages approximate the likelihood using the Laplace method (Bates, 2010), whereas the `MCMCglmm` package approximates it via MCMC, a simulation-based approach that is very flexible but not very fast. The more parameters there are in the model, the more difficult it will be to approximate the likelihood, irrespective of the fitting technique used. Problems fitting models usually show up in software like `lme4` and `glmmTMB` as convergence issues, with error or warnings, whereas problems may be more subtle in MCMC software like `MCMCglmm`, with the Markov chains needing to be interrogated for signs of non-convergence (Gelman et al., 2013, Section 11.4).

Code Box 11.6: Fitting a hierarchical GLM to spider data on `glmmTMB`

```
> library(glmmTMB)
> spid_glmm = glmmTMB(abundance~genus+soil.dry:genus+moss:genus
 +(0+genus|rows), family="poisson",data=spiderGeneraLong)
> summary(spid_glmm)
Random effects:
 Groups Name      Variance Std.Dev. Corr
 rows   colsAlop 1.3576   1.1652
        colsPard 1.5900   1.2610   0.75
        colsTroc 0.7682   0.8765   0.59 0.82
Number of obs: 84, groups:  rows, 28

Conditional model:
                   Estimate Std. Error z value Pr(>|z|)
(Intercept)          1.9843     0.2463   8.058 7.76e-16 ***
colsPard             1.0846     0.2100   5.165 2.41e-07 ***
colsTroc             0.6072     0.2335   2.600  0.00932 **
colsAlop:soil.dry   -0.6012     0.3284  -1.831  0.06716 .
colsPard:soil.dry    1.4041     0.3704   3.791  0.00015 ***
colsTroc:soil.dry    1.4108     0.2950   4.782 1.74e-06 ***
colsAlop:moss        0.3435     0.3322   1.034  0.30103
colsPard:moss        0.7361     0.3547   2.075  0.03796 *
colsTroc:moss       -0.2033     0.2648  -0.768  0.44264
Signif. codes:  0 `***' 0.001 `**' 0.01 `*' 0.05 `.' 0.1 ` ' 1
```

Can you see any differences in the response of different spider genera to environmental conditions?

Notice that in Code Box 11.6, the formula was specified as

```
abundance~genus+soil.dry:genus+moss:genus+(0+genus|rows)
```

where genus refers to genera (which are stored in different columns when data are in short format) and rows refers to different sampling events (the rows when in short format). Thus, the model was fitted without main effects for environmental variables and without an intercept term in the random effects model. Neither of these decisions was necessary, but they simplified the interpretation of parameters. The main effects for environmental variables were left out so that for each genus, separate estimates were returned of the effect of each environmental variable (Code Box 11.6). The intercept was left out of the random effects model so that a separate random effect was estimated for each spider genus, so the correlations between random effects in Code Box 11.6 correspond to the estimated correlations in ϵ_{ij} between genera. This would not have happened by default, which would have involved the use of indicator variables along the lines of Sect. 3.2. The correlations are quite large (in one case, 0.82); note that these are for process model errors ϵ_{ij} and not for the raw counts y_{ij}, which have smaller correlations, dampened by the Poisson noise in the model. This is a caveat of hierarchical modelling—the meaning (in terms of observed data) of parameters is not always clear when they are in the process model, because their effect gets distorted by the data model. Put another way, the model for data y_{ij} has been specified conditionally, and it is hard to interpret what this means for y_{ij} marginally.

In `MCMCglmm` the formula is similar (Code Box 11.7), except that columns are called `trait`, rows are called `units`, and we fit an unstructured variance–covariance matrix via `us`. The output from `MCMCglmm` is a sequence (or *chain*) of plausible values for each model parameter, which are a function only of the previous value (a *Markov* chain) plus some random noise (making it Markov chain *Monte Carlo*). Decisions need to be made about what prior distribution to sample from (`prior` argument), how long to run the MCMC chain for (`itt`) and which samples from it to keep (`burnin`, `thin`), but default specifications were used in Code Box 11.7.

Petrus's results are reassuringly similar for `glmmTMB` and `MCMCglmm` (Code Boxes 11.6–11.7). The random effect variances tend to be larger for `MCMCglmm` than `glmmTMB`, but this is mostly because of how results are reported rather than how the model is fitted. Specifically, `MCMCglmm` reports a mean of the chain of sampled parameter values, whereas `glmmTMB` reports a maximum likelihood estimator, similar to finding the highest point (*mode*) of the distribution. The mode and mean are usually similar (because most parameters have symmetric distributions), but for variance parameters the distribution tends to be right-skewed, and the mean is much larger than the mode.

Code Box 11.7: `MCMCglmm` fit to Petrus's spider genus data

```
> library(MCMCglmm)
> set.seed(1)
```

```
> ft_MCMC = MCMCglmm(cbind(Alop,Pard,Troc)~trait+soil.dry:trait+
   moss:trait, rcov=~us(trait):units, data=spidGeneraWide,
   family=rep("poisson",3))
> summary(ft_MCMC)

                            post.mean    l-95% CI   u-95% CI eff.samp
traitAlop:traitAlop.units 1.9673797 0.6895887 3.395750 171.8871
traitPard:traitAlop.units 1.5216329 0.5111319 2.618777 750.4314
traitTroc:traitAlop.units 0.8476355 0.1973054 1.620263 735.4733
traitAlop:traitPard.units 1.5216329 0.5111319 2.618777 750.4314
traitPard:traitPard.units 2.1080892 1.0426738 3.557205 624.7579
traitTroc:traitPard.units 1.2399104 0.5336700 2.169283 738.6973
traitAlop:traitTroc.units 0.8476355 0.1973054 1.620263 735.4733
traitPard:traitTroc.units 1.2399104 0.5336700 2.169283 738.6973
traitTroc:traitTroc.units 1.0876851 0.4375083 1.836845 545.7719

 Location effects: cbind(Alop, Pard, Troc) ~ trait + soil.dry:trait
                   + moss:trait

                    post.mean   l-95% CI   u-95% CI eff.samp  pMCMC
(Intercept)           1.96106  1.40677  2.56137     564.0 <0.001 ***
traitPard             1.10316  0.62128  1.55902     307.0 <0.001 ***
traitTroc             0.60433  0.07406  1.12209     314.2  0.028 *
traitAlop:soil.dry   -0.60527 -1.38983  0.14573     797.5  0.102
traitPard:soil.dry    1.44406  0.61769  2.22122     741.5 <0.001 ***
traitTroc:soil.dry    1.46411  0.76420  2.04801     353.2 <0.001 ***
traitAlop:moss        0.35515 -0.34662  1.14501     683.7  0.332
traitPard:moss        0.76450 -0.01952  1.60829     902.7  0.064 .
traitTroc:moss       -0.19314 -0.76568  0.37943     646.0  0.498
---
Signif. codes:  0 `***' 0.001 `**' 0.01 `*' 0.05 `.' 0.1 ` ' 1
```

How do results compare to the model fitted using `glmmTMB`?

Exercise 11.6: Different effects on different spider genera?
Petrus would like to know if there is evidence against the claim that all spider genera respond to their environment in the same way.

Use `spiderGeneraLong` (Code Box 11.5) to fit a model that assumes all spiders respond in the same way to their environment. Now use `anova` to compare this model to `spid_glmm` (Code Box 11.6).

Is there evidence that different spider genera respond in different ways to their environment?

As previously, we should plot the raw data to try to confirm the patterns we see in analyses (Code Box 11.6). The main story in the data is that different genera have different responses to environment (Fig. 11.4). We could also construct scatterplots of counts (or of Dunn-Smyth residuals, to control for environmental effects) to study the nature of correlation of abundances.

11.3.3 Hierarchical GLMs for Presence–Absence Data

You can use `MCMCglmm` to fit a hierarchical GLM to presence–absence data, but not `lme4`. A complication is that while the process model errors in a hierarchical GLM introduce overdispersion, presence–absence data carry no information about overdispersion. This means that variance parameters of the process model errors cannot be estimated (the variances are "unidentifiable"), and to fit the model, the variances need to be fixed to a constant (commonly fixed to one). The precise value the variances are fixed to does not matter, as long as it is larger than the covariances in the data. The `MCMCglmm` function can fix the variance on all random effects using `fix=1` in the `prior` argument for the multivariate random effect (Hadfield et al., 2010, Section 3.6).[2]

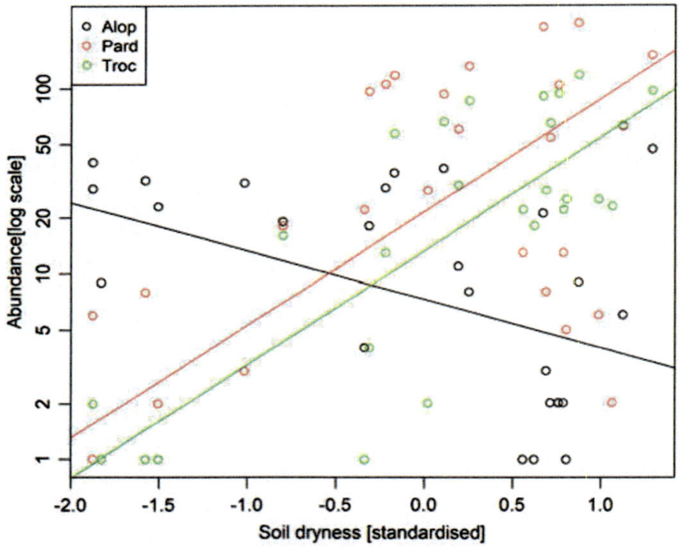

Fig. 11.4: Scatterplot of abundance against soil dryness for each genus of Petrus's spider data. Fitted lines are taken from `spid_glmm` (Code Box 11.6). Note that *Alop* has a flatter association with soil dryness than the other genera, as indicated in the `summary` output of Code Box 11.6

[2] Note that `fix=1` does not mean "fix the variance to one"; it means "fix the variances for all random effects from the first one onwards". If a model has three different types of random effects in it, `fix=2` would fix the variance of the second and third random effects but not the first.

Table 11.1: Common research questions you can answer using a hierarchical GLM

Research question	Model components of interest	Key tools
How do taxa respond to their environment?	$\boldsymbol{\beta}_j$	`summary`, `confint`
Is there an association between a treatment/environmental variable and a community of taxa?	$\boldsymbol{\beta}_j (= 0?)$	`anova`
Which environmental variables are associated with community abundance?	Choosing the $\mathbf{x}_i$	Model selection tools (e.g. `BIC`)
Which species co-occur? To what extent is this explained by a common response to the environment?	$\boldsymbol{\Sigma}$	`VarCorr`

Maths Box 11.5: No overdispersion in binary data

The sample mean–variance relationship of binary data has a special form. To see this, consider a sample of n binary values, for which the proportion of values that are one is $\hat{\mu}$. The sample mean is $\frac{1}{n}\sum_i y_i = \hat{\mu}$. We will calculate the sample variance, but using n in the denominator of the variance formula (rather than $n-1$), because it gives a slightly neater answer:

$$\hat{V}(\hat{\mu}) = \frac{1}{n}\sum_{i=1}^{n}(y_i - \hat{\mu})^2 = \hat{\mu}(1-\hat{\mu})^2 + (1-\hat{\mu})(0-\hat{\mu})^2$$

$$= \hat{\mu}(1-\hat{\mu})(1-\hat{\mu}+\hat{\mu}) = \hat{\mu}(1-\hat{\mu})$$

So the sample mean–variance relationship of a binary response is always $\hat{V}(\mu) = \mu(1-\mu)$ (as in Fig. 10.1), irrespective of what the true model for the data is. This means there is no capacity to use binary data to estimate overdispersion. We can still estimate covariance (from the extent to which ones and zeros of different responses occur at the same time), but not variances. Hence, in a hierarchical model for binary data, the variance–covariance matrix $\boldsymbol{\Sigma}$ must have the variances all fixed to constant values, although the covariances need not be.

11.3.4 Mind Your Ps and Qs

Recall that we always need to check that our analysis is aligned with the research question (*Qs*) and data properties (*Ps*).

In terms of research questions, a hierarchical GLM is a flexible tool that can be used for many different things, some of which are summarised in Table 11.1. Like any regression model, it can be used to test for and estimate associations, for variable selection, or for prediction.[3] Important additional questions can also be answered using covariances (in Σ), which can be used to study co-occurrence of taxa and the extent to which co-occurrence can be explained by environmental variables. This idea has attracted a lot of attention recently in ecology—using estimates of Σ from models, including environmental variables, to tease apart competing explanations for species co-occurrence (as in Letten et al., 2015, for example).

In terms of data properties, as usual, the distributional assumptions are not key; the focus is instead on assumptions of independence, mean–variance, and linearity. These assumptions should be considered in the usual way. For example, independence as always depends on the design, although you can use the data to try to diagnose special types of dependence, such as spatial or temporal autocorrelation (Chap. 7). Other assumptions are checked on residual plots, as in Code Box 11.8. Note, as in Chap. 6, we use the `DHARMa` package (Hartig, 2020) to get Dunn-Smyth residuals, given their absence from `glmmTMB` (our usual sources of Dunn-Smyth residuals, `statmod` and `mvabund`, currently lack mixed model functionality). As in Chap. 6, matters are complicated by having two sources of randomness in the model—the y_{ij} and the ϵ_{ij} (data and process models, respectively). These work together in model fitting, and if you look at one without the other, you may not see the full story. It is a good idea to use fixed effect fitted values (`re.form=NA`) in residual plots, since random effects influence the residuals, so they shouldn't feature in the fitted values also.

Code Box 11.8: Diagnostic plots for a hierarchical GLM of Petrus's spider data

```
library(DHARMa)
spidFits = predict(spid_glmm, re.form=NA)
res_spid = qnorm( simulateResiduals(spid_glmm)$scaledResiduals )
plot(spidFits, res_spid, col=spiderGeneraLong$genus,
 xlab="Fitted values", ylab="Dunn-Smyth residuals")
abline(h=0, col="olivedrab")
addSmooth(spidFits,res_spid)
qqenvelope(res_spid, col=spiderGeneraLong$genus, main="")
```

The `addSmooth` function, provided in the `ecostats` package, adds a GAM smoother to the data and confidence bands. Note these confidence bands give pointwise rather than global coverage (meaning they do not control for multiple testing), unlike the output from `plotenvelope` or `qqenvelope` (below right)

[3] Although if the purpose of the study is prediction, additional features that are known to often help are shrinking parameters (using a LASSO or assuming regression coefficients are drawn from a common distribution), or in large datasets, using flexible regression tools (like additive models) to handle non-linearity.

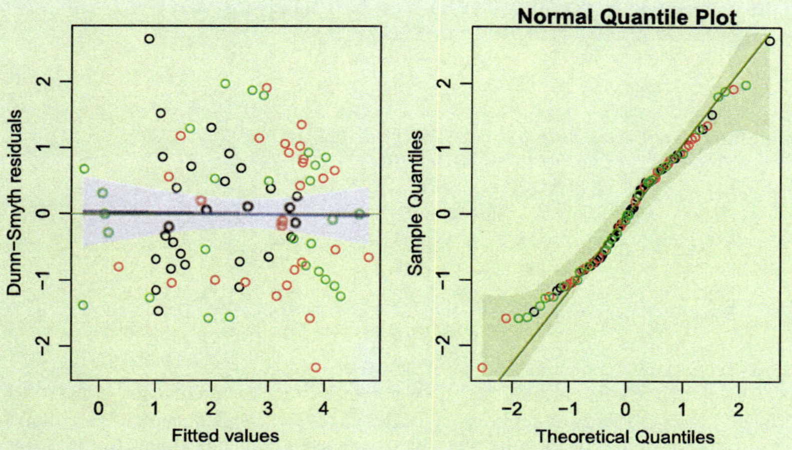

What do these plots tell us about model assumptions?

When fitting models using MCMC, an important thing to check is whether the Markov chain converged. To check MCMC convergence, we can run multiple chains and check that they have similar properties (Code Box 11.9). This is especially important for hierarchical models like Petrus's, which is parameter rich. If there are too many response variables compared to the number of observations (where too many is actually a small number, let's say $\sqrt{n/2}$), it is unlikely that the chain will converge to the posterior distribution (Exercise 11.7). One option is to change the priors to be *informative*, essentially telling the model the approximate values of some of the parameters, in the event that they are hard to estimate from data. This has parallels to penalised estimation (Sect. 5.6). A better idea usually is to consider a different model that has fewer parameters (along the lines of Chap. 12).

Code Box 11.9: Diagnosing convergence in a `MCMCglmm` fit

To check if our `MCMCglmm` chain converged, we will repeat the analysis three times and compare chains:

```
> set.seed(2)
> ft_MCMC2 = MCMCglmm(cbind(Alop,Pard,Troc)~trait+soil.dry:trait+
    moss:trait, rcov=~us(trait):units, data=spidGenera,
    family=rep("poisson",3))
> set.seed(3)
> ft_MCMC3 = MCMCglmm(cbind(Alop,Pard,Troc)~trait+soil.dry:trait+
    moss:trait, rcov=~us(trait):units, data=spidGenera,
    family=rep("poisson",3))
> whichPlot=c(1:3,5:6,9) # indices of unique variance-covar parameters
> par(mfrow=c(length(whichPlot),1),mar=c(2,0.5,1.5,0))
> for(iPlot in whichPlot)
{
```

```
    plot.default(ft_MCMC$VCV[,iPlot],type="l",lwd=0.3,yaxt="n")
    lines(ft_MCMC2$VCV[,iPlot],col=2,lwd=0.3)
    lines(ft_MCMC3$VCV[,iPlot],col=3,lwd=0.3)
    mtext(colnames(ft_MCMC$VCV)[iPlot])
}
> gelman.diag(mcmc.list(ft_MCMC$VCV[,whichPlot],ft_MCMC2$VCV
    [,whichPlot],ft_MCMC3$VCV[,whichPlot]))
Potential scale reduction factors:
                             Point est. Upper C.I.
traitAlop:traitAlop.units          1.00       1.01
traitPard:traitAlop.units          1.00       1.01
traitTroc:traitAlop.units          1.00       1.00
traitPard:traitPard.units          1.01       1.02
traitTroc:traitPard.units          1.00       1.00
traitTroc:traitTroc.units          1.01       1.04
```

A trace plot and Gelman-Rubin statistic is constructed for each parameter in the 3×3 covariance matrix (`whichPlot` removes duplicates of covariance parameters). We want the colours to be all mixed up on the trace plots, with parameters bouncing around the full range of values relatively quickly. We want the Gelman-Rubin statistic to be close to one, which seems to be the case here.

Exercise 11.7: Non-converging model for Petrus's *Alopecosa* species

Consider again Petrus's data, but now consider counts from just the first three species of the dataset (the species from the *Alopecosa* genus):

```
data(spider)
spider3Wide = data.frame(rows=1:28, scale(spider$x[,c(1,4)]),
                spider$abund[,1:3])
```

Try to fit a hierarchical GLM, along the lines of Code Boxes 11.6 or 11.7, to the three Alopecosa *species.*

Whether using `glmmTMB` or `MCMCglmm`, the model does not converge. But this is less obvious when using `MCMCglmm`, as some runs don't produce a non-convergence error.

For a `MCMCglmm` *run that does not return an error, construct a trace plot, along the lines of Code Box 11.9. Look in particular at the traces of covariance parameters (rows 2, 3, and 5). Are there any signs of non-convergence in the trace plots or in the Gelman-Rubin statistics?*

This model does not converge because there are many zeros in the species data—almost half this dataset are absences, whereas there were only a couple of absences in the genus-level dataset of Exercise 11.3. Lots of absence is especially problematic for estimating covariance. Covariances try to estimate co-occurrence patterns, which can't be done without occurrences!

11.4 Other Approaches to Multivariate Analysis

The introduction to multivariate analysis in this chapter has been relatively narrow, focusing on simple multivariate regression models. There are many other types of research question for which different types of multivariate analysis are appropriate, examples of which appear in Table 11.2. Most of these can be answered using multivariate regression models, with some extra bells and whistles, as we will see over the coming chapters.

Table 11.2: Some common research questions you can't answer using the multivariate regression techniques of this chapter

Research question	Method	Reference
How do I plot the data?	Ordination and related methods	Chap. 12
What are the main axes of variation in the data?	Ordination	Chaps. 12–13
Which combination of response variables best captures how treatments differ from each other?	Discriminant analysis	Hastie et al. (2009, Chapter 4)
Is there an effect of a predictor on multivariate response? What is it? Oh, and I have heaps of responses	High-dimensional regression	Chap. 14
Is there a relative change in responses (i.e. ratios of mean response matter, not the mean *per se*)?	Compositional analysis	Chap. 14
Can responses/observations be assigned to groups with similar values?	Classification	Chap. 16
Why do different variables have different responses to predictors?	"Fourth corner" approaches	Chap. 16

Chapter 12
Visualising Many Responses

A key step in any analysis is data visualisation. This is a challenging topic for multivariate data because (if we have more than two responses) it is not possible to jointly visualise the data in a way that captures correlation across responses or how they relate to predictors. In this chapter, we will discuss a few key techniques to try.

Exercise 12.1: Flower Size and Shape
Edgar (Anderson, 1935) was studying *Iris* flowers. For 50 flowers on each of three *Iris* species, he took four measurements—length and width of petals and length and width of sepals. He wants to know:

> How do these three species of Iris differ in terms of flower size and shape?

What sort of graph should Edgar produce to visualise how species differ in flower size and shape?

Exercise 12.2: Bush Regeneration and Invertebrate Counts
In his revegetation study (Exercise 10.3), Anthony classified anything that fell into his pitfall traps into orders and thus counted the abundance of each of 24 invertebrate orders across 10 sites. He wants to know:

> Is there evidence of a change in invertebrate communities due to bush regeneration efforts?

What sort of graph should Anthony produce to visualise the effects of bush regeneration on invertebrate communities?

D. I. Warton, *Eco-Stats: Data Analysis in Ecology*, Methods in Statistical Ecology,
https://doi.org/10.1007/978-3-030-88443-7_12

12.1 One at a Time: Visualising Marginal Response

A good starting point is to plot each response variable separately against predictors. The `ggplot2` package (Wickham, 2016) makes this quite easy to do, or you could try the `plot` function in the `mvabund` package (Code Box 12.1). Note that by default, `plot.mvabund` only shows the 12 response variables with the highest total abundance. This can be changed using `n.vars` and `var.subset` to control how many taxa, and which, are plotted. Even with just a subset of 12 response variables, the plot looks quite busy—as it should because there is a lot of data and (potentially) a lot of information in it.

Code Box 12.1: Plotting the Bush Regeneration Data of Exercise 12.2 Using mvabund

```
library(mvabund)
library(ecostats)
data(reveg)
reveg$abundMV=mvabund(reveg$abund) #to treat data as multivariate
plot(abundMV~treatment, data=reveg)
```

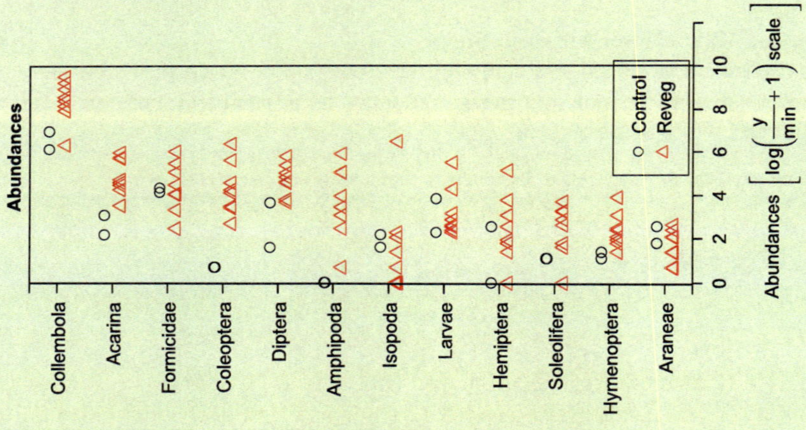

Can you see any taxa that seem to be associated with bush regeneration?

You can learn a fair bit about the relationship between a multivariate y and x from separate plots of y against x for each y variable. For example, in Code Box 12.1 we see that the two control sites often had lower invertebrate abundances (especially for *Collembola*, *Acarina*, *Coleoptera*, and *Amphipoda*), but this effect wasn't seen in all orders (e.g. *Formicidae*). Note, however, that this one-at-a-time approach, plotting the *marginal effect* on y, will not detect more complex relationships between x and combinations of y. As in Fig. 11.1, you can't see joint effects on y of x without jointly plotting y variables against x in some way. Another option is to plot responses two at a time in a scatterplot matrix (e.g. using `pairs` in R). But too many variables would be a nightmare—for 100 response variables, there are *4950* possible pairwise scatterplots! Another option is *ordination*, discussed below.

12.2 Ordination for Multivariate Normal Data

The intention of ordination is to summarise data as best as can be done on just a few axes (usually two). This requires some form of *data reduction*, reducing p response variables to just a few. There are many approaches to doing this. For (approximately) multivariate normal data, we use principal component analysis (PCA) or factor analysis.

12.2.1 Principal Components Analysis

PCA is the simplest of ordination tools. It tries to rotate data to identify axes that (sequentially) explain as much of the variation as possible, based on just the variance–covariance matrix or the correlation matrix (Maths Box 12.1). In R you can use the `princomp` function, as in Code Box 12.2.

PCA will report the rotation of the data that sequentially maximised sample variation, often referred to as the loadings, as well as the amount of variation explained by each axis. For example, in Code Box 12.2, the first component was constructed as 0.521 × `Sepal.Length` − 0.269 × `Sepal.Width` + 0.580 × `Petal.Length` + 0.565 × `Petal.Width`, which had a standard deviation of 1.71, explaining 73% of the variation in the data. The direction defined by this linear combination of the four responses gives the largest possible standard deviation for this dataset. The second component explains the most variation in the data beyond that explained by the first component and captures an additional 23% of the variation. Together, the first two principal components explain 96% of the variation in the data, so most of what is happening in the data can be captured by just focusing on these two principal component scores instead of looking at all four responses.

Maths Box 12.1: Principal Component Analysis as Eigendecomposition

Any square symmetric matrix **A** can be written

$$\mathbf{A} = \mathbf{P}'\boldsymbol{\Lambda}\mathbf{P}$$

where **P** is an orthonormal matrix, i.e. a matrix that satisfies $\mathbf{P}\mathbf{P}' = \mathbf{P}'\mathbf{P} = \mathbf{I}$, and $\boldsymbol{\Lambda}$ is diagonal. An orthonormal matrix can be interpreted geometrically as a rotation. The above expression is known as the *eigendecomposition* of **A**, and **P** is a matrix of *eigenvectors*, each of which has a corresponding *eigenvalue* in $\boldsymbol{\Lambda}$.

Applying eigendecomposition to the sample variance–covariance matrix $\widehat{\boldsymbol{\Sigma}}$, we can re-express in the form

$$\widehat{\boldsymbol{\Lambda}} = \mathbf{P}\widehat{\boldsymbol{\Sigma}}\mathbf{P}'$$

which looks like Eq. 11.1 of Maths Box 11.1. Specifically, $\mathbf{P}$ is a matrix of loadings describing a rotation that can be applied to data, $\mathbf{z} = \mathbf{Py}$, which will give uncorrelated values (since the variance–covariance matrix of $\mathbf{z}$ is $\widehat{\boldsymbol{\Lambda}}$, which is diagonal, i.e. it has zeros in all off-diagonal elements). A bit of further work shows that the largest eigenvalue in $\widehat{\boldsymbol{\Lambda}}$ is the biggest possible variance that could be obtained from a rotation of data $\mathbf{y}$. The corresponding eigenvector gives us the loadings for this first principal component. The next largest eigenvalue is the largest variance obtainable by a rotation uncorrelated with the first axis, so we use this to construct the second principal component, and so forth.

The trace of a matrix, $\text{tr}(\mathbf{A})$, is the sum of its diagonal elements. The trace of $\widehat{\boldsymbol{\Sigma}}$ is a measure of the total variation in the data. Now a special rule for traces of products is that $\text{tr}(\mathbf{AB}) = \text{tr}(\mathbf{BA})$, meaning that

$$\text{tr}(\widehat{\boldsymbol{\Lambda}}) = \text{tr}(\mathbf{P}\widehat{\boldsymbol{\Sigma}}\mathbf{P}') = \text{tr}(\mathbf{P}'\mathbf{P}\widehat{\boldsymbol{\Sigma}}) = \text{tr}(\widehat{\boldsymbol{\Sigma}})$$

so the sum of the variances of the principal component scores $\mathbf{Py}$ is the same as the sum of the variances of $\mathbf{y}$, and we can look at the relative sizes of the values in $\widehat{\boldsymbol{\Lambda}}$ as telling us the proportion of variance in $\mathbf{y}$ explained by each principal component.

The sample correlation matrix $\mathbf{R}$ can be understood as the variance–covariance matrix of standardised data. So applying an eigendecomposition to the sample correlation matrix gives us a PCA of standardised data. Now the diagonals of $\mathbf{R}$ are all one, so $\text{tr}(\mathbf{R}) = p$, and the eigenvalues from a PCA on standardised data will sum to p.

An important step in PCA is to look at the loadings and think about what they mean in order to try to interpret principal components. For example, the loadings for the first principal component of the *Iris* data (labelled `Comp.1` as in Code Box 12.2) are large and positive for most responses, so this can be interpreted as a size variable, a measure of how big flowers are. The second principal component has a very negative loading for `Sepal.Width` and so can be understood as (predominantly) characterising sepal narrowness. Given that 96% of the variation in the data is explained by these two principal components, it seems that most of what is going on is that the *Iris* flowers vary in size, but some flowers of a given size will have narrower sepals than others.

Code Box 12.2: A PCA of Edgar's *Iris* Data

```
> data("iris")
> pc = princomp(iris[,1:4],cor=TRUE)
> pc
Call:
princomp(x = iris[, 1:4], cor = TRUE)

Standard deviations:
   Comp.1    Comp.2    Comp.3    Comp.4
1.7083611 0.9560494 0.3830886 0.1439265

 4  variables and  150 observations.
```

Variances (square of standard deviation) of the principal components can be understood as partitioning the total variance in the data (total variance equals four, the number of response variables, when doing a PCA on standardised data). So most of the variation in the data is explained by the first two principal components ($1.7^2/4 = 73\%$ and $0.96^2/4 = 23\%$).

We can look at the loadings to see what these principal components mean:

```
> loadings(pc)
Loadings:
             Comp.1 Comp.2 Comp.3 Comp.4
Sepal.Length  0.521  0.377  0.720  0.261
Sepal.Width  -0.269  0.923 -0.244 -0.124
Petal.Length  0.580        -0.142 -0.801
Petal.Width   0.565        -0.634  0.524
```

Any loading less than 0.1 is not shown. The first principal component is mostly about flower size (especially petals, higher values for bigger petals). The second principal component is mostly about sepal width. For a biplot as in Fig. 12.1

```
> biplot( pc, xlabs=rep("\u00B0",dim(iris)[1]) )
```

The `xlabs` argument has been used to plot points as circles, instead of individually labelling each observation.

Exercise 12.3: Making Sense of the *Iris* Biplot

Consider the biplot of Edgar's *Iris* data in Fig. 12.1.

Can you distinguish the three species from each other on this plot? Which variables seem most associated with cross-species differences?

While a lot of information about how the principal components were constructed can be seen from looking at their loadings and standard deviations, usually we really want a plot of the data. We can't readily plot the original data jointly for all four responses (might need four-dimensional paper!), but we can plot the scores of each observation along the first two principal components to visualise the main patterns in the data. As before, this captures most (96%) of the variation, but it is harder to interpret than when plotting raw data, because the axes are based, not on measured variables, but linear combinations of them. We can make things a little easier to understand by adding axes for the originally measured variables to the plot, based

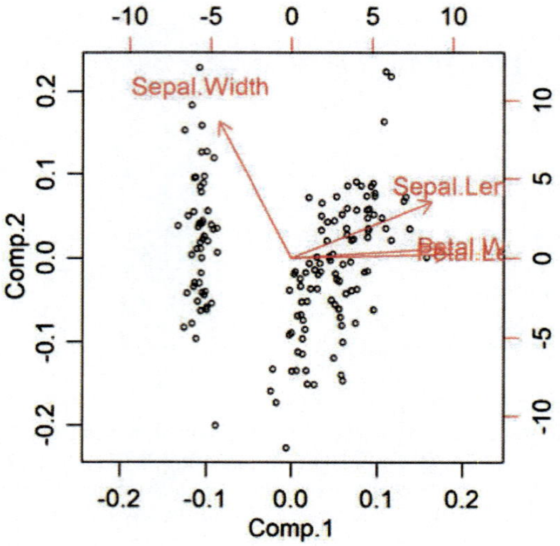

Fig. 12.1: PCA biplot of Edgar's *Iris* data. Points are scores of individual flowers on the first two principal components; arrows are the loadings of each variable on these principal components. The loadings suggest (as in Code Box 12.2) that bigger flowers are towards the right and flowers with wider sepals are towards the top

on their principal component loadings, to make a *biplot* (Fig. 12.1). As discussed previously, most measured variables run along the first principal component axis ("size"), meaning they go (roughly) from left to right on the biplot. Looking at Fig. 12.1, it seems that the three species seem to cluster into different groups, mostly along the first principal component axis, i.e. the species differ from each other primarily in size.

A PCA can be conducted based on either a covariance matrix or a correlation matrix, and the results will be slightly different for each. So which one should you use? If variables are measured on the same scale you can do a PCA using the covariance matrix (the default on `princomp`), but if there are large differences in variation across responses, or if they are measured using different scales, it would make more sense to standardise data before analysing them, which is equivalent to doing PCA on a correlation matrix (`cor=TRUE`). If data have been log-transformed, you could argue that this has put responses on the same scale (a proportional scale) and so standardising is not needed, unless there are substantial differences in variation across responses on a log scale.

Being based on the variances and correlations, PCA is sensitive to outliers and skewed variables. So to do a PCA, data should be fairly symmetric and not too far from normally distributed. For data with outliers you could do a PCA on covariances as estimated using robust methods, e.g. using `cov.rob` from the `MASS` package, but data should still be roughly elliptically distributed for this to be meaningful. As such, PCA is a good option for morphological (size/shape) data like Edgar's

(Exercise 12.1) but a bad option for discrete data with lots of zeros like Anthony's (Exercise 12.2).

12.2.2 Factor Analysis

Factor analysis (often called exploratory factor analysis) is an ordination technique that models the data as multivariate normal with all correlation across responses determined by a few key underlying *factors* (or ordination axes). While not very widely used in ecology, it is the platform for more general methods that are increasingly being used in ecology, in particular, generalised latent variable models (GLVMs) (discussed in the following section) and structural equation models (Grace, 2006).

The goal of factor analysis sometimes is not ordination; rather it is to describe the underlying ordination axes or *factors*. The technique is used a lot in the social sciences, e.g. in questionnaire analysis, where each question is treated as a response variable, and we are trying to quantify a small number of underlying factors that characterise questionnaire responses. In this setting, rather than trying to visualise the main ways people differ in response, the primary goal might be to quantify some intangible quantity (or "construct") like intelligence or anxiety.

A key point of difference from PCA is that factor analysis specifies a statistical model for the data. PCA, in contrast, is a transformation (specifically, a rotation) of the data, so is more of an exploratory technique rather than a modelling technique. The mathematics turn out to be very similar, however, and the two methods often return qualitatively similar results, as in Code Boxes 12.2 and 12.3.

Code Box 12.3: Factor Analysis of *Iris* Data

We will use the fa function in the psych package to do a factor analysis by maximum likelihood using a varimax rotation.

```
> library(psych)
> fa.iris <- fa(iris[,1:4], nfactors=2, fm="ml", rotate="varimax")
> loadings(fa.iris)

Loadings:
             ML1    ML2
Sepal.Length  0.997
Sepal.Width  -0.115 -0.665
Petal.Length  0.871  0.486
Petal.Width   0.818  0.514

                 ML1   ML2
SS loadings    2.436 0.942
Proportion Var 0.609 0.236
Cumulative Var 0.609 0.844
```

How do the results compare to the PCA?

Factor analysis, with K factors, assumes that the ith observation of the jth response variable comes from the following model:

$$y_{ij} = \mathbf{z}_i^T \boldsymbol{\lambda}_j + \epsilon_{ij} \tag{12.1}$$

where the errors ϵ_{ij} are assumed to be independent and normal with variance σ_j^2 (which is constant across *observations* but may differ across responses). The K factor scores are also assumed to be independently normally distributed with common variance and are stored in $\mathbf{z}_i$. These assumptions on ϵ_{ij} and $\mathbf{z}_i$ imply that y_{ij} is assumed to be multivariate normal. Covariation is introduced across the responses by assuming they all respond to some relatively small number (K) of common factors. This covariation is captured by the loadings $\boldsymbol{\lambda}_j$—if data are highly correlated, the loadings will be large, otherwise they will be small (although they are often standardised in output, in which case look for a statement of proportion of variance explained, as in Code Box 12.3). If a pair of response variables is highly correlated, then their loadings will be similar; if a pair of response variables is independent, no pattern will be apparent on comparing their loadings. The other parameters of interest are the variances of the errors (ϵ_{ij}), often known as *communalities*, which are smaller when the data are more strongly correlated.

The factor analysis model looks a lot like a multivariate linear regression model, and it is. The main difference is that now we don't have measured predictors (no x_i) but instead use the axes of covariation in the data to estimate these factor scores (z_i). You can think of these factor scores as unmeasured predictors. If you have measured predictors as well, it is possible to extend the factor analysis model to account for them.

It seems weird, kind of like voodoo magic, that we can assume there are some predictors in our model that are important but unmeasured and then fit a model that goes and estimates them. But it turns out that if there were some underlying axis that all responses were related to, it can actually be seen in the pattern of responses. Consider, for example, the bivariate case, with two response variables that are linearly related. A factor analysis model assumes that the sole cause of this correlation is that both variables are responding to some underlying factor, which must then be proportional to scores along some line of best fit through the data (Fig. 12.2). The errors ϵ_{ij} capture the variation in the data around this line of best fit, along each axis.

12.2.2.1 Mind Your Ps and Qs—Factor Analysis

A factor analysis assumes data are a sample from a multivariate normal variable, but where the variance–covariance matrix has a special structure that can be summarised using a small number of factors. We can break down the assumptions as follows:

1. The observed y-values are *independent* given z.
2. The y-values are *normally distributed*, and within each response variable they have *constant variance*.
3. The mean of each y is a *linear* function of K factor scores.

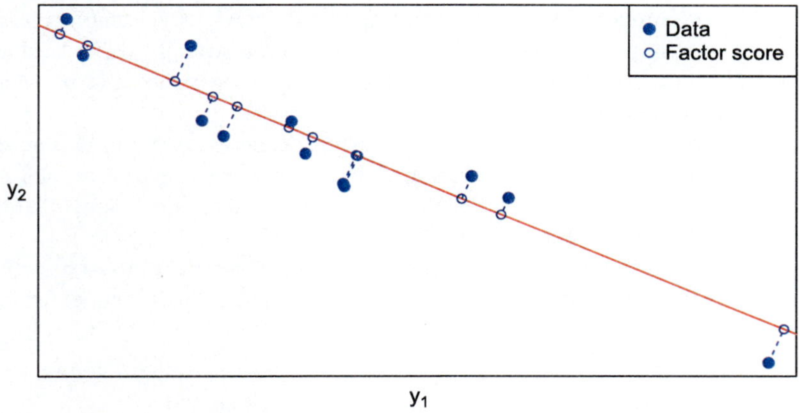

Fig. 12.2: An illustration of how factor analysis works. The two response variables here are (negatively) correlated with each other, and factor analysis assumes this is because they both respond to some underlying, unmeasured factor. Our estimates of the scores for this factor fall along a line of best fit, chosen in such a way that errors from the line are uncorrelated noise, and so that all the covariation is captured by the factor scores

4. The factor scores $\mathbf{z}$ are *independent* (across samples and factors) and are *normally distributed* with *constant variance*.

These assumptions look a lot like linear model assumptions—because, as previously, a factor analysis is like a linear model, but with unmeasured predictors. More specifically, these look like mixed model assumptions, and this model can in fact be fitted as a type of mixed model.

So we have four assumptions. How robust is this model to failure of assumptions?

The independence assumptions are critical when making inferences from a factor analysis model. Note it is independence across observations that is required, not across response variables; in fact, the point of factor analysis is to model the dependence across responses. When using data to determine the number of factors to use in the model, then note this is a type of model selection (an inference technique), and, as always, independence of observations is needed for model selection methods to work reliably.

Despite the foregoing discussion, often we use a factor analysis for descriptive purposes (ordination, trying to identify underlying factors), and we don't want to make inferences from it. If we aren't going to make inferences, then the independence assumption across observations is not critical. You could even argue that the factors estimate and account for dependence across observations, indirectly, e.g. if you collect spatially structured data, that spatial structure will (to some extent) be captured by the factors. However, if you have a known source of structured dependence in your data, you could expect better results when modelling it directly, which is possible using more advanced methods (Thorson et al., 2015; Ovaskainen et al., 2016).

Normality assumptions here don't really matter, as previously, except in terms of efficiency. It is important that we don't have outliers or strongly skewed values. These analyses rely heavily on estimates of the variance–covariance matrix, which can be quite sensitive to outlying values.

The equal variance assumption, like the independence assumption, is required for inferences to be valid. If using factor analysis for descriptive purposes, and not making inferences from the model, then this assumption is not critical, but violations of it will affect how well the method works.

The most important assumption in factor analysis is the linearity of **y** (Assumption 3). Linearity is important because factor analysis, as a type of linear model, is only able to capture linear relationships across response variables.

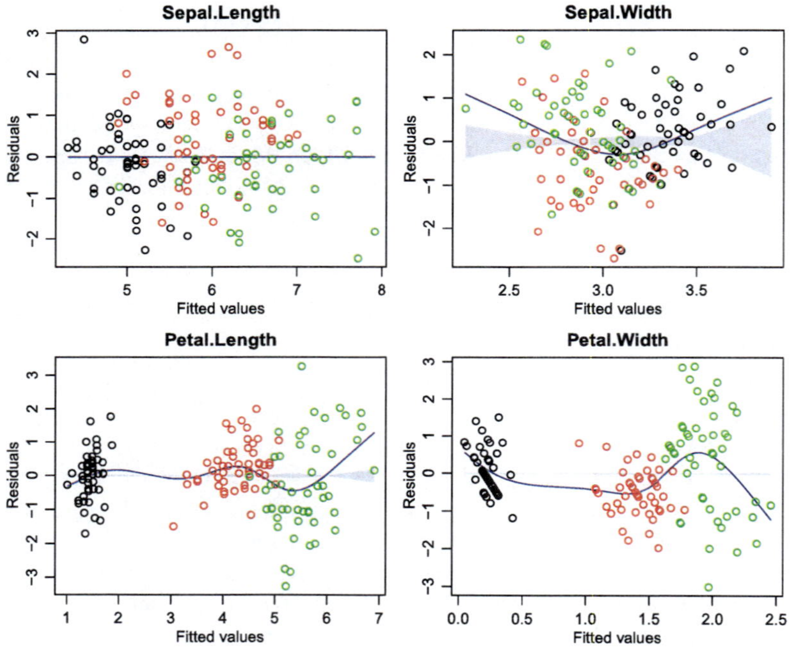

Fig. 12.3: One way to check the assumptions of a factor analysis is by fitting a simple linear regression for each response variable as a function of factor scores, as done here for the *Iris* data. Different species have been labelled with different colours. Note that assumptions do not appear to be satisfied; in particular, there appears to be a fan shape in the petal variables

One way to check factor analysis assumptions is to fit a linear model for each response variable as a function of the factors (Code Box 12.4). If the number of response variables is small enough to permit it, it might also be a good idea to construct a scatterplot matrix.

Code Box 12.4: Assumption Checking for a Factor Analysis of *Iris* Data

To check linearity and equal variance assumptions, we will check the assumptions of a linear model for each response as a function of each factor (Fig. 12.3):

```
for(iVar in 1:4)
  plotenvelope(lm(iris[,iVar]~fa.iris$scores), which=1, col=iris$
  Species)
```

For a scatterplot matrix, as follows:

```
plot(iris[,1:4],col=iris$Species)
```

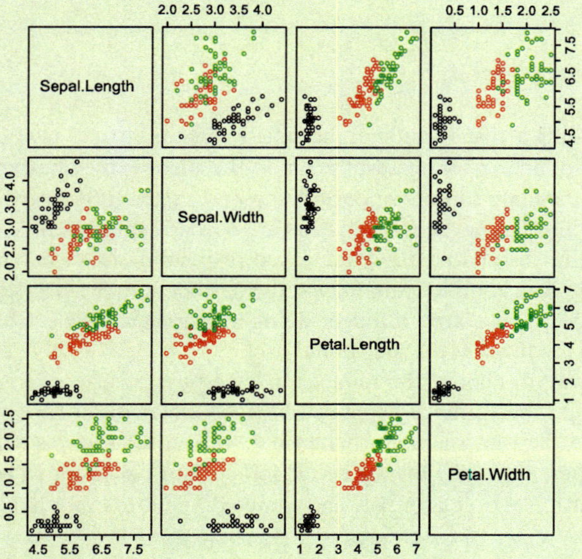

The assumptions don't look good, so inferences based on this model would be questionable. Within each species, linearity and equal variance look reasonable, so separate factor analyses by species (or one with a species term added to the model) would be more reasonable.

Although the latent variables were assumed to be independent, this doesn't actually impose any constraints on the model, because the loadings that $\mathbf{z}$ are multiplied by induce dependence. Similarly, the equal variance assumption on $\mathbf{z}$ imposes no constraints because the loadings handle how each factor scales against each response. The variance of $\mathbf{z}$ is often set to one, without loss of generality.

It is worth keeping in mind that it only makes sense to do a factor analysis if you expect some of your response variables to be correlated. The reason for this is that the factors are introduced to capture correlation patterns, so if there are none there, then the factors will have nothing to explain.

Exercise 12.4: A Factor Analysis for Anthony's Data(?)

Load Anthony's revegetation data (stored as `reveg` in the `ecostats` package), and do a factor analysis (with two factors). You might not be able to get a solution when using maximum likelihood estimation (`fm="ml"`), in which case, try fitting without specifying the `fm` argument (which tries to minimise residuals).

Check some of the assumptions by fitting a linear model to some of the response variables as a function of factor scores.

Can you see any issues with factor analysis assumptions?

12.2.2.2 How Many Factors?

Often, when seeking a visualisation of the data, we choose to use a two-factor model so that we can plot one factor against the other. But there is the question of whether this captures all the main sources of covariation in the data, or if more factors would be needed to do this. Or maybe fewer factors are needed—maybe there is only really one underlying dimension that all variables are responding to, or maybe there is no covariation in the data at all to explain, and the correct number of factors to choose is zero (in which case we are no longer doing a factor analysis!). So how do you work out how many factors (K) you should use?

A common way to choose the number of factors is to use a *scree plot* (Code Box 12.5)—to plot the variance explained by each factor in the model and choose a solution where the plot seems to flatten out, with a relatively small amount of additional variation explained by any extra factors. This can work when the answer is quite clear, but it is a little arbitrary and can be difficult to use reliably.

Code Box 12.5: Choosing the Number of Factors for the *Iris* Data

An informal way to choose the number of factors is to look for a kink on a scree plot:

```
> plot(fa.iris$values,type="l")
```

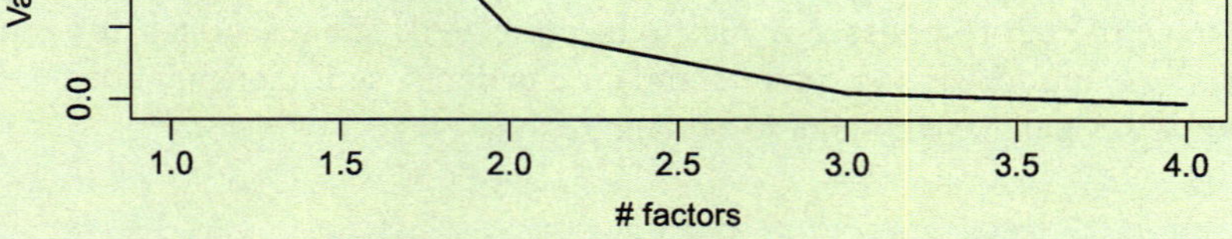

which suggests choosing one or two factors (maybe one, since the variance explained is less than one for the second factor).

A more formal approach is to compare BICs for models with one, two, or three factors. Unfortunately, these are not given in `fa` output, so we need to compute them manually. This can be done inside a loop to avoid the repetition of calculations across the three models:

```
> nFactors=3 # to compare models with up to 3 factors
> BICs = rep(NA,nFactors) # define the vector that BIC values go in
> names(BICs) = 1:nFactors # name its values according to #factors
> for(iFactors in 1:nFactors)  {
  fa.iris <- fa(iris[,1:4], nfactors=iFactors, fm="ml",
rotate="varimax")
  BICs[iFactors] = fa.iris$objective - log(fa.iris$nh) * fa.iris$dof
}
> BICs # correct up to a constant, which is ignorable
        1         2         3
-9.436629  5.171006 15.031906
```

How many factors are supported by the data?

A better approach is to exploit the model-based nature of factor analysis and use model-based approaches to choose K. For example, we could choose the value of K that minimises BIC, as in Code Box 12.5. For Edgar's *Iris* data, this led us to a model with one factor, representing size. The second factor, which was essentially petal width, is not a joint factor across multiple responses, so it can readily be characterised by the error term ϵ_{ij} for petal width.

For PCA, we cannot use model-based approaches to choose the number of principal components to focus on. The reason is that PCA is basically a transformation of data, not a model for it, and we are stuck using more arbitrary rules like scree plots or taking all components that explain more than $1/p$ of the variation in the data (which, when performed on a correlation matrix, means choosing all components with a variance greater than one). For the *Iris* data, this would again lead us to a model with just one component, similar to factor analysis.

12.2.2.3 Rotations

Factor analysis solutions are invariant under rotation (Maths Box 12.2), which essentially means that results are the same when you look at a biplot from any orientation. This issue doesn't actually matter when it comes to interpreting an ordination because the relative positions of points and arrows are unaffected by the plot's orientation.

But if you wish to study factor loadings and try to attach an interpretation to axes, a decision needs to be made as to which orientation of the solution to use. The most common choice is a *varimax* rotation, which tends to push loadings towards zero or one (Maths Box 12.2). This has no biological or statistical justification, it is done to try and make it easier to interpret factors because they will tend to be a function of fewer responses.

Maths Box 12.2: Factor Rotation and Identifiability
In a factor analysis model, the rotation of the solution that is presented is arbitrary.

To see this, recall that a rotation of values $\mathbf{y}$ can be written $\mathbf{Py}$, where $\mathbf{P}$ is a "rotation matrix" satisfying $\mathbf{P'P} = \mathbf{I}$. Now, if a factor analysis solution gives factor scores z_i and loadings $\boldsymbol{\lambda}_j$, then another equivalent solution is to use factor scores $\mathbf{P}z_i$ and loadings $\mathbf{P}\boldsymbol{\lambda}_j$ for any rotation matrix $\mathbf{P}$ since

$$(\mathbf{P}z_i)'(\mathbf{P}\boldsymbol{\lambda}_j) = z_i'\mathbf{P}'\mathbf{P}\boldsymbol{\lambda}_j = z_i'\boldsymbol{\lambda}_j$$

So which rotation should we use? Well, I guess it doesn't really matter! It is the relative position of points on the ordination that matters, not their orientation.

In fitting the model, often we assume (without loss of generality) that the matrix of loadings is lower triangular, meaning that the first response has loadings falling along the first axis, the second response is somewhere on the plane specified by the first two axes, and so forth. Once the model has been fitted, the solution can be rotated any way you like to obtain an equivalent solution.

In viewing the factor analysis solution, a *varimax* rotation is often used, intended to simplify interpretation of the fitted model. A varimax rotation aims to maximise the variance of the squared factor loadings, which tends to move them either towards zero or one, so each involves fewer variables, simplifying interpretation.

12.3 Generalised Latent Variable Models

Recall Anthony's data are discrete and are not going to satisfy the all-important linearity and equal variance assumptions. In Exercise 12.4, you may have found instances where each assumption was violated. You might also have noticed some big outliers and some predicted abundances that were negative—signs we are modelling mean response on the wrong scale! What Anthony needs is a technique that combines the ideas of factor analysis with something like generalised linear models to handle the mean–variance relationship in his count data and to put mean response on a more appropriate scale. Such techniques exist and are often called generalised latent variable models (Skrondal & Rabe-Hesketh, 2004). They were initially developed in the social sciences for questionnaire analysis, since questionnaire responses are rarely normally distributed. They have been extended to ecology relatively recently (Walker & Jackson, 2011; Hui et al., 2015).

Assumptions of a GLVM for ordination:

1. The observed **y**-values are *independent*, after conditioning on latent variables **z**.
2. The **y**-values come from *a known distribution* (from the exponential family) with known *mean–variance relationship* $V(\mu)$.
3. A *linear relationship* between some known function of the mean of **y** and each latent variable **z**

$$g(\mu_{ij}) = \beta_{0j} + \mathbf{z}_i^T \boldsymbol{\Lambda}_j \tag{12.2}$$

4. The latent variables **z** are independent (across samples and each other) and are standard normal in distribution

$$z_{ik} \sim N(0, 1)$$

Basically, this is a GLM with latent variables instead of measured predictor variables. As previously, **z** are factor scores, which can be used as ordination axes, and the coefficients the latent variables are multiplied by ($\boldsymbol{\Lambda}_j$) are often called *loadings*.

Although the latent variables were assumed to be independent with mean zero and variance one, this doesn't actually impose any constraints on the model. This is because the loadings that **z** are multiplied by induce dependence and different sized effects for different responses, and the intercept term β_{0j} sets the overall mean for each response.

A GLVM is a type of hierarchical model, as in Sect. 11.3. In fact, the factor model $\mathbf{z}_i^T \boldsymbol{\Lambda}_j$ can be rewritten as a multivariate normal random intercept ϵ_{ij}, which has a reduced rank variance–covariance matrix, $\boldsymbol{\Sigma} = \boldsymbol{\Lambda}\boldsymbol{\Lambda}'$. This basically means that a GLVM is like the hierarchical GLM in Sect. 11.3, but with fewer parameters, because we are assuming a simpler form for $\boldsymbol{\Sigma}$.

When it comes to fitting GLVMs, there are a few different options around. In R, the fastest (at the time of writing) and one of the easiest to use is `gllvm` (Code Box 12.6). This package uses maximum likelihood estimation, but for this sort of model, the likelihood function has a weird shape and it can be hard to find its maximum, especially for a dataset with many responses that are mostly zeros. So it is advisable to run a few times and check that the maximised value for the likelihood doesn't change (preferably, using `jitter.var=0.2` or something similar, to jitter starting guesses of ordination scores so as to start estimation from slightly different places). If it does change, you might need to run it about 10 times and keep the solution with the highest likelihood. For Anthony's data, on five runs I got −689.3 on every run, with a small amount of variation on the second decimal place, which is ignorable.

Assumptions can be checked in the usual way (Fig. 12.4). Residuals vs fitted value plots are constructed essentially using tools for fitting a GLM as a function of factor scores, analogously to what is done in Fig. 12.3.

As previously, biplots can be readily constructed by overlaying factor loadings on top of plots of factor scores (Code Box 12.6). Note from the biplot in Code Box 12.6 that sites 1 and 5 appear towards the left of the plot—these are the two control sites. Notice also that *Blattodea* scores highly near these sites, while *Amphipoda* and *Coleoptera* have high loadings away from these sites. This suggests that there are more cockroaches at control sites and more amphipods and beetles away from them, as an inspection of the data confirms (Table 12.1).

Table 12.1: Counts for key orders from Anthony's revegetation data, with control sites indicated in black and revegetated sites in red. Notice how closely these correspond to the information on the biplot in Code Box 12.6

Site	*Blattodea*	*Amphipoda*	*Coleoptera*
1	3	0	1
2	0	11	61
3	0	21	609
4	0	159	279
5	4	0	1
6	1	52	14
7	0	31	30
8	0	1	32
9	0	156	97
10	0	376	69

The `gllvm` package can handle some family arguments commonly used in ecology; for more options see the `boral` package (Hui, 2016) or `Hmsc` (Ovaskainen et al., 2017b; Ovaskainen & Abrego, 2020, Section 7.3). The `glmmTMB` package was recently extended to fit latent variable models (version 1.1.2.2 and later), via the `rr` correlation structure, which stands for "reduced rank". This is an exciting development because it nests the capacity to fit latent variable models in software that can flexibly fit mixed models, with a range of different distribution options.

Code Box 12.6: A Generalised Latent Variable Model for Anthony's Revegetation Data

```
> data(reveg)
> library(gllvm)
> reveg_LVM = gllvm(reveg$abund, num.lv=2, family="negative.binomial",
    trace=TRUE, jitter.var=0.2)
> logLik(reveg_LVM)
'log Lik.' -689.3072 (df=95)
```

Repeating this several times seems to return an answer within about 0.01 of this value, so we can be confident this is (close to) the maximum likelihood solution. To get a biplot of this solution, labelling only the 12 responses with highest loadings (to reduce clutter):

```
> ordiplot(reveg_LVM, col=as.numeric(reveg$treatment), biplot=TRUE,
    ind.spp=12)
```

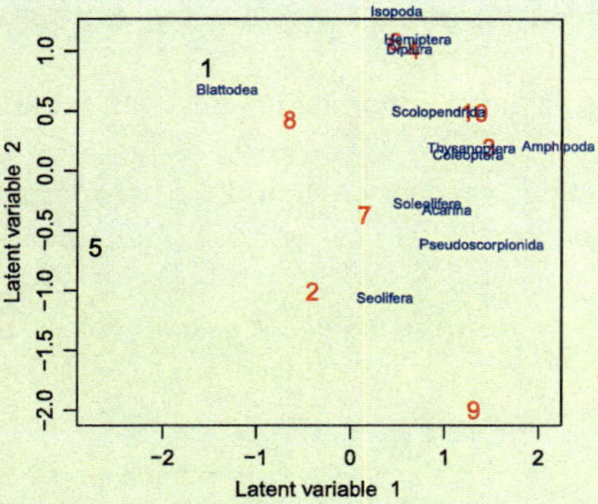

You may get a different ordination, which is more or less a rotation or reflection of this one, which should have a similar interpretation. Checking assumptions:

```
> par(mfrow=c(1,3))
> plot(reveg_LVM, which=c(1,2,5))
```

which returns the plot in Fig. 12.4.

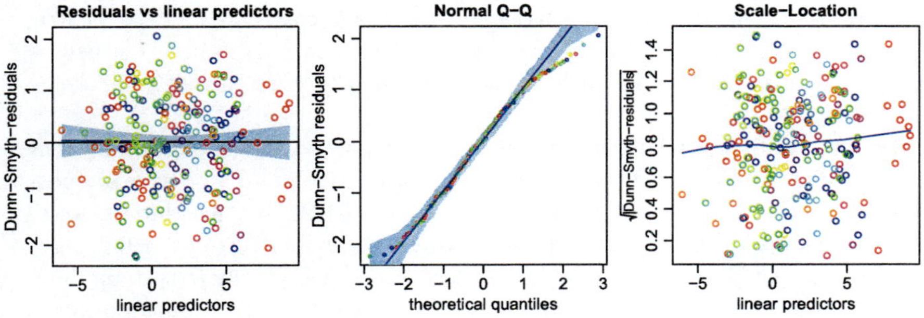

Fig. 12.4: Assumption checks of the GLVM fitted to Anthony's revegetation data in Code Box 12.6. There is no appreciable pattern in plots that would lead us to worry about our variance or linearity assumptions, with no fan or U shape in the residuals vs fits plot. There is a slight trend in the quantile plot, which remained there on repeat runs. This suggests a possible violation of distributional assumptions, with large residuals being slightly smaller than they were expected to be. This means the distribution of counts was not quite as strongly right-skewed as expected by the model, which is not a critical issue, since the mean–variance trend seems to have been accounted for adequately

Exercise 12.5: Checking Analysis Decisions for Anthony's Revegetation Data
In Code Box 12.6, a negative binomial model was fitted using two latent variables.

Are two latent variables needed, or should we use more, or less? *Fit a few models varying the number of latent variables.* Which model fits the data best, according to BIC?

Fit a Poisson model to the data and check assumptions. Are there any signs of overdispersion?

12.4 Multi-Dimensional Scaling and Algorithms Using Pairwise Dissimilarities

Another ordination method, historically very commonly used in ecology, is (non-metric) multi-dimensional scaling ("MDS" or "nMDS", Kruskal & Wish, 1978). A MDS can be fitted on `R` using the `vegan` package (Oksanen et al., 2017) as in Code Box 12.7. The general idea is to capture the "main characteristics" of the data by arranging points so that straight-line distances between pairs of points as best as possible reflect *dissimilarities* that have been computed between pairs of multivariate observations.

The main difficulty in MDS is the decision on how to measure dissimilarities between pairs of multivariate observations, for which there are a great many possibilities (e.g. see the list in Legendre & Legendre, 2012), and whether or not to transform data prior to computing dissimilarities. Different decisions at this stage can lead to strikingly different results (Exercise 12.6). A difficulty in making these decisions is that MDS is an *ad hoc* algorithm for producing an ordination, not a statistical model for data, so model selection tools are not available and we resort largely to *ad hoc* decisions about dissimilarity measures and data transformation, which seem to be quite hard to get right.

A feature of abundance data that can cause problems for MDS is the typically strong mean–variance relationship, with more abundant taxa tending to be much more variable than rare taxa—by more than a factor of a million in the case of Anthony's data (Fig. 10.3b, left). Transformation can reduce this change in variability, but it cannot remove it for rarer taxa (Warton, 2018, also see Fig. 10.3b, right). The use of different distance metrics and data transformations can be understood as making quite different implicit assumptions about the mean–variance relationship in data, and when there is a disconnect between the distance measure used and properties of the data being analysed, this can introduce artefacts into analyses (Warton et al., 2012b). This is the main source of the problems in Exercise 12.6, which can be avoided using a GLVM with the appropriate distributional assumptions for your data.

Code Box 12.7: A Non-Metric Multi-Dimensional Scaling Ordination of Anthony's Data

```
> library(vegan)
> ord_mds=metaMDS(reveg$abund)
Square-root transformation
Wisconsin double standardisation
Run 0 stress 0.1611237
Run 1 stress 0.1680773
Run 2 stress 0.1934608
... New best solution
... Procrustes: rmse 5.511554e-05  max resid 0.0001003286
*** Solution reached
> plot(ord_mds$points, pch=as.numeric(reveg$treatment),
        col=reveg$treatment)
```

By default, the vegan package uses the Bray-Curtis distances and a bunch of fairly arbitrary transformations.

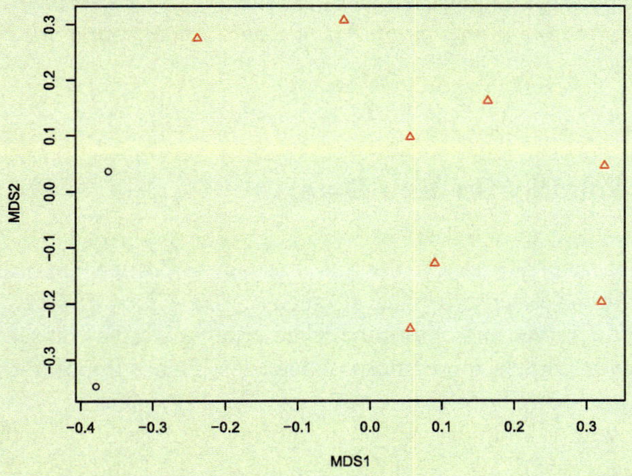

The separation of the two control plots from the others suggests to us that there is an effect of bush regeneration treatment on the invertebrate community, although it doesn't give us much sense for what the nature of this effect is.

Exercise 12.6: MDS Ordinations of Coral Data
Warwick et al. (1990) measured coral cover along 10 transects in the Tikus Islands (Indonesia) at different times—we will focus on 1981 and 1983 data, before and after an El Niño event, respectively. Warwick et al. (1990) used this dataset and MDS ordinations to argue that stress increases dispersion in coral communities.

The data can be found in the mvabund package and can be loaded using:

```
library(mvabund)
data(tikus)
tikus20 = tikus$abund[1:20,] # for 1981 and 1983 data only
tikusAbund = tikus20[,apply(tikus20,2,sum)>0] # remove
zerotons
```

Construct a MDS plot of the data using the Bray-Curtis distance (default) and colour-code symbols by year of sampling. *Does this plot agree with the Warwick et al. (1990) interpretation?*

Construct another MDS plot using the Euclidean distance on log(y + 1)-transformed data. *Does this plot agree with the Warwick et al. (1990) interpretation?*

Use the `plot.mvabund` function to plot each coral response variable as a function of time. *What is the main pattern you see?*

Convert the data into presence–absence as follows:

```
tikusPA = tikusAbund
tikusPA[tikusPA>1]=1
```

and use the `gllvm` package to construct an ordination.

Do the assumptions appear reasonable? How would you interpret this plot?

12.5 Make Sure You Plot the Raw Data!

Visualising multivariate data is tricky, and while it is easy to produce an ordination plot, it is not as easy to make sense of what it means. *Always look at plots of your raw data* as well as ordinations, and determine whether any patterns you see can be reproduced on plots with simpler, more interpretable axes. This is important for two reasons.

Firstly, it may provide a simpler way of presenting your results, so they are easier for readers to understand. For example, compare Fig. 12.1 and Fig. 12.5. The seasoned biplot reader can see the main patterns in Fig. 12.1, but the boxplots of Fig. 12.5 are more readable and have meaningful scales on their axes. It is clear from Fig. 12.5, for example, that *Iris setosa* has much smaller petals than the other species, with petal lengths and widths tending to be about an order of magnitude smaller, and no overlap with the other species on either of these variables. It also tends to have the widest sepals, despite these also tending to be the shortest.

A second reason for not relying on ordinations alone is that sometimes you can get artefacts in your results due to bad choices of ordination—patterns that don't really mean anything, that are instead due to unfortunate decisions you made in your analysis. For example, using MDS, you could make an unfortunate choice of dissimilarity (such as either of the options in Exercise 12.6), or using a latent variable model you could assume the wrong mean–variance relationship (mind your Ps and Qs!). If you see a result in an ordination that you can't reproduce on the originally measured data, then you should question whether the pattern is really there and maybe think harder about your choice of ordination approach.

Code Box 12.8: Studying Each Observation Separately for the *Iris* Data

For a summary of sample means for each species and each response variable:

```
> by(iris, iris$Species, function(dat){ apply(dat[,1:4],2,mean) } )
iris$Species: setosa
Sepal.Length  Sepal.Width Petal.Length  Petal.Width
       5.006        3.428        1.462        0.246
------------------------------------------------------------
iris$Species: versicolor
Sepal.Length  Sepal.Width Petal.Length  Petal.Width
       5.936        2.770        4.260        1.326
------------------------------------------------------------
iris$Species: virginica
Sepal.Length  Sepal.Width Petal.Length  Petal.Width
       6.588        2.974        5.552        2.026
```

Note that *I. verginica* tends to have the largest flowers (in petal size and sepal length), and *I. setosa* tends to have the smallest flowers (but the widest sepals). Using boxplots to visualise this:

```
> par(mfrow=c(2,2))
> plot(Sepal.Length~Species,data=iris,xlab="")
> plot(Sepal.Width~Species,data=iris,xlab="")
> plot(Petal.Length~Species,data=iris,xlab="")
> plot(Petal.Width~Species,data=iris,xlab="")
```

produces Fig. 12.5.

Key Point

Interpretation is always difficult in multivariate analysis, and mistakes are easily made. The best safeguard against this is to try to visualise key results using your *raw data* to complement more abstract tools like ordination.

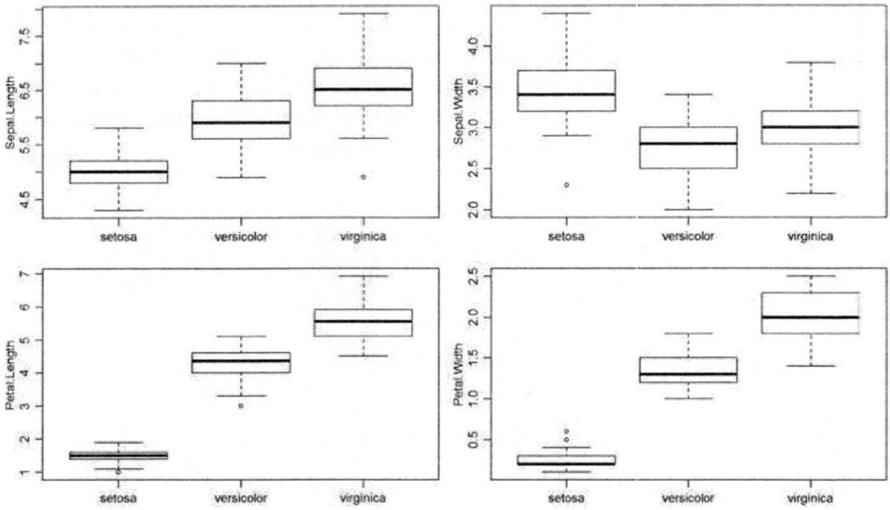

Fig. 12.5: Boxplots of *Iris* flower data for each species. Note the species differ in size, especially in petal variables, although something different seems to be happening with sepal width, with the widest sepals for the smallest flowers (*I. setosa*). This graph is arguably more informative than the biplot, showing the value of looking for ways to visualise patterns using the originally measured variables

Chapter 13
Allometric Line Fitting

Exercise 13.1: Brain Size–Body Size Relationships
Body size and brain size were recorded for 28 mammals. Data are available as `Animals` in the `MASS` package and look as follows:

Species	Mountain beaver	Cow	Grey wolf	Goat	Guinea Pig	...
Body weight (kg)	1.35	465	36.33	27.66	1.04	...
Brain weight (g)	8.1	423	199.5	115	5.5	...

We would like to know:

Does brain size scale as the 2/3 power of body size?

(A crude argument for the "2/3 power law" is that a major function of the brain is to interpret signals from the skin, so they should scale proportionately to surface area rather than body mass, Rensch, 1954; Gould, 1966).

How should we analyse the data to answer this research question?

Exercise 13.2: Leaf Economics and Environment
Ian has noticed a "leaf economics spectrum" where long-lived leaves tend to have larger leaf mass per area. He would like to know if this relationship varies with environment. In particular:

How steep is the slope of the line representing the leaf economics spectrum? Is it steeper than one? Does the steepness of the line vary across environments?

What method should he use to do this?

D. I. Warton, *Eco-Stats: Data Analysis in Ecology*, Methods in Statistical Ecology,
https://doi.org/10.1007/978-3-030-88443-7_13

Consider Exercises 13.1–13.2. In both cases, we have two variables, but we do not have a response and a predictor, so we should not be thinking about univariate methods of analysis. Instead we have a multivariate problem, well, a bivariate problem (with two responses), and there is interest in how these responses covary. This looks like a job for factor analysis or principal component analysis (PCA)—trying to estimate some underlying factor that characterises how these responses covary. But there is an important difference in the way the research question has been phrased, specifically, the focus here is on making inferences about the slope of the line of best fit. Factor loadings store this information, but factor analysis output will not tell you if the loadings on y_1 and y_2 are significantly different from each other, which is what we would want to know if trying to determine whether the slope is one ("isometry"), and it will not help us see if there is a significant departure from the $2/3$ power law of Exercise 13.1 (Fig. 13.1).

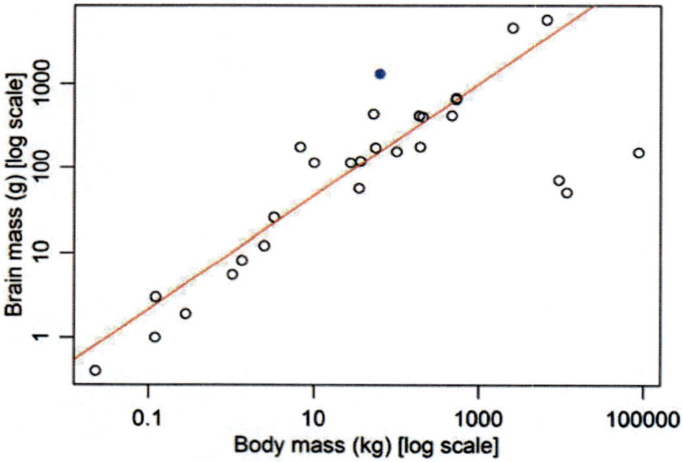

Fig. 13.1: Brain size vs body size for different species of vertebrate, with a $2/3$ power law overlaid (red line). All points correspond to different mammal species, except the three outliers, which are dinosaurs. Should brain size go on the y-axis, the x-axis, or either? Or put another way—why do humans (solid blue point) not lie on the line—because they have large brain mass for their body size or a small body for their brain size? This is a matter of perspective, and the ambiguities here complicate analysis and lead us to think about multivariate techniques rather than standard linear regression

Exercises 13.1–13.2 come from studies of *allometry*, the study of size-correlated variation (Niklas, 2004). Methods of analysis for problems like these are sometimes referred to as allometric line-fitting (Warton et al., 2006) because allometry is the main application area we have in mind. Allometric line fitting makes most sense when thought of in a multivariate way. In fact, common analysis methods used in allometry are variations on PCA.

13.1 Why Not Just Use a Linear Model?

Because we are talking about fitting a line to relate two variables, it is tempting to think of this as a linear modelling problem and apply simple linear regression as in Chap. 2. However, linear regression is fitted using least squares, minimising errors predicting y. This certainly makes sense if your goal is to predict or explain y, but that is not our goal in Exercises 13.1–13.2. Rather than having a single response variable to explain, we have two y variables, and we are trying to estimate some underlying factor that best explains why our responses covary. Another key distinguishing feature here is that we are interested in specific values of the slope—in Exercise 13.1 we want to know if there is evidence that the slope is not $2/3$, and in Exercise 13.2 we want to know if the slope differs from one. Because there is no obvious choice of predictor and response, we could equally well do the analysis with either variable as the predictor (Code Box 13.1), but this gives completely different answers!

Code Box 13.1: Linear Models of Brain Size–Body Size Data

```
> library(MASS)
> data(Animals)
> ftBrainBody=lm(log(brain)~log(body),data=Animals)
> confint(ftBrainBody)
                2.5 %    97.5 %
(Intercept) 1.7056829 3.4041133
log(body)   0.3353152 0.6566742
```

From a linear regression of brain size against body size, a 95% confidence interval (CI) for the true slope does not quite cover $2/3$, so we would conclude that there is some evidence that the relationship is flatter than $2/3$ scaling. But why do things this way around—what if we regressed body size against brain size?

```
> ftBodyBrain=lm(log(body)~log(brain),data=Animals)
> confint(ftBodyBrain)
                 2.5 %    97.5 %
(Intercept) -3.6396580 0.3396307
log(brain)   0.8281789 1.6218881
```

Reversing axes, we might expect the slopes to be the inverse of what we saw before—so we are now interested in whether or not the slope is $3/2$. Confusingly, this time around the CI does cover $3/2$, so we have no evidence against the $2/3$ scaling law.

The reason these lines give different answers is that they do different jobs—one tries to predict y from x, while the other tries to predict x from y. Neither is what we are really after when trying to characterise the main axis along which brain mass and body mass covary.

The problem that can arise using regression in allometry is known as *regression to the mean*, with regression slopes being flatter than you might naïvely expect.

The term "regression" comes from the idea that y predictions tend to be closer to their mean (in standardised units) than the values of x from which their predictions were made, hence the variable seems to "regress" towards its mean. The idea was first discussed by Galton (1886) in relation to predicting sons' heights from fathers' heights, where sons were always predicted to be of more average stature than their father, with tall fathers having shorter sons and short fathers having taller sons. Regression to the mean is entirely appropriate when predicting y from x, because we do expect things to become more average. For example, if you do a test and get your best score ever, topping the class, do you expect to do just as well on your next test for that subject? You will probably do well again, but you should not expect to do quite as well.

In allometry, we often do not want regression to the mean. In Exercise 13.1, when studying the $2/3$ power law, regression to the mean makes the slope of the line relating brain mass to body mass flatter than we might expect, and if we regress body mass against brain mass, this line is then steeper than we might expect (hence the contradictory results of Exercise 13.1). So instead of minimising error predicting y (or x), maybe we should minimise something in between, like the straight-line distance of points from the line.

13.2 The (Standardised) Major Axis

The most common tools used for allometric line fitting are variants on PCA. One option is to fit a "major axis" (MA), so named because it the major axis of the ellipse that best fits the data. It can be understood as the line that minimises the sum of squared straight line distances of each point from the line. The major axis turns out to have the same slope as the first principal component fitted to a variance–covariance matrix and can be interpreted as a PCA of the data—it estimates the axis or dimension that characterises most of the (co)variation in the data.

If response variables are on different scales, it might make sense to standardise prior to fitting, but then to rescale to the original axes for interpretation, known as a "standardised major axis" or "reduced major axis" (hereafter SMA). This can be understood as a PCA on the correlation matrix, but rescaled to the original axes. The main difference between these methods is that MA is invariant under rotation (i.e. if you rotate your data, the fitted line will be rotated by the same amount), whereas SMA is invariant under changes of scale (i.e. if you change height from metres to centimetres, the slope will change by a factor of 100). SMA is more commonly used, but if the estimated slope is near one, then the methods will give quite similar answers.

(S)MA methods are sometimes referred to as *Model II regression* (Sokal & Rohlf, 1995) and have some equivalences with models that assume there is measurement error in the x variable (Carroll & Ruppert, 1996, "measurement error models"). It is important, however, to recognise that in allometry we typically do not have a regression problem because we do not seek to predict y from x, rather we want to study how y_1 and y_2 covary. Measurement error models are a form of regression model that estimate how much error there is in measuring x and adjust predictions for y accordingly. As such it is best not to think too deeply about relations between allometric line-fitting methods and measurement error models, as their intentions are quite different. An issue applying measurement error approaches in allometry is that typically the main form of error is not due to measurement of x but to a lack of fit of the model (Carroll & Ruppert, 1996, "equation error").

13.2.1 Inference About (S)MA Lines Using `smatr` Package

Methods for estimating and making inferences about MA or SMA slopes can be found in the `smatr` package (Warton et al., 2012a). Different types of hypotheses that are often of interest are illustrated in Fig. 13.2.

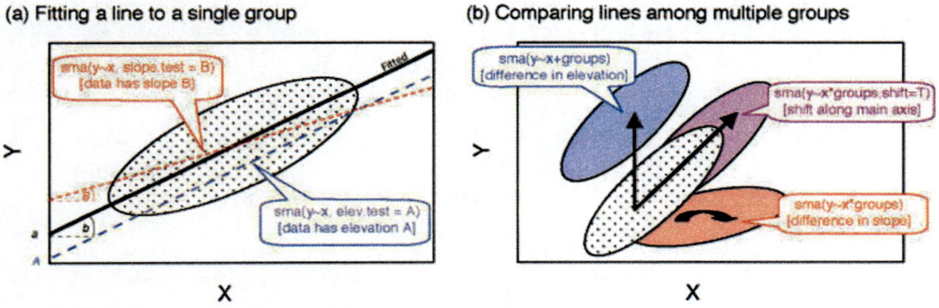

Fig. 13.2: Schematic diagram of different types of hypothesis tests in `smatr` package, and code to implement them, from Warton et al. (2012a). (**a**) Testing if the true SMA slope is B, based on a single sample. (**b**) Comparing slopes of several samples, or if slopes can be assumed to be common, comparing the location of samples, looking either for a change in elevation or a shift along the common axis

For a single sample, as in Exercise 13.1, interest is most commonly in the slope of the (S)MA. Methods to test the slope of the line have been around for a long time (Pitman, 1939; Creasy, 1957, Code Box 13.2) and can be understood as testing for correlation between fitted values and residuals, where residuals are measured in different directions depending on which line-fitting method is used (hence, the test statistic in Code Box 13.2 is a correlation coefficient). Confidence intervals for slope are constructed by inverting this test—finding the range of values for the slope such that the fitted values and residuals are not significantly correlated. When testing if the slope is one, which often coincides with a test for "isometry" (Gould, 1966), tests for MA and SMA are equivalent.

Code Box 13.2: Testing if the Brain–Body Mass Slope is 2/3

```
> library(smatr)
> sma_brainBody = sma(brain~body, data=Animals,log="xy",slope.test=2/3)
> sma_brainBody
Coefficients:
             elevation     slope
estimate     0.8797718 0.6363038
lower limit 0.4999123 0.4955982
upper limit 1.2596314 0.8169572
------------------------------------------------------------
H0 : slope not different from 0.6666667
Test statistic : r= -0.07424 with 26 degrees of freedom under H0
P-value : 0.70734
```

What happens if you reverse the axes? Do you get the same answer?

```
> sma(body~brain, data=Animals,log="xy",slope.test=3/2)
Coefficients:
              elevation    slope
estimate    -1.3826286 1.571576
lower limit -2.2584635 1.224054
upper limit -0.5067936 2.017763
------------------------------------------------------------
H0 : slope not different from 1.5
Test statistic : r= 0.07424 with 26 degrees of freedom under H0
P-value : 0.70734
```

Is this what you would have expected? Is there evidence against the $2/3$ *power law?*

When several lines correspond to independent samples that we want to compare, as in Exercise 13.2, several different types of questions could be asked. One can test for a common MA or SMA slope using a likelihood ratio statistic (Code Box 13.3) based on methods drawn from the principal component literature (Flury, 1984; Warton & Weber, 2002). If slopes can be assumed to be common, then one can compare locations of points across samples, to see if there is a change in elevation (Code Box 13.4), or possibly a shift in location along a common axis. These procedures make use of Wald tests (Warton et al., 2006) and can be understood as being a lot like MANOVA. A distinction from MANOVA is that the shift in means is being decomposed into shifts along the (standardised) major axis or otherwise. A secondary distinction from MANOVA is that we do not assume equal variance–covariance matrices, only a common slope, which is a weaker assumption. In the case of comparing more than two samples, it may be of interest to identify pairs of samples for which we have evidence of differences. Multiple comparison procedures are available in `smatr` via a `multcomp` argument.

Code Box 13.3: Comparing Allometric Slopes for Ian's Data Using smatr

First we will test for a common slope, log-transforming both variables:

```
> data(leaflife)
> leafSlopes = sma(longev~lma*site, log="xy", data=leaflife)
> summary(leafSlopes)
------------------------------------------------------------
Results of comparing lines among groups.

H0 : slopes are equal.
Likelihood ratio statistic : 9.382 with 3 degrees of freedom
P-value : 0.02462
------------------------------------------------------------
Coefficients by group in variable "site"

Group: 1
            elevation    slope
estimate    -4.218236 2.119823
lower limit -5.903527 1.451816
upper limit -2.532946 3.095192
...
> plot(leafSlopes)
```

Output for groups 2–4 has been excluded due to space considerations. The plot appears in Fig. 13.3.

The output suggests there is some evidence ($P = 0.02$) that SMA slopes for the so-called leaf economics spectrum are different across the four sites that were sampled. Confidence intervals for SMA slopes for each group are also reported, and it can be seen above that the slope does not cover one for group 1 (high rainfall and high soil nutrients). This is also the case for group 4 (low rainfall and low soil nutrients). So we can conclude that the relationship is not isometric, with evidence that (at least in some cases) leaf longevity changes by more than leaf mass per area (on a proportional scale) as you move along the leaf economics spectrum.

Code Box 13.4: Comparing Elevations of Allometric Lines for Ian's Low Soil Nutrient Data Using smatr

We subset data to just sites with low soil nutrients, for which SMA slopes were quite similar (it makes little sense to compare elevations when slopes are different).

```
> leaf_low_soilp = subset(leaflife, soilp == "low")
> leafElev = sma(longev~lma+rain, log="xy", data=leaf_low_soilp)
> leafElev
------------------------------------------------------------
Results of comparing lines among groups.

H0 : slopes are equal.
Likelihood ratio statistic : 2.367 with 1 degree of freedom
P-value : 0.12395
------------------------------------------------------------

H0 : no difference in elevation.
Wald statistic: 6.566 with 1 degree of freedom
```

```
P-value : 0.010393
------------------------------------------------------------
```

The results suggest some evidence of a difference in elevation of the leaf economics spectrum between high- and low-rainfall sites. Looking at elevation estimates (using the `summary` function), we would see that elevation is higher at high-rainfall sites, meaning that at high-rainfall sites, leaves could have lower leaf mass per area without a cost in leaf longevity.

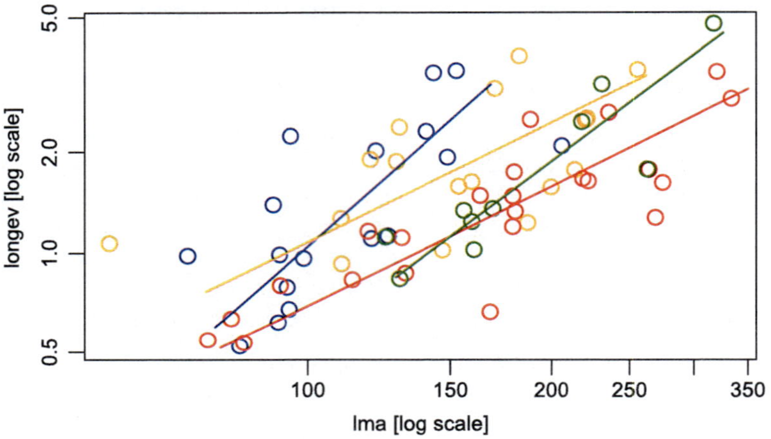

Fig. 13.3: Ian's leaf economics data from Exercise 13.2, with separately fitted lines for each treatment combination, as produced by `plot(leafSlopes)` in Code Box 13.3. There is a suggestion here that some lines differ from each other in slope, and others may differ in elevation, explored in Code Boxes 13.3–13.4

13.2.2 Mind Your Ps and Qs for (S)MA

Inference procedures about (S)MA, as in Code Boxes 13.2–13.4, essentially assume observations are independent and bivariate normal. As previously, the most important parts of the bivariate normality assumption are linearity and equal variance. As usual, we can diagnose these assumptions graphically using a plot of residuals against fitted values (Code Box 13.5).

Code Box 13.5: Residual Plots for Brain–Body Size Relationship

```
plot(sma_brainBody,which="residual") # residual plot
abline(a=0,b=0,col="red")
qqnorm(residuals(sma_brainBody))
qqline(residuals(sma_brainBody), col="red")
```

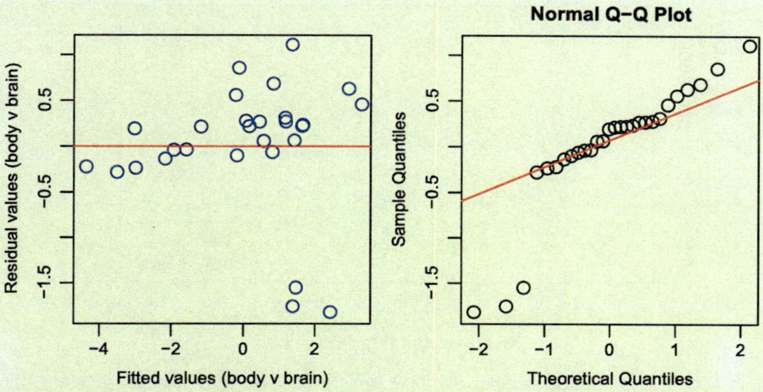

Do the assumptions look reasonable?

Being based on variances and covariances, (S)MA methods are quite sensitive to outliers. Robust extensions have been developed (Taskinen & Warton, 2011, using Huber's M estimation) and are available in the `smatr` package (Code Box 13.6). These methods are advisable if there are any outliers in the data. Outliers can have undue influence on fits, substantially reducing efficiency and making it harder to see the signal in data.

Code Box 13.6: Robust SMA for Brain–Body Size Relationship

```
> sma_brainBodyRobust = sma(brain~body, data=Animals,log="xy",
    slope.test=2/3,robust=TRUE)
> sma_brainBodyRobust
Confidence intervals (CI) are at 95%
------------------------------------------------------------
Coefficients:
            elevation     slope
estimate    0.8367912 0.7541820
lower limit 0.6056688 0.6452409
upper limit 1.0679136 0.8815164
------------------------------------------------------------
H0 : slope not different from 0.6666667
Test statistic : r= 0.3458 with 26 degrees of freedom under H0
P-value : 0.11673

> plot(brain~body,data=Animals,log="xy")
> abline(sma_brainBody, col="red")
> abline(sma_brainBodyRobust, col="blue")
```

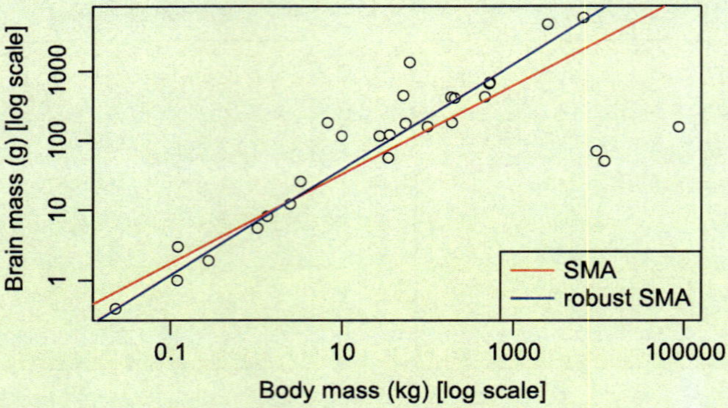

Notice that this line is slightly steeper than previously because it is less sensitive to the outliers below the line; also note that the confidence interval is narrower. *Is this what you expected to happen?*

Obvious limitations of the `smatr` software are that it is designed for problems with only two size variables and for linear relationships. For problems with more than two responses, PCA techniques could be used, but inference procedures might be harder to get a handle on. There is a literature on non-linear approaches to principal components (starting with Hastie & Stuetzle, 1989) that could be exploited to develop non-linear extensions of (standardised) major axis techniques.

Exercise 13.3: Outlier Sensitivity for the Brain–Body Mass Data
The plot in Fig. 13.1 has three outlying values towards the right. These correspond to dinosaurs, whereas all other measurements were taken on mammals. Maybe these values don't belong in this dataset at all.

Repeat the analyses of the brain–body mass data, in Code Boxes 13.2 and 13.6, excluding the three dinosaur species from analysis using `AnimalsSnipped=Animals[-c(6,16,26),]`.

Is robust SMA less sensitive to the dinosaur outliers? Is this what you expected?

Exercise 13.4: Robust Allometric Line Fitting for Ian's Leaf Data
The plot in Fig. 13.3 seems to have an outlying value towards the left, suggesting that maybe we should think about using robust methods of analysis here, too.

Repeat the analysis of Ian's leaf economics data, as in Code Boxes 13.3–13.4, using `robust=TRUE`. *Do the results work out differently?*

13.3 Controversies in the Allometry Literature

The literature on allometric line fitting has a history of controversy, making it difficult for users to navigate. Some key points of contention are summarised below.

While it was argued in this chapter that allometry should be treated as a multivariate problem, with two y variables, there is the question of when to actually do this or whether one of the variables could serve as a predictor, such that we have a linear model (as in Chap. 2). This would considerably simplify things and so should always be considered. Smith (2009) argued that a good way to check if (standardised) major axis techniques are warranted is to see if a problem makes as much sense if you flip the axes around—this is a way to check if you have two responses (y_1 and y_2, as in this chapter) or a response and a predictor (y and x, as in Chap. 2). For example, is the problem described in Code Box 13.1 really an issue, or was one or other of the regressions fitted there the correct one.

Allometric data, being size measurements, are commonly log-transformed prior to analysis. Packard (2013, and elsewhere) argues against transformation for a few reasons, including interpretability, whereas Kerkhoff and Enquist (2009) argue that if a variable is best understood as being the outcome of a set of multiplicative processes (as size variables often are), then the log scale is more natural. Xiao et al. (2011) make the obvious but important point that the decision to transform or not can be informed by your data, by checking assumptions of the fitted model.

There was a fair bit of discussion in the fishery literature, going back to at least the 1970s (Ricker, 1973; Jolicoeur, 1975), about whether a major axis or standardised major axis approach should be preferred. This is directly analogous to the question of whether to do a PCA on a covariance or a correlation matrix, although, interestingly, the problem has not really courted controversy when phrased this way. Inference about slope is rarely of interest in PCA, but it is central to allometry, which may be why the issue of how to estimate slope has taken on greater importance in the allometric literature. A sensible way forward may, however, be to take advice from the principal component literature—to standardise if variables are on different scales, or perhaps if they differ appreciably in their variability. An alternative way to make the decision is to think about the properties of the different lines, in particular whether it is more desirable to fit a line that is invariant under changes of scale (SMA) or under rotation (MA). Most would argue the former. Fortunately, because MA and SMA tests for isometry are equivalent, in many situations it doesn't actually matter which is used.

It has been argued (Hansen & Bartoszek, 2012, for example) that because you can also derive (S)MA methods from measurement error models, we should instead estimate how much measurement error there is in y_1 and y_2 and use this to choose the method of line fitting used. The problem here is that there are other sources of error beyond measurement error, most importantly, due to a lack of model fit, and studying measurement error is uninformative about how to measure lack of fit. For example, humans lie above the line in Fig. 13.1, and the reason they do so is not because of inaccuracies measuring body and brain size; it is because humans actually have large brains for their body size. (Or should we say small bodies for their brain size?) Looking at how accurately body and brain size have been measured should not tell us how we allocate error from the line due to lack of fit; it should be treated and adjusted for as a separate issue (Warton et al., 2006).

A lot of the ambiguity in the literature arises because when finding a line of best fit there is no single correct answer, with different options available depending on your objective. Technically, at least in the case of multivariate normal data, if we wish to attribute error to both y_1 and y_2, then a single line characterising the data is unidentifiable (Moran, 1971) without additional information on the relative errors in y_1 vs y_2. That is, a line cannot be estimated without a decision first being made concerning how to measure error from the line (e.g. in the vertical direction for linear regression, perpendicular to the line for a major axis). Some authors have advised on line-fitting methods using simulation studies, generating data from a particular line, with errors in y_1 and y_2, and looking at how well different methods reproduce that line. But the unidentifiability issue means that even in simulated data there is no single best line, hence no single right answer—it is more helpful to think of the data as coming from some true (bivariate) distribution, with a true variance–covariance matrix, which could be characterised in a number of different ways (in terms of its SMA or MA, for example). So one could argue against using simulation studies to decide which of SMA, MA, or other to use, because they all characterise different aspects of the same data.

Recall that in this text it has been emphasised that analysis methods are informed by data properties and research questions, so we always need to *mind our Ps and Qs*. A key point to recognise, which can help navigate the conflicting views in the allometry literature, is that the decision to use MA, SMA, linear regression, or another type of line-fitting method is more about the Qs than the Ps.

Part III
Regression Analysis for Multivariate Abundances

Chapter 14
Multivariate Abundances—Inference About Environmental Associations

The most common type of multivariate data collected in ecology is also one of the most challenging types to analyse—when some abundance-related measure (e.g. counts, presence–absence, biomass) is simultaneously collected for all taxa or species encountered in a sample, as in Exercises 14.1–14.3. The rest of the book will focus on the analysis of these *multivariate abundances*.

Exercise 14.1: Revegetation and Invertebrate Counts
In his revegetation study (Exercise 10.3), Anthony classified anything that fell into his pitfall traps into orders, and thus counted the abundance of each of 24 invertebrate orders across 10 sites. He wants to know:

Is there evidence of a change in invertebrate communities due to revegetation efforts?

What type of response variable(s) does he have? How should Anthony analyse his data?

Exercise 14.2: Invertebrates Settling on Seaweed
In Exercise 1.13, David and Alistair looked at invertebrate epifauna settling on algal beds (seaweed) with different levels of isolation (0, 2, or 10 m buffer) from each other, at two sampling times (5 and 10 weeks). They observed presence–absence patterns of 16 different types of invertebrate (across 10 replicates).

They would like to know if there is any evidence of a difference in invertebrate presence–absence patterns with distance of isolation. *How should they analyse the data?*

D. I. Warton, *Eco-Stats: Data Analysis in Ecology*, Methods in Statistical Ecology,
https://doi.org/10.1007/978-3-030-88443-7_14

Exercise 14.3: Do Offshore Wind Farms Affect Fish Communities?
As in Exercise 10.2, Lena studied the effects of an offshore wind farm on fish communities by collecting paired data before and after wind farm construction, at 36 stations in each of 3 zones (wind farm, north, and south). She counted how many fish were caught at each station, classified into 16 different taxa.

Lena wants to know if there is any evidence of a change in fish communities at wind farm stations, compared to others, following construction of the wind farm. *How should she analyse the data?*

This type of data goes by lots of other names—"species by site data", "community composition data", even sometimes "multivariate ecological data", which sounds a bit too broad, given that there are other types of multivariate data used in ecology (such as allometric data, see Chap. 13). The term multivariate abundances is intended to put the focus on the following key statistical properties.

Multivariate: There are many correlated response variables, sometimes more variables than there are observations:

- In Exercise 14.1, Anthony has 10 observations and 24 variables.
- In Exercise 14.2, David and Alistair have 57 observations and 16 variables.
- In Exercise 14.3, Lena has 179 observations and 16 variables.

Abundance: Abundance or presence–absence data usually exhibits a strong mean–variance relationship, as in Fig. 14.1.

You need to account for both properties in your analysis.

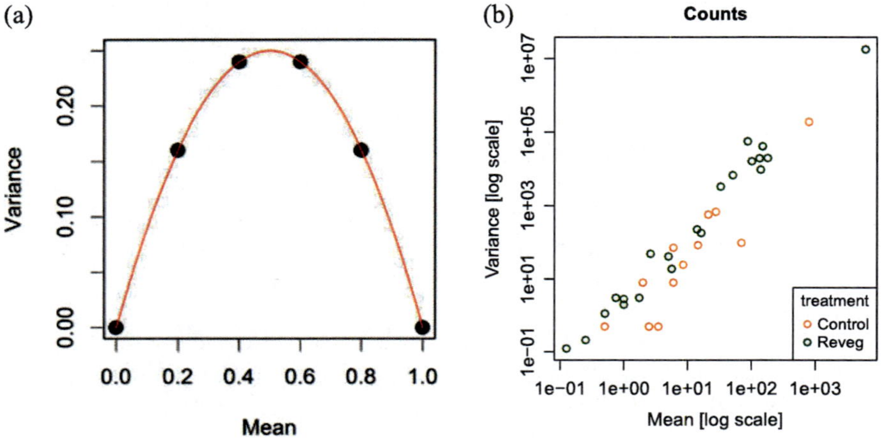

Fig. 14.1: Mean–variance relationships for (**a**) David's and Alistair's data of Exercise 14.2 and (**b**) the revegetation study of Exercise 14.1

Multivariate abundance data are especially common in ecology, probably for two reasons. Firstly, it is often of interest to say something collectively about a community, e.g. in environmental impact assessment, we want to know if there is any impact of some event on the ecological community. Secondly, this sort of data arises naturally in sampling—even when you're interested in some target species, others will often be collected incidentally along the way, e.g. pitfall traps set specifically for ants will inevitably capture a range of other types of invertebrate also. So even when they are interested in something else, many ecologists end up with multivariate abundances and feel like they should do something with them. In this second case we do not have a good reason to analyse multivariate abundances. Only bother with multivariate analysis if the primary research question of interest is multivariate, i.e. if a community or assemblage of species is of primary interest. Don't go multivariate just because you have the data.

There are a few different types of questions one might wish to answer using multivariate abundances. The most common type of question, as in each of Exercises 14.1–14.3, is whether or not the community is associated with some predictor (or set of predictors) characterising aspects of the environment—whether looking at the effect on a community of an experimental treatment, testing for environmental impact (Exercise 14.3), or something else again.

In Chap. 11 some multivariate regression techniques were introduced, and model-based inference was used to study the effects of predictors on response. If there were only a few taxa in the community, those methods would be applicable. But (as flagged in Table 11.2) a key challenge with multivariate abundances is that typically there are many responses. It's called biodiversity for a reason! There are lots of different types of organisms out there. The methods discussed in this chapter are types of *high-dimensional regression*, intended for when you have many responses, but if you only have a few responses, you might be better off back in Chap. 11. High-dimensional regression is technically difficult and is currently a fast-moving field.

In this chapter we will use *design-based inference* (as in Chap. 9). Design-based inference has been common in ecology for this sort of problem for a long time as a way to handle the *multivariate* property, and the focus in this chapter will be on applying design-based inference to models that appropriately account for the mean–variance relationship in data (to also handle the *abundance* property). There are some potential analysis options beyond design-based inference, which we will discuss later.

Key Point

Multivariate abundance data (also "species by site data", "community composition data", and so forth) has two key properties: a *multivariate* property, that there are many correlated response variables, and an *abundance* property, a strong mean–variance relationship. It is important to account for *both* properties in your analysis.

14.1 Generalised Estimating Equations

Generalised estimating equations (GEEs, Liang & Zeger, 1986; Zeger & Liang, 1986) are a fast way to fit a model to correlated counts, compared to hierarchical models (Chaps. 11–12). Design-based inference techniques like the bootstrap (Chap. 9) tend to be computationally intensive, especially when applied to many correlated response variables, GEEs are a better choice when planning to use design-based inference. Parameters from GEEs are also slightly easier to interpret than those of a hierarchical model, because they specify marginal rather than conditional models, so parameters in the mean model have direct implications for mean abundance (see Maths Box 11.4 for problems with marginal interpretation of hierarchical parameters).

GEEs are *ad hoc* extensions of equations used to estimate parameters in a GLM, defined by taking the estimating equations from a GLM, forcing them to be multivariate (Maths Box 14.1), and hoping for the best. An assumption about the correlation structure of the data is required for GEEs. Independence of responses is commonly assumed, sometimes called *independence estimating equations*, which simplifies estimation to a GLM problem, and then correlation in the data is adjusted for later (using "sandwich estimators" for standard errors, Hardin & Hilbe, 2002).

Maths Box 14.1: Generalised Estimating Equations

As in Maths Box 10.2, maximum likelihood is typically used to estimate parameters in a GLM, which ends up meaning that we need to find the values of parameters that solve the following *score equations*:

$$\mathbf{0} = \sum_{i=1}^{n} \boldsymbol{d}_i V(\mu_i)^{-1}(y_i - \mu_i)$$

(where $\boldsymbol{d}_i = \frac{\partial \mu_i}{\partial \boldsymbol{\beta}} = \frac{x_i}{g'(\mu_i)}$, as in Eq. 10.4). The GEE approach involves taking these *estimating equations* and making them multivariate, by replacing the response with a vector of correlated responses, replacing the variance with a variance–covariance matrix, and hoping for the best:

$$\mathbf{0} = \sum_{i=1}^{n} \boldsymbol{D}_i \boldsymbol{V}(\boldsymbol{\mu}_i)^{-1}(\boldsymbol{y}_i - \boldsymbol{\mu}_i) \tag{14.1}$$

$\boldsymbol{V}(\boldsymbol{\mu}_i)$ is now a variance–covariance matrix, requiring a "working correlation" structure to be specified.

A similar fitting algorithm is used as for GLMs, which means that GEEs are typically relatively quick to fit.

Notice that whereas the score equations for GLMs are derived as the gradient of the log-likelihood function, that is not how GEEs are derived. In fact, unless responses are assumed to be normally distributed, or they are assumed to be independent of each other, *there is no GEE likelihood function*. This complicates inference, because standard likelihood-based tools such as AIC, BIC, and likelihood ratio tests cannot be used because we cannot calculate a GEE likelihood.

Some difficulties arise when using GEEs, because of the fact that they are motivated from equations for estimating parameters, rather than from a parametric model for the data. GEEs define marginal models for data, but (usually) not a joint model, with the estimating equations no longer corresponding to the derivative of any known likelihood function. One difficulty that this creates is that we cannot simulate data under a GEE "model". A second difficulty is that without a likelihood, a likelihood ratio statistic can't be constructed. Instead, another member of the "Holy Trinity of Statistics" (Rao, 1973) could be used for inference, a Wald or score statistic. Maybe this should be called the Destiny's Child of Statistics (Maths Box 14.2), because while the Wald and score statistics are good performers in their own right, the likelihood ratio statistic is the main star (the Beyoncé). Wald statistics have been met previously, with the output from `summary` for most `R` objects returning Wald statistics. These statistics are based on parameter estimates under the alternative hypothesis, by testing if parameter estimates are significantly different from what is expected under the null hypothesis. A score statistic (or Rao's score statistic) is based on the estimating equations themselves, exploiting the fact that plausible estimates of parameters should give values of the estimating equations that are close to zero. Specifically, parameter estimates under the null hypothesis are plugged into the estimating equations under the alternative hypothesis, and a statistic constructed to test for evidence that the expression on the right-hand side of Eq. 14.1 is significantly different from zero.

Maths Box 14.2: The Destiny's Child of Statistics
Consider testing for evidence against a null model ($\mathcal{M}_0$) with parameter θ_0, in favour of a more general alternative model ($\mathcal{M}_1$) with parameter $\hat{\theta}_1$. There are three main types of likelihood-based test statistics, the *Destiny's Child of Statistics* (or the *Holy Trinity of Statistics*, according to Rao, 1973). These can be visualised in a plot of the log-likelihood function against θ:

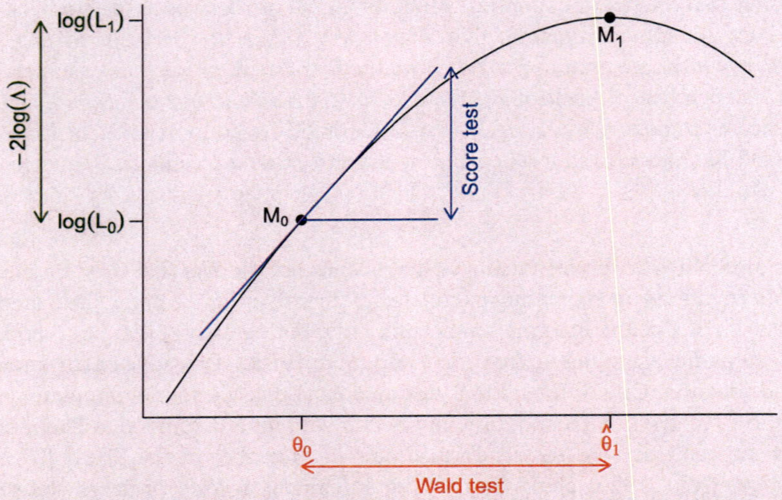

The *likelihood ratio test* $-2\log\Lambda(\mathcal{M}_0,\mathcal{M}_1) = 2\ell_{\mathcal{M}_1}(\hat{\theta}_1;\mathbf{y}) - 2\ell_{\mathcal{M}_0}(\theta_0;\mathbf{y})$ focuses on whether the likelihoods of the two models are significantly different (vertical axis).

The *Wald statistic* focuses on the parameter (horizontal axis) of $\mathcal{M}_1$ and whether $\hat{\theta}_1$ is significantly far from what would be expected under $\mathcal{M}_0$, using $\frac{\hat{\theta}_1-\theta_0}{\hat{\sigma}_{\hat{\theta}_1}}$.

The *score statistic* focuses on the score equation $u(\theta)$, the gradient of the log-likelihood at $\mathcal{M}_0$. The likelihood should be nearly flat for a model that fits the data well. So if $\mathcal{M}_0$ is the correct model, $u(\theta_0)$ should be near zero, and we can use as a test statistic $\frac{u(\theta_0)}{\hat{\sigma}_{u(\theta_0)}}$.

In GEEs, $u(\theta)$ is defined, hence θ can be estimated, but the likelihood is not defined (unless assuming all variables are independent). So for correlated counts we can use GEEs to calculate a Wald or score statistic, but not a likelihood ratio statistic. Sorry, no Beyoncé!

14.2 Design-Based Inference Using GEEs

A simple GEE model for abundance at site i of taxon j is

$$\begin{aligned} y_{ij} &\sim F(\mu_{ij}, \phi_i) \text{ such that } \sigma^2_{y_{ij}} = V(\mu_{ij}, \phi_j) \\ g(\mu_{ij}) &= \beta_{0j} + \boldsymbol{x}'_i \boldsymbol{\beta}_j \\ \text{cor}(r_{ij}, r_{ij'}) &= \mathbf{R}_{jj'} \quad \text{where } r_{ij} = \frac{y_{ij} - \mu_{ij}}{\sigma_{y_{ij}}} \end{aligned} \tag{14.2}$$

An offset or a row effect term can be added to account for variation in sampling intensity, which is useful for diversity partitioning, as discussed later (Sect. 14.3).

A working correlation matrix (**R**) is needed, specifying how abundances are associated with each other across taxa. The simplest approach is to use *independence estimating equations*, ignoring correlation for the purposes of estimation (assuming **R** = **I**, a diagonal matrix of ones, with all correlations equal to zero), so that the model simplifies to fitting a GLM separately to each response variable. This is pretty much the simplest possible model that will account for the *abundance* property, and by choosing a simple model, we hope that resampling won't be computationally prohibitive.

We need to handle the *multivariate* property of the data to make valid multivariate inferences about the effects of predictors (environmental associations), and this can be done by resampling *rows* of data. Resampling rows keeps site abundances for all taxa together in resamples, to preserve the correlation between taxa. This is a form of block resampling (Sect. 9.7.1). Correlation can also be accounted for in constructing the test statistic.

The `manyglm` function in the `mvabund` package was written to carry out the preceding operation, and it behaves a lot like `glm`, so it is relatively easy to use if you are familiar with the methods of Chap. 10 (Code Box 14.1). It does, however, take longer to run (for `anova` or `summary`), so sometimes you have to be patient. Unlike the `glm` function, `manyglm` defaults to `family="negative.binomial"`. This is done because the package was designed to analyse multivariate abundances (hence the name), and these are most commonly available as overdispersed counts.

Code Box 14.1: Using mvabund to Test for an Effect of Revegetation in Exercise 12.2

```
> library(ecostats)
> library(mvabund)
> data(reveg)
> reveg$abundMV=mvabund(reveg$abund)
> ft_reveg=manyglm(abundMV~treatment+offset(log(pitfalls)),
family="negative.binomial", data=reveg) # offset included as in
        Ex 10.9
> anova(ft_reveg)
Time elapsed: 0 hr 0 min 9 sec
Analysis of Deviance Table

Model: manyglm(formula = abundRe ~ treatment + offset(log(pitfalls)),
Model:     family = "negative.binomial")
```

```
Multivariate test:
            Res.Df Df.diff   Dev Pr(>Dev)
(Intercept)      9
treatment        8       1 78.25    0.024 *
---
Signif. codes:  0 `***' 0.001 `**' 0.01 `*' 0.05 `.' 0.1 ` ' 1
Arguments:
 Test statistics calculated assuming uncorrelated response (for faster
 computation)P-value calculated using 999 iterations via PIT-trap
 resampling.
```

Exercise 14.4: Testing for an Effect of Isolation on Invertebrates in Seaweed

Consider David and Alistair's study of invertebrate epifauna settling on algal beds with different levels of isolation (0, 2, or 10 m buffer) at different sampling times (5 and 10 weeks), with varying seaweed biomass in each patch.

What sort of model is appropriate for this dataset? Fit this model and call it `ft_epiAlt` and run `anova(ft_epiAlt)`. (This might take a couple of minutes to run.)

Now fit a model under the null hypothesis that there is no effect of distance of isolation, and call it `ft_epiNull`. Run `anova(ft_epiNull, ft_epiAlt)`. *This second* `anova` *took much less time to fit—why?*

Is there evidence of an effect of distance of isolation on presence–absence patterns in the invertebrate community?

14.2.1 Mind Your Ps and Qs

The `manyglm` function makes the same assumptions as for GLMs, plus a correlation assumption:

1. The observed y_{ij}-values are *independent* across observations (across i), after conditioning on x_i.
2. The y_{ij}-values come from *a known distribution* (from the exponential family) with known *mean–variance relationship* $V(\mu_{ij})$.
3. There is a *straight-line relationship* between some known function of the mean of $\mathbf{y}_j$ and $\mathbf{x}$
$$g(\mu_{ij}) = \beta_{0j} + \mathbf{x}_i^T \boldsymbol{\beta}_j$$
4. Residuals have a *constant correlation matrix* across observations.

Check assumptions 2 and 3 as for GLMs—using Dunn-Smyth residual plots (Code Box 14.2). As usual, the `plot` function could be used to construct residual plots for `mvabund` objects, or the `plotenvelope` function could be used to add simulation envelopes capturing the range of variation to expect if assumptions were satisfied. For large datasets, `plotenvelope` could take a long time to run, unless using `sim.method="stand.norm"` (as in Code Box 14.3) to simulate standard normal random variables, instead of simulating new responses and refitting the model for each. As usual, we want no trend in the residuals vs fits plot and would be particularly worried by a U shape (non-linearity) or a fan shape (problems with the mean–variance assumption, as in Code Box 14.3), and in the normal quantile plot we expect residuals to stay close to the trend line. The `meanvar.plot` function can also be used to plot sample variances against sample means, by taxon and optionally by treatment (Code Box 14.3).

Code Box 14.2: Checking Assumptions for the Revegetation Model of Code Box 14.1

```
par(mfrow=c(1,3))
ft_reveg=manyglm(abundMV~treatment,offset=log(pitfalls),
        family="negative.binomial", data=reveg)
plotenvelope(ft_reveg, which=1:3)
```

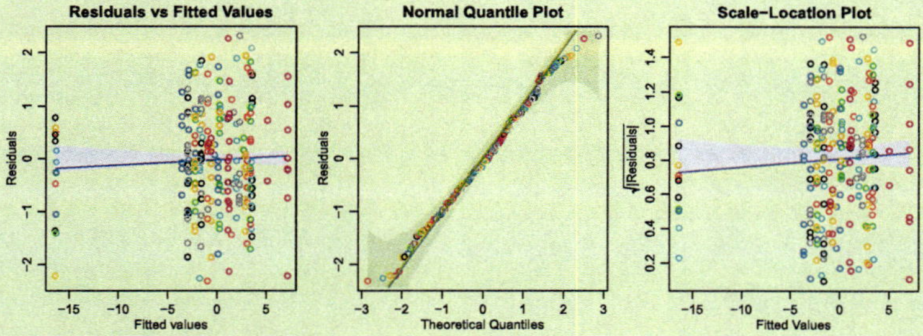

What do you reckon?

Code Box 14.3: Checking Mean–Variance Assumptions for a Poisson Revegetation Model

```
ft_revegP=manyglm(abundMV~treatment, offset=log(pitfalls),
          family="poisson", data=reveg)
par(mfrow=c(1,3))
plotenvelope(ft_revegP, which=1:3, sim.method="stand.norm")
```

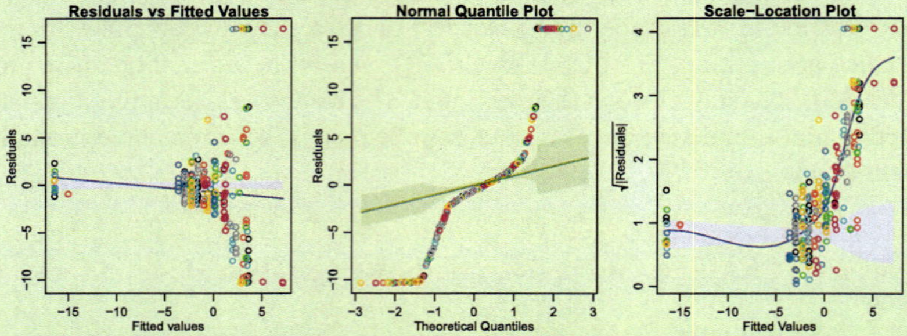

Plotting sample variances against sample means for each taxon and treatment:

```
meanvar.plot(reveg$abundMV~reveg$treatment)
abline(a=0,b=1,col="darkgreen")
```

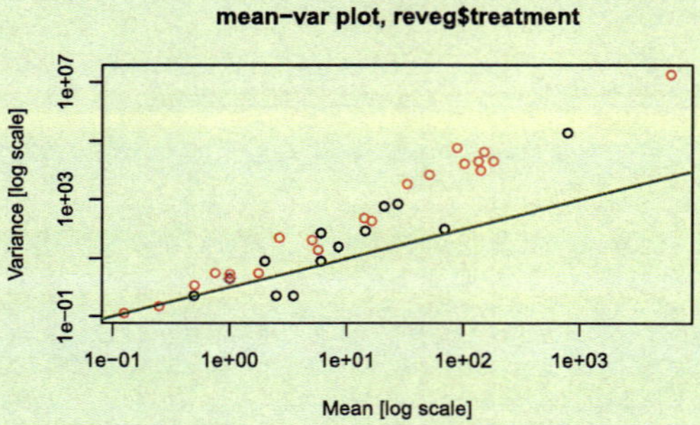

How's the Poisson assumption looking?

Exercise 14.5: Checking Assumptions for the Habitat Configuration Data
Consider the multivariate analysis of the habitat configuration study (Exercise 14.4).

What assumptions were made?

Where possible, check these assumptions.

Do the assumptions seem reasonable?

Exercise 14.6: Checking Assumptions for Wind Farm Data
Consider Lena's offshore wind farm study (Exercise 14.3). Fit an appropriate model to the data. Make sure you include a `Station` main effect (to account for the paired sampling design).

What assumptions were made?

Where possible, check these assumptions.

Do the assumptions seem reasonable? In particular, think about whether there is evidence that the counts are overdispersed compared to the Poisson.

14.2.2 Test Statistics Accounting for Correlation

When using `anova` on a `manyglm` object, a "sum-of-LR" statistic (Warton et al., 2012b) is the default—a likelihood ratio statistic computed separately for each taxon, then summed across taxa for a community-level measure. By summing across taxa, the sum-of-LR statistic is calculated assuming independent responses (and the job of accounting for the *multivariate* property is left to row resampling).

If you want to account for correlation between variables in the test statistic, you need to change both the type of test statistic (via the `test` argument) and the assumed correlation structure (via the `cor.type` argument). The test statistic to use is a score (`test="score"`) or Wald (`test="wald"`) statistic, as described previously. The type of correlation structure to assume is controlled by the `cor.type` argument. Options currently available include the following:

`cor.type="I"` (Default) Assumes independence for test statistic calculation—sums across taxa for a faster fit.

`cor.type="R"` Assumes unstructured correlation between all variables, i.e. estimates a separate correlation coefficient between each pair of responses. Not recommended if there are many variables compared to the number of observations, because it will become numerically unstable (Warton, 2008).

`cor.type="shrink"` A middle option between the previous two. Use this to account for correlation unless you have only a few variables. This method shrinks an unstructured correlation matrix towards the matrix you would use if assuming independence, using the data to work out how far to move towards independence (Warton, 2011, as in Code Box 14.4).

Note that even if you ignore correlation when constructing a test statistic, it is accounted for in the P-value because rows of observations are resampled. This means the procedure is valid even if the independence assumption used in constructing the test statistic is wrong. But recall that *valid≠efficient*—while this procedure is valid, the main risk when using this statistic is that if there are correlated variables you can miss structure in the data (the scenario depicted in Fig. 11.1). So one approach, as when making inferences from multivariate linear models (Sect. 11.2), is to try a couple of different test statistics, in case the structure in the data is captured by one of these but not another. This approach would be especially advisable if using Wald statistics, because they can be insensitive when many predicted values for a taxon are zero (as in Chap. 10).

Code Box 14.4: A manyglm Analysis of Revegetation Data Using a Statistic Accounting for Correlation

```
> anova(ft_reveg,test="wald",cor.type="shrink")
Time elapsed: 0 hr 0 min 6 sec
Analysis of Variance Table

Model: abundMV ~ treatment + offset(log(pitfalls))

Multivariate test:
            Res.Df Df.diff  wald Pr(>wald)
(Intercept)      9
treatment        8       1 8.698     0.039 *
---
Signif. codes:  0 `***' 0.001 `**' 0.01 `*' 0.05 `.' 0.1 ` ' 1
Arguments:
 Test statistics calculated assuming correlated response via ridge
 regularization P-value calculated using 999 iterations via PIT-trap
  resampling.
```

You can also use the summary function for manyglm objects, but the results aren't quite as trustworthy as for anova. The reason is that resamples are taken under the alternative hypothesis for summary, where there is a greater chance of fitted values being zero, especially for rarer taxa (e.g. if there is a treatment combination in which a taxon is never present). Abundances don't resample well if their predicted mean is zero.

14.2.3 Computation Time

One major difference between glm and manyglm is in computation time. Analysing your data using glm is near instantaneous, unless you have a very large dataset. But in Code Box 14.1, an anova call to a manyglm object took almost 10 s, on a small dataset with 24 responses variables. Bigger datasets will take minutes, hours, or sometimes days! The main problem is that *resampling is computationally intensive*—by default, this function will fit a glm to each response variable 1000 times, so there are 24,000 GLMs in total. If an individual GLM were to take 1s to fit, then fitting 24,000 of them would take almost 7 h (fortunately, that would only happen for a pretty large dataset).

For large datasets, try setting nBoot=49 or 99 to get a faster but less precise answer. Then scale it up to around 999 when you need a final answer for publication.

You can also use the `show.time="all"` argument to get updates every 100 bootstrap samples, e.g. `anova(ft_reveg,nBoot=499,show.time="all")`.

If you are dealing with long computation times, parallel computing is a solution to this problem—if you have 4 computing cores to run an analysis on, you could send 250 resamples to each core then combine, to cut computation down four-fold. If you have access to a computational cluster, you could even send 1000 separate jobs to 1000 nodes, each consisting of just one resample, and reduce a hard problem from days to minutes. By default, `mvabund` will split operations up across however many nodes are available to it at the time.

Another issue to consider with long computation times is whether some of the taxa can be removed from the analysis. Most datasets contain many taxa that are observed very few times (e.g. *singletons*, seen only once, and doubletons or tripletons), and these typically provide very little information to the analysis, while slowing computation times. The slowdown due to rarer taxa can be considerable because they are more difficult to fit models to. So an obvious approach to consider is removing rarer taxa from the analysis—this rarely results in loss of signal from the data but removes a lot of noise, so typically you will get faster (and better) results from removing rare species (as in Exercise 14.7). It is worth exploring this idea for yourself and seeing what effect removing rarer taxa has on results. Removing species seen three or fewer times is usually a pretty safe bet.

Exercise 14.7: Testing for an Effect of Offshore Wind Farms (Slowly)
Consider Lena's offshore wind farm study (Exercise 14.3). The data contain a total of 179 rows of data and a `Station` main effect (to account for the paired sampling) that has lots of terms in it. Analysis will take a while.

Fit models under the null and alternative hypotheses of interest. Run an `anova` to compare these 2 models, with just 19 bootstrap resamples, to estimate computation time.

Remove zerotons and singletons from the dataset using

```
windMV = mvabund(windFarms$abund[,colSums(windFarms$abund>0)>1])
```

Now fit a model to this new response variable, again with just 19 bootstrap resamples. *Did this run take less time? How do the results compare? How long do you think it would take to fit a model with 999 bootstrap resamples, for an accurate P-value?*

14.2.4 The `manyany` Function

The `manyglm` function is currently limited to just a few choices of family and link function to do with count or presence–absence data, focusing on distributions like the negative binomial, Poisson, and binomial. An extension of it is the `manyany` function (Code Box 14.5), which allows you to fit (in principle) any univariate function to each column of data and use anova to resample rows to compare two competing models. The cost of this added flexibility is that this function is very slow—manyglm was coded in C (which is much faster than R) and optimised for speed, but manyany was not.

Code Box 14.5: Analysing Ordinal Data from Habitat Configuration Study Using manyany

Regression of ordinal data is not currently available in the manyglm function, but it can be achieved using manyany:

```
> habOrd = counts = as.matrix( round(seaweed[,6:21]*seaweed$Wmass))
> habOrd[counts>0 & counts<10] = 1
> habOrd[counts>=10] = 2
> library(ordinal)
> summary(habOrd) # Amphipods are all "2" which would return an
  error in clm
> habOrd=habOrd[,-1] #remove Amphipods
> manyOrd=manyany(habOrd~Dist*Time*Size,"clm",data=seaweed)
> manyOrdNull=manyany(habOrd~Time*Size,"clm",data=seaweed)
> anova(manyOrdNull, manyOrd)

              LR Pr(>LR)
sum-of-LR 101.1     0.12
---
Signif. codes:  0 `***' 0.001 `**' 0.01 `*' 0.05 `.' 0.1 ` ' 1
```

What hypothesis has been tested here? Is there any evidence against it?

14.3 Compositional Change and Partitioning Effects on α- and β-Diversity

In the foregoing analyses, the focus was on modelling mean abundance, but sometimes we wish to focus on relative abundance or *composition*. The main reason for wanting to do this is if there are changes in sampling intensity for reasons that can't be directly measured. For example, pitfall traps are often set in terrestrial systems to catch insects, but some will be more effective than others because of factors unrelated to the abundance of invertebrates, such as how well pitfall traps were placed and the extent to which ground vegetation impedes movement in the vicinity of the trap (Greenslade, 1964). A key point here is that some of the variation in abundance measurements is due to changes in the way the sample was taken rather than being due to changes in the study organisms—variation is explained by the sampling mechanism as well as by ecological mechanisms. In this situation, only *relative abundance* across taxa is of interest, after controlling for variation in sampling intensity. In principle, it is straightforward to study relative abundances using a model-based approach—we simply add a term to the model to account for variation in abundance across samples. So the model for abundance at site i of taxon j becomes

$$
\begin{aligned}
y_{ij} &\sim F(\mu_{ij}, \phi_i) \text{ such that } Var(y_{ij}) = V(\mu_{ij}, \phi_j) \\
g(\mu_{ij}) &= \boldsymbol{x}_i'\boldsymbol{\alpha} + \alpha_{0i} + \beta_{0j} + \boldsymbol{x}_i'\boldsymbol{\beta}_j
\end{aligned} \tag{14.3}
$$

The new term in the model, α_{0i}, accounts for variation across samples in total abundance, so that remaining terms in the model can focus on change in relative abundance. The optional term $\boldsymbol{x}_i'\boldsymbol{\alpha}$ quantifies how much of this variation in total abundance can be explained by environmental variables. The terms in Eq. 14.3 have thus been partitioned into those studying total abundance (the α) and those studying relative abundance (the β). Put another way, the effects of environmental variables have been split into main effects (the α) and their interactions with taxa (the β, which take different values for different taxa). The model needs additional constraints for all the terms to be estimable, which R handles automatically (e.g. by setting $\alpha_{01} = 0$).

Key Point
Often the primary research interest is in studying the effects of environmental variables on community composition or species turnover. This is especially useful if some variation in abundance is explained by the sampling mechanism, as well as ecological mechanisms. This can be accounted for in a multivariate analysis by adding a "row effect" to the model, a term that takes a different value for each sample according to its total abundance. Thus, all remaining terms in the model estimate compositional effects (β-diversity), after controlling for effects on total abundance (α-diversity).

A classic paper by Whittaker (1972) described the idea of partitioning species diversity into α-diversity, "the community's richness of species", and β-diversity, the "extent of differentiation in communities along habitat gradients".[1] The parameters of Eq. 14.3 have been written as either α or β to emphasise their connections to α-diversity and β-diversity. Specifically, larger α coefficients in Eq. 14.3 correspond to samples or environmental variables that have larger effects on abundance of all species (and hence on species richness), whereas larger β coefficients in Eq. 14.3 correspond to taxa that differ from the overall α-trend in terms of how their abundance relates to the environmental variable, implying greater species turnover along the gradient. The use of statistical models to tease apart effects of environmental variables on α- and β-diversity is a new idea that has a lot of potential.

A model along the lines of Eq. 14.3 can be readily fitted via `manyglm` using the `composition` argument, as in Code Box 14.6. The `manyany` function also has a `composition` argument, which behaves similarly.

[1] He also defined γ-diversity, the richness of species in a region, but this is of less interest to us here.

Code Box 14.6: A Compositional Analysis of Anthony's Revegetation Data

```
> ft_comp=manyglm(abundMV~treatment+offset(log(pitfalls)),
data=reveg, composition=TRUE)
> anova(ft_comp,nBoot=99)
Time elapsed: 0 hr 0 min 21 sec
Model: abundMV ~ cols + treatment + offset(log(pitfalls)) + rows
                    + cols:(treatment + offset(log(pitfalls)))

                Res.Df Df.diff   Dev Pr(>Dev)
(Intercept)        239
cols               216      23 361.2     0.01 **
treatment          215       1  14.1     0.01 **
rows               206       9  25.5     0.02 *
cols:treatment     184      23  56.7     0.01 **
```

In this model, coefficients of `cols`, `treatment`, `rows`, and `cols:treatment` correspond in Eq. 14.3 to β_{0j}, α, α_{0i}, and $\boldsymbol{\beta}_j$, respectively.

Which term measures the effect of treatment on relative abundance? Is there evidence of an effect on relative abundance?

Fitting models using `composition=TRUE` is currently computationally slow. Data are re-expressed in long format (along the lines of Code Box 11.5) to fit a single GLM as in Eq. 14.3, with abundance treated as a univariate response and row and column factors used as predictors to distinguish different samples and responses (respectively). This is fitted using the `manyglm` computational machinery, but keeping all observations from the same site together in resamples, as previously. A limitation of this approach is that the model is much slower to fit than in short format and may not fit at all for large datasets because of substantial inefficiencies that a long format introduces. In particular, the design matrix storing x variables has p times as many rows in it and nearly p times as many columns! Computation times for `anova` can be reduced by using it to compare just the null and alternative models for the test of interest, manually fitted in long format, as in Code Box 14.7. Further limitations, related to treating the response as univariate, are that the model is unable to handle correlation across responses (so `cor.type="I"` is the only option for compositional analyses), multiple testing across responses (see following section) is unavailable, and any overdispersion parameters in the model are assumed constant across responses (i.e. for each j, we assumed $\phi_j = \phi$ in Code Box 14.6). Writing faster and more flexible algorithms for this sort of model is possible and a worthwhile avenue for future research, whether using sparse design matrices (Bates & Maechler, 2015) or short format.

Code Box 14.7: A Faster Compositional Analysis of Anthony's Revegetation Data

In Code Box 14.6, every term in `ft_comp` was tested, even though only the last term was of interest. The data used to fit this model are stored in long format in `ft_comp$data`, so we can use this data frame to specify the precise null and alternative models we want to test so as to save computation time:

```
> ft_null = manyglm(abundMV~cols+rows+offset(log(pitfalls)),
data=ft_comp$data)
> ft_alt = manyglm(abundMV~cols+rows+treatment:cols
+offset(log(pitfalls)), data=ft_comp$data)
> anova(ft_null, ft_alt, nBoot=99, block=ft_comp$rows)
Time elapsed: 0 hr 0 min 5 sec

ft_null: abundMV ~ cols + rows + offset(log(pitfalls))
ft_alt: abundMV ~ cols + rows + treatment:cols + offset(log(pitfalls))

        Res.Df Df.diff   Dev Pr(>Dev)
ft_null    207
ft_alt     184      23 56.74     0.01 **
---
Signif. codes:  0 `***' 0.001 `**' 0.01 `*' 0.05 `.' 0.1 ` ' 1
Arguments: P-value calculated using 99 iterations via PIT-trap
resampling.
```

The results are the same (same test statistic and similar P-value) but take about a quarter of the computation time. Notice also that a main effect for the `treatment` term was left out of the formulas. *Why didn't exclusion of the `treatment` term change the answer?*

14.3.1 Quick-and-Dirty Approach Using Offsets

For large datasets it may not be practical to convert data to long format, in which case the `composition=TRUE` argument is not a practical option. In this situation a so-called quick-and-dirty alternative for count data is to calculate the quantity

$$s_i = \log \sum_{j=1}^{p} y_{ij} - \log \sum_{j=1}^{p} \hat{\mu}_{ij} \qquad (14.4)$$

and use this as an offset (Code Box 14.8). The term $\hat{\mu}$ refers to the predicted value for y_{ij} from the model that would be fitted if you were to exclude the compositional term. The s_i estimate the row effect for observation i as the difference in log-row sums between the data and what would be expected for a model without row effects. The best way to use this approach would be to calculate a separate offset for each model being compared, as in Code Box 14.8. If there was already an offset in the model, it stays there, and we now add a second offset as well (Code Box 14.8).

Code Box 14.8: Quick-and-Dirty Compositional Analysis of Anthony's Revegetation Data

We can approximate the row effects α_{0i} using the expression in Eq. 14.4:

```
> # calculate null model offset and fit quick-and-dirty null model
> ft_reveg0 = manyglm(abundMV~1+offset(log(pitfalls)), data=reveg)
> QDrows0 = log(rowSums(reveg$abundMV)) - log(rowSums(fitted(ft_reveg0)))
> ft_row0 = manyglm(abundMV~1+offset(log(pitfalls))+
              offset(QDrows0), data=reveg)
> # calculate alt model offset and fit quick-and-dirty alt model
> ft_reveg = manyglm(abundMV~treatment+offset(log(pitfalls)),
              data=reveg)
> QDrows = log(rowSums(reveg$abundMV)) - log(rowSums(fitted(ft_reveg)))
> ft_row = manyglm(abundMV~treatment+offset(log(pitfalls))+
             offset(QDrows), data=reveg)
> anova(ft_row0,ft_row)
Time elapsed: 0 hr 0 min 7 sec
Analysis of Deviance Table

ft_row0: abundMV ~ 1 + offset(QDrows0)
ft_row: abundMV ~ treatment + offset(QDrows)

Multivariate test:
        Res.Df Df.diff   Dev Pr(>Dev)
ft_row0      9
ft_row       8       1 50.26    0.048 *
---
Signif. codes:  0 `***' 0.001 `**' 0.01 `*' 0.05 `.' 0.1 ` ' 1
Arguments:
 Test statistics calculated assuming uncorrelated response (for faster
 computation)P-value calculated using 999 iterations via PIT-trap
 resampling.
```

This was over 10 times quicker than Code Box 14.7 (note that it used 10 times as many resamples), but the results are slightly different—the test statistic is slightly smaller and the P-value larger. *Why do you think this might be the case?*

The approach of Code Box 14.8 is quick (because it uses short format), and we will call it quick-and-dirty for two reasons. Firstly, unless counts are Poisson, it does not use the maximum likelihood estimator of α_{0i}, so in this sense it can be considered sub-optimal. Secondly, when resampling, the offset is not re-estimated for each resample when it should be, so P-values become more approximate. Simulations suggest this approach is conservative, so perhaps the main cost of using the quick-and-dirty approach is loss of power—test statistics are typically slightly smaller and less significant, as in Code Box 14.8. Thus the `composition` argument should be preferred, where practical.

Key Point

A key challenge in any multivariate analysis is understanding what the main story is and communicating it in a simple way. Some tools that can help with this include the following:

- Identifying a short list of *indicator taxa* that capture most of the effect.
- Visualisation tools—maybe an ordination, but especially looking for ways to see the key results using the raw data.

14.4 In Which Taxa Is There an Effect?

In Code Box 14.1, Anthony established that revegetation does affect invertebrate communities. This means that somewhere in the invertebrate community, there is evidence that some invertebrates responded to revegetation—maybe all taxa responded, or maybe just one did. The next step is to think about which of the response variables most strongly express the revegetation effect. This can be done by adding a `p.uni` argument as in Code Box 14.9.

Code Box 14.9: Posthoc Testing for Bush Regeneration Data

```
> an_reveg = anova(ft_reveg,p.uni="adjusted")
> an_reveg

Analysis of Deviance Table

Model: manyglm(formula = dat ~ treatment + offset(log(pitfalls)),
                    family = "negative.binomial")

Multivariate test:
            Res.Df Df.diff   Dev Pr(>Dev)
(Intercept)      9
treatment        8       1 78.25    0.022 *

Univariate Tests:
            Acarina          Amphipoda          Araneae           Blattodea
                Dev Pr(>Dev)       Dev Pr(>Dev)     Dev Pr(>Dev)        Dev
(Intercept)
treatment     8.538    0.208     9.363    0.172   0.493    0.979     10.679
                       Coleoptera          Collembola          Dermaptera
            Pr(>Dev)         Dev Pr(>Dev)        Dev Pr(>Dev)         Dev
(Intercept)
treatment      0.117       9.741    0.151      6.786    0.307       0.196
⋮
```

The `p.uni` argument allows univariate test statistics to be stored for each response variable, with *P*-values from separate tests reported for each response, to identify taxa in which there is statistical evidence of an association with predictors. The `p.uni="adjusted"` argument uses *multiple testing*, adjusting *P*-values to control family-wise Type I error, so that the chance of a false positive is controlled jointly across all responses (e.g. for each term in the model, if there were no effect of that term, there would be a 10% chance, at most, of at least one response having a *P*-value less than 0.1). The more response variables there are, the bigger the *P*-value adjustment and the harder it is to get significant *P*-values, after adjusting for multiple testing. It is not uncommon to get global significance but no univariate significance—e.g. Anthony has good evidence of an effect on invertebrate communities but can't point to any individual taxon as being significant. This comes back to one of the original arguments for why we do multivariate analysis (Chap. 11, introduction)—it is more efficient statistically than separately analysing each response one at a time.

The size of univariate test statistics can be used as a guide to *indicator taxa*, those that contribute most to a significant multivariate result. A test statistic constructed assuming independence (`cor.type="I"`) is a sum of univariate test statistics for each response, so it is straightforward to work out what fraction of it is due to any given subset of taxa. For example, the "top 5" taxa from Anthony's revegetation study account for more than half of the treatment effect (Code Box 14.10). This type of approach offers a short list of taxa to focus on when studying the nature of a treatment effect, which can be done by studying their coefficients (Code Box 14.10), plotting the subset, and (especially for smaller datasets) looking at the raw data.

Code Box 14.10: Exploring Indicator Taxa Most Strongly Associated with Treatment Effect in Anthony's Revegetation Data

Firstly, sorting univariate test statistics and viewing the top 5:

```
> sortedRevegStats = sort(an_reveg$uni.test[2,],decreasing=T,
                      index.return=T)
> sortedRevegStats$x[1:5]
 Blattodea Coleoptera  Amphipoda    Acarina Collembola
 10.679374   9.741038   9.362519   8.537903   6.785946
```

How much of the overall treatment effect is due to these five orders of invertebrates? The multivariate test statistic across all invertebrates, stored in `an$table[2,3]`, is 78.25. Thus, the proportion of the difference in deviance due to the top 5 taxa is

```
> sum(sortedRevegStats$x[1:5])/an_reveg$table[2,3]
[1] 0.5764636
```

So about 58% of the change in deviance is due to these five orders.

The model coefficients and corresponding standard errors for these five orders are as follows:

```
> coef(ft_reveg)[,sortedRevegStats$ix[1:5]]
                 Blattodea Coleoptera Amphipoda  Acarina Collembola
(Intercept)     -0.3566749  -1.609438 -16.42495 1.064711   5.056246
treatmentReveg -3.3068867   5.009950  19.42990 2.518570   2.045361
> ft_reveg$stderr[,sortedRevegStats$ix[1:5]]
                Blattodea Coleoptera Amphipoda   Acarina Collembola
(Intercept)     0.3779645   1.004969  707.1068 0.5171539  0.4879159
```

```
treatmentReveg 1.0690450   1.066918  707.1069 0.5713194  0.5453801
```
Note that a log-linear model was fitted, so the exponent of coefficients tells us the proportional change when moving from the control to the treatment group. For example, cockroach abundance (`Blattodea`) decreased by a factor of about $e^{3.3} = 27$ on revegetation, while the other four orders increased in abundance with revegetation. *Can you construct an approximate 95% confidence interval for the change in abundance of beetles (`Coleoptera`) on revegetation?*

Exercise 14.8: Indicator Species for Offshore Wind Farms?
Which fish species are most strongly associated with offshore wind farms in Lena's study?

Reanalyse the data to obtain univariate test statistics and univariate *P*-values that have been adjusted for multiple testing. Recall that the key term of interest, in terms of measuring the effects of offshore wind farms on fish communities, is the interaction between `Zone` and `Year`. *Is there evidence that any species clearly have a `Zone:Year` interaction, after adjusting for multiple testing? What proportion of the total `Zone:Year` effect is attributable to these potential indicator species?*

Plot the abundance of each potential indicator species against `Zone` and `Year`. *What is the nature of the wind farm effect for each species? Do you think these species are good indicators of an effect of wind farms?*

14.5 Random Factors

One limitation of design-based inference approaches like `mvabund` is computation time; another is difficulties dealing with *mixed models* to account for random factors (as in Chap. 6). There are additional technical challenges associated with constructing resampling schemes for mixed models, but the main obstacle is that resampling mixed models is computationally intensive, to the point that resampling a "manyglmm" function would not be practical for most datasets. A model-based approach, making use of hierarchical models, holds some promise, and we hope that this can address the issue in the near future.

14.6 Other Frameworks for Making Inferences About Community–Environment Associations

This chapter has focused on design-based inference using GEEs. What other options are there for making inferences about community–environment associations? A few alternative frameworks could be used; their key features are summarised in

Table 14.1. Copula models are mentioned in the table and will be discussed in more detail later (Chap. 17).

Table 14.1: Summary of the main differences in functionality of four frameworks for modelling multivariate abundances

Framework	Fast?	Ordination?	Composition?[a]	Co-occurrence?[b]
Hierarchical GLMs	×	✓	✓	✓
GEEs	✓✓	×	✓	×
Copulas	✓	✓	✓	✓
Dissimilarity-based algorithms	✓✓✓	✓	×	×

[a]That is, can they appropriately account for changes in sampling intensity, for valid inferences about community composition?
[b]That is, can they be used to study patterns in co-occurrence of taxa, e.g. by quantifying correlation

14.6.1 Problems with Dissimilarity-Based Algorithms

Multivariate analysis in ecology has a history dating back to the 1950s (Bray & Curtis, 1957, for example), whereas the other techniques mentioned in Table 14.1 are modern advances using technology not available in most of the twentieth century, and only actually introduced to ecology in the 2010s (Walker and Jackson, 2011; Wang et al., 2012; Popovic et al., 2019). In the intervening years, ecologists developed some algorithms to answer research questions using multivariate abundance data, which were quite clever considering the computational and technological constraints of the time. These methods are still available and widely used in software like PRIMER (Anderson et al., 2008), CANOCO (ter Braak & Smilauer, 1998), and free versions such as in the `ade4` (Dray et al., 2007) or `vegan` (Oksanen et al., 2017) packages.

The methods in those packages (Clarke, 1993; Anderson, 2001, for example) tend to be stand-alone algorithms that are not motivated by an underlying statistical model for abundance,[2] in contrast to GEEs and all other methods in this book (so-called model-based approaches). The algorithmic methods are typically faster than those using a model-based framework because they were developed a couple of decades ago to deal with computational constraints that were much more inhibiting than they are now. However, these computational gains come at potentially high cost in terms of statistical performance, and algorithmic approaches are difficult to reconcile conceptually with conventional regression approaches used elsewhere in ecology (Chaps. 2–11). So while at the time of writing many algorithmic techniques are still widely used and taught to ecologists, a movement has been gathering pace

[2] Although the methods in CANOCO have some connections to Poisson regression.

in recent years towards model-based approaches to multivariate analysis in ecology, whether a GEE approach or another model-based framework for analysis. A few of the key issues that arise when using the algorithmic approach are briefly reviewed below.

Recall that a particular issue for multivariate abundances is the *abundance* property, with strong mean–variance patterns being the rule rather than the exception and a million-fold range of variances across taxa not being uncommon (e.g. Fig. 10.3). Algorithmic methods were not constructed in a way that can account for the abundance property; instead, data are typically transformed or standardised in pre-processing steps to try to address this issue, rather than addressing it in the analysis method itself. However, this approach is known to address the issue ineffectively (Warton, 2018) and can lead to undesirable and potentially misleading artefacts in ensuing analyses (for example Warton et al., 2012b, or Fig. 14.2).

Another issue with algorithmic approaches is that because they lack an explicit mean model, they have difficulty capturing important processes affecting the mean, such as variation in sampling intensity. Adjusting for changes in sampling intensity is essential to making valid inferences about changes in community composition. It is relatively easy to do in a statistical model using an offset term (as in Sect. 10.5) or a row effect (Sect. 14.3), but algorithmic approaches instead try to adjust for this using data pre-processing steps like row standardisation. This can be problematic and can lead to counter-intuitive results (Warton & Hui, 2017), because the effects of changing sampling intensity can differ across datasets and depend on data properties (e.g. the effects of sampling intensity on variance are governed by the mean–variance relationship). Related difficulties are encountered when algorithmic approaches are used to try to capture interactions or random factors.

A final issue worthy of mention is that we always need to *mind our Ps and Qs*. But without a model for abundance, the assumptions of algorithmic approaches are not made explicit, making it harder to understand what data *p*roperties we should be checking. This also makes it more difficult to study how these methods behave for data with different types of data properties, but the results we do have are less than encouraging (Warton et al., 2012b; Warton & Hui, 2017).

14.6.2 Why Not Model-Based Inference?

Design-based inference was used in this chapter to make inferences from models about community–environment associations. As in Chap. 9, design-based inference is often used in place of model-based inference when the sampling distribution of a statistic cannot be derived without making assumptions that are considered unrealistic or when it is not possible to derive the sampling distribution at all. A bit of both is happening here, with high dimensionality making it difficult to specify good models for multivariate abundances and to work out the relevant distribution theory. However, progress is being made on both fronts.

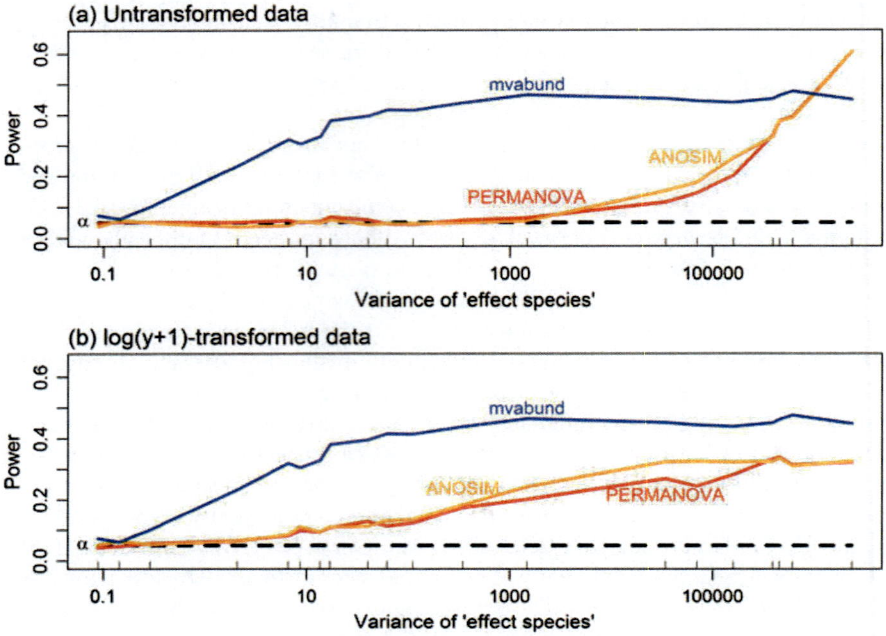

Fig. 14.2: Simulation results showing that while algorithmic approaches may be valid, they are not necessarily efficient when testing no-effect null hypotheses. In this simulation there were counts in two independent groups of observations (as in Anthony's revegetation study, Exercise 12.2), with identical means for all response variables except for one, which had a large (10-fold) change in mean. Power (at the 0.05 significance level) is plotted against the variance of this one "effect variable" when analysing (**a**) untransformed counts; (**b**) $\log(y + 1)$-transformed counts using dissimilarity-based approaches, compared to a model-based approach ("mvabund"). A good method will have the power to detect a range of types of effects, but the dissimilarity-based approaches only detect differences when expressed in responses with high variance

A specific challenge for model-based inference procedures is that the number of parameters of interest needs to be small relative to the size of the dataset. For a model that has different parameters for each taxon (as in Eq. 14.2), this means that the number of taxa would need to be small. For example, Exercise 11.6 took Petrus's data and analysed just the three most abundant genera of hunting spiders. This model used six parameters (in $\boldsymbol{\beta}$) to capture community–environment associations. If such an approach were applied to all 12 hunting spider species, 24 parameters would be needed to make inferences about community–environment associations, and standard approaches for doing this would not be reliable (especially considering that there are only 28 observations). A parametric bootstrap could be used instead, but as previously, this is computationally intensive and not a good match for a hierarchical GLM, unless you are very patient.

There are a few ways forward to deal with this issue that involve simplifying the regression parameters, $\boldsymbol{\beta}$, e.g. assuming they come from a common distribution (Ovaskainen & Soininen, 2011) or have reduced rank (Yee, 2006). Using hierarchical GLMs for inference about community–environment associations is an area of active research, but at the time of writing, issues with model-based inference had not been adequately resolved. Although they may well be soon!

Chapter 15
Predicting Multivariate Abundances

Exercise 15.1: Predicting Fish Communities at Wind Farms?
Consider again Lena's wind farm study (Exercise 14.3). She would like to predict what fish occur where.

What type of model should she use to get the best predictions?

Exercise 15.2: Which Environmental Variables Predict Hunting Spider Communities?
Petrus set up 28 pitfall traps on sand dunes in different environmental conditions (e.g. in the open, under shrubs) and classified all hunting spiders that fell into the trap to species. He measured six environmental variables characterising each site. He would like to know:

Which environmental variables best predict hunting spider communities?

What analysis approach should he use?

In Chap. 14 the focus was on using design-based inference to test hypotheses about environment–community associations and estimate model parameters with confidence. But recall that sometimes the primary goal is not inference about the effect of predictors. Other possible primary objectives are *model selection*, for example, to study which type of model best predicts abundances in a community (Exercise 15.1), or *variable importance*, studying which environmental variables are most useful for predicting abundance (Exercise 15.2). In these cases, the problem of interest can be framed in terms of prediction, and the techniques of Chap. 5 are applicable. However, there is a lot that can go wrong, so the model selection technique and the predictive model to be used should be chosen with care.

D. I. Warton, *Eco-Stats: Data Analysis in Ecology*, Methods in Statistical Ecology,
https://doi.org/10.1007/978-3-030-88443-7_15

15.1 Special Considerations for Multivariate Abundances

As usual, a suitable basis for choosing between competing models for data is to use an information criterion (such as AIC and BIC) or cross-validation, on a suitably chosen objective function. Like the models being fitted, the model selection method should be chosen in a way that accounts for the key properties of the data, in particular, the *abundance* and *multivariate* properties.

The *abundance* property of data can be accounted for using as the objective function the log-likelihood of a model that specifies an appropriate mean–variance relationship. If the log-likelihood is used in combination with (cross-)validation, this is known as *predictive likelihood*. In predictive likelihood, we estimate model parameters from training data, then use these to compute the likelihood for test data (Code Box 15.1). This method is related to AIC, having been shown to estimate the same quantity (Stone, 1977). The main distinction is that information criteria are a form of model-based inference, requiring the model to be (close to) correct, whereas cross-validation is a form of design-based inference, requiring only that training and test datasets be independent. While mean squared error was used previously (Sect. 5.2) to compare linear models, this does not account for the mean–variance relationship of abundance data, which can be very strong. If using mean squared error on multivariate abundances, the model would put undue weight on how well models predicted abundant (highly variable) taxa, at the expense of rare taxa. (And mean squared error on the scale of the linear predictor might do the opposite.)

Model selection can account for the *multivariate* property of data in different ways, depending on the approach taken. Information criteria, being a form of model-based inference, require a correlation to be accounted for in model specification. But correlation is not accounted for in the `manyglm` function, so while `AIC` can be computed in this case, it should only be considered as a very rough guide. Cross-validation, being a form of design-based inference, can handle the multivariate property by assigning *rows* of observations to training and test samples, keeping correlated observations at a site together in training or test sets. If there is additional dependence structure in the rows, correlated rows should be kept together also (as in Code Box 15.1). Cross-validation could be applied to `manyglm` or (in principle) any function capable of predicting new values.

A final important consideration when applying model selection to multivariate abundances is how to make predictions for rare taxa. There is little information in a rare taxon, so one should not expect good predictions from a model that separately estimates parameters for each response variable, as in Eq. 14.2. In fact, model selection approaches can return some strange answers for rare taxa if they aren't modelled carefully. For example, in cross-validation (Code Box 15.1, Exercise 15.3), if a taxon is completely absent from a training sample (or from a factor level in the training sample), `manyglm` would have a predicted mean of zero and near-infinite regression parameters (as in Maths Box 10.6). If such a taxon is then observed in the test data, it will have near-infinite predictive likelihood. This problem is a type of overfitting, and it is common when modelling multivariate abundances, especially with factor predictors. Information criteria also can behave strangely when modelling many responses if a model with many parameters is being used, for the reasons raised previously when discussing model-based inference in a different context (Sect. 14.6.2). The solution to these issues is to use a simpler model that *borrows strength* across taxa, as discussed in what follows.

Code Box 15.1: Predictive Likelihood for Wind Farm Data

To calculate a predictive likelihood on test data, we need to decide how many rows should be assigned to the training set and assign them in such a way that the training and test sets are (approximately) independent. Notice that because Lena collected the data in a paired design, with two sampling times at each station, random *stations* need to be chosen as the test sample, not random rows of data. The pair of rows in the dataset corresponding to each station are jointly allocated to test our training data at random, giving independent training and test sets if fish communities at different stations are independent. Half of the stations will be assigned to test data and half to training.

```
> library(ecostats)
> data(windFarms)
> set.seed(5) # use this seed to get the same results as below:
> nStations = length(levels(as.factor(windFarms$X$Station)))
> isTestStn = sample(nStations,nStations/2)
> isTest = windFarms$X$Station %in%
             levels(windFarms$X$Station)[isTestStn]
> library(mvabund)
> windMV = mvabund(windFarms$abund)
> windFt_Train=manyglm(windMV[isTest==FALSE,]~Year+Zone,
      data=windFarms$X[isTest==FALSE,],family="poisson")
> windFt_Int_Train=manyglm(windMV[isTest==FALSE,]~Year*Zone,
      data=windFarms$X[isTest==FALSE,],family="poisson")
> prWind_Test = predict(windFt_Train,newdata=windFarms$X[isTest,],
      type="response")
> prWind_Int_Test = predict(windFt_Int_Train,
      newdata=windFarms$X[isTest,], type="response")
> predLogL = dpois(windMV[isTest,],lambda=prWind_Test,log=TRUE)
> predLogL_Int = dpois(windMV[isTest,],lambda=prWind_Int_Test,
      log=TRUE)
> c(sum(predLogL),sum(predLogL_Int))
[1] -931.5643 -928.3885
```

What do these results tell us about which model is preferred?

Exercise 15.3: Cross-Validation for Wind Farm Data and Rare Species

Repeat the analyses of Code Box 15.1, but after removing rare species (observed less than 10 times), using the following code:

```
notRare=colSums(windMV>0)>10
windMVnotRare=mvabund(windFarms$abund[,notRare])
```

Did you get a similar answer?

Note that so far we have only considered one test sample, and there is randomness in the choice of training/test split. Repeat the analyses of Code

Box 15.1 as well as those you have done here, with and without rare species, multiple times (which is a form of *cross-validation*).

Which set of results tends to be more reliable (less variable)—the ones with or without the rare species? Why do you think this happened?

Key Point

Sometimes it is of primary interest to predict communities, or some community-level property, from multivariate abundance data. A good predictive model will include the following features:

- The model will account for the *multivariate* and *abundance* properties.
- It will *borrow strength* across taxa to use information in more abundant taxa to help guide predictions for rare taxa.
- It will be capable of handling *non-linear* responses of taxa to their environment.

15.2 Borrowing Strength Across Taxa

It is hard to reliably estimate environmental responses for rare taxa because there is little information in them that can be used for estimation. However, if model coefficients were estimated in a way that *borrowed strength* across responses, predictions for rarer taxa could be guided in part by other taxa, for which more information might be available.

When predicting multivariate abundances, it is a good idea to borrow strength across response variables to improve predictions for rare taxa. This is done by imposing some structure across taxa in some way (e.g. assuming they come from a common distribution) to shrink parameters for rare taxa towards values commonly seen for other taxa. Incidentally, a model that imposes structure across taxa will have fewer parameters in it, making it easier to interpret, and sometimes it can be used to better understand why different taxa respond to environment differently (Chap. 16).

One way to borrow strength is to shrink parameters towards a common value. Fitting a mixed model, with a random effect for each environmental coefficient that takes different values across taxa (Ovaskainen & Soininen, 2011, or see Code Box 15.2), will shrink environmental coefficients towards each other. Rarer taxa will tend to have their coefficients shrunk more strongly towards others. A LASSO is an alternative way to shrink parameters (Code Box 15.3, using `glmnet`), which will shrink parameters to exactly zero if they have little effect on response. An interesting prospect, little studied for multivariate abundance data, is the group LASSO (Yuan & Lin, 2006, Code Box 15.4, using the `grplasso` package), which groups parameters into different types and shrinks the whole group towards zero jointly. This is quite

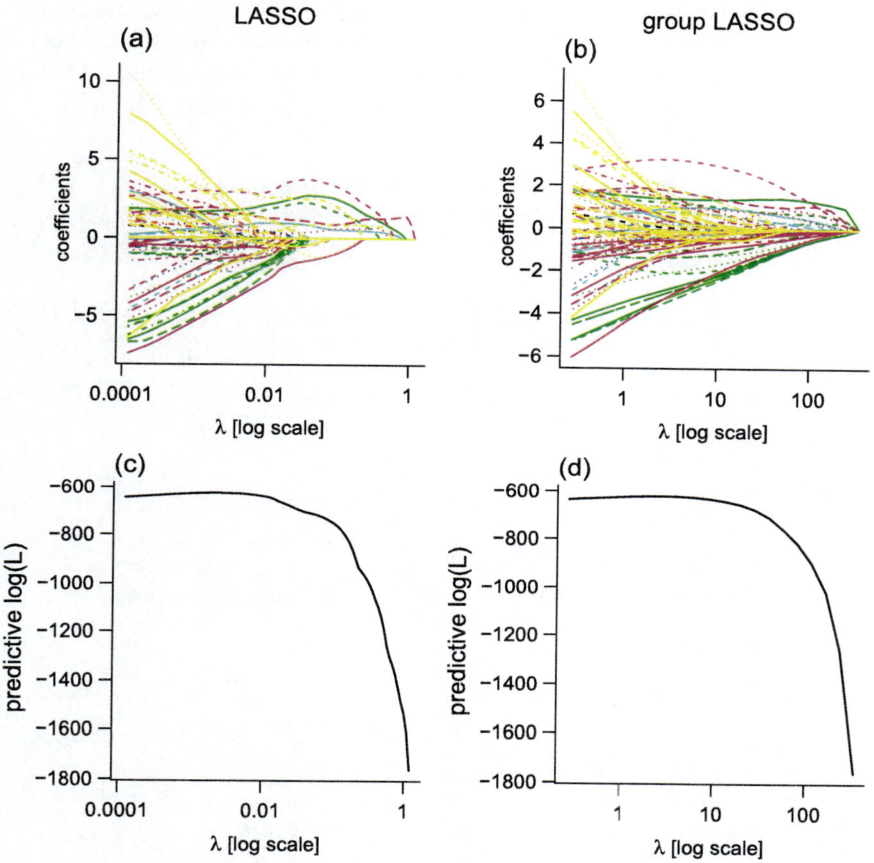

Fig. 15.1: LASSO analyses of Lena's wind farm data, including the regularisation path for (**a**) a LASSO fit (Code Box 15.3) and (**b**) a group LASSO (Code Box 15.4), and predictive likelihood for (**c**) the LASSO fit and (**d**) the group LASSO fit. Note in (**a**) and (**b**) that all slope parameters start at zero for large values of the LASSO penalty (λ) and fan out towards their unpenalised values (towards $\lambda = 0$). Different types of model terms (e.g. for `Zone`, `Year`, and their interaction) have been colour-coded; note that all parameters of the same type of model term enter at the same time in the group LASSO fit because they are penalised as a group. Note that (**c**) and (**d**) follow a J curve, but the curve is upside down (because we maximise likelihood, not minimise error) and flipped horizontally (because smaller models have larger λ). The optimal value for the LASSO penalty is $\lambda \approx 0.002$ and for the group LASSO $\lambda \approx 2$. These values are not on the same scale so are not directly comparable

a natural approach to variable selection in multivariate regression because it allows coefficients of less important environmental variables to be set to zero simultaneously for all taxa (Fig. 15.1b). The `glmnet` and `grplasso` packages, at the time of writing, could easily fit penalised logistic regression or Poisson regression. The `glm1path` function in the `mvabund` package was written to behave similarly to `glmpath`, but it can fit a negative binomial regression with a LASSO penalty. For a group LASSO, new family functions can be written into `grplasso` using the `grpl.model` function.

Code Box 15.2: Fitting a Mixed Model to Wind Farm Data

We first need data in long format, which can easily be constructed from `manyglm(..., comp=TRUE)`:

```
> windComp = manyglm(windMV~Zone*Year, data=windFarms$X,
                composition=TRUE)
```

and `windComp$data` stores the data in long format, with species in `cols`. We will fit a model where each term has an uncorrelated (hence `diag`) random effect across species:

```
> library(glmmTMB)
> wind_glmm = glmmTMB(windMV~Year*Zone+diag(Year*Zone|cols),
                 family=poisson(), data=windComp$data)
> summary(wind_glmm)
 Groups Name           Variance Std.Dev. Corr
cols    (Intercept)    5.3421   2.3113
        Year2010       1.6302   1.2768   0.00
        ZoneN          0.6807   0.8251   0.00 0.00
        ZoneS          3.9012   1.9751   0.00 0.00 0.00
        Year2010:ZoneN 0.8519   0.9230   0.00 0.00 0.00 0.00
        Year2010:ZoneS 0.3777   0.6146   0.00 0.00 0.00 0.00 0.00
Number of obs: 2864, groups:  cols, 16

               Estimate Std. Error z value Pr(>|z|)
(Intercept)    -3.18913    0.65414  -4.875 1.09e-06 ***
Year2010       -0.26989    0.45008  -0.600    0.549
ZoneN          -0.15541    0.35209  -0.441    0.659
ZoneS          -0.16619    0.61670  -0.269    0.788
Year2010:ZoneN  0.08998    0.41305   0.218    0.828
Year2010:ZoneS  0.56211    0.37247   1.509    0.131
```

The first table gives the random effect variances, and the second gives the fixed effects (the random effect means), e.g. the intercept terms across taxa are assumed to follow a normal distribution with mean −3.19 and standard deviation 2.31.

Exercise 15.4: Predictive Likelihood for Wind Farm Mixed Model

Calculate the predictive likelihood for the mixed model fitted to the wind farm data (Code Box 15.2), following the method in Code Box 15.1.

Did this have better predictive performance? Why do you think this happened?

Code Box 15.3: Fitting LASSO to Wind Farm Data via `glmnet`

We will use the data in long format from `windComp$data` (Code Box 15.2). Note, however, that the `glmnet` package currently doesn't take formula inputs—it requires a design matrix of predictors and a vector of responses as the first two arguments:

```
library(glmnet)
X = model.matrix(windMV~Year*Zone*cols,data=windComp$data)
y = windComp$data$windMV
windLasso = glmnet(X,y, family="poisson")
```

This fits a whole path of models, varying `lambda` (Fig. 15.1a). But which of these models has the best fit? We can use validation to look at this using the same test stations as in Code Box 15.1 (stored in `isTestStn`):

```
isTest = windComp$data$Station %in%
        levels(windComp$data$Station)[isTestStn]
windLassoTrain = glmnet(X[isTest==FALSE,],y[isTest==FALSE],
        family="poisson")
prLassoTest = predict(windLassoTrain,X[isTest,],type="response")
predLLlasso=colSums(dpois(windComp$data$windMV[isTest],prLassoTest,
        log=TRUE))
plot(windLassoTrain$lambda,predLLlasso,type="l",log="x")
isBestLambda = which(predLLlasso==max(predLLlasso))
```

Results are in Fig. 15.1b. The value of λ for the best-fitting model is stored in `windLassoTrain$lambda[isBestLambda]`, and coefficients for this model are stored in `coef(windLassoTrain)[,isBestLambda]`. This model includes 58 non-zero parameters, across main effects and interactions for different species.

Code Box 15.4: Fitting a Group LASSO to Wind Farm Data

As before, we will use data in long format stored in `windComp$data`. Using the `grplasso` package, we need to manually construct a range of values for the LASSO penalty parameter (`lambda`) to use in fitting. `lambdamax` finds the biggest possible penalty, corresponding to an intercept model, which we will multiply by `0.7^(0:19)`, to get 20 values reducing λ a thousand-fold, to fit a path all the way up to a full model:

```
library(grplasso)
windLambdaTrain = lambdamax(windMV~Year*Zone*cols,
   data=windComp$data, subset=isTest==FALSE,
   model = PoissReg()) * 0.7^(0:19)
windGrplasso = grplasso(windMV~Year*Zone*cols, data=windComp$data,
               lambda=windLambdaTrain, subset=isTest==FALSE,
               model = PoissReg())
```

This fits a whole path of models, varying `lambda`. Terms for `Year`, `Zone`, and their interaction are each grouped across species, so each enters the model for all species at the same time (Fig. 15.1c). Again, we can use validation to choose a model that predicts well using new data, with the same test stations as in Code Box 15.1 (stored in `isTestStn`):

```
prGrpTest = predict(windGrplasso,newdata=windComp$data[isTest,],
        type="response")
predLLgrplasso = colSums(dpois(windComp$data$windMV[isTest],
        prGrpTest,log=TRUE))
plot(windGrplasso$lambda,predLLgrplasso,log="x",type="l")
isBestLambdaGrplasso = which(predLLgrplasso==max(predLLgrplasso))
```

and the best-fitting value of `lambda` is `windLambdaTrain[isBestLambdaGrplasso]`, which ends up being about 2 (Fig. 15.1d). This model includes species-specific terms for all of `Year`, `Zone`, and their interaction.

Exercise 15.5: Comparing Predictive Likelihoods for Wind Farm Data
Compare the predictive log-likelihoods from Code Box 15.1 and Fig. 15.1c, d. These models were all applied using the same test dataset. *Which model seems to fit the data better? Why do you think this is?*

A very different way to borrow strength is classification—specifically, classifying taxa based on their environmental response (Chap. 16), an approach first applied in Dunstan et al. (2011). It is often referred to as a species archetype model (Hui et al., 2013), because it is assumed that there are a small number of so-called archetypal environmental responses, and each taxon follows one of these archetypes. These archetypal responses are estimated from the data, and all taxa are classified into one of these archetypes. This approach borrows strength across taxa because all taxa are used to estimate the archetypal responses—although abundant taxa, being more informative, will have more say in the shape of these estimated environmental responses. This has been shown to be an effective way to borrow strength for rare taxa (Hui et al., 2013)—we are not asking for a lot from rare taxa, because we don't need to estimate their environmental response, we only need to use them to classify rare taxa into an archetype.

If there are several environmental terms in the model, then another way to borrow strength is to use *reduced rank regression*—assuming the matrix of regression coefficients has low rank. One way to do this is to use the `rrvglm` function in `VGAM` (Code Box 15.5 Yee, 2006); it can also be done using the `Hmsc` package (Ovaskainen & Abrego, 2020, Section 9.5). This method is related to generalised linear latent variable models, except that the reduced rank assumption is on $\boldsymbol{\beta}$ rather than on $\boldsymbol{\Sigma}$ (in fact there are typically no random effects in this model, making it much faster to fit). Conceptually, this approach is also somewhat related to species archetype models because it also involves estimating a small number of archetypal environmental responses. The key difference is that it is then assumed that the linear predictor for each taxon will be a linear combination of archetypal predictions. As before, most of the work estimating archetypal responses is done by abundant taxa, so rarer taxa borrow strength from these. Reduced rank regression isn't useful if the number of parameters used to model environmental response is small because then the rank (which is the number of parameters per taxon) is already small.

Reduced rank regression can be used to construct a *constrained ordination*, which is an ordination in which axes are a linear function of predictors (environmental variables) rather than factors derived from responses (abundances). A popular multivariate tool in ecology for several decades has been canonical correspondence analysis (ter Braak, 1986, as in the CANOCO software), which enables constrained ordination using a weighted least-squares approximation to reduced rank regression. CANOCO ordination software has, however, been superseded by maximum likelihood approaches to reduced rank regression for count data (Yee, 2010).

Code Box 15.5: Reduced Rank Regression for Wind Farm Data

`VGAM` accepts data in short format. To fit a rank two Poisson reduced rank regression:

```
> library(VGAM)
> wind_RR2=rrvglm(as.matrix(windFarms$abund)~Year*Zone,
                  family=poissonff, data=windFarms$X, Rank=2)
```

We could compare this to a model fitted via `manyglm`:

```
> wind_manyglm = manyglm(windMV~Year*Zone, data=windFarms$X,
                                        family=poisson())
> c( BIC(wind_RR2), sum(BIC(wind_manyglm)))
[1] 2626.993 2742.920
```

Which model fits better, according to BIC?

Linear predictors for each "archetype", characterising how fish species typically respond, are obtained using `latvar`:

```
zoneyear = interaction(windFarms$X$Zone,windFarms$X$Year)
matplot(as.numeric(zoneyear),latvar(wind_RR2),pch=c(1,19))
```

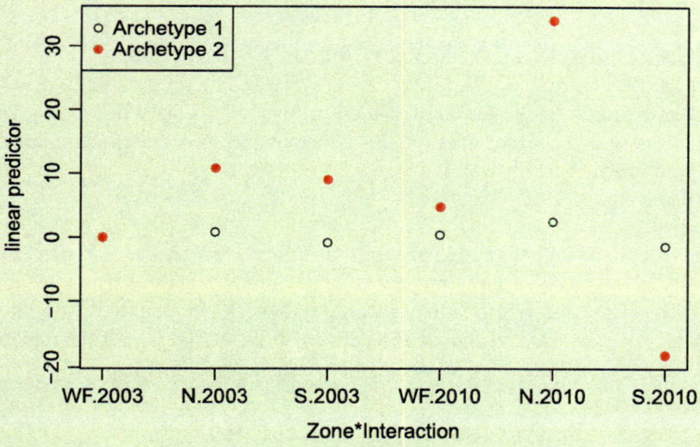

Note that this model does not detect a strong effect of wind farms and that some linear predictors take very high or very low values, suggesting overfitting.

15.3 Non-Linearity of Environmental Response and Interactions

Some studies (Elith et al., 2006, for example) suggest models predict better if they can handle non-linearity and interactions between environmental variables. Non-linearity is important to think about because many species have an optimal value for environmental variables, at which they can be found in high abundance, and they reduce in abundance as you move away from this optimum in either direction (a *uni-modal* response, ter Braak & Prentice, 1988). For example, thinking about the response of a species to temperature; every organism has a point where it is too hot and a point where it is too cold. This idea cannot be captured using linear terms alone, and at the very least a quadratic term would be needed to account for this, or some other method to handle non-linearity (such as in Chap. 8). One could argue that, as a rule, quantitative environmental variables should be entered into models for species response in a manner that can handle non-linear responses, and uni-modal responses in particular (ter Braak & Prentice, 1988).

Generalised additive models (Chap. 8) are another possible way to handle non-linearity, but this method is quite data hungry and doesn't always hold up well in comparisons of predictive performance for this sort of data (Norberg et al., 2019). Alternatives include the `mistnet` package (Harris, 2015), which can fit a hierarchical model via artificial neural networks to presence–absence data, and the `marge` package (Stoklosa & Warton, 2018), a generalised estimating equation (GEE) extension of multivariate adaptive regression splines ("MARS", Friedman, 1991, also see Elith & Leathwick, 2007). There has been a boom in machine learning methods recently, and many of these are capable of flexibly handling non-linear responses, such as artificial neural networks (Olden et al., 2008) and deep learning (Christin et al., 2019, 2021).

Code Box 15.6: Using the LASSO for Petrus's Spider Data

We will fit a negative binomial regression to Petrus's spider data, but with a LASSO penalty to force parameters to zero if they are not related to response. This will be done using the `traitglm` function, which coerces data into long format and stores coefficients in a matrix:

```
library(mvabund)
data(spider)
# fit model:
spid.trait = traitglm(spider$abund,spider$x,method="cv.glm1path")
```

The `method="cv.glm1path"` argument meant that the model was fitted using the `glm1path` function, a LASSO algorithm written into `mvabund` that works for negative binomial regression. Cross-validation was used to choose the value of the LASSO penalty parameter.

To plot coefficients in a heat map where white means zero (Fig. 15.2):

```
library(lattice)
a         = max( abs(spid.trait$fourth.corner) )
colort    = colorRampPalette(c("blue","white","red"))
plot.4th = levelplot(t(as.matrix(spid.trait$fourth.corner)),
                xlab="Environmental Variables", ylab="Species",
                col.regions=colort(100), at=seq(-a, a, length=100),
                scales = list( x= list(rot = 45)) )
print(plot.4th)
```

15.4 Relative Importance of Predictors

Which predictors tend to have the strongest effects on community abundance? This question asks us to quantify *variable importance*, which can be addressed as previously (Sect. 5.7), e.g. using a leave-one-out change in deviance (via the `drop1` function) or looking at the size of standardised coefficients (as in Code Box 15.6). Other options are to use a mixed model, with a random effect on environmental coefficients that takes a different value for each taxon, and to look at the size of variance components for standardised predictors (Code Box 15.7). The larger the variance

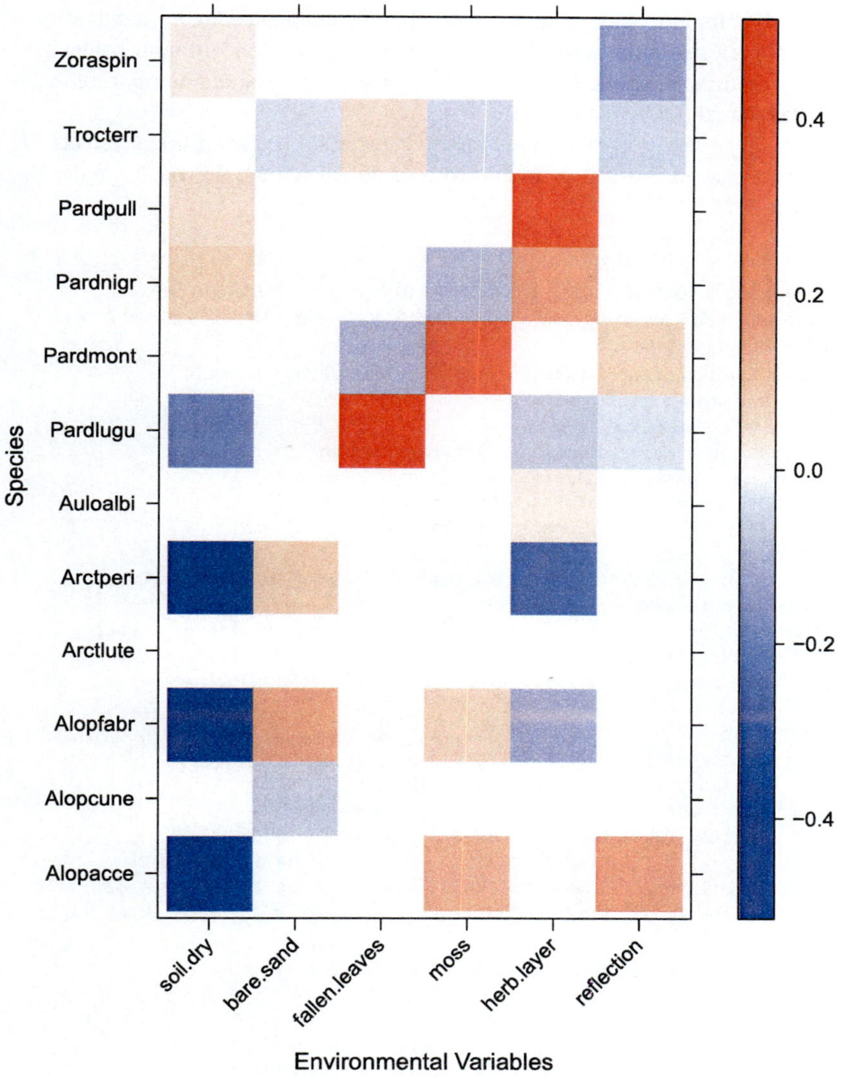

Fig. 15.2: Heat map of standardised coefficients from negative binomial LASSO regression of Code Box 15.6. Darker coefficients correspond to stronger effects. *Which environmental variables seem to have the strongest effect on spider abundances?*

component, the more the standardised coefficients vary across taxa, hence the greater the variation in community composition due to that predictor. In Code Box 15.7, this approach would suggest that soil dryness is the most important predictor of changes in hunting spider composition, with a variance component of 1.84, almost three times larger than for any other predictor. Note that predictors need to be standardised prior to this analysis—otherwise the slope coefficients are on different scales for the different predictors, so their variance components are, too, and comparing their relative sizes would not be meaningful.

Note that all of the aforementioned methods estimate the conditional effects of variables, and careful thought is required to establish if that is what is really of interest (Sect. 5.7).

Code Box 15.7: Mixed Model Prediction of Spider Abundances

To fit a mixed model, shrinking coefficients together across taxa, we will first put data in long format, then use glmmTMB:

```
> spidXY = data.frame(scale(spider$x),spider$abund) # scale
  standardises data!
> library(reshape2)
> spiderLong = melt(id=1:6,spidXY,variable.name="cols")
> Xformula = paste(colnames(spider$x),collapse="+")
> fullFormula = formula(paste0("value~cols+",Xformula,
                 "+(",Xformula,"|cols)"))
> library(glmmTMB)
> spid_glmm = glmmTMB(fullFormula,family=nbinom2(),data=spiderLong)
> summary(spid_glmm)
Formula: value ~ cols + soil.dry + bare.sand + fallen.leaves + moss +
    herb.layer + reflection + (soil.dry + bare.sand + fallen.leaves +
    moss + herb.layer + reflection | cols)

 Groups Name          Variance Std.Dev. Corr
 cols   (Intercept)   0.6198   0.7873
        soil.dry      1.8397   1.3564   -0.88
        bare.sand     0.1567   0.3959   -0.10 -0.08
        fallen.leaves 0.4022   0.6342    0.67 -0.67 -0.68
        moss          0.2354   0.4852   -0.70  0.40  0.51 -0.59
        herb.layer    0.6003   0.7748   -0.91  0.82 -0.20 -0.40  0.47
        reflection    0.6677   0.8172    0.15 -0.52  0.33  0.20  0.40 -0.33

Overdispersion parameter for nbinom2 family (): 1.51

              Estimate Std. Error z value Pr(>|z|)
...
soil.dry       0.65262    0.42430   1.538 0.124020
bare.sand      0.06150    0.15994   0.385 0.700601
fallen.leaves -0.35286    0.26457  -1.334 0.182309
moss           0.13402    0.18688   0.717 0.473289
herb.layer     1.10662    0.27995   3.953 7.72e-05 ***
reflection    -0.03521    0.29688  -0.119 0.905585
```

Which predictors seem to be most important to spider community composition?

Chapter 16
Understanding Variation in Environmental Response Across Taxa

When studying how a community responds to its environment, it is typically the case that different taxa will respond in different ways. An important challenge for the ecologist is to go deeper (Shipley, 2010; McGill et al., 2006), to look for patterns in environmental responses, and, where possible, to *capture the mechanisms* by which taxa vary in their environmental response (as in Exercise 16.1). What are the main types of response to environmental gradients? Why do taxa differ in their environmental response?

In this chapter, we will study two types of techniques intended to better characterise variation in environmental response across taxa—using classification to group taxa by environmental response, and functional traits to look for predictors that explain variation in response.

Exercise 16.1: Understanding How Spiders Vary in Environmental Response
Consider again Petrus's hunting spider data of Exercise 15.2. He established (e.g. in Code Box 15.7) that different species respond differently to environmental gradients, especially so for soil dryness, herb layer, and reflectance. He would like to know what the main types of environmental response were, across species.

What approach should he use to answer this question?

16.1 Classifying Species by Environmental Response

When looking to develop a deeper understanding of how taxa vary in environmental response (Exercise 16.1), one possible approach is to start with an assumption that

D. I. Warton, *Eco-Stats: Data Analysis in Ecology*, Methods in Statistical Ecology,
https://doi.org/10.1007/978-3-030-88443-7_16

there are a (relatively) small number of different types of environmental response and to classify each taxon into one of these response types. This is a type of *classification* problem. Classification has a long history in multivariate analysis (Lance & Williams, 1967), especially in ecology (Williams & Lambert, 1966), although the older algorithmic approaches are not really suited to the problem of classifying species based on environmental response. Instead, we will use a type of *mixture model*, pioneered by Dunstan et al. (2011), often referred to as a species archetype model (after Hui et al., 2013).

Key Point
To study *how* taxa differ in environmental response, a used tool is to classify them into *archetypes* based on their environmental response. This can help characterise the main ways taxa differ in their environmental responses.

16.1.1 A Brief Introduction to Mixture Models

A finite mixture model with G components assumes that each observation comes from one of G distinct component distributions, but we don't know in advance which observation comes from which component, and estimate this from the data. We make some assumptions about the form of the component distributions (e.g. normal, Poisson, . . .), but estimate their parameters from the data, as well as estimating the overall ("prior") proportion of observations falling in each component and the "posterior" probability that each observation belongs to any given component. Hence, mixture models are a form of *soft classification*, with observations assigned a probability of belonging to each of the G components, rather than being simply assigned to one component (hard classification). Those familiar with Bayesian analysis will know the terms "prior" and "posterior", but the term is being used in a different context here, and a finite mixture model is not necessarily fitted using Bayesian techniques. In fact, mixture models are most often fitted by maximum likelihood.

We will use mixture models primarily for *classification*, where we wish to classify observations according to how well they fit into each of several component distributions. In particular, we want to classify taxa based on how well their environmental response falls into one of several categories of response type.

There are other reasons, beyond classification, you might want to fit mixture models. In particular, they are a flexible method of *fitting distributions* to data, capable of generating weird distributions for weird data. The most familiar example of this in ecology is *zero-inflated distributions* (Welsh et al., 1996), count distributions with extra zeros in them, often fitted as a two-component mixture of a count distribution and a distribution that is guaranteed to take the value zero (*degenerate* at zero). Another important reason for fitting a mixture model is as a simple way to model *heterogeneity*, assuming all observations come from one of a few distinct "types". An example of this is in capture–recapture modelling (Pledger, 2000), to account for differences in the capture probability of different individuals (e.g. due to differences in behaviour). To read more about mixture models and other ways they are used, see McLachlan and Peel (2000).

"Mixture model" sounds a lot like "mixed model", but it is important not to confuse the two; they are quite different models.[1]

Maths Box 16.1: Mixture Models Have Weird Likelihood Surfaces
A two-component species archetype model, as a function of one predictor x_i at site i, assumes the abundance of taxon j (y_{ij}) is drawn from an exponential family (or related) distribution with mean either m_{ij1}, if it comes from component 1, or m_{ij2}, if it comes from component 2, where

$$g(m_{ijk}) = \beta_{0j} + x_i\beta_k$$

for some link function $g(\cdot)$ and $k \in \{1, 2\}$. Taxa are assumed to be independently drawn from component 1 (with probability π) or 2 (with probability $1 - \pi$). If we also assume that abundances are independent (conditional on group membership), the log-likelihood function is

$$\ell(\boldsymbol{\beta}) = \sum_{i=1}^{p} \log\left(\pi \prod_{i=1}^{N} f(y_{ij}; m_{ij1}) + (1-\pi)\prod_{i=1}^{N} f(y_{ij}; m_{ij2})\right)$$

which tends to be a weird function that is difficult to maximise.

One difficulty is *label switching*—the component labels are interchangeable. That is, if (β_1, β_2) is a solution, with probability π of being in component 1, then (β_2, β_1) is also a solution, with probability $1-\pi$ of being in component 1. So the likelihood surface will have two maxima, meaning it also has an intervening minimum, as below (for the spider data, with *local maxima* at -2.3 and -1.8).

[1] Although technically they are actually related, a mixture model is a type of mixed model where the random effect is not normally distributed; instead it has a multinomial distribution that takes G different values.

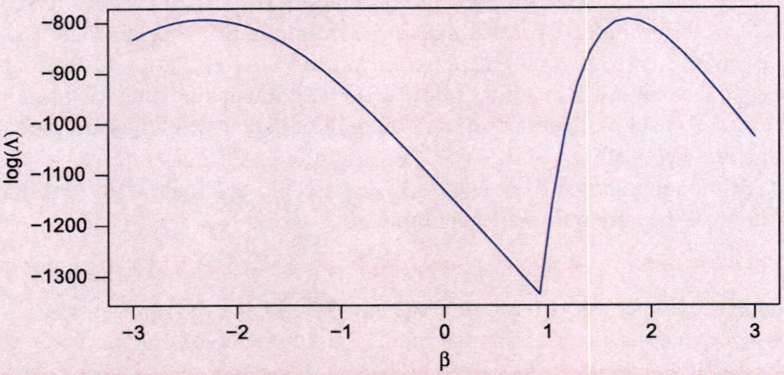

Previously we maximised the likelihood function by finding its stationary point and solving the subsequent score equations (Maths Box 10.2). But now there is more than one stationary point, and some may not be maxima. To have a better chance of finding the *global maximum*, we should give the search algorithm thoughtful initial guesses for parameters (*starting values*) or try multiple runs with different starting values and keep the solution with the largest likelihood. Or try a bit of both.

16.1.2 Species Archetype Models to Classify Taxa

A *species archetype model* fits a mixture model to classify taxa based on their environmental response (Dunstan et al., 2011), as in Fig. 16.1. This is a special type of mixture model for the regression setting, known as a *finite mixture of regressions*, where we assume that observations come from one of a few distinct regression lines. Instead of having to estimate and study a different environmental response for every taxon, we can focus our efforts on estimating and characterising a smaller number of so-called archetypal environmental responses.

A species archetype model is a mixture of generalised linear models (GLMs), where we mix on the regression parameters, assuming the following:

- Each taxon is independently drawn from one of G archetypes. The (prior) probability for the kth component is π_k.
- The observed y_{ij}-values (i for observation/replicate, j for taxon) are *independent*, conditional on the mean m_{ijk} for archetype k.
- Conditional on their mean m_{ijk}, the y-values come from *a known distribution* (from the exponential family) with known *mean–variance relationship* $V(m_{ijk})$.
- Within each archetype, there is a straight-line relationship between *some known function of the mean* of y and each x, with a separate intercept for each taxon β_{0j}:

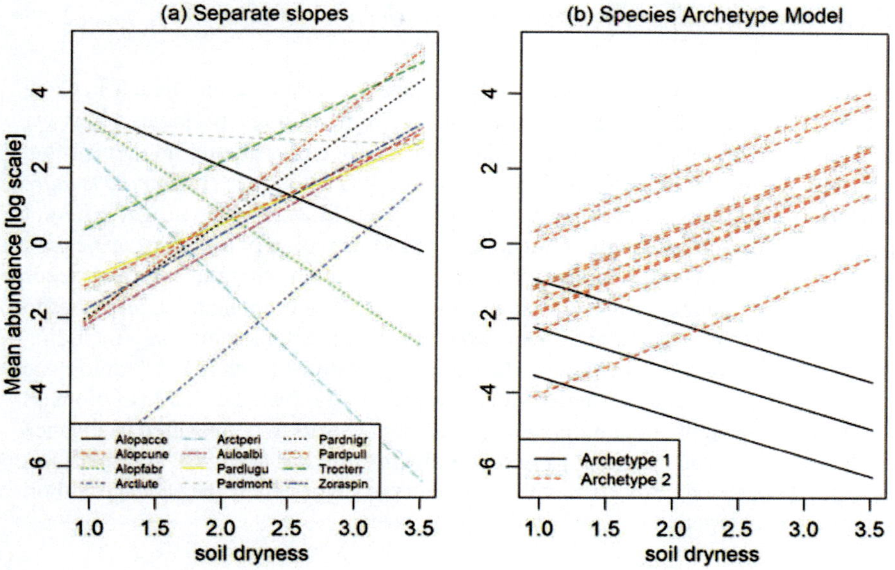

Fig. 16.1: Understanding variation in environmental response using a mixture model for Petrus's hunting spider data. (**a**) A separate regression model for each species gives 12 different lines to interpret. (**b**) A species archetype model (fitted using `SpeciesMix`) assumes there is a small number of archetypal responses to soil dryness (in this case, two), and each species is classified into one of these archetypes. Notice results are roughly concordant across the two plots, but much simplified in (**b**), where spider species have essentially been classified as "increasers" or "decreasers" in response to soil dryness

$$g(m_{ijk}) = \beta_{0j} + \mathbf{x}_i^T \boldsymbol{\beta}_k \tag{16.1}$$

It is important to pay attention to the subscripts in the preceding terms—note in particular that there is a subscript k in the $\boldsymbol{\beta}_k$ of Eq. 16.1, which means that regression slopes are being estimated separately for each archetype (k), not for each taxon (j). Hence we assume that there are only a small number of forms of environmental response, "archetypes", and each taxon has a response that tends to follow one (or possibly a mixture) of these. In contrast, the subscript j in β_{0j} indicates that the intercept term is allowed to take different values for different taxa. So we have a model along the lines of Fig. 16.1b, with common slopes across groups of taxa sharing the same archetype, but different intercepts. Note also that in Eq. 16.1, the mean for the ith observation from the jth taxon is referred to as m_{ijk}, because it refers to the mean under the assumption that this taxon belongs in archetype k. (The actual mean is a weighted average of the m_{ijk}, weighted by the probability that taxon j belongs in archetype k.)

16.1.3 Fitting a Species Archetype Model

Example code for fitting a species archetype model to Petrus's spider data is in Code Box 16.1. This uses the `species_mix` function in the `ecomix` package, which fits models by maximum likelihood. The likelihood surface of a mixture model is often quite bumpy, making it hard to find the maximum. It is a good idea to *fit models more than once* from different starting points (as in Sect. 12.3), and if you get different answers, use the one with highest likelihood. The `species_mix` function can be used as for most R regression functions; the only trick to note is that the response needs to be entered in matrix format. As usual, a family for the component GLMs needs to be specified via the `family` argument, which takes character vectors only including `"negative.binomial"` for negative binomial regression, `"bernoulli"` for logistic regression of presence-absence data, and `"tweedie"` for Tweedie GLMs of biomass data. Finally, the number of component archetypes G needs to be specified in advance (via the `nArchetypes` argument) and defaults to three. A `control` argument has been specified in Code Box 16.1, with parameters as suggested by the package author, to improve convergence for small and noisy datasets.

Code Box 16.1: Fitting a Species Archetype Model to Petrus's Spider Data

The `species_mix` function from the `ecomix` package will be used to fit a species archetype model to Petrus's spider data to classify species by environmental response:

```
> library(mvabund)
> library(ecomix)
> data(spider)
> SpiderDF=data.frame(spider$x)
> SpiderDF$abund=as.matrix(spider$abund)
> spiderFormula = abund ~ soil.dry + bare.sand + fallen.leaves + moss +
                  herb.layer + reflection
> ft_Mix = species_mix(spiderFormula, data=SpiderDF,
    family="negative.binomial", nArchetypes=2,
    control=list(init_method='kmeans',ecm_refit=5, ecm_steps=2) )

SAM modelling
There are 2 archetypes to group the species into.
There are 28 site observations for 12 species.
...
iter  60 value 777.946121
iter  70 value 777.933389
final  value 777.933345
converged
```

The values reported are negative log-likelihood (so the smaller, the better). This function doesn't always converge to a global maximum, so try several times and stick with the fit with the lowest likelihood. You can ignore errors earlier in the output if the final model converges.

```
> coef(ft_Mix)$beta
              soil.dry   bare.sand fallen.leaves       moss herb.layer reflection
Archetype1  1.5627792 -0.05472319    -0.2784647 -0.1450093  0.7081597 -0.4730827
Archetype2 -0.1070417  0.22425755    -0.2854891  0.4187408  0.5896159  0.4760607
```

Which predictors vary the most across archetypes (hence across taxa)? Is this what you saw in Code Box 15.7?

Key parts of the output to look at are the model coefficients (available as usual via the `coef` function), in particular the regression coefficients for archetypes (stored as `coef(ftMix)$beta` for a fitted model `ftMix`), the posterior probabilities of group membership (`ftMix$tau`, which are often mostly zero and one), and the intercept terms (`coef(ftMix)$alpha`).

16.1.4 Mind Your Ps and Qs

In terms of research *q*uestions, species archetype models are best used for exploratory work, understanding the reasons why environmental response varies across taxa, as for Exercise 16.1. They can also be used for prediction, given their ability to borrow strength across taxa to improve predictions for rare taxa (Hui et al., 2013).

Recall that multivariate abundances always have two key data *p*roperties, a multivariate assumption (abundances are correlated across taxa) and an abundance assumption (mean–variance relationship). A GLM approach is used to handle the abundance assumption, and standard GLM tools can be used to check this assumption, and linearity, as in Code Box 16.2. However, note that species archetype models lack any correlation in the model and instead assume (conditional) independence of abundances across taxa. This assumption is rarely reasonable, so community-level inferences from these models should be treated with caution. But there is no issue if using the method for exploratory purposes or for prediction of abundances in each taxon since the prediction of individual response variables is usually little affected by correlation across responses.

Code Box 16.2: Minding Your Ps and Qs for Petrus's Species Archetype Model

Taking the fit from Code Box 16.1 we can just ask for a `plot`:

```
> plot(ft_Mix, fitted.scale="log")
```

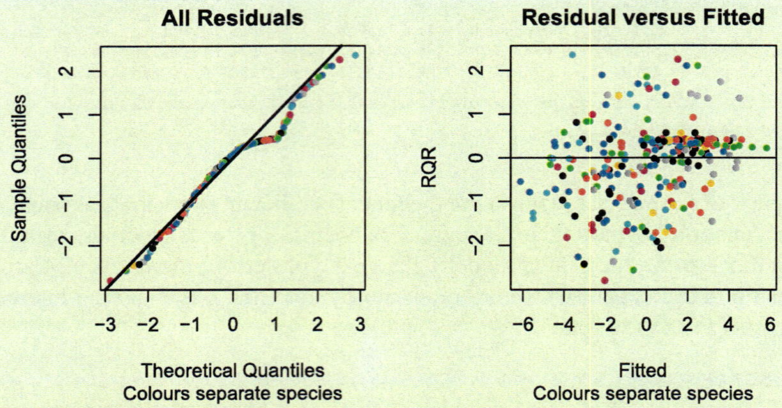

Do model assumptions seem reasonable here?

Exercise 16.2: Archetypal Revegetation Response
Consider again Anthony's revegetation study. We would like to better characterise how different invertebrate orders respond to the revegetation treatment. Fit a species archetype model with two archetypes. *Explain what the main two types of response to revegetation were, across invertebrate taxa.*

Look at the posterior probabilities for each order, identify a few taxa that characterise each response type, and plot the raw data for these taxa. *Does the species archetype model seem to capture the main trends in response to revegetation treatment?*

16.1.5 Choosing the Number of Archetypes

To fit a finite mixture model, the number of archetypes G needs to be specified in advance. What value of G should be used? This is a model selection problem, so the approaches of Chap. 5 apply. AIC, however, should not be used to guide the choice of G because some of the theory that is needed to derive it is not satisfied for mixture models (McLachlan & Peel, 2000). BIC is a good option and has been shown to consistently estimate G (Keribin, 2000), meaning that as the sample size gets larger, the chance that BIC will choose the correct value of G goes to 100%, if mixture model assumptions are satisfied. But as mentioned previously, species archetype models make an assumption of independence of responses that is unlikely to be satisfied, a possible effect of which is to overfit models. Design-based approaches such as cross-validation could be used as an alternative for inferences robust to violation of this assumption.

Code Box 16.3 fits models with different numbers of archetypes to Petrus's spider data and plots BIC as a function of number of archetypes.

Code Box 16.3: Choosing the Number of Archetypes for Petrus's Spider Data

```
nClust=rep(2:6,3)
bics   = rep(NA, length(nClust))
for(iClust in 1:length(nClust))
{
>  fti_Mix = species_mix(spiderFormula, data=SpiderDF,
      family="negative.binomial", nArchetypes=nClust[iClust],
      control=list(init_method='kmeans',ecm_refit=5, ecm_steps=2))
}
plot(bics~nClust, ylab="BIC", xlab="# archetypes")
```

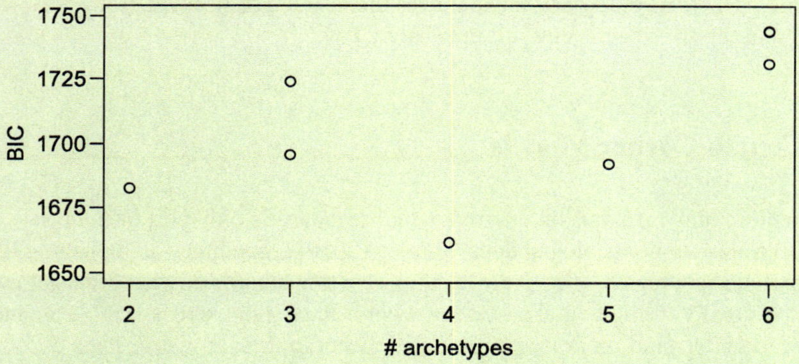

How many archetypes would you use?

Note that repeat runs didn't always give the same BIC. *Why did this happen?*

Exercise 16.3: How Many Revegetation Archetypes?
Consider again a species archetype model for Anthony's revegetation data. In Exercise 16.2, we assumed there were only two archetypes. *How many archetypes should be used for Anthony's data?* Answer this question using the model selection technique of your choice.

16.1.6 Other Multivariate Mixture Models

The `species_mix` function classifies taxa based on their environmental response. You can also cluster observations based on their composition (Foster et al., 2017) using the `regional_mix` function in the same package. This function fits a so-called mixture-of-experts model (McLachlan & Peel, 2000), which uses environmental variables as predictors on the posterior probabilities of group membership. This approach was designed for mapping bioregions, such as vegetation types. The process of vegetation mapping typically involves

- collecting multivariate abundances of vegetation from many different sites and using them to define vegetation types according to which sites are classified
- mapping these vegetation types out at the landscape scale, usually informed by maps of environmental variables that are thought to be strongly associated with different vegetation types.

A mixture-of-experts approach can build a model that does both of these steps at once!

Shirley Pledger (Victoria University of Wellington) and collaborators have developed some software for clustering species and sites jointly ("biclustering", Pledger & Arnold, 2014), currently downloadable as the `clustglm` package from `http://homepages.ecs.vuw.ac.nz/~shirley/`.

16.2 Fourth Corner Models

Documenting how taxa vary in environmental response is only part of the battle. If that is all a community ecologist does, Shipley (2010) argued they are behaving like a "demented accountant", effectively keeping separate records of what different taxa do, without really trying to make sense of what is happening across taxa. In science we want to understand processes, meaning we want to look at *why* species differ in their environmental response (as in Exercise 16.4).

Exercise 16.4: Understanding Why Spiders Vary in Environmental Response

Consider again Petrus's hunting spider data of Exercise 15.2. We would like to understand why species differ in their environmental responses. Data are available on body size and colour. He wants to know the extent to which species traits explain interspecific variation in environmental response.

What approach should he use to answer this question?

Regression is a key tool for answering *why* questions, because you can introduce predictors to try to explain why a response varies. Here we need predictors across taxa, the columns of the multivariate abundance dataset, in order to explain why taxa have different patterns of environmental response. These are commonly referred to as species traits or functional traits, and it has been argued that they should have a much greater role in community ecology (McGill et al., 2006; Shipley, 2010). Because we are using traits to understand changes in environmental response, we are specifically interested in whether species traits *interact* with environmental variables in predicting abundance.

Key Point
To study *why* taxa differ in environmental response, a useful tool is to measure functional traits for different taxa and predict abundance as a function of these traits, environmental variables, and their interaction (a *fourth corner model*). Of primary interest is the fourth corner, the matrix characterising how environment and traits interact, which is simply the matrix of environment–trait interaction coefficients.

Shipley et al. (2006) proposed a maximum entropy technique, later referred to as community assembly by trait selection (CATS, Shipley 2010), to understand why abundance varies across taxa as a function of their traits, at a single site. Warton et al. (2015) showed this is equivalent to a Poisson GLM to predict abundance at a site as a function of traits. But what is needed in Exercise 16.4 are methods to predict abundance across multiple sites simultaneously, as a function of environment as well as traits.

An early approach to this problem labelled it the *fourth corner problem* (Legendre et al., 1997), seeing the issue as being that we have three matrices (environmental variables, abundance, and traits), and what we care about is a fourth matrix, relating environment to traits, as in Fig. 16.2a. The `fourthcorner` function in the `vegan` package implements the methods of Legendre et al. (1997), which do not deal explicitly with the abundance property, although they have some connections to model-based methods (Brown et al., 2014; ter Braak, 2017).

An especially helpful way of viewing this problem is to formulate it as a regression model for multivariate abundance (Pollock et al., 2012; Jamil et al., 2013; Brown et al., 2014). We can write a model much like what we saw previously, but now including traits ($\mathbf{z}$) as an additional predictor in the model and, importantly, their interaction with environmental variables ($\mathbf{x}$) in predicting the mean abundance at site i of taxon j:

$$g(\mu_{ij}) = \beta_{0j} + \boldsymbol{x}_i'\boldsymbol{\beta}_{(x)} + \boldsymbol{z}_j'\boldsymbol{\beta}_{(z)} + \left(\boldsymbol{x}_i \otimes \boldsymbol{z}_j\right)'\boldsymbol{\beta}_{(x\times z)} + \boldsymbol{x}_i'\boldsymbol{b}_j \qquad (16.2)$$

The term $\mathbf{x} \otimes \mathbf{z}$ denotes the interaction, which captures the idea that environmental response varies across taxa due to their traits. Importantly, the matrix of interaction coefficients $\boldsymbol{\beta}_{(x\times z)}$ can be understood as the fourth corner (Fig. 16.2b). Thus, this type of regression model could be described as a *fourth corner model*. In principle, any regression framework could be used to construct a fourth corner model by incorporating trait variables and their interaction with other predictors. For example, a fourth corner generalised latent variable model could be constructed by adding latent variables to the mean model in Eq. 16.2. To date, software for resampling-

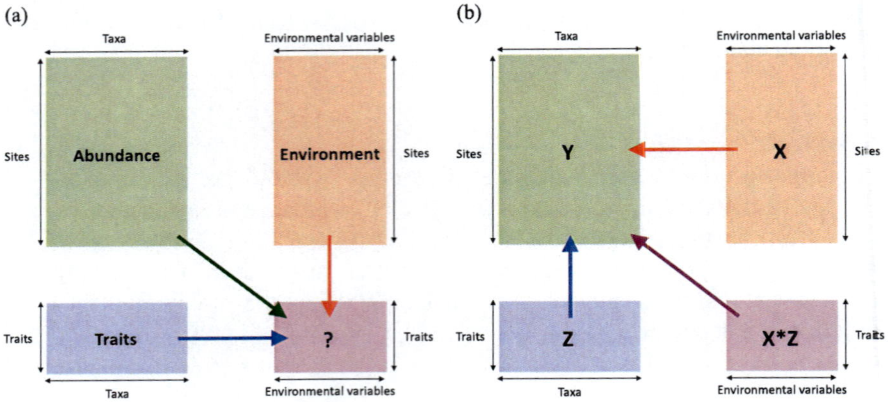

Fig. 16.2: Schematic diagrams illustrating approaches to the fourth corner problem. (**a**) The problem as presented in Legendre et al. (1997), as matrices for multivariate abundance, environmental variables, and functional traits, with the goal being to estimate a matrix connecting environment and traits (labelled with a question mark). (**b**) A regression model for this problem views the multivariate abundances as the response (**Y**), environmental variables (**X**), and species traits (**Z**) as predictors, and the matrix of interaction coefficients for $\mathbf{X} \times \mathbf{Z}$ as the fourth corner matrix

based testing (Wang et al., 2012) and ordination (Hui, 2016; Ovaskainen et al., 2017b; Niku et al., 2019) has fourth corner functionality.

Figure 16.3 illustrates conceptually the relationship between fourth corner models and some other models described in this text that treat abundance as the response (as in Warton et al., 2015). In particular, the essential distinction from multivariate regression models introduced in Chap. 14 is the use of predictors on columns of abundance (traits) as well as on rows (environment).

The main effect term for traits $\boldsymbol{z}_j' \boldsymbol{\beta}_{(z)}$ in Eq. 16.2 can usually be omitted. This main effect term estimates changes in total abundance across responses due to traits, but β_{0j} already ensures that responses can differ from each other in total abundance. In practice, the main effect for traits is usually left out of the model, although it could remain if a random effect were put on the β_{0j}. If a row effect were included in the model (as in Sect. 14.3), then the main effect for environment could similarly be omitted from the model.

In Eq. 16.2, the $\boldsymbol{x}_i' \boldsymbol{b}_j$, where $\boldsymbol{b}_j$ are typically drawn from a (multivariate) normal distribution, allow taxon j to vary in environmental response for reasons not explained by its traits ($\mathbf{z}_j$). It is important to include this term when making inferences about environment–trait associations, because it is unrealistic to expect all variation in environmental response across taxa to be explained by the traits included in the model. Incorporating this term requires the use of regression software with mixed model functionality or perhaps penalised likelihood estimation (e.g. using a LASSO).

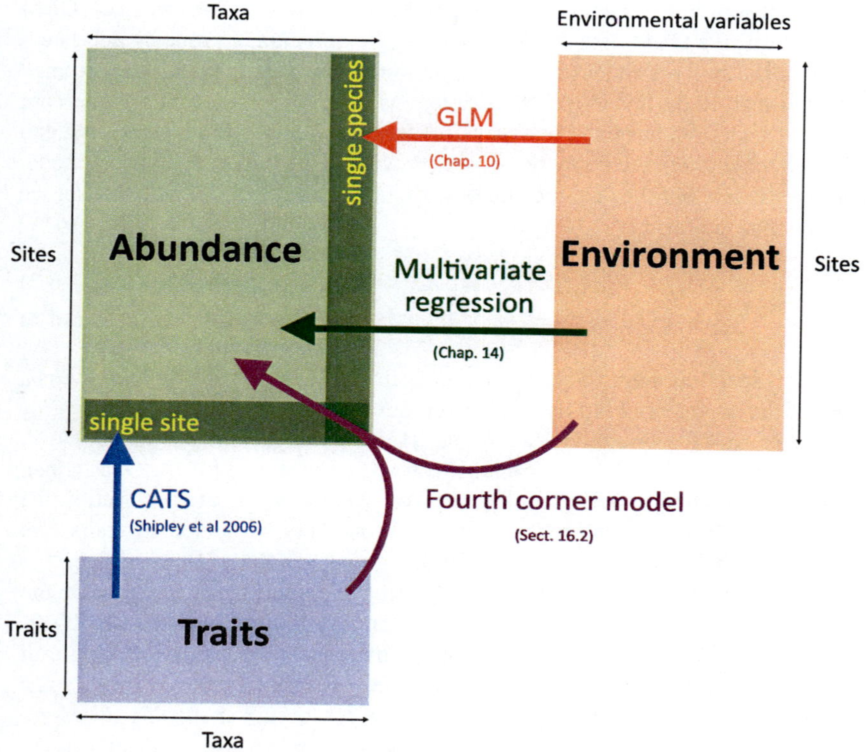

Fig. 16.3: Diagram showing interrelationship between fourth corner models and other methods we have seen. A GLM, like for David and Alistair's crab data in Chap. 10) and multivariate regression (as for Anthony's data in Chap. 14) both relate abundance to environmental variables. Now we are adding a new type of predictor, *Traits*, which acts across columns (*Taxa*) rather than across rows (*Sites*). The interaction between environmental and trait variables explains differences in environmental response across taxa

16.2.1 Fitting Fourth Corner Models

A fourth corner model can be fitted using any software suitable for analysing multivariate abundance data, capable of including predictors that operate across columns (taxa) as well as down rows (sites). While some of the original fourth corner models were fitted using standard mixed modelling software (`lme4` was used in Jamil et al., 2013), it is important to also account for the *multivariate* property (Chap. 14), that abundances are correlated across taxa. In the following we will focus on using a generalised linear latent variable model (as in Chap. 12) for this reason.

A generalised linear latent variable model is a type of hierarchical GLM (Sect. 11.3) using a GLM data model to handle the *abundance* property and latent variables in the process model to handle the *multivariate* property. As mentioned previously (Sect. 12.3), this model more or less uses a multivariate random intercept to capture correlation, but fitted in a way that uses much fewer parameters, so it can still fit even when there are many responses. Previously we fitted this type of model without predictors, but they can be added without extra difficulty. So the model we will fit now has the form

$$y_{ij} \sim F(m_{ij}, \phi_i) \text{ such that } Var(y_{ij}) = V(m_{ij}, \phi_j) \tag{16.3}$$

$$g(m_{ij}) = g(\mu_{ij}) + \mathbf{z}_i'\boldsymbol{\Lambda}_j \tag{16.4}$$

where $g(\mu_{ij})$ specifies a fourth corner model as in Eq. 16.2, and as previously F is a member of the exponential family of distributions (Maths Box 10.1).

Code Box 16.4 fits this model using the `gllvm` package (Niku et al., 2019) in R; another candidate is the `Hmsc` package (Ovaskainen et al., 2017b; Ovaskainen & Abrego, 2020, Section 6.3). Recall that taxon-specific terms are required in the model to handle variation in environmental response not explained by traits (the $\boldsymbol{x}_i'\boldsymbol{b}_j$ from Eq. 16.2), done using the `randomX` argument in the `gllvm` package.

Fourth corner models can also be fitted using the `traitglm` function of the `mvabund` package, but it is best used as an exploratory tool only. ter Braak (2017) showed that `traitglm` has problems making inferences about environment–trait associations, which arises because taxon-specific terms are not included in `traitglm` when a trait argument (`Q`) has been specified. The reason taxon-specific terms are omitted is that `anova` calls in the `mvabund` package make use of design-based inference (as in Chap. 14), which has difficulty dealing with random effects.

Code Box 16.4: A Fourth Corner Model for Spider Data Using `traitglm`

We will fit a fourth corner model using the `gllvm` package. The `randomX` argument is included to capture species-specific environmental responses not explained by other predictors. Only soil dryness and herb cover are included as predictors to capture the main environmental trends. Note that with only 12 species, there is not enough information in the data to estimate random effects across many traits.

```
library(gllvm)
data(spider)
X = spider$x[,c("soil.dry","herb.layer")]
ft_trait = gllvm(spider$abund, X, spider$trait,
   randomX=~soil.dry+herb.layer, family="negative.binomial")
logLik(ft_trait)
```

The log-likelihood bounces around a little, but for a good fit it should be greater than −721.

Fourth corner coefficients, stored in `ft_trait$fourth.corner`, capture interactions between environmental and trait variables. They can be plotted to study patterns in environmental response across species that can be explained by traits.

```
library(lattice)
a        = max( abs(ft_trait$fourth.corner) )
colort   = colorRampPalette(c("blue","white","red"))
```

```
plot.4th = levelplot(ft_trait$fourth.corner, col.regions=colort(100),
           at=seq(-a, a, length=100), scales = list( x=list(rot = 45)) )
print(plot_4th)
coefplot(ft_trait)
```
which returns a plot along the lines of Fig. 16.4.

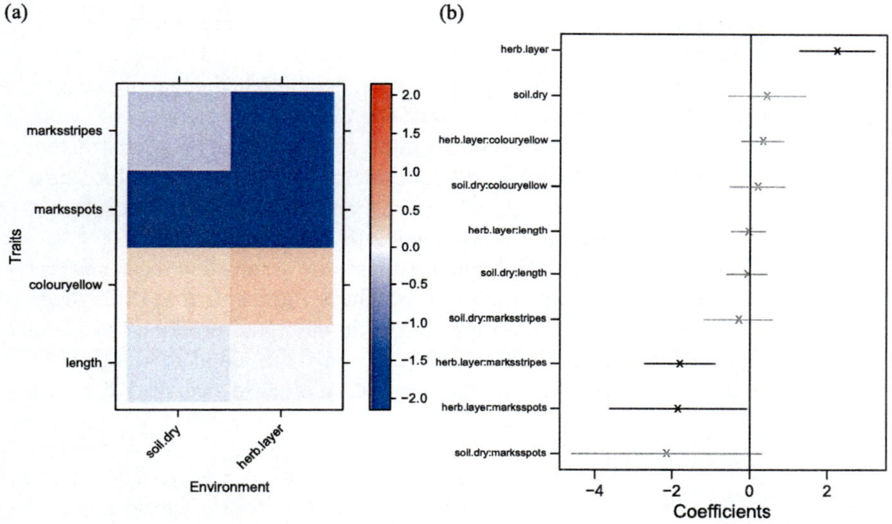

Fig. 16.4: Fourth corner coefficients from model fitted in Code Box 16.4, plotted as follows: (**a**) heat map of fourth corner; (**b**) confidence intervals for all coefficients (Eq. 16.2). Fourth corner coefficients explain how the environmental response of different taxa varies with their traits. For example, the negative interaction coefficient between soil dryness and spots suggests that as soil gets drier, spiders with spots on them are less likely to be found

Exercise 16.5: Heloise's Ants

Heloise collected counts of 41 species of ants at 30 sites across south-eastern Australia, stored as `antTraits` in the `mvabund` package. She also recorded five environmental variables at each site and five functional traits for each species. She would like to know if these traits explain why ant species differ in environmental response.

Fit a fourth corner model to this dataset. Don't forget to mind your Ps and Qs!

What are the key traits that capture why (and how) ants differ in environmental response?

16.2.2 Visualising Fourth Corner Interactions

Two ways to visualise fourth corner coefficients are presented in Fig. 16.4. However, often it is hard to understand an interaction from the coefficients alone, and it is advisable to look for ways to plot how abundance varies with key combinations of environmental variables and traits. A standard tool for this purpose, if at least one of the predictors is categorical, is to use an interaction plot (Code Box 16.5). Currently these plots must be produced manually using the `predict` function. Looking at this plot, we see clearly that, when moving along a gradient towards increasing soil dryness, we expect to see fewer spiders with spots and more spiders with stripes. There were only two spiders in the dataset with spots, and both had negative species-specific slopes (Fig. 16.1a, `Alopacce`, and `Arctperi`).

If both predictors are quantitative, then an interaction plot can't be constructed, unless one of these is binned into categories. An alternative would be to plot a heat map of predicted values as a function of the trait and environmental variable of interest (over the range of these values that was jointly observed).

An issue to keep in mind is that if different responses are assumed to have different intercepts, predictions for different responses will have intercepts that differ from each other in fairly arbitrary ways. So, for example, in the figure produced by Code Box 16.5, each line is for a different response, so only the slope of the lines is meaningful, but the positions of the lines relative to each other in a vertical direction are not.

Code Box 16.5: A Fourth Corner Interaction Plot for Petrus's Spider Data

To manually construct a fourth corner interaction plot, to study how the response of spider abundance to soil dryness varies for spiders with different markings on them:

```
nVars = dim(spider$abund)[2]
newTraits = spider$trait
# set factors not of interest here to be a constant value
newTraits$length= mean(spider$trait$length) #set length to its mean
newTraits$colour=factor(rep(levels(spider$trait$colour)[1],nVars),
   levels=levels(spider$trait$colour)) #set to first level of factor
# set starting rows of 'marks' to take all possible values
nMarks = nlevels(spider$trait$marks)
newTraits$marks[1:nMarks]=levels(spider$trait$marks)
# create a new env dataset where the only thing that varies is soil.dry:
newEnv = spider$x[1:2,c("soil.dry","herb.layer")]
newEnv[,"soil.dry"]=range(scale(spider$x[,"soil.dry"]))
newEnv[,"herb.layer"]=0
#make predictions and plot:
newPreds = predict(ft_trait,newX=newEnv,newTR=newTraits,type="response")
matplot(newEnv[,1], newPreds[,1:nMarks],type="l", log="y")
legend("topright",levels(newTraits$marks),lty=1:nMarks,col=1:nMarks)
```

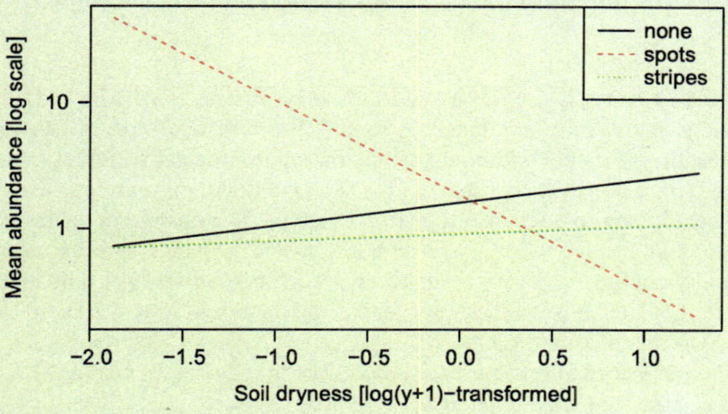

Intercept terms are arbitrary (species-specific), and only the slopes on this plot are meaningful. Spider species with spots tend to decrease in abundance on drier soil, whereas there seems to be little change in response to soil dryness otherwise.

Exercise 16.6: A Fourth Corner Interaction Plot for Heloise's Ants
Recall the fourth corner model you fitted to Heloise's data in Exercise 16.5.

For an interaction of your choice, construct a plot to visualise how the environment–abundance association changes for different values of the functional trait.

16.2.3 Quantifying the Importance of Traits in a Fourth Corner Model

In Exercise 16.4, Petrus wants to quantify the extent to which traits explain variation in environmental response. This can be done by fitting three models—a *main effects model*, capturing effects of environmental variables on total abundance, a fourth corner model, and a *species-specific model*, which lets different species have different environmental responses. Following the ideas in Sect. 14.3, the main effects model can be understood as capturing effects of environmental variables on α-diversity. The additional terms in the species-specific model capture β-diversity along these environmental gradients, and the fourth corner model attempts to capture this β-diversity using traits. A simple way to quantify how effectively it does this is to fit all three models and compare their deviances, e.g. using the `anova` function in the `gllvm` package (Code Box 16.6).

The proportion of β-diversity deviance explained, as previously, is a type of R^2 measure, and, like any such measures, it does not account for model complexity. This means that as more traits are added to the model, the proportion of deviance explained will increase. To account for this, one could use cross-validation to instead compute the proportion of β-diversity deviance explained *at new sites*.

Code Box 16.6: Quantifying How Effectively Traits Explain β-Diversity

To what extent is β-diversity explained by traits? We will use three `gllvm` models to look at this question—a model with main effects for environmental variables (capturing α-diversity, `ft_main`), a fourth corner model (`ft_trait` but without `randomX`), and a model with species-specific environmental responses (`ft_spp`). The change in deviance between the main effects and species models measures the extent to which species response and, hence, community composition vary along these gradients (β-diversity). The fourth corner model is an intermediate between these two, capturing the extent to which these β-diversity patterns are due to the measured traits.

```
> ft_spp = gllvm(spider$abund, X, family="negative.binomial")
> ft_trait = gllvm(spider$abund, X, spider$trait,
                                   family="negative.binomial")
> ft_main = gllvm(spider$abund, X, spider$trait,
              family="negative.binomial", formula=~soil.dry+herb.layer)
> an_spider4th = anova(ft_main, ft_trait, ft_spp)
Model  1 :  ~ soil.dry + herb.layer
Model  2 :  y ~ soil.dry + herb.layer + (soil.dry + herb.layer):(length
              + colour + marks)
Model  3 :  y ~ X
> an_spider4th
  Resid.Df         D Df.diff     P.value
1      287  0.00000       0
2      279 40.12997       8 3.02998e-06
3      265 20.84413      14    0.105694
> an_spider4th$D[2]/sum(an_spider4th$D)
[1] 0.6581477
```

So about two-thirds of the variation in environmental response ($40.1/(40.1 + 20.8) \approx 66\%$) could be explained by these traits. This was achieved using about a third of the available parameters (8 rather than $8 + 14 = 22$). The non-significant P-value when comparing the fourth corner model to the species model suggests that there is no evidence of β-diversity across environments that remains unexplained by traits.

Exercise 16.7: Variation Explained by Traits for Heloise's Ants

Consider again Heloise's ant data.

What proportion of the variation in environmental response is explained by the measured species traits?

Chapter 17
Studying Co-occurrence Patterns

While the previous two chapters focused on studying community-environment associations, here we will focus on characterising associations between taxa within communities, as in Exercise 17.1.

When the focus is on co-occurrence, attention moves to a different part of the fitted model. Irrespective of the precise modelling framework, the key components of the model are the parameters characterising community–environment associations, which we have called $\boldsymbol{\beta}$, and parameters characterising associations between taxa within communities, which we have called $\boldsymbol{\Sigma}$. For example, the mean model of a hierarchical GLM has the form

$$g(m_{ij}) = \beta_{0j} + \mathbf{x}_i^T \boldsymbol{\beta}_j + \epsilon_{ij}$$

where the process model errors $\boldsymbol{\epsilon}_i = (\epsilon_{i1}, \ldots, \epsilon_{ip})$ satisfy

$$\boldsymbol{\epsilon}_i \sim MVN(\mathbf{0}, \boldsymbol{\Sigma})$$

The previous two chapters focused on the regression coefficients $\boldsymbol{\beta}$. In this chapter, we will focus on the variance–covariance matrix $\boldsymbol{\Sigma}$.

The variance–covariance matrix $\boldsymbol{\Sigma}$ can be understood as capturing co-occurrence patterns—positive correlations across taxa indicate that they co-occur more often than expected by chance. Negative correlations across taxa indicate that the taxa co-occur less often than expected by chance.

Exercise 17.1: Co-occurrence in Hunting Spider Data
Petrus would like to think about the co-occurrence of hunting spider species. He wants to know which species co-occur and the extent to which these co-occurrence patterns can be explained by joint response to environmental gradients.

What approach should he used to look at these questions?

D. I. Warton, *Eco-Stats: Data Analysis in Ecology*, Methods in Statistical Ecology,
https://doi.org/10.1007/978-3-030-88443-7_17

There are multiple reasons taxa may co-occur more or less often than expected by chance. Broadly, this may be due to interactions across taxa (e.g. predation, facilitation), or it may be because of shared responses to another variable. We will further break down these potential causes of co-occurrence as follows:

1. The taxa may interact directly.
2. The taxa may both respond to a common *mediator* taxon.
3. Both taxa may respond to a common environmental variable.

The modelling tools presented in this chapter can isolate some of the effects of (3) or potentially (2), but only in the case where the environmental variable of interest has been included in the model and its effect has been estimated correctly. A model fitted to observational data can never tell the difference between species interaction (1) and shared response to some unmeasured or imperfectly modelled predictor (3). It is important to keep this qualification in mind when studying co-occurrence, because a couple of things we know for sure in ecology are that organisms respond to their environment and that our models are never quite right. There should always be an expectation that some taxa will respond to common aspects of their environment in ways that are not perfectly captured by the model, which induces a form of correlation across taxa that we cannot distinguish from direct interaction across taxa.

Key Point
There are three reasons taxa may co-occur:

1. They may interact directly.
2. They may both respond to some mediator taxon.
3. They may respond to a common environmental variable.

We can model species correlation to tease apart these three sources, and quantify the extent to which (2) and (3) drive co-occurrence. However, a challenge is that models aren't perfect and there will probably be missing environmental variables, or predictors whose effect is not estimated correctly by the model. This means that we can't use models to fully account for (2) and (3), so we can never really tell from an observational study whether co-occurrence is due to species interaction or missing predictors (or other sources of model misspecification).

A distinction was made in the preceding discussion between a common response to another taxon (2) vs an environmental variable (3) because these sources of co-occurrence are stored in different parts of the model, and different techniques are required to quantify each of them. Specifically, because abundance of taxa is the response ($\mathbf{y}$), if the mediator taxon is one of the responses, then information about its role in driving co-occurrence patterns is captured in the structure of the variance–covariance matrix of the response ($\boldsymbol{\Sigma}$). However, if taxa are correlated because they both respond to the same environmental variable, this can be accounted for by regressing responses directly against the relevant environmental predictors (and is captured in $\boldsymbol{\beta}$).

In this chapter, two main modelling tools will be considered—latent variable models, which can tease apart the effects of environmental variables (3), and graphical models, which can additionally identify the effects of (2).

17.1 Copula Frameworks for Modelling Co-occurrence

Table 14.1 mentioned four main frameworks for analysing multivariate abundances. Algorithmic distance-based approaches to analysis, used widely in the past, compromise performance in search of short computation times and can no longer be recommended. Generalised estimating equations (GEEs) are useful for fitting models quickly, for example, in combination with computationally intensive techniques for design-based inference (Chap. 14). Hierarchical GLMs (Chap. 11) have been used for ordination (Sect. 12.3) and fourth corner modelling (Sect. 16.2). They could also be used here, but we will instead use another modelling framework—copulas—whose main advantage over hierarchical approaches is in computation time.

A *copula model* is a type of marginal model, like generalised estimating equations (GEEs). The word *copula* comes from the idea that the model *couples* together a marginal model for data with a covariance model, as explained below. Compared to GEEs, a copula has the advantage that it is a fully parametric model, meaning likelihood-based or Bayesian inference is an option for a copula model. This comes at some cost in terms of computation time, so copula models tend to be slower to fit than GEEs, but they are typically much faster than hierarchical models. They are used in this chapter because flexible tools for modelling co-occurrence have been proposed in a Gaussian copula framework (Popovic et al., 2018).

The idea of a Gaussian copula model is to map abundances from their marginal distribution to a standard normal *copula variable* and then analyse the copula variables assuming they are multivariate normal. If data were continuous, we would construct copula values from a marginal model for the jth response at the ith site by solving

$$\Phi(z_{ij}) = F(y_{ij})$$

where $F(\cdot)$ is the cumulative distribution function of y_{ij}, and $\Phi(\cdot)$ is the cumulative distribution function of the standard normal distribution. Abundances are not continuous, so we will need to use a discrete copula model, defined as follows:

$$F(y_{ij}^{-}) < \Phi(z_{ij}) \leq F(y_{ij}) \tag{17.1}$$

$$\mathbf{z}_i \sim MVN(\mathbf{0}, \boldsymbol{\Sigma}) \tag{17.2}$$

where y_{ij}^- is the previous value of y_{ij}. The copula values z_{ij} are not observed, so they are treated as random effects. Mathematically, this means the likelihood function has an integral in it, which slightly complicates estimation.

The *covariance model* tells us the form of $\boldsymbol{\Sigma}$ to be used in Eq. 17.2. This is like an error term, capturing co-occurrence patterns that are not explained elsewhere in the model. In this chapter we will use two covariance modelling approaches: a latent variable model, which assumes $\boldsymbol{\Sigma}$ has reduced rank; and a graphical model, which assumes many of the responses are conditionally independent of each other. Because copula values have been mapped to the standard normal, with variance one, $\boldsymbol{\Sigma}$ is actually a correlation matrix.

A *marginal model* is needed to determine the form of cumulative distribution function $F(y_{ij})$ to be used in Eq. 17.1. This is the part of the model where environmental variables can be introduced, to capture their effects on abundance. In principle, any marginal model can be used; the approach taken here will be to assume each response follows a generalised linear model (or related method) and to check these assumptions as appropriate, along the lines of Chaps. 10 and 14. So an example model for the abundance of taxon j at site i is

$$y_{ij} \sim F(\mu_{ij}, \phi_j) \text{ such that } Var(y_{ij}) = V(\mu_{ij})$$
$$g(\mu_{ij}) = \beta_0 + \boldsymbol{x}_i'\boldsymbol{\beta}_j$$

The marginal model is not the main focus of the chapter, but it is important as always to specify it appropriately in order to make meaningful inferences about co-occurrence.

It is fairly common to estimate the marginal model for y separately from the covariance model for z, as a two-step process. This is an approximation, which ignores any interplay between the two parts of the model. However, algorithms that estimate both models jointly are much more complicated to work with and tend to perform only slightly better (if at all). In this chapter we will make use of the `ecoCopula` package, which uses the two-step process to maximum likelihood estimation but phrases it as a type of Monte Carlo expectation-maximisation (EM) algorithm (Popovic et al., 2018) that could be applied to combine (in principle) any parametric marginal model with any covariance modelling tool designed for multivariate normal data. This algorithm uses Dunn-Smyth residuals to map observed data onto the standard normal distribution, but using multiple sets of Dunn-Smyth residuals (which vary across runs due to jittering) and weighting them according to how well each residual fits a multivariate normal distribution.

17.1.1 *Mind Your Ps and Qs*

Each of Eqs. 17.1–17.2 involves making an assumption. Firstly, Eq. 17.1 requires an assumption that the marginal model for abundances is correct in order to be able to transform abundances from the marginal model to standard normal copula values

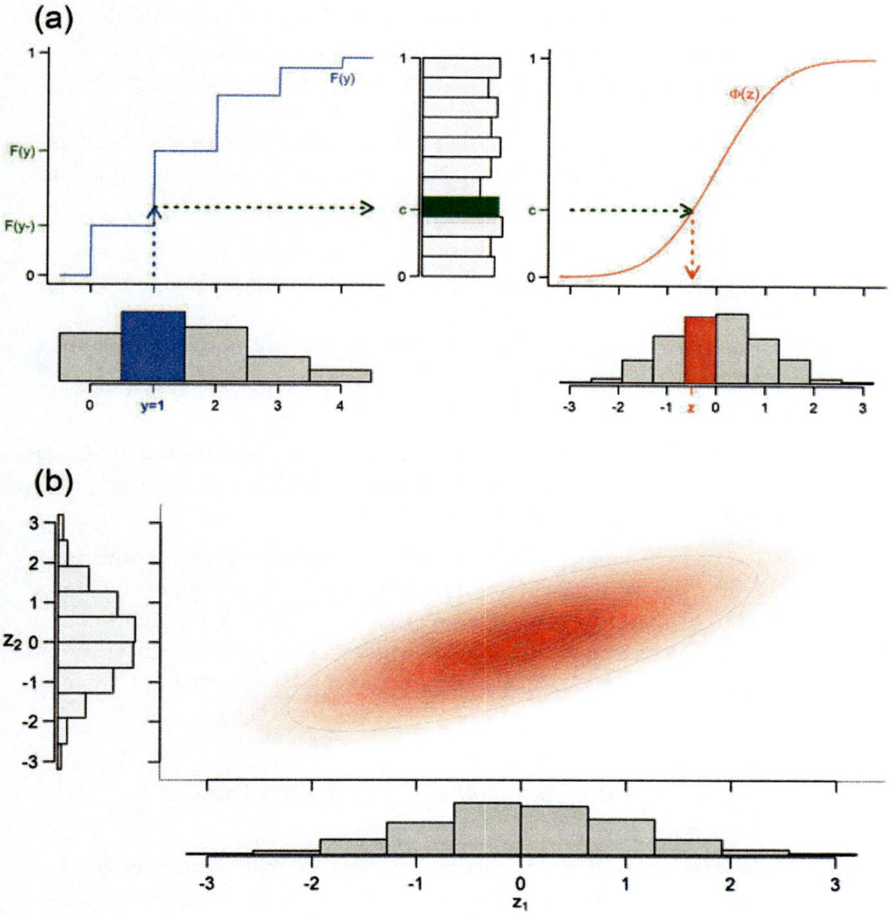

Fig. 17.1: Schematic of how a Gaussian copula model is constructed. (**a**) In a Gaussian copula model, copula values (z, *right*) are computed by equating values from the distribution function of abundance y under the assumed marginal model ($F(y)$, blue step curve, *left*) with the standard normal distribution function for copula values ($\Phi(z)$, red curve, *right*). (**b**) Copula values are then assumed to be multivariate normal. If data are discrete, the cumulative distribution function of the data (a, left) is a step function, and we need to choose a value (c) along the relevant step to use in generating z. While an initial guess could be chosen at random, as for Dunn-Smyth residuals (Fig. 10.7), the copula model actually suggests some values are better than others, which needs to be accounted for in model fitting. If the assumed marginal model is correct, the values of $F(y)$ (c, *centre*) are uniformly distributed between zero and one, and values of z will be standard normal

z_{ij}, e.g. by computing Dunn-Smyth residuals as in Fig. 17.1. Secondly, it is not necessarily the case that marginally normal variables are multivariate normal. So asserting the multivariate normality of the z_{ij}, as in Eq. 17.2, is itself an assumption.

The precise assumptions that are made in a copula model thus depend on the type of marginal model that is fitted and the type of covariance model that is fitted. For example, if we use `manyglm` to fit a marginal model, then we have made the following assumptions:

1. The observed multivariate y-values are *independent*, after conditioning on x.
2. The y-values come from *a known distribution* (from the exponential family) with known *mean–variance relationship* $V(\mu)$.
3. There is a *straight-line relationship* between some known function of the mean of y and each x:
$$g(\mu_y) = \beta_0 + \mathbf{x}^T \boldsymbol{\beta}$$
4. Copula values z, which satisfy $F(y^-) < \Phi(z) \leq F(y)$, come from a multivariate normal distribution with variance–covariance matrix $\boldsymbol{\Sigma}$ (which is actually a correlation matrix, with all variances one).

Assumptions 1–3 come from the marginal model, assumption 4 from the covariance model. The details of what is meant by assumption 4 depend on the type of covariance model that is fitted, which determines the precise form of $\boldsymbol{\Sigma}$. It is difficult to check assumption 4 directly from data because it involves unobservable values z, but competing covariance models could be compared via model selection.

17.2 Inferring Co-occurrence Using Latent Variables

Chapter 11 introduced the notion of a variance–covariance matrix and methods that can be used to estimate an unstructured matrix $\boldsymbol{\Sigma}$, with separate parameters for each variance and covariance (an *unstructured* variance–covariance matrix). A hierarchical GLM with an unstructured variance–covariance matrix, as in Sect. 11.3, has been proposed to study co-occurrence (Pollock et al., 2014). However, multivariate abundance data typically involve many taxa and data that are information-poor, with many absences. A model that requires separate estimates of every pairwise covariance is not a good match to that situation.

Latent variable models can be used to estimate a variance–covariance matrix $\boldsymbol{\Sigma}$, but do so in a parsimonious fashion, as explained in Maths Box 17.1. A hierarchical modelling approach was used previously to fit these models (Sects. 12.3 and 16.2), but a copula model is used instead in Code Box 17.1, via the `ecoCopula` package. This package uses a Gaussian copula model to map data from, in principle, any parametric distribution to the multivariate normal, then applies a factor analysis to the copula values. This is very fast, relative to `gllvm`, and is even competitive with dissimilarity-based algorithms (as in Sect. 12.4) for large datasets (Popovic et al., 2022).

Maths Box 17.1: Latent Variables as Reduced Rank Covariance Estimators

A factor analysis model as in Eq. 12.1, but written in vector notation and potentially including predictors, is

$$\mathbf{y}_i = \boldsymbol{\beta}_0 + \boldsymbol{x}_i'\mathbf{B} + \boldsymbol{\Lambda}'\mathbf{z}_i + \boldsymbol{\epsilon}_i$$

where $\boldsymbol{\Lambda}$ is a matrix whose jth column is Λ_j, $\boldsymbol{\beta}_0$ and $\mathbf{B}$ are defined similarly, and $\boldsymbol{\epsilon}_i$ is a vector of independent error terms.

This model can capture correlation across responses via the latent variables. The first two terms are not random, so they do not feature in the variance–covariance matrix of $\mathbf{y}_i$, which can be written

$$\boldsymbol{\Sigma}_{\mathbf{y}_i} = \boldsymbol{\Lambda}'\boldsymbol{\Sigma}_{\mathbf{z}_i}\boldsymbol{\Lambda} + \boldsymbol{\Sigma}_{\boldsymbol{\epsilon}_i} = \boldsymbol{\Lambda}'\boldsymbol{\Lambda} + \boldsymbol{\Sigma}_{\boldsymbol{\epsilon}}$$

Because the elements of $\boldsymbol{\epsilon}_i$ are independent, $\boldsymbol{\Sigma}_{\boldsymbol{\epsilon}}$ is diagonal, and so all information about covariances (the off-diagonal elements) of $\boldsymbol{\Sigma}_{\mathbf{y}_i}$ is stored in the factor loadings, $\boldsymbol{\Lambda}$. If a model has $d < p$ factors, this matrix has only d rows, and it is upper triangular (Maths Box 12.2), with the $\binom{d}{2}$ lower triangular terms set to zero. Hence, the number of parameters in the model for covariances of $\boldsymbol{\Sigma}_{\mathbf{y}_i}$ is $dp - \binom{d}{2}$. The number of parameters in an unstructured estimate of the covariances of $\boldsymbol{\Sigma}_{\mathbf{y}_i}$ is $\binom{p}{2}$, so latent variables can lead to major savings in parameters:

parameters in:

p	Unstructured $\widehat{\boldsymbol{\Sigma}}$	2-factor model
5	10	9
10	45	19
20	190	39
100	4950	199

Notice that the number of parameters needed to model $\boldsymbol{\Sigma}$ increases much more slowly with p in a factor analytic model—it increases *linearly* with p rather than *quadratically*.

This idea applies to latent variable models irrespective of whether they are constructed in a hierarchical GLM framework or using a Gaussian copula.

Code Box 17.1: Estimating Co-occurrence Patterns Using a Copula Model

The `cord` function in the `ecoCopula` package will be used to estimate a variance–covariance matrix capturing co-occurrence information for Petrus's spider data.

```
library(ecoCopula)
library(mvabund)
data(spider)
spiderAbund = mvabund(spider$abund)
spider_glmInt = manyglm(spiderAbund~1,data=spider$x)
```

```
ord_spiderInt = cord(spider_glmInt)
plot(ord_spiderInt, biplot=TRUE) #for a biplot

# now plot a correlation matrix
library(gclus)
sortVars = order.single(ord_spiderInt$sigma)
library(corrplot)
corrplot(ord_spiderInt$sigma[sortVars, sortVars], type="lower",
         diag=FALSE, method="square")
```

Corresponding plots can be found in Fig. 17.2a, b.

The idea is illustrated in Fig. 17.2a, b for Petrus's spider data (Exercise 17.1). Figure 17.2a presents a biplot of factor scores and loadings using a two-factor Gaussian copula, and Fig. 17.2b is the estimated correlation matrix this implies. Species at opposite sides of a biplot have negative estimated correlations, and species close together have positive estimated correlations. The strength of these correlations increases the further the loadings are from the origin. This close correspondence between the two plots is expected because the correlations are computed as a function of factor loadings (Maths Box 17.1).

The two-factor Gaussian copula model of Fig. 17.2a, b was constructed using the `cord` function of the `ecoCopula` package (Code Box 17.1). This software makes explicit its two-step approach to estimation by first requiring a marginal model to be fitted (e.g. using the `mvabund` package); then a covariance model is fitted by applying the `cord` function to this marginal model object.

17.3 Co-occurrence Induced by Environmental Variables

The correlations in $\boldsymbol{\Sigma}$ capture patterns not accounted for by predictors already included in the model. The correlation patterns seen in Fig. 17.2a, b are based on a model with no predictors in it, and some of these correlations are potentially explained by shared (or contrasting) response to environmental variables. To study the extent to which measured environmental variables explain these patterns, we can refit the model with predictors included and compare results, as in Fig. 17.2c, d.

Note that correlation patterns seem to be substantially weaker after removing effects due to environmental variables (Fig. 17.2d), with most correlations being smaller than they were previously (Fig. 17.2b). In particular, there were strongly negative correlations between the three species on the bottom rows of Fig. 17.2b and the remaining nine species, which all disappeared on inclusion of environmental variables. These three species had a negative response to soil dryness (as in Fig. 16.1), whereas the remaining nine species had a positive response. These contrasting responses to soil dryness are the likely reason for the negative co-occurrence patterns of these species (Fig. 17.2b).

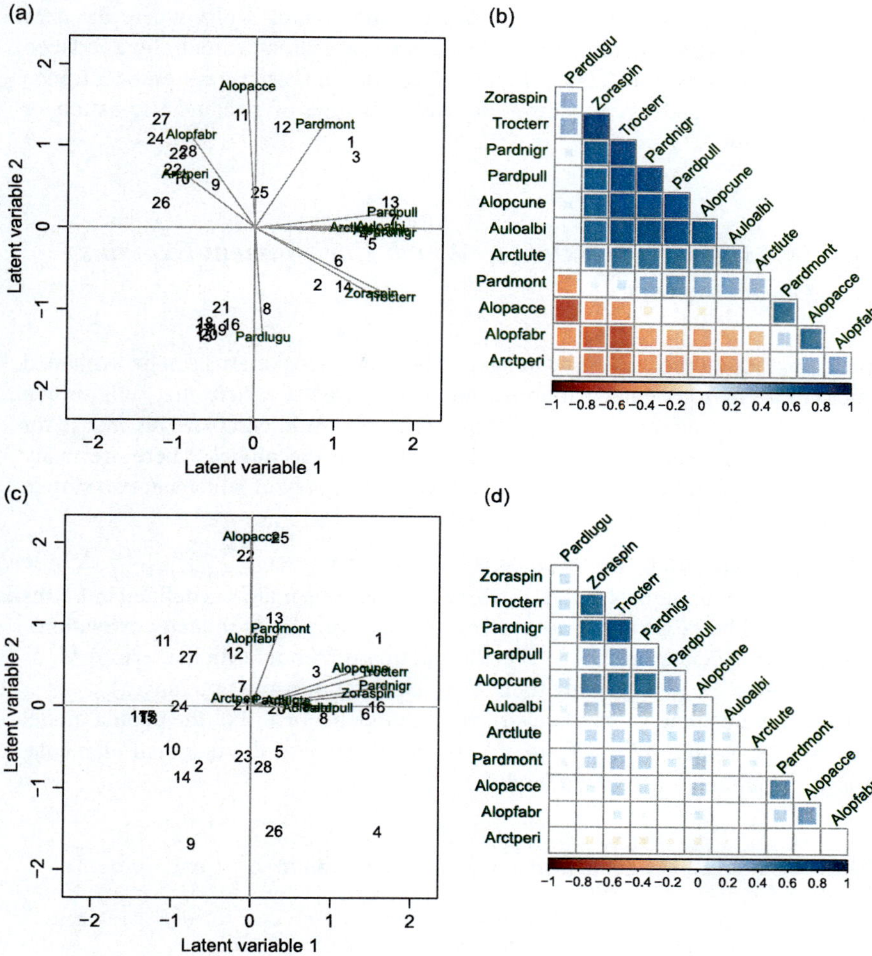

Fig. 17.2: Exploring co-occurrence in Petrus's spider data: (**a**) unconstrained biplot of a two-factor Gaussian copula model with no predictors; (**b**) the correlation matrix this model implies; (**c**) a "residual" biplot of a two-factor Gaussian copula model after including environmental variables; (**d**) the correlation matrix this implies. Notice that species far apart in a biplot have strongly negative estimated correlations (e.g. *Alopecosa accentuata* and *Pardosa lugubris* in **a** and **b**), and species close together and far from the origin have strongly positive estimated correlations (e.g. *Trochosa terricola* and *Zora spinimana* in **a** and **b**). Notice also that after controlling for environmental variables, factor loadings have changed (**c**), and correlation estimates tend to be closer to zero (**d**)

ter Braak and Prentice (1988) introduced the term *unconstrained ordination* to describe an ordination without any predictors in it, as in Fig. 17.2a. ter Braak and

Prentice (1988) further defined a constrained ordination as a plot where the axes were derived as a linear combination of measured predictors, essentially a reduced rank regression as in Yee (2006). Figure 17.2c, in contrast, shows co-occurrence patterns not explained by measured covariates, a type of residual ordination or *partial ordination*.

17.3.1 Quantifying the Extent to Which Environment Explains Co-occurrence

To put a number on precisely how much of co-occurrence patterns can be explained by measured environmental variables, we would need a metric that tells us the strength of (estimated) correlation patterns. Then we could compare this metric for models with and without environmental variables in the model. There are many potential approaches here, and little in the literature by way of guidance; two simple measures that could be used are as follows:

- The sum of squared factor loadings, which can be written $\sum_{i=1}^{p} \sum_{j=1}^{d} \lambda_{ij}^2$. A little algebra shows this equals $\text{tr}(\mathbf{\Lambda}\mathbf{\Lambda}')$ (where the trace of a matrix is defined in Maths Box 12.1). The larger the loadings are, the stronger the estimated correlations, since covariance parameters in $\mathbf{\Sigma}$ are a function of $\mathbf{\Lambda}$ only (via the term $\mathbf{\Lambda}\mathbf{\Lambda}'$, as in Maths Box 17.1, suggesting $\text{tr}(\mathbf{\Lambda}\mathbf{\Lambda}')$ as a natural measure of strength).
- The sum of absolute value of covariance coefficients in $\mathbf{\Sigma}$. For the copula model this can be written $\sum_{i=1}^{p} \sum_{i=1}^{p} |\sigma_{ij}| - p$ (the p removes the diagonal elements, which are all one for a copula model).

Code Box 17.2: Co-occurrence Patterns Explained by Environmental Variables

We can repeat the procedure of Code Box 17.1 to study the extent to which measured environmental variables explain the co-occurrence patterns of Fig. 17.2:

```
spider_glmX = manyglm(spiderAbund~.,data=spider$x)
ord_spiderX = cord(spider_glmX)
plot(ord_spiderX,biplot=TRUE)
corrplot(ord_spiderX$sigma[sortVars,sortVars],type="lower",
                    diag=FALSE, method="square")
```

Corresponding plots can be found in Fig. 17.2c, d.

To what extent do measured environmental variables explain patterns in co-occurrence? One way to look at this is via the size of the factor loadings:

```
> ss = c(sum(ord_spiderInt$loadings^2), sum(ord_spiderX$loadings^2))
> c(ss, 1-ss[2]/ss[1])
[1] 7.4976904 4.1602363 0.4451309
```

So including environmental variables substantially reduced the amount of covariation in the factor analysis—the sum of squared loadings is 45% smaller, after including predictors in the model. Another way to look at this is to look at the size of correlations directly:

```
> absCor = c( sum(abs(ord_spiderInt$sigma)),
             sum( abs(ord_spiderX$sigma)) ) - ncol(spider$abund)
```

```
> c(absCor, 1-absCor[2]/absCor[1])
[1] 52.6900602 21.9502380  0.5834084
```

The sum of absolute values of correlation coefficients reduced from 53 to 22 upon inclusion of environmental variables, a reduction of 58%. This can be seen visually in Fig. 17.2d, where correlations are all much weaker than in Fig. 17.2b.

Both the aforementioned measures are calculated in Code Box 17.2, for models with and without environmental variables. Both measures were roughly halved on inclusion of environmental predictors, meaning that about half of co-occurrence patterns can be explained by measured environmental variables. Note the measures did not return exactly the same answer (45% vs 58%)—as always, different ways of measuring things can lead to different measurements!

Exercise 17.2: Co-occurrence Patterns in Spider Presence–Absence
Repeat the analyses of Code Box 17.1 on presence–absence data, which you can construct using `spiderPA=pmin(spiderAbund,1)`. Estimate the correlation matrix. *How does the correlation matrix compare to Fig. 17.2b?*

There is a lot less information in presence–absence data, and the weaker signals in the marginal distribution typically present as much weaker correlations.

Calculate the sum of squared loadings, with and without predictors, as in Code Box 17.2. *Are the values smaller for the presence–absence data? Is this what you expected?*

Are your conclusions any different to Code Box 17.2, in terms of the extent to which patterns in co-occurrence can be explained by environmental variables?

Exercise 17.3: Co-occurrence in Bird Communities
Françoise visited 51 sites near Lyon, France, and estimated the abundance of 40 different species of birds (by listening to bird songs for 15 min) (Tatibouët, 1981). He would like to understand bird co-occurrence and the extent to which it is explained by predictors related to urbanisation. Load the data and fit ordinal regressions to each response:

```
library(ade4)
data(aviurba)
abund=mvabund(aviurba$fau)
library(ordinal)
ft_birdsInt=manyany(abund~1, "clm", family="ordinal",
 data=aviurba$mil)
```

Use `cord` to fit a latent variable model and estimate the correlation matrix.

Add `fields` to the model to study the extent to which presence or absence of fields explains co-occurrence patterns.

Calculate the sum of squared loadings for latent variable models with and without `fields` as a predictor. Notice that these are a lot smaller than in Code Box 17.2.

What can you conclude about the co-occurrence patterns of these birds and the extent to which they are explained by presence or absence of fields?

17.4 Co-occurrence Induced by Mediator Taxa

Recall that another possible reason two taxa may co-occur, beyond the scenario where both taxa respond to the same environmental variable, is that they are both related to some *mediator taxon*.

Graphical modelling is a technique designed to tease these two ideas apart, which identifies pairs of *conditionally dependent* taxa—taxa that remain correlated with each other even after accounting for all others in the model. Popovic et al. (2019) wrote a helpful introduction to the idea of conditional dependence and how to visualise conditional relationships between a set of response variables.

Recall from Chap. 3 that a key idea in multiple regression is that coefficients are interpreted conditionally, after controlling for the effects of all others in the model. Thus, multiple regression is a tool for finding conditionally dependent associations. So one approach to finding pairs of responses that are conditionally dependent would be to apply multiple regression to each response, as a function of all others, and look for slope coefficients that are clearly non-zero. Graphical modelling takes a slightly different (but related) approach, estimating the inverse of the variance–covariance matrix and looking for values that are clearly non-zero.

If two variables are conditionally independent, we do not necessarily get a zero in the variance–covariance matrix ($\mathbf{\Sigma}$) itself because correlation might be induced by a third *mediator* taxon. We would, however, get a zero in the appropriate cell of the inverse of the variance–covariance matrix ($\mathbf{\Sigma}^{-1}$, sometimes called a *precision matrix*), if data are multivariate normal. Maths Box 17.2 uses multiple regression results from Chap. 3, and some matrix algebra, to show why this is the case. Figure 17.3 illustrates the idea for a hypothetical variance–covariance matrix, and its precision matrix, for *snakes*, *mice*, and *lions*. *Snakes* and *lions* are conditionally independent but positively correlated because of the mediator *mice*, which is positively correlated with each of them.

Maths Box 17.2: Why the precision matrix captures conditional dependence

Consider a linear model for one response $\mathbf{y}_p$ in terms of other responses $\mathbf{Y}_X$, $\mu_p = \beta_0 + \mathbf{Y}_X'\boldsymbol{\beta}_1$, where $(\mathbf{Y}_X, \mathbf{y}_p)$ are multivariate normal with variance-covariance matrix:

$$\boldsymbol{\Sigma} = \begin{bmatrix} \boldsymbol{\Sigma}_{XX} & \boldsymbol{\sigma}_{Xp} \\ \boldsymbol{\sigma}_{pX} & \sigma_{pp} \end{bmatrix}$$

If we have are more than two responses, $\boldsymbol{\Sigma}_{XX}$ is a matrix, whereas $\boldsymbol{\sigma}_{Xp}$ is a vector and σ_{pp} is scalar. It turns out that the true values for slope coefficients are:

$$\boldsymbol{\beta}_1 = \boldsymbol{\Sigma}_{XX}^{-1}\boldsymbol{\sigma}_{Xp}$$

This expression is essentially the least squares estimator from Maths Box 3.1, but with the relevant (cross-)products replaced with (co-)variances.

If $\mathbf{y}_p$ is conditionally independent of the jth variable in $\mathbf{Y}_X$, then the jth regression coefficient is zero (as in Chapter 3), which we will write as $[\boldsymbol{\beta}_1]_j = 0$ or $\left[\boldsymbol{\Sigma}_{XX}^{-1}\boldsymbol{\sigma}_{Xp}\right]_j = 0$. In matrix algebra, $\left[\boldsymbol{\Sigma}_{XX}^{-1}\boldsymbol{\sigma}_{Xp}\right]_j = 0$ does not imply that $[\boldsymbol{\sigma}_{Xp}]_j = 0$, so conditional independence does not imply zero covariance, nor zero correlation.

But consider $\boldsymbol{\phi}_{Xp}$, the corresponding term in the precision matrix:

$$\boldsymbol{\Sigma}^{-1} = \begin{bmatrix} \boldsymbol{\Phi}_{XX} & \boldsymbol{\phi}_{Xp} \\ \boldsymbol{\phi}_{pX} & \phi_{pp} \end{bmatrix}$$

We can find an expression for $\boldsymbol{\phi}_{Xp}$ in terms of $\boldsymbol{\Sigma}$ using the equation $\boldsymbol{\Sigma}\boldsymbol{\Sigma}^{-1} = \mathbf{I}$:

$$\begin{bmatrix} \boldsymbol{\Sigma}_{XX} & \boldsymbol{\sigma}_{Xp} \\ \boldsymbol{\sigma}_{pX} & \sigma_{pp} \end{bmatrix} \begin{bmatrix} \boldsymbol{\Phi}_{XX} & \boldsymbol{\phi}_{Xp} \\ \boldsymbol{\phi}_{pX} & \phi_{pp} \end{bmatrix} = \begin{bmatrix} \boldsymbol{I} & \mathbf{0} \\ \mathbf{0} & 1 \end{bmatrix}$$

Taking the top-right partition of this equation and solving for $\boldsymbol{\phi}_{Xp}$:

$$\boldsymbol{\Sigma}_{XX}\boldsymbol{\phi}_{Xp} + \boldsymbol{\sigma}_{Xp}\phi_{pp} = \mathbf{0}$$
$$\text{so } \boldsymbol{\phi}_{Xp} = -\boldsymbol{\Sigma}_{XX}^{-1}\boldsymbol{\sigma}_{Xp}\phi_{pp} = -\boldsymbol{\beta}_1\phi_{pp}$$

and since ϕ_{pp} is scalar, $[\boldsymbol{\beta}_1]_j = 0$ implies that $[\boldsymbol{\beta}_1\phi_{pp}]_j = 0$, i.e. $[\boldsymbol{\phi}_{Xp}]_j = 0$. So if $\mathbf{y}_p$ is conditionally independent of a variable, the corresponding element of $\boldsymbol{\Sigma}^{-1}$ is zero.

It turns out that this argument works both ways, for multivariate normal variables – conditional independence implies a zero in the precision matrix, and a zero in the precision matrix implies conditional independence.

Conditional dependence relationships can be demonstrated graphically by plotting a point for each taxon, with lines joining pairs of taxa that are conditionally dependent (Fig. 17.3). Mediator taxa are those that indirectly link taxa to each other, inducing a correlation in them. In Fig. 17.3, there is no line joining *snakes* and *lions*, implying conditional independence. As previously, these taxa are linked to each other via the mediator *mice* and are correlated with each other because of this connection.

$$\boldsymbol{\Sigma} = \begin{array}{c} \text{snakes mice lions} \\ \begin{pmatrix} 1 & 1 & 1 \\ 1 & 2 & 2 \\ 1 & 2 & 2.5 \end{pmatrix} \end{array} \begin{array}{l} \\ \text{snakes} \\ \text{mice} \\ \text{lions} \end{array} \qquad \boldsymbol{\Sigma}^{-1} = \begin{array}{c} \text{snakes mice lions} \\ \begin{pmatrix} 2 & -1 & 0 \\ -1 & 3 & -2 \\ 0 & -2 & 3 \end{pmatrix} \end{array} \begin{array}{l} \\ \text{snakes} \\ \text{mice} \\ \text{lions} \end{array}$$

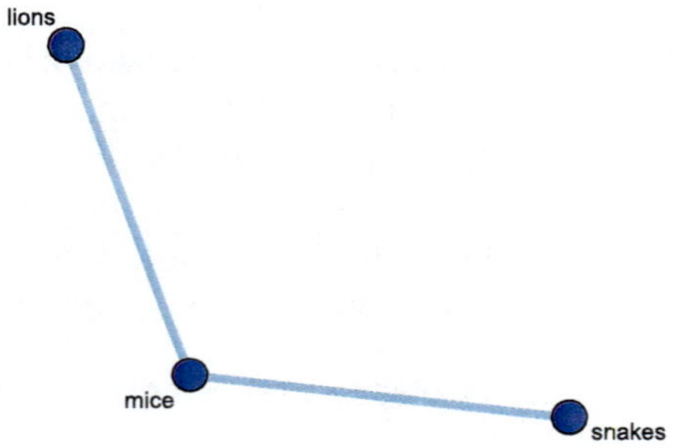

Fig. 17.3: A hypothetical variance–covariance matrix ($\boldsymbol{\Sigma}$) from abundance data for snakes, mice, and lions and its corresponding precision matrix ($\boldsymbol{\Sigma}^{-1}$). While all taxa are positively correlated ($\boldsymbol{\Sigma}$), there is a zero in $\boldsymbol{\Sigma}^{-1}$ for *snakes* and *lions*, meaning they are conditionally independent of each other. A graph of conditional dependence relationships connects *mice* with each of *snakes* and *lions*, but *snakes* and *lions* are not connected because they are conditionally dependent and their correlation is induced by shared positive associations with the mediator taxon, *mice*

17.5 The Graphical LASSO for Multivariate Abundances

There are some nice computational tools for graphical modelling of multivariate normal data, most notably the *graphical LASSO* (Friedman et al., 2008, the `glasso` package in R). This method estimates a precision matrix using a LASSO approach via penalised maximum likelihood:

$$\log \mathcal{L}(\boldsymbol{\beta}, \boldsymbol{\Sigma}; \mathbf{y}) - \lambda ||\boldsymbol{\Sigma}^{-1}||_{1,1} \qquad (17.3)$$

where $||\Sigma^{-1}||_{1,1}$ is the element-wise L_1 norm, the sum of absolute values of parameters. The key difference from the LASSO as introduced in Chap. 5 is that the penalty is applied to a precision matrix rather than to a vector of regression coefficients, so terms in the precision matrix are shrunk towards zero, rather than regression coefficients. The end result is (usually) a precision matrix with lots of zeros in it, which considerably simplifies interpretation, by putting our focus on the taxa that drive key dependence patterns in the data.

A difficulty we have applying this idea to multivariate abundance data is that the method assumes multivariate normality, which can be overcome by mapping values to the multivariate normal using a Gaussian copula (Popovic et al., 2018). Other graphical modelling methods have been applied in ecology, using related modelling frameworks, in the special case of presence–absence data (Harris, 2016; Clark et al., 2018).

Graphical models can be fitted to multivariate abundance data using the `cgr` function of the `ecoCopula` package (Code Box 17.3). As previously, models can be fitted with and without environmental variables to study the extent to which conditional dependence relationships can be explained by environmental variables, leading to the graphs of Fig. 17.4. As before, most of the co-occurrence patterns in the data seem to be explainable by response to common environmental variables because the relationships are weaker in Fig. 17.4b and there are slightly fewer of them (reduced from 37 to 31). This effect does, however, look less dramatic on Fig. 17.4 than in the correlation plots of Fig. 17.2, although it seems comparable in scale (Code Box 17.3 suggests that in absolute terms, correlations reduced by about 60%, as previously).

Results from graphical models are very sensitive to the GLASSO penalty parameter, the value of λ in Eq. 17.3. In a regular LASSO regression (Sect. 5.6), when λ is large enough, all terms are left out of the model, and when it is small enough, all predictors get included. In much the same way, in GLASSO, if λ is large enough, then no correlations are included in the model, and if it is small enough, then all of them are included. When comparing different graphical models to look at the effects of predictors on co-occurrence patterns (Code Box 17.3), it is a good idea to use the same value of λ to put the two models on an even footing. When reporting a final graphical model, it is a good idea to re-estimate λ, but when specifically looking at the extent to which predictors explain co-occurrence, λ should be fixed to a common value.

The two species with the most conditional dependence relationships detected (Fig. 17.4) were the two most abundant species (*Pardosa pullata* and *Trochosa terricola*). The three rarest species, *Arctosa lutetiana*, *Arctosa perita*, and *Alopecosa fabrilis*, tended to have fewer and weaker conditional dependence relationships. This pattern suggests that the conditional dependence relationships for a species are in part a function of how much information we have on that species. We need a lot of information on occurrence patterns if we are to detect co-occurrence patterns!

Code Box 17.3: A Copula Graphical Model for Petrus's Spider Data

The `cgr` function in the `ecoCopula` package will be used to construct a Gaussian copula graphical model from negative binomial regressions of each hunting spider species:

```
# fit an intercept model (for an "unconstrained" graph)
graph_spiderInt = cgr(spider_glmInt)
plot(graph_spiderInt, vary.edge.lwd=TRUE)

# repeat with all environmental variables included
graph_spiderX = cgr(spider_glmX, graph_spiderInt$all_graphs$lambda.opt)
plot(graph_spiderX, vary.edge.lwd=TRUE)
```

This results in the plots of Fig. 17.4. For both models, the same GLASSO penalty parameter was used (`graph_spiderInt$all_graphs$lambda.opt`), so that the models are comparable in terms of the strength of shrinkage applied.

```
> absCor = c( sum(abs(graph_spiderInt$best_graph$cov)),
       sum( abs(graph_spiderX$best_graph$cov)) ) - ncol(spider$abund)
> c(absCor, 1-absCor[2]/absCor[1])
[1] 45.6394291 17.2511592  0.6220119
```

So environmental variables explained a considerable proportion (about 62%) of estimated covariance patterns in the data.

Exercise 17.4: Does Soil Dryness Explain Co-occurrence Patterns in Petrus's Data?

Recall that in Fig. 16.1a, *Alopecosa accentuata*, *Alopecosa fabrilis*, and *Arctosa perita* decreased in response to soil dryness, while all other species increased. Note that the "unconstrained" correlation matrix of Fig. 17.2b found negative correlation patterns between these species and most others.

To what extent do contrasting responses to soil dryness explain the negative correlations of Fig. 17.2b?

Answer this question by fitting a covariance model of your choice to the spider data with and without soil dryness as a predictor.

17.5.1 Graphical Modelling as Covariance Estimation

Notice that the number of links in the graphical models is relatively small—Fig. 17.4b was constructed from a precision matrix with 31 non-zero values in it, whereas an unstructured matrix would have 66 non-zero values. A smaller number of parameters makes interpretation easier but also often gives a better estimate of $\mathbf{\Sigma}$ by choosing a more appropriate point on the bias–variance trade-off.

Recall that the number of parameters in an unstructured variance–covariance matrix increases quadratically with the number of responses (p), and unless p is small, it is not practical to estimate $\mathbf{\Sigma}$ without assuming some structure to it. Using a generalised latent variable models is one way to impose structure, assuming $\mathbf{\Sigma}$ has

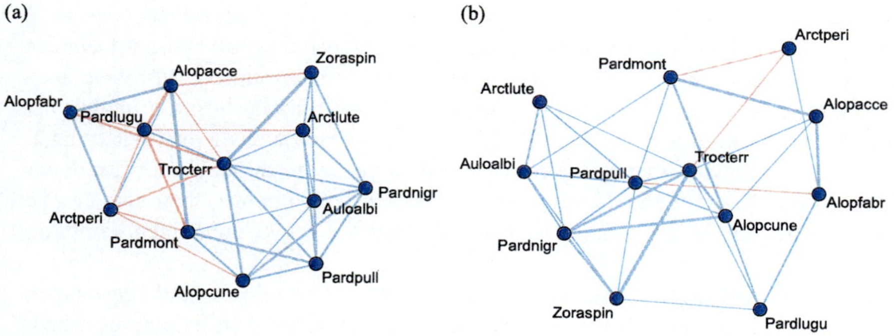

Fig. 17.4: Graphs of conditional dependence relationships between species in Petrus's spider data (**a**) without and (**b**) with environmental variables included in the model. Negative conditional dependence is coloured pink, positive dependence is blue, stronger dependence is indicated by a thicker line. Note that after controlling for environmental variables, there are fewer estimated conditional dependence relationships (31 vs 37), they tend to be weaker, and most of the negative dependence has disappeared. Hence most of the negative co-occurrence patterns exhibited in (**a**) could be explained by patterns in environmental response across species

reduced rank (Maths Box 17.1). Graphical modelling is an alternative way to impose structure, assuming many pairs of responses are conditionally independent.

17.6 Other Models for Co-occurrence

There are many other options for ways to model correlation patterns and, hence, quantify the various potential sources of co-occurrence patterns. One option not considered above is correlation from phylogenetic sources (Sect. 7.3). Closely related taxa might co-occur more often than expected—Li and Ives (2017) explain that one potential reason for this is missing predictors, specifically, missing functional traits that may explain some co-occurrence. Traits often carry a phylogenetic signal, and so traits that are omitted from a model might induce phylogenetic correlation. ter Braak (2019) proposed adding random slopes for traits to a model to specifically account for this. However, any method that accounts for the *multivariate* property, such as those in Chaps. 11 and 14 and this chapter, are capable of accounting for phylogenetic correlation, among other causes of correlated responses. If the research question of interest is to quantify the phylogenetic signal in co-occurrence, then we need to directly quantify it, which can be done using the `phyr` package (Li et al., 2020) or using `Hmsc` (Ovaskainen & Abrego, 2020, Section 6.4).

Another idea not touched on in this chapter is that co-occurrence patterns may change as the environment changes—there is ample empirical evidence that this happens (Tylianakis & Morris, 2017). But a fixed variance–covariance matrix $\mathbf{\Sigma}$ was assumed in this chapter, meaning that co-occurrence from all sources has been assumed to occur to the same extent irrespective of environment. We have seen that there are many parameters to estimate in $\mathbf{\Sigma}$, and reliably fitting a model that can be used for inference in the case of fixed $\mathbf{\Sigma}$ is already a challenge. Allowing $\mathbf{\Sigma}$ to vary across environments makes the problem yet more difficult and introduces the question of how it should vary. If this question is of interest, then data can be collected to quantify variation in $\mathbf{\Sigma}$ in controlled ways, e.g. in a hierarchical sampling design (Tikhonov et al., 2017).

While the methods of this chapter can tease apart some potential reasons for co-occurrence, they cannot distinguish species interactions from situations where species are responding to some unmeasured environmental variable (or some other form of model misspecification). The only certain way to rule out confounders is to move beyond observational studies and design a controlled experiment to look for direct interaction.

Note also that species interactions may be asymmetric—species A may be highly dependent on species B (e.g. requiring it as a food source), while species B has much weaker dependence on A. This idea is not captured at all by the models discussed here, which estimate correlation, a asymmetric measure of co-occurrence. Blanchet et al. (2020) caution strongly against the use of co-occurrence to infer species interactions—their most compelling arguments being asymmetry and (as previously) the possibility that co-occurrence can arise due to model misspecification (e.g. unmeasured environmental variables). The approaches in this chapter can be most helpfully thought of as a starting point—these are exploratory methods, which could lead to further work to demonstrate species interaction. Short of designed experiments, one way to get closer to diagnosing species interaction is to collect time series data across multiple species and look for a delayed response of one species to another (Ovaskainen et al., 2017a, for example, using a multivariate autoregression model).

Chapter 18
Closing Advice

This book has been a long journey! But this is *not* the end. Something I've learned over my career so far is that the more I know, the more I realise that I don't know. I am regularly finding out about methods and problems I wasn't previously aware of or new techniques that have been developed recently.

One way to understand the breadth of statistics is to recall that the way data should be analysed primarily depends on the **P**s and **Q**s:

P Data **p**roperties. In regression, the properties of the response variable are what matters.

Q The research **q**uestion guides us in finding a target quantity or technique for analysis.

When you look across ecology, or any other discipline, there are lots of different ways people can collect data, and there are lots of different types of research questions to ask. So there is a need for a whole lot of different analysis methods! And new ones all the time, as technology brings new methods of data collection, and as people think of new types of questions they can ask. So one thing we know for sure is that you will not find in this book all the analysis methods you are going to need over the course of your career.

So to close things out, let's first summarise the main lessons in a common framework. Then we will discuss what to do when you come across a new method not covered in this book, as you inevitably will!

18.1 A Framework for Data Analysis—Mind Your Ps and Qs

A framework for data analysis is presented in Fig. 18.1—it's all about minding your Ps and Qs. As with the remainder of this book, Fig. 18.1 and the following discussion primarily focus on problems that can be posed as a regression model—the situation where we want to relate a response, or set of responses, to predictors.

D. I. Warton, *Eco-Stats: Data Analysis in Ecology*, Methods in Statistical Ecology,
https://doi.org/10.1007/978-3-030-88443-7_18

Key Point
Always mind your **P**s and **Q**s!

P Data **p**roperties. In regression, the properties of the response variable are what matters. The main properties we have considered in this book are number of responses, response type, dependence across observations, and form of response to predictors.
Q The research **q**uestion guides us in finding a target quantity or technique for analysis. In particular, it tells us if we should focus on exploratory data analysis only, inferences about key parameters, hypothesis testing, model selection, or predictive modelling.

For more details see Fig. 18.1.

18.1.1 Ps—Data Properties

Some key data properties to consider when choosing an analysis method:

- How many response variables are of interest to the research question? One response leads to a univariate analysis (Chaps. 4–10), a few leads to multivariate analysis using an unstructured correlation matrix (Chap. 11), many responses is the tricky one (Chaps. 14–17).
- What type of response variables are they—what is needed in specifying a model for them? Quantitative variables can often be handled using linear models (Chap. 4), although the equal variance assumption must be checked. Discrete data can often be handled using generalised linear models (Chap. 10), and with only a few exceptions, these models should include a term to handle *overdispersion*.
- Do we expect correlation or clustering across responses? This depends largely on how data were collected—repeated measures or longitudinal data, spatially structured sampling, or comparative studies across multiple species often imply the need to use methods designed for dependent data (Chap. 7), and multi-level or hierarchical sampling designs can be handled using random effects (Chap. 6).
- How do we expect response variables to relate to predictors? If the response is expected to be non-linear, e.g. response of species to environmental gradients, additive models (Chap. 8) or, if there aren't much data available, quadratic terms can be used.

Note that while most of the chapters in this book dealt with one or two of these issues, in principle, you might need to deal with complexities introduced by a few of the foregoing issues all at once. For example, one might require a multivariate analysis of discrete data that are spatially correlated, with non-linear responses to predictors!

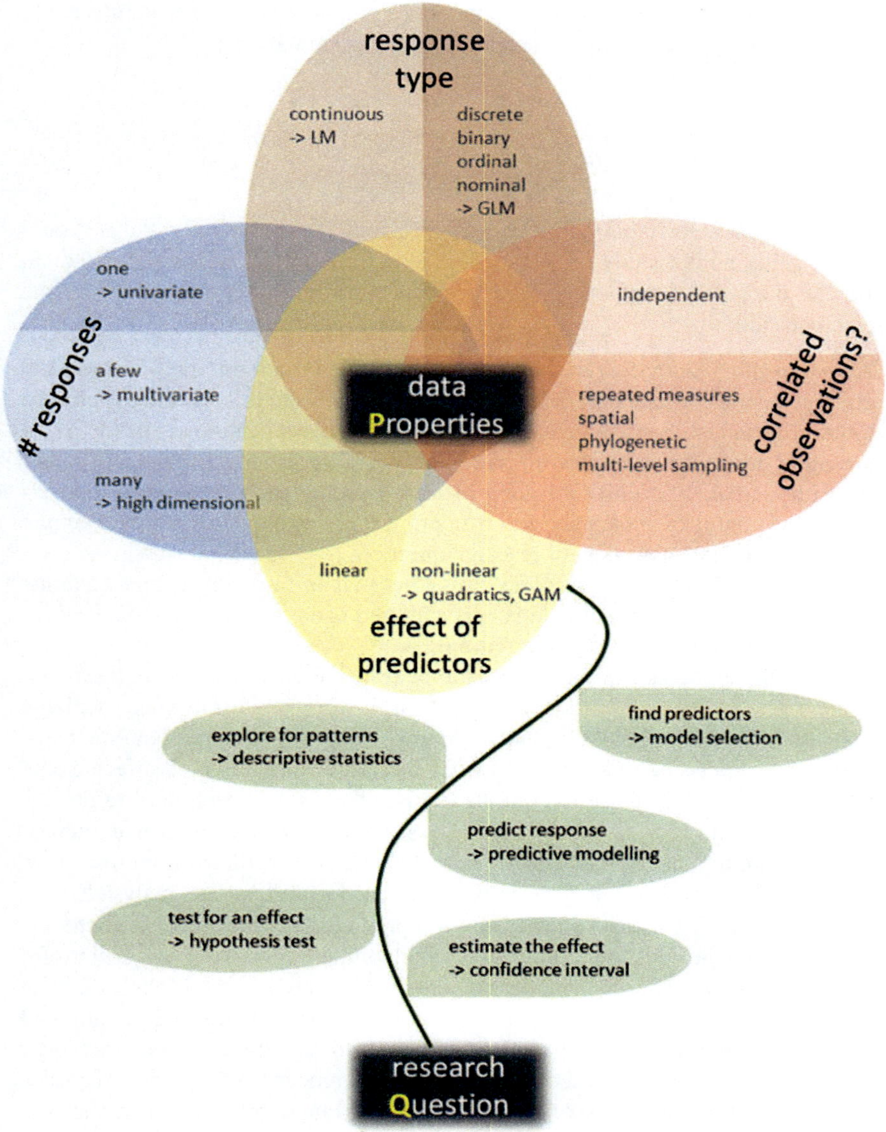

Fig. 18.1: Schematic diagram of approaches to data analysis. In data analysis you have to mind your Ps and Qs—the analysis method needs to be aligned with data **p**roperties and the research **q**uestion. Some key data properties to consider are the number and type of response variables, potential sources of correlation in response, and how they respond to predictors. Any combination of these properties could be encountered in practice. The research question informs what type of inference procedure (if any) is needed, for a suitably chosen regression model

(The first three of these issues could be handled along the lines of Ovaskainen et al., 2016; non-linearity would require quadratic terms or smoothers.)

18.1.2 Qs—Research Question

In analysis it is always important to keep the research question in mind—this should not only guide how a study is designed, it also feeds into how it is analysed. In particular, the type of method of inference that is of interest is determined by the research question:

- An *exploratory data analysis*, looking for patterns ("data mining"), should not attempt to make inferences. Such a study should be used to generate hypotheses rather than to test them. Ordination (Chap. 12) and classification (as in Chap. 16) are more advanced examples of procedures that are usually exploratory rather than inferential. It is *always* a good idea to do some exploratory data analysis where the goal is to understand key data properties, or to try to produce a simple visualisation that answers your research question.
- When a specific quantity is of interest, the goal is to estimate this quantity. Uncertainty around the estimate is best captured using a *confidence interval* (or prediction interval, in a Bayesian setting).
- If a specific hypothesis or claim needs to be tested, then a formal *hypothesis test* is appropriate. The most common example of this is when testing if one variable is related to another. Where possible, it is also helpful to think about effect size and to construct a confidence interval for the size of the observed effect. Some prefer to do this instead of hypothesis testing. This is, however, hard to do in a multivariate setting, where many parameters contribute to the question of interest, and it is hard to formulate an effect size that captures everything of interest. If we did have such a measure, maybe we wouldn't need a multivariate analysis!
- Rather than asking if there is an effect of a predictor, often a set of predictors is available, and the question of interest is which of these predictors are most useful in predicting the response, a *model selection* problem (Chap. 5). Model selection should be thought of as a type of inference, distinct from hypothesis testing and confidence interval estimation, and not to be mixed with the two (because model selection messes with the validity of any later inferences from the same dataset).
- When there is a target variable of interest that we want to predict, the emphasis is on *predictive modelling* (e.g. Chap. 15)—finding a model that predicts new values as accurately as possible. Such models can be quite complex and have parameters that are difficult to interpret, but interpretation and simplicity are not the goal. The literature on machine learning, which has been booming this century, is founded largely on the concept of predictive modelling (Hastie et al., 2009).

18.1.3 Your Model Is Wrong—Check Assumptions and Build in "Model Error"

Always remember that while we should try to capture all key data properties, we are unlikely to do so successfully. The working assumption should always be that your model is wrong.

Where possible, assumptions should always be checked, whether using residual plots, model selection tools to compare models with competing assumptions, or related approaches, to try and detect mistakes that were made in building the model. But assumption checks are not perfect, and in most cases we can expect missing predictors or other forms of *model misspecification* that have not been fixed in the analysis process. Fortunately, terms can be added to a model to give some robustness to violations of assumptions, including the following:

- When modelling discrete data, *include an overdispersion term*. This term is needed to capture any variation in response not explained by the model and account for it when making inferences. Linear models have a built-in error term that performs this role (as in Maths Box 4.1), but Poisson regression does not and will go badly wrong if there is overdispersion that was not accounted for.
- When modelling multiple responses simultaneously, account for correlation across responses (e.g. using latent variables, or an unstructured correlation matrix). Even if responses are not expected to directly interact, correlation will be induced by missing predictors that affect multiple responses.

Taking these steps will not prevent confounding variables from biasing coefficients (as in Maths Box 4.1), but it will help ensure standard errors and predictions are reasonable in the presence of missing information.

18.1.4 Does Your Analysis Pass the Common Sense Test?

Sometimes statistical procedures return surprising answers. Sometimes analysis reveals patterns in your data you didn't see before, because it looks at the data in a different way. But maybe the results look weird because something is actually wrong with the analysis! This could happen if an assumption about data properties was strongly violated but went undetected (e.g. overdispersion was not accounted for so lots of effects falsely appear significant), or maybe some computational issues went unnoticed (e.g. the model didn't converge, or you are using new software that hasn't been tested well and didn't work in your case). So if you get surprising results, should you trust them?

It is always worthwhile to look at your data in multiple ways. It is key for effective communication to produce intuitive and accessible ways to see your results, but it also offers a safeguard against an analysis that went wrong. If you can't produce a graph of the raw data that visualise the key pattern your analysis tells you is there, maybe there is a problem with your analysis!

For example, in early drafts of this book, I fitted the hierarchical GLM in Section 11.3 using the `lme4` package, and eventually found a solution that didn't report convergence issues. What puzzled me was that when I constructed a plot for this fit corresponding to Fig. 11.4, the lines didn't really go through the data, some seemed to miss the data entirely. I eventually tried refitting the model using `glmmTMB`, which gave a qualitatively different fit that fitted the data much more closely (Fig. 11.4). Clearly my earlier `lme4` fit had not converged (at least, not to a sensible answer), and if I hadn't tried plotting the data and fitting trend lines to it I would never have noticed the issue.

18.2 Beyond the Methods Discussed in This Book

There are heaps of things we haven't talked about, some of which you will need to know at some stage. Further, new methods are being developed all the time, so there is always the possibility that some new method will come up that you will want to learn how to use. So let's talk briefly about the next steps beyond this book.

18.2.1 A Few Particular Topics We Missed

There are still plenty of other data considerations, and methods of data analysis, that we have not had time to explore in this book. Let's run through a few ideas.

- Whereas a regression model involves splitting variables into response and predictor variables, more complex networks are possible. A variable that is a predictor may itself respond to another predictor, as in a food web. It is possible to build competing hypothesised networks of causal pathways and test which are more consistent with the data, using *structural equation models* (Grace, 2006) or *path analysis* (Shipley, 2016), although the development of such methods for non-normal data is a work in progress. There is a lot of related literature developing techniques intended for *causal inference* from observational studies (Morgan & Winship, 2015), the first step of which is to think through all possible alternate causes of an association you see in practice, and what can be measured about each of these to account for them and, hence, estimate the causal effect of primary interest. This is a really helpful step to work through as you progress towards a better understanding of your study system, irrespective of intent to make causal inferences. It may then be possible to make causal statements about the effect of interest, under the very strong assumption that your model is correct, an assumption that invites some scepticism.
- Predictor variables may not be measured perfectly and may come with measurement error. If the size of this error can be estimated, it can (if desired) be corrected for in order to estimate the response to the true value of the predictor using *mea-*

surement error models, otherwise known as *errors-in-variables models* (Carroll et al., 2006).

- Not all individuals that are present at a site may be observed on any given visit—especially cryptic species, or animals concerned with predator avoidance. This *imperfect detection* can be estimated from independent repeated visits or by noting time to detection (MacKenzie et al., 2017; Guillera-Arroita, 2017). Imperfect detection needs to be corrected for if the absolute number of individuals is of interest or in cases where one is studying the response of a species to the environment and expecting detection rates to vary with the environment (e.g. because it is harder to see study organisms in thicker vegetation).
- Sometimes the data of interest aren't the values of a response variable but rather the times or locations where an event happened, e.g. places where a species has been sighted. In these cases, a *point process model* may be appropriate, modelling the likelihood of observing new points across space or time. The mathematics of these methods at first look quite different to the regression models seen here but are actually quite closely related to Poisson GLMs (Renner et al., 2015).
- Animals move. Sometimes the pattern of movement is of interest, so we track animals over time and ask questions about how and why they move as they do. This is a special type of spatiotemporal model—one where we might be interested not just in the location of animals but also in the rate and direction of movement (Hooten et al., 2017).
- Sometimes the total number of individuals in a population is of interest, in particular when establishing threatened species status or monitoring an endangered population. In this case, the goal is to estimate not a mean but a total number of individuals, currently done through capture-recapture (Borchers et al., 2002) and related techniques.
- Some issues arise in connection with taking the methods of this book and scaling them up to large datasets. *Big data* can mean different things and has different implications depending on the context. Data can be big in the sense that they are long (many observations) or wide (many responses or predictors), or both. Long data tend to be easier to deal with, but they can require different solutions regarding storage (Lumley, 2020; Jordan et al., 2013) and visualisation (Cook et al., 2016). An important exception is the analysis of dependent observations (Chap. 7), which gets hard quickly as sample size increases and requires specialist solutions (Hefley et al., 2017). Wide data are more of a challenge, but the methods of Chaps. 12–17 were all designed with many responses in mind, and most of the software in these chapters has been used with some success when there are thousands of response variables. The presence of many predictors is often handled using sparse penalties (Hastie et al., 2015) like the LASSO (Sect. 5.6) using algorithms designed to scale well to big data.

This is not an exhaustive list!

18.2.2 Should You Even Use Some New Analysis Method?

So what steps should you go through in working out how to use some new method?

Before we talk about the steps to work through when using some new method, first reconsider whether you really need or want to use it at all. Potential costs of using some new method include the following: a longer start-up time; if the method is not well known to readers of your research, then it will be harder for them to follow; it is probably not as tried-and-tested as existing methods and so could have problems not yet recognised. Something else worth considering is whether you are looking for a fancy new method for the right reasons—do you need a better analysis or a better question? Sometimes researchers use fancy new or fashionable methods to dress up their research and make it look more attractive to good journals. But if the underlying science is not that interesting, then no amount of bells and whistles (sometimes called *statistical machismo* after a blog post by Brian McGill) can compensate, and the hard truth is that in that situation you may be better off going back to the drawing board.

Having said that, there are plenty of potential gains to using a fancy new method. New methods:

- Might work better or better answer your research question than existing methods. They are usually improvements on older methods in one of these two respects.
- Could add novelty to your research. Sometimes you can even write a paper about the methodology as well as about the results!
- Can be useful as a learning experience.
- Can go over better with reviewers, assessors, and examiners, depending on your audience. This should mostly be the case when the new methods are better suited to your research question.

18.2.3 Things to Look for When Studying Up on Some New Analysis Method

When learning about a new method of analysis, the most important question to answer is how do you *mind your Ps and Qs*? Specifically, what are the key data properties for which this method was intended? What are the main research questions it was designed to answer? And what procedures can you use to check you did the right analysis for your data?

Another important consideration is *performance*—has this method been demonstrated to do what it claims to, and if there are alternative approaches, has it been shown that this new method works better than those alternatives? The main tool for demonstrating that a method works is a simulation experiment—data are generated repeatedly under a particular, controlled scenario, and we look at how effectively the new method recovers the correct answer. A paper proposing new software or a new method should include a simulation experiment, unless the proposed method is an application of an established method but in a new context, in which case its performance has already been established.

A final consideration is how you actually use the new method. Many researchers jump to this step first, e.g. if the method was suggested to them by a colleague, but really it should be much lower on the priorities, because without good answers to the foregoing questions, there isn't much value in using the software. Software typically comes with reproducible worked examples in the documentation, and familiarising yourself with one or two worked examples, and stepping through the analysis yourself, can help you get a feel for how to apply the method to your data.

In looking for answers to the preceding questions, you pretty much use the same research skills as in any literature search—you are looking for sources (software papers, software documentation, or literature about software) that makes well-reasoned arguments, addressing the aforementioned issues, supported by evidence. In principle, such software or literature could be found in any outlet, but it is more likely to come from reputable sources—well-known authors, institutions, or journals. But judging a book by its cover is a very dangerous game, and software from a reputable person or journal is no substitute for clear arguments for why the software should be used and how to mind your Ps and Qs while using it. A clear difficulty, if you are not highly trained in statistics, is that it would be hard to judge the quality of software, or an article describing software, for yourself, and so again I strongly suggest that you see a *statistical consultant*. When unsure about something, you can also write directly to the software developer—you may be surprised how often they respond!

Exercise 18.1: Learn a Funky New Method!
Sometimes you might be talking to a supervisor or colleague about a project you are working on and they forward you a paper and tell you it would be a good idea to use that method for analysis. Let's practice this process.

First, choose one of the following papers:

- Schaub and Abadi (2011), a review of the use of integrated population models in ecology.
- Michelot et al. (2016) or McClintock and Michelot (2018), describing software for animal movement modelling using hidden Markov models.
- Hefley et al. (2017), a tutorial on basis functions (as in Chap. 8) and how to use them to fit spatiotemporal models to big data.
- Renner et al. (2015), a tutorial on how to use point process models to analyse a list of species sightings.

Now see if you can use your chosen paper, and related resources it directs you to, to answer the following questions:

- Qs—what sort of research questions can it be used to answer?
- Ps—what sort of data properties is it intended for?
- Mind your Ps and Qs—how could you check assumptions?
- Has this new method been shown to work well?

- Is there some existing, competing method or software? If so, why use the proposed methods instead?
- Is there software available with a worked example? See if you can work through this example by yourself.

Key Point

Some key things to look for in a document describing how to use some new software:

- What are the Ps—what data properties did developers have in mind when designing the method? This should be explicitly stated! It should also be implicit in their motivating examples.
- What are the Qs—what sorts of research questions was the method designed to answer? This again may not be explicitly stated and again can sometimes be inferred from the worked examples.
- How do you mind your Ps and Qs—how do you check assumptions? Does the paper suggest any particular model diagnostics that might be useful?
- Has it been shown (usually by simulation experiment) that this method actually works?
- What other methods are available, and why is this software proposed instead?
- Is there a worked example you can follow yourself?

Remember it would do no harm at all to *see a statistical consultant*.

Yeah, so that's it! I wish you the best for your journey—may your research questions be rippers.

References

Akaike, H. (1972). Information theory as an extension of the maximum likelihood principle. In: B. N. Petrov & F. Csaki (Eds.), *Second International Symposium on Information Theory* (pp. 267–281). Budapest: Akademiai Kiado.

Akaike, H. (1974). A new look at the statistical model identification. *IEEE Transactions on Automatic Control, AC-19*, 716–723. System identification and time-series analysis.

Anderson, E. (1935). The irises of the Gaspe Peninsula. *Bulletin of the American Iris Society, 59*, 2–5.

Anderson, M. J. (2001). A new method for non-parametric multivariate analysis of variance. *Austral Ecology, 26*, 32–46.

Anderson, M. J., Gorley, R. N., & Clarke, K. R. (2008). *PERMANOVA+for PRIMER: Guide to Software and Statistical Methods*. Plymouth: PRIMER-E.

Anderson, T. W. (2003). *An introduction to multivariate statistical analysis* (3rd ed.). New York: Wiley.

Austin, M. (2002). Spatial prediction of species distribution: An interface between ecological theory and statistical modelling. *Ecological Modelling, 157*, 101–118.

Banerjee, S., Gelfand, A. E., Finley, A. O., & Sang, H. (2008). Gaussian predictive process models for large spatial data sets. *Journal of the Royal Statistical Society: Series B (Statistical Methodology), 70*, 825–848.

Bates, D., Mächler, M., Bolker, B., & Walker, S. (2015). Fitting linear mixed-effects models using lme4. *Journal of Statistical Software, 67*, 1–48.

Bates, D., & Maechler, M. (2015). *MatrixModels: Modelling with sparse and dense matrices*. R package version 0.4-1.

Bates, D. M. (2010). *lme4: Mixed-effects modeling with R*. New York: Springer.

Bear, R., Rintoul, D., Snyder, B., Smith-Caldas, M., Herren, C., & Horne, E. (2017). *Principles of biology*. OpenStax CNX, Rice University.

Bergström, L., Sundqvist, F., & Bergström, U. (2013). Effects of an offshore wind farm on temporal and spatial patterns in the demersal fish community. *Marine Ecology Progress Series, 485*, 199–210.

D. I. Warton, *Eco-Stats: Data Analysis in Ecology*, Methods in Statistical Ecology,
https://doi.org/10.1007/978-3-030-88443-7

Blanchet, F. G., Cazelles, K., & Gravel, D. (2020). Co-occurrence is not evidence of ecological interactions. *Ecology Letters, 23*, 1050–1063.

Blomberg, S. P., Lefevre, J. G., Wells, J. A., & Waterhouse, M. (2012). Independent contrasts and PGLS regression estimators are equivalent. *Systematic Biology, 61*, 382–391.

Borchers, D. L., Buckland, S. T., Stephens, W., Zucchini, W., et al. (2002). *Estimating animal abundance: Closed populations* (Vol. 13). Springer.

Bray, J. R., & Curtis, J. T. (1957). An ordination of the upland forest communities of southern Wisconsin. *Ecological Monographs, 27*, 325–349.

Brooks, M. E., Kristensen, K., van Benthem, K. J., Magnusson, A., Berg, C. W., Nielsen, A., Skaug, H. J., Machler, M., & Bolker, B. M. (2017). glmmTMB balances speed and flexibility among packages for zero-inflated generalized linear mixed modeling. *The R journal, 9*, 378–400.

Brown, A. M., Warton, D. I., Andrew, N. R., Binns, M., Cassis, G., & Gibb, H. (2014). The fourth-corner solution—using predictive models to understand how species traits interact with the environment. *Methods in Ecology and Evolution, 5*, 344–352.

Brown, B. M., & Maritz, J. S. (1982). Distribution-free methods in regression. *Australian Journal of Statistics, 24*, 318–331.

Carroll, R. J., & Ruppert, D. (1996). The use and misuse of orthogonal regression in linear errors-in-variables models. *American Statistician, 50*, 1–6.

Carroll, R. J., Ruppert, D., Stefanski, L. A., & Crainiceanu, C. M. (2006). *Measurement error in nonlinear models: A modern perspective*. CRC Press.

Chevan, A., & Sutherland, M. (1991). Hierarchical partitioning. *The American Statistician, 45*, 90–96.

Christin, S., Hervet, É., & Lecomte, N. (2019). Applications for deep learning in ecology. *Methods in Ecology and Evolution, 10*, 1632–1644.

Christin, S., Hervet, É., & Lecomte, N. (2021). Going further with model verification and deep learning. *Methods in Ecology and Evolution, 12*, 130–134.

Clark, G. F., Kelaher, B. P., Dafforn, K. A., Coleman, M. A., Knott, N. A., Marzinelli, E. M., & Johnston, E. L. (2015). What does impacted look like? High diversity and abundance of epibiota in modified estuaries. *Environmental Pollution, 196*, 12–20.

Clark, J. S. (2007). *Models for ecological data*. Princeton, NJ: Princeton University Press.

Clark, J. S., Gelfand, A. E., Woodall, C. W., & Zhu, K. (2014). More than the sum of the parts: Forest climate response from joint species distribution models. *Ecological Applications, 24*, 990–999.

Clark, N. J., Wells, K., & Lindberg, O. (2018). Unravelling changing interspecific interactions across environmental gradients using Markov random fields. *Ecology, 99*, 1277–1283.

Clarke, K. R. (1993). Non-parametric multivariate analyses of changes in community structure. *Australian Journal of Ecology, 18*, 117–143.

Cook, D., Lee, E.-K., & Majumder, M. (2016). Data visualization and statistical graphics in big data analysis. *Annual Review of Statistics and Its Application, 3*, 133–159.

Cooper, V. S., Bennett, A. F., & Lenski, R. E. (2001). Evolution of thermal dependence of growth rate of escherichia coli populations during 20,000 generations in a constant environment. *Evolution, 55*, 889–896.

Corder, G. W., & Foreman, D. I. (2009). *Nonparametric statistics for non-statisticians*. Wiley. ISBN: 9781118165881.

Creasy, M. A. (1957). Confidence limits for the gradient in the linear functional relationship. *Journal of the Royal Statistical Society B, 18*, 65–69.

Cressie, N. (2015). *Statistics for spatial data*. Wiley.

Cressie, N., Calder, C. A., Clark, J. S., Hoef, J. M. V., & Wikle, C. K. (2009). Accounting for uncertainty in ecological analysis: The strengths and limitations of hierarchical statistical modeling. *Ecological Applications, 19*, 553–570.

Cressie, N., & Johannesson, G. (2008). Fixed rank kriging for very large spatial data sets. *Journal of the Royal Statistical Society: Series B (Statistical Methodology), 70*, 209–226.

Davis, P. J., & Rabinowitz, P. (2007). *Methods of numerical integration*. Courier Corporation.

Davison, A. C., & Hinkley, D. V. (1997). *Bootstrap methods and their application*. Cambridge: Cambridge University Press.

Diggle, P. J., Heagerty, P., Liang, K.-Y &, Zeger, S. (2002). *Analysis of longitudinal data* (2nd ed.). Oxford University Press.

Dormann, C. F., M. McPherson, J., B. Araújo, M., Bivand, R., Bolliger, J., Carl, G., G. Davies, R., Hirzel, A., Jetz, W., Kissling, W. D., Kühn, I., Ohlemüller, R., R. Peres-Neto, P., Reineking, B., Schröder, B., M. Schurr, F., & Wilson, R. (2007). Methods to account for spatial autocorrelation in the analysis of species distributional data: A review. *Ecography, 30*, 609–628.

Dray, S., Dufour, A.-B., et al. (2007). The ade4 package: Implementing the duality diagram for ecologists. *Journal of Statistical Software, 22*, 1–20.

Duan, N. (1983). Smearing estimate: A nonparametric retransformation method. *Journal of the American Statistical Association, 78*, 605–610.

Dunn, P., & Smyth, G. (1996). Randomized quantile residuals. *Journal of Computational and Graphical Statistics, 5*, 236–244.

Dunstan, P. K., Foster, S. D., & Darnell, R. (2011). Model based grouping of species across environmental gradients. *Ecological Modelling, 222*, 955–963.

Edgington, E. S. (1995). *Randomization tests*. (3rd ed.). New York: Marcel Dekker.

Efron, B. (2004). The estimation of prediction error: Covariance penalties and cross-validation. *Journal of the American Statistical Association, 99*, 619–642.

Efron, B., & Tibshirani, R. (1997). Improvements on cross-validation: The 632+ bootstrap method. *Journal of the American Statistical Association, 92*, 548–560.

Elith, J., Graham, C., Anderson, R., Dudik, M., Ferrier, S., Guisan, A., Hijmans, R., Huettmann, F., Leathwick, J., Lehmann, A., Li, J., Lohmann, L., Loiselle, B., Manion, G., Moritz, C., Nakamura, M., Nakazawa, Y., Overton, J., Peterson, A., Phillips, S., Richardson, K., Scachetti-Pereira, R., Schapire, R., Soberon, J.,

Williams, S., Wisz, M., & Zimmermann, N. (2006). Novel methods improve prediction of species' distributions from occurrence data. *Ecography, 29*, 129–151.

Elith, J., & Leathwick, J. (2007). Predicting species distributions from museum and herbarium records using multiresponse models fitted with multivariate adaptive regression splines. *Diversity and Distributions, 13*, 265–275.

Ellison, A. M., Gotelli, N. J., Inouye, B. D., & Strong, D. R. (2014). *P* values, hypothesis testing, and model selection: It's déjà vu all over again. *Ecology, 95*, 609–610.

Evans, M., & Swartz, T. (2000). *Approximating integrals via Monte Carlo and deterministic methods*. Oxford statistical science series. OUP Oxford. ISBN: 9780191589874.

Fan, J., & Li, R. (2001). Variable selection via nonconcave penalized likelihood and its oracle properties. *Journal of the American Statistical Association, 96*, 1348–1360.

Felsenstein, J. (1985). Phylogenies and the comparative method. *The American Naturalist, 125*, 1–15.

Ferrari, S., & Cribari-Neto, F. (2004). Beta regression for modelling rates and proportions. *Journal of Applied Statistics, 31*, 799–815.

Finley, A. O., Sang, H., Banerjee, S., & Gelfand, A. E. (2009). Improving the performance of predictive process modeling for large datasets. *Computational Statistics and Data Analysis, 53*, 2873–2884.

Fisher, R. A. (1922). On the mathematical foundations of theoretical statistics. *Philosophical Transactions of the Royal Society of London. Series A, Containing Papers of a Mathematical or Physical Character, 222*, 309–368.

Flury, B. N. (1984). Common principal components in *k* groups. *Journal of the American Statistical Association, 79*, 892–898.

Foster, S. D., Hill, N. A., & Lyons, M. (2017). Ecological grouping of survey sites when sampling artefacts are present. *Journal of the Royal Statistical Society: Series C (Applied Statistics), 66*, 1031–1047.

Freckleton, R., Harvey, P., & Pagel, M. (2002). Phylogenetic analysis and comparative data: A test and review of evidence. *The American Naturalist, 160*, 712–726.

Friedman, J. (1991). Multivariate adaptive regression splines. *Annals of Statistics, 19*, 1–67.

Friedman, J., Hastie, T., & Tibshirani, R. (2008). Sparse inverse covariance estimation with the graphical lasso. *Biostatistics, 9*, 432–441.

Friedman, J., Hastie, T., & Tibshirani, R. (2010). Regularization paths for generalized linear models via coordinate descent. *Journal of Statistical Software, 33*, 1–22.

Galton, F. (1886). Regression towards mediocrity in hereditary stature. *Journal of the Anthropological Institute, 15*, 246–263.

Gelman, A., Stern, H. S., Carlin, J. B., Dunson, D. B., Vehtari, A., & Rubin, D. B. (2013). *Bayesian data analysis*. Chapman and Hall/CRC.

Geman, S., Bienenstock, E., & Doursat, R. (1992). Neural networks and the bias/variance dilemma. *Neural computation, 4*, 1–58.

Gould, S. J. (1966). Allometry and size in ontogeny and phylogeny. *Biological Reviews, 41*, 587–638.

Grace, J. B. (2006). *Structural equation modeling and natural systems*. Cambridge: Cambridge University Press.

Grafen, A. (1989). The phylogenetic regression. *Philosophical Transactions of the Royal Society of London. Series B, Biological Sciences, 326*, 119–157.

Grass, I., Lehmann, K., Thies, C., & Tscharntke, T. (2017). Insectivorous birds disrupt biological control of cereal aphids. *Ecology, 98*, 1583–1590.

Greenslade, P. (1964). Pitfall trapping as a method for studying populations of carabidae (coleoptera). *The Journal of Animal Ecology, 33*, 301–310.

Guillera-Arroita, G. (2017). Modelling of species distributions, range dynamics and communities under imperfect detection: Advances, challenges and opportunities. *Ecography, 40*, 281–295.

Guisan, A., & Zimmerman, N. E. (2000). Predictive habitat distribution models in ecology. *Ecological modelling, 135*, 147–186.

Hadfield, J. D. et al. (2010). MCMC methods for multi-response generalized linear mixed models: The MCMCglmm R package. *Journal of Statistical Software, 33*, 1–22.

Hall, P., & Wilson, S. R. (1991). Two guidelines for bootstrap hypothesis testing. *Biometrics, 47*, 757–762.

Hansen, T. F., & Bartoszek, K. (2012). Interpreting the evolutionary regression: The interplay between observational and biological errors in phylogenetic comparative studies. *Systematic Biology, 61*, 413–425.

Hardin, J. W., & Hilbe, J. M. (2002). *Generalized estimating equations*. Boca Raton: Chapman & Hall.

Harris, D. J. (2015). Generating realistic assemblages with a joint species distribution model. *Methods in Ecology and Evolution, 6*, 465–473.

Harris, D. J. (2016). Inferring species interactions from co-occurrence data with Markov networks. *Ecology, 97*, 3308–3314.

Hartig, F. (2020). *DHARMa: Residual diagnostics for hierarchical (multi-level/mixed) regression models*. R package version 0.3.2.0.

Harvey, P. H., Read, A. F., & Nee, S. (1995). Why ecologists need to be phylogenetically challenged. *Journal of Ecology, 83*, 535–536.

Hastie, T., & Stuetzle, W. (1989). Principal curves. *Journal of the American Statistical Association, 84*, 502–516.

Hastie, T., Tibshirani, R., & Friedman, J. (2009). *The elements of statistical learning* (2nd ed.). New York: Springer.

Hastie, T., Tibshirani, R., & Wainwright, M. (2015). *Statistical learning with sparsity: The lasso and generalizations*. CRC Press.

Hastie, T. J., & Tibshirani, R. J. (1990). *Generalized additive models*. London: Chapman & Hall.

Hefley, T. J., Broms, K. M., Brost, B. M., Buderman, F. E., Kay, S. L., Scharf, H. R., Tipton, J. R., Williams, P. J., & Hooten, M. B. (2017). The basis function approach for modeling autocorrelation in ecological data. *Ecology, 98*, 632–646.

Hodges, J. S., & Reich, B. J. (2010). Adding spatially-correlated errors can mess up the fixed effect you love. *The American Statistician, 64*, 325–334.
Hooten, M. B., Johnson, D. S., McClintock, B. T., & Morales, J. M. (2017). *Animal movement: Statistical models for telemetry data*. CRC Press.
Hui, F. K. C. (2016). boral—Bayesian ordination and regression analysis of multivariate abundance data in R. *Methods in Ecology and Evolution, 7*, 744–750.
Hui, F. K. C., Taskinen, S., Pledger, S., Foster, S. D., & Warton, D. I. (2015). Model-based approaches to unconstrained ordination. *Methods in Ecology and Evolution, 6*, 399–411.
Hui, F. K. C., Warton, D. I., Foster, S., & Dunstan, P. (2013). To mix or not to mix: Comparing the predictive performance of mixture models versus separate species distribution models. *Ecology, 94*, 1913–1919.
Hurlbert, S. H. (1984). Pseudoreplication and the design of ecological field experiments. *Ecological Monographs, 54*, 187–211.
Hyndman, R. J., & Athanasopoulos, G. (2014). *Forecasting: Principles and practice*. OTexts.
Ives, A. R. (2015). For testing the significance of regression coefficients, go ahead and log-transform count data. *Methods in Ecology and Evolution, 6*, 828–835.
Jamil, T., Ozinga, W. A., Kleyer, M., & ter Braak, C. J. F. (2013). Selecting traits that explain species–environment relationships: A generalized linear mixed model approach. *Journal of Vegetation Science, 24*, 988–1000.
Johns, J. M., Walters, P. A., & Zimmerman, L. I. (1993). The effects of chronic prenatal exposure to nicotine on the behavior of guinea pigs (Cavia porcellus). *The Journal of General Psychology, 120*, 49–63.
Johnson, D. H. (1999). The insignificance of statistical significance testing. *The Journal of Wildlife Management, 63*, 763–772.
Johnson, N., Kotz, S., & Balakrishnan, N. (1997). *Discrete multivariate distributions*. Wiley series in probability and statistics. Wiley. ISBN: 9780471128441.
Jolicoeur, P. (1975). Linear regression in fishery research: Some comments. *Journal of the Fisheries Research Board of Canada, 32*, 1491–1494.
Jordan, M. I. et al. (2013). On statistics, computation and scalability. *Bernoulli, 19*, 1378–1390.
Keck, F., Rimet, F., Bouchez, A., & Franc, A. (2016). phylosignal: An R package to measure, test, and explore the phylogenetic signal. *Ecology and Evolution, 6*, 2774–2780.
Keribin, C. (2000). Consistent estimation of the order of mixture models. *SankhyāĄ: The Indian Journal of Statistics, Series A, 62*, 49–66.
Kerkhoff, A., & Enquist, B. (2009). Multiplicative by nature: Why logarithmic transformation is necessary in allometry. *Journal of Theoretical Biology, 257*, 519–521.
Koenker, R., & Hallock, K. F. (2001). Quantile regression. *Journal of Economic Perspectives, 15*, 143–156.
Kruskal, J. B., & Wish, M. (1978). *Multidimensional scaling*. Beverley Hills: Sage Publications.

Kuchibhotla, A. K., Kolassa, J. E. & Kuffner, T. A. (2022) Post-Selection Inference. *Annual Review of Statistics and Its Application, 9*, 505–527.

Lance, G. N., & Williams, W. T. (1967). A general theory of classificatory sorting strategies: 1. Hierarchical systems. *The Computer Journal, 9*, 373–380.

Legendre, P., Galzin, R., & Harmelin-Vivien, M. L. (1997). Relating behavior to habitat: Solutions to the fourth-corner problem. *Ecology, 78*, 547–562.

Legendre, P., & Legendre, L. (2012). *Numerical ecology*. Third English edition. Amsterdam: Elsevier Science.

Letten, A. D., Keith, D. A., Tozer, M. G., & Hui, F. K. (2015). Fine-scale hydrological niche differentiation through the lens of multi-species co-occurrence models. *Journal of Ecology, 103*, 1264–1275.

Li, D., Dinnage, R., Nell, L., Helmus, M. R., & Ives, A. (2020). phyr: An R package for phylogenetic species-distribution modelling in ecological communities. *Methods in Ecology and Evolution, 11*, 1455–1463.

Li, D., & Ives, A. R. (2017). The statistical need to include phylogeny in trait-based analyses of community composition. *Methods in Ecology and Evolution, 8*, 1192–1199.

Liang, K. Y., & Zeger, S. L. (1986). Longitudinal data analysis using generalized linear models. *Biometrika, 73*, 13–22.

Lislevand, T., & Thomas, G. H. (2006). Limited male incubation ability and the evolution of egg size in shorebirds. *Biology Letters, 2*, 206–208.

Ludbrook, J., & Dudley, H. (1998). Why permutation tests are superior to *t* and *F* tests in biomedical research. *The American Statistician, 52*, 127–132.

Lumley, T. (2020). *biglm: Bounded memory linear and generalized linear models*. R package version 0.9-2.

MacKenzie, D. I., Nichols, J. D., Royle, J. A., Pollock, K. H., Bailey, L., & Hines, J. E. (2017). *Occupancy estimation and modeling: Inferring patterns and dynamics of species occurrence*. Elsevier.

Maronna, R., Martin, R. D., & Yohai, V. (2006). *Robust statistics*. Chichester: Wiley.

McClintock, B. T., & Michelot, T. (2018). momentuHMM: R package for generalized hidden Markov models of animal movement. *Methods in Ecology and Evolution, 9*, 1518–1530.

McCullagh, P., & Nelder, J. A. (1989). *Generalized linear models*. (2nd ed.). London: Chapman & Hall.

McDonald, T. L., Erickson, W. P., & McDonald, L. L. (2000). Analysis of count data from before-after control-impact studies. *Journal of Agricultural, Biological, and Environmental Statistics, 5*, 262–279.

McGill, B. J., Enquist, B. J., Weiher, E., & Westoby, M. (2006). Rebuilding community ecology from functional traits. *Trends in Ecology and Evolution, 21*, 178–185.

McLachlan, G., & Peel, D. (2000). *Finite mixture models*. Wiley.

Michelot, T., Langrock, R., & Patterson, T. A. (2016). moveHMM: An R package for the statistical modelling of animal movement data using hidden Markov models. *Methods in Ecology and Evolution, 7*, 1308–1315.

Miller Jr., R. G. (1997). *Beyond ANOVA: Basics of applied statistics*. Boca Raton, FL: Chapman & Hall.

Moles, A. T., Warton, D. I., Warman, L., Swenson, N. G., Laffan, S. W., Zanne, A. E., Pitman, A., Hemmings, F. A., & Leishman, M. R. (2009). Global patterns in plant height. *Journal of Ecology, 97*, 923–932.

Moore, D. S., McCabe, G. P., & Craig, B. A. (2014). Introduction to the practice of statistics.

Moran, P. A. P. (1971). Estimating structural and functional relationships. *Journal of Multivariate Analysis, 1*, 232–255.

Morgan, S. L., & Winship, C. (2015). *Counterfactuals and causal inference*. Cambridge University Press.

Morrisey, D., Underwood, A., Howitt, L., & Stark, J. (1992). Temporal variation in soft-sediment benthos. *Journal of Experimental Marine Biology and Ecology, 164*, 233–245.

Murtaugh, P. A. (2007). Simplicity and complexity in ecological data analysis. *Ecology, 88*, 56–62.

Nakagawa, S., & Cuthill, I. C. (2007). Effect size, confidence interval and statistical significance: A practical guide for biologists. *Biological Reviews, 82*, 591–605.

Nelder, J. A., & Wedderburn, R. W. M. (1972). Generalized linear models. *Journal of the Royal Statistical Society. Series A (General), 135*, 370–384.

Niklas, K. J. (2004). Plant allometry: Is there a grand unifying theory? *Biological Reviews, 79*, 871–889.

Niku, J., Hui, F. K. C., Taskinen, S., & Warton, D. I. (2019). gllvm: Fast analysis of multivariate abundance data with generalized linear latent variable models in R. *Methods in Ecology and Evolution, 10*, 2173–2182.

Nishii, R. (1984). Asymptotic properties of criteria for selection of variables in multiple regression. *The Annals of Statistics, 12*, 758–765.

Norberg, A., Abrego, N., Blanchet, F. G., Adler, F. R., Anderson, B. J., Anttila, J., Araújo, M. B., Dallas, T., Dunson, D., Elith, J., Foster, S. D., Fox, R., Franklin, J., Godsoe, W., Guisan, A., O'Hara, B., Hill, N. A., Holt, R. D., Hui, F. K. C., Husby, M., Kålås, J. A., Lehikoinen, A., Luoto, M., Mod, H. K., Newell, G., Renner, I., Roslin, T., Soininen, J., Thuiller, W., Vanhatalo, J., Warton, D., White, M., Zimmermann, N. E., Gravel, D., & Ovaskainen, O. (2019). A comprehensive evaluation of predictive performance of 33 species distribution models at species and community levels. *Ecological Monographs, 89*, e01370.

Oberdorff, T., & Hughes, R. M. (1992). Modification of an index of biotic integrity based on fish assemblages to characterize rivers of the seine basin, France. *Hydrobiologia, 228*, 117–130.

Oksanen, J., Blanchet, F. G., Friendly, M., Kindt, R., Legendre, P., McGlinn, D., Minchin, P. R., O'Hara, R. B., Simpson, G. L., Solymos, P., Stevens, M. H. H., Szoecs, E., & Wagner, H. (2017). *vegan: Community ecology package*. R package version 2.4-3.

Olden, J., Lawler, J., & Poff, N. (2008). Machine learning methods without tears: A primer for ecologists. *The Quarterly Review of Biology, 83*, 171–193.

Ord, T. J., Charles, G. K., Palmer, M., & Stamps, J. A. (2016). Plasticity in social communication and its implications for the colonization of novel habitats. *Behavioral Ecology, 27*, 341–351.

Ovaskainen, O., & Abrego, N. (2020). *Joint species distribution modelling: With applications in R*. Cambridge: Cambridge University Press.

Ovaskainen, O., Roy, D. B., Fox, R., & Anderson, B. J. (2016). Uncovering hidden spatial structure in species communities with spatially explicit joint species distribution models. *Methods in Ecology and Evolution, 7*, 428–436.

Ovaskainen, O., & Soininen, J. (2011). Making more out of sparse data: Hierarchical modeling of species communities. *Ecology, 92*, 289–295.

Ovaskainen, O., Tikhonov, G., Dunson, D., Grótan, V., Engen, S., Sæther, B.-E., & Abrego, N. (2017a). How are species interactions structured in species-rich communities? A new method for analysing time-series data. *Proceedings of the Royal Society B: Biological Sciences, 284*, 20170768.

Ovaskainen, O., Tikhonov, G., Norberg, A., Guillaume Blanchet, F., Duan, L., Dunson, D., Roslin, T., & Abrego, N. (2017b). How to make more out of community data? A conceptual framework and its implementation as models and software. *Ecology Letters, 20*, 561–576.

Packard, G. C. (2013). Is logarithmic transformation necessary in allometry? *Biological Journal of the Linnean Society, 109*, 476–486.

Pagel, M. (1997). Inferring evolutionary processes from phylogenies. *Zoologica Scripta, 26*, 331–348.

Pagel, M. (1999). Inferring the historical patterns of biological evolution. *Nature, 401*, 877–84.

Phillips, S. J., Anderson, R. P., & Schapire, R. E. (2006). Maximum entropy modeling of species geographic distributions. *Ecological Modelling, 190*, 231–259.

Pitman, E. T. G. (1939). A note on normal correlation. *Biometrika, 31*, 9–12.

Pledger, S. (2000). Unified maximum likelihood estimates for closed capture–recapture models using mixtures. *Biometrics, 56*, 434–442.

Pledger, S., & Arnold, R. (2014). Multivariate methods using mixtures: Correspondence analysis, scaling and pattern-detection. *Computational Statistics & Data Analysis, 71*, 241–261.

Pollock, L. J., Morris, W. K., & Vesk, P. A. (2012). The role of functional traits in species distributions revealed through a hierarchical model. *Ecography, 35*, 716–725.

Pollock, L. J., Tingley, R., Morris, W. K., Golding, N., O'Hara, R. B., Parris, K. M., Vesk, P. A., & McCarthy, M. A. (2014). Understanding co-occurrence by modelling species simultaneously with a Joint Species Distribution Model (JSDM). *Methods in Ecology and Evolution, 5*, 397–406.

Popovic, G. C., Hui, F. K., & Warton, D. I. (2018). A general algorithm for covariance modeling of discrete data. *Journal of Multivariate Analysis, 165*, 86–100.

Popovic, G. C., Hui, F. K. C., & Warton, D. I. (2022). Fast model-based ordination with copulas. *Methods in Ecology and Evolution*, 13, 194–202

Popovic, G. C., Warton, D. I., Thomson, F. J., Hui, F. K. C., & Moles, A. T. (2019). Untangling direct species associations from indirect mediator species effects with graphical models. *Methods in Ecology and Evolution, 10*, 1571–1583.

Rao, C. R. (1973). *Linear statistical inference and its applications* (2nd ed.). New York: Wiley.

Renner, I. W., Elith, J., Baddeley, A., Fithian, W., Hastie, T., Phillips, S. J., Popovic, G., & Warton, D. I. (2015). Point process models for presence-only analysis. *Methods in Ecology and Evolution, 6*, 366–379.

Renner, I. W., & Warton, D. I. (2013). Equivalence of MAXENT and Poisson point process models for species distribution modeling in ecology. *Biometrics, 69*, 274–281.

Rensch, B. (1954). The relation between the evolution of central nervous functions and the body size of animals. *Evolution as a Process* 181–200.

Rice, K. (2004). Sprint research runs into a credibility gap. *Nature, 432*, 147.

Ricker, W. E. (1973). Linear regressions in fishery research. *Journal of the Fisheries Research Board of Canada, 30*, 409–434.

Robert, C., & Casella, G. (2013). *Monte Carlo statistical methods*. Springer.

Roberts, D. A., & Poore, A. G. (2006). Habitat configuration affects colonisation of epifauna in a marine algal bed. *Biological Conservation, 127*, 18–26.

Roberts, D. R., Bahn, V., Ciuti, S., Boyce, M. S., Elith, J., Guillera-Arroita, G., Hauenstein, S., Lahoz-Monfort, J. J., Schröder, B., Thuiller, W., Warton, D. I., Wintle, B. A., Hartig, F., & Dormann, C. F. (2017). Cross-validation strategies for data with temporal, spatial, hierarchical, or phylogenetic structure. *Ecography, 40*, 913–929.

Schaub, M., & Abadi, F. (2011). Integrated population models: A novel analysis framework for deeper insights into population dynamics. *Journal of Ornithology, 152*, 227–237.

Schielzeth, H., Dingemanse, N. J., Nakagawa, S., Westneat, D. F., Allegue, H., Teplitsky, C., Réale, D., Dochtermann, N. A., Garamszegi, L. Z., & Araya-Ajoy, Y. G. (2020). Robustness of linear mixed-effects models to violations of distributional assumptions. *Methods in Ecology and Evolution, 11*, 1141–1152.

Schwarz, G. (1978). Estimating the dimension of a model. *The Annals of Statistics, 6*, 461–464.

Sentinella, A. T., Warton, D. I., Sherwin, W. B., Offord, C. A., & Moles, A. T. (2020). Tropical plants do not have narrower temperature tolerances, but are more at risk from warming because they are close to their upper thermal limits. *Global Ecology and Biogeography, 29*, 1387–1398.

Shao, J. (1993). Linear model selection by cross-validation. *Journal of the American Statistical Association, 88*, 486–494.

Shermer, M. (2012). *The believing brain: From spiritual faiths to political convictions–how we construct beliefs and reinforce them as truths.* Hachette, UK.

Shipley, B. (2010). *From plant traits to vegetation structure: Chance and selection in the assembly of ecological communities*. Cambridge University Press.

Shipley, B. (2016). *Cause and correlation in biology: A user's guide to path analysis, structural equations and causal inference with R.* Cambridge University Press.

Shipley, B., Vile, D., & Garnier, E. (2006). From plant traits to plant communities: A statistical mechanistic approach to biodiversity. *Science, 314*, 812–814.

Simpson, G. L. (2016). *Permute: Functions for generating restricted permutations of data*. R package version 0.9-4.

Skrondal, A., & Rabe-Hesketh, S. (2004). *Generalized latent variable modeling: Multilevel, longitudinal and structural equation models*. Boca Raton: Chapman & Hall.

Smith, R. J. (2009). Use and misuse of the reduced major axis for line-fitting. *American Journal of Physical Anthropology, 140*, 476–486.

Sokal, R. R., & Rohlf, F. J. (1995). *Biometry: The principles and practice of statistics in biological research*, third edition. New York: Freeman.

Stephens, P., Hunter, C., Bignell, G., Edkins, S., Davies, H., Teague, J., Stevens, C., O'Meara, S., Smith, R., Parker, A., et al. (2004). Athletics: Momentous sprint at the 2156 olympics? *Nature, 431*, 525.

Stoklosa, J., & Warton, D. I. (2018). A generalized estimating equation approach to multivariate adaptive regression splines. *Journal of Computational and Graphical Statistics, 27*, 245–253.

Stone, M. (1977). An asymptotic equivalence of choice of model by cross-validation and Akaike's criterion. *Journal of the Royal Statistical Society: Series B (Methodological), 39*, 44–47.

Student (1931). The Lanarkshire milk experiment. *Biometrika, 23*, 398–406.

Symonds, M. R. E., & Blomberg, S. P. (2014). *A primer on phylogenetic generalised least squares* (pp. 105–130). Springer: Heidelberg.

Taskinen, S., & Warton, D. I. (2011). Robust estimation and inference for bivariate line-fitting in allometry. *Biometrical Journal, 53*, 652–672.

Tatibouët, F. (1981). *Approche écologique d'un établissement humain (environnement et structure): exemple de la Communauté Urbaine de Lyon*. Ph.D. thesis, University of Lyon 1.

Taylor, L. R. (1961). Aggregation, variance and mean. *Nature, 189*, 732–735.

ter Braak, C. J. F. (1986). Canonical correspondence analysis: A new eigenvector technique for multivariate direct gradient analysis. *Ecology, 67*, 1167–1179.

ter Braak, C. J. F. (2017). Fourth-corner correlation is a score test statistic in a log-linear trait–environment model that is useful in permutation testing. *Environmental and Ecological Statistics, 24*, 219–242.

ter Braak, C. J. F. (2019). New robust weighted averaging- and model-based methods for assessing trait–environment relationships. *Methods in Ecology and Evolution, 10*, 1962–1971.

ter Braak, C. J. F., & Prentice, I. C. (1988). A theory of gradient analysis. *Advances in Ecological Research, 18*, 271–317.

ter Braak, C. J. F., & Smilauer, P. (1998). *CANOCO reference manual and user's guide to CANOCO for Windows: Software for canonical community ordination (version 4)*. New York: Microcomputer Power.

Thorson, J. T., Scheuerell, M. D., Shelton, A. O., See, K. E., Skaug, H. J., & Kristensen, K. (2015). Spatial factor analysis: A new tool for estimating joint species distributions and correlations in species range. *Methods in Ecology and Evolution, 6*, 627–637.

Tibshirani, R. (1996). Regression shrinkage and selection via the lasso. *Journal of the Royal Statistical Society. Series B (Methodological), 58*, 267–288.

Tikhonov, G., Abrego, N., Dunson, D., & Ovaskainen, O. (2017). Using joint species distribution models for evaluating how species-to-species associations depend on the environmental context. *Methods in Ecology and Evolution, 8*, 443–452.

Tylianakis, J. M., & Morris, R. J. (2017). Ecological networks across environmental gradients. *Annual Review of Ecology, Evolution, and Systematics, 48*, 25–48.

Væth, M. (1985). On the use of Wald's test in exponential families. *International Statistical Review, 53*, 199–214.

Walker, S. C., & Jackson, D. A. (2011). Random-effects ordination: Describing and predicting multivariate correlations and co-occurrences. *Ecological Monographs, 81*, 635–663.

Wang, Y., Naumann, U., Wright, S. T., & Warton, D. I. (2012). `mvabund`—an R package for model-based analysis of multivariate abundance data. *Methods in Ecology and Evolution, 3*, 471–474.

Warton, D. I. (2005). Many zeros does not mean zero inflation: Comparing the goodness-of-fit of parametric models to multivariate abundance data. *Environmetrics, 16*, 275–289.

Warton, D. I. (2007). Robustness to failure of assumptions of tests for a common slope amongst several allometric lines—a simulation study. *Biometrical Journal, 49*, 286–299.

Warton, D. I. (2008). Penalized normal likelihood and ridge regularization of correlation and covariance matrices. *Journal of the American Statistical Association, 103*, 340–349.

Warton, D. I. (2011). Regularized sandwich estimators for analysis of high dimensional data using generalized estimating equations. *Biometrics, 67*, 116–123.

Warton, D. I. (2018). Why you cannot transform your way out of trouble for small counts. *Biometrics, 74*, 362–368

Warton, D. I., & Aarts, G. (2013). Advancing our thinking in presence-only and used-available analysis. *Journal of Animal Ecology, 82*, 1125–1134.

Warton, D. I., Duursma, R. A., Falster, D. S., & Taskinen, S. (2012a). `smatr 3`—an R package for estimation and inference about allometric lines. *Methods in Ecology and Evolution, 3*, 257–259.

Warton, D. I., & Hui, F. K. C. (2011). The arcsine is asinine: The analysis of proportions in ecology. *Ecology, 92*, 3–10.

Warton, D. I., & Hui, F. K. C. (2017). The central role of mean-variance relationships in the analysis of multivariate abundance data: A response to Roberts (2017). *Methods in Ecology and Evolution, 8*, 1408–1414.

Warton, D. I., Lyons, M., Stoklosa, J., & Ives, A. R. (2016). Three points to consider when choosing a LM or GLM test for count data. *Methods in Ecology and Evolution, 7*, 882–890.

Warton, D. I., Shipley, B., & Hastie, T. (2015). CATS regression—a model-based approach to studying trait-based community assembly. *Methods in Ecology and Evolution, 6*, 389–398.

Warton, D. I., Thibaut, L., & Wang, Y. A. (2017). The PIT-trap—a "model-free" bootstrap procedure for inference about regression models with discrete, multivariate responses. *PLoS One, 12*, e0181790.

Warton, D. I., & Weber, N. C. (2002). Common slope tests for errors-in-variables models. *Biometrical Journal, 44*, 161–174.

Warton, D. I., Wright, I. J., Falster, D. S., & Westoby, M. (2006). Bivariate line-fitting methods for allometry. *Biological Reviews, 81*, 259–291.

Warton, D. I., Wright, S. T., & Wang, Y. (2012b). Distance-based multivariate analyses confound location and dispersion effects. *Methods in Ecology and Evolution, 3*, 89–101.

Warwick, R. M., Clarke, K. R., & Suharsono (1990). A statistical analysis of coral community responses to the 1982–1983 El Niño in the Thousand Islands, Indonesia. *Coral Reefs, 8*, 171–179.

Wasserstein, R. L., & Lazar, N. A. (2016). The ASA's statement on p-values: Context, process, and purpose. *The American Statistician, 70*, 129–133.

Weisbecker, V., & Warton, D. I. (2006). Evidence at hand: Diversity, functional implications, and locomotor prediction in intrinsic hand proportions of diprotodontian marsupials. *Journal of Morphology, 267*, 1469–1485.

Welsh, A. H., Cunningham, R. B., Donnelly, C. F., & Lindenmeyer, D. B. (1996). Modelling the abundance of rare species: Statistical methods for counts with extra zeros. *Ecological Modelling, 88*, 297–308.

Westoby, M., Leishman, M. R., & Lord, J. M. (1995). On misinterpreting the 'phylogenetic correction'. *Journal of Ecology, 83*, 531–534.

Wheeler, J. A., Cortés, A. J., Sedlacek, J., Karrenberg, S., van Kleunen, M., Wipf, S., Hoch, G., Bossdorf, O., & Rixen, C. (2016). The snow and the willows: Earlier spring snowmelt reduces performance in the low-lying alpine shrub salix herbacea. *Journal of Ecology, 104*, 1041–1050.

White, C. (2005). Hunters ring dinner bell for ravens: Experimental evidence of a unique foraging strategy. *Ecology, 86*, 1057–1060.

Whittaker, R. H. (1972). Evolution and measurement of species diversity. *Taxon, 21*, 213–251.

Wickham, H. (2016). *ggplot2: Elegant graphics for data analysis*. New York: Springer. ISBN: 978-3-319-24277-4.

Williams, W., & Lambert, J. T. (1966). Multivariate methods in plant ecology: V. Similarity analyses and information-analysis. *The Journal of Ecology, 54*, 427–445.

Wood, S. N. (2011). Fast stable restricted maximum likelihood and marginal likelihood estimation of semiparametric generalized linear models. *Journal of the Royal Statistical Society: Series B (Statistical Methodology), 73*, 3–36.

Wood, S. N. (2017). *Generalized additive models: An introduction with R* (2nd edn.). CRC Press.

Xiao, X., White, E. P., Hooten, M. B., & Durham, S. L. (2011). On the use of log-transformation vs. nonlinear regression for analyzing biological power laws. *Ecology, 92*, 1887–1894.

Yee, T. (2006). Constrained additive ordination. *Ecology, 87*, 203–213.

Yee, T. W. (2010). The VGAM package for categorical data analysis. *Journal of Statistical Software, 32*, 1–34.

Yee, T. W., & Mitchell, N. D. (1991). Generalized additive models in plant ecology. *Journal of Vegetation Science, 2*, 587–602.

Yuan, M., & Lin, Y. (2006). Model selection and estimation in regression with grouped variables. *Journal of the Royal Statistical Society: Series B (Statistical Methodology), 68*, 49–67.

Zeger, S. L., & Liang, K. Y. (1986). Longitudinal data analysis for discrete and continuous outcomes. *Biometrics, 42*, 121–130.

Zou, H. (2006). The adaptive lasso and its oracle properties. *Journal of the American Statistical Association, 101*, 1418–1429.

Zuur, A., Ieno, E., Walker, N., Saveliev, A., & Smith, G. (2009). *Mixed effects models and extensions in ecology with R*. New York: Springer.

Index

D. I. Warton, *Eco-Stats: Data Analysis in Ecology*, Methods in Statistical Ecology,
https://doi.org/10.1007/978-3-030-88443-7